Springer Series in Geomechanics and Geoengineering

Series editor

Wei Wu, Universität für Bodenkultur, Vienna, Austria
e-mail: wei.wu@boku.ac.at

About this Series

Geomechanics deals with the application of the principle of mechanics to geomaterials including experimental, analytical and numerical investigations into the mechanical, physical, hydraulic and thermal properties of geomaterials as multiphase media. Geoengineering covers a wide range of engineering disciplines related to geomaterials from traditional to emerging areas.

The objective of the book series is to publish monographs, handbooks, workshop proceedings and textbooks. The book series is intended to cover both the state-of-the-art and the recent developments in geomechanics and geoengineering. Besides researchers, the series provides valuable references for engineering practitioners and graduate students.

More information about this series at http://www.springer.com/series/8069

Renpeng Chen · Gang Zheng · Changyu Ou
Editors

Proceedings of the 2nd International Symposium on Asia Urban GeoEngineering

ZHEJIANG UNIVERSITY PRESS
浙江大学出版社

Springer

Editors
Renpeng Chen
Hunan University
Changsha, Hunan
China

Gang Zheng
Tianjin University
Tianjin
China

Changyu Ou
National Taiwan University of Science and Technology
Taipei
China

ISSN 1866-8755 ISSN 1866-8763 (electronic)
Springer Series in Geomechanics and Geoengineering
ISBN 978-981-13-4919-5 ISBN 978-981-10-6632-0 (eBook)
https://doi.org/10.1007/978-981-10-6632-0

Jointly published with Zhejiang University Press, Hangzhou, China

The print edition is not for sale in China Mainland. Customers from China Mainland please order the print book from: Zhejiang University Press

Printed on acid-free paper

This Springer imprint is published by Springer Nature
The registered company is Springer Nature Singapore Pte Ltd.
The registered company address is: 152 Beach Road, #21-01/04 Gateway East, Singapore 189721, Singapore

Preface

The 2nd International Symposium on Asia Urban GeoEngineering held from November 24–27, 2017, in Changsha, a historical and modern city in Hunan, China. In 2015, the first symposium was held in Fukuoka, Japan. The aim of this symposium is to share ideas and experiences about urban geoengineering in Asian countries among engineers, researchers, and academic professors. The emphasis will be on improving our knowledge in meeting geoengineering requirements for a long-term sustainable urban development and the need to protect and preserve our environment.

Extensively, urban construction is going on in Asian countries, and environmental friendliness and sustainability are emphasized in modern urbanization processes. Environmental friendliness addresses the control of ground movement, mitigation measures, and risk management. Those control or mitigation measures should be sustainable. Moreover, the performance of deep excavation and tunnels, mitigation measures, and risk management are often site dependent and vary from country to country. The ground conditions, construction practices, codes, and standards are very different among Asian countries. The aim of this symposium is to share knowledge and experiences of the analysis, design, construction, and maintenance of urban geoengineering among engineers, researchers, and academic professors in Asian countries.

Totally, more than 70 papers have been received by the symposium organizer. After technical review, more than 40 papers have been accepted to be presented at the symposium and also included in the conference proceedings co-published by Springer and the Zhejiang University Press. These papers have covered a wide range of issues concerning on urban geoengineering problems in Asian countries. Showcasing diversity and quality, these papers report the state of the art and point to future directions of research and development in this exciting area. We are also honored to have outstanding speakers to give invited keynote speeches at the symposium.

The success of the symposium is due to the contributions from every participants and organizations. On behalf of the Organizing Committee, I really appreciate all authors for your careful preparation of submissions and all speakers for sharing your knowledge, experience, and research at the symposium. We are grateful to all the members of the Scientific Committee and the Organizing Committee for your important contributions to the symposium.

Part of researches and the publication of this book received supports from the National Key Research and Development Program of China (2016YFC0800207), the National Natural Science Foundation of China (51508181), the Provincial Key Research and Development Program of Hunan (0105679005), the Industrial Technology and Development Program of Zhongjian Tunnel Construction Co., Ltd. (17430102000417) and the Research Program of Construction Engineering of Shenzhen (20151118003B). We would like to thank all the financial supports from these projects.

November 2017

Renpeng Chen
Chairman of the Organizing Committee 2nd Asia UGE

The original version of the book was revised: New chapters have been inserted and misspelt university name has been corrected. The erratum to the book is available at https://doi.org/10.1007/978-981-10-6632-0_54

Proceedings of the 2nd International Symposium on Asia Urban GeoEngineering

November 24–27, 2017, Hunan University, Changsha, China

Edited by

Renpeng Chen
Gang Zheng
Changyu Ou

Organized by

Hunan University, Changsha, China
Tianjin University, Tianjin, China

Host

International Society for Soil Mechanics and Geotechnical Engineering ATC6

Scientific Committee

Chairman

Changyu Ou — National Taiwan University of Science and Technology, Taipei

Members

Benson Hsiung	National Kaohsiung University of Applied Sciences, Taipei
Anthony Goh	Nanyang Technology University, Singapore
Askar Zhussupbekov	ISSMGE, Kazakhstan
ChiaRay Seng	CECI Engineering Consultants, Inc., Taipei
ChungRen Huang	Sinotech Engineering Consultants, LTD, Taipei
DerWen Chang	Tamkang University, Taipei
Duc Long Phung	WSP Vietnam Company, Vietnam
FookHou Lee	National University of Singapore, Singapore
Gang Zheng	Tianjin University, China
Hirokazu Akagi	Waseda University, Japan
Hongwei Huang	Tongji University, China
Hsien C. Juang	Clemson University, Taipei
HsiiSheng Hsieh	Trinity Consultant Co., Taipei
Jian Hua Wang	Shanghai Jiao Tong University, China
Jiro Takemura	Tokyo Institute of Technology, Japan
Juneja Ashish	Indian Institute of Technology, India
Kenny Yee	AGSSEA, Malaysia
KuoHsin Yang	National Taiwan University of Science and Technology, Taipei
Mamoru Kikumoto	Yokohama National University, Japan
Masyhur Irsyam	Institut Teknologi Bandung, Indonesia
Mitsutaka Sugimoto	Nagaoka University of Technology, Japan
Ng Tiong Guan	Golder Associates (Singapore), Singapore
Phienwej Noppadol	Asian Institute of Technology, Thailand
Pornpot Tanseng	Suranaree University of Technology, Thailand
Rahardjo Paulus	Parahyangan Catholic University, Indonesia
Renpeng Chen	Hunan University, China
Seungrae Lee	Korea Advanced Institute of Science and Technology, Korea
ShawShong Liew	G & P Professional, Malaysia
ShunMin Lee	CECI Engineering Consultants, Inc., Taipei
Victor C.W. Ong	ONE SMART Engineering, Singapore
Wanhuan Zhou	University of Macau, China
William Cheang	PLAXIS Asia, Singapore
YeanChin Tan	G & P Professional, Malaysia
YungShow Fang	National Chiao Tung University, Taipei

Organizing Committee

Chairman

Renpeng Chen	Hunan University, China
Gang Zheng	Tianjin University, China

Members

Changfu Chen	Hunan University, China
Huayang Lei	Tianjin University, China
Ling Zhang	Hunan University, China
Minghua Huang	Hunan University, China
Minghua Zhao	Hunan University, China
Minghui Yang	Hunan University, China
Qijian Liu	Hunan University, China
Suhua Zhou	Hunan University, China
Wengui Cao	Hunan University, China
Xiaomin Liu	Hunan University, China
Xin Kang	Hunan University, China
Xin Tan	Hunan University, China
Xinjun Zou	Hunan University, China
Xuesong Cheng	Tianjin University, China
Yonghua Su	Hunan University, China
Yu Diao	Tianjin University, China

Contents

Excavation and Slope Engineering

Numerical Methods in Geotechnical Engineering

Fundamental Behavior and Constitutive Model of Geomaterials

Compaction Characteristics of Aeolian Sand with Different Sediment Percentage

Xiu-kun Dong and Li-ying Liu(✉)

School of Civil Engineering and Architecture, Chongqing University of Science and Technology, Chongqing, China
zhangyiluu@126.com

Abstract. In order to research compaction characteristic of aeolian sand with different sediment percentage, and get suitable roadbed filling, seven kinds of aeolian sand with different sediment percentage is made by taking representative sample of aeolian sand. Aeolian sand samples are compacted by heavy compaction method. The sediment percentage of aeolian sand is greater than 8%, the roadbed filling is suitable for compacted by using heavy compaction method to determine the maximum dry density of aeolian sand with different sediment percentage. When the sediment percentage of aeolian sand is range of 16.5%–30.2%, the dry density of aeolian sand is close to maximum dry density and has the best compaction effect, the aeolian sand is a kind of suitable material for roadbed filling.

Keywords: Aeolian sand with different sediment percentage · Compaction characteristics · Heavy compaction method · Roadbed filling

1 Introduction

With the planning and implementation of the western development, Highway construction project in desert area of China will be more and more. Aeolian sand is the most abundant and cheapest material in desert area. It's unknown whether aeolian sand can be used subgrade and roadbed filling material. Chen Zhongda, Zhang Dengliang and other scholars carried out the study of the engineering properties of aeolian sand and achieved some results. But at present the compaction characteristics of aeolian sand with different sediment percentage are seldom studied. Combined with the engineering practice the typical aeolian sand samples have been taken, and seven kinds of aeolian sand with different sediment percentage were prepared (Zhang and Wang 2013). Compaction characteristics of aeolian sand with different sediment percentage has been studied experimentally. Then through comparative analysis, the paper finds the best compaction effect of aeolian sand which under different sediment percentage. Research results will promote aeolian sand with different sediment percentage widely used in highway construction project of desert area. Research results has very important economic significance in highway construction.

R. Chen et al. (eds.), *Proceedings of the 2nd International Symposium on Asia Urban GeoEngineering*, Springer Series in Geomechanics and Geoengineering,
https://doi.org/10.1007/978-981-10-6632-0_1

2 Basic Characteristics of Aeolian Sand

2.1 Basic Characteristics of Natural Aeolian Sand

Through the experiment the dry density of natural Aeolian sand samples is in range of 1.56–1.69 g/cm^3, and relative density of natural Aeolian sand samples is in range of 2.65–2.72 g/cm^3. Its natural water content is under 3.80% generally. The particle size of Aeolian sand samples for more than 90% is about 0.074–0.60 mm. The non-uniform coefficient of Aeolian sand samples is Cu = 2.13, and the gradation coefficient Cc = 1.01. By analyzing the mechanical composition of Aeolian sand particles, it was found that the gradation of aeolian sand was bad, with even fine grains with diameters mostly ranged from 0.074 mm to 0.60 mm.

2.2 Grading Index of Aeolian Sand with Different Sediment Percentage

Aeolian sand samples were prepared with different sediment percentage respectively at 5%, 15%, 30%, 40%, 50%, 60% and 70%. Grading Indexes of prepared Aeolian sand samples were listed in Table 1.

Table 1. Grading indexes of samples

Samples	Effective particle size (mm)	Non-uniform coefficient	Curvature coefficient	Diameter < 0.074
Sample1	0.090	3.500	1.125	5.200
Sample2	0.050	5.400	1.250	16.500
Sample3	0.031	5.150	1.050	30.200
Sample4	0.030	4.670	1.040	41.300
Sample5	0.025	3.640	1.800	49.500
Sample6	0.020	3.600	2.500	59.400
Sample7	0.018	3.700	2.640	68.600

The Table 1 shows that non-uniform coefficient increase first and then decrease with the increasing of sediment percentage. The effective particle size decreased with the increasing of sediment percentage. The values of curvature coefficient in the range of 1–3. The result indicates that mostly Aeolian sand samples were bad Gradation.

3 Compaction Test

3.1 Standard Heavy Compaction Test

3.1.1 Test Equipment and Scheme

Test equipment: heavy compaction instrument, balance System, Mixing tools, Oven and dryer, measuring cylinder, aluminum box, repair knife, flat ruler, and etc.

Test scheme: According to the Ministry of Transport of the People's Republic of China standards "Test Method of Soils for Highway Engineering (JTG E40-2007)", the soils of compaction test has three layers and each layer was impacted 98 times. The compaction hammer falls according to 45 cm drop height. Seven kinds of Aeolian sand with different sediment percentage samples were selected and tested under different water contents.

3.1.2 Experiment Results and Analyses

Seven kinds of Aeolian sand with different sediment percentage samples were prepared to do standard heavy compaction test. The compaction curves were shown in Fig. 1. The relationship of maximum dry densities and the quantity sediment shows in Fig. 2.

As can be seen from Fig. 1, the dry density of aeolian sand with sediment appeared a peak value when water content was zero. The dry density decreases with the increasing of water content. When water content was in range of 4%–6%, the dry density begin to increase until appears to another peak value as water content was about 10%. The compaction characteristics of aeolian sand with sediment is like to aeolian sand. It's typical property of aeolian sand. When sediment percentage was in range of 30.2%–59.4%, the compaction characteristics of aeolian sand with sediment was in the transition zone. The compaction curves show the samples are mixing of aeolian sand and sediments. The dry densities are lower than aeolian sand. When sediment percentage is over 59.4%, the compaction characteristics of the samples obviously are properties of fine grained soils. As water content of the samples is below the optimum water content, the dry density increases gradually with the increasing of water content (Li et al. 2006; Zheng 2014). When water content reaches the optimum water content the dry density reaches maximum. If water content is more than the optimum water content, the dry density decreases. We can also kwon the dry density will decrease with the increasing of sediment percentage. The optimum water content also increase with the increasing of sediment percentage.

In Fig. 2, the maximum dry density increase with the increasing of sediment percentage of aeolian sand. When sediment percentage was about 20%, the maximum dry density reaches the maximum value. Then the dry density gradually decreases with the increasing of sediment percentage. When sediment percentage of aeolian sand is in range of 20%–40%, the large change rate of the maximum dry density is relatively obvious (Yang et al. 2011; Zhang et al. 2015). Sediment percentage of aeolian sand is over 40% then the maximum dry density reducing rate is relatively slow.

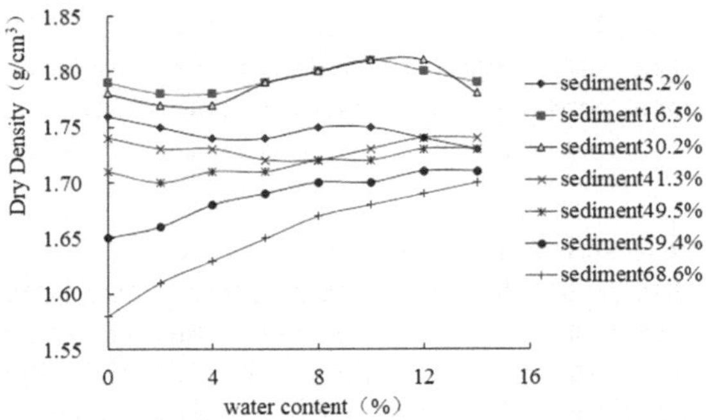

Fig. 1. Relations of maximum dry density and the quantity of sediment

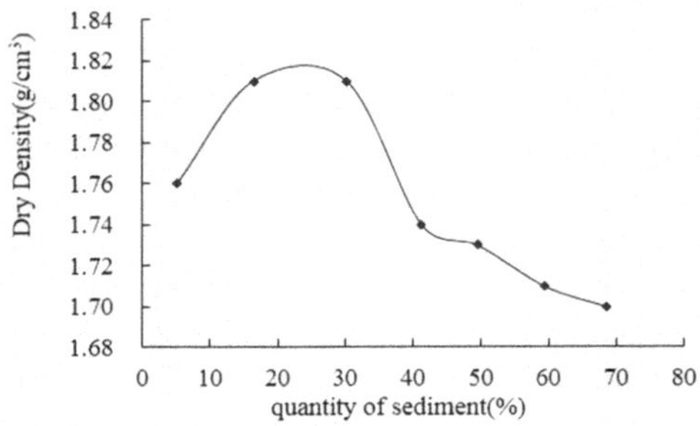

Fig. 2. Heavy compaction curve of seven samples

4 Conclusion

When sediment percentage of aeolian sand is below 30.2%, the compaction properties behave the characteristics of aeolian sand. When sediment percentage is over 59.4%, the compaction properties of the samples obviously show characteristics of fine grained soils. When sediment percentage was in range of 30.2%–59.4%, the compaction characteristics of aeolian sand with sediment showed a mixture of aeolian and sediment. The dry density reaches to the maximum value when sediment percentage of aeolian sand is 20%.

It's sediment percentage of aeolian sand is greater than 8%, the roadbed filling is suitable for compacted by using heavy compaction method to determine the maximum dry density of. When the sediment percentage of aeolian sand is range of 16.5%–30.2%, the dry density of aeolian sand is close to maximum dry density and has the best compaction effect, the aeolian sand is a kind of suitable material for roadbed filling.

Acknowledgement. This work is funded by Science and technology research project of Chongqing Education Commission (KJ1501332).

References

Zhang, H., Wang, Z.: Research on dynamic compaction characteristics of Aeolian sand in desert region of Keerqin. Rock Soil Mech. **25**(2), 45–49 (2013)

Li, Z., Cao, Y., Liang, N.: Compaction mechanism of aeolian sand. China J. Highw. Transp. **9**(5), 6–11 (2006)

Zheng, C.: Expansive soil slope stability analysis affected by many factors. J. Railw. Sci. Eng. **11**(1), 82–87 (2014)

Yang, R., Zeng, J., Lin, D.: Compacting mechanism and features of aeolian sands. J. Chang'an Univ. (Natural Science Edition) **15**(4), 105–110 (2011)

Zhang, H., Li, Z., She, X.: Compaction and compression characteristics of different levels of aeolian sand in Maowusu area. J. Arid Land Resour. Environ. **29**(8), 167–172 (2015)

Dynamic Effect in Capillary Pressure – Saturation Relationship Using Lattice Boltzmann Simulation

Guanxi Yan[1(✉)], Zi Li[1], Thierry Bore[1], Sergio Galindo-Torres[2], Alexander Scheuermann[1], and Ling Li[2]

[1] Geotechnical Engineering Centre, School of Civil Engineering, The University of Queensland, St Lucia, QLD, Australia
g.yan@uq.edu.au
[2] National Centre for Groundwater Research and Training, The University of Queensland, St Lucia, QLD, Australia

Abstract. Soil water retention curve (SWRC) as the constitutive relationship of hydro-mechanical coupling bridge of unsaturated soil has been experimentally investigated decades regarding impacts from dynamic effects in sandy soil and deformation in soft soil. However, due to inaccessibility of observation of microscale dynamic capillary behavior in geotechnical testing scale, most of the experimental methods can only provide the deviation between static SWRC and dynamic SWRC on drainage and imbibition. With the development of Computational Fluid Dynamic (CFD) simulation, several numerical methods so far can be utilized to investigate the soil water retention behavior in both micro- and macro-scale and upscaling between them, such as pore network model, Navier-Stokes integrated with Volume of Fluid and Level Set Method. Nevertheless, none of them provide a vision of interaction between fundamental fluid fractions in order to replicate the physical behavior of fluid tension from their mathematical expression. Compared to those CFD methods, Lattice Boltzmann Methods (LBM) is formulated on microscale for simulation of fluid dynamics. In addition to the interaction forces between fluid-fluid and fluid-solid phases, it fundamentally replicates the physical meaning of immiscible multiphase flow behavior in porous media. Therefore, LBM is selected to investigate the dynamic effect in soil water retention behavior in this study. The aim is to investigate the dynamic capillary pressure (soil suction in soil mechanics) varying with the saturation of each phase in a Representative Elementary Volume (REV) domain.

Keywords: Multiphase flow · Porous media · Dynamic effect · Capillary pressure · Saturation

1 Introduction

The dynamic two-phase flow seepage in porous media was theoretically extrapolated based on mass conservation (continuity equation), momentum conservation (Darcy's law) and fundamental constitutive relationships (capillary pressure/relative

R. Chen et al. (eds.), *Proceedings of the 2nd International Symposium on Asia Urban GeoEngineering*, Springer Series in Geomechanics and Geoengineering,
https://doi.org/10.1007/978-981-10-6632-0_2

permeability-saturation relations) in describing fluid-fluid retention behaviour for a Representative Elementary Volume (REV) (Buckley and Leverett 1942; Jacob 1972; Richards 1931). Therefore, the prediction of transient two-phase flow in porous media is dependent on the experimental determination of fluid-fluid retention behavior and hydraulic boundaries applied on simulating domain. Few experimental and theoretical studies from previous literatures demonstrated the velocity-dependence of capillary pressure-saturation (Abidoye and Das 2014; Barenblatt and Vinnichenko 1980; Chen 2006; Das and Mirzaei 2012; Hassanizadeh et al. 2002; Kalaydjian 1987; Mirzaei and Das 2007; O'Carroll et al. 2005; Sakaki et al. 2010; Topp et al. 1967), but those works were based on the framework of continuum mechanics without additionally understanding of multiphase physics in micro-scale. With an aim to further discover two-phase transient flow in porous media based on the microscale Computational Fluid Dynamic (CFD) simulation, the Multiphase Multicomponent Lattice Boltzmann Method (LBM) is selected as an effective approach (Galindo-Torres et al. 2013; Galindo-Torres et al. 2013; Porter et al. 2009; Schaap et al. 2007; Scheuermann et al. 2014). In comparison with other CFD methods solving Naiver-Stokes equation (Ferrari 2014; Ferrari et al. 2015; Ferrari and Lunati 2013; Ferrari and Lunati 2014), Multiphase LBM own fundamental physical meanings on the fluid-fluid interface (Shan and Chen 1993).

In this study, the Shan-Chen LBM were implemented for simulating the dynamic two-phase flow in porous media and interrelationships between each pair of macroscale state variables are analyzed for further interpreting dynamic effects. According to the findings, Multiphase LBM was demonstrated as a powerful tool for investigating dynamic multiphase flow seepage. Furthermore, the research orientation for future is proposed with discussions on a few limitations in LBM model.

2 Numerical Method

2.1 Shan-Chen Multiphase Lattice Boltzmann Model

Lattice Boltzmann Method (LBM) is a method to numerically approximating the solution of Boltzmann equation by assuming statistical mechanics of microscale particles inside the single lattice node. Compared to computational fluid dynamics, which solves the problem in control volume (Naiver-Stoke equation), LBM is more fundamental because it directly provides a solution from a level of particle clouds described by statistics. LBM cannot clarify the density and momentum of a single fluid particle (the fluid molecule in most case) but assign a statistical distribution between a central node and surrounding nodes to describe the probability of the density of a cloud of particles inside a specified lattice cell (Sukop 2006). A D2Q9 lattice cell is shown in Fig. 1. The microscale velocity of a single node is shown in the center of each lattice. This is like upscaling molecule scale to the specified lattice scale, and then approximate Boltzmann equation on a domain made up of these lattices, using computational iteration to determine the density and velocity of the central node of the single lattice cell for each time step. To reconstruct the particles streaming and collision mechanism in a domain, the interaction between each lattice is given by a relaxation behavior towards an equilibrium distribution of particles (Eq. (3)).

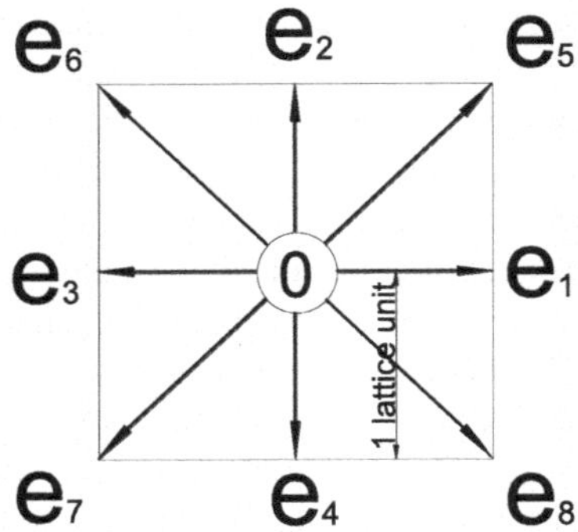

Fig. 1. D2Q9 scheme of LBM

In this experimental study, LBM is just taken as a numerical experiment to explore the dynamic effect. The D2Q9 SCLBM engine is available in Mechsys (Multi-physics Simulation Library, developed by Dr. Sergio Galindo Torres at School of Civil Engineering, the University of Queensland). The construction of this Shan-Chen LBM (SCLBM) engine is listed in Table 1. The explanation of LBM in details can be sourced from Sukop (2006).

Equations (1)–(4) give the standard LBM using BGK equilibrium distribution function. Shan-Chen LBM added inter-particle pseudo forces (Eq. (5)), multiphase repulsive force (Eq. (7)) and particle-solid interaction force (Eq. (8)) to simulate compressibility of each fluid phase and interface (Shan and Chen 1993). Due to introducing these lattice interaction forces, velocity for equilibrium distribution function has to be locally averaged using Eq. (9). In the single component multiphase system, there is only in need of interparticle attractive force (Eq. (5)). For the multi-component, multiphase system, Eq. (5) is modified to Eq. (8) to conduct attraction between wetting phase and solid particles and repulsion between non-wetting phase and solid particles. When the lattice cell contains solid nodes on one side of four edges, to generate the solid adsorption behavior, an imbalance force between fluid-fluid and fluid-solid is given by setting function distributions to unity instead of density.

Table 1. The construction of D2Q9 Shan-Chen LBM simulation engine

D2Q9 Cell	Lattice Fluid Density: $\rho = \sum_{i=0}^{8} f_i$ (1) Lattice Fluid Velocity: $\mathbf{u} = \frac{\sum_{i=0}^{8} f_i \mathbf{e_i}}{\rho}$ (2)
Governing Equation	$\overbrace{f_i(\mathbf{x}+\mathbf{e}_i\Delta t, t+\Delta t) - f_i(\mathbf{x}, t)}^{\text{Streaming}} = \overbrace{\frac{\left[\overbrace{w_i\rho(\mathbf{x})\left(1+3\frac{\mathbf{e_i}\cdot\mathbf{u}}{C}+4.5\frac{(\mathbf{e}\cdot\mathbf{u})^2}{C^4}-1.5\frac{\mathbf{u}^2}{C^2}\right)}^{\text{Bhatnagar-Gross-Krook(BGK)approximation of } f_i^{eq}} - f_i(\mathbf{x}, t)\right]}{\underbrace{3\frac{\upsilon}{c_s}+\frac{1}{2}}_{\text{Relaxiation timer} > 0.5}}}^{\text{Collision}}$ (3)
Body Force Equation	Newton Second Law: $\mathbf{u^{eq}} = \mathbf{u} + \frac{\tau\mathbf{F}}{\rho}$ (4)
Particles interaction Force	$F_a = -G_a\psi_j(\mathbf{x}, t)\sum_{i=1}^{8} w_i\psi_j(\mathbf{x}+\mathbf{e_i}\Delta t, t)\mathbf{e_i}$ (5)
LBM EOS	$P = \frac{\rho}{3} + \frac{G}{6}\psi_0^2\exp(\frac{-2\rho_0}{\rho})$ (6)
Immiscible Multiphase Equations	Immiscible fluids repulsive force: $F_r = -G_r\rho_j(\mathbf{x}, t)\sum_{i=1}^{8} w_i\rho_{j'}(\mathbf{x}+\mathbf{e_i}\Delta t, t)\mathbf{e_i}$ (7) Solid-Fluids interaction force: $F_s = -G_s\psi_j(\mathbf{x}, t)\sum_{i=1}^{8} w_i s(\mathbf{x}+\mathbf{e_i}\Delta t, t)\mathbf{e_i}$ (8)
Correction Equation for f_i^{eq}	Velocity Correction: $\mathbf{u} = \frac{\sum_\sigma \frac{1}{\tau_\sigma}\sum_{i=0}^{8} f_i^\sigma \mathbf{e_i}}{\sum_\sigma \frac{\rho_\sigma}{\tau_\sigma}}$ (9) Density for each phase: $\rho_\sigma = \sum_{i=0}^{8} f_i^\sigma$ (10)
Pressure of single cell	$P(x) = \frac{\rho_j(x) + \rho_{j'}(x) + G_r\rho_j(x)\rho_{j'}(x)}{3}$ (11)
Young-Laplace Equation for SCLBM	Capillary pressure: $P_c = P_{j'} - P_j = \frac{2\sigma_s}{R}$ (12) Contact angle: $\cos\theta = \frac{G_s^{j'} - G_s^{j}}{G_r}$ (13)

where f_i = distribution function for direction i projecting from central node of single lattice, e_i = discretised velocity shooting from central node to surrounding nodes of single lattice, f_i^{eq} = equilibrium distribution function, x = the position vector of single node, t = time, τ = relaxation time (better close to 1), $C = \Delta t/\Delta x$ (Δt = 1 *ts*, Δx = 1 *lu*, $v = \mu/\rho$ kinematic viscosity, c_s = speed of sound of lattice ($C/3^{0.5}$), u^{eq} = macroscale fluid Velocity in equilibrium, F_a = body force (it could be Gravity only if there is no other action), G_a = attractive intensity between fluid particles of single phase, ψ_j = interaction potential of phase j, w_j = weight factor of phase j for each node in single cell ($w_0 = 4/9$, $w_{1\text{-}4} = 1/9$, $w_{5\text{-}8} = 1/36$), s = a 'switch' value usually to be density but is switched to 1 if zero velocity solid boundary are encountered, ψ_0 = constant initial interaction potential, ρ_0 = constant initial macroscale density, F_r = immiscible phase repulsive force, G_r = two phase repulsive intensity (positive value), F_s = fluid-solid interaction force (attraction for wetting phase j and repulsion for non-wetting phase j'), G_s = fluid-solid attractive/repulsive intensity, ρ_j = macroscale density of wetting phase j, $\rho_{j'}$ = macroscale density of non-wetting phase j', σ_s = surface tension, Θ = contact angle, G_s^j = wetting phase fluid-solid attractive intensity (negative value), $G_s^{j'}$ = non-wetting phase fluid-solid repulsive intensity (positive value).

2.2 Simulation Setup

For simulating two-phase flow in 2D porous media, a package of perfect circular disk is filled into a 500*500 square of lattice 2D LBM domain as shown in Fig. 2. The cyan background is the void space, where the two-phase flow is percolating through the colorful disks representing the granular soil particles in this domain. This setup aims to replicate the phase displacement phenomenon, so the right and left boundary conditions are fixed to solid as same as the outline of each grain using bounce back boundary conditions in Sukop and Thorne (2006)

$$f_i = f_{-i} \tag{14}$$

where $-i$ represents the distribution function reverse its direction after a fluid collision with a solid boundary. The top and bottom boundary conditions are directly controlled by density difference without any buffer zone, which represents pressure difference using LBM EOS transformation (Eq. (11)). Wetting or non-wetting phase fluid is fully filled into void space as the initial condition for drainage or imbibition simulation.

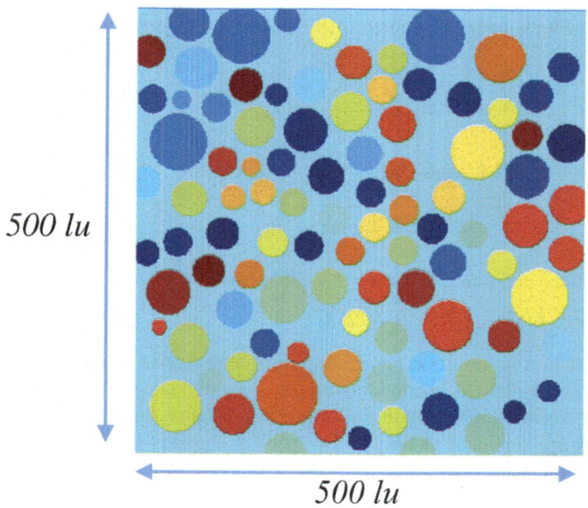

Fig. 2. A package of circular disk in 500*500 lu^2 2D lattice domain

Grain Size Distribution (GSD) is measured using the Image Toolbox built in MATLAB© R2014b. Pore Size Distribution (PSD) is measured using the image processing algorithm developed by Rabbani et al. (2014) The GSD and PSD of virtual soil are shown in Fig. 3. According to Fig. 3, the PSD is wider than GSD, and mean pore size is larger than the mean grain size, so the packing condition is not identical to natural soil but similar to glass beads package with a lot of interconnected pore channels.

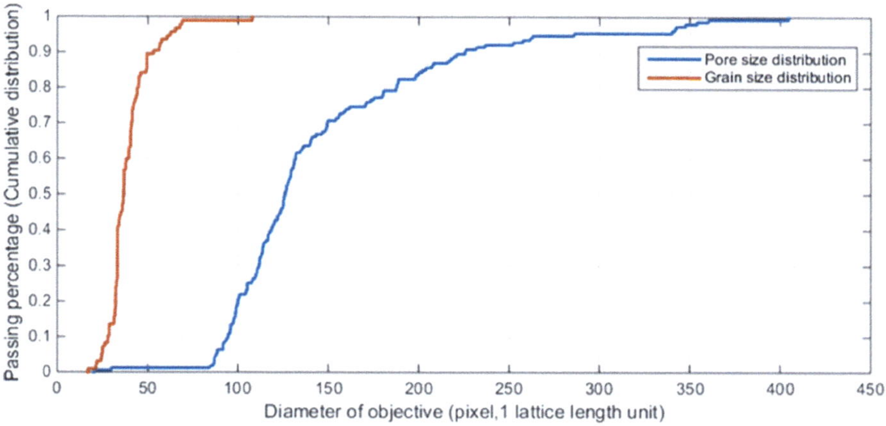

Fig. 3. Pore size distribution and Grain size distribution of virtual grains in 2D LBM simulation

The parameters for this simulation setup is given in Table 2. According to this set of parameters, the lattice surface tension is determined at 0.1818 mu/Δts^2 using bubble simulation, and the contact angle is set to zero calculated by Eq. (13). Once the simulation initiates, non-wetting phase or wetting phase drives into void space by density difference boundary between top and bottom, until there is an equilibrium achieved between such density difference and interfacial meniscus. In this state, the single data point on static Capillary Pressure – Saturation (P_c-S_w) is derived. By applying density difference from zero to a value that can drain each phase out of the void domain, various capillary equilibrium condition can be produced to the corresponding saturation of wetting phase measured by counting pixels occupied by wetting phase fluid.

Table 2. Lattice-scaled parameter values used in numerical experiment

Lattice-scaled Parameters	Value
Wetting phase fluid density	2.0
Non-wetting phase fluid density	2.0
Dynamic viscosity of wetting phase fluid	0.16
Dynamic viscosity of non-wetting phase fluid	0.16
Lattice unit length	1.0
Lattice time step	1.0
Two phase repulsive intensity	1.0
Wetting fluid attractive intensity	−0.5
Non-wetting fluid attractive intensity	0.5
Gross simulation time	500000
Time interval	5000
Density difference/Initial density	0% – 80%
Initial density for two-phase fluids	2.0
Infinitesimal density for each other phase for mixing	0.0001

As for post data analysis, the results of the simulation are given in four arrays: wetting phase density, wetting phase velocity, non-wetting phase density, and non-wetting phase velocity. The macroscale pressure of each fluid phase is determined using Eq. (11) in consideration of a small amount of the other phase mixing inside (Galindo-Torres et al. 2013). Finally, the capillary pressure (P_c) can be given by the difference between pressures of two mixing areas (two fluid phases) using Eq. (12). The saturation of wetting fluid phase (S_w) can be determined by counting the lattice nodes (Eq. (15)) when local density is larger than half of initial density. This density threshold rule has been applied by Porter et al. (2009) and Galindo-Torres et al. (2013). As for the measurement of the specific interfacial area (a_{nw}), the area of each fluid phase and solid phase can be given by counting the pixel occupied by interfaces between each pair of phases. Finally, the specific interfacial area is calculated using Eq. (16). With known capillary pressure, saturation for both two-phase fluids and interfacial area of the two-phase meniscus on the end of each time interval, the static $Pc\text{-}S_w\text{-}a_{nw}$ or dynamic $Pc\text{-}S_w\text{-}a_{nw}$ can be extracted from equilibrium approached and transient flow conditions (Table 3).

Table 3. Upscaling post analysis for determining the microscale variables

Macroscale state variables (REV scale)	Upscaling equations
Saturation (S_w)	wetting phase: $S_w = \frac{\sum_{i=1}^{N_w} A_{i,w}}{\sum_{i=1}^{n_T} A_i}$ (15)
Interfacial area (A_{nw})	Specific interfacial area: $a_{nw} = \frac{\sum_{i=1}^{N_w} A_{i,w} + \sum_{i=1}^{N_{nw}} A_{i,n} - \sum_{i=1}^{n_{nw}} A_{i,s}}{2\sum_{i=1}^{n_T} A_i}$ (16)

3 Result and Analysis

3.1 Performance of SC-LBM Simulation for One-Step in/Outflow

According to Table 4, there is an observation that with the pressure difference increase between the top and bottom boundary, more wetting phase is pushed out the beads package, and more non-wetting phase invaded the domain. When the density difference on two boundaries achieves 0.4 mu/lu^2, a preferential path is generated in the center of virtual soil. The wetting phase fluid is separated into two portions on two corners of bottom separately. In this simulation, to reduce computational effort, the initial density and viscosity ratio are both set at one as listed in Table 2. Therefore, such a preferential path is not caused by different flow patterns, but geometrical determined, where the larger pore throat, where the non-wetting phase breaks through the entry pressure first.

According to the Table 5, the imbibition process shows more preferential paths on the left side of the domain because the pore throats on the left are smaller than the right

Table 4. Visualization of 2D LBM simulation of drainage

Density difference boundary	Equilibrium condition of wetting phase	Equilibrium condition of non-wetting phase
Initial condition	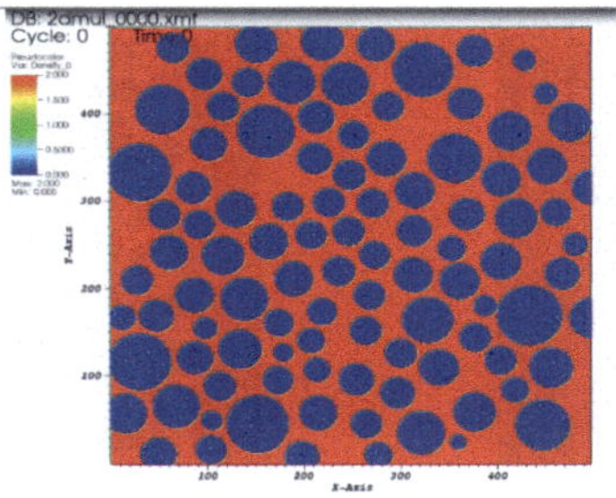	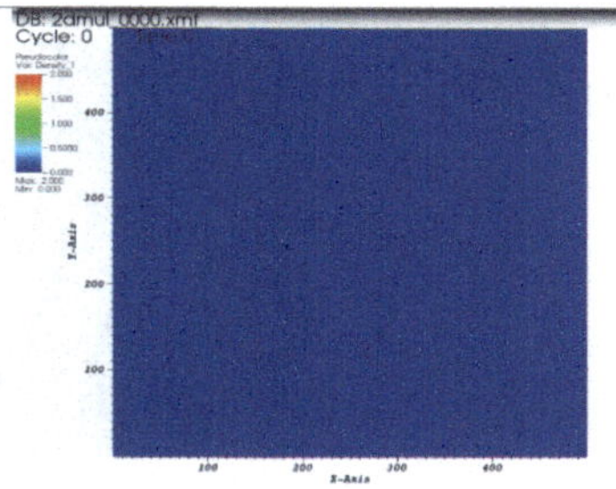
0	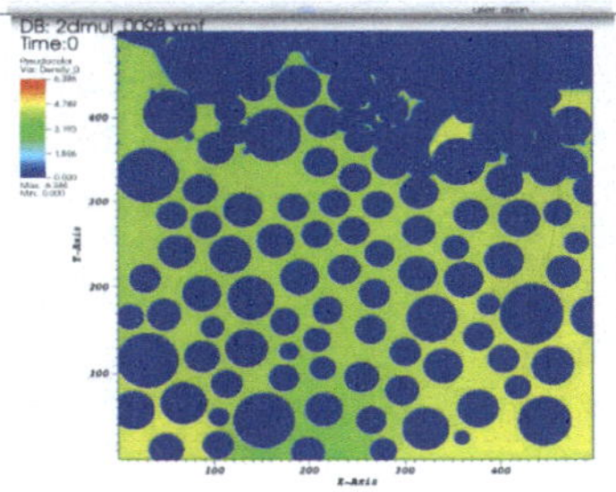	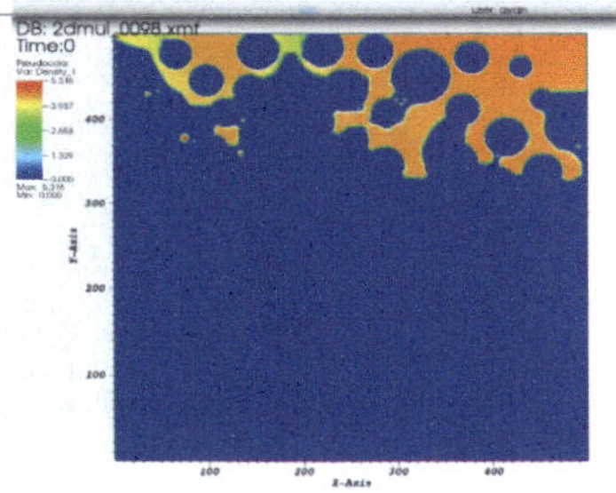
0.4	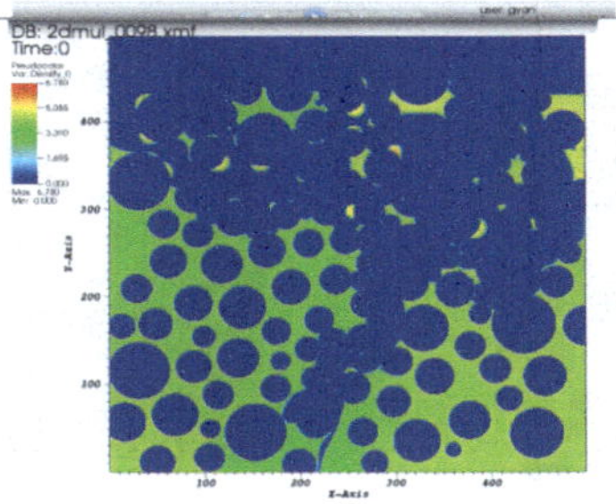	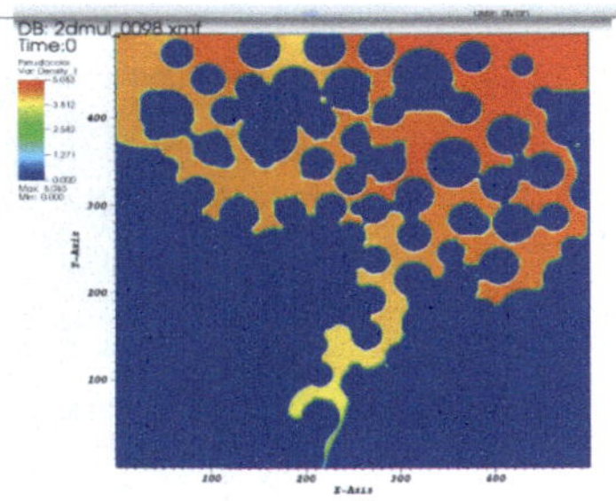
1.2	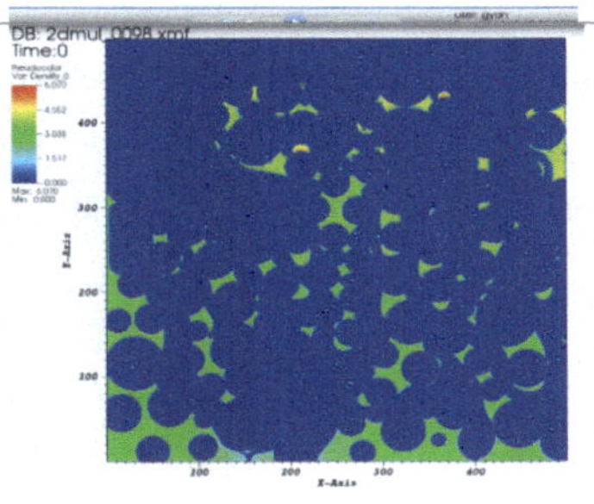	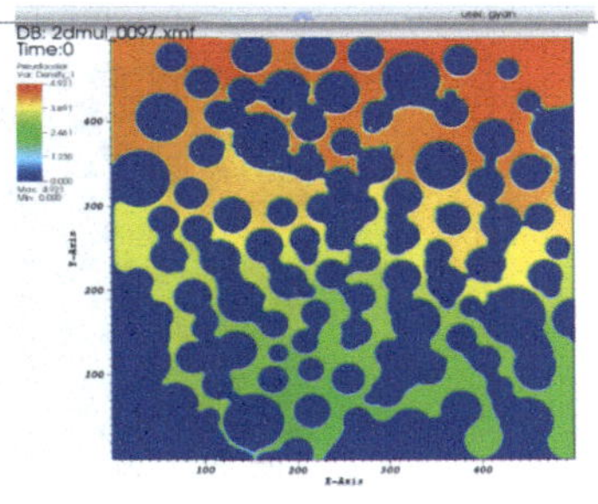

Table 5. Visualization of 2D LBM simulation of imbibition

Density difference boundary	Equilibrium condition of wetting phase	Equilibrium condition of non-wetting phase
Initial condition	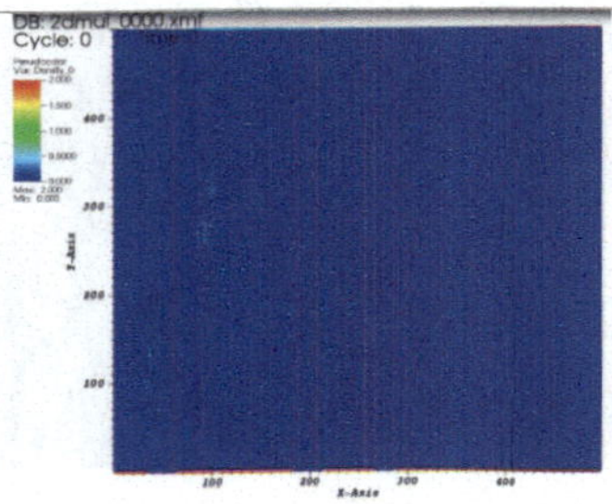	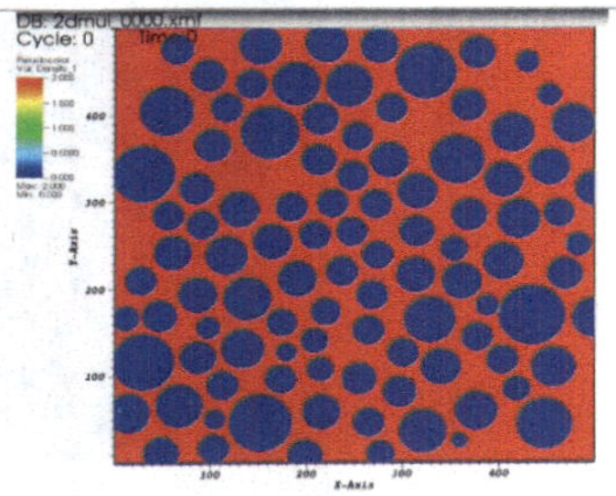
0	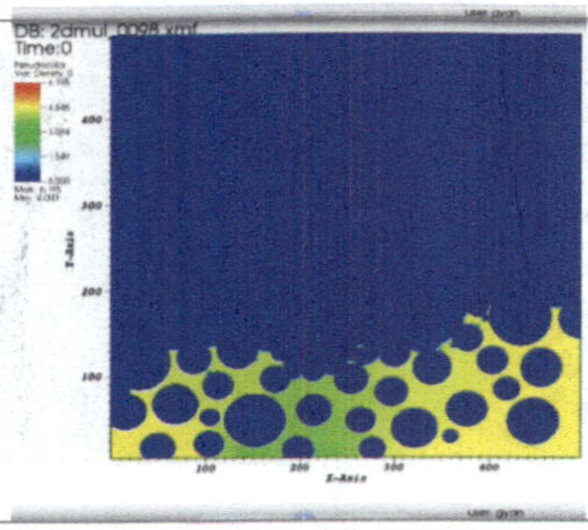	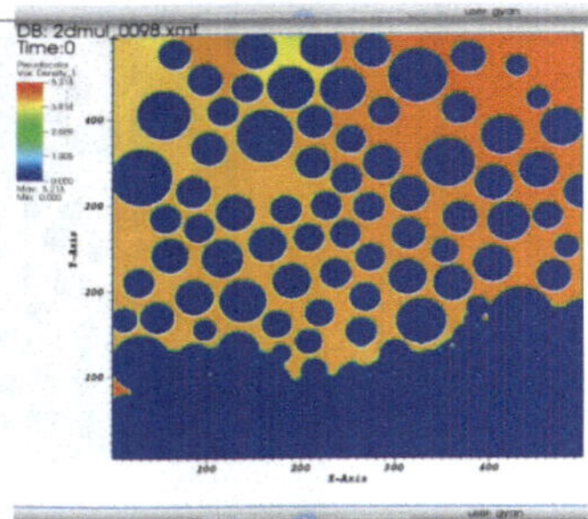
0.2	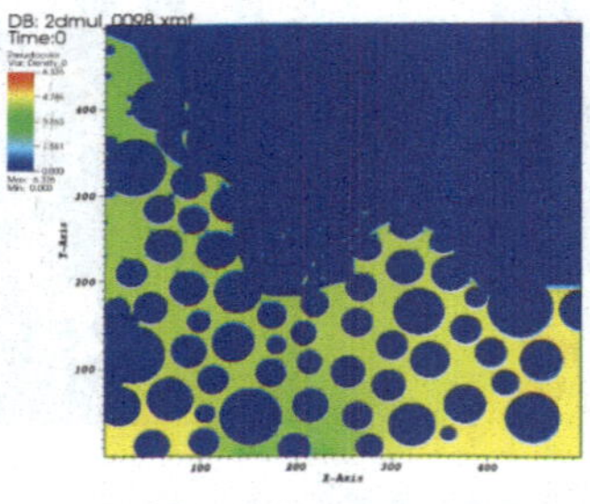	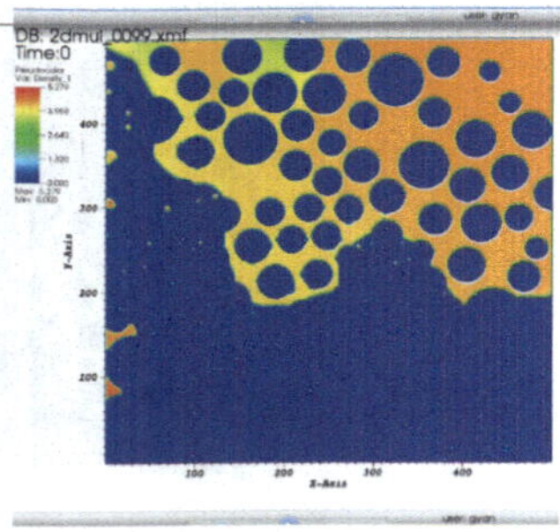
1.6	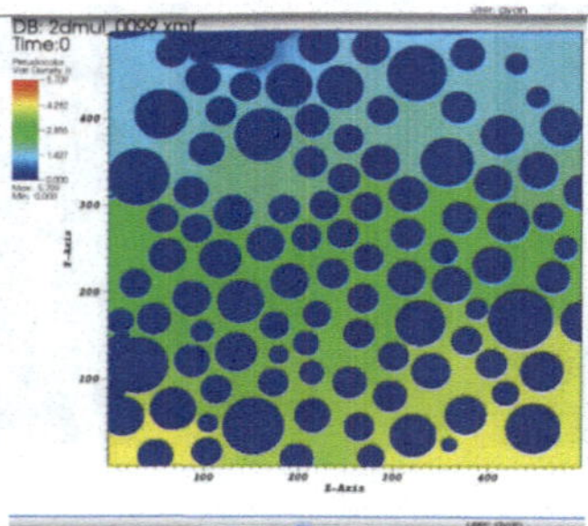	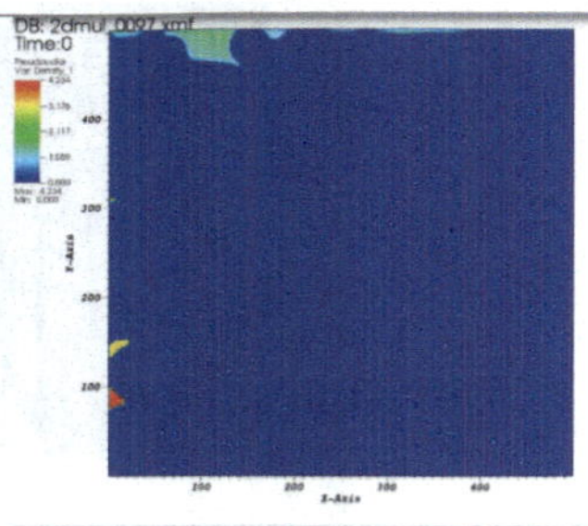

side. Also, a small amount of wetting phase trapping and non-wetting phase trapping can be found. However, this trapping is only a little in the domain. When the final equilibrium achieved, most of the trapping phase slowly evaporates, causing a slight decrease of each trapping phase fluid. This simulating performance is consistent with Schaap et al. (2007). Schaap et al. (2007) claimed their confidence on their results having such issues. Here, we follow their recommendations on neglecting the disappearance of phase trapping. The visualization in Tables 4 and 5 demonstrates that the SCLBM physically replicates capillary flow in a granular porous media, and the observation is consistent with two-phase displacement in natural porous media.

3.2 Dynamic Effects in Capillary Pressure-Saturation Curves

The static and dynamic P_c-S_w simulated using SCLBM are plotted in Fig. 5(a). Each data point is taken for plotting static P_c-S_w after equilibrium achievement. The equilibrium of static P_c-S_w is ensured by checking the S_w varying with lattice time step shown in Fig. 4. After $2*10^5$ time steps, the equilibrium is approached already with a

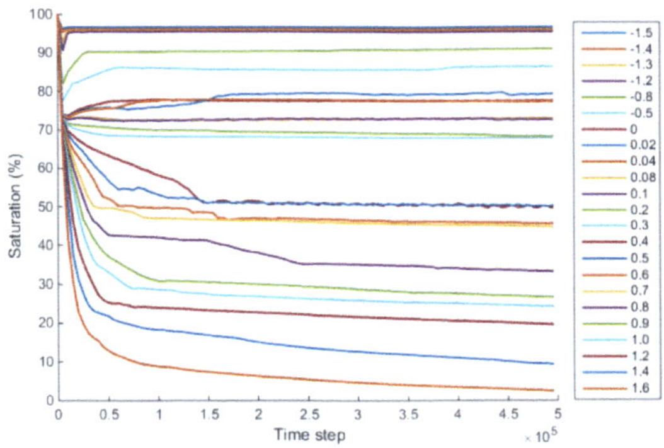

(a)

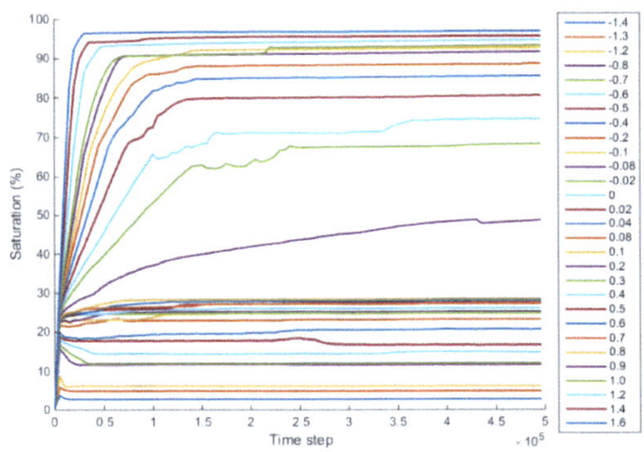

(b)

Fig. 4. Historical logging of Saturation for various density boundary conditions listed in legend: (a) Drainage, (b) Imbibition slight fluctuation caused by numerical artifact on meniscus instability of wetting front. The dynamic responses under transient flow conditions are more significantly within a first $0.5*10^5$ time step. Within this period, the dynamic P_c-S_w-a_{nw} can be plotted, as can be checked from Fig. 5(c).

Except for some anomaly for bubble point and residual trapping saturation of drainage P_c-S_w, the results shown in Fig. 5 replicate the primary static drainage and imbibition P_c-S_w highlighted in yellow zone in Fig. 5, and the area of dynamic P_c-S_w highlighted in dark blue zone in Fig. 5. Also, having the plot of P_c-S_w corresponding to each time frame of simulation, the dynamic P_c-S_w can be checked through a simulation

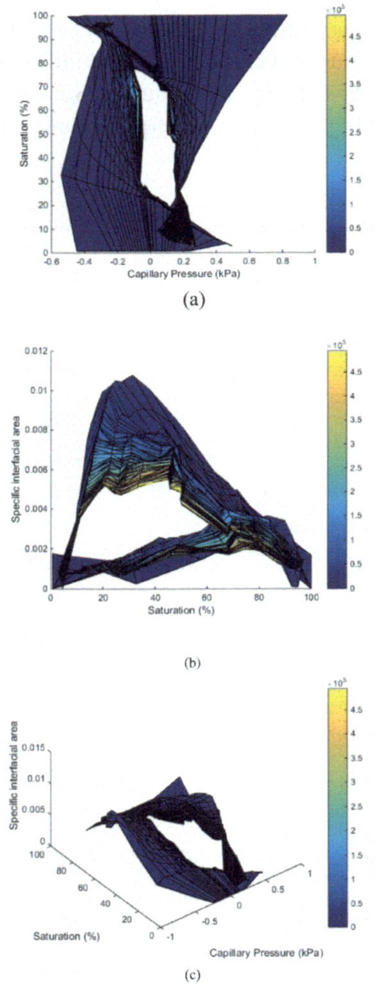

Fig. 5. Dynamic Pc-Sw-anw relationship with contour color bar marking the time step: (a) the projection of 3D contour plot of Pc-Sw-anw on 2D Pc-Sw plane, (b) the projection of 3D contour plot of Pc-Sw-anw on 2D Sw-anw plane, (c) Dynamic Pc-Sw-anw relationship

of transient two-phase displacement in the virtual soil. This is a demonstration of dynamic capillary effect in P_c-S_w for a transient two-phase seepage, which is missed in our conventional theory just requiring achievement of equilibrium on every time step instantaneously.

These results are the first effort on studying dynamic multiphase flow in porous media using 2D SCLBM. Although the model setup is simple, it is still a good beginning for applying SCLBM to investigate dynamic effects in immiscible fluid retention behavior under transient flow conditions. Due to the dimensional constraints, the simulation result temporarily can only be seen as a straightforward proof. The 2D setup leads to less phase trapping, and large pore throats as shown in Tables 4 and 5 additionally result in low capillary entry pressure or even positive pore pressure for imbibition process as shown in Fig. 5(a). In the future, quasi-2D and 3D setup are worth to be implemented for a better replication of physical testing conditions.

3.3 Dynamic Effects in Capillary Pressure-Saturation-Specific Interfacial Area

The other interesting theoretical framework in the advanced theory of multiphase flow in porous media is the uniqueness of P_c-S_w-a_{nw} constitutive surface proposed by Hassanizadeh and Gray (1993b). Porter et al. (2009) confirmed the uniqueness of the static P_c-S_w-a_{nw} surface using 3D SCLBM. However, there is no any information about the dynamic effect for this constitutive surface under transient flow using SCLBM. Hence, the 2D SCLBM is applied here to discover the uniqueness of surface under transient condition. Karadimitriou et al. (2014) experimentally investigated this constitutive surface under transient condition using an elongated PDMS micromodel, and find that this surface is not unique under transient flow condition. Also, it is Capillary number dependent, which means that it depends on seepage velocity. Joekar-Niasar et al. (2010) also studied this surface under transient condition using numerical dynamic pore network model and concluded that due to the each phase fluid losing continuity in a transient condition, there is a significant discrepancy between equilibrium and nonequilibrium constitutive surface. For this numerical experiment, the evolution of these three state variables is plotted in Fig. 5.

From Fig. 5(b), there is a clear temporal evolution of wetting phase saturation-interfacial area. Checking through the color bar representing the time step, it is straightforward to find that there is less variation of saturation-interfacial area relationship after $1*10^5$ time steps. Beforehand, there is a different saturation-interfacial area plot within each time step. In Fig. 5(b), the curves for drainage is located on the upper side while the curves for imbibition is located downside. The saturation-interfacial area relationship follows different paths for hydraulic loading history. This also replicates the physical phenomenon that meniscus area is pulled larger for drying than for wetting process in which the curvature of the meniscus alleviated by fluid inertial (Hoffman 1983). It is also an explanation that there is no relationship between the dynamic contact angle of the meniscus and static hysteresis of SWRC because the dynamic contact angle contributes to the capillary pressure difference between dynamic and static capillary pressure (Hassanizadeh and Gray 1993a), but the data points of static hysteresis SWRC is taken on equilibrium state.

The Fig. 5(c) shows the variation of the 3D capillary pressure-saturation-interfacial area with time forwarding to the next equilibrium state for each density difference boundary condition. The temporal evolution of imbibition is on the left side while the other temporal evolution of drainage is on the right side. As can be seen from the color bar of temporal evolution in Fig. 5(c), it is evident that the equilibrium can only be achieved when there is no variation of the specific interfacial area. Most of the dynamic variation of capillary pressure and the interfacial area is located within $0–1*10^5$ time steps. Only within this time scale, there is significant saturation varying with time, indicating that there is real flux seepage through this assumed Darcy scale REV. Once the equilibrium achieved, the static SWRC can be plotted, but there is no flow at all. When the dynamic process is ongoing from one state to next state, the capillary pressure-saturation-interfacial area is continuously changing with time steps. There is no any uniqueness of SWRC and the 3D constitutive surface for the evolution of the state.

4 Conclusion and Reflection

In this study, the Shan-Chen Lattice Boltzmann simulation is conducted to model two-phase flow displacement in a granular porous media. According to the simulation performance, it can conclude that SCLBM is a powerful tool to investigate multiphase hydrodynamic seepage problems for unsaturated flow in the vadose zone and oil reservoir enhancement. The results manifest that the SCLBM can replicate the dynamic effects in fluid-fluid retention behavior. Also, the capillary pressure–saturation relationship shows a transitional behavior between initial states and next equilibrium state under transient flow conditions. However, according to the conventional theory, the transition from one state to the next is assumed to follow the capillary pressure-saturation curve only in the static state. Also, the larger the density boundary (pressure boundary of LBM) applied, the larger the dynamic capillary pressure response and larger interfacial area variation appear during a transient process. These findings might lead to a further question on how the effective hydraulic conductivity can be evaluated based on the static P_c-S_w using any empirical and statistical Hydraulic Conductivity Function (HCF) models. According to the nonlinear diffusion theory, which is the mathematical form of Richards' equation, the temporal evolution of the volumetric water content should be equal to the divergence of the flux in 3D space. If the flux for each dimension cannot be approximated using the hydraulic conductivity predicted by static P_c-S_w, the purpose of using such theory somehow becomes questionable. More interestingly, so far the HCF models are all correlated to only the static drainage P_c-S_w by adopting phase fraction network assumptions. Under the transient flow condition, there is no guarantee for this continuous phase flowing through the fraction they occupied. The dynamic P_c-S_w is an indication of the dynamic response of two-phase displacement in porous media. The application on the estimation of seepage flux so far has not been investigated yet. Each dynamic P_c-S_w only exists within evolution, and finally, return to static one. Incorporating dynamic relationships into Richard's equation, therefore, seems vague on explaining the physical meaning of HCF. For further investigating the dynamic multiphase flow, flow governing equation in regards to relative permeability concept for calculating two-phase seepage flux should be further studied using SCLBM.

Acknowledgements. The first author wants to acknowledge the support from the University of Queensland International Scholarship (UQI). The simulations were implemented using the Mechsys open source library on the Macondo high-performance cluster, currently merged into Goliath supercluster, at the University of Queensland.

References

Abidoye, L.K., Das, D.B.: Scale dependent dynamic capillary pressure effect for two-phase flow in porous media. Adv. Water Resour. **74**, 212–230 (2014)

Barenblatt, G.I., Vinnichenko, A.: Non-equilibrium seepage of immiscible fluids. Adv. Mech. **3** (3), 35–50 (1980)

Buckley, S.E., Leverett, M.: Mechanism of fluid displacement in sands, 146(01), 107–116 (1942)

Chen, L.: Hysteresis and dynamic effects in the relationship between capillary pressure, saturation, and air-water interfacial area in porous media (2006)

Das, D.B., Mirzaei, M.: Dynamic effects in capillary pressure relationships for two-phase flow in porous media: Experiments and numerical analyses. AIChE J. **58**(12), 3891–3903 (2012)

Ferrari, A.: Pore-scale modeling of two-phase flow instabilities in porous media. University of Torino (Italy) (2014)

Ferrari, A., Jimenez-Martinez, J., Borgne, T.L., Méheust, Y., Lunati, I.: Challenges in modeling unstable two-phase flow experiments in porous micromodels. Water Resour. Res. **51**(3), 1381–1400 (2015)

Ferrari, A., Lunati, I.: Direct numerical simulations of interface dynamics to link capillary pressure and total surface energy. Adv. Water Resour. **57**, 19–31 (2013)

Ferrari, A., Lunati, I.: Inertial effects during irreversible meniscus reconfiguration in angular pores. Adv. Water Resour. **74**, 1–13 (2014)

Galindo-Torres, S., Scheuermann, A., Li, L., Pedroso, D., Williams, D.: A lattice boltzmann model for studying transient effects during imbibition–drainage cycles in unsaturated soils. Comput. Phys. Commun. **184**(4), 1086–1093 (2013)

Galindo-Torres, S., Scheuermann, A., Pedroso, D., Li, L.: Effect of boundary conditions on measured water retention behavior within soils. In: AGU Fall Meeting Abstracts, p. 1516 (2013)

Hassanizadeh, S.M., Celia, M.A., Dahle, H.K.: Dynamic effect in the capillary pressure–saturation relationship and its impacts on unsaturated flow. Vadose Zone J. **1**(1), 38–57 (2002)

Hassanizadeh, S.M., Gray, W.G.: Thermodynamic basis of capillary pressure in porous media. Water Resour. Res. **29**(10), 3389–3405 (1993a)

Hassanizadeh, S.M., Gray, W.G.: Toward an improved description of the physics of two-phase flow. Adv. Water Resour. **16**(1), 53–67 (1993b)

Hoffman, R.L.: A study of the advancing interface: II. Theoretical prediction of the dynamic contact angle in liquid-gas systems. J. Colloid Interface Sci. **94**(2), 470–486 (1983)

Jacob, B.: Dynamics of Fluids in Porous Media. Dover Publications, New York (1972)

Kalaydjian, F.: A macroscopic description of multiphase flow in porous media involving spacetime evolution of fluid/fluid interface. Transp. Porous Media **2**(6), 537–552 (1987)

Karadimitriou, N., Hassanizadeh, S., Joekar-Niasar, V., Kleingeld, P.: Micromodel study of two-phase flow under transient conditions: Quantifying effects of specific interfacial area. Water Resour. Res. **50**(10), 8125–8140 (2014)

Mirzaei, M., Das, D.B.: Dynamic effects in capillary pressure–saturations relationships for two-phase flow in 3D porous media: Implications of micro-heterogeneities. Chem. Eng. Sci. **62**(7), 1927–1947 (2007)

O'Carroll, D.M., Phelan, T.J., Abriola, L.M.: Exploring dynamic effects in capillary pressure in multistep outflow experiments 41(11) (2005)

Porter, M.L., Schaap, M.G., Wildenschild, D.: Lattice-Boltzmann simulations of the capillary pressure–saturation–interfacial area relationship for porous media. Adv. Water Resour. **32** (11), 1632–1640 (2009)

Rabbani, A., Jamshidi, S., Salehi, S.: An automated simple algorithm for realistic pore network extraction from micro-tomography images. J. Petrol. Sci. Eng. **123**, 164–171 (2014)

Richards, L.A.: Capillary conduction of liquids through porous mediums. J. Appl. Phys. **1**(5), 318–333 (1931)

Sakaki, T., O'Carroll, D.M., Illangasekare, T.H.: Direct quantification of dynamic effects in capillary pressure for drainage–wetting cycles. Vadose Zone J. **9**(2), 424–437 (2010)

Schaap, M.G., Porter, M.L., Christensen, B.S., Wildenschild, D.: Comparison of pressure-saturation characteristics derived from computed tomography and lattice Boltzmann simulations 43(12) (2007)

Scheuermann, A., Galindo-Torres, S., Pedroso, D., Williams, D., Li, L.: Dynamics of water movements with reversals in unsaturated soils. In: 6th International Conference on Unsaturated Soils, UNSAT 2014, pp. 1053–1059 (2014)

Shan, X., Chen, H.: Lattice Boltzmann model for simulating flows with multiple phases and components 47(3), 1815 (1993)

Sukop, M., Thorne, D.T.: Lattice Boltzmann Modeling. Springer, Heidelberg (2006)

Topp, G., Klute, A., Peters, D.: Comparison of water content-pressure head data obtained by equilibrium steady-state and unsteady-state methods. Soil Sci. Soc. Am. J. **31**(3), 312–314 (1967)

Effect of Water Content on the Strength of Bio-Cemented Sand in Various Drying Process

Chen Zeng, Jun-jie Zheng(✉), Ming-juan Cui, and Xiang-chuan Yao

School of Civil Engineering, Huazhong University of Science and Technology, Wuhan, China
zhengjj@hust.edu.cn

Abstract. Microbially induced calcium carbonate precipitation (MICP) is a novel ground improvement technique which has promising applications in geotechnical engineering. The research of this technique commonly focuses on calcium carbonate precipitation and geotechnical properties of bio-cemented soil in fully saturated or dry states. However, the natural soil may contain different amount of water, which will have influences on the geotechnical properties of soil. The objective of the present study is to investigate the mechanical properties of bio-cemented sand with different water contents and drying temperatures. During MICP treatment, all the bio-cemented sand samples were prepared at the same cementation level. Meanwhile, two drying temperatures (i.e., 30 °C and 60 °C) were adopted to obtain different water contents. Calcium carbonate content and water distribution were tested, and unconfined compression test was conducted to measure the unconfined compressive strength. The experimental results show that, as the water content increases, the strength decreases firstly and then increases at the same drying process. This phenomenon may be caused by the interaction of pore water and calcium carbonate crystals during the drying period. However, under the same water content, drying in 30 °C has more remarkable influences on the strength in comparison with the drying in 60 °C. The much higher strength obtained by the 30 °C drying in dried state implies that different temperatures may change the crystal forms of calcium carbonate.

Keywords: Bio-cemented sand · Water content · Temperature · Unconfined compression test · Calcium carbonate crystal

1 Introduction

Nowadays, due to the high-speed development in infrastructure, the land that is suitable for construction is limited. Traditional ground improvement techniques, such as chemical grouting, are usually adopted to improve the soil properties to satisfy the engineering demands. However, these ground improvement techniques would cause environmental problems. Hence, the use of human-made chemical materials should be reduced or banned (DeJong et al. 2010).

Currently, microbially induced calcium carbonate precipitation (MICP) has been introduced into the geotechnical engineering. It is a natural biomineralization process that the bacteria hydrolyse the urea into ammonia and carbonate ions to induce calcium

R. Chen et al. (eds.), *Proceedings of the 2nd International Symposium on Asia Urban GeoEngineering*, Springer Series in Geomechanics and Geoengineering,
https://doi.org/10.1007/978-981-10-6632-0_3

carbonate crystals at the presence of calcium ions. The reactions involved in this process are as follows:

$$CO(NH_2)_2 + 2H_2O \rightarrow 2NH_4^+ + CO_3^{2-} \tag{1}$$

$$Ca^{2+} + CO_3^{2-} \rightarrow CaCO_3 \downarrow \tag{2}$$

The calcium carbonate crystals precipitate in the void to fill the pore space and/or bond the soil particles and then the soil properties can be improved (Ivanov and Chu 2008). It can be found from the reactions that four key parameters may influence the precipitation of calcium carbonate crystals, including calcium concentration, carbonate concentration, pH and the availability of nucleation sites (Whiffin et al. 2007; Hammes and Verstraete 2002). Many researchers have conducted investigations to study the effects of MICP treatment, such as the type and activity of bacteria (Okwadha and Li 2010; Hata et al. 2011), component of chemical solutions (Al Qabany and Soga 2013; Zhang et al. 2015), injection methods of bacterial suspension or chemical solutions (Rong et al. 2012; Rahim et al. 2015), and chemical reaction conditions (pH, temperature and catalyzer) (Stocks-Fischer et al. 1999; Bachemeier and Ringuelet 2002; Cheng and Shahin 2016).

However, the factors that affect mechanical properties of bio-cemented soil are more than these. The circumstance of soils after solidified treatment (i.e. MICP treatment) plays an important role in soil properties, such as water content and temperature of drying process.

There are many reports about the effects of water content and temperature on properties of different type of soils. With the increasing of water content, cohesion of soil would first increase to the peak and then decrease, while internal friction angle would decrease to nearly zero (Chen et al. 2007). For cemented soils, there is an optimum water content that can get the maximum strength of soil (Ribeiro et al. 2016). This may be caused by the optimum amount of bonds which can provide better junctions for soil particles. What's more, with the variation of water content, pores between soil particles would be affected that lead to the strength changes (Zuo et al. 2016). Although many researches have demonstrated strong connection between water content and soil properties, the studies about the influence of water content on the properties of bio-cemented soil are very few. This can cause the misestimate of bio-cemented soil properties. Besides, in many experiments of bio-cemented soil, temperature of drying process in the oven after MICP treatment are usually 30 °C (Li 2015) or 60 °C (Truong et al. 2012). But the comparison between these two temperatures of drying process are limited. The changing of water content and temperature of soils are common in the field which could be caused by weather changes, flow changes and so on. Therefore, it is very essential to investigate the effect of water content and drying temperature on mechanical properties of bio-cemented soil in drying process.

Nevertheless, salinity of bio-cemented soil after MICP treatment is an important factor which should be noticed first before the study of influences on drying process. Truong et al. (2010) found that a very low content of salt in soil can change the small strain stiffness greatly. In addition, this result was consistent with the study of Li (2015) that an overestimation strength of soil would be caused by the residual of calcium chloride. The salinity of samples is crucial for the estimation of factors on soil properties and some measures should be taken to eliminate this impact.

The objective of this study is to investigate the effect of water content on the strength of bio-cemented sand in two kinds of drying process. A series of bio-cemented sand specimens were prepared and then drying processes (30 °C and 60 °C) were conducted after the specimens were washed by tap water. After fully saturated and dried for different time, the unconfined compression tests were conducted to analyse the strength of bio-cemented sand. The results indicate that both water content and temperature correlate to the strength of bio-cemented sand. The overestimation of strength may occur without taking into account water content and drying temperature.

2 Materials and Methods

2.1 Bacterial Suspension and Urease Activity

Sporosarcina pasteurii [American type culture collection (ATCC 11859)] was used in this study. According to the NH_4-YE medium (ATCC 1376), 0.13 mol l^{-1} Tris buffer (pH = 9.0), 10 g l^{-1} $(NH_4)_2SO_4$, and 20 g l^{-1} yeast extract were included, and individual ingredients were autoclaved separately before mixing together. After inoculated into the medium, the bacterium was incubated at 30 °C with a shake rate of 150 r/min for 24 h. As the concentration of bacterial suspension was proportional to the optical density (OD_{600}), the visible light spectrophotometer was used to determine the OD_{600} of bacterial suspension, which was 1.66 in this study.

Urease activity could present the activity of bacteria that was vital for MICP process. To obtain the urease activity of the bacteria, conductivity method was used, which was similar to the research of Whiffin (2004). The rate of conductivity increase was 0.121 mS/cm/min, urease activity was 1.344 mM/min, and specific urease activity was 0.81 mM/min.

2.2 Sand and Specimen Preparation

PVC tube (140 mm in length and 37 mm in diameter) was used in this study to prepare the specimens (37 mm in diameter and 80 mm in length). The specimen preparation was similar to that of Cui et al. (2015). The tube was positioned vertically with a rubber plug in the bottom and packed with ISO standard sand for 80 mm in height, the grain size of ISO was in Fig. 1. The dry mass of each sand specimen was 160 g. A porous stone with thickness of 5 mm was put on the top and bottom of the sand sample to facilitate the solution diffusion. The dry density and initial porous ratio of these samples were 1.86 g/cm^3 and 0.30–0.32. Prior to the injection of bacterial suspension, each sand specimen was saturated for 24 h to exhaust air.

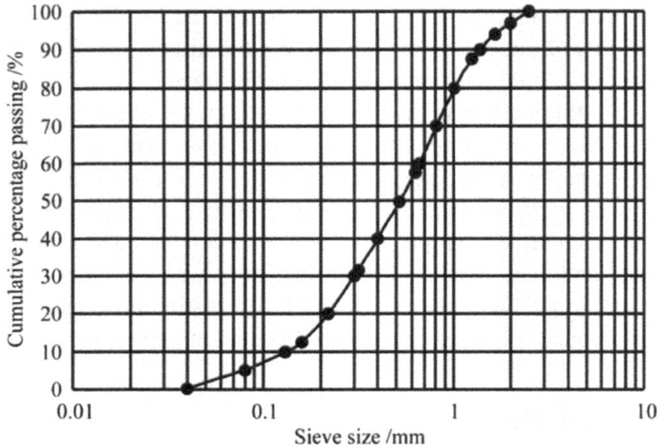

Fig. 1. Grain size sieve analysis curve of ISO sand

2.3 MICP Treatment

In the MICP treatment, bacterial suspension and cementation reagents were injected into specimens to induce calcium carbonate precipitation. The concentration of the cementation reagents (calcium chloride-urea solution) was 0.5 mol/L. The MICP treatment was conducted in the following two steps:

(1) Injection of bacterial suspension: One pore volume of bacterial suspension was injected into each specimen through a peristaltic pump with an input rate of 5 ml/min. Then, the injection was stopped for 6 h for the adequate adsorption of bacteria on sand particles.
(2) Injection of cementation reagents: One pore volume of cementation solution was injected into each specimen through the peristaltic pump with an input rate of 10 ml/min for every 24 h. All the sand specimens were treated for seven times in total.

2.4 Drying Process and Unconfined Compression Test

After the completion of MICP treatment, all specimens were treated by following steps:

(1) Specimens were flushed by at least 10 pore volume water to cut off the reaction of remained reagents and eliminate the effect of salt on strength in sand samples (He and Chu 2016).
(2) Specimens were saturated for 24 h using vacuum saturation method. Then, the weight of the bio-cemented specimens would be measured quickly after the rubber plug was removed and water on the tube wall was wiped. The weight of samples was termed as m_1.
(3) Specimens were divided into three series as shown in Table 1. The samples of Series I would be conducted the unconfined compression test (UCT) immediately after saturation and removal of PVC tube. The samples of series II and III were dried in

the oven for different temperatures and time before UCT. The weight of specimens after different drying time was m_2, while the weight of PVC tube removed before UCT was m_{pvc}. The water loss of each specimen was decided by m_1-m_2. And there were five groups in each series depending on the average water loss of each group.

Table 1. Drying process of three series

Series	Drying process	Average water loss (g)				
		A	B	C	D	E
I	Saturation	–	–	–	–	–
II	60 °C	7.7	10.6	17.8	29.2	32.5
III	30 °C	7.7	10.6	17.8	29.2	32.5

Note: Specimens in group A-C were cut into three parts after UCT

(4) After the unconfined compression test, samples of group A-C were cut into three parts to measure the water and calcium carbonate distribution of each group. Then samples were dried in the oven of 105 °C until there was no weight change.

2.5 Measurement of Calcium Carbonate Content

The content of calcium carbonate was measured by acid treatment (Montoya and DeJong 2015). The dry weight before and after acid treatment by 2 M HCl was m_3 and m_4. The calcium carbonate mass (m_c) was determined as follows:

$$m_c = m_3 - m_4 \tag{3}$$

In this study, the water content (w_w) and saturation degree (S_r) of each sample were used to measure the water volume of samples by the following equations:

$$w_w = \frac{m_2 - m_4 - m_{pvc}}{m_4} \times 100\% \tag{4}$$

$$S_r = \frac{V_{water}}{V_{void}} \times 100\% = \frac{m_2 - m_4 - m_{pvc}}{m_1 - m_4 - m_{pvc}} \times 100\% \tag{5}$$

The V_{water} was the volume of water in samples after different drying process, and V_{void} was the volume of water of samples in saturation state.

3 Results

3.1 Mass of Calcium Carbonate of Bio-Cemented Sand

The precipitation of calcium carbonate is the key point of properties improvement of bio-cemented sand. In many researches, mass of calcium carbonate is a measurement for cementation level which can strongly affect the strength of samples (Whiffin and van

Paassen 2007; Cheng et al. 2013; Soon et al. 2013). In this study, the average calcium carbonate mass of each group is shown in Fig. 2. It can be seen that all the bio-cemented sand samples have similar calcium carbonate mass after the same MICP treatment, indicating that the amount of precipitated $CaCO_3$ in different groups is relatively uniform. This confirms that the influence of calcium carbonate content on the strength of bio-cemented sand can be ignored.

In the same level of calcium carbonate mass, differences of distribution in samples may have effects on the strength of bio-cemented sand (Harkes et al. 2010). In MICP process, calcium ions are attached to the bacteria due to the negative charge of cell wall. Hence, calcium carbonate will precipitate around the bacteria cell that is absorbed on sand grains (Muynck et al. 2010). In other words, the sites of calcium carbonate are fixed after MICP treatment and the distribution of calcium carbonate has no effects on strength of samples in this test.

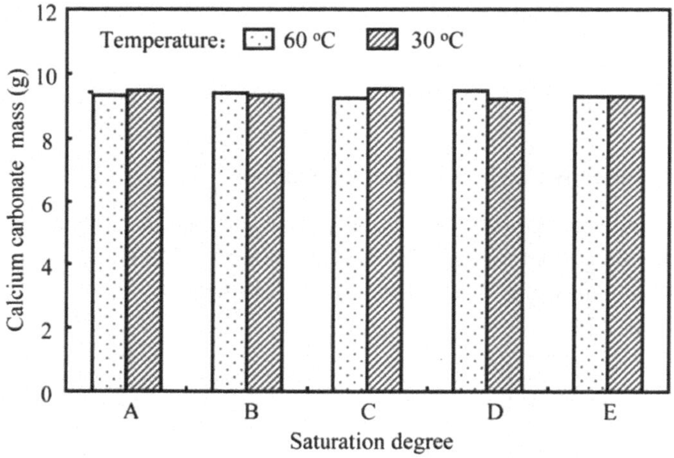

Fig. 2. The average calcium carbonate mass of each group

3.2 Effect of Water Content on Strength of Bio-Cemented Sand

The relationship between water content and unconfined compressive strength (UCS, termed as strength hereon) is presented in Figs. 3 and 4. The saturation degree of 100% and 0% represents the fully saturated and fully dried states, respectively. It can be found that strength has a strong correlation to the water content of bio-cemented sand samples. As the saturation degree decreases, the strength decreases slightly first and then increases quickly up to the peak strength of the fully dried samples. The variation trends of water content versus strength are both in concave shape in two figures. The strength of fully dried bio-cemented sand samples is higher than that of fully saturated ones, which is similar to the findings of Cheng (2013) and Li (2015). However, the enhancement of strength in fully dried state is not as high as the reported strength of Li (2015) due to the salt washing of samples before saturation.

During the drying process, the rate of water loss in different parts is various with the approach to the central of samples. In this case, water content of different parts in samples (termed as water distribution hereon) is measured to study the effect of it on the strength. The result of water distribution in different water loss is shown in Fig. 5. It can be seen that the total water content of samples decreases with the increase of water loss. Nevertheless, there is no distinct relationship between water distribution and water loss along samples. It suggests that water distribution is similar under the same drying temperature.

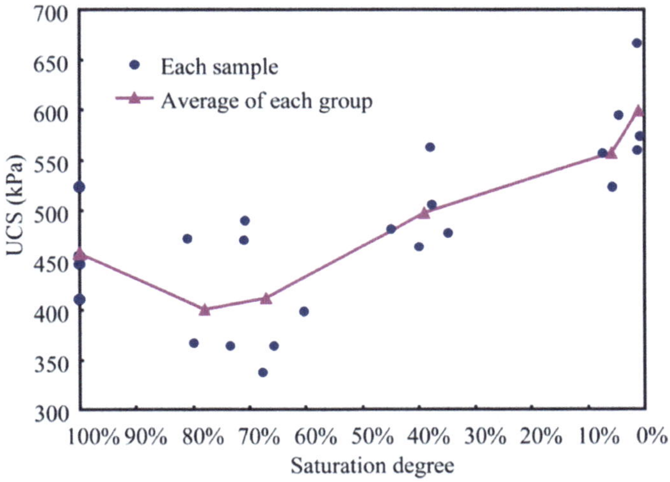

Fig. 3. Effect of water content on strength of bio-cemented sand in 60 °C drying

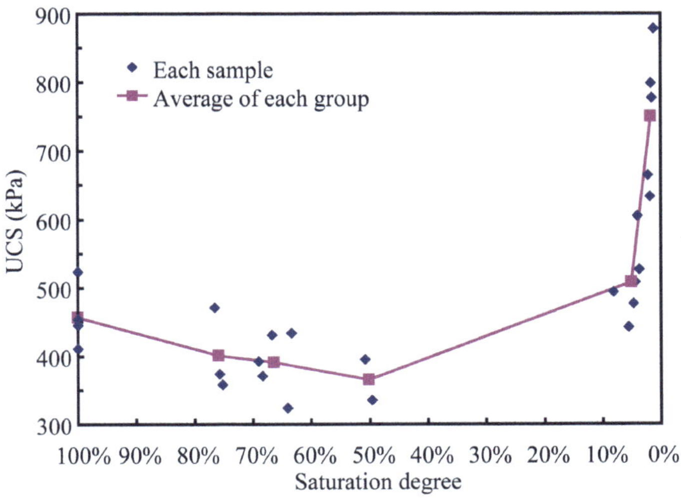

Fig. 4. Effect of water content on strength of bio-cemented sand in 30 °C drying

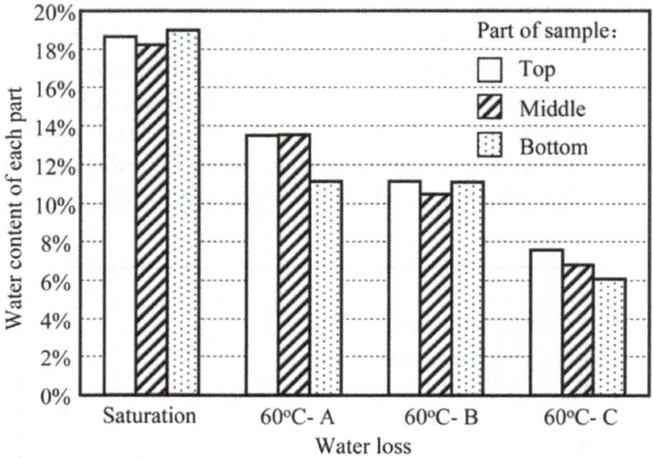

Fig. 5. Effect of water loss on water distribution of samples in 60 °C drying

3.3 Effect of Temperature on Strength of Bio-Cemented Soil

The influence of drying temperature on the strength of bio-cemented sand samples is shown in Fig. 6. Overall, the variation of strength is highly dependent on the temperature, presenting that higher strength can be obtained at higher temperature when the saturation degree is not too high. In the present study, when the saturation degree is larger than 80%, the strength is barely influenced by the drying temperature. However, if the bio-cemented sand is fully dried (i.e., saturation degree is zero), a relatively higher strength

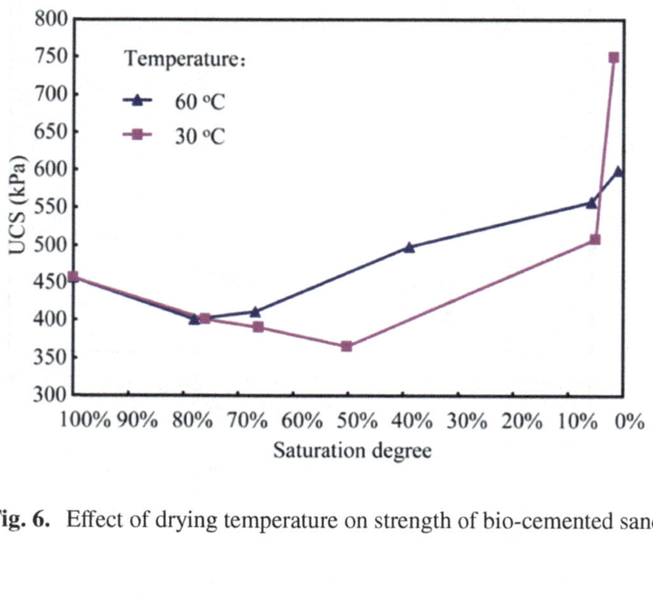

Fig. 6. Effect of drying temperature on strength of bio-cemented sand

is achieved. The considerable differences may be related to the variation of crystal structure caused by the different drying temperatures.

Similarly, the stiffness of bio-cemented sand samples is denoted by the average secant modulus, E_{50}. The value of E_{50} is equal to the ratio of 50% strength and its corresponding axial strain. The variation of E_{50} is shown in Fig. 7. The variation trend of stiffness is similar to that of strength, presenting a concave shape. That is to say, the stiffness of bio-cemented sand samples is also influenced by the drying temperature.

As a conceivable factor that affecting strength of samples, the water distribution of samples under different drying process is showed in Fig. 8. Although the average water content of top-middle-bottom parts is similar under the same water loss, the variation range is different, which represents the water distribution of samples. In the same water loss, the variation range of 30 °C drying is higher than that of 60 °C. In addition, the difference of variation range increases with the rise of water loss, indicating the non-uniform water distribution in 30 °C drying. In other words, the distribution of water content in 30 °C drying samples is inferior, which can cause the reduction of strength. It is consistent with the result of strength comparison under different drying temperatures in Fig. 6.

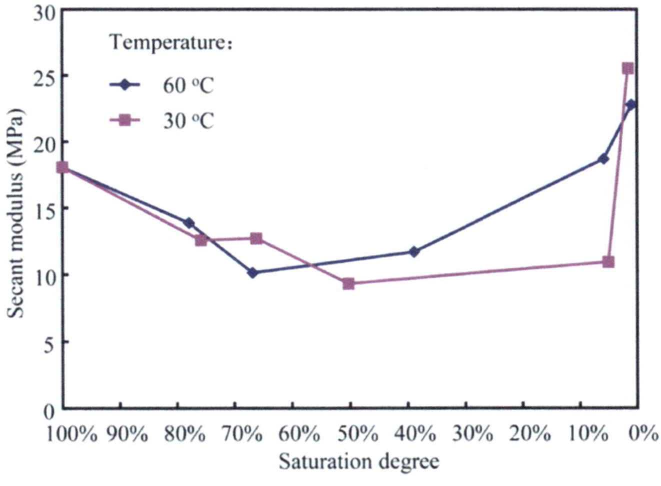

Fig. 7. Effect of drying temperature and water content on stiffness of bio-cemented sand

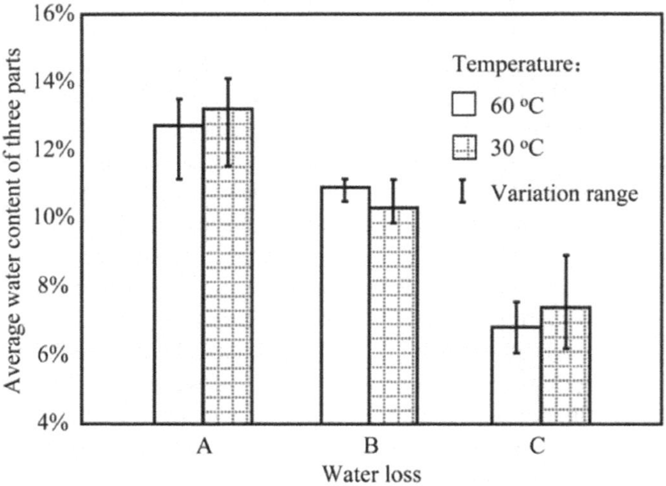

Fig. 8. Effect of water loss on average water content of top-middle-bottom parts in 60 °C and 30 °C drying

4 Discussion

From the theory of MICP process, calcium carbonate content, distribution, and crystal forms are three important factors of strength changing in the perspective of precipitated $CaCO_3$. According to the analysis above, it verifies that the content and distribution would not change in drying process after the same MICP treatment. Thus, it can infer that the crystal forms are responsible for the strength changing in drying process.

There are three crystal forms that have been detected in bio-cemented soil including calcite, aragonite and vaterite (Zhang et al. 2015). The differences between crystal forms can lead to the changing of cementation properties. It has been proved in aquatic environment that pure crystal of calcite, vaterite and aragonite is found with the temperature of 10 °C, 55 °C and 70 °C, respectively. In addition, mixture of vaterite and calcite is gotten in 25 °C (Huang et al. 2017). It suggests that the crystal forms of the precipitated calcium carbonate are various in different temperature of drying process which can cause different strength of samples as shown in Fig. 4. Moreover, the increase of supersaturation can accelerate the rate of nucleation and crystal growth (Han et al. 2006). In this study, with the water loss of samples, supersaturation can rise up to affect crystal forms of precipitated calcium carbonate. In that way, the strength of samples is various with different water content as shown in Figs. 2 and 3. This suggests that the influences of water content may be caused by the interaction of pore water and calcium carbonate crystals during the drying period. However, the crystal forms of calcium carbonate in different drying process should be tested to survey the polymorph of them, which is a valuable study in prospective research.

This study shows the relationship between strength and the temperature and water content in drying process which is meaningful for experimental work and engineering applications. In the lab, many studies commonly focus on the influences in material and MICP treatment and use fully saturated or dried samples for strength measurement but neglect the factors of water content and temperature in drying process. This can lead to the overestimation of strength. Compared to lab environment, the temperature and water content are different in the field, especially in seashore or riverside. The overestimation of strength could cause potential safety risk, while the strength changing may lead to damage of structure. The effect of these factors should be noticed to promote the application of MICP techniques.

5 Conclusion

In this study, the mechanical properties of bio-cemented sand with various water content are investigated under two drying temperatures. All samples are treated with the same MICP treatment and dealt with different drying process after salt washing. The following conclusions are drawn:

(1) In the same drying temperature, the strength of samples is strongly related to the water content. With the saturation decreases, the strength of samples is decreased slightly first and then increases quickly up to the peak strength in fully dried condition. And the interaction of pore water and calcium carbonate crystals in drying period is responsible for that.
(2) In the same water content, the strength of samples is various in different drying temperatures. The strength of samples in 30 °C drying is lower than that of 60 °C under the saturation degree of 5%–80% which may be caused by the inferior water distribution. However, when the saturation degree is lower than 5%, the strength of samples in 30 °C drying is higher than that of 60 °C, especially in the fully dried condition. And the stiffness of samples is influenced by temperature in the similar way of strength. These phenomena are mainly caused by the crystal forms of calcium carbonate which are transformational in different temperatures.

The results of this study indicate that the mechanical properties of bio-cemented sand after MICP treatment can be affected by the water content and drying temperature. This will cause the overestimation of strength in lab work or filed applications which should be noticed in the prospective studies.

References

Al Qabany, A., Soga, K.: Effect of chemical treatment used in MICP on engineering properties of cemented soils. Géotechnique **63**(4), 331–339 (2013)

Bachemeier, O.A., Ringuelet, A.: Kinetics of soil nitrogen mineralization from undisturbed and disturbed soil. Commun. Soil Sci. Plant Anal. **33**(19–20), 3703–3721 (2002)

Chen, H.X., Li, F.H., Hao, S.L., Zhang, X.P.: Effects of soil water content and soil sodicity on soil shearing strength. Trans. CSAE **23**(2), 21–25 (2007). (in Chinese)

Cheng, L., Crod-Ruwisch, R., Shahim, M.A.: Cementation of sand soil by microbially induced calcite precipitation at various degrees of saturation. Can. Geotech. J. **50**(1), 81–90 (2013)

Cheng, L., Shahin, M.A.: Urease active bioslurry: a novel soil improvement approach based on microbially induced carbonate precipitation. Can. Geotech. J. **53**(9), 1376–1385 (2016)
Cui, M.J., Zheng, J.J., Zhang, R.J., Miao, C.X., Zhang, J.J.: Study of effect of chemical treatment on strength of bio-cemented sand. Rocks Soil Mech. **36**(S1), 392–396 (2015). (in Chinese)
DeJong, J.T., Mortensen, B.M., Martinez, B.C., Nelson, D.C.: Bio-mediated improvement. Ecol. Eng. **36**(2), 197–210 (2010)
Hammes, F., Verstraete, W.: Key roles of pH metabolism in microbial carbonate precipitation. Rev. Environ. Sci. Bio/Technol. **1**(1), 3–7 (2002)
Han, Y.S., Hadiko, G., Fuji, M.: Factors affecting the phase and morphology of $CaCO_3$ prepared by a bubbling method. J. Eur. Ceram. Soc. **26**(4), 843–847 (2006)
Harekes, M.P., van Paassen, L.A., Booster, J.L., Whiffin, V.S., van Loosdrecht, M.C.M.: Fixation and distribution of bacterial activity in sand to induce carbonate precipitation for ground reinforcement. Ecol. Eng. **36**(2), 112–117 (2010)
Hata, T., Tsukamoto, M., Inagaki, Y., Mori, H., Kuwano, R., Gourc, J.P.: Evaluation of multiple soil improvement techniques based on microbial functions. In: Geo-frontiers Congress, pp. 3945–3955 (2011)
He, J., Chu, J.: Cementation of sand due to salt precipitation in drying process. Mar. Georesour. Geotechnol. **35**, 441–445 (2016)
Hu, Y.B.: The study of the formation and stability of ikaite ($CaCO_3{\cdot}6H_2O$) in the sea ice. MS Thesis, Ocean University of China, China (2010). (in Chinese)
Huang, L.L., Ma, X.D., Liu, W.K., Xue, B.D., Liu, D.F.: Study on the factors of crystal polymorphs controlling. Shandong Ind. Technol. **6**, 41 (2017). (in Chinese)
Ivanov, V., Chu, J.: Applications of microorganism to geotechnical engineering for bioclogging and biocementation of soil in situ. Rev. Environ. Sci. Bio/Technol. **7**(2), 139–153 (2008)
Li, B.: Geotechnical properties of biocement treated sand and clay. Ph.D. Thesis, Nanyang Technological University, Singapore (2015)
Montoya, B.M., DeJong, J.T.: Stress-strain behavior of sands cemented by microbially induced calcite precipitation. J. Geotech. Geoenviron. Eng. **141**(6), 04015019 (2015)
Muynck, W.D., Belie, N.D., Verstraete, W.: Microbial carbonate precipitation in construction materials: a review. Ecol. Eng. **36**(2), 118–136 (2010)
Okwadha, G.D.O., Li, J.: Optimum conditions for microbial carbonate precipitation. Chemosphere **81**(9), 1143–1148 (2010)
Rahim, S.S., Brendan, C.O., Ali, N., O'Kelly, B.C.: Improving sand with microbial-induced carbonate precipitation. Proc. Inst. Civ. Eng. Gr. Improv. **168**(GI3), 217–230 (2015)
Ribeiro, D., Néri, R., Cardoso, R.: Influence of water content in the UCS of soil-cement mixtures for different cement dosages. Procedia Eng. **143**, 59–66 (2016)
Rong, H., Qian, C.X., Li, L.Z.: Influence of molding process on mechanical properties of sandstone cemented by microbe cement. Constr. Build. Mater. **28**(1), 238–243 (2012)
Stocks-Fischer, S., Galinat, J.K., Bang, S.S.: Microbiological precipitation of $CaCO_3$. Soil Biol. Biochem. **31**(11), 1563–1571 (1999)
Soon, N.W., Lee, L.M., Khun, T.C., Ling, H.S.: Improvements in engineering properties of soils through microbial-induced calcite precipitation. J. Civ. Eng. **17**(4), 718–728 (2013)
Truong, Q.H., Eom, Y.H., Byun, Y.H., Lee, J.S.: Characteristic of elastic waves according to cementation of dissolved salt. Vadose Zone J. **9**(9), 662–669 (2010)
Truong, Q.H., Lee, C., Kim, Y.U., Lee, J.S.: Small strain stiffness of salt-cemented granular media under low confinement. Géotechnique **62**(10), 949–953 (2012)
Whiffin, V.S.: Microbial $CaCO_3$ Precipitation for the production of Biocement. Ph.D. Thesis, Murdoch University, Western Australia (2004)

Whiffin, V.S., van Paassen, L.A., Harkes, M.P.: Microbial carbonate precipitation as a soil improvement technique. Geomicrobiol. J. **24**(5), 417–423 (2007)

Zhang, Y., Guo, X.H., Cheng, X.H.: Influences of calcium sources on microbially induced carbonate precipitation in porous media. Mater. Res. Innov. **18**(S2), S2-79–S2-84 (2015)

Zuo, C.Q., Liu, D., Ding, S., Chen, J.P.: Micro-characteristics of strength reduction of tuff residual soil with different moisture. KSCE J. Civ. Eng. **20**(2), 639–646 (2016)

Electro-Osmotic Chemical Treatment of Marine Clayey Soil Using $CaCl_2$ Solution

Lei Zhang[1], Liping Jing[1(✉)], Ningwei Wang[1,2], Chen Fang[3], Zhendong Shan[1], and Yongqiang Li[1]

[1] Key Laboratory of Earthquake Engineering and Engineering Vibration, Institute of Engineering Mechanics, China Earthquake Administration, Harbin, People's Republic of China
{ray_iem, jlp_iem}@163.com
[2] School of Civil Engineering, Shenyang Jianzhu University, Shenyang, People's Republic of China
[3] Department of Civil Engineering, University of Nebraska-Lincoln, Lincoln, USA

Abstract. This paper performed an experimental study on the electro-osmotic chemical treatment (ECT) for marine clayey soil using $CaCl_2$ solution. The soil samples were collected from the southern region in Liaodong Peninsula, Dalian city. The study focused on the use of ECT to improve mechanical and physical properties of soil. A series of laboratory experiments with the self-designed equipment were conducted to analyze the feasibility of ECT using $CaCl_2$ solution. The experimental results demonstrated that the ECT is an effective method to decrease the water content and increase the bearing capacity of the marine clayey soil. Furthermore, the average undrained shear strength of the marine clayey soil increased with the ETC technique. The bearing capacity of the treated soil was improved up to $450\% \sim 700\%$. The energy consumption and the electrode corrosion were reduced through adding Polyvinylchlorid (PVC) pipe to the electrodes. The electro-osmotic chemical treatment technique using $CaCl_2$ solution proposed in this paper has the potential to strengthen marine clayey soil foundations and to increase the economic efficiency and feasibility of improvement of marine clayey soil.

Keywords: Marine clayey soil · Electro-osmotic chemical treatment · Charging time · Bearing capacity · Consolidation

1 Introduction

Marine clayey soils along the seashore are characterized with high water content and high compressibility, low bearing capacity and low shear strength. These poor properties of marine clayey soils often cause foundation problems to the building structures. Due to the construction activities in these regions, it is necessary to improve the bearing capacity and shear strength of the soil. Dalian city is located in the southern region of Liaodong Peninsula, and a new campus of Dalian Ocean University is building in the region. The reclaimed soils will be used for the construction of a series of buildings, and the construction project cannot be performed before the soils are consolidated. It

R. Chen et al. (eds.), *Proceedings of the 2nd International Symposium on Asia Urban GeoEngineering*, Springer Series in Geomechanics and Geoengineering,
https://doi.org/10.1007/978-981-10-6632-0_4

takes a long time to use the traditional techniques for the soil consolidation due to the 10-meter-thickness marine clayey soil layer. Based on various studies, electro-osmosis chemical treatment (ECT) method has the potential to accelerate consolidation of the marine clayey soil with large water content and high compressibility. However, the high energy consumption and the corrosion of electrodes raise economic constraints and reduce its feasibility. Therefore, several foundation treatment methods including the surcharge preloading, vacuum preloading, electro-osmosis consolidation, and ECT were investigated and evaluated. Results show that ECT is an effective technique to consolidate the marine clayey soil in Dalian city relative to other techniques.

The ECT is a useful technique for consolidation. In ECT, the current is provided to pass through the soil, and the electrolyte solution is added in the anode. An electric field produced in the soil leads to the electro-osmosis, electrophoresis, and ionic migration. The consolidation mechanism consists of two components: (1) electro-osmosis: the hydrated ions combined with the cationic ions adsorbed in the soil and the water molecules move to the cathode under the action of electric field; (2) soil colloids: the chemical reactions are conducted between the soil particles and the ions provided by the electrolyte solution.

The electro-osmosis was discovered and proposed by Reuss in 1809, and then many scholars started to develop the related researches. It has been proved as a useful method for improving soil strength. The physical and mechanical properties of the strengthened soil are improved to meet the requirements of construction. Electro-osmotic consolidation (EC) has been used for the stabilization of various soils to improve their strength and durability [1–6]. Lo conducted some experiments of soft soils from Gloucester in the UK using self-designed apparatus, and result in decreasing the water content and increasing the average undrained shear strength after treatment [7, 8]. Katti [9] was the first person to consolidate the marine clayey soils using ECT, and then many scholars discussed the improvement of the soils using different chemical reagents. Esrig [10] added calcium chloride in the illite to investigate the mechanism of ECT. Gray [11] used the solution of aluminum chloride to strengthen four different groups of synthetic soil and study the relationship between ECT and pH value. Ozkan [12] applied aluminum ion and phosphate ion to conduct the ECT on the kaolin clays. The research showed that the strength of soil was five times improved and the liquid limit was increased by 30%. Alshawabkeh [13] conducted a series of tests by adding nitronic acid and nitric acid respectively in the soil, and results demonstrated that the bearing capacity of the soil with nitric acid increased 5.6 times. Asavadorndeja [14] developed a theory that the cementation only appeared in the cathode when the calcium chloride solution was added. The study demonstrated that the alkaline solution increased the area of the alkaline zone in the soil and extended the cemented zone to the anode. In this research, the relationship between calcium ions and the cementation of soil was developed. Zeng [15] was the first scholar in China to conduct the investigation of electro-osmosis consolidation on soft clay. Ou and Chien [16–20] made a series of experimental testing based on ECT to consolidate the precipitated clay in Taipei using self-designed testing apparatus to investigate the effects of different salt solutions and the concentrations of solutions on strengthening soft clay. Based on the studies of Ou and Chien, Chang [21] investigated the effect of different concentrations of $CaCl_2$ and different charging times on strengthening process using ECT.

Zhuang [22, 23] analyzed the interfacial resistance during electro-osmosis consolidation and studied the electric electrode materials in the strengthening period. Hu [24] provided a reference basis for designing electro-osmosis consolidation by establishing a control equation of multi-field coupling problems and developing the finite element software.

This study investigated the feasibility of ETC to consolidate the marine clayey soils collected from Dalian city. The objectives of the study were to examine: (1) the effect of an electric field parameter, (2) the effect of ECT, (3) the role of electrochemical reactions. The objectives were originated from practical engineering problems in Dalian city. Several laboratory tests were performed to strengthen these marine clayey soils, and results provided the theoretical basis for the future field tests.

2 Experimental Apparatus

A schematic of the experimental apparatus used for the ECT tests is depicted in Fig. 1. The apparatus consisted of an electrical circuit, a DC power supply (the measuring range of ammeter is 0 ~ 30A, and the range of voltage meter is 0 ~ 100 V), metal electrodes, polyvinylchloride (PVC) pipes for grouting, soil sample test tank, ampere meters. Furthermore, several auxiliary apparatuses including a multi-meter,

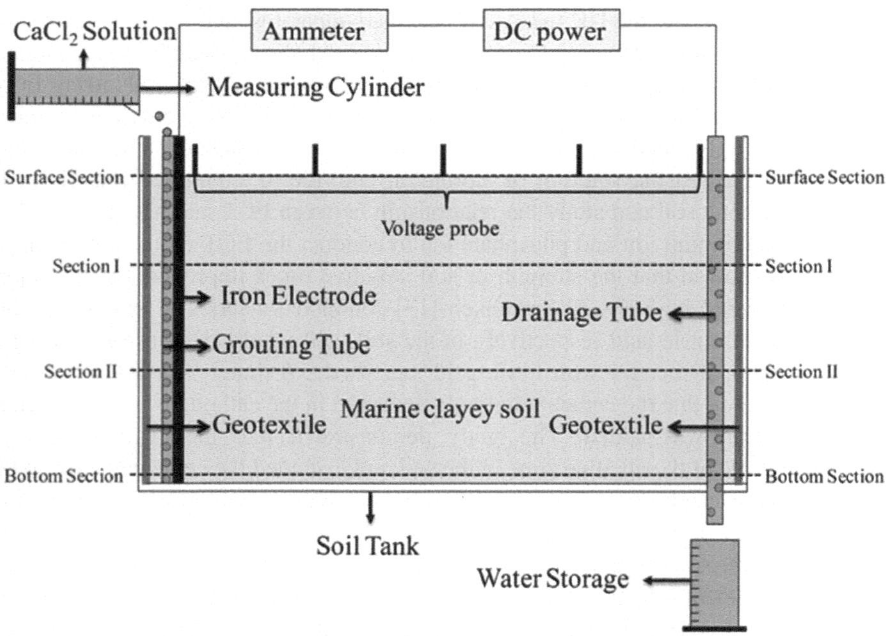

Fig. 1. Schematic of the laboratory set up for electro-osmosis chemical grouting

thermometers, a soil conductivity meter and a pocket penetrometer were prepared for the experiment.

The soil sample test tank was made of polyethylene with the dimension of 300 mm length by 220 mm width by 250 mm depth. The depth of soil sample in the test tank was 220 mm. The ammeter and voltmeter were equipped to measure the electric current and voltage of the soil during consolidation. A measuring cylinder was designed to control the injection of the solution. PVC pipes were selected to reduce the corrosion of these electrodes. A cylinder was used to storage the discharged water from the soil. Therefore, the amount of the discharged water from the soil sample was measured.

The electrodes were connected to the regulated power supply for the generation of an electric field in the soil sample. Two mild steel tubes which had a diameter of 8 mm and a length of 300 mm were inserted into the soil to a depth of 200 mm and served as the electrodes. Some holes were drilled in the steel wall, allowing the salt solutions to move freely from the tube to the soil. A PVC pipe was bundled with the anode to inject the salt solution. The electrodes were connected to the regulated power supply, and the soils were subjected to direct current under normal temperature and humidity conditions. The soil samples were covered with a geotextile in the left and right sides of the tank to prevent evaporation. The voltage of the soil sample was measured using a voltage probe, as shown in Fig. 1. A voltage probe was made of several copper wires and fixed on the bottom of the test tank.

In the study, several parameters including electric field intensity, current density, energy consumption, and dewatering efficiency were calculated and analyzed to investigate the feasibility of ECT on the marine clayey soils. The relationship between the electric field intensity and the voltage was formulated as Eq. (1).

$$E = \frac{dV}{dl} \tag{1}$$

where E is the electric field intensity (V/m), V is the voltage (V), and l is the distance from the anode to the cathode (m). The current density is defined as the ratio of the current to the cross-sectional area, formulated as Eq. (2).

$$i = \frac{I}{A} \tag{2}$$

where i is the current density (A/m^2), I is the current of the soil sample (A), and A is the cross-sectional area of the soil sample (m^2). Energy consumption is an important parameter to demonstrate the feasibility of ETC on the marine clayey soil, formulated as Eq. (3).

$$C_{t_2} = C_{t_1} + \int_{t_1}^{t_2} U_t I_t dt \tag{3}$$

In the equation, U_t and I_t represents the effective voltage and the current from t_1 to t_2, respectively. C_{t1}, C_{t2} represents the energy consumption at the time of t_1 and t_2, respectively. The dewatering efficiency is also considered in the study and expressed as Eq. (4).

$$\eta = Q_j / C \tag{4}$$

In the equation, η represents the dewatering efficiency (mL/W·h). Q_j (mL) is the amount of discharged water from the soil sample. C (W·h) is the electro-osmotic energy consumption.

The marine clayey soil was collected in Dalian city. The soils were obtained at the site at a depth of 10 m and then transported to the laboratory at Shenyang Jianzhu University. The samples of marine clayey soils with gray-black color had low bearing capacities. The shear strengths of the untreated soil samples were determined from 15 kPa to 20 kPa. The water content of the untreated soil sample was 55.72%. The basic parameters of untreated soil samples are summarized in Table 1.

Table 1. Basic parameters of untreated soil

Index Name	Unit	Index value			
		Average	Maximum	Minimum	Standard
Water content w	%	55.72	63.7	46.4	56.6
Specific gravity γ	kN/m^3	17.6	18.5	16.8	17.5
Dry gravity γ_d	kN/m^3	10.3	11.4	9.7	10.4
Porosity ratio e	–	1.75	2.04	1.53	1.77
Saturation S_r	%	98.91	100.1	95.1	99.08
Liquid limit w_L	%	40.2	49.2	28.6	39.8
Plastic limit w_P	%	21.5	30.7	16.1	22.4
Plastic index I_L	–	1.64	2.07	1.01	1.69
Liquid index I_P	–	15.9	16.8	14.6	15.8
Cohesive force c_q	kPa	5.7	7.8	3.1	5.4
Frictional angle φ_q	°	2.6	4.1	1.5	2.43
Unconfined compressive strength q_u	kPa	16.4	21.3	4.6	15.9
Compression modulus E_s	MPa	2.08	2.61	1.51	2.06
Salinity	g/L	35	–	–	–

According to the testing requirements, the untreated marine clayey soils were dehydrated, dried, triturated, and then mixed with sufficient amounts of deionized water to make remolded soils sample. The water content of the remolded samples reached up to 52%. The salinity of the interstitial water was 35 g/L which was obtained from electrical conductivity measurements. The pH of the interstitial water was determined as 7.3 using a pH electrode. The pumping test indicated that the permeability coefficient (K) of the soil layer was 0.4 m/d and the radius of influence (R) was 150.30 m.

Four different tests were designed and performed in the study, and the experimental conditions are summarized in Table 2. In these tests, the water content, voltage, electrode spacing and volume of the marine clayey soils were kept constant during the process of testing. The $CaCl_2$ solution was selected as the electrolyte for the experiment, and the concentration of the CaCl2 solution injected in the anode was an influencing factor for improving the strength of the marine clayey soil. Then, the electrodes were passed through the plastic membrane and inserted into the soil. These metal electrodes were divided into two groups as the anodes and the cathode. The power supply was connected to these electrodes, and the control parameters of voltage were set before testing.

Table 2. Test conditions of the laboratory experiment

Test NO.	Water content (%)	Electric field (V/m)	Porosity ratio	Salt solution	Concentration (%)	Total amount of solution (mL)	Charging time (h)
C1-1	52.42	1	1.579	Water	–	20	16.5
C1-2	51.61	1	1.579	$CaCl_2$	5	20	16.5
C1-3	52.77	1	1.579	$CaCl_2$	10	20	16.5
C1-4	52.64	1	1.579	$CaCl_2$	15	20	16.5

3 Results and Discussions

The current and the amount of discharged water were recorded during the process of testing. When testing had been terminated, the ultimate energy consumption and the amount of discharged water of unit energy consumption were obtained based on the recorded data and Eqs. (1)–(4).

3.1 Amount of Discharged Water

The amount of discharged water versus time curves were received according to the testing data, as shown in Fig. 2. It was concluded from Fig. 2 that, the amounts of discharged water in four tests were identical before $CaCl_2$ was added into the tests. The slopes of three curves increased due to adding $CaCl_2$, but the slope of the C1-1 curve did not change because the pure water was added in the test C1-1. The results demonstrated that adding $CaCl_2$ can increase the amount of the discharged water. In these four tests, the accumulated amount of discharged water was 61 mL, 61 mL, 61 mL, and 61 mL, respectively, before adding the solution. With the solutions, the rate of dewatering was increased, and therefore the accumulated amount of discharged water was 76 mL, 86 mL, 90 mL, and 95 mL, respectively. Furthermore, the amount of discharged water in three tests adding $CaCl_2$ solution was larger 1.13, 1.2, and 1.26 times than that in the C1-1 test, respectively. As a result, the increase of the concentration of added solution resulted in the increase of the amount of discharged water.

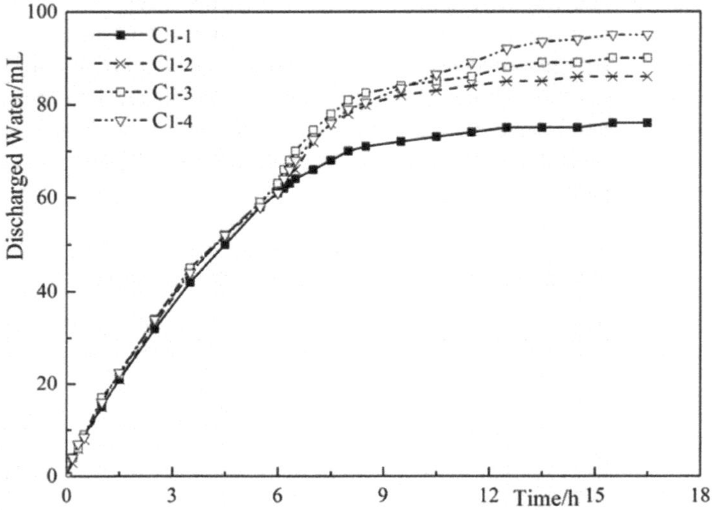

Fig. 2. Dewatering amount versus time

3.2 Current

In Fig. 3, the changing trends of four tests were same. In the first 0.5 h, the current increased from 1.1A to 1.3A, and then the current reduced until the solution was added into the tests. At the initial stage, the slight increase of the current was shown owing to the potential change of the soil and the polarization of metal electrodes. As the time increased, free water in the soil pores moved to the cathode and was drained out of the

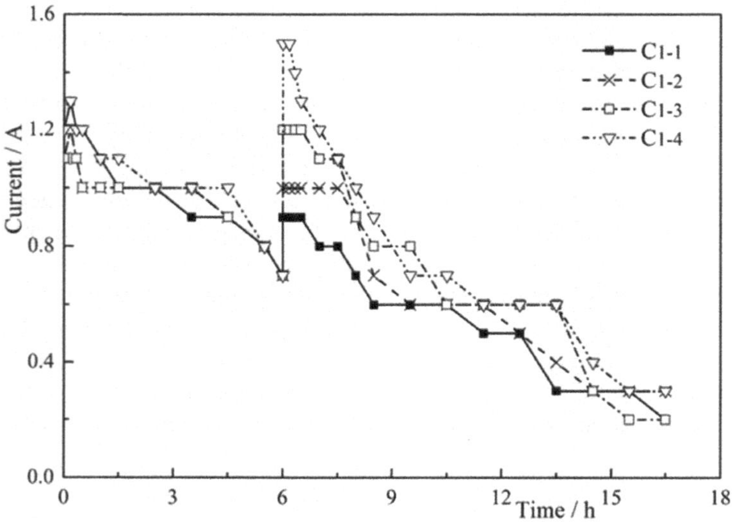

Fig. 3. Current density versus time

soils. The water content in the anode decreased to result in the decrease of the conductivity of the soil. As a result, the current of the soil reduced. The identical volumes of solutions were added to increase the conductivity of the soils in the anode. The current increased with the increase of the concentration of the added solution. The currents in four tests are 0.7A before adding the solutions. The currents increased to 0.90A, 1.00A, 1.20A, and 1.50A, respectively, after adding the solutions. The percentage increase of current in four tests was 28.6%, 42.8%, 71.4%, and 114.3%, respectively. After the peak current, the current decreased gradually. The current directly reflects the movement of ions. The decrease of the current demonstrated the decrease in the number of ions and the occurrence of the stable status of the ions. The decrease of the current is caused by the ion exchange, ion precipitation, and dewatering in the cathode.

Based on the electro-osmosis theory proposed by Helmoholtz, the positive ions are combined with the hydro ones owing to hydration reaction. Free waters and some bound waters moved to the cathode under the action of the electric field. In Fig. 3, owing to adding the solution, the number of ions in the soils was increased to enhance the speed of ions movement. Therefore, the current in the soils was increased by the solutions. According to Fig. 2 and 3, the amount of discharged water was increased with the increase of the current. As the concentration of the solution increased, the accumulated amount of discharged water also increased. This demonstrated that the increase of the current has an improvement effect on the dewatering.

3.3 Energy Consumption

The accumulated energy consumption and the dewatering efficiency were calculated and determined using Eqs. 3 and 4. Figures 4 and 5 shows the accumulated energy consumption-time curves and the dewatering efficiency-time curves, respectively.

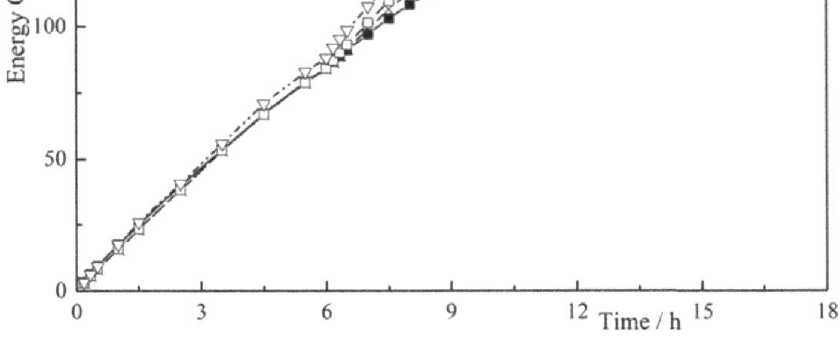

Fig. 4. Energy consumption versus time

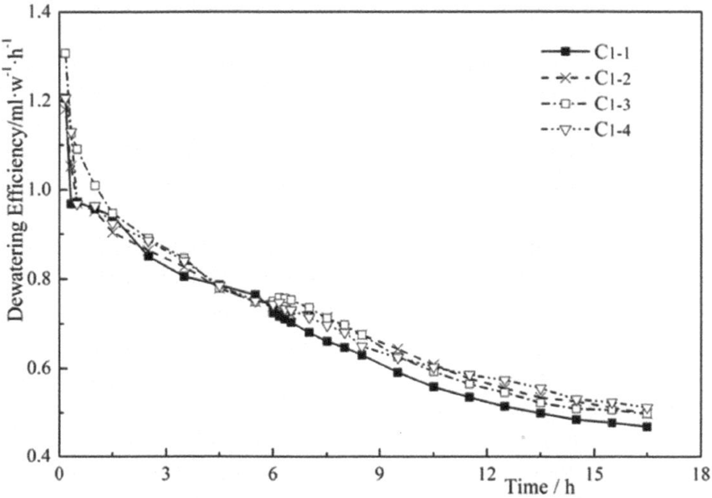

Fig. 5. Dewatering efficiency versus time curves

It was concluded from Fig. 4 that, the energy consumption was changed adding the solutions, and the energy consumption increased with the increase of the concentration of the solution. The current and the voltage are the primary parameters to affect the accumulated energy consumption. In the test, the voltage is a constant, and therefore the current is the only parameter to change the accumulated energy consumption. As a result, the energy consumption presented the same changing trend as the current. The increase of the concentration of the solution resulted in the energy consumption.

In Fig. 5, dewatering efficiency in each test decreased before the solutions were added owing to the decrease of water content and the increase of the conductivity of the soils. After the solutions were added in four tests, vibrations were not shown in the dewatering efficiency-time curve of C1-1, and the curves present obvious vibrations in the other three tests. After adding the solutions thirty minutes, the dewatering efficiencies decreased with the increase of the time, and the dewatering efficiency in the test C1-4 was higher than other three tests. The solutions are effective to improve the dewatering efficiency of the soils, and the increase of the concentration of the solution can result in the increase of the dewatering efficiency. According to Figs. 4 and 5, as the concentration of the solution increased, the accumulated energy consumption and the dewatering efficiency increased. This demonstrated that the increase of the concentration of the solution brings benefits to the effective use of electronics for dewatering, and it is more economical to use the solution of low concentration for dewatering.

3.4 Bearing Capacity

In the initial stage of testing, some bubbles and water appeared in the anodes. After charging 30 min up to 1 h, an electrolytic reaction of the metal electrode was taken place in the anode to produce some reddish substances. As the charging time increased, the region where the reddish substances were produced expanded. Based on the various studies, the reddish substances were preliminarily identified as several ferruginous colloids consisting of FeO, Fe_2O_3, $Fe(OH)_2$, $Fe(OH)_3$, and so forth. With the increase in the charging time, the temperature of the soil increased gradually, which demonstrated the exothermic reaction in the soils during the process of testing. After charging 6 h, the solutions were injected into these tests, and the data was recorded in the ampere meter within a short time, which means the currents can be increased quickly in a short time owing to the injected solutions. Some gases with the pungent smell were also produced in the soils during the testing. The gases were composed of the chlorine gas created from an oxidation reaction on chlorine ions of the soil and some other gases, such as SO_2 and H_2S, which were produced from organic materials in the soils. After charging 6 h to 7 h, the white substances appeared in the cathode in the test C1-2~C1-4. These substances were determined as products from ion precipitations and chemical reactions between solutions and salt contents in the soils under the action of the electronic field. And the regions where the white substances appeared were developed from the cathode with the increase of the charging time. After charging 4 h, several small cracks occurred near the anode, and these cracks increased and developed to the cathode with the increase of the charging time. The development of the crack was related to the dry-wet status procedure and the distribution of the potential in the soils. Since the electro-osmotic compaction occurred during the process of dewatering, the settlements of the soils were detected as the water was discharged out of the soils. When the testing was terminated, the soils were placed for three days, and then soils

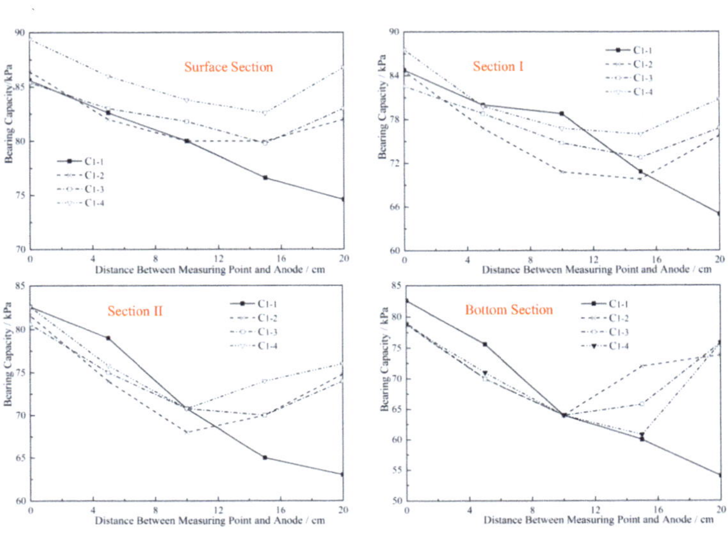

Fig. 6. Bearing Capacity of the Soil

were cut horizontally into different layers, as shown in Fig. 1. The bearing capacities of the soil at different depths were determined using a pocket penetrometer, and the vibration owing to cutting was ignored in the testing. Figure 6 shows the bearing capacities of the soils at different depths.

The same remolded soils were used in these tests, and the initial bearing capacities of the soils were 25 kPa. According to Fig. 6, the bearing capacity of the soil was improved when the testing was terminated. It is clear that the bearing capacities of the soils in tests C1-2 ~ C1-4 were improved, and the bearing capacity of the soil in test C1-1 was not changed. During the process of charging, free waters in the soils and weakly bound waters moved to the cathode and were discharged out of the soil. The water content of the soil was reduced owing to dewatering to result in the electro-osmotic compaction effect on the soils. Therefore, the improvement of the bearing capacity of the soil was mainly due to dewatering. The $CaCl_2$ solution was injected into the testing to increase the current, enhance the speed of ion movement, improve the exchanging abilities of ions, and promote the dewatering effect on the soils. Ca^{2+} ions were moved and precipitated to the cathode under the action of electronic field. When charging was terminated, some ions in the soils moved freely to improve the bearing capacity of the soil near the cathode. Furthermore, electrolysis was performed in the anode to produce ferric hydroxide colloids, and the ferric hydroxide colloids possessed the good cementation on the soils to consolidate the soils near the anode.

4 Conclusions

According to analysis on the amount of discharged water, current, energy consumption, and dewatering efficiency, the concentration of the injected solution had obvious effects on the electro-osmotic chemical consolidation on the soils. The increase of the concentration of the solution resulted in the increase of the amount of discharged water and the increase of the current. In addition, the speed of dewatering and the dewatering efficiency were influenced by the concentration of the injected solution. The dewatering efficiency increased with the increase of the concentration of the injected solution, and the energy consumption in the testing increased with the increase of the concentration of the solution. The higher concentration of the solution provided more economic efficiency to the electro-osmotic chemical treatment on the soils. According to the analysis on the bearing capacity of the treated soil, the bearing capacities of soils in four tests were improved. Water was injected in the test C1-1, and the bearing capacity of the soils presented the decreasing trend from the anode to the cathode. In the test C1-2 ~ C1-4, the bearing capacity of the soil near the cathode also increased, and the improvement effect in the tests with the salt solutions was better than the test with the water.

Acknowledgement. This project was supported by the Scientific Research Fund of Institute of Engineering Mechanics, China Earthquake Administration (Grant No. 2014B03), the National Key Research and Development Program of China (Grant NO.2016YFC0800205), and the National Natural Science Foundation of China (Grant NO.51438004 and NO.51408566).

References

1. Casagrande, L.: Electro-osmosis in soils. Géotechnique **1**(3), 159–177 (1949)
2. Casagrande, L.: Electro-osmotic stabilization of soils. J. Boston Soc. Civil Eng. **39**(1), 51–83 (1952)
3. Casagrande, L.: Stabilization of soils by means of electro-osmotic state-of-art. J. Boston Soc. Civil Eng. **69**(3), 255–302 (1983)
4. Chew, S.H., Karunaratne, G.P., Kuma, V.M., et al.: A field trial for soft clay consolidation using electric vertical drains. Geotext. Geomembr. **22**(1), 17–35 (2004)
5. Shang, J.Q., Lo, K.Y., Huang, K.M.: On factors influencing electro-osmotic consolidation. Geotech. Eng. **27**(4), 23–26 (1996)
6. Fetzer, C.A.: Electro-osmotic stabilization of west branch dam. J. Soil Mech. Found. Div. **93** (4), 85–106 (1967)
7. Lo, K.Y., Ho, K.S., Inculet, I.I.: Field test of electro-osmotic strengthening of soft sensitive clays. Can. Geotech. J. **28**(l), 74–83 (1991a)
8. Lo, K., Inculet, I.I., Ho, K.: The effects of electroosmotic field treatment on the soil properties of a soft sensitive clay. Can. Geotech. J. **28**(6), 763–770 (1991)
9. Katti, R.K.: Stabilization of Bombay marine clay with various inorganic additives. In: Proceedings of the 2nd South East Asian Conference on Soil Engineering, vol. 27(1), pp. 589–599 (1967)
10. Esrig, M.I.: Pore pressure, consolidation and electrokinetics. J. Soil Mech. Found. Div. ASCE **94**(4), 899–922 (1968)
11. Gray, D.H.: Electrochemical hardening of clay soils. Géotechnique **20**(1), 81–93 (1970)
12. Ozkan, S., Gale, R.J., Seals, R.K.: Electrokinetic stabilization of kaolinite by injection of Al and PO_4^{-3} ions. Proc. ICE-Ground Improv. **3**, 135–144 (1999)
13. Alshawabkeh, A.N., Thomas, C.S.: Soft soil stabilization by ionic injection under electric fields. Proc. ICE-Ground Improv. **7**(4), 177–185 (2003)
14. Asavadorndeja, P., Glawe, U.: Electrokinetic strengthening of soft clay using the anode depolarization method. Bull. Eng. Geol. Env. **64**(3), 237–245 (2005)
15. Zeng, G.X., Gao, Y.C.: Electrochemical reinforcement of soft clay. J. Zhejiang Univ. **8**(1), 12–35 (1956)
16. Ou, C.Y., Chien, S.C., Wang, Y.G.: On the enhancement of electro-osmotic soil improvement by the injection of saline solutions. Appl. Clay Sci. **44**(1–2), 130–136 (2009)
17. Chien, S.C.: A study of ground improvement using electroosmotic chemical grouting. Ph.D. thesis, National Taiwan University of Science and Technology, Taipei, Taiwan (2003)
18. Chien, S.C., Ou, C.Y., Wang, Y.H.: Soil improvement using electroosmosis with the injection of chemical solutions: laboratory tests. J. Chin. Inst. Eng. **34**(7), 863–875 (2011)
19. Chien, S.C., Ou, C.Y., Wang, M.K.: Injection of saline solutions to improve the electro-osmotic pressure and consolidation of foundation soil. Appl. Clay Sci. **44**(3–4), 218–224 (2009)
20. Ou, C.Y., Chien, S.C., Chang, H.H.: Soil improvement using electroosmosis with the injection of chemical solutions: field tests. Can. Geotech. J. **46**(6), 727–733 (2009)
21. Chang, H.W., Krishna, P.G., Chien, S.C., Wang, M.K.: Electro-osmotic chemical treatments: Effects of Ca^{2+} concentration on the mechanical strength and pH of kaolin. Clays Clay Miner. **58**(2), 154–163 (2010)
22. Zhuang, Y.F., Li, X., Wang, Z.: Application of electro-kinetic geosynthetics in reinforced slope. In: GeoAsia 2004 Proceedings of the 3rd Asian Regional Conference on Geosynthetics, pp. 1042–1047 (2004)

23. Zhuang, Y.F., Wang, Z.: Interface electric resistance of electroosmotic consolidation. J. Geotech. Geoenviron. Eng. **133**(12), 1617–1621 (2007)
24. Hu, L., Wu, W., Wu, H.: Numerical model of electro-osmotic consolidation in clay. Géotechnique **62**(6), 537–541 (2012)

Experimental Study of Fiber-Reinforced Sand Subject to Seepage

K.-H. Yang[1(✉)], W.M. Adilehou[2], S.-T. Jian[2], and B.-C. Hsiung[3]

[1] Department of Civil Engineering, National Taiwan University, Taipei, Taiwan
khyang@ntu.edu.tw
[2] Department of Civil and Construction Engineering,
National Taiwan University of Science and Technology, Taipei, Taiwan
[3] Department of Civil Engineering,
National Kaohsiung University of Applied Science, Kaohsoung, Taiwan

Abstract. This paper presents an experimental study on the hydraulic responses of fiber-reinforced soil (FRS) subject to seepage. A series of upward seepage tests on unreinforced and reinforced sand was conducted to investigate the influence of soil density and fiber parameters (i.e., fiber contents and lengths) on the piping failure mode, hydraulic conductivity k, and critical hydraulic gradient i_{cr} of FRS. Direct shear tests were also performed to establish the relationships between soil shear strength and critical hydraulic gradient of FRS. The seepage test results revealed that k decreases and i_{cr} increases as the fiber content increases. Short fiber appears to reduce k; however, the fiber length has only a minor influence on the i_{cr}. The fiber has a greater effect on dense specimens than it does on loose specimens. The test results also indicate the i_{cr} of FRS is strongly correlated to its soil shear strength. The findings in this study suggest that the use of FRS as backfill in hydraulic structures can effectively delay the advance of seepage, reduce soil piping potential, and improve system stability against seepage. The results and discussion in this study provide insightful information for the application of FRS to hydraulic structures.

Keywords: Fiber-reinforced sand · Critical hydraulic gradient · Hydraulic conductivity · Soil shear strength

1 Introduction

Hydraulic failures of geotechnical earth structures are caused by impacts from wave forces, toe scour, overtopping, and soil piping and erosion (Brandl 2011). As taller hydraulic structures are constructed because of rising flood levels due to the influence of global warming and extreme weather, the increasing hazards of soil piping and erosion have gained much attention (Danka and Zhang 2015).

Past case studies have reported that failures of many manmade earth fill structures, natural soil masses, and rock deposits (e.g., landslide dams) have been associated with seepage-induced piping and erosion (Foster et al. 2000). Consequently, mitigating seepage-induced adverse impacts and enhancing the stability of the earth structures has become an urgent and challenging issue in waterfront protection.

R. Chen et al. (eds.), *Proceedings of the 2nd International Symposium on Asia Urban GeoEngineering*, Springer Series in Geomechanics and Geoengineering,
https://doi.org/10.1007/978-981-10-6632-0_5

Effective countermeasures against soil piping involve the use of cut-off walls, impervious blankets, and pressure relief wells to reduce the hydraulic gradient within the soil or the use of soil improvement and filter layers to increase soil piping resistance. Among these measures, soil improvement through fiber reinforcement is the focus of this study.

The majority of previous studies have focused on the mechanical behavior of fiber-reinforced soil (FRS) and demonstrated that mixing fiber with soil can effectively enhance peak shear strength and reduce the loss of postpeak shear strength (Consoli et al. 2009; Michalowski and Čermák 2003; Zornberg 2002). Soil type, fiber type, length, and content, as well as cement content and compaction conditions, are the key parameters evaluated in these studies (Chou et al. 2016).

In contrast to the studies on mechanical behavior of FRS, studies on the hydraulic performance of FRS are relatively limited. Only a few studies of experimental seepage tests on FRS (Das et al. 2009; Das and Viswanadham 2010; Estabragh et al. 2014, 2016; Furumoto et al. 2002; Sivakumar Babu and Vasudevan 2008) and model tests for levees reinforced with short fibres (Furumoto et al. 2002) have been reported in the literature. In summary, these studies have found that fiber reinforcement can effectively enhance the piping resistance of soil by deferring the occurrence of soil piping at a high hydraulic gradient. Little attention has focused on the effect of fiber parameters on the hydraulic conductivity of FRS, which is another crucial hydraulic parameter of FSR. The preceding discussion is the basis of this study, which involved conducting a series of upward seepage tests on unreinforced and reinforced sands. The objectives of this study are as follows: (1) to evaluate the influence of soil density and fiber parameters (i.e., fiber length and content) on the hydraulic responses (i.e., piping failure mode, hydraulic conductivity, and critical hydraulic gradient) of FRS; and (2) to establish the relationship between the critical hydraulic gradient and the shear strength parameter of FRS. The results and discussion in this study provide insightful information for the application of FRS to hydraulic structures.

2 Experimental Program

2.1 Test System

A series of seepage tests were conducted to evaluate the influences of soil density and fiber parameters on the hydraulic conductivity and critical hydraulic gradient of FRS. For this purpose, an upward seepage test system (Fig. 1), consisting of a constant head device, a permeameter, and measuring systems, was developed in this study and is described in this section.

The permeameter consists of a cylindrical cell (10.5 cm in diameter and 38 cm in height) and a bottom pedestal. To avoid a scale effect, the ratios of the specimen diameter to the mean grain diameter of sand in this test were 90, which is larger than the values (8–12) specified in ASTM (D2434). Additionally, the diameter of the specimen (i.e., 10.5 cm) also satisfies the ASTM requirement (i.e., cylinder diameter >7.6 cm) with respect to the grain sizes of the used soils. The pedestal, filled with marbles and covered with porous screens was used to distribute the upward seepage

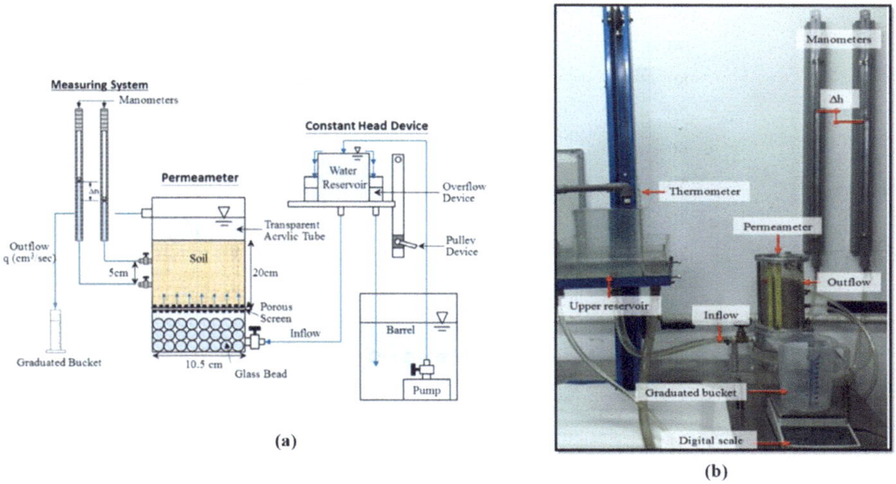

Fig. 1. Upward seepage test system: (a) schematic illustration; (b) overview photo

evenly across the soil specimen. The porous screens comprised two perforated metal plates and a nonwoven geotextile. The perforated metal plates, with numerous punched holes, were used to support the overburden pressure from soil specimens. The nonwoven geotextile was placed between the two perforated metal plates and served as a filter to prevent the loss of soil.

The water flow from the top of the specimen was measured and then discharged to the barrel. Discharge velocity v at a given hydraulic gradient was calculated by dividing the collected volume of discharge at a certain time period by the cross-sectional area of the soil specimen. The permeameter cylinder was perforated at distances of 7 and 12 cm from the bottom of the specimen (Fig. 1) and connected to graduated manometers to measure hydraulic head difference (i.e., head loss) at a given distance of the seepage path. The corresponding hydraulic gradient i can then be calculated at each stage of the test using the following equation:

$$i = \frac{\Delta h}{L} \tag{1}$$

where Δh is the head difference between two manometers and L (=5 cm) is the distance between the two measuring valves connected to the manometers.

2.2 Test Material and Test Program

Uniform quartz sand was tested in this study. Figure 2 presents the grain size distribution curve of the test soil. Table 1 summarizes the soil properties. The minimum and maximum dry unit weights of sand, conducted in accordance with ASTM (D4254, D4253) at the two target relative densities (D_r = 50% and 70%) to represent soil loose and dense conditions, were $\gamma_{d,min}$ = 13.92 kN/m^3 and $\gamma_{d,max}$ = 15.35 kN/m^3. Based on

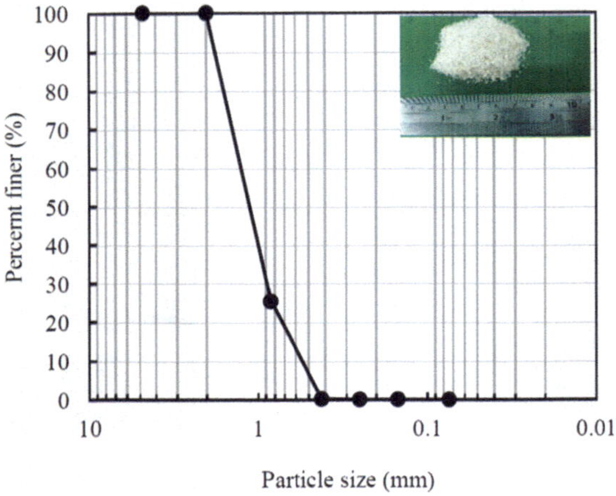

Fig. 2. Grain size distribution curve and photo of the tested sand

Table 1. Summary of test soil properties

Soil properties	Value
Specific gravity G_s	2.65
Effective particle size d_{10} (mm)	0.61
Mean particle size d_{50} (mm)	1.12
Uniformity coefficient C_u	2.02
Coefficient of curvature C_c	1.08
Soil classification (USCS)	SP
Maximum dry unit weight $\gamma_{d,max}$ (kN/m^3)	15.35
Minimum dry unit weight $\gamma_{d,min}$ (kN/m^3)	13.92
Maximum void ratio e_{max}	0.87
Minimum void ratio e_{min}	0.63
[a]Hydraulic conductivity of soil k (m/s)	8.2×10^{-3} and 6.1×10^{-3}
[a]Critical hydraulic gradient of soil i_{cr}	0.86 and 0.92

[a] Values are for soil density D_r = 50% and 70%, respectively.

the constant head test results presented later, the saturated hydraulic conductivity of the soil was $k = 8.2 \times 10^{-3}$ and 6.1×10^{-3} m/s, and the critical hydraulic gradient of the soil was i_{cr} = 0.86 and 0.92 for soil at D_r = 50% and 70%, respectively.

Polypropylene (PP) fiber was used in this study. PP fiber is the most widely adopted synthetic fiber for soil reinforcement (Yetimoglu et al. 2003; Hejazi et al. 2012). Das and Viswanadham (2010) reported that PP fiber performed better than polyester (PET) fiber in increasing seepage resistance. It is because the PET fiber has a specific gravity higher than the PP fiber. For the same fiber content, a larger specific gravity implies a lower fiber volume and a less number of fibers, and hence reduces the benefit of improving the piping resistance of a soil. The PP fiber tested in this study has a circular cross-section with an average diameter of 0.0577 mm. The specific gravity of the fiber is $G_{sf} = 0.91$, slightly lower than that of water. Table 2 summarizes the physical and mechanical properties of the test fiber. A total of 20 seepage tests were conducted on both unreinforced and reinforced soil specimens. The test variables are

Table 2. Physical and mechanical properties of fiber

Properties	Value
Type	Polypropylene (PP) fiber
Cross-section shape	Circular
Equivalent diameter d_f (mm)	0.0557
Length L_f (mm)	6, 12, 19
Specific gravity G_{sf}	0.91
Denier (g/9000 m)	20
Ultimate tensile strength (MPa)	250
Melting point (°C)	160–170
Water absorption	No

Table 3. Summary of seepage test conditions and results

Test	ω_f (%)	L_f (mm)	i_{cr}	i_{cr} ratio	k (m/s)	k ratio
Test Series L (Loose Specimen)						
U-50	0	0	0.86	1	8.2×10^{-3}	1
R-50-0.5-6	0.5	6	1.16	1.35	4.3×10^{-3}	0.52
R-50-1-6	1	6	1.2	1.40	3.7×10^{-3}	0.45
R-50-1.5-6	1.5	6	1.52	1.77	2.2×10^{-3}	0.27
R-50-0.5-12	0.5	12	1.08	1.26	5.3×10^{-3}	0.65
R-50-1-12	1	12	1.3	1.51	4.4×10^{-3}	0.54
R-50-1.5-12	1.5	12	1.52	1.77	3.2×10^{-3}	0.39
R-50-0.5-19	0.5	19	1.14	1.33	6.5×10^{-3}	0.79
R-50-1-19	1	19	1.34	1.56	3.9×10^{-3}	0.48
R-50-1.5-19	1.5	19	1.36	1.58	3.5×10^{-3}	0.43
Test Series D (Dense Specimen)						
U-70	0	0	0.92	1	6.1×10^{-3}	1
R-70-0.5-6	0.5	6	1.28	1.39	4.1×10^{-3}	0.68
R-70-1-6	1	6	1.40	1.52	2.8×10^{-3}	0.46
R-70-1.5-6	1.5	6	1.94	2.11	1.4×10^{-3}	0.24
R-70-0.5-12	0.5	12	1.22	1.33	3.3×10^{-3}	0.54
R-70-1-12	1	12	1.50	1.63	3.1×10^{-3}	0.51
R-70-1.5-12	1.5	12	1.74	1.89	2.5×10^{-3}	0.41
R-70-0.5-19	0.5	19	1.28	1.39	4.1×10^{-3}	0.67
R-70-1-19	1	19	1.36	1.48	3.3×10^{-3}	0.54
R-70-1.5-19	1.5	19	1.62	1.76	2.5×10^{-3}	0.41

soil relative density (D_r = 50% and 70%), fiber content (ω_f = 0%, 0.5%, 1.0%, and 1.5%), and fiber length (L_f = 6, 12, and 19 mm).

Table 3 lists the seepage test program. The test numbering was defined as follows. The first part, a letter "R" or "U," indicates a reinforced or unreinforced specimen, respectively. The second, third, and fourth parts denote soil relative density, fiber content, and fiber length, respectively. For example, U-50 indicates an unreinforced specimen with soil relative density D_r = 50% and R-70-0.5-6 indicates a reinforced specimen with soil relative density D_r = 70%, fiber content ω_f = 0.5% and fiber length L_f = 6 mm.

2.3 Specimen Preparation and Test Procedure

The specimens were prepared at loose and dense conditions (i.e., Test Series L and D), corresponding to soil relative density D_r = 50% and 70%. Each specimen was carefully prepared to ensure that its soil had a uniform density and full saturation. The required weight of dry soil for the target relative density was determined using the relative density equation:

$$D_r = \frac{e_{\max} - e}{e_{\max} - e_{\min}} \tag{2}$$

where e_{max}, e_{min}, and e are the maximum, minimum, and target void ratios of soil. The desired weight of fiber for reinforced specimens was determined considering the dry weight of the soil and the desired gravimetric fiber content, as expressed in Eq. (3):

$$\omega_f = \frac{W_f}{W_s} \tag{3}$$

where W_f and W_s are the dry weight of fiber and soil, respectively.

A known quantity of soil and fiber was carefully mixed by hand. The hand mixing method has been commonly adopted by various researchers (Yetimoglu and Salbas 2003; Das et al. 2009; Estabragh et al. 2014). The soil–fiber mixture was moisturized by adding water (10% of total weight) to avoid soil–fiber segregation before being spread into the permeameter. The permeameter was carefully filled with the wet soil–fiber mixture in five layers (4-cm thick for each layer). Each layer was slightly compacted using a metal rod to control its height. This procedure was repeated until the desired specimen height (H = 20 cm) was reached. Visual inspection showed that good uniformity was achieved. The repeatability and consistency of the test results were verified by examining test results performed under the same conditions.

Figure 3 shows the scanning electron microscope (SEM) photographs of unreinforced specimens at different relative densities (Figs. 3a and b) and the FRS specimen of R-70-1-6 (Fig. 3c). The interaction between fibers and soil grains can be clearly observed from Fig. 3c. Compared with unreinforced sand, some soil pore spaces (as highlighted by red dashed lines in Fig. 3c) of the reinforced sand were partially filled by the fibers. Consequently, the fibers could block some pore channels for seepage and restrict the seepage flow within these pore channels.

Fig. 3. Scanning electron microscope (SEM) photo: (a) unreinforced sand at D_r = 50% (U-50); (b) unreinforced sand at D_r = 70% (U-70); (c) fiber-reinforced sand (R-70-1-6)

After specimen preparation, the specimen was submerged in water and subjected to a constant seepage flow under a low hydraulic head that did not affect the specimen stability for 24 h to ensure the full saturation of specimens. Afterward, the seepage test began by applying a series of incrementally increased hydraulic heads to the specimen until soil piping failure occurred. The applied hydraulic head was increased by 2 cm (approximately Δi = 0.1) for each increment and maintained for at least 10 min until the hydraulic heads in the manometers stabilized, indicating that equilibrium was reached. The hydraulic gradient i and corresponding discharge velocity v were recorded in each stage of the test.

3 Results and Discussion

In this section, the test results including seepage failure mode and the i–v relations were analyzed and discussed. The i–v plots were used to determine the k and i_{cr} of FRS. The influence of soil density and fiber parameters on the hydraulic responses of FRS is quantitatively evaluated and discussed, and the relationship between the critical hydraulic gradient and the shear strength of FRS is established.

3.1 Failure Modes

Figure 4 shows the typical failure modes of unreinforced and reinforced specimens at and after the critical hydraulic gradient i_{cr}. For the unreinforced specimen at i_{cr} (Fig. 4a), the specimen (U-70) exhibited an expansion (≈0.8 cm). At this stage, the soil seemed to have liquefied (the author's finger can easily penetrate into the specimen without feeling much resistance). When the next hydraulic head increment after i_{cr} was applied, the specimen showed a sudden and notable heave (≈2 cm), following by the sand piping/boiling phenomenon (Fig. 4b). The soil lost it overall stability. The heave and boiling phenomenon are strong evidence of soil failure subjected to seepage. The soil boiling happened globally within the unreinforced specimen in which soil particles were forced to migrate with the upward seepage. Vigorous soil boiling on top of the specimen can also be clearly observed.

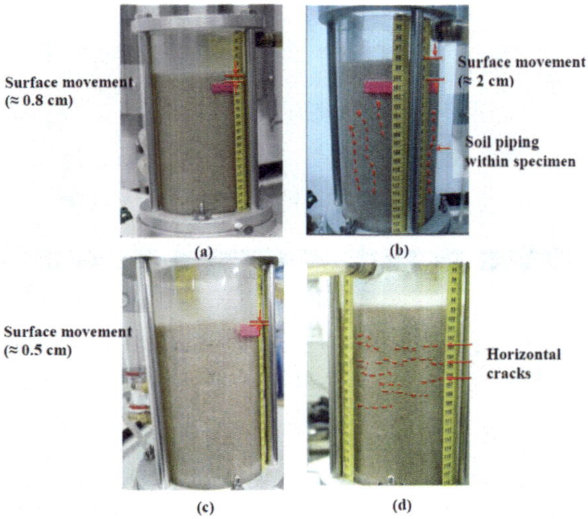

Fig. 4. Failure mode of unreinforced and reinforced specimens: (a) U-70 at i_{cr} = 0.92; (b) U-70 at the next applied hydraulic head increment after i_{cr}; (c) R-70-1.5-19 at i_{cr} = 1.62; (d) development of horizontal cracks of R-70-1.5-19 at the next applied hydraulic head increment after i_{cr}

For the reinforced specimen at i_{cr} (Fig. 4c), the specimen (R-70-1.5-19) displayed an isotropic failure mode: soil experienced a uniformly slight heave (≈0.5 cm). At this stage (onset of soil seepage failure), the surface movement of the reinforced specimen is smaller than that of the unreinforced specimen even though the reinforced specimen was subject to a seepage force (i_{cr} = 1.62) higher than that of the unreinforced specimen (i_{cr} = 0.92). Figure 4d shows the failure mode of the reinforced specimen when the next hydraulic head increment after i_{cr} was applied. Several horizontal micro cracks developed within the specimen at this stage.

The development of horizontal cracks within the reinforced specimen indicates significant distress on the specimen upon seepage force. The observed horizontal cracks also reveal that the seepage force induced tensile force could have exceeded the tensile resistance provided by fibers at local areas within the specimen; consequently, soil particles at these areas tend to separate and then the cracks develop. Unlike the failure mode of the unreinforced specimen, the reinforced specimen experienced neither global soil piping within the specimen nor vigorous soil boiling on top of the specimen. This observation demonstrates that fiber can effectively bind soil together against seepage and convert the vigorous soil erosion and piping in the unreinforced soil to a global and isotropic soil expansion.

3.2 *I–V* Relations

Figures 5 and 6 show the *i–v* plots for different soil densities, fiber contents, and fiber lengths. The *i–v* curves consist of two parts. At the first part of the curves, the *v* value increases linearly with *i*. The flow is laminar and the hydraulic conductivity *k* can be obtained according to Darcy's law (i.e., $v = ki$). At the second part of the curves, the i–v curves exhibit either a drastic increase in the discharge velocity or a decrease in the measured hydraulic gradient. Both phenomena indicate a change in the hydraulic behavior of specimens, suggesting the occurrence of the soil seepage failure. The critical hydraulic gradient was determined at one stage prior to this stage. Table 3 summarizes the test results of k and i_{cr}.

Although the system total head was increased by elevating the upper water reservoir, the drop in the hydraulic gradient after i_{cr} suggests the relief of accumulated pore water pressure within specimens after soil piping failure (Fig. 4a) and the development of horizontal cracks (Fig. 4d). The drop can also be explained by the fact that the head

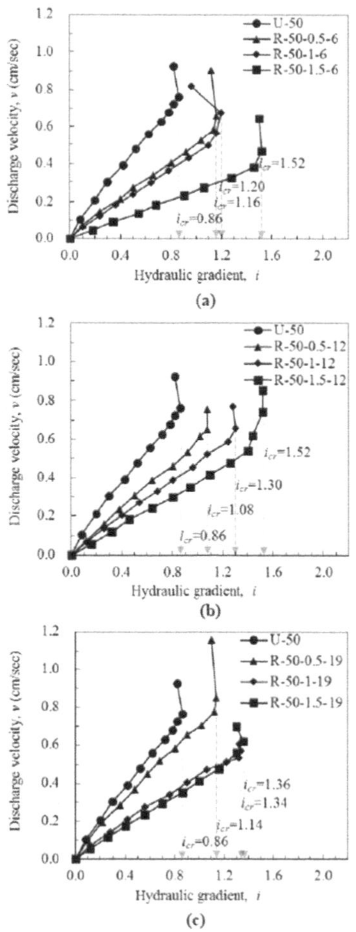

Fig. 5. Results for Test Series L:(a) $L_f = 6$ mm; (b) $L_f = 12$ mm; (c) $L_f = 19$ mm

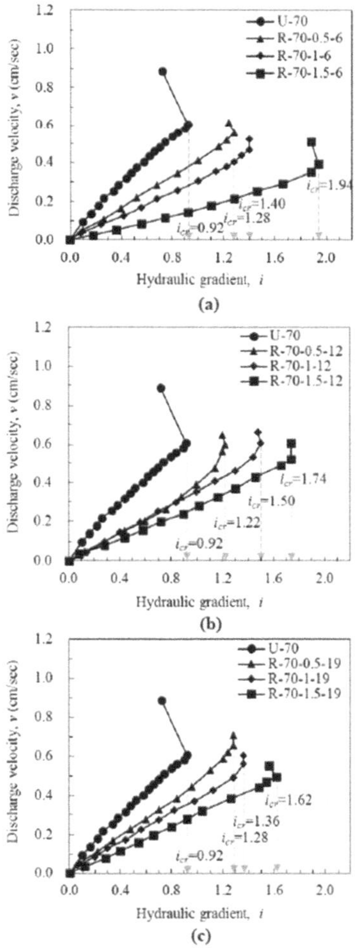

Fig. 6. Results for Test Series D:(a) $L_f = 6$ mm; (b) $L_f = 12$ mm; (c) $L_f = 19$ mm

loss of seepage through specimens decreased because of the loosened soil packing state after the soil seepage failure. A pressure drop has also been observed in experimental and field tests (Parekh et al. 2016). Consequently, the water pressure measured at this stage cannot represent the real hydraulic conditions.

Figures 5 and 6 clearly reveal that for specimens at the same relative density and fiber length, the *i*–*v* curves shift to the right as the fiber content increases, signifying an increase in i_{cr} and a decrease in *k*. The experimental results likely result from the fiber inclusion providing tensile resistance against soil piping and erosion, and restricting the seepage flow within some pore channels which were partially blocked by fibers. An increase in the critical hydraulic gradient of FRS with fiber content was also reported in Das et al. (2009), Das and Viswanadham (2010), Estabragh et al. (2014, 2016), and Sivakumar Babu and Vasudevan (2008). A detailed and quantitative evaluation of the effect of fiber content and length follows in the next section.

3.3 Effect of Fiber Parameters

This section discusses the effect of fiber parameters (i.e., fiber content and length) on the hydraulic response of FRS. The variation of *k* and i_{cr} with fiber parameters was quantitatively evaluated using the *k* and i_{cr} ratios, defined as the ratios of the hydraulic conductivity and critical hydraulic gradient of FRS to those of unreinforced sand. The *k* and i_{cr} ratios serve as indices for assessing the reduction of seepage velocity and improvement of soil piping resistance.

Figure 7 shows the influence of fiber content on *k* and i_{cr} and Table 3 lists the associated *k* and i_{cr} ratio values. A clear trend of increasing i_{cr} and decreasing *k* with increasing fiber content can be observed, regardless of soil density. For example, for FRS with $L_f = 6$ mm in Test Series L, the *k* decreases from 8.2×10^{-3} to 2.2×10^{-3} m/s (*k* ratio decreases from 1.0 to 0.27) and the i_{cr} increases from 0.86 to 1.52 (i_{cr} ratio increases from 1.0 to 1.77) as the fiber content increases from 0 to 1.5%. These results suggest that soil piping resistance increases to 177% of, and hydraulic conductivity decreases to 27% of that of unreinforced sand after adding 1.5% of fiber to the soil. Similarly, for FRS with $L_f = 6$ mm in Test Series D, the *k* decreases from 6.1×10^{-3} to 1.4×10^{-3} m/s (*k* ratio decreases from 1.0 to 0.24) and the i_{cr} increases from 0.92 to 1.94 (i_{cr} ratio increases from 1.0 to 2.11) as the fiber content increases from 0 to 1.5%.

The test results also reveal that the fiber has a greater effect in dense specimens (Test Series D) than in loose specimens (Test Series L); at a given fiber length and content, the larger i_{cr} and smaller *k* values were measured for dense specimens. The high i_{cr} improvement likely results from increased soil–fiber interaction in dense soil states, and high *k* reduction is attributable to the fact that the pore space of soil in a dense state, which is less abundant and smaller than that of soil in a loose state, could be easily blocked or filled with a given amount of fiber.

Figure 8 shows the influence of fiber length on *k* and i_{cr} and Table 3 lists the associated *k* and i_{cr} ratio values. In general, the *k* values appear to decrease with a decrease in fiber length for specimens at both densities (Fig. 8a and b). That is, compared with long fiber, the short fiber can produce higher *k* reduction. For example, for FRS with $\omega_f = 1.5\%$, the *k* increases from 2.2×10^{-3} to 3.5×10^{-3} m/s (*k* ratio increases from 0.27 to 0.43) in Test Series L and from 1.4×10^{-3} to 2.5×10^{-3} m/s

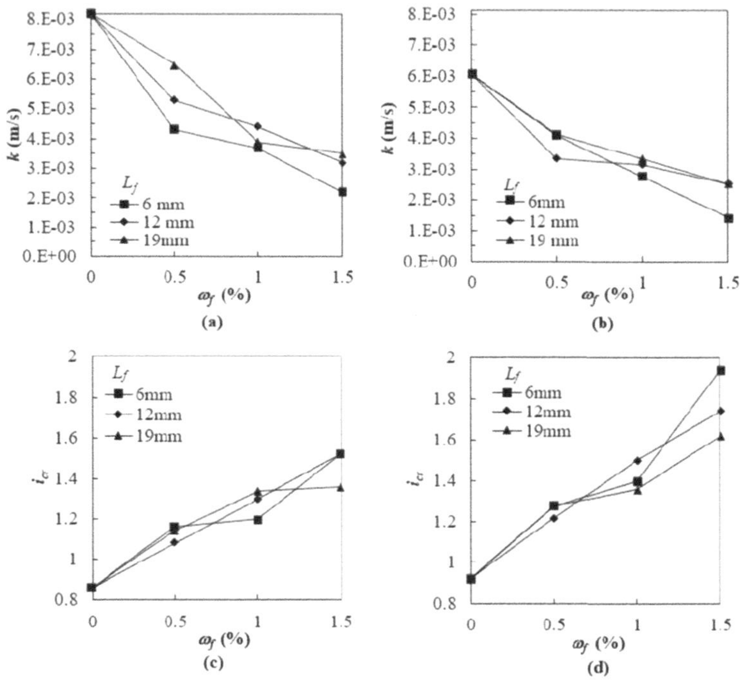

Fig. 7. Effect of fiber content on: hydraulic conductivity (a) Test Series L: (b) Test Series D, and critical hydraulic gradient (c) Test Series L, (d) Test Series D

(*k* ratio increases from 0.24 to 0.41) in Test Series D, respectively, as the fiber length increases from 6 to 19 mm.

Better performance in short fiber (i.e., producing a low k value) is likely because the total amount of short fiber is greater than that of long fiber at the given fiber content, resulting in short fiber possibly being able to fill more pore space than long fiber can. Finally, the variation of i_{cr} with fiber length does not show a clear trend (Figs. 8c and d), suggesting that the fiber length has only a minor influence on the i_{cr}.

In summary, test results suggest that FRS, prepared with high fiber content and short fiber length, and compacted into a dense soil state, has a superior hydraulic performance for the improvement of soil piping resistance and the reduction of seepage velocity.

3.4 Relationship Between Soil Shear Strength and Critical Hydraulic Gradient

A series of direct shear tests was performed to establish the relationship between the soil shear strength and the critical hydraulic gradient. The direct shear test was conducted in accordance with ASTM (D3080). The specimen size was 6.33 cm in diameter of and 4.66 cm in height, subjected to normal pressures of 100, 200, and 400 kPa, under a shearing rate of 1.5 mm/min. The specimens were prepared in the

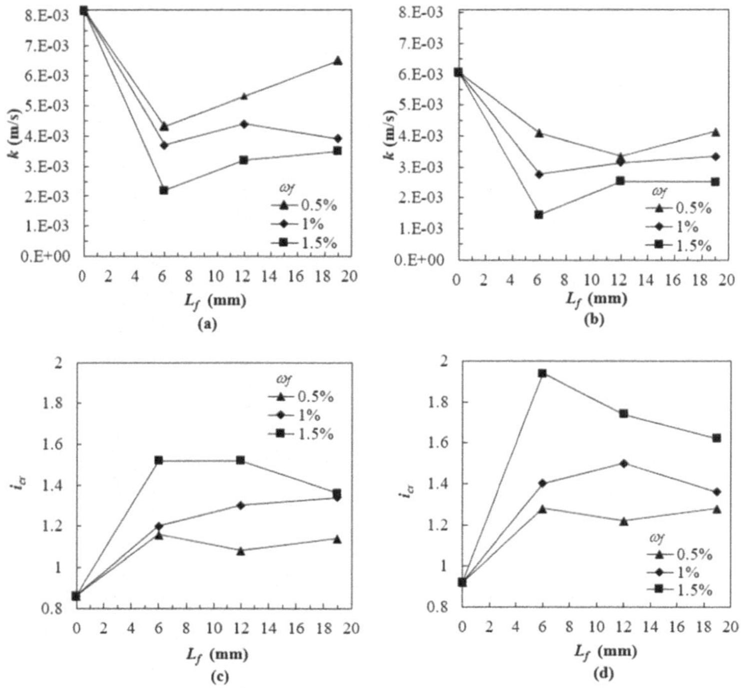

Fig. 8. Effect of fiber length on: hydraulic conductivity (a) Test Series L: (b) Test Series D, and critical hydraulic gradient (c) Test Series L, (d) Test Series D

same manner as the ones used for the seepage tests. The failure was determined corresponding to the maximum shear stress measured, or the shear stress at the relative displacement of 10 mm for the cases that no definite peak was noticed on the stress-displacement curve.

The peak soil shear strength τ_f for the specimens in the seepage test were determined from the Mohr–Coulomb failure envelopes using normal pressure in accordance with the overburden pressure at the bottom of specimens ($\approx$4 kPa). After obtaining the peak shear strength of reinforced and unreinforced sand, the relationship between the soil shear strength and critical hydraulic gradient can be established (Fig. 9). As shown in Fig. 9, the i_{cr} was strongly correlated with soil τ_f. The data from specimens at different soil densities and fiber parameters fell into a unique linear relationship. This linear relationship demonstrated that the soil shear strength improvement from the fiber inclusion directly contributed to the piping resistance of FRS. This strong correlation between the soil shear strength and critical hydraulic gradient could be also related to the global and isotropic expansion failure mode of FRS. It should be noted that the established relationship cannot be extrapolated because it was established based on the results of the soil tested under low overburden pressures, representing the cases of soils at a shallow depth underneath the surface of geotechnical earth structures. Further study is required to investigate the influence of the overburden pressure on the τ_f and i_{cr} relationship.

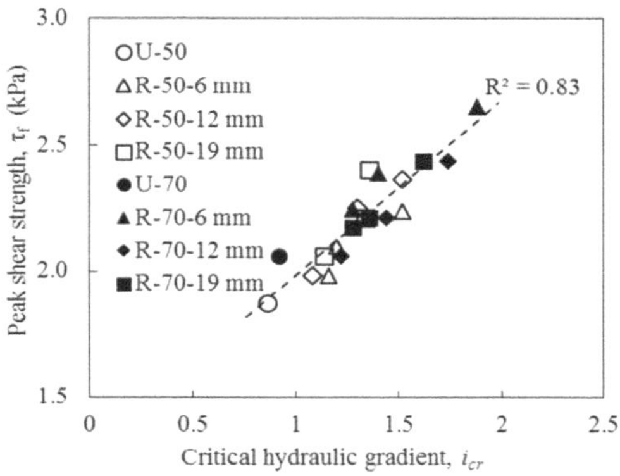

Fig. 9. Relationships between critical hydraulic gradient and soil shear strength

4 Conclusion

In this study, experimental seepage tests were conducted to investigate the hydraulic responses (i.e., piping failure mode, hydraulic conductivity, and critical hydraulic gradient) of FRS subject to seepage. Based on the test results, the following conclusions can be drawn:

(1) Unreinforced specimens had a failure mode associated with a significant soil heave and vigorous soil piping/boiling, whereas the reinforced specimens exhibited an isotropic failure mode (a uniform soil expansion) with several horizontal micro cracks developed within the specimen.
(2) Seepage test results revealed that i_{cr} increases and k decreases as the fiber content increases. The experimental results likely result from the fiber inclusion providing tensile resistance against soil piping and erosion, and restricting the seepage flow within some pore channels which were partially blocked by fibers.
(3) Test results suggested that FRS, prepared with high fiber content and short fiber length, and compacted into a dense soil state, has superior hydraulic performance in the improvement of soil piping resistance and reduction of seepage velocity.
(4) A unique linear relationship exists between i_{cr} and τ_f, indicating that soil shear strength improvement from fiber inclusion directly contributed to the piping resistance of FRS. This strong correlation between the soil shear strength and the critical hydraulic gradient is also related to the global and isotropic expansion failure mode of FRS as observed from the experimental tests.

References

ASTM D2434: Standard test method for permeability of granular soils (constant head). ASTM International, West Conshohocken

ASTM D3080: Standard test method for direct shear test of soils under consolidated drained condtions. ASTM International, West Conshohocken

ASTM D4253: Standard test methods for maximum index density and unit weight of soils using a vibratory table. ASTM International, West Conshohocken

ASTM D4254: Standard test methods for minimum index density and unit wieight of soils and calculation of relative density. ASTM International, West Conshohocken

Brandl, H.: Geosynthetics applications for the mitigation of natural disasters and for environmental protection. Geosynth. Int. **18**(6), 340–390 (2011)

Chou, J.-S., Yang, K.-H., Lin, J.-Y.: Peak shear strength of discrete fiber-reinforced soils computed by machine learning and metaensemble methods. J. Comput. Civ. Eng. **30**(6), 1–17 (2016). doi:10.1061/(ASCE)CP.1943-5487.0000595. ASCE

Consoli, N.C., Vendruscolo, M.A., Fonini, A., Rosa, F.D.: Fiber reinforcement effects on sand considering a wide cementation range. Geotext. Geomembr. **27**(3), 196–203 (2009)

Danka, J., Zhang, L.M.: Dike failure mechanisms and breaching parameters. J. Geotech. Geoenviron. Eng. **141**(9), 04015039 (2015). ASCE

Das, A., Jayashree, C., Viswanadham, B.V.S.: Effect of randomly distributed geofibers on the piping behaviour of embankments constructed with fly ash as a fill material. Geotext. Geomembr. **27**(5), 341–349 (2009)

Das, A., Viswanadham, B.V.S.: Experiments on the piping behavior of geofiber-reinforced soil. Geosynth. Int. **17**(4), 171–182 (2010)

Estabragh, A.R., Soltannajad, K., Javadi, A.A.: Improving piping resistance using randomly distributed fibers. Geotext. Geomembr. **42**(1), 15–24 (2014)

Estabragh, A.R., Soltanijad, A., Javadi, A.A.: Models for predicting the seepage velocity and seepage force in a fiber reinforced silty soil. Comput. Geotech. **75**, 174–181 (2016)

Foster, M., Fell, R., Spannagle, M.: The statistics of embankment dam failures and accidents. Can. Geotech. J. **37**(5), 1000–1024 (2000)

Furumoto, K., Miki, H., Tsuneoka, N., Obata, T.: Model test on the piping resistance of short fiber reinforced soil and its application to river levee. In: Proceeding of the 7th International Conference on Geosynthetics, Nice, France, pp. 1241–1244 (2002)

Hejazi, S.M., Sheikhzadeh, M., Abtahi, S.M., Zadhoush, A.: A simple review of soil reinforcement by using natural and synthetic fibers. Constr. Build. Mater. **30**, 100–116 (2012)

Michalowski, R., Čermák, J.: Triaxial compression of sand reinforced with fibers. J. Geotech. Geoenviron. Eng. **129**(2), 125–136 (2003). ASCE

Parekh, M., Kanning, W., Bocovich, C., Mooney, M.A., Koelewijn, A.R.: Backward erosion monitored by spatial-temporal pore pressure changes during field experiments. J. Geotech. Geoenviron. Eng. **142**(10), 1–10 (2016). doi:10.1061/(ASCE)GT.1943-5606.0001528. ASCE

Sivakumar Babu, G.L., Vasudevan, A.K.: Seepage velocity and piping resistance of coir fiber mixed soils. J. Irrig. Drain. Eng. **134**(5), 668–680 (2008). ASCE

Yetimoglu, T., Salbas, O.: A study on shear strength of sands reinforced with randomly distributed discrete fibers. Geotext. Geomembr. **21**(2), 103–110 (2003)

Zornberg, J.: Discrete framework for limit equilibrium analysis of fiber-reinforced soil. Géotechnique **52**(8), 593–604 (2002)

One Dimensional Consolidation Theory Considering the Strain Rate Effect

Quan Liu, Yue-bao Deng(✉), Fei Chen, and Sen-jie Tong

Institute of Geotechnical Engineering, Ningbo University, Ningbo, Zhejiang, China
dengyuebao@nbu.edu.cn

Abstract. The traditional Terzaghi's one dimensional consolidation theory assumes a linear stress-strain relationship for soil. But in reality, the stress-strain relationship together with the permeability behavior of soft soil is usually nonlinear. Following Davis's infiltration hypothesis, an assumption about the permeability index and the compression index is made. The effect of the strain rate on the compression behavior soft soil is also taken into considering. Based on these assumptions, a new one dimensional consolidation theory considering the strain rate is established. The effective stress-strain-strain rate model is thus set up. The effect of the strain rate on the consolidation process is analyzed. The conclusion shows that the consolidation rate considering the strain rate is faster than it by traditional one-dimensional consolidation method. Also, it found that the higher the strain rate is, the faster the consolidation rate will be.

Keywords: Soft clay · Consolidation · Settlement · Strain rate effect · Analytical solution

1 Introduction

It has been generally accepted that the stress-strain relationship of saturated clay is strongly influenced by the strain rate. As early as 1936, Buisman found the soil deformation would develop slowly over time. In the early 1960s, Craw-ford (1964) considering the effect of strain rate during consolidation and found the strain rate had a significant effect on the soil deformation. By summing up a large group of one-dimensional compression tests of different types of clay, Leroueil [5, 6] further explored the effect of strain rate and found the pre-consolidation stress σ'_p increased with the strain rate, especially in the double logarithmic coordinates the pre-consolidation stress and strain rate showed a linear relationship. In addition, many triaxial test results show that the strain rate effect can be extended to the whole initial yield surface, the larger the initial yield surface as the strain rate increased [4, 8]. At the same time Leroueil [7] pointed out that in the last 20 years, people had done lot of consolidation tests on natural clay with different strain rates (CRS) (for eastern clay in Canada; Crawford1963, Vailetal1979; Leroueil et al. 1983, etc.) All the experiments showed a similar result that the strain rate affects the behaviour of clay [1, 12, 13]. Jarrett (1967) found that for a given strain the higher the strain rate, the higher the effective stress would be. After that, Burghignoli (1979) and Larsson (1981) also

R. Chen et al. (eds.), *Proceedings of the 2nd International Symposium on Asia Urban GeoEngineering*, Springer Series in Geomechanics and Geoengineering,
https://doi.org/10.1007/978-981-10-6632-0_6

verified the validity of this view according to different types of consolidation tests. Duncan [3] also pointed out that both indoor and field tests indicated that the compression of the clay depended on the change of the strain rate. And the excessive strain rate of indoor tests will prevent the compression process while the relatively low strain rate of site tests will accelerate the compression process.

Many scholars have improved the one-dimensional consolidation theory of Terzaghi, and developed into a new one-dimensional consolidation theory considering the nonlinearity of compression and permeability. For example, based on the hypothesis that the permeability coefficient and the compression coefficient are changed synchronously during the consolidation, Davis et al. [2] (1965) first got the analytical solution of saturated clay under the self-stress was uniform. Barden et al. (1965), Mesri et al. (1974) used the currently accepted e-logσ' and e-logk_v linear models, obtained the corresponding compression curve by using finite difference method. In domestic, Xie Kanghe et al. [9–11] (1996, 2002) based on Davis's assumption, obtained the one-dimensional nonlinear consolidation analysis under the condition of single and double layers foundation.

In summary, many scholars have studied the impact of the strain rate, but quantitative study of the impact of strain rate is still blank, and how to introduce strain rate factor into consolidation theory is still important. Based on available cases this paper proposed a new consolidation model with considering of stress rate. Finally numerical analysis is used to verify the correctness of the model.

2 Constitutive Model of Soil

Leroueil (2006) summed up the behavior of soft soil. Professor Šuklje (1957) presented a test result about soil samples, which describes the relation between the pore ratio to the effective stress curve; as Fig. 1 shows.

The abscissa is the effective stress, and the ordinate is the pore ratio. The strain rates in the test are between 10^{-4} 1/s to 10^{-13} 1/s. It can be seen from the figure that the compression behaviour of soil is corresponding to the strain rate. A new constitutive model of soft soil considering the strain rate can be gained according to fit these measured points. Figure 2 shows the part of measured value and fitting results using the new expressions, i.e., Eq. (1).

The relationship between the fitting curve and the measured point is shown in Fig. 2. The different curves are the stress-strain relation at different strain rates. When the abscissa is 0, the intersection with the ordinate is 0.98. It thus can be gained that the initial porosity ratio $e_0 = 0.98$.

The strain rate in Fig. 1 is 10–4 1/s. It can be expressed as the following equation:

$$e - 0.98 = -0.1223 \lg \sigma' \tag{1}$$

In Eq. 1, the initial pore ratio is 0.98. It can be referred from Eq. (1) that the compression index a is corresponding to the strain rate. Figure 3 shows the fitting results between compression index a and strain rate.

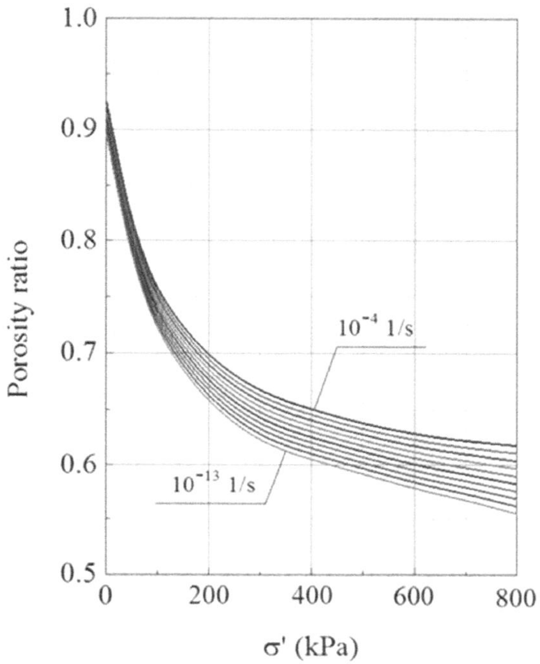

Fig. 1. Compression test result (from Šuklje 1957)

According to Fig. 3, it can be gained that

$$a = A + \alpha \lg \dot{\varepsilon} \tag{2}$$

Parameters A and α are the values relate the type of soil. According to the test result of Šuklje (1957), there are A = 0.114, and α = −0.002.

Substituting Eq. (2) to Eq. (1), then

$$e - 0.98 = -(A + \alpha \lg \dot{\varepsilon}) \cdot \lg \sigma' \tag{3}$$

Equation (3) is gained from test result of Šuklje (1957). Extending this equation to the general soft soil, it can be obtained that

$$e = e_0 - (A + \alpha \lg \dot{\varepsilon}) \cdot \lg(\sigma'/\sigma'_0) \tag{4}$$

where e_0 is initial porosity ratio, σ'_0 is the initial effective stress. Equation (4) is the proposed constitute model for soft soil considering the strain rate effect in the present paper.

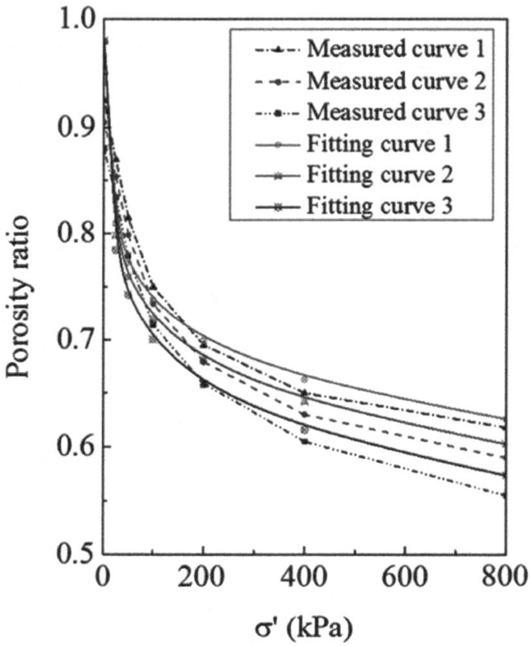

Fig. 2. Measured and fitting result for the compression test result from Šuklje (1957)

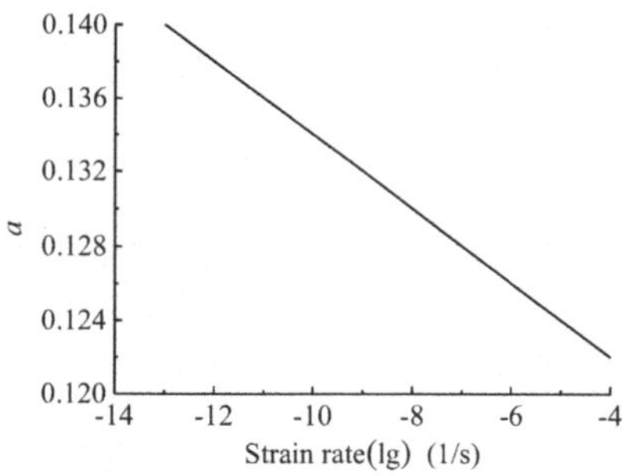

Fig. 3. The relationship between strain rate and slope

3 Consolidation Theory Considering Strain Rate

3.1 Basic Assumption

In order to solve the deformation of saturated soil during infiltration consolidation, it is usually calculated by the one-dimensional consolidation theory proposed by Terzaghi. Here, the basic assumptions are that: (1) the load area is much larger than the thickness of the compressible soil; (2) the pore water in the foundation mainly flows along the vertical direction.

On the other hand, it has been found that the compressive modulus and the permeability coefficient change with time. It means the soil has a non-linear characteristic during the consolidation process. On the basis of the existing research, a new one-dimensional nonlinear consolidation equation is established by using the newly accepted e $\sim$ logσ' value and e $\sim$ logk_v models:

$$e - e_0 = C_c \log\left(\sigma'_0/\sigma'\right) \tag{5a}$$

$$e - e_0 = C_k \log(k_v/k_{v0}) \tag{5b}$$

Where C_k is permeability index, k_{v0} is initial permeability coefficient.

According to the Eqs. (5a) and (5b), there is

$$k_v = k_{v0}\left(\sigma'_0/\sigma'\right)^{C_c/C_k} \tag{6a}$$

It can be deduced if $C_c/C_k = 1$ that

$$k_v = k_{v0}\frac{\sigma'_0}{\sigma'} \tag{6b}$$

Therefore, the basic assumption of the one-dimensional consolidation theory considering the strain rate is as follows: (1) the soil is homogeneous, isotropic and fully saturated; (2) soil particles and pore water are incompressible; (3) the additional stress in the soil along the horizontal plane is infinite distribution, so the soil deformation and water flow are vertical; (4) the seepage of water in the soil is subject to Darcy's law; (5) The consolidation deformation is small, that is, the deformation is completely caused by the dissipation of excess pore water pressure in the soil; (6) the permeability coefficient k_v of the soil is changed during the consolidation process; (7) the compression index C_c is change during the consolidation process.

For the e $\sim$ logσ' value model, the slope of the line in Fig. 3 is the compression index C_c. The compression index is varied with strain rate. However, it can be seen from Table 1 that the variation range is small. Thus, in the following deducing, value C_c is fixed as a constant.

3.2 Determination of the Varied Consolidation Coefficient

The consolidation coefficient is calculated as

$$c_v = \frac{k_v}{m_v \gamma_w} \tag{7a}$$

where m_v is volume compression coefficient. Here it should be noted that m_v is a changed value; it can be deduced from the following equation.

$$m_{\mathrm{v}} = -\frac{1}{1+e_1}\cdot\frac{\partial e}{\partial \sigma'} = \frac{0.434C_{\mathrm{c}}}{(1+e_1)\sigma'} \tag{7b}$$

Where e_1 is porosity ratio under natural condition. Combining Eqs. (7a) and (7b), there is

$$c_{\mathrm{v}} = \frac{k_{\mathrm{v}}}{m_{\mathrm{v}}\gamma_{\mathrm{w}}} = \frac{k_{\mathrm{v0}}\sigma'_0(1+e_1)}{0.434\gamma_{\mathrm{w}}}\cdot\frac{1}{C_{\mathrm{c}}} \tag{7c}$$

It should be noted that c_{v} in Eq. (7c) is varied with permeability coefficient, porosity ratio and strain rate. Equation (7c) can be reformed as

$$c_{\mathrm{v}} = \frac{C}{C_{\mathrm{c}}} \tag{7d}$$

Where

$$C = \frac{k_{\mathrm{v0}}\sigma'_0(1+e_1)}{0.434\gamma_{\mathrm{w}}} \tag{7e}$$

3.3 Controlling Equation

According the Darcy's law, the average seepage velocity of any section is

$$v = k_{\mathrm{v}}i = -\frac{k_{\mathrm{v}}}{\gamma_{\mathrm{w}}}\cdot\frac{\partial u}{\partial z} \tag{8}$$

then

$$\frac{\partial v}{\partial z}\mathrm{d}z = \frac{\partial}{\partial z}\left(-\frac{k_{\mathrm{v}}}{\gamma_{\mathrm{w}}}\cdot\frac{\partial u}{\partial z}\right)\mathrm{d}z \tag{9}$$

Substituting Eqs. (5b) and (7c) to the above equation, there is

$$\frac{\partial v}{\partial z}\mathrm{d}z = -Bc_{\mathrm{v}}C_{\mathrm{c}}\left[\frac{1}{\sigma'}\cdot\frac{\partial^2 u}{\partial z^2} - \frac{1}{\sigma'^2}\frac{\partial u}{\partial z}\cdot\frac{\partial \sigma'}{\partial z}\right]\mathrm{d}z \tag{10}$$

where $B = 0.434/(1 + e_1)$.

According to the basic assumption (1), there is $V_{\mathrm{v}} = V_{\mathrm{w}}$, then

$$\frac{\partial V_{\mathrm{w}}}{\partial t} = -\frac{\partial}{\partial t}\left(\frac{e}{1+e_1}\mathrm{d}x\mathrm{d}y\mathrm{d}z\right) \tag{11}$$

Further, based on the relationship between strain and stress, there is

$$\frac{\partial e}{\partial t} = -\frac{0.434C_c}{\sigma'} \cdot \frac{\partial \sigma'}{\partial t} \tag{12}$$

Substituting Eq. (12) to Eq. (11), there is

$$\frac{\partial V_w}{\partial t} = \frac{0.434C_c}{(1+e_1)\sigma'} \cdot \frac{\partial \sigma'}{\partial t} \mathrm{d}x\mathrm{d}y\mathrm{d}z = \frac{BC_c}{\sigma'} \cdot \frac{\partial \sigma'}{\partial t} \mathrm{d}x\mathrm{d}y\mathrm{d}z \tag{13}$$

According to continuous conditions, that the reduction in pore volume in the soil unit should be equal to the amount of water flowing out of the unit in the same period, then

$$\frac{\partial V_w}{\partial t} \mathrm{d}t = \frac{\partial q}{\partial z} \mathrm{d}z\mathrm{d}t \tag{14a}$$

Because of $q = v\mathrm{d}x\mathrm{d}y$, it can be further deduced that

$$\frac{\partial V_w}{\partial t} \mathrm{d}t = \frac{\partial v}{\partial z} \mathrm{d}z\mathrm{d}x\mathrm{d}y\mathrm{d}t \tag{14b}$$

Combining Eqs. (13) and (14b), then

$$\frac{BC_c}{\sigma'} \cdot \frac{\partial \sigma'}{\partial t} = \frac{\partial v}{\partial z} \tag{15}$$

Substituting Eq. (15) to Eq. (10), there is

$$\frac{1}{\sigma'} \cdot \frac{\partial \sigma'}{\partial t} = -c_v \left[\frac{1}{\sigma'} \cdot \frac{\partial^2 u}{\partial z^2} - \frac{1}{\sigma'^2} \frac{\partial u}{\partial z} \cdot \frac{\partial \sigma'}{\partial z} \right] \tag{16}$$

According to the principle of effective stress, $\sigma' = \sigma\text{-}u$, then

$$\frac{1}{\sigma'} \cdot \frac{\partial \sigma'}{\partial t} = -\frac{C}{C_c} \left[\frac{1}{\sigma'} \cdot \frac{\partial^2 u}{\partial z^2} + \frac{1}{\sigma'^2} \left(\frac{\partial u}{\partial z} \right)^2 \right] \tag{17}$$

The above equation is the one-dimensional consolidation differential equation considering the strain rate for the saturated soil.

4 Analytical Solutions

It can be seen from Eq. (17) that the newly present consolidation equation is a non-linear equation. Here, we make the following assumption for further deducing.

$$\omega = \ln\frac{\sigma'}{\sigma'_{\mathrm{f}}} \tag{18a}$$

where σ'_{f} is the final effective stress. Then, there are

$$\frac{\partial\omega}{\partial t} = \frac{1}{\sigma'}\frac{\partial\sigma'}{\partial t} \tag{18b}$$

$$\frac{\partial\omega}{\partial z} = \frac{1}{\sigma'}\frac{\partial\sigma'}{\partial z} \tag{18c}$$

$$\frac{\partial^2\omega}{\partial z^2} = -\frac{1}{\sigma'^2}\cdot\left(\frac{\partial\sigma'}{\partial z}\right)^2 + \frac{1}{\sigma'}\cdot\frac{\partial^2\sigma'}{\partial z^2} \tag{18d}$$

Substituting the above equations into Eq. (17), and combining the principle of effective stress, there is

$$\frac{C}{C_{\mathrm{c}}}\cdot\frac{\partial^2\omega}{\partial z^2} = \frac{\partial\omega}{\partial t} \tag{19}$$

To solve the above equation, the initial conditions and boundary conditions are needed. According to the classical Terzaghi's one dimensional consolidation theory, there are:

The initial condition: $t = 0$, $u = p_0$; thus $\omega = \ln(1-p_0/\sigma'_{\mathrm{f}}) = \ln(\sigma'_0/\sigma'_{\mathrm{f}})$;

The boundary conditions:

(1) permeable boundary for the up boundary, that is $z = 0$, $u = 0$, thus $\omega = 0$;
(2) impermeable boundary for the base boundary, that $z = H$, $\partial u/\partial z = 0$, and $\partial\omega/\partial z = 0$;

Additionally, when t tends to infinity, there is $u = 0$ and $\omega = 0$.

Combining the above equation and conditions, there is

$$\begin{cases} \frac{\partial\omega}{\partial t} = \frac{C}{C_{\mathrm{c}}}\cdot\frac{\partial^2\omega}{\partial z^2} \\ \omega(0,t) = 0, \frac{\partial\omega}{\partial z}|_{z=H} = 0 \quad (0\leq z\leq H) \\ \omega(z,0) = \ln\frac{\sigma'_{\mathrm{f}}-p_0}{\sigma'_{\mathrm{f}}} \end{cases} \tag{20}$$

Equation (20) can be solved by the separation variable method. Here, it is assumed that:

$$\omega(z,t) = X(z)T(t) \tag{21}$$

Then, Eq. (20) can be reformed to the following two equations:

$$\begin{aligned} & X''(z) + \lambda X(z) = 0 \\ & T'(t) + \lambda \frac{C}{C_{\mathrm{c}}} T(t) = 0 \end{aligned} \tag{22}$$

Firstly, function $X(z)$ can be solved as:

$$\begin{cases} X''(z) + \lambda X(z) = 0 \\ X(0) = X_z(H) = 0 \end{cases} \tag{23}$$

The following three cases λ about will be analyzed:

(1) if $\lambda < 0$, nontrivial solution cannot be gained for Eq. (23). In fact, it is known from the ordinary differential equation that the general solution of the equation is

$$X(z) = Ae^{\sqrt{-\lambda z}} + Be^{-\sqrt{-\lambda z}} \tag{24a}$$

It can be gained by the boundary condition ($A = B = 0$) that $X(z)$ will be fixed as zero.

(2) if $\lambda = 0$, it also cannot be gained a nontrivial solution. The general solution for this condition is:

$$X(z) = Az + B \tag{24b}$$

According to the boundary conditions (i.e., $A = B = 0$), it can be deduced that $X(z)$ will be fixed as zero.

(3) if $\lambda > 0$, the general solution is

$$X(z) = A\cos\sqrt{\lambda}z + B\sin\sqrt{\lambda}z \tag{24c}$$

According to the boundary conditions, there is

$$X(0) = A = 0, X_z(H) = B\sqrt{\lambda}\cos\sqrt{\lambda}H = 0 \tag{24d}$$

It can thus be gained that

$$\lambda = \lambda_n = \left(\frac{n\pi}{2H}\right)^2 \tag{25}$$

where n = 1,3,5,……. Thus, a family of non-zero solution can be gained as

$$X_n(z) = \sin\frac{n\pi z}{2H} \quad (n = 1, 3, 5.\ldots.) \tag{26}$$

Substituting the above equations to the Eq. (22) about $T(t)$, then

$$T' = -\left(\frac{n\pi}{2H}\right)^2 \frac{C}{C_c} T \tag{27a}$$

The general solution for this equation is

$$T_n(t) = C_n e^{-\left(\frac{n^2\pi^2 C}{4H^2 c_c}t\right)} \quad (n = 1, 3, 5.\ldots.) \tag{27b}$$

then

$$\omega(z,t) = \sum_{n=1}^{\infty} X_n(z)T_n(t) = \sum_{n=1}^{\infty} C_n \sin\frac{n\pi z}{2H} \cdot e^{-\left(\frac{n^2\pi^2 C}{4H^2 C_c}t\right)} \quad (n = 1, 3, 5.\ldots.) \tag{28a}$$

Where

$$C_n = \frac{2}{H}\int_0^H \phi(t)\sin\frac{n\pi z}{2H}dz = \frac{2}{H}\cdot \ln\frac{\sigma'_0}{\sigma'_f}\cdot \int_0^H \sin\frac{n\pi z}{2H}dz = \frac{4}{n\pi}\cdot \ln\frac{\sigma'_0}{\sigma'_f} \tag{28b}$$

Suming up the above equations and subsitituing them into Eq. (21), there is

$$\omega = \ln\frac{\sigma'_0}{\sigma'_f}\cdot \sum_{m=0}^{\infty}\left(\frac{2}{M}\sin\frac{Mz}{H}\right)\exp\left(-\frac{M^2 C}{H^2 C_c}t\right) \quad (m = 0, 1, 2.\ldots.) \tag{28c}$$

Where $m = (n-1)/2$, $M = \pi(2m + 1)/2$.
Substituting Eq. (18a) into Eq. (28c), there is

$$\frac{\sigma'_f - u}{\sigma'_f} = \left(\frac{\sigma'_0}{\sigma'_f}\right)^{\sum_{m=0}^{\infty}\left(\frac{2}{M}\sin\frac{Mz}{H}\right)\exp\left(-\frac{M^2 C}{H^2 C_c}t\right)} \tag{29}$$

$$u = \sigma'_f\left(1 - \left(\frac{\sigma'_0}{\sigma'_f}\right)^{\sum_{m=0}^{\infty}\left(\frac{2}{M}\sin\frac{Mz}{H}\right)\exp\left(-\frac{M^2 C}{H^2 C_c}t\right)}\right) \tag{30a}$$

Lastly, it can be reformed as

$$u = \sigma'_{\mathrm{f}}\left(1 - \left(\frac{\sigma'_0}{\sigma'_{\mathrm{f}}}\right)^{\sum_{m=0}^{\infty}\left(\frac{2}{M}\sin\frac{Mz}{H}\right)\exp\left(-\frac{M^2C}{H^2(A+\alpha\lg\dot{\varepsilon})}t\right)}\right) \tag{30b}$$

This is the one-dimensional consolidation solution considering the strain rate about the excess pore water pressure.

Additionally, the average degree of consolidation can be calculated as:

$$U = \frac{\int_0^H (e_0 - e)\mathrm{d}z}{\int_0^H (e_0 - e_{\mathrm{f}})\mathrm{d}z} \tag{31a}$$

Where e_0 is the initial Porosity ratio; e_{f} is the final porosity ratio. According to the relationship of porosity ratio and effective stress, the following equation can be gained from Eq. (31b).

$$U = \frac{\int_0^H \left(\lg\frac{\sigma'_{\mathrm{f}}-u}{\sigma'_0}\right)\mathrm{d}z}{\int_0^H \left(\lg\frac{\sigma'_{\mathrm{f}}}{\sigma'_0}\right)\mathrm{d}z} \tag{31b}$$

Substituting Eq. (30b) into Eq. (31b), then

$$\begin{aligned} U &= \frac{\int_0^H \left(\lg\frac{\sigma'_{\mathrm{f}}}{\sigma'_0}\cdot\left(\frac{\sigma'_0}{\sigma'_{\mathrm{f}}}\right)^{\sum_{m=0}^{\infty}\left(\frac{2}{M}\sin\frac{Mz}{H}\right)\exp\left(-\frac{M^2C}{H^2(A+\alpha\lg\dot{\varepsilon})}t\right)}\right)\mathrm{d}z}{\int_0^H \left(\lg\frac{\sigma'_{\mathrm{f}}}{\sigma'_0}\right)\mathrm{d}z} \\ &= 1 - \sum_{m=0}^{\infty}\frac{2}{M^2}\exp\left(-\frac{M^2C}{H^2(A+\alpha\lg\dot{\varepsilon})}t\right) \end{aligned} \tag{31c}$$

Finally, the average degree of consolidation considering strain rate effect is gained as

$$U = 1 - \sum_{m=0}^{\infty}\frac{2}{M^2}\exp\left(-\frac{M^2C}{H^2(A+\alpha\lg\dot{\varepsilon})}t\right) \tag{31d}$$

5 Computation and Analysis

Comparison between the classical Terzaghi's consolidation theory and the newly present theory is made in the following. In the computation, the constitute model of soil is gained from the test result by Šuklje (1957). Other conditions are listed here:

$H = 1$ m; $k_{v0} = 10^{-8}$ m/s; $\sigma'_0 = 1$ kPa; $\sigma'_f = 101$ kPa; $e_1 = 1.2$. The strain rate is assumed changed from 10^{-4} 1/s to 10^{-13} 1/s.

Figure 4 shows the pore water pressure at depth $z = 0.8$ m. Figure 4(a) is the result by Terzaghi's consolidation theory, while Fig. 4 (b) is the result by the present method. It can be seen from the two figures that (1) under the same total stress, initial permeability coefficient and natural porosity ratio, the pore water pressure dissipation rate considering the strain rate is faster than the dissipation rate by the classical one-dimensional consolidation theory; (2) the dissipation for the pore water pressure at depth $z = 0.8$ m is about 2.5 years when considering the strain rate, while it needs 4.4 years for the same condition by the classical method; (3) the higher the strain rate is, the faster the pore water pressure dissipates; (4) a strain rate of 10^{-4} is about 0.3 years faster than it under strain rate 10^{-13} for 90% dissipation of pore water pressure.

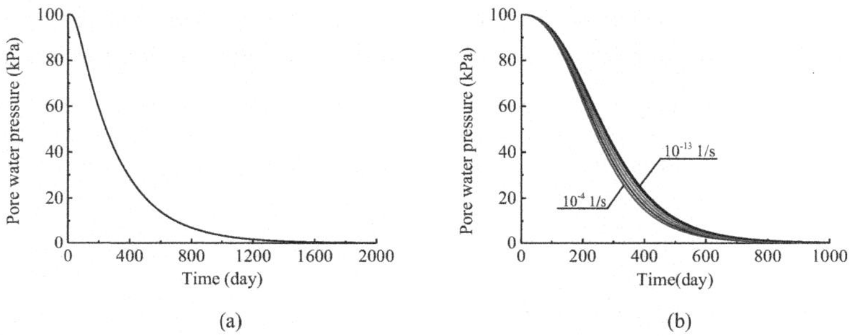

Fig. 4. The relationship between pore water pressure and time at $z = 0.8$ m (a) by Terzaghi's classical consolidation theory; (b) by the present method

Figure 5 shows the average degree of consolidation with time. Figure 5(a) is the result by Terzaghi's consolidation theory; Fig. 5 (b) is the result by the present method. It can be seen from the two figures that (1) under the same permeability coefficient and natural porosity ratio, the consolidation time considering the strain rate is shorter than the consolidation time by the classical one-dimensional consolidation theory; (2) the

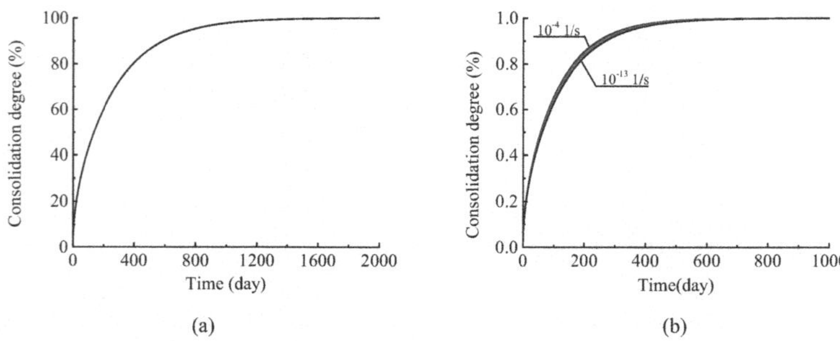

Fig. 5. The relationship between average degree of consolidation and time (a) by Terzaghi's classical consolidation theory; (b) by the present method

time to 99% of degree of consolidation is 1.9 years for the present method, while it is 4.1 years for the same condition by Terzaghi's method; (3) the higher the strain rate is, the faster the consolidation rate will be; (4) comparing to the consolidation under 10^{-3} 1/s, it is 0.16 years faster to reach 90% of degree of consolidation for 10^{-4} 1/s strain rate condition.

6 Conclusions

(1) Under the same conditions, the dissipation rate of pore water pressure considering the strain rate effect is faster than that with an average constant strain rate by classical consolidation theory.
(2) The greater the strain rate is, the faster the dissipation of pore water pressure will be.
(3) Under the same conditions, the consolidation rate considering the strain rate by the newly proposed method is faster than that by the classical one-dimensional consolidation theory.
(4) The greater the strain rate is, the faster the consolidation rate will be.

References

1. Cheng, C.M., Yin, J.H.: Strain-rate dependent stress-strain behavior of undisturbed Hong Kong Marine deposits under oedometric and triaxial stress states. Mar. Georesouces Geotechnol. **23**, 61–92 (2005)
2. Davis, E.H., Raymond, G.P.: A nonlinear theory of consolidation. Geotechnique **15**(2), 161–173 (1965)
3. Duncan, J.M.: Limitations of conventional analysis of consolidation settlement. J. Geotech. Eng. **119**, 1333–1359 (1993)
4. Gao, Y.B., Wang, Z.W.: Effect of strain rate on undrained shear strength of clays. Chin. J. Rock Mech. Eng. **24**(2), 5779–5783 (2005). (In Chinese)
5. Graham, J., Crooks, J.H.A., Bell, A.L.: Time effects on the stress - strain behaviour of natural soft clays. Geotechnique **33**(3), 327–340 (1983)
6. Leroueil, S., Kabbaj, M., Tavenas, F., Bouchard, R.: Stress-strain-strainl rate relation for the compressibility of sensitive natural clays. Geotechnique **35**(2), 159–180 (1985)
7. Leroueil, S.: Recent developments in consolidation of natural clays. Can. Geotech. J. **25**, 85–107 (1988)
8. Leroueil, S., Marques, M.E.S.: Importance of strain rate and temperature effects in geotechnical engineering. In: Sheaharl, T.C. (ed.) Measuring and Modeling Time Dependent Soil Behavior, No. 61, pp. 1–60. Geotechnical Special Publication, New York (1996)
9. Xie, K.H., Pan, Q.Y.: One-dimensional consolidation theory of arbitrary layers under time-dependent loading. Chin. J. Geotech. Eng. **17**(5), 80–85 (1995). (In Chinese)
10. Xie, K.H., Li, B.H., Li, Q.L.: A nonlinear theory of consolidation under time-dependent loading. In: Proceedings of the 2nd International Conference on Soft Soil Engineering, [s.l.]: [s. n.], pp. 193–198 (1996)

11. Xie, K.H., Zheng, H., Li, B.H., et al.: Analysis of one-dimensional nonlinear consolidation of layered soils under time-dependent loading. J. Zhejiang Univ. **37**(4), 426–431 (2003). (In Chinese)
12. Yin, J.H., Cheng, C.M.: Comparison of strain-rate dependent stress-strain behavior from K_0-consolidated compression and extension tests on natural Hong Kong marine deposits. Mar. Georesouces Geotech. **24**, 119–147 (2006)
13. Zhu, J.G., Yin, J.H.: Strain-rate dependent stress-strain behaviour of over consolidated Hong Kong marine clay. Can. Geotech. J. **37**(6), 1272–1282 (2000)

Statistical Damage Constitutive Model for Warm Frozen Clay Considering Effect of Damage Threshold

Hongfa Xu, Hansheng Geng(✉), Xiao Chen, and Zhun Bai

State Key Laboratory of Disaster Prevention
and Mitigation of Explosion and Impact,
PLA University of Science and Technology,
Nanjing 210007, Jiangsu, China
genghansheng@163.com

Abstract. Like most geotechnical materials, frozen clay is a kind of multiphase particulate composite, which makes the mechanical properties and behaviors show great uncertainty and randomness, for that reason, it is more rational to investigate constitutive relationship of frozen soil using a probabilistic approach rather than a deterministic approach. This research cites previous uniaxial experimental data and theoretical results of two kinds of permafrost for the three temperature conditions. Based on Mohr-Coulomb failure criterion of geotechnical materials and rock damage model, a new statistical damage constitutive model considering the effect of damage threshold is introduced. Finally, through a series of fitting analysis and compared with previous models, it turns out that the new model proposed by this paper can be better close to the test results and reflect the characteristics of deformation for both the warm frozen clay and warm ice-rich frozen clay. Particularly in the elastic stage and the post-peak stage, the experimental and theoretical curves are much closer.

Keywords: Warm frozen clay · Constitutive model · Statistical damage · Damage threshold

1 Introduction

Frozen clay naturally formed in the cold region such as Tibetan, Greenland and Northern Canada, which refers to all kinds of rocks and soils contains ice below 0 °C and contains at least the following components: soil, ice, liquids, and gases. As is well known, the mechanical properties of most of the geotechnical materials and artificial composites change very little by the temperature. But, the frozen clay discussed in this paper differs from other similar materials, because of the different mechanical properties between ice and liquid and they convert to each other easily with the temperature changes (Ladanyi 2004). Many important projects are being built in such environment, for example, the Qinghai-Tibetan Railway, almost half of which is built on warm frozen permafrost (Cheng 2003). Here, 'warm frozen permafrost ' indicates that the temperature of ground is a little high, the range is about from 0 to −1.5 °C yearly. 'ice-rich' means that the ice content is a little high, it is about 25% or more. Due to the

R. Chen et al. (eds.), *Proceedings of the 2nd International Symposium on Asia Urban GeoEngineering*, Springer Series in Geomechanics and Geoengineering,
https://doi.org/10.1007/978-981-10-6632-0_7

need of economic construction, more and more building project will be built in cold area. Therefore, the engineering mechanical properties of frozen clay causes many researchers' attention, and some experiments were taken (Ladanyi 2004).

As the melting point of ice is 0 °C, it make the production of frozen soil samples very difficult. But, there still exist some successful experiments on this subject. E.g, by field experiment, Ma discovered that a small pressure applied on warm ice-rich frozen soil would result in huge deformation (Ma 2006), the explanation of this phenomenon can also explain the major reason of uneven subsidence setting and lots of fissures along the roadbed of the QTR for the past decade. Besides, the relationship between strengths and water contents is also studied, the compressive strength of this kind of clay decrease linearly with the temperature increase. As a matter of fact, there are many defects randomly distributed in these samples of water-bearing clay. Any deterministic approach may be improper to be adopt in the research of constitutive relationships of them. In the study of Lai et al. (Lai 2008), a mass of uniaxial tests were taken, at three temperatures. Large numbers of stress-strain curves and experiment data were obtained. Then, they introduced the statistical theory and continuous damage theory into the research. Three distribution functions were substituted into the constitutive model to fit the experiment data. After comparison, this model proved to be well in describing the whole stress-strain process, and the Weibull distribution was best in three distribution functions. Li, et al. put forward an optimized model (Li 2009), which was first introduced by Krajcinovic et al. (Krajcinovic and Silva 2012), which regarded the micro-element strength as stochastic variable. In their study, Failure criterions are used to decide if the micro-element is damaged. It turned out that Li's theory results is more suitable to the fit the test data than Lai's. But in their models, damage produced as soon as load began to increase. Both of them did not take the linear elastic stage of deformation into account. Li, et al. introduced the damage threshold for the first time when they study the stress-strain curves of rock which has not been used in soil, results better than previous research were obtained (Li 2012). In this study on frozen clay, the influences of soil damage threshold is also taken into account. A new measurement method more suitable to the micro-element strength of soil is established. Through further research, this research aimed at reflecting the linear elastic deformation characteristics of warm frozen soil in the low stress levels. Results shows that the new model is closer to the test results than models of by Lai et al. and Li et al. (Lai 2008; Li 2009).

2 The Establishment of Statistical Damage Constituted Model

In this paper, based on the results of Lai et al. and Li et al., a new model is induced by considering the damage threshold (Lai 2008; Li 2009).

At present, the statistical damage theory has been successfully applied to the simulation of rock and soil deformation process, one of the key is to establishment of damage model. In accordance with the strain equivalence assumption of Lemaitre (Lemaitre and Chaboche 1970), the stress-strain relationship can be written as:

$$[\sigma'] = [\sigma]/(1 - \mathrm{D}) = [\mathrm{C}][\varepsilon]/(1 - \mathrm{D}) \tag{1}$$

The above expression can also be written as:

$$\sigma_1 = (1 - D)E\varepsilon_1 + 2\mu\sigma_3 \tag{2}$$

Where σ_i is the macro nominal positive stress; σ_i' is microscopic stress on undamaged elements or effective normal stress;E' and E denote the macro nominal elastic tensor and undamaged part; D is the damage factor, its definition is:

$$D = \frac{N_t}{N} \tag{3}$$

where N denotes all element, and N_t denotes the number of damaged element.

Equation (2) can also be written as the following form for the uniaxial compression case:

$$\sigma_i = (1 - D)E\varepsilon_i \tag{4}$$

2.1 The Change Rule of the Damage Factor

For the purpose of getting the better stress-strain model of warm frozen clay (WFC, frozen content below 25% in permafrost) and warm ice-rich frozen clay (WIRFC, ice content of more than 25% of frozen soil), it is necessary to discuss the change rule of rock damage variable based on the experimental data, so the value of damage factor can be understand. First of all, Eq. (2) can be transformed in the form of the expression below:

$$D = 1 - \frac{\sigma_1 - 2\mu\sigma_3}{E\varepsilon_1} \tag{5}$$

Then, substitute the experiment data given by previous experiment of Lai et al. into Eq. (5). So it is easy to calculate the corresponding damage factor D with any mathematical software. Thus, the $D - \varepsilon_1$ curves of different specimens can be plot, it can be seen from Fig. 1 that:

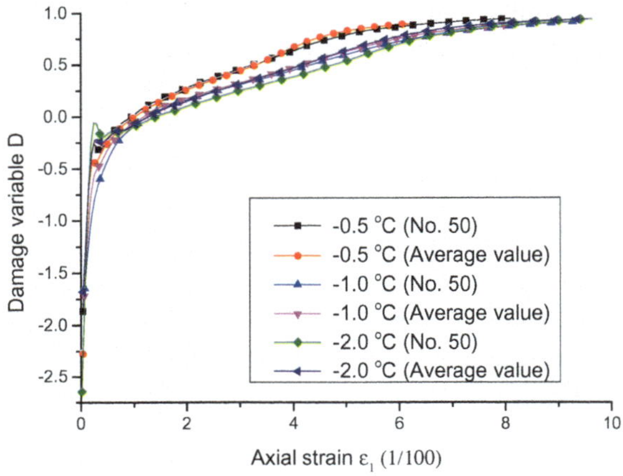

a. WFC

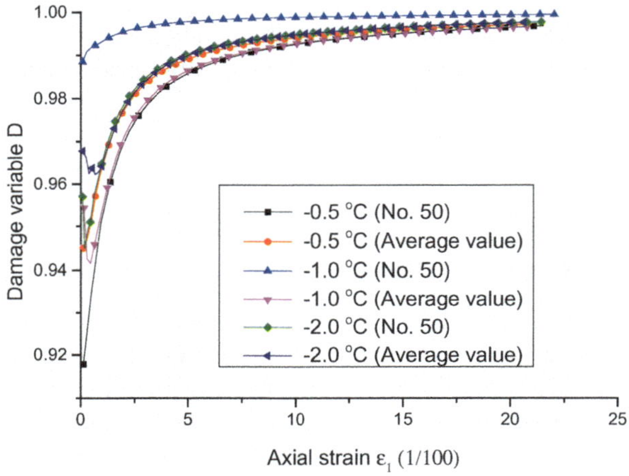

b. WIRFC

Fig. 1. Theoretical curves of $D - \varepsilon_1$

(1) When the axial deformation is small, the damage factor changes in the lack of regularity, sometimes D is less than 0, this is obviously not in the reasonable value range ($0 \ll D \ll 1$). When the deformation is quite small, the specimen is in a linear elastic state and doesn't produce damage. In this stage, D should be constant to 0. This suggests that the effect of damage threshold has very important practical significance.

(2) Followed by deformation increases, the damage factor or stress levels increase from 0 to nearly 1. Because when the deformation or stress exceeds a certain value, the material begins to yield to produce damage, and increases with deformation or stress levels, the damage degree of WFC increase either.

(3) For specimens under different temperature, the damage threshold is also different, and different ice and water content has also big influence. Of course, these difference nature is caused by the role of water and ice, besides, temperature is an important external cause.

In a word, damage threshold is exist here. Before the threshold (i.e. the linear elastic phase), no damage is produced with deformation, the damage does not begin to produce until the threshold, and it is closely related to temperature. Hence, the new damage evolution model of WFC must be able to reflect the above-mentioned damage evolution rules and characteristics.

2.2 The Establishment of Model Considering the Effect of Damage Threshold

There are different ways to establish the damage evolution model, among them the statistical theory is a relatively successful method, and the key lies in the reasonable measurement of micro-element strength. Lai et al. (Lai 2008) proposed the measurement of micro-element strength by axial strain, good results have been achieved. However, the micro-element strength of soil is not simply determined by the deformation size of a certain direction, but rather with the stress state. Li et al. adopt Mohr - Coulomb failure criterion to describe the micro-element strength (Li 2009), which is able to reflect the effect of stress state, and is more reasonable. The Mohr-Coulomb failure criterion is;

$$F'' = (1 - \sin\varphi_f)\sigma'_1 - (1 + \sin\varphi_f)\sigma'_3 = 2c_f \cos\varphi_f \tag{6}$$

where c_f and φ_f denotes the cohesion and internal friction angle corresponding to the yield stress. 'F''' is a more rational choice to express the microcosmic element strength than ε_1, but still flawed. Equation (6) indicates that the damage begins to occur at the moment of loading. In fact, the damage of WFC is not the case, it begins happen only when the stress reaches a certain level. For this reason, a definition of F is proposed in this paper:

$$F' = (1 - \sin\varphi_f)\sigma'_1 - (1 + \sin\varphi_f)\sigma'_3 - k'_0 \tag{7}$$

where k'_0 means the threshold of a sample begins to produce damage.

Compare Eq. (6) with Eq. (7), it indicates that an initial constant must be deducted by 'F''. The damage does not start to produce until $F' > 0$. When $F' \leq 0$, material is in the linear elastic deformation stage, and the damage variable is 0. The new definition of F not only reflects the damage threshold problem, but also gives a reasonable point for the starting point of damage. For the convenience of calculation, both sides of Eq. (7) are divide by '$1 - \sin\varphi$'. The final soil failure strength is obtained.

$$F = F'/(1 - \sin\varphi) = \sigma'_1 - \alpha\sigma'_3 - k_0 \tag{8}$$

where:$\alpha = \frac{1 + \sin\varphi_f}{1 - \sin\varphi_f}$, $k_0 = k'_0/(1 - \sin\varphi)$ is constant.

The following equations are obtained by simultaneous Eqs. (2) and (8):

$$F = \frac{E\varepsilon_1(\sigma_1 - \alpha\sigma_3)}{\sigma_1 - 2\mu\sigma_3} - k_0 \tag{9}$$

Assuming that F obeys Weibull distribution, the new model is established:

$$D = \begin{cases} 1 - \exp\left[-\left(\frac{F}{F_0}\right)^m\right] & F > 0 \\ 0 & F \leq 0 \end{cases} \tag{10}$$

where F_0 and m are both unknown parameters.

Substituting Eq. (10) into (2), we have:

$$\sigma_1 = \begin{cases} E\varepsilon_1 \exp\left[-\left(\frac{F}{F_0}\right)^m\right] + 2\mu\sigma_3 & F > 0 \\ E\varepsilon_1 + 2\mu\sigma_3 & F \leq 0 \end{cases} \tag{11}$$

Equation (11) represents the new constitutive model under triaxial loading condition. In the case of uniaxial loading, Eq. (11) can be written as:

$$\sigma_1 = \begin{cases} E\varepsilon_1 \exp\left[-\left(\frac{F}{F_0}\right)^m\right] & F > 0 \\ E\varepsilon_1 & F \leq 0 \end{cases} \tag{12}$$

2.3 The Parameter Determination Method

Determination of parameters (E, F_0, m and k_0) is one of the critical process to obtain complete model. There exists two main methods. Below is the regression method to determine these parameters by the stress-strain curves of tests.

(1) For the determination of E, the linear regression method is used, based on the data on the elastic stage of tests. This step is relatively simple to operate, and the calculation process is omitted for space limitation.

(2) Substitute E, σ_1, σ_3 and ε_1 into Eq. (5), the damage variable D is obtained. when $D = 0$ and at the moment begins to increase (D value of the inflection point), substitute the corresponding strain $\varepsilon_{1,D=0}$ and stress $\sigma_{1,D=0}$, σ_3 into Eq. (9), the solution of k_0 is as follows:

$$k_0 = \frac{E\varepsilon_{1,D=0}(\sigma_{1,D=0} - \alpha\sigma_3)}{\sigma_{1,D=0} - 2\mu\sigma_3} \tag{13}$$

where $\varepsilon_{1,D=0}$ and $\sigma_{1,D=0}$ are the corresponding strain and stress at the inflection point of D.

(3) Using the nonlinear regression method, the experiment data corresponding to $D > 0$ are fitted by the first formula of Eq. (11) to obtain F_0 and m.

Through the above analysis, the model parameters E, F_0, m and k_0 are determined.

3 Case Analysis and Discussion

This paper cites the experimental data of Lai et al. (Lai 2008), In their study, two groups of specimens (as displayed in Fig. 1 (a, b), one group is WFC, the other one is WIRFC) were made from clay and different amount of water, respectively 17.6% and 30.0% of total mass, which was randomly distributed in the specimens. In three different conditions (the temperature were, respectively −0.5 °C, −1.0 °C and − 2.0 °C), 50 samples for each group were prepared for uniaxial compression test by a MTS low temperature testing machine. Detailed experimental information can be obtained by consulting their paper.

Based on the above experiments data, the parameters E, F_0, m and k_0 are got through numerical calculation. Substitute these parameters in to Eqs. (11) or (12), so the new model considering the damage threshold for specimen under each condition is

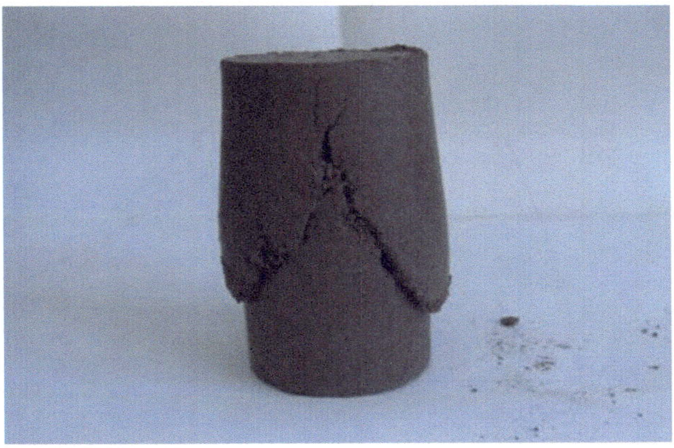

(a) Failure form of WFC

(b) Failure form of WIRFC

Fig. 2. The experimental specimens

able to be obtained. To prove consistency and exactness of the new model of this study, this section select 12 representative test curves to be compared to the new model and previous models used in frozen clays.

3.1 Analysis for WFC

Figures 3, 4 and 5 show the test data at three different temperature. After comparison through numerical analysis, the new model shows a big advantage, especially in the elastic stage, the experimental and new theoretical curves are much closed. At the peak of the curves, the new model does not perform as well as the original model, but errors are in the acceptable range. In the stage after the peak, the new model is much better, and is significantly closer to the experimental data. Overall, the new model considering

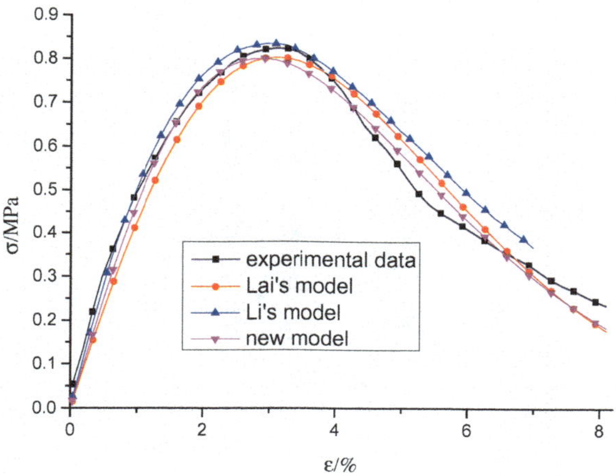

(a) Specimen of No. 50

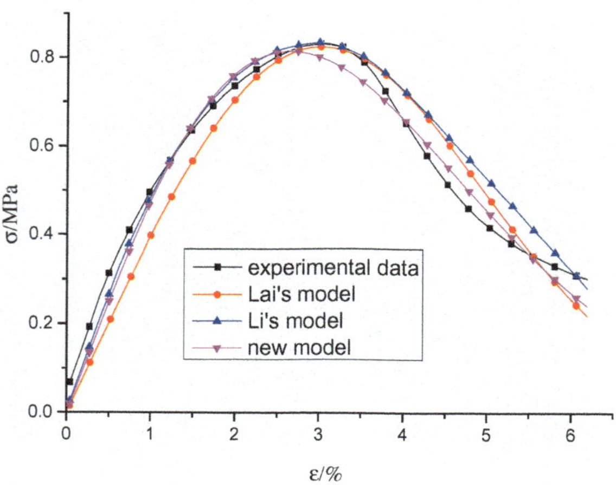

(b) Average value

Fig. 3. Test and fit data of WFC in the case of T = − 0.5 °C

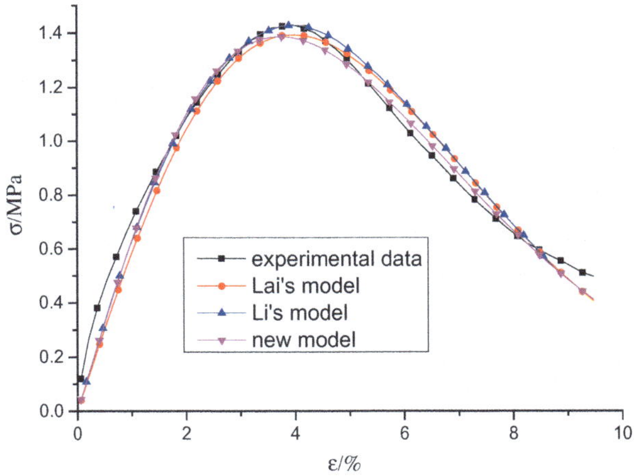

(a) Specimen of No. 50

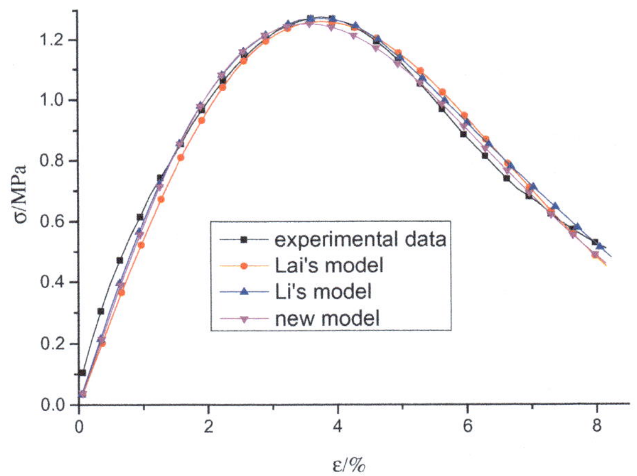

(b) Average value

Fig. 4. Test and fit data of WFC in the case of T = −1.0 °C

damage threshold is a finer choice to describe the constitutive model of warm frozen clay than other original theories. Moreover, we also discover that for warm frozen clay, the higher the temperature, the smaller the elastic modulus and strength, which has been found by some previous researchers (Haynes 1978). The parameters of three models are listed in Table 1.

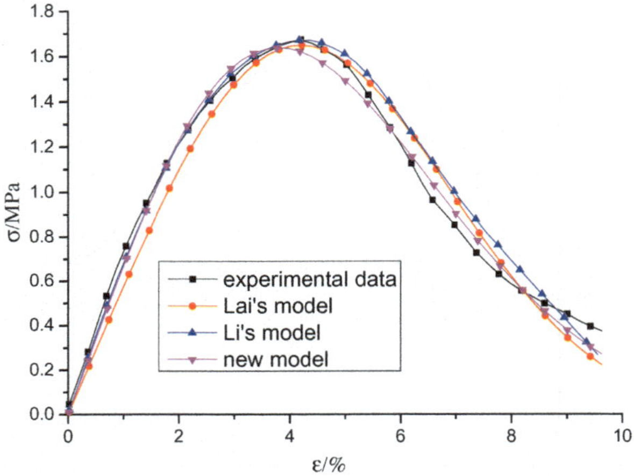

(a) Specimen of No. 50

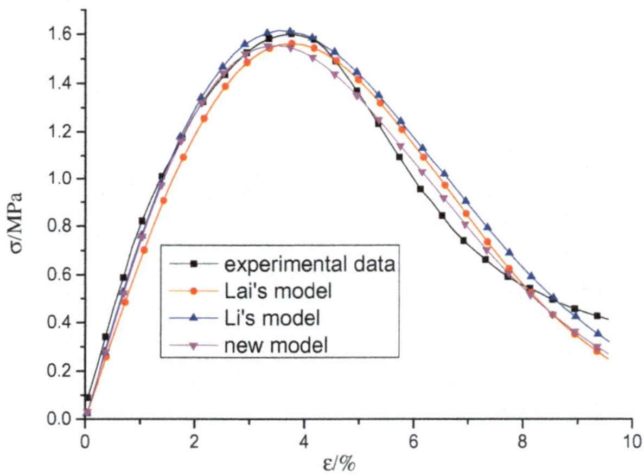

(b) Average value

Fig. 5. Test and fit data of WFC when T = − 2.0 °C

Table 1. Parameters of the average test values

Group	Temperature	E/MPa			F_0			m		
		Lai's model	Li's model	New model	Lai's model	Li's model	New model	Lai's model	Li's model	New model
WFC	−0.5 °C	43.78	56.765	50.43	0.04439	3.120	3.423	2.217	1.406	1.766
	−1.0 °C	57.19	65.958	60.72	0.05323	4.904	4.525	1.890	1.483	1.631
	−2.0 °C	66.99	79.189	75.33	0.05439	6.345	4.400	2.203	1.754	1.648
WIRFC	−0.5 °C	894.71	849.711	1167	0.00004809	0.04336	0.01913	0.203	0.199	0.187
	−1.0 °C	1042.62	1042.624	1334	0.0001003	0.09173	0.17906	0.216	0.208	0.200
	−2.0 °C	1375.81	1371.813	2895	0.0002231	0.23572	0.04063	0.237	0.221	0.193

3.2 Analysis for WIRFC

Figures 6, 7 and 8 shows the experimental and theoretical constitutive curves at three different temperature. We can see that each model is able to describe constitutive relationship of them under uniaxial condition, reason for that is the damage threshold is zero in these six cases. However, Figs. 7(b) and 8(a, b) demonstrates that the test curves is little different from those curves of original models, the new one can be much more closed to the test data on the curve. All these curves illustrate that the constitutive model of this kind of clay is ideal elastic-plastic, we can also come to this conclusion from Fig. 2(b). In general, the new model is a finer choice to present constitutive model

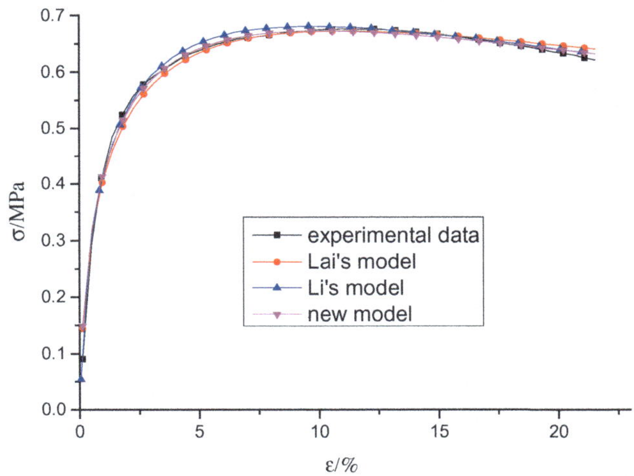

(a) Specimen of No. 50

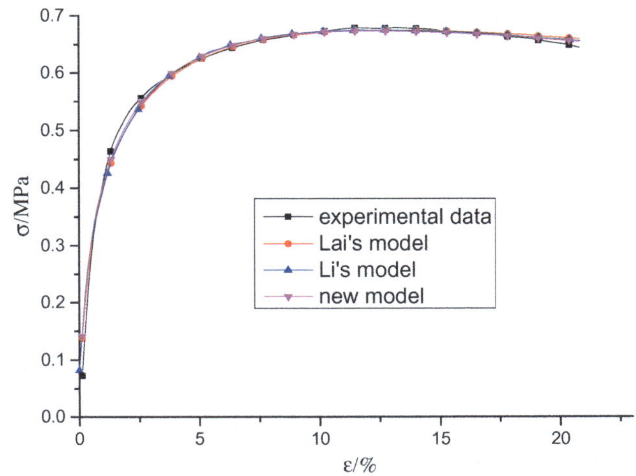

(b) Average value

Fig. 6. Test and fit data of WIRFC in the case of T = − 0.5 °C

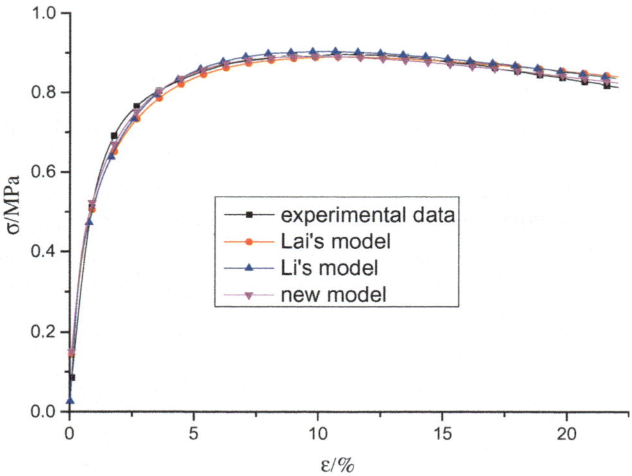

(a) Specimen of No. 50

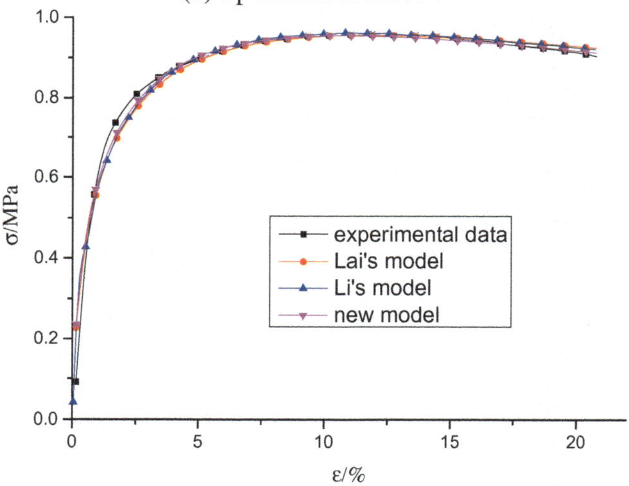

(b) Average value

Fig. 7. Test and fit data of WIRFC in the case of T = − 1.0 °C

of WIRFC than other original theories, although the advantage is not obvious. The parameters of the three models are listed in Table 1.

Considering the threshold value for the rock damage problem is more realistic than the method without considering the threshold value, because when the load is very small, the specimen is only elastically deformed without damage. Only when the load reaches a certain degree, the crack begins to develop and develop. Through the above analysis, it is not difficult to find that the model of this paper is not only realistic in the damage mechanism, but also fit the experimental data better for both WFC and WIRFC.

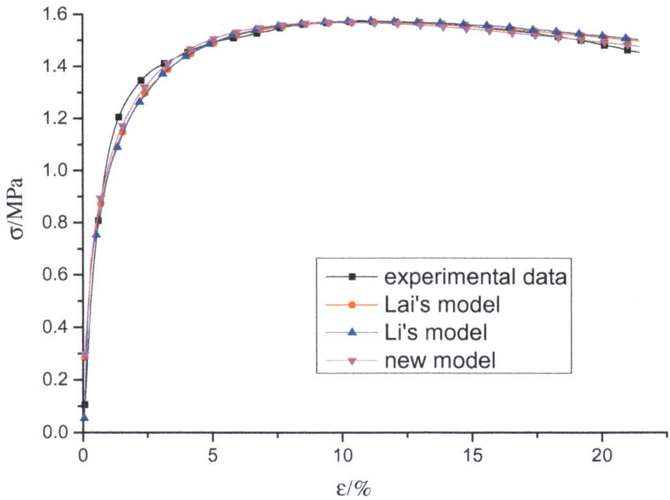

(a) Specimen of No. 50

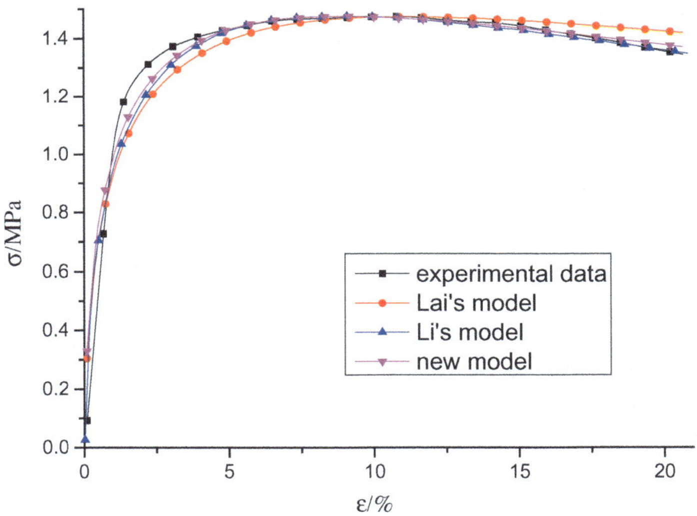

(b) Average value

Fig. 8. Test and fit data of WIRFC in the case of T = − 2.0 °C

4 Conclusion and Discussions

Due to the complex structures and components of WFC and WIRFC, their mechanical properties and behavior show great undeterminancy, the probability method is more logical to explore the constitutive relationship of them. This paper establishes a new model considering effect of damage threshold. In order to test the correctness of this new model, a mass of test data of different condition for the two kinds of frozen clay obtained by previous researchers and two theoretical models deduced by previous scholar are cited (Lai 2008; Li 2009). Firstly, the thresholds were calculated by analyzing the experimental data, and then through the non-linear fitting method, we can see that the theoretical curve agrees well with the experimental data after taking the damage threshold into consideration. After comparison with the previous studies, the new model of this paper can better fit the experimental data. Particularly in the elastic stage and post-peak stage, the experimental and theoretical curves are much closed. Besides, the model established in this paper can be applied to triaxial condition theoretically, but further experiments are needed to support it.

More or less, the statistical distribution function has its limitation, it can't fit the post-yield stage of experiment curves well. The new model presented in this study is not able to performance the constitutive behavior of the frozen clay at this stage precisely either and can only be closer to the experiment value numerically, the results shows better than previous models. It still deserves further study to explain the post-yield mechanical properties of this clay. In summary, despite the shortcomings, results shows that the model proposed by this paper is more appropriate for warm frozen soil.

Acknowledgements. The authors would like to thank Natural Science Foundation of Jiangsu Province of China (BK20141067) for their financial support.

References

Cheng, G.D.: Construction of Qinghai-Tibetan railway with cooled roadbed. China Railway Sci. **24**(3), 1–4 (2003). doi:10.3321/j.issn:1001-4632.2003.03.001

Haynes, F.D.: Strength and deformation of frozen silt. In: Proceedings 3rd International Conference on Permafrost, Edmonton, Alberta, vol. 1, pp. 656–716. National Research Council of Canada, Ottawa (1978)

Krajcinovic, D., Silva, M.A.G.: Statistical aspects of the continuous damage theory. Int. J. Solids Struct. **18**(7), 551–562 (2012). doi:10.1016/0020-7683(82)90039-7

Ladanyi, B., Andersland, O.B.: Frozen Ground Engineering, 2nd edn. ASCE, pp. 656–716 (2004)

Lai, Y.M., Li, S.Y., Qi, J.L., Chang, X.X.: Strength distribution of warm frozen clay and its stochastic damage constitutive model. Cold Reg. Sci. Technol. **53**(2), 200–215 (2008). doi:10.1016/j.coldregions.2007.11.001

Lemaitre, J., Chaboche, J.L.: Mechanics of Solid Materials. Cambridge University Press, London (1970)

Li, S.Y., Lai, Y.M., Zhang, S.J.: An improved statistical damage constitutive model for warm frozen clay based on Mohr-Coulomb criterion. Cold Reg. Sci. Technol. **57**(2), 154–159 (2009). doi:10.1016/j.coldregions.2009.02.010

Li, X., Cao, W.G., Su, Y.H.: A statistical damage constitutive model for softening behavior of rocks. Eng. Geol. **143–144**(2012), 1–17 (2012). doi:10.1016/j.enggeo.2012.05.005

Ma, X.J.: Research on strength and creep of warm and ice-rich frozen soil (in Chinese), A Dissertation Submitted to Graduate School of Chinese Academy of Sciences for the Degree of Master, Lanzhou (2006)

Excavation and Slope Engineering

A Simplified Expression for Ground Movements Induced by Excavations in Soft Clay

Fuchen Teng(✉)

Department of Civil and Construction Engineering,
National Taiwan University of Science and Technology, Taipei, Taiwan
fteng@mail.ntust.edu.tw

Abstract. Clays are generally known not to be isotropic materials. The inherent anisotropy of clays results from the deposition process, which tends to induce a horizontal bedding plane in the soil layer. Soil movements induced by excavation activities are a crucial issue in urban geotechnical engineering with soft clays. However, empirical and semi-empirical methods for evaluation of lateral soil movements induced by excavations with consideration of inherent anisotropy have not been well-established. This study aims to establish a simplified evaluation for predicting ground movements reasonably based on a soil model which considered small-strain properties and anisotropic behavior. Parametric study by numerical analyses with an anisotropic soil model was used in this study. The anisotropic ratio for soil moduli was one of the factors in the simplified expression and ranged from 1.0 to 1.4 which agreed well with test data on many soft clays. A set of simplified expressions for predicting ground movements is proposed. Only the excavation geometry, the subsoil properties, stiffness of braces, and inherent anisotropy are required to evaluate the ground movements.

Keywords: Excavation · Inherent anisotropy · Soil movement · Simplified expression

1 Introduction

Ground movements induced by excavations generally involve the lateral deflection of retaining structure and settlement of surrounding soils. Adjacent buildings may be damaged when the ground movement is greater than the tolerable value. Thus, the prediction of ground movement near an excavation becomes an important topic in the design of excavation. Finite-element method (FEM) is generally employed to estimate the wall deflections and ground surface settlements. The accuracy of estimation highly depends on the soil constitutive model used in the FEM. Although soil models that considered the small-strain behavior (Whittle and Kavvads 1994; Benz 2007; Hsieh and Ou 2011) and stiffness anisotropy of soil (Teng et al. 2014) are available in recent years. However, observation and selection of soil characteristic such as behaviors at small strain and properties of anisotropy is generally difficult and costly because those properties can only be obtained through specialized tests, which are not

R. Chen et al. (eds.), *Proceedings of the 2nd International Symposium on Asia Urban GeoEngineering*, Springer Series in Geomechanics and Geoengineering,
https://doi.org/10.1007/978-981-10-6632-0_8

available to most of consultants and designers. Therefore, a simple and reasonable assessment on ground movements induced by excavation is desirable and required to examine the design of excavation.

Several empirical and simplified methods (Clough and O'Rourke 1990; Kung et al. 2007; Hsieh 2001) have been developed to predict the maximum wall deflection induced by deep excavation through case histories. Results of those estimations are satisfactory even though the deflection of wall involves complex soil-structure interactions. Nevertheless, the empirical and simplified predictions of ground settlement are relatively unreliable because case histories with observations of ground settlements are very limited. Observation of ground surface settlements adjacent to excavations is often influenced by construction vehicles, dewatering, existing structures nearby, and accuracy of instruments. In addition to case histories, numerical parametric studies can also be used to establish simplified methods for predicting ground surface settlements. To accurately predict the ground surface settlement behind the wall using numerical methods, the constitutive model of soil should be able to reproduce the soil behavior at various strain level and include the non-linear and anisotropic behavior (Pestana and Whittle 1999; Ng et al. 2004; Teng et al. 2014). However, advanced models are seldom used in parametric studies to examine the characteristic of ground settlement.

For determining the maximum ground surface settlement, evaluation methods proposed in several works (Peck 1969; Bowles 1986; Clough and O'Rourke 1990; Ou et al. 1993; Hsieh and Ou 1998; Kung et al. 2007) are often used for estimating the ground surface settlement. Evaluation methods based on field observations, i.e., empirical equations, were highly affected by the accuracy of measurements.

In addition to ground surface settlement, the prediction of ground settlement below the ground surface is also of interest to engineers. As shown in Fig. 1, footings of adjacent buildings are usually placed underground rather than placed on the ground surface. However, the behavior of subsurface settlement, i.e. the shape and magnitude of settlement troughs, is not well investigated. The evaluation of damage potential of buildings adjacent to a braced excavation is dependent on the subsurface settlement profile. Therefore, the ground settlement, which includes subsurface settlement and surface settlement, needs to be properly investigated.

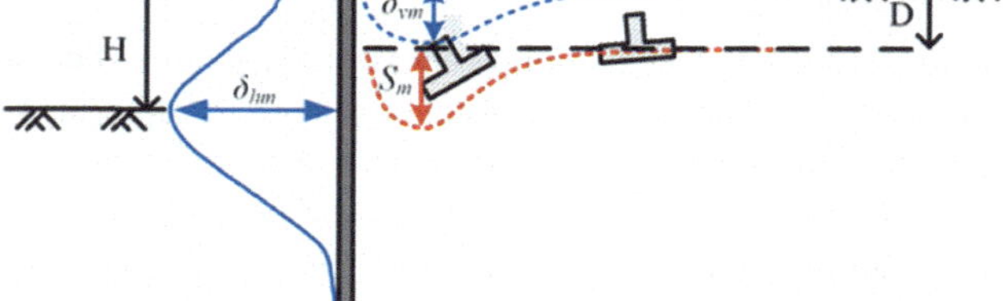

Fig. 1. Schematic diagram for ground movements induced by excavation

This study aims to establish a simplified method for predicting the excavation induced movement, including the maximum wall deflection (δ_{hm} in Fig. 1), the maximum ground settlement (δ_{hm} in Fig. 1), and the subsurface ground settlements (S_m in

Fig. 1), induced by deep excavation. Parametric study with an advanced soil model which considers the small-strain behavior and stiffness anisotropy of soils (Teng et al. 2014) was carried out in this study. A large number of FEM analyses of hypothetical cases with different configurations of influence factors such as excavation dimensions, soil strength, structure stiffness, and stiffness anisotropy of soils were employed. Finally, the proposed simplified evaluation methods are validated through comparison with data from 35 excavation case histories.

2 Influence Factors to Ground Movement

Factors that influence ground movements induced by excavation have been studied by various researchers (e.g. Wong and Broms 1989; Clough and O'Rourke 1990; Hashash and Whittle 1996; Hsieh 2001; Kung et al. 2007). These factors include the excavation depth (H), the excavation width (B), the undrained shear strength of the subsoil (s_u) or the factor of safety against basal heave, the system stiffness of the wall (S_w), and the axial stiffness of wall bracing (S_a). According to Clough and O'Rourke (1990), the system stiffness of wall is dimensionless and defined as

$$S_w = \frac{E_w I_w}{\gamma_{wat} h_{avg}^4} \tag{1}$$

where E_w is the Young's modulus of the wall; I_w is the moment of inertia per unit length of the wall; γ_{wat} is the unit weight of water; h_{avg} is the average vertical spacing of lateral struts.

The axial stiffness of the lateral strut per unit length of the wall is defined as

$$S_a = \frac{A_{st} E_{st}}{s} \tag{2}$$

where A_{st} is the average cross sectional area of the lateral struts for each level; E_{st} is the Young's modulus of the lateral strut; s is the horizontal spacing between the lateral struts.

In addition to factors mentioned above, the stiffness anisotropy of soils could affect the ground movement surrounding an excavation. The inherent anisotropy of stiffness was observed in multi-directional bender element tests on several soft clays as listed in Table 1 (Ng et al. 2004; Cho and Finno 2010; Teng et al. 2014). The stiffness anisotropy of clays was generally assumed to be cross-anisotropic since there is a horizontal plane of isotropy, i.e., the bedding plane. The cross-anisotropic constitutive equation is usually used to simulate the anisotropy of soils. Cross-anisotropic soil models have been applied in numerical analyses to predict ground settlements due to tunneling (Lee and Rowe 1989; Simpson and Ng 1995). It has been shown that the use of cross-anisotropic elastic parameters improves the accuracy of predictions of settlement profiles, including the shape and magnitude of settlement troughs, and the predictions agreed well with field observations (Teng et al. 2014). The anisotropic ratio of stiffness is defined as

Table 1. Anisotropy of stiffness for different clays

	Ng et al. (2004)	Cho and Finno (2010)	Teng et al. (2014)
Soil type	Completely decomposed tuff	Chicago glacial clay	Taipei silty clay
Sample type	Intact sample	Intact sample	Intact sample
Anisotropic ratio	1.15–1.36	1.0–1.4	1.15–1.42

$$AR = \frac{G_{hh}}{G_{vh}} \tag{3}$$

where G_{vh} is the shear modulus in the vertical plane; G_{hh} is the shear modulus in the horizontal plane.

Six major factors, i.e., H, B, s_u, S_w, S_a, and AR, that affect ground movements caused by excavation were employed in the parametric study presented herein. The configuration of the parametric study is introduced in the following section.

3 The Fem Parametric Study

3.1 Introduction and Validation of FEM with An-USC

The parametric studies were performed in a finite-element program (PLAXIS 2D) with an advanced soil constitutive model, the anisotropic undrained soft clay moel (An-USC model). The An-USC model which originated from the undrained soft clay model (Hsieh and Ou 2011), is a stress path dependent undrained soil model and based on the concept of effective stress. Both the anisotropy of soil strength, the anisotropy of stiffness, and small-strain behaviors of soils are simulated by the An-USC model. Only eight parameters are required to fully describe the An-USC model: undrained shear strength in axial compression (s_u), Young's modulus at small strain (E_i), failure ratio (R_f), Poisson's ratio (υ), anisotropic ratio of undrained strength (K_s), anisotropic ratio of stiffness (AR), and degradation parameters (m and n). Among these parameters, s_u and R_f are exactly the same as those in the original Duncan-Chang model (Duncan and Chang 1970). The Young's modulus at small strain (E_i) can be obtained from small strain tests, bender element tests or empirical correlations. The anisotropic strength factor (K_s) is obtained from triaxial compression tests and extension tests. The anisotropic ratio of stiffness can be obtained from multi-directional bender element test. Degradation parameters (m and n) can be obtained from unloading-reloading triaxial tests.

The FEM analyses with An-USC model are validated by predicting the excavation-induced ground movements. A well-documented excavation case with field monitoring data and soil testing data (Ou et al. 1998, 2000a, and 2000b) is used to calibrate the FEM solution. The geometry of the excavation site was trapezoid and show in Fig. 2a. The depth of the excavation was 19.7 m, and dimension of diaphragm wall was 0.9 m thick and 35 m deep. The basement was completed by the top-down construction method, in which the wall was supported by solid concrete floor slabs of 0.15 m thick. Figure 2b shows the comparison of wall deflection and ground surface settlement between field observations and predictions, respectively. Based on the FEM results, An-USC model gave predictions close to field observations on both the maximum wall displacement and the shape of wall deformation curves in all stages. For the ground surface settlement, the prediction was not only accurate on the maximum settlement but also the settlement beyond the location of the maximum settlement. In addition, the location of maximum ground surface settlement was also well predicted. Thus, the accuracy of the FEM with An-USC model is considered satisfactory for the following parametric study.

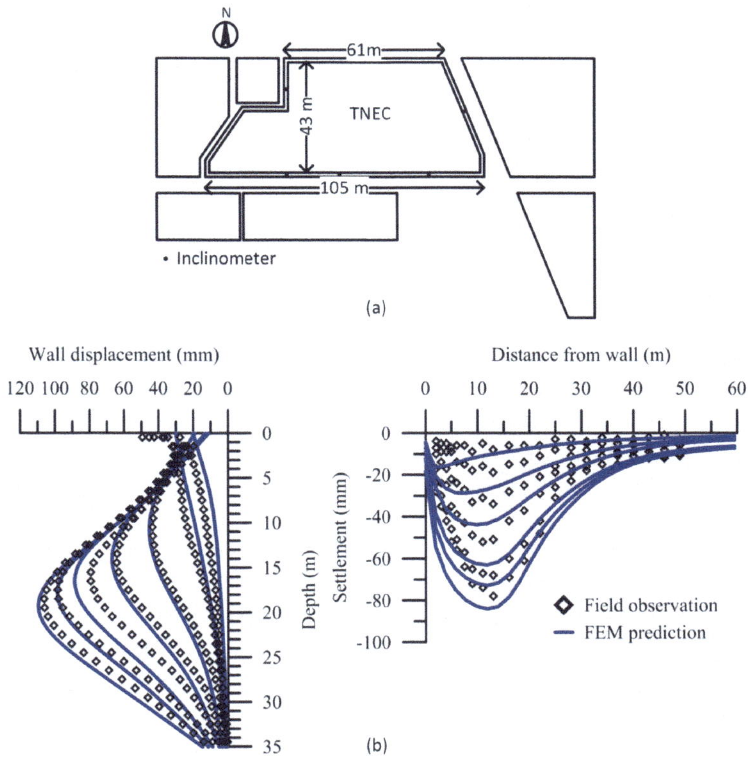

Fig. 2. (a) Geometry of the calibration case, (b) prediction of wall deflection and ground surface settlement by FEM with An-USC model

3.2 Configuration of Hypothetic Excavations

The mesh of hypothetical case used in the parametric study is shown in Fig. 3, in which the excavation was assumed symmetric to the centerline of the excavation width.

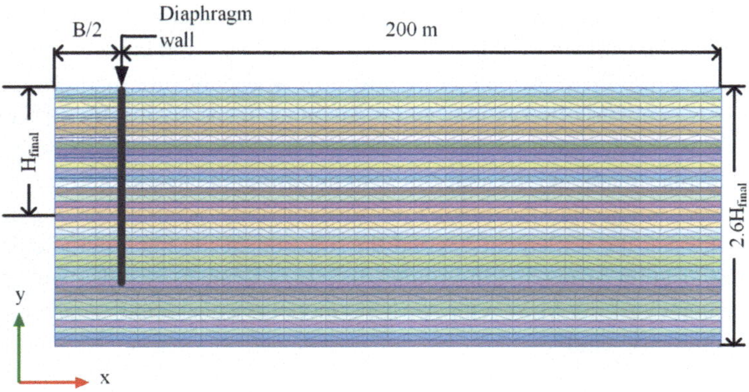

Fig. 3. Mesh in numerical analysis

As shown in the Fig. 3, the left boundary of the finite element mesh is at the center of excavation and the right boundary is at the distance of 200 m from the excavation center, which is sufficiently long for all hypothetic cases. The bottom boundary is at a depth of 2.6 times of the final excavation depth. The length of the wall is equal to twice of the final excavation depth. For the boundary conditions, the boundary of the bottom was restrained from both horizontal and vertical displacements while the left and right boundaries were restrained from the horizontal displacement.

The subsoil was assumed to be a uniform clay layer with a unit weight of 18.24 kN/m^3. The groundwater level was located at the ground surface and in a hydrostatic distribution with the depth. The normalized Young's modulus of the clay at small strain (E_i/σ'_v, where σ'_v is the effective vertical stress) is within the range of 588 to 714. The coefficient of earth pressure at rest (K_O) was assumed to be 0.5. All structures in the analysis such as walls and struts were simulated by linear elastic model. Plate elements were used to model the wall while anchors were used to simulate the lateral struts.

The values of factors that influenced ground movements in the parametric study, i.e., H, B, s_u, S_w, S_a, and AR, were varied as listed in Table 2. Three different final excavation depths, 10 m, 20 m, and 30 m, were employed in the parametric study. Detailed construction sequence of hypothetic excavations is shown in Fig. 4. The lateral strut of each level was installed at a depth of 1.0 m above the current excavation surface. The variation of the factors listed in Table 2 generally covered most of practical excavation cases. Ranges of excavation dimensions, i.e., H and B, are 10 m to 30 m and 20 m to 80 m, respectively. The system stiffness of wall, S_w, started from the flexible to stiff structures. The S_a coved the axial stiffness of steel struts to that of concrete floor slabs. As observed in Table 1, the anisotropic stiffness ratios for soft clays are in the range of 1.1 to 1.4. Therefore, anisotropic ratios of soil stiffness were set as 1.0 (isotropic) to 1.4 herein. A total number of 729 hypothetic excavations were generated based on the variation of the factors in Table 2. Among these 729 artificial cases, some had unrealistic excavation conditions, such as a deep and wide excavation (H : 30 m $\times$ B : 80 m) in soft clays but with a soft wall and soft lateral supports.

Table 2. Variation of the factors that used in the parametric study

Factor	Values
H (m)	10, 20, 30
B (m)	20, 40, 80
s_u/σ'_v	0.28, 0.3, 0.34
S_w	352, 1760, 5280
S_a (MN/m)	550, 2960, 11830
AR	1.0, 1.2, 1.4

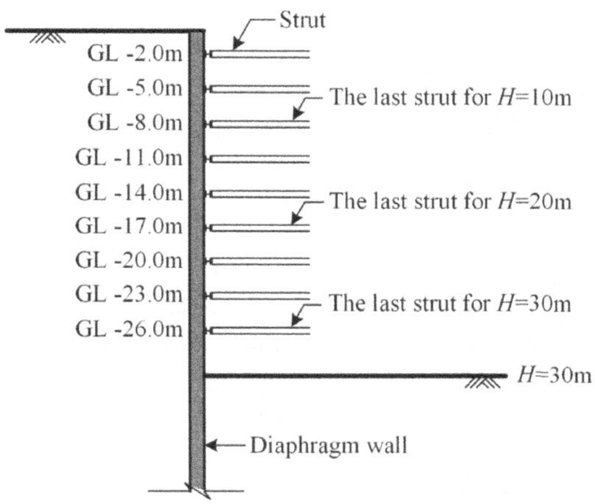

Fig. 4. Construction sequence of the hypothetical excavation

By excluding such unreasonable cases, the total number of FEM analyses is thus reduced to 180, which will be used to establish a simplified evaluation method for excavation-induced ground movements.

4 Simplified Methods for Predicting Ground Movements

4.1 Determination of the Maximum Wall Deflection

Results of FEM numerical analyses of 180 artificial excavations were collected and used to obtain the relationship between maximum wall deflections (δ_{hm}) and factors of influence (H, B, s_u, S_w, S_a, and AR). By applying the multivariate regression analysis on the results and the parameters, the normalized maximum wall deflection, δ_{hm}/H, could be calculated by equations below:

$$\delta_{hm}/H = 0.09\exp(78.03X_{e,hm}) \quad (4)$$

$$X_{e,hm} = 3.05 \times 10^{-4}H + 1.53 \times 10^{-4}B - 0.124 s_u/\sigma'_v - 1.49 \times 10^{-6}S_w - 2.63 \times 10^{-7}S_a - 2.25 \times 10^{-3}AR \quad (5)$$

where $X_{e,hm}$ is a synthetic parameter for determining the maximum wall deflection which takes all factors of influence such as excavation geometry, subsoil properties, and stiffness of braces into consideration. The units of parameters in Eq. (5): H and B are in meter; S_a is in MN/m; s_u/σ'_v, S_w, and AR are dimensionless. δ_{hm} from FEM analyses, $X_{e,hm}$ calculated by Eq. (5) based on parameters listed in Table 2, and the regression curve are shown in Fig. 5. The coefficient of determination, R^2, of the regression curve in Fig. 5 is 0.81.

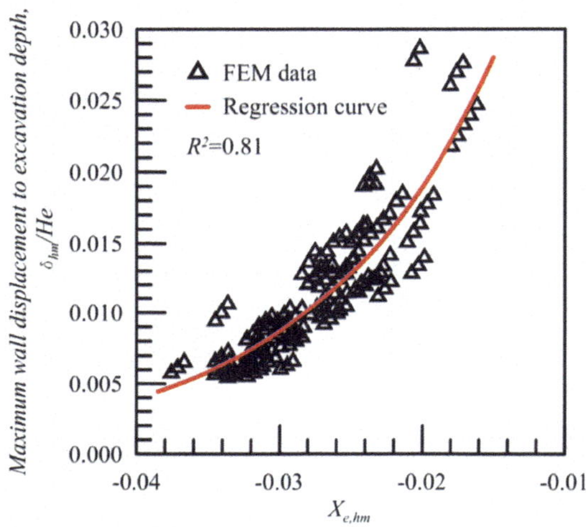

Fig. 5. Relationship between the ratio of the maximum wall deflection to the excavation depth and $X_{e,hm}$

According to the work by Hsieh (2001), the wall deflection will increase with the increase of the depth of a hard stratum, H_g, measured from the ground surface. When the H_g/H is greater than or equal to 1.8 which implies the hard stratum is sufficiently deep, the wall deflection is not affected by the existence of the hard stratum. However, for the situation that H_g/H is less than 1.8, the hard stratum will restrain the wall deflection induced by excavation. A modification factor, α_g, is proposed as

$$\alpha_g = 0.91\frac{H_g}{H} - 0.64 \leq 1.0 \quad (6)$$

The proposed simplified equation for estimating the wall deflection is based on the parametric study on FEM in which the bottom boundary, i.e., the hard stratum, is located at a depth of 2.6H. Therefore, the Eq. (4) is applied to excavations with $H_g/H \geq 1.8$ only.

For excavations with shallower hard stratum, i.e., $H_g/H < 1.8$, the maximum wall deflection calculated by Eq. (4) should multiply the modification factor α_g.

The system stiffness of the wall, S_w, in Eq. (5) is usually related to the final excavation depth. In practice, a stiffer wall is designed for an excavation with deeper final excavation depth. Hsieh (2001) suggested the relationship between the system stiffness of the wall (S_w) and the final excavation depth (H_{ef}) by observing field data as

$$S_w = 98.02H_{ef} - 32.39 \tag{7}$$

The stiffness of the strut, S_a, is also related to the final excavation depth. Usually, a stiffer strut is designed for a deeper excavation. The relationship between S_a and H_{ef} was obtained on several case histories as shown in Fig. 6. The relationship is as follows:

$$S_a = 91.45H_{ef} - 515.9 \tag{8}$$

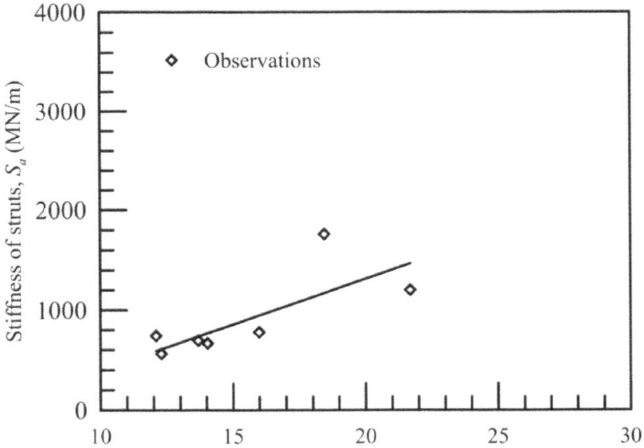

Fig. 6. Relationship between S_a and H

By substituting Eqs. (7) and (8) into Eq. (5), the calculation of parameter $X_{e,hm}$ could be reduced as an equation with only four parameters, i.e., H_{ef}, B, s_u, and AR. For most of normally consolidated clay, the AR can be assumed to be 1.2. Therefore, a quick evaluation method for the maximum wall deflection is proposed. Figure 7 shows such a relationship for soft clay ($s_u/\sigma'_v = 0.28$) and medium clay ($s_u/\sigma'_v = 0.34$). Each of the curve in the figure, namely linear, corresponding to a value of B can be approximated by a specific ratio of δ_{hm} to H_e. For a specific area of soils, for example,

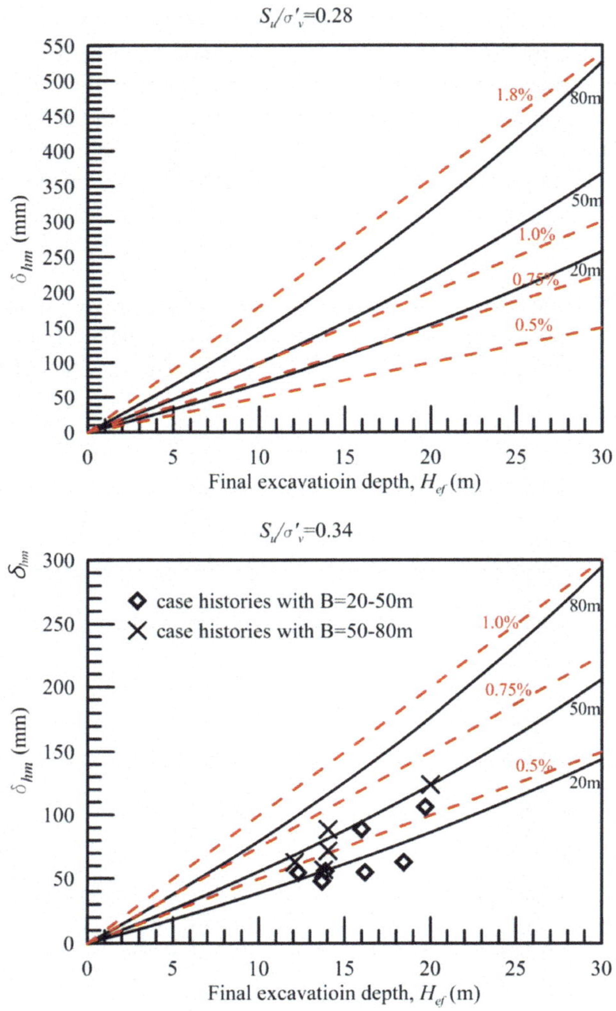

Fig. 7. Evaluation for the maximum wall deflection based on H_{ef}

$s_u/\sigma'_v = 0.28$, with a known B = 50 m and H_e = 20 m, an approximate line of $\delta_{hm}/H_e \cong 1.1\%$ can be selected as a first assessment, and therefore, the initial estimation for δ_{hm} would be 220 mm.

4.2 Determination of the Maximum Ground Surface Settlement

The maximum ground surface settlements, δ_{vm}, obtained from the FEM analyses were also used to derive the simplified evaluation equation. The relationship between the maximum ground surface settlement (δ_{vm}) and factors of influence (H, B, s_u, S_w, S_a, and AR) is obtained by applying the multivariate regression analysis on the results and the parameters. The normalized maximum ground surface settlement, δ_{vm}/H, could be calculated by

$$\delta_{vm}/H = 0.102\exp(106.19X_{e,vm}) \tag{9}$$

$$\begin{aligned} X_{e,vm} = {} & 1.78 \times 10^{-4}H + 1.14 \times 10^{-4}B - 0.102 s_u/\sigma'_v - 1.1 \times 10^{-6}S_w \\ & - 2.23 \times 10^{-7}S_a - 1.54 \times 10^{-3}AR \end{aligned} \tag{10}$$

where $X_{e,vm}$ is a synthetic parameter for determining the maximum ground surface settlement, inclusive of excavation geometry, soil properties, and stiffness of braces. Figure 8 shows the numerical data of δ_{vm} and $X_{e,vm}$ and the regression curve where $R^2 = 0.8$.

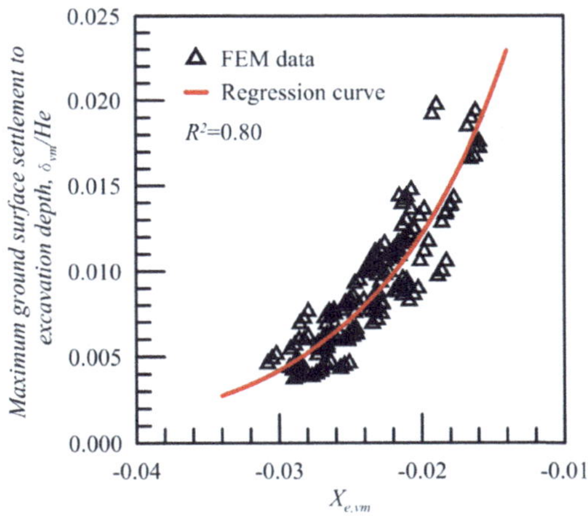

Fig. 8. Relationship between the ratio of the maximum ground surface settlement to the excavation depth and $X_{e,vm}$

The Eq. (10) could be also reduced by Eqs. (7) and (8), and resulted in the quick evaluation for estimating the maximum ground surface settlement without the parameter of S_w and S_a. Figure 9 shows the relationship between δ_{vm} and the H_e for soft clay ($s_u/\sigma'_v = 0.28$) and medium clay ($s_u/\sigma'_v = 0.34$). Similarly, for $s_u/\sigma'_v = 0.28$, with the known B = 50 m and H_e = 20 m, an approximate line of $\delta_{vm}/H_e \cong 0.8\%$ can

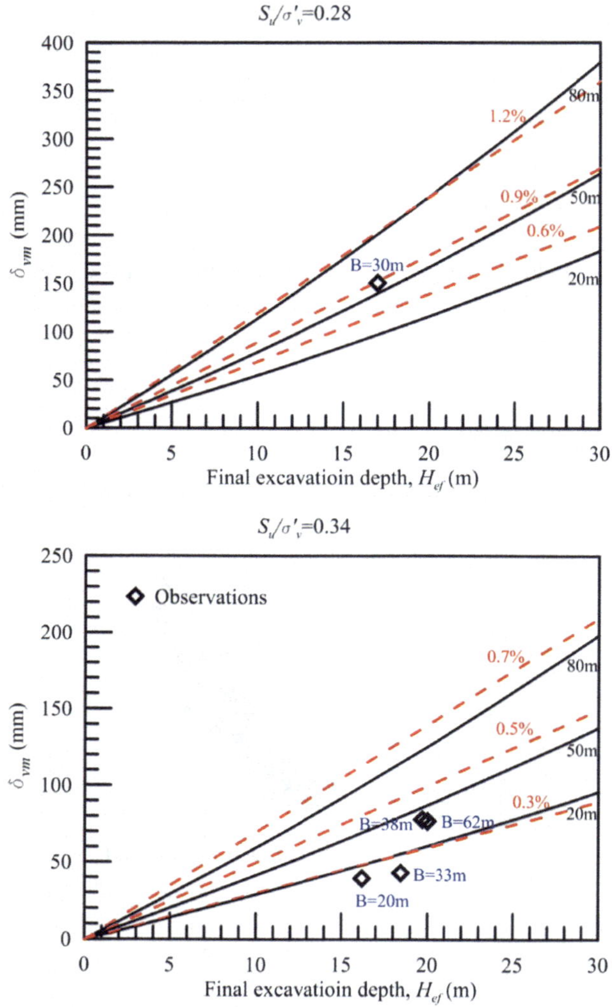

Fig. 9. Evaluation for the maximum ground surface settlement based on H_{ef}

be selected as a first assessment, and therefore, the initial estimation for δ_{vm} would be 160 mm.

The ground surface settlement can be determined directly from Eqs. (9) and (10), which based on the influence factors. However, for those situations that reliable maximum wall deflections were known, the ground surface settlement can also be determined by the relationship between δ_{vm} and δ_{hm}. Results from numerical analyses on hypothetic excavations were used to establish the relationship between δ_{vm} and δ_{hm}. Figure 10 shows the maximum ground surface settlement against the maximum wall

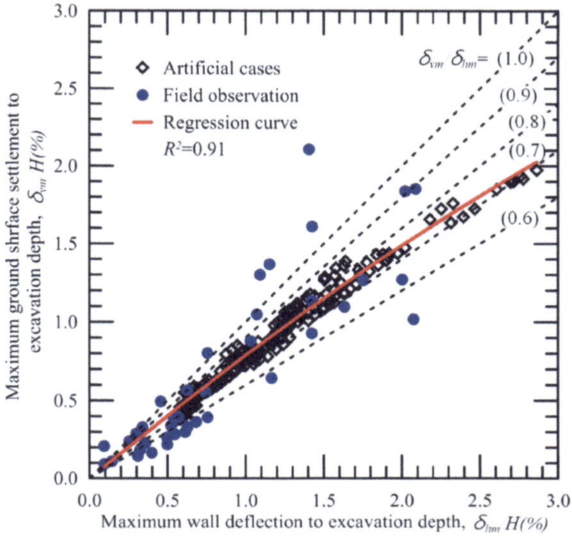

Fig. 10. Relationship between max ground surface settlement and max wall deflection

deflection either from FEM analyses or field observations. For the generated excavation data, the ratios of the maximum ground surface settlement to the maximum wall deflection were generally located within the range of 0.7 to 0.9. While the δ_{vm}/δ_{hm} is more scattered for field observations. The regression equation shown in Fig. 10 is as follows.

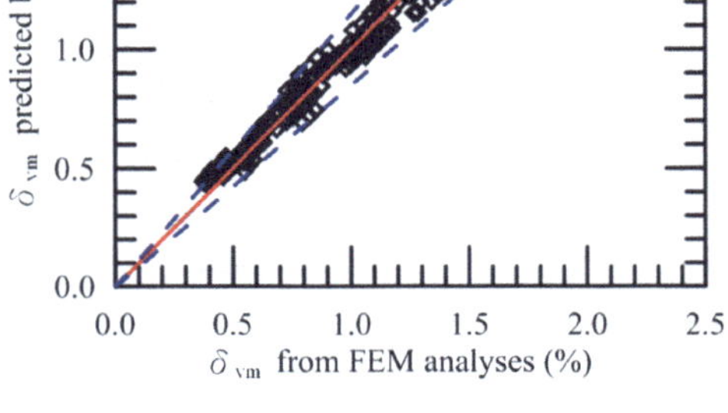

Fig. 11. Performance of Eq. (11) in predicting the ground surface settlement for hypothetical cases

$$\frac{\delta_{vm}}{H} = -0.04\left(\frac{\delta_{hm}}{H}\right)^2 + 0.82\frac{\delta_{hm}}{H} \quad (11)$$

Figure 11 shows the performance of Eq. (11) in reproducing the results of FEM analyses. As observed in the figure, the prediction made by Eq. (11) is consistent with results calculated from FEM analyses. The high R^2 value (0.98) and low coefficient of variance (COV = 0.05) illustrates that the accurate maximum ground surface settlement could be obtained if a reliable maximum wall deflection is known. In addition, most of data points shown in the figure were located in the range of $\pm 15\%$ of the line where predicted δ_{vm} is equal to δ_{vm} from FEM. Further validations of Eq. (11) on case histories will be included in the section of validation.

4.3 Determination of the Subsurface Settlement

In addition to the maximum ground surface settlement, subsurface settlements at different depths were collected from results of FEM analyses on artificial excavations. Both the shape and the magnitude of subsurface settlement troughs were rarely discussed before.

Amounts of the maximum subsurface settlements, S_m, at various depths were firstly investigated in the hypothetic cases. Ratios of the maximum subsurface settlement to the maximum ground surface settlement, S_m/δ_{vm}, are plotted against the normalized depth, D/H, in Fig. 12. Only data points from hypothetic cases with AR = 1.2 were plotted since data of different AR are almost coincide with each other. The ratios of S_m to δ_{vm} from results of FEM on the calibration case were also put in the figure. As observed in the figure, the S_m at depths between 0.1H to 0.4H is about 10% greater than the δ_{vm}. For conservative, ratio of S_m to δ_{vm} could be assumed to be 1.1 for depths

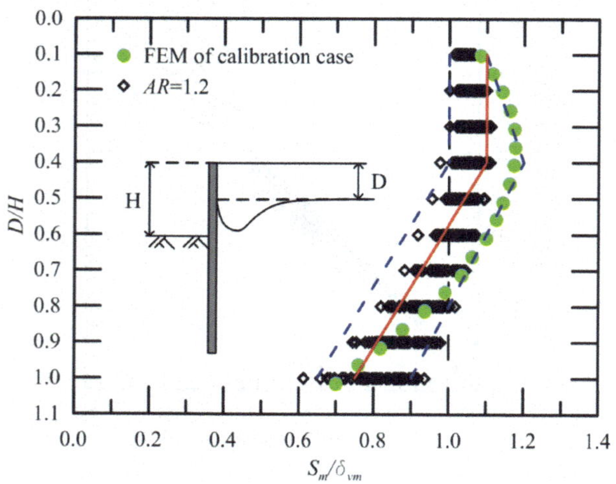

Fig. 12. Normalized subsurface settlement against normalized depth

above 0.4H. The ratio then reduced gradually to a final value of 0.75 at the depth of 1.0H. The upper/lower limits for the ratio of S_m to δ_{vm} were also plotted in the figure.

The profile of the subsurface settlement is as important as the determination of the maximum subsurface settlement. Hsieh and Ou (1998) proposed a concept of influence zones on the ground surface settlement due to excavation and a procedure for estimating the ground surface settlement profile. The profile is divided into two segments, i.e., the primary influence zone (*PIZ*) and secondary influence zone (*SIZ*), and could be determined once the δ_{vm} is known. The estimation of ground surface profile and determination of *PIZ* is modified by Ou and Hsieh (2011), which is based on the excavation depth (*H*), the excavation width (*B*), the depth of soft clay bottom (H_f), and the depth of the hard stratum (H_g). The proposed *PIZ* is shown in Fig. 13. The subsurface settlement profiles in selected representative hypothetic cases, which included various excavation depths, excavation widths, soil properties and structural stiffness, were plotted in Fig. 14. The subsurface settlement (*S*) at any distance behind the wall (*d*) is normalized by the maximum subsurface settlement (S_m) at the selected depth. For each of the selected hypothetic cases, the *PIZ* is calculated and used as a factor to normalize the distance (*d*). By observing the relationship between S/S_m and *d/PIZ* at any normalized depth (*D/H*), the ground settlement profiles are generally similar. The location of maximum settlements varied slightly from 1/3 to 1/4 of *d/PIZ* with the increase of depths; moreover, the settlement ratio (S/S_m) at position of $d = PIZ$ is reduced from 1/6 when D = 0 to 1/10 when D = 1.0H. However, in general, the settlement profile proposed by Ou and Hsieh (2011) still works quite well for those on the surface and subsurface. The reason could be that the PIZ is related to the strain induced by the excavation. For soils within the range of PIZ, both the subsurface settlement and the ground surface settlement were affected by the same mechanism.

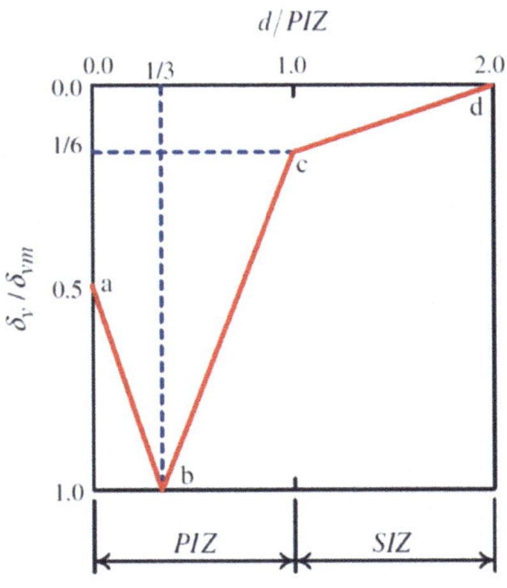

Fig. 13. Primary influence zone (reprint after Ou and Hsieh 2011)

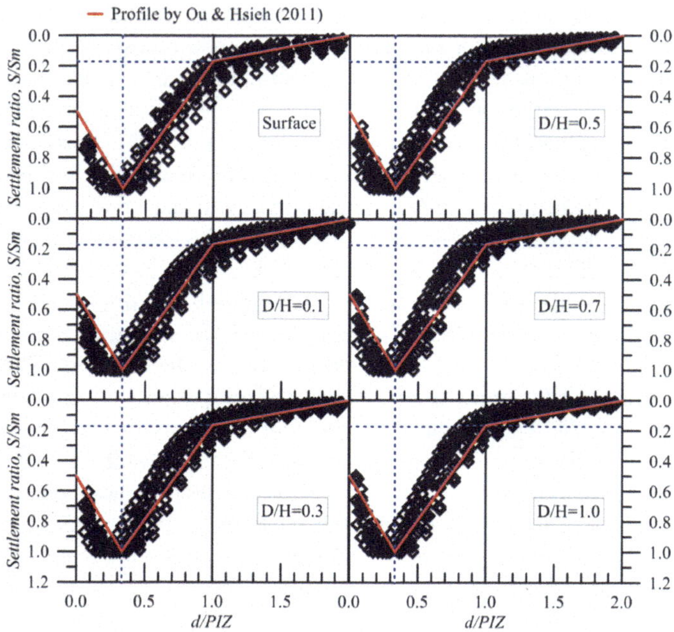

Fig. 14. Ground settlement profiles at different depths

Therefore, the subsurface settlement at any depth and distance behind the wall can be determined by applying Figs. 12, 13 and 14.

4.4 Summary of the Simplified Evaluation Method

The flowchart of the proposed simplified evaluation for ground movements is shown in Fig. 15 which included the following parts:

1. The simplified equation for determination of the maximum wall deflection based on parameters of H, B, s_u, S_w, S_a, and AR, i.e., Eqs. (4) and (5);
2. The simplified equation for determination of the maximum ground surface settlement based on parameters of H, B, s_u, S_w, S_a, and AR, i.e., Eqs. (9) and (10);
3. A quick assessment for the maximum wall deflection and the maximum ground surface settlement by Figs. 7 and 9
4. An additional evaluation equation for the maximum ground surface settlement based on the relationship between the maximum wall deflection and the maximum ground surface settlement, i.e., Eq. (11);
5. Simplified evaluation for the maximum value and profiles of the subsurface settlement, i.e., Figs. 12 and 14.

The simplified evaluation was obtained from multivariate regression analysis of numerical analyses on hypothetic excavations with: H = 10 to 30 m, B = 20 to 80 m, s_u/σ'_v = 0.28 to 0.34, S_w of diaphragm walls to sheet piles, S_a of steel struts and

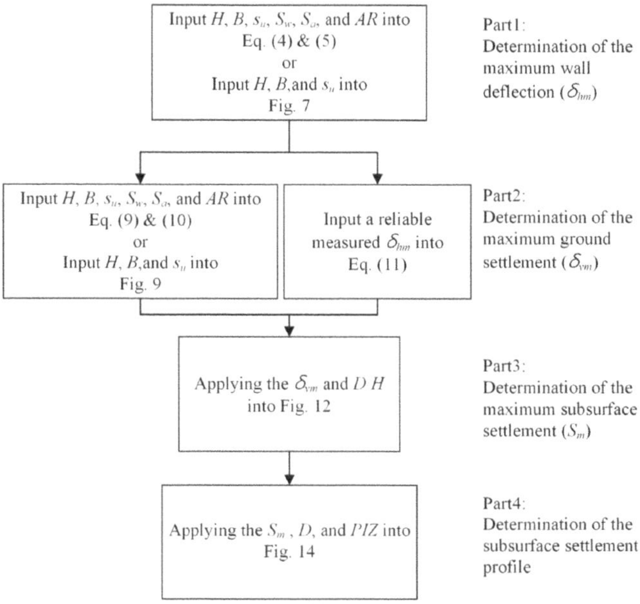

Fig. 15. Flowchart of the proposed simplified method

concrete floor slabs, and *AR* of isotropy and anisotropy. It should be noted the undrained shear strength used in the evaluations are that of axial compression under K_0-consolidation (CK_0U-AC). The average undrained shear strength used in isotropic analysis, similar to that obtained from the unconsolidated-undrained test (UU) or the FV test (field van shear), could be calculated by the following equation.

$$s_{u,avg} = \frac{s_{u,AC} + s_{u,AE}}{2} = \frac{s_{u,AC} + K_s s_{u,Ac}}{2} = \frac{(1 + K_s)s_{u,AC}}{2} \tag{12}$$

where $s_{u,iso}$ is the undrained strength in Mohr-Coulomb model; $s_{u,AC}$ is the undrained strength of axial compression; $s_{u,AE}$ is the undrained strength of axial extension; K_s is the anisotropic ratio of undrained strength.

Thus, the proposed simplified method should better be applied to the excavation which the dimension, soil properties, and stiffness of braces are in the above range.

5 Validation

Four case histories (Ou et al. 1998; Hsieh and Ou 2012) with detailed construction information were employed to validate the proposed simplified evaluation method. Table 3 lists the excavation geometry and subsoil properties used in the simplified evaluation. The anisotropic ratio of soil stiffness in case 1 was determined by test results (Teng et al. 2014), and ratios in other three cases were assumed to be equal to the average value of Taipei clay shown in Table 1. Intermediate stages in four cases

Table 3. Summary of excavation case histories

Case No.	B (m)	$\frac{s_u}{\sigma'_v}$	H_g (m)	Stage No.	H (m)	S_w	S_a (MN/m)	AR	α_g	observed δ_{hm} (mm)	observed δ_{vm} (mm)
1	38.2	0.33	46.0	4	11.8	638	2625	1.4	1.00	65	34
				5	15.2	732	2893	1.4	1.00	80	52
				6	17.3	1065	3054	1.4	1.00	99	68
				7	19.7	1313	2760	1.4	1.00	106	78
2	20.0	0.33	51.0	4	9.2	3065	3443	1.3	1.00	32	–
				5	11.7	3573	3017	1.3	1.00	41	–
				6	13.9	4495	2854	1.3	1.00	47	–
				7	16.2	4952	2745	1.3	1.00	55	–
3	62.4	0.34	32.0	3	12.4	179	2197	1.3	1.00	91	55
				4	15.4	285	2697	1.3	1.00	102	61
				5	16.9	562	2947	1.3	1.00	114	70
				6	20.0	598	3097	1.3	0.82	124	77
4	33.4	0.34	31.0	4	10.2	1239	1039	1.3	1.00	42	–
				5	13.2	1239	1307	1.3	1.00	49	–
				6	16.2	1255	1579	1.3	1.00	60	–
				7	18.5	1498	1761	1.3	0.89	63	43

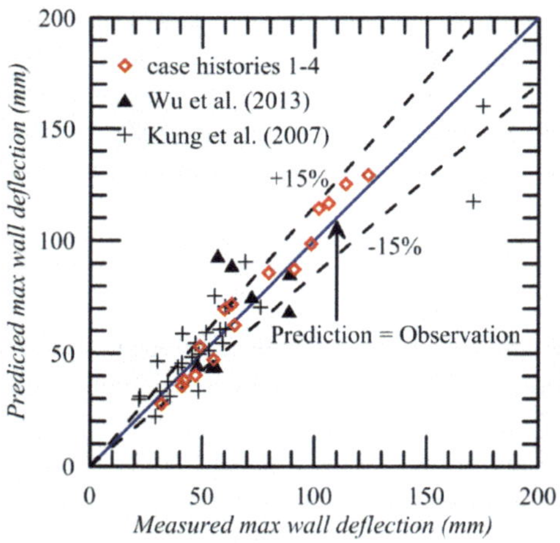

Fig. 16. Comparison between measured and calculated max wall deflections for case history 1–4

were treated as individual case histories, and the maximum wall deflection could be estimated by Eqs. (4) and (5). Further validations were made on additional 8 case histories (Wu et al. 2013) and 23 case histories (Kung et al. 2007).

Figure 16 shows the comparison of the maximum wall deflection between observations and predictions made by using Eq. (5). For δ_{hm} at intermediate stages and the final stage predicted by Eq. (5) are generally close to the line of prediction = observation. This result shows that the Eq. (5) is applicable to predict δ_{vm} not only at the final stage but intermediate stages. The result proves that Eq. (5) can yield a satisfactory prediction on δ_{hm}, which generally located within a range of $\pm 15\%$ of observed δ_{hm}.

In addition to the validation of prediction on δ_{hm}, the observed maximum ground surface settlements in case 1, 3 and 4 as listed in Table 3, and those in additional four case histories collected by Kung et al. (2007) were used for the validation of predicting the δ_{vm}. There are three evaluation methods presented hereinbefore, i.e., by Eqs. (10) and (11). Equation (10) was employed in calculating the maximum ground surface settlement at all stages. Equation (11) calculated the δ_{vm} based on the observed δ_{hm}/H. The comparison between observed and calculated δ_{hm} by three different methods are shown in Fig. 17. The performance of Eq. (10) is better than that of Eq. (11), since more factors such as excavation geometry and subsoil properties were considered in Eq. (10). However, the prediction results of Eqs. (10) and (11) are generally rational

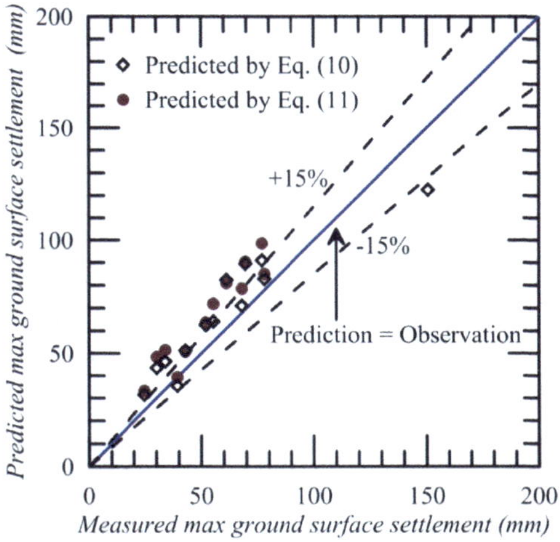

Fig. 17. Comparison between measured and calculated max ground surface settlement

Table 4. Estimated maximum subsurface settlement for case history 1

D (m)	D/H	S_m/δ_{vm}	δ_{vm} (mm)	S_m (mm)	d/PIZ		
8.0	0.41	1.10	78	85.8	0.05	0.20	0.41
12.0	0.61	1.00	78	78.0			
16.5	0.84	0.85	78	66.3			
20.0	1.02	0.75	78	58.5			

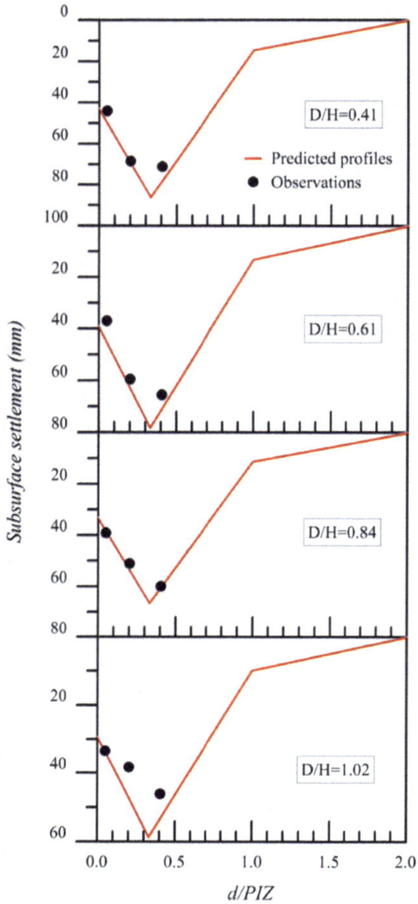

Fig. 18. Prediction for subsurface settlements in case history 1

and close to the range of 0.85 to 1.15 times observed value. The proposed evaluation methods for the maximum ground surface settlement are satisfactory with the above validation.

Validations of quick evaluation method for the maximum wall deflection and the maximum ground surface settlement are shown in Figs. 7 and 9. For the quick assessment of the maximum wall deflection, case histories with B = 20–80 m were employed. Although the prediction is not as accurate as that by Eq. (5), the results in Fig. 7 still demonstrate a satisfactory quick assessment prior to detail simulations on FEM. For the quick evaluation of the maximum ground surface settlement, five case histories with different undrained strength and excavation width are applied. The prediction results are generally sufficient for the initial assessment in the field when the engineer has limited information.

The prediction for subsurface settlement was validated through the case history 1 where the subsurface settlements were measured. The subsurface settlement was monitored by extensometers, placed at 2 m, 8 m, and 16 m behind the wall, at depths of 8 m, 12 m, 17.5 m, and 20 m. The primary influence zone (*PIZ*) for case history 1 was calculated to be 39.4 m (Ou and Hsieh 2011). To estimate the subsurface settlement, the magnitude of the maximum subsurface settlements should be determined firstly by applying the *D/H* into Fig. 12. The estimated maximum subsurface settlements at different depths are listed in Table 4. Once the S_m is determined, the subsurface settlement profile can be plotted. The comparison of subsurface settlement between field observations and prediction profiles is shown in Fig. 18. The predicted settlement profiles matched the observations quite well at all depths and positions. It demonstrates the accuracy of the proposed evaluation method for determining the subsurface settlement.

6 Conclusion

35 case histories were employed to validate the prediction of maximum wall deflection. Of these case histories, four are well-documented case histories with detail construction information at all stages. The simplified equation for evaluating the maximum ground surface settlement was validated through seven case histories. The computed maximum wall deflections and maximum ground surface settlements were generally in good agreement with the field observations. The proposed simplified evaluation provided a quick and reliable assessment of ground movements induced by excavation. In addition, a quick evaluation method for both wall deflections and ground surface settlements based on soil strength and excavation geometry were proposed and validated. It provides an initial assessment for engineers in the field with limited information.

The evaluation for damage potential of buildings adjacent to a braced excavation should base on the subsurface settlement profile. An evaluation method for the subsurface settlement, including the magnitude and shape, was suggested herein. The maximum subsurface settlement at a depth less than 0.4H was possibly greater than the ground surface settlement for 10% to 20%. The maximum subsurface settlement at depth deeper than 0.4H decreased gradually to an average value of $0.75S_m$ at the depth of 1.0H. For the shape of the subsurface settlement, the location of maximum settlements varied slightly from 1/3 to 1/4 of *d/PIZ* with the increase of depths, and the settlement ratio (S/S_m) at position of $d = PIZ$ is reduced from 1/6 to 1/10 at depth of 1.0H. However, the subsurface settlement profiles are generally similar to that on the ground surface. The settlement profile proposed by Ou and Hsieh (2012) is thus suggested. A case history that monitored the subsurface settlements at various depths and distances was employed to validate the proposed method. The result shows a good agreement between predictions and observations which implied the subsurface settlement could be well predicted by the proposed evaluation method.

References

Atkinson, J.H., Sallfors, G.: Experimental determination of soil properties. In: Proceedings of 10th ECSMFE, Florence, vol. 3, pp. 915–956 (1991)
Benz, T.: Small-strain stiffness of soils and its numerical consequences. Ph.D. dissertation, University Stuttgart, Germany (2007)
Bowles, J.E.: Foundation Analysis and Design, 4th edn. McGraw-Hill, New York (1986)
Burland, J.B.: Ninth Laurits Bjerrum Memorial Lecture. 'Small is beautiful'. The stiffness of soils at small strains. Can. Geotech. J. **26**(4), 499–516 (1989)
Cho, W., Finno, R.J.: Stress-strain responses of block samples of compressible chicago glacial clays. J. Geotech. Geoenviron. Eng. **136**(1), 178–188 (2010)
Clough, G.W., O'Rourke, T.D.: Construction-induced movements of in-situ walls. In: Proceedings, Design and Performance of Earth Retaining Structure, ASCE Special Conference, Ithaca, New York, pp. 439–470 (1990)
Duncan, J.M., Chang, C.Y.: Nonlinear analysis of stress and strain in soils. J. Soil Mech. Found. Div. ASCE **96**(5), 637–659 (1970)
Hashash, Y.M.A., Whittle, A.J.: Ground movement prediction for deep excavations in soft clay. J. Geotechn. Geoenviron. Eng. **122**(6), 474–486 (1996)
Hashash, Y.M.A., Whittle, A.J.: Mechanisms of load transfer and arching for braced excavations in clay. J. Geotech. Geoenviron. Eng. **128**(3), 187–197 (2002)
Hsieh, P.G., Ou, C.Y.: Shape of ground surface settlement profiles caused by excavation. Can. Geotech. J. **35**(6), 1004–1017 (1998)
Hsieh, P.G.: Prediction of maximum ground movements induced by deep excavation in clay. J. Chin. Inst. Civil Hydraul. Eng. **13**(3), 489–498 (2001). (in Chinese)
Hsieh, P.G., Ou, C.Y.: Analysis of nonlinear stress and strain in clay under the undrained condition. J. Mech. **27**(2), 201–213 (2011)
Hsieh, P.G., Ou, C.Y.: Analysis of deep excavations in clay under the undrained condition with small strain characteristics. J. Chin. Inst. Eng. **35**(5), 601–616 (2012)
Jardine, R.J., Potts, D.M., Fourie, A.B., Burland, J.B.: Studies of the influence of non-linear characteristics in soil-structure interaction. Geotechnique **36**(3), 377–396 (1986)
Kung, T.C., Juang, C.H., Hsiao, C.L., Hashash, Y.M.A.: Simplified model for wall deflection and ground surface settlement caused by braced excavation in clays. J. Geotech. Geoenviron. Eng. **133**(6), 731–747 (2007)
Lee, K.M., Rowe, R.K.: Deformations caused by surface loading and tunneling: the role of elastic anisotropy. Geotechnique **39**(1), 125–140 (1989)
Mana, A.I., Clough, G.W.: Prediction of movements for braced cuts in clay. J. Geotech. Eng. ASCE **107**(6), 759–777 (1981)
Mesri, G., Huvaj, N.: Shear strength mobilized in undrained failure of soft clay and silt deposits. ASCE Geotechnical Special Publication 173, Geo-Denver (2007)
Ng, C.W.W., Leung, E.H.Y., Lau, C.K.: Inherent anisotropic stiffness of weathered geomaterial and its influence on ground deformations around deep excavations. Can. Geotech. J. **41**(1), 12–24 (2004)
Ou, C.Y., Heish, P.G., Chiou, D.C.: Characteristics of ground surface settlement during excavation. Can. Geotech. J. **30**(5), 758–767 (1993)
Ou, C.Y., Hsieh, P.G.: A simplified method for predicting ground settlement profiles induced by excavation in soft clay. Comput. Geotech. **38**(8), 987–997 (2011)
Peck, R.B.: Deep excavation and tunneling in soft ground. In: Proceedings of the 7th International Conference on Soil Mechanics and Foundation Engineering, State-of-the-art volume, Mexico City, pp. 225–290 (1969)

Simpson, B., Ng, C.W.W.: Anisotropic analysis of settlement trough due to tunnelling in an overconsolidated clay. In: Proceedings of the 10th Asian Regional Conference on Soil Mechanics and Foundation Engineering, vol. 2, pp. 166–167. International Academic Publisher, Beijing (1995)

Teng, F.C., Ou, C.Y., Hsieh, P.G.: Measurements and numerical simulations of the inherent stiffness anisotropy in soft Taipei clay. J. Geotech. Geoenviron. Eng. **140**(1), 237–250 (2014)

Whittle, A.J., Kavvadas, M.J.: Formulation of the MIT-E3 constitutive model for overconsolidated clays. J. Geotech. Eng. ASCE **120**(1), 173–198 (1994)

Whittle, A.J., Degroot, D.J., Ladd, C.C., Seah, T.H.: Model prediction of anisotropic behavior of Boston blue clay. J. Geotech. Eng. ASCE **120**(1), 199–224 (1994)

Wong, K.S., Broms, B.B.: Lateral wall deflections of braced excavations in clay. J. Geotech. Eng. ASCE **115**(6), 853–870 (1989)

Wu, S.H., Ching, J., Ou, C.Y.: Predicting wall displacements for excavations with cross walls in soft clay. J. Geotech. Geoenviron. Eng. **139**(6), 914–927 (2013)

A Three-Dimensional Axisymmetric Arc Sliding Method for Checking Basal Heave Stability of Circular Foundation Pits

Ming-ju Zhang, Zhen-bo Zhang(✉), and Meng Yang

Beijing Collaborative Innovation Center for Metropolitan Transportation, Beijing University of Technology, Beijing, China
zzb759619047@126.com

Abstract. Nowadays, a large number of foundation pits in circular plane shape have appeared during urban construction in China. In the design process of the circular foundation pits, it is necessary to check their basal heave stability. Based on the circular arc sliding model of limit equilibrium method, a Three-dimensional Axisymmetric Arc Sliding Method (TAASM), in which the stiffness of the enclosure structure and the spatial effects are considered, was proposed. Furthermore, the TAASM was applied to check a practical engineering and its result was compared with those of other methods. The main conclusions are as follows: (a) The TAASM considers not only the effects of the enclosure structure stiffness and deformation but also the spatial effects of circular foundation pits on the sliding surface; (b) Compared with results of other plane algorithm methods, the result of TAASM is reasonable relatively; And (c) By TAASM, the calculated safety factor of basal heave stability for circular foundation pits will be larger than that by any other existing plane algorithm, so the embedded depth of enclosure structure may be optimized to lower the enclosure structure cost.

Keywords: Circular foundation pits · Basal heave stability · Spatial effects · Arc sliding model · Enclosure structure

1 Introduction

Nowadays, a large number of foundation pits in circular plane have appeared during urban construction in China. For example, the foundation pit of Shanghai Tower located in the Shanghai Lujiazui is in circular plane with diameter 123.4 m and depth 31.2 m (Zhai et al. 2010), the foundation pit of Shanghai Expo 500 kV underground substation is in circular plane with diameter 130 m and depth 34.0 m (Weng et al. 2010), and the south anchorage foundation pit of Wuhan Yangluo Changjiang river bridge is in circular plane with diameter 70 m and depth 45 m (Ruan et al. 2005). The circular foundation pits need to be designed for many cylindrical buildings as well, such as a circular revolving sedimentation tank of hot-rolled steel engineering in circular foundation pit with diameter 33 m and depth of 39.3 m (Li et al. 2006), a circular pool in circular foundation pit with diameter 101 m and depth 16 m (Li et al. 2011),

R. Chen et al. (eds.), *Proceedings of the 2nd International Symposium on Asia Urban GeoEngineering*, Springer Series in Geomechanics and Geoengineering,
https://doi.org/10.1007/978-981-10-6632-0_9

and a underground cylindrical 3D garage with diameter 22 ~ 36 m and depth 24.5 m (Zhang et al. 2016), and so on.

The process of foundation pit design involves the basal heave stability. There are generally two methods for checking the basal heave stability of foundation pit: the Bearing Capacity Theory Method (BCTM) and the Circular Sliding Method (CSM). With regard to BCTM, Terzaghi had previously given formula for the checking calculations based on the bearing capacity theory, but this method is limited to shallow excavation and clay (Terzaghi 1943). Afterwards, based on the Terzaghi method, Bjerrum had also given formula for the checking calculations, but this method is limited to clay (Bjerrum and Eide 1956). With regard to CSM, many researchers studied how to get the reasonable parameters and analyzed the effects of parameters, such as soil parameters, the width of excavation, and the embedded depth of diaphragm wall, on basal heave stability (Hu et al. 2001; Huang et al. 2008; Zheng et al. Zheng and Cheng 2012). Also, both BCTM and CSM belong to the plane algorithm. However, the circular foundation pits are of self-stability performance because it has the spatial effects during excavation.

In the circular foundation pits, the hoop stress was introduced to modify the earth pressure, for example, hoop stress σ_θ was considered as first main stress. The 3D axisymmetric limit equilibrium equations were established by the slip line method, then the equations were solved to obtain the earth pressures (Березанцев 1952). Afterwards, the solved results were taken to check the basal heave stability of circular foundation pit (Wang 2013). Meanwhile, a new method, in which the definition of the basal heave stability of safety is the soil limit state unload quantity and actual unload quantity ratio, was put forward, but its results are conservative (Wang 2013).

Recently, the Shear Strength Reduction Method (SSRM) has been used to check the basal heave stability of circular foundation pits. In this method, elastoplastic constitutive relationship was adopted for the soil, and the basal heave failure curves of circular foundation pit were gotten (Wang and Xu 2010; Kong et al. 2010). Meanwhile, Finite Element Method (FEM) was applied to analyze enclosure structure of small diameter circular foundation pits for their mechanical properties and the effects of the clay and sand on the basal heave deformation (Huang 2013). However, both the SSRM and FEM highly depend on the parameters and the constitutive model, so they will usually not be recommended for checking the basal heave stability in design process.

Although many researchers used plenty of methods to check the basal heave stability, the spatial effects of foundation pit and enclosure structure are often ignored. Especially for the circular foundation pits, the constraints of the adjacent soil are very strong because the hoop stress affects on the soil. So the circular foundation pits have significant spatial effects and they are of self-stability performance. When the excavation depths are the same, the lateral displacements of enclosure structure in circular foundation pit will be smaller than those in rectangular foundation pit. If a plane algorithm is chosen to calculate the basal heave stability of a circular foundation pit, the results of the embedded depth of enclosure structure will be so large that the enclosure structure will cost much more.

In this paper, based on the circular arc sliding model of limit equilibrium method, a Three-dimensional Axisymmetric Arc Sliding Method (TAASM) was proposed to check the basal heave stability of the circular foundation pits. In TAASM, stiffness and deformation of the enclosure structure were considered by assuming parabolic deformation of the enclosure structure and the hoop stress effects on sliding face were considered by the elastic foundation beam method and the circular ring axisymmetric method.

2 Definitions and Assumptions

Here,the basal heave stability safety factor is also defined as the anti-moment divided by and the sliding moment:

$$k_s = \frac{M_r}{M_s} \tag{1}$$

Where, M_r is the anti-moment; M_s is the sliding moment; and k_s is the safety factor of basal heave stability.

The calculating diagram is shown as Fig. 1.

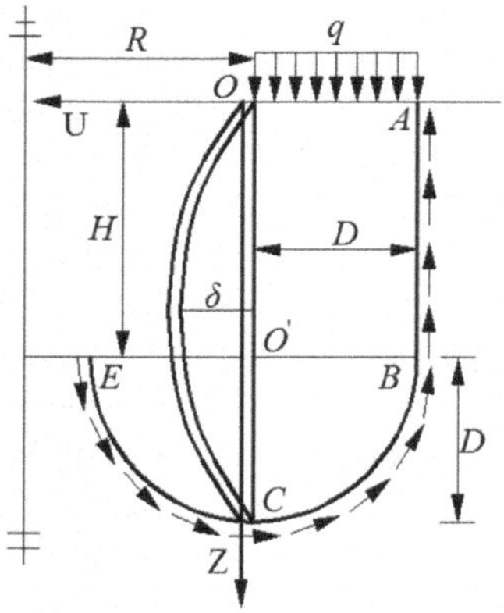

Fig. 1. The calculating diagram

In Fig. 1, H is the depth of excavation; D is the embedded depth of enclosure structure; q is the ground overload; O' is the center of sliding circle; R is the radius of foundation pit; and δ is the maximum lateral displacement of enclosure structure.

The assumptions are made as follows:

(a) Soil slides alone the sliding surface ABCE, on which the shear stress provides the anti-moment;
(b) The constitutive relationship of soil is modeled by Mohr-Coulomb model;
(c) The function $2c\tan(\pi/4-\varphi/2)$ is ignored in the active earth pressure and the function $2c\tan(\pi/4 + \varphi/2)$ is ignored in the passive earth pressure;
(d) Parabolic deformation is assumed in the enclosure structure;

And (e) The spatial effects on the soil below the bottom are ignored.

According to the assumption (a), the anti-moment is generated by the shear stress on the sliding surface AB, BC, CE and internal support, so

$$M_r = M_1 + M_2 + M_3 + M_h \tag{2}$$

Where, M_1 is anti-moment generated by shear stress on the sliding surface AB;

M_2 is anti-moment generated by shear stress on the sliding surface BC;
M_3 is anti-moment generated by shear stress on the sliding surface CE;
M_h is anti-moment generated by internal support, M_h = 800kN·m for concrete internal support, and M_h = 800kN·m for steel internal support (Wang et al. 2007).

According to the assumption (b), the shear stress can be expressed as:

$$\tau = \sigma \tan \varphi + c \tag{3}$$

Where, σ is the normal stress, τ is the shear stress, c is the cohesive, and φ is the friction angle.

According to the assumption (d), taking the coordinate system in the Fig. 1, the lateral displacement of any point in the enclosure structure can be given by

$$u = -\frac{4\delta}{(D+H)^2}z^2 + \frac{4\delta}{(D+H)}z \tag{4}$$

Where, u and z are the abscissa and the ordinate in UOZ coordinate system, respectively.

3 Formula Derivations

The radial stress and hoop stress generate the shear stress on the sliding surface AB together in the different plane. When the sliding force greater than or equal to the maximum value of the anti-sliding force generated by the radial stress and hoop stress, the sliding body start to slip. And the hoop stress is greater than the radial stress in this condition. So take the anti-moment generated by the hoop stress as the anti-moment generated by sliding surface AB. Furthermore, the hoop stress of hollow cylinder is in Eq. (3). The stress calculation diagram of hollow cylinder is shown as Fig. 2.

Where, R_1 is the internal radius; ρ is the radius of any cross section; R_2 is the external radius; q_1 is the pressure on the internal circular arc; and q_2 is the pressure on the external circular arc.

The stress solution to the hollow cylinder can be obtained through elastic mechanics (Xu 2002) and the hoop stress of hollow cylinder on any cross section is

$$\sigma_\varphi = \frac{\frac{R_2^2}{\rho^2} + 1}{\frac{R_2^2}{R_1^2} - 1}q_1 - \frac{1 + \frac{R_1^2}{\rho^2}}{1 - \frac{R_1^2}{R_2^2}}q_2 \tag{5}$$

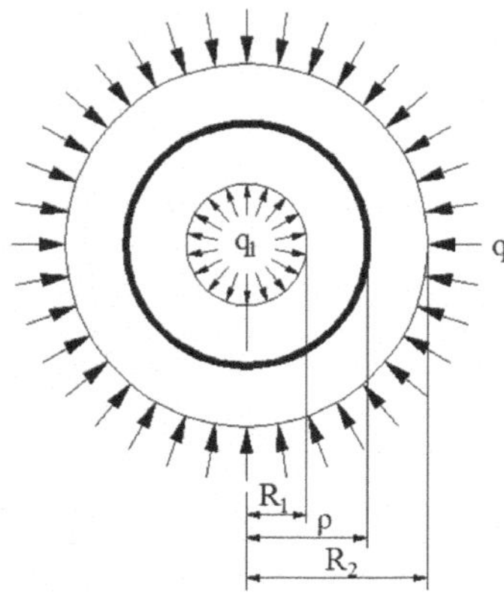

Fig. 2. The stress calculation diagram of hollow cylinder

Where, σ_{φ} is the hoop stress of hollow cylinder and the tensile stress is stipulated to be positive value.

The effects of circular foundation pit excavation on surrounding soil are limited to some space. If it is assumed that R_2 equals $3R_1$, the pressure on the external circular arc can be considered to be the static earth pressure, i.e.

$$q_2 = k_0 \gamma H \tag{6}$$

Where, k_0 is the coefficient of static lateral earth pressure and here is $k_0 = 1 - \sin\varphi$; γ is the soil unit weight.

Taking the deformation of the enclosure structure into consideration, the pressure q_1 on the internal circular arc can be given by the elastic foundation beam method as follows:

$$q_1 = k_d u \tag{7}$$

$$k_d = \frac{E_d b}{R_0^2} \tag{8}$$

Where, k_d is the equivalent stiffness of the diaphragm wall;

b is the thickness of the diaphragm wall;
R_0 is the radius of the diaphragm wall;
E_d is the circumferential comprehensive compression modulus of the diaphragm wall. Here, $E_d = 0.5 \sim 0.7E$, where E is the elastic modulus of the diaphragm wall and E_d should be taken smaller value when the R_0 is larger or the panel segment of the diaphragm wall is more.

The anti-moment generated on the sliding surface AB can be expressed by the following integral formula:

$$M_1 = \int_0^H \tau D dz \tag{9}$$

Substituting the expressions of Eqs. (3) ~ (8) into Eq. (9) and operation leads to

$$\begin{aligned} M_1 &= k_1(1-\sin\varphi)(\frac{\gamma H^2}{2}+qH)D\tan\varphi \\ &- k_2[\frac{2\delta}{(D+H)}H^2 - \frac{4\delta}{3(D+H)^2}H^3]Dk_d\tan\varphi + cDH \end{aligned} \tag{10}$$

Where, $k_1 = (\frac{9}{8}+\frac{9R^2}{8(R+D)^2}), k_2 = [\frac{9R^2}{8(R+D)^2}+\frac{1}{8}]$.

The calculating diagram of the anti-moment generated on the sliding surface BC and CE are shown as Fig. 3, respectively.

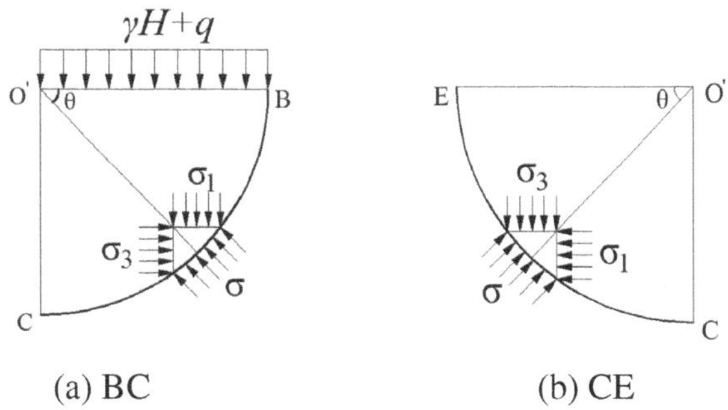

Fig. 3. The calculating diagram of anti-moment on sliding surface

The pressure on the sliding surface BC is considered to be active earth pressure, so the normal stress on the sliding surface BC can be expressed as:

$$\begin{cases} \sigma = \sigma_1 \sin\theta + \sigma_3 \cos\theta \\ \sigma_1 = D\gamma \sin\theta + q + \gamma H \\ \sigma_3 = k_a \sigma_1 \end{cases} \tag{11}$$

k_a is the active earth pressure factor, k_a = tan2(π/4-φ/2).

Accordingly, the anti-moment M_2 on sliding surface BC can be derived by the following integral calculus:

$$M_2 = \int_0^{\frac{\pi}{2}} \tau D^2 d\theta = [(\frac{\pi}{4}D\gamma + q + \gamma H) + (\frac{D\gamma}{2} + q + \gamma H)k_a]D^2\tan\varphi + \frac{\pi}{2}cD^2 \tag{12}$$

The pressure on the sliding surface CE is considered to be passive earth pressure, so the normal stress on the sliding surface CE can be expressed as:

$$\begin{cases} \sigma = \sigma_1 \cos\theta + \sigma_3 \sin\theta \\ \sigma_1 = k_p \sigma_3 \\ \sigma_3 = D\gamma \sin\theta \end{cases} \tag{13}$$

k_p is the passive earth pressure factor, $k_p = \tan^2(\pi/4 + \varphi/2)$.

Similarly, the anti-moment M_3 on sliding surface CE can be derived by the following integral calculus:

$$M_3 = \int_0^{\frac{\pi}{2}} \tau D^2 d\theta = \frac{k_p D^3 \gamma}{2} \tan\varphi + \frac{\pi}{4} D^3 \gamma \tan\varphi + \frac{\pi}{2} c D^2 \tag{14}$$

Where, M_h = 800kN·m for the concrete internal support, and M_h = 800kN·m for the steel internal support.

The sliding moment M_s is generated by sliding body ABO'O in Fig. 1. And the formula of M_s is expressed:

$$M_s = \frac{(\gamma H + q) D^2}{2} \tag{15}$$

4 An Example for Practical Engineering Checking

The basal heave stability of the foundation pit of Shanghai Expo 500 kV Underground Substation (Wang and Xu et al. 2010) will be checked as an example. The pit is in circular plane with diameter 130 m and depth 34.0 m. Its enclosure structure parameters are: thickness of diaphragm wall 1.2 m, embedded depth of diaphragm wall 23.5 m, and the elastic module of concrete 25 GPa. The field geological compositions and their physical and mechanical parameters are shown in Table 1.

Table 1. The geological compositions and their physical and mechanical parameters

Geological compositions	H/m	γ/kN·m^{-3}	c/kPa	φ/°
Silt plain fill	3.1	18.8	10	20
Mucky silty clay	7.0	17.8	11	26
Mucky silty clay	7.0	17.2	12	18
Clay	9.6	18.2	12	21
Silty clay	4.0	19.6	20	22
Sandy silt	6.5	19.1	2	27
Silty sand	8.3	19.3	2	28
Silty clay	27.8	18.5	10	21
Silty clay with silty sand interbed	4.0	19.4	7	24
Medium sand	4.0	18.6	2	30
Coarse sand	27.8	16.7	1	32

The basal heave stability of the pit was analyzed by the TAASM and its result was compared with those by other method from the reference (Wang and Xu 2010), as shown in Table 2.

Table 2. The safety factors of the basal heave stability for the pit by different methods

Method	BCTM	CSM	TAASM	SSRM
Safety factor	2.50	3.32	3.62	4.90

Because the spatial effects on the sliding surface is considered in the TAASM, the safety factor obtained by this method is larger than that by BCTM or by CSM. However, the mutual effects of soil and enclosure structure and the anti-sliding effects of adjacent soil on the sliding soil are ignored, so the safety factor by TAASM is lower than that by SSRM. The SSRM highly depends on parameters and constitutive model, it is a good method for qualitative and comparative analysis, but it will usually be not recommended for checking basal heave stability in design process.

5 Discussions an Conclusions

The new method (TAASM) is based on the circular arc sliding model, which is limited to the case that the ratio of embedded depth to excavation depth is larger than 0.5. When the ratio of embedded depth to excavation depth is smaller than 0.5, further studies which consider spatial effects should be conduct in the future. The main conclusions are as follows: (a) The TAASM combines the stiffness with the deformation of enclosure structure to check basal heave stability of circular foundation pits and it also takes the spatial effects of circular foundation pit into consideration; (b) The calculation result of TAASM is larger than that of any other plane algorithm method, but it is smaller than that of SSRM. Through the checking of an engineering example, it is proved that the result of TAASM is reasonable; and (c) The calculation result of TAASM is larger than that of the CSM. Under the circumstances of circular foundation pits, if basal heave stability safety factors are the same, the embedded depth of enclosure structure may be optimized to lower the cost of enclosure structure.

Acknowledgment. This work was carried on the Beijing University of Technology. The authors would like to acknowledge the financial support from the Collaborative Innovation Project of Chaoyang District, Beijing(XC1402), National Natural Science Foundation of China(51538001) and Beijing Natural Science Foundation (8161001), and technical support from Innovative Research Team of Beijing University of Technology.

References

Zhai, J.Q., Xie, X.L., Jia, J.: Design method for large-deep circular foundation pit of Shanghai Tower. Chin. J. Geotech. Eng. **32**(S1), 392–396 (2010)

Weng, Q.P., Wang, W.D., Zhou, J.L.: Design of diaphragm wall of an ultra_deep cylindrical excavation constructed by top_down method. J. Build. Struct. **31**(5), 188–194 (2010)

Ruan, W.J., He, C.R.: Bracing technique of a super round pit excavation. Chin. J. Undergr. Space Eng. **1**(1), 125–128+132 (2005)

Li, F.: Studies on circle-style retaining structures. Chin. J. Geotech. Eng. **28**(S1), 1522–1524 (2006)

Li, J., Li, H.M., Wu, H., Jin, G.L.: Study on design method of great cylindrical foundation pits. Chin. J. Undergr. Space Eng. **7**(5), 938–944+967 (2011)

Zhang, M.J., Zhang, Z.B., Du, X.L.: Architectural scheme design and equipment selection for the underground cylindrical 3D garage. In: China Civil Engineering Society. Symposium of China Civil Engineering Society Academic Conference in 2016. China City Press, pp. 183–197 (2016)

Terzaghi, K.: Theoretical soil mechanics, pp. 118–143. Willey, New York (1943)

Bjerrum, L., Eide, O.: Stability of strutted excavations in clay. Geotechnique **6**(1), 32–47 (1956)

Hu, Z.F., Zhou, J., Yang, L.D.: Study on subsoil stability under deep excavation. China Civil Engineering Journal **34**(02), 84–88+95 (2001)

Huang, M.S., Qin, H.L.: Basal stability analysis methods for braced foundation pit. Chin. J. Geotech. Eng. **30**(S1), 182–186 (2008)

Zheng, G., Cheng, X.S.: Basal stability analysis method considering arc length and normal stress correction. Chinese Journal of Geotechnical Engineering **34**(5), 781–789 (2012)

Ъерезанцев, В.Г.: ОСЕСИММЕТРИЧНАЯ ЗАДАЧА ТЕОРИИ ПРЕДЕЛЬНОГО РАВНОВЕСНЯ СЫПУЧЕЙ СРЕДЫ, pp. 40–64 (1952)

Wang, H.X.: Safety factor of heave-resistant stability considering two- and three- dimensional size effects of foundation pits. Chin. J. Geotech. Eng. **35**(11), 2144–2152 (2013)

Wang, W.D., Xu, Z.H.: Strength reduction approach for analyzing safety against basal heave of circular deep excavations. J. Build. Struct. **31**(05), 195–201 (2010)

Kong, D.S., Men, Y.Q., WW, Zhang, et al.: Numerical analysis of basal heave stability for deep foundation pits in soft soils. Chin. J. Geotech. Eng. **32**(S1), 186–190 (2010)

Huang, X.G.: Structural design and analysis of the mechanical properties of the small diameter circular pit. MS thesis, China University of Geosciences, Beijng (2013)

Wang, H.X., Chen, J.J., Liu, J.S.: Practical calculation analysis and application of safety factor of heave-resistant stability of foundation pit. Chin. J. Rock Mechan. Eng. **26**(S1), 3223–3230 (2007)

Xu, Z.L.: A Concise Course in Elasticity, 3rd edn., pp. 63–64. Higher Education Press, Beijing (2002)

Analysis of Sand Boiling Failure in Deep Excavations

Ignatius Tommy Pratama(✉) and Chang-Yu Ou

National Taiwan University of Science and Technology, Taipei, Taiwan
tommyignatius@gmail.com

Abstract. Sand boiling is an issue of the stability of excavations in the sand under high groundwater level condition. Many approaches have been investigated in recent years to quantify the safety against sand boiling failure. Nonetheless, some conventional methods such as Terzaghi, Harza, Marsland, Harr, and the simplified one-dimensional techniques are still commonly used to analyze it due to their simplicity. Consequently, it may lead to a misinterpretation in predicting the failure condition because each method gives different values of the factor of safety. The objectives of this study are to predict the failure mechanism caused by groundwater flow and to perceive the factor of safety values in representing the safety against sand boiling. A series of the parametric study was carried out using the conventional method of flow net and numerical analysis to obtain the hydraulic gradient and the effective stresses in the excavation zone during groundwater flow. The numerical models involved 12 cases in a homogeneous dense sand and 11 cases in a homogenous loose sand with various excavation geometry. According to the analysed results, the excavation geometry and the soil density have some significant effects in governing the hydraulic gradient and the failure mechanism. It shows that the critical width of the soil prism is smaller than that is postulated by Terzaghi. Moreover, the most likely possible failure mechanism in loose sand is bulk heave of the soil prism, whereas the expulsion of the sand-water mixture on the excavation surface will be indicated in the dense sand cases.

Keywords: Sand boiling · Factor of safety · Failure mechanism · Excavation geometry · Sand density

1 Introduction

In the recent decades, the number of deep excavation projects for basements and tunnels has been widely increasing along with the growth of infrastructure needs in many urban areas. The limitation of open lands for further development is one of the reasons behind that trend. Nonetheless, deep excavation is a high-risk construction process to construct infrastructures because of the high-level uncertainty of soils and natural conditions. According to Ou (2006), in the design of deep excavation, there are several issues related to the stability of an excavation such as overall shear failure, upheaval failure, and sand boiling failure. Those concerns are relevant for layered cohesive and cohesionless soils cases, except sand boiling failure. It merely occurs in cohesionless soils under high groundwater level condition.

R. Chen et al. (eds.), *Proceedings of the 2nd International Symposium on Asia Urban GeoEngineering*, Springer Series in Geomechanics and Geoengineering,
https://doi.org/10.1007/978-981-10-6632-0_10

Sand boiling catastrophe in either excavation case or cofferdam case involves seepage flow in the mechanism failure. The movement of water inside the soil body can decrease or increase the effective stress of soil based on the direction of flow. It generates the unbalance forces along the sheet pile which leads to the disturbance of the equilibrium condition. Once the failure occurs, it becomes an unstoppable incidence because of the rapid occurrence. A proper analysis should be conveyed to estimate the safety of an excavation against sand boiling. Various conventional approaches such as Terzaghi (1922), Harza (1935), Marsland (1953), Harr (1962), and Ou (2006) methods have been studied. Those methods derived from either experimental models or mathematical approaches to express the safety against sand boiling failure in an excavation and cofferdam systems. Due to the development of advanced technology in computing some challenging and time-consuming mathematical formulas, numerical analysis becomes a promising and reliable approach to capturing and observing the behaviour of sand boiling failure. Nonetheless, the conventional methods are still widely used by engineers because of their simplicity.

The major issue related to the traditional approaches is the number of techniques in which it can offer a reasonable and reliable factor of safety and simulate well the real condition. Besides the number of methods, analysing the same case of excavation with different formulas will provide different values of the factor of safety. Different assumptions in each technique are suspected as one of the causes behind that phenomenon. Numerous researchers and geotechnical codes also have already suggested a different minimum standard of the factor of safety as a safe adequacy condition. As a matter of effect, it may lead to a misrepresentative information of the safety against sand boiling. Therefore, the objective of this research is to investigate several factors of safety formulas against sand boiling failure and other factors that govern its value in explaining the boiling failure phenomena by the two-dimensional numerical analysis and the conventional approaches. Moreover, the interpretation of effective stress distribution in the influence of groundwater flow to forecast the failure mechanism is also one of the interests of this work.

2 Philosophy of the Factor of Safety Against Sand Boiling

Terzaghi (1922) conducted model tests to perceive the mechanism of piping or sand boiling, and he found that the piping failure due to heave is initiated by the rise of the sand prism as shown in Fig. 1. The failure zone has a width, S about equal to the half of the wall penetration depth, $H_p/2$, and a depth of approximately equal to $H_{p,0}$ below excavation surface (i.e. $0 \leq H_{p,0} \leq H_p$). The piping failure condition attains when the effective weight of the prism, W' is less than or equal to the excess hydrostatic pressure, U_e which acts on the base of the prism. Simultaneously, the effective vertical stress at any section in the prism and horizontal stress on the sides of the prism are approximately nil. Furthermore, Terzaghi (1943) mentioned that the critical segment was represented by a depth of $H_{p,0}$ that the value nearly equals to H_p. Therefore, the factor of safety against piping is

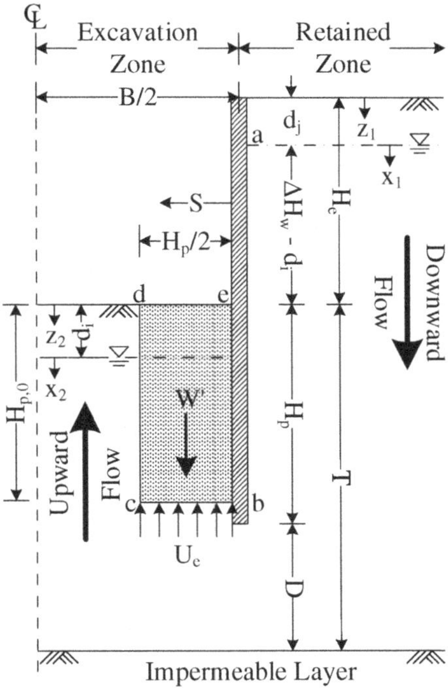

Fig. 1. The potential heave zone under seepage condition

$$FS = \frac{W'}{U_e} = \frac{\gamma' H_p}{\gamma_w h_a} = \frac{\gamma'}{\gamma_w i_{prism}} = \frac{i_c}{i_{prism}} \tag{1}$$

where γ' is the effective unit weight of soil, γ_w is the unit weight of water, h_a is the average hydrostatic head on the base of the prism, i_c is the critical hydraulic gradient, and i_{prism} is the average hydraulic gradient of the submerged prism. The i_c/i_{prism} formula is only applicable to cofferdam case (i.e. having no excavation) and to the excavation case that the groundwater level in the excavation zone is above the excavation level or at the same level as the excavation level. When the groundwater water level is below the excavation level, the Terzaghi (1922) factor of safety can be modified as follows:

$$FS = \frac{W'}{U_e} = \frac{\gamma_n d_i + \gamma'(H_p - d_i)}{\gamma_w i_{prism}(H_p - d_i)} \tag{2}$$

in which γ_n denotes the wet unit weight, and d_i is the distance between the groundwater level and the excavation level. According to the previous condition, the i_c value becomes a function of depth that is measured from the groundwater level.

Harza (1935) describes a factor of safety formula against sand boiling as a ratio between critical hydraulic gradient, i_c and maximum exit gradient, i_e on the downstream side. The maximum exit gradient is most likely occur at the downstream side where the water flows and leaves the soil mass by the shortest flow path. He defined the critical condition against sand boiling is attained when the ratio between i_c and i_e equals to one. Then, according to Harza (1935), the factor of safety against sand boiling can be expressed as

$$FS = i_c / i_e \tag{3}$$

The exit gradient, i_e of Harza (1935) definition can be obtained either from the conventional method of flow net or mathematical derivations. Harr (1962) investigated several circumstances of groundwater flow by the conformal mapping technique and proposed an analytical solution to estimate i_e for every particular groundwater flow case. According to Harr (1962), the critical exit point is located near the sheet pile or at point e as shown in Fig. 1. To make the equations to be more applicable for practice, Fig. 2 exhibits chart solutions to estimate the factor of safety against sand boiling in two distinct cases. In Fig. 2, the factor of safety trends were calculated by Harza (1935) formula with the i_c value equals unity, while the i_e values were estimated by Harr (1962) equation. If the i_c does not equal to one, multiply the $H_p/\Delta H_w$ ratio by the i_c value. Linear interpolation or logarithmic interpolation is allowable to be adopted in those charts. Harr (1962) also provided an exit gradient equation for a simple cofferdam case without impermeable layer beneath the toe of retaining wall as follows:

$$i_e = \frac{\Delta H_w}{H_p \pi} \tag{4}$$

Harr (1962) suggested that the minimum factor of safety value is greater than or equal to 4.0 to 5.0 to be considered adequate on the safety of excavation or dam against piping or boiling.

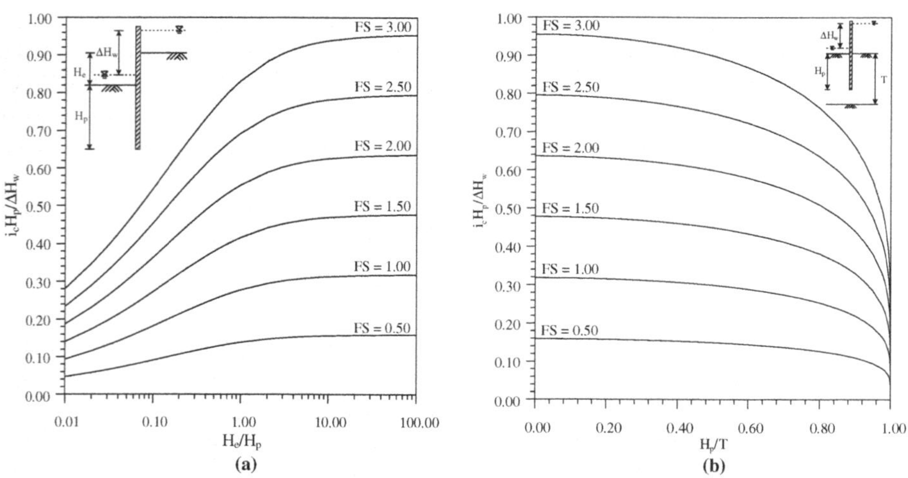

Fig. 2. Harr (1962) factor of safety against sand boiling

Harr (1962) implicitly showed that the geometry of a retaining system and the existence of impermeable layer affect the magnitude of exit gradient. His equations demonstrate that every equation is only applicable for a particular condition. To observe the influence of excavation geometry and the existence of impermeable layer, Fig. 3 depicts the reduction factors of impermeable layer, RF_{imp} and the reduction factor of excavation height, RF_{exc}. Those factors were obtained by comparing the exit gradient equations in which the exit gradient equations consider and do not consider the factors (i.e. the existence of impermeable layer and the excavation height). The charts can be applied to estimate the exit gradient of the related case by multiplying the reduction factor by Eq. (4). Figure 3a illustrates the closer the distance between impermeable layer and the wall toe, the larger the decrease in the exit gradient value. Whereas, Fig. 3b shows that the increase of excavation height also reduces the magnitude of exit gradient. Those tendencies imply that a larger factor of safety against sand boiling can be attained by the decrease of the distance between wall toe and impermeable layer and the increment of the excavation height.

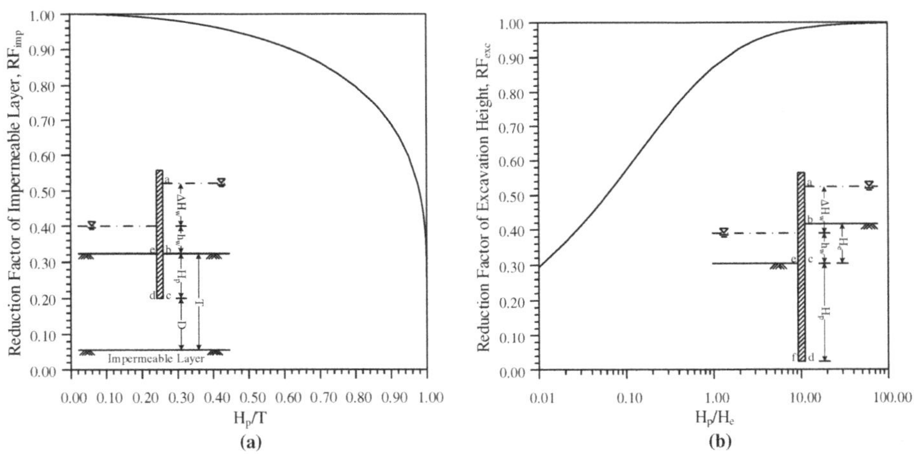

Fig. 3. Reduction factors of (a) the existence of impermeable layer and (b) the excavation height

Marsland (1953) conducted seepage model experiments on strutted excavations in either homogeneous coarse-grained soils or different soil strata to verify the results of the model test and those obtained mathematically. Figure 4 shows the Marsland (1953) factor of safety charts for various conditions. The factor of safety was estimated by taking an assumption that the groundwater level is at the same level as the ground level either at the downstream or upstream sides. The previous condition equals to heavy rainfall or flooding around the excavation. If the i_c value does not equal to one, the H_p/H_e component should be divided by i_c.

Ou (2006) suggested a simplified approach to estimating the factor of safety against sand boiling. By assuming the seepage flow is one-dimensional flow, and the hydraulic gradient is uniform along the flow path, the factor of safety equation is expressed as follows:

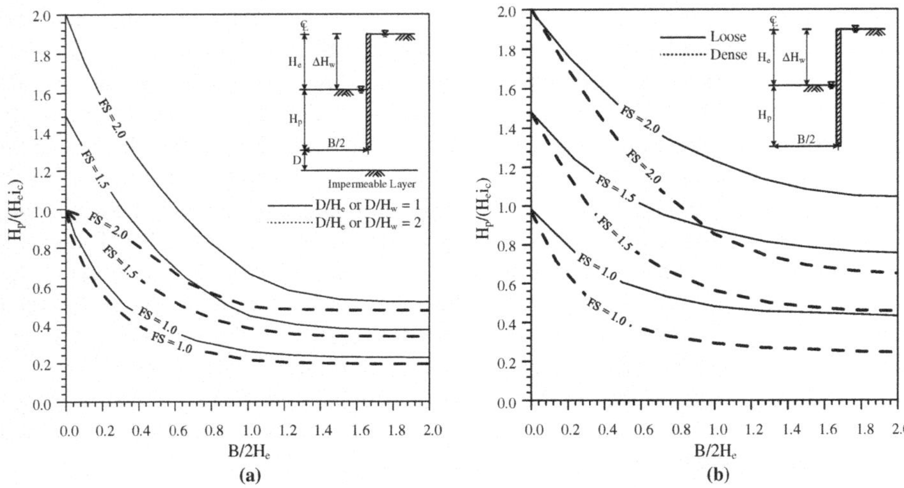

Fig. 4. Marsland (1953) factor of safety against sand boiling

$$FS = \frac{i_c}{i_{avg}} = \frac{\gamma' \left(H_e + 2H_p - d_i - d_j \right)}{\gamma_w \Delta H_w} \tag{5}$$

in which i_{avg} is the average hydraulic gradient along the flow path, H_e is the excavation height, and d_j stands for the distance between groundwater level and ground surface in the retained zone. According to Ou (2006), the factor of safety against sand boiling should be greater than or equal to 1.5.

After the overviews, it appears that the leading factor in designating the sand boiling failure due to seepage condition is the hydraulic gradient. Hydraulic gradient in groundwater flow case affects the magnitude of effective stresses acting on the soil particles. The movement of water inside the soil drags the soil particles in the direction of the current, and it produces additional forces. The upward flow as shown in Fig. 1 yields the decrease of the effective stress, whereas the downward flow will escalate the effective stress of the soil. Those manners imply the modification of the effective stress equation as shown in Eq. (6) for downward flow and Eq. (7) for upward flow.

$$\sigma'_v = \gamma_m z_1 + \gamma' x_1 + \gamma_w i_{down} x_1 \tag{6}$$

$$\sigma'_v = \gamma_m z_2 + \gamma' x_2 - \gamma_w i_{up} x_2 \tag{7}$$

in which σ'_v is the effective vertical stress of the soil, γ_m is the moist unit weight, γ' is the effective unit weight, z_i is the depth which is measured from the ground surface ($0 \leq z_1 \leq d_j$; $0 \leq z_2 \leq d_i$), x_i is the depth which is measured from the ground water level, i_{up} is the upward hydraulic gradient, and i_{down} is the downward hydraulic gradient. Consequently, the effective stress of soil and the hydraulic gradient are very fundamental in analysing the failure against seepage because they govern failure mechanism of the soil. In the further analyses, it associates in controlling volume change and soil shear strength. Nonetheless, many features are still influencing the magnitude of hydraulic gradient in practice such as the geometry of cofferdam or excavation, the ratio of wall penetration to total soil thickness, wall behaviour, particle transport mechanism, and many other aspects.

3 Analysis Procedure

Marsland (1953) model tests in the homogeneous cohesionless soil were adopted in this study to investigate the failure mechanism of sand boiling and to observe the factor of safety value tendency in analysing failure case. The investigation involved numerical analysis using SEEP/W program and conventional methods of flow net to associate the results and to achieve a better understanding of the seepage failure mechanism. The cohesionless soil model is Ham River sand which is categorized into two different densities; relatively loose sand (D_r = 59.35%) and dense sand (D_r = 96.97%). The simulation consisted of 12 cases of homogeneous dense sand and 11 cases of

Table 1. Ham river sand parameter

Parameter	Unit	Loose sand	Dense sand	Reference(s)
Maximum void ratio, e_{max}		0.92		Bishop and Green (1965)
Minimum void ratio, e_{min}		0.59		Bishop and Green (1965)
Diameter at 60% passing, d_{60}	mm	0.25		Bishop and Green (1965)
Effective diameter, d_{10}	mm	0.16		Bishop and Green (1965)
Porosity, n		0.42	0.37	Marsland (1953)
Void ratio, e		0.72	0.59	
Relative density, D_r	%	59.35	96.97	
Critical hydraulic gradient, i_c		0.97	1.05	Marsland (1953)
Soil permeability, k	m/min	0.068	0.018	Chapuis (2004), Marsland (1953)
Saturated unit weight, γ_{sat}	kN/m^3	19.32	20.10	
Effective friction angle, ϕ'	degree	32.68	36.95	Kulhawy and Mayne (1990)
Effective Young's modulus, E	kPa	29221.59	37401.70	Seed and Idriss (1970)
Modulus of compressibility, m_v	1/kPa	3.42×10^{-5}	3.42×10^{-5}	

homogeneous loose sand under steady state flow type with the properties as shown in Table 1. The permeability of dense sand was estimated using Chapuis (2004) equation that is presented as follows:

$$k = 2.4622\left[d_{10}^{2}\frac{e^{3}}{1+e}\right]^{0.7825} \tag{8}$$

wherein k stands for soil permeability in cm/s, d_{10} is the effective diameter, and e is the void ratio. Even though the permeability of dense sand was evaluated using Chapuis (2004) equation, the permeability of loose sand was obtained directly from Marsland (1953) data. Moreover, for the reason that the material model under the isotropic homogeneous condition, the permeability of sand does not vary in any direction of flow and the magnitude is constant. Besides the soil parameter that is available in Marsland (1953), Bishop and Green (1965) in his study also provided some soil parameters of Ham River sand in which their values are also presented in Table 1.

Kulhawy and Mayne (1990) proposed a chart relationship between effective friction angle and the normalized dry unit weight of clean coarse-grained soils. Figure 5 depicts the friction angle of loose sand and dense sand in which the friction angle for loose sand in this simulation equals to 32.68°, and it equals to 36.95° for dense sand. With the purpose of estimating the modulus of compressibility of the Ham river sand, the following equation was used to determine the modulus of compressibility, m_v value:

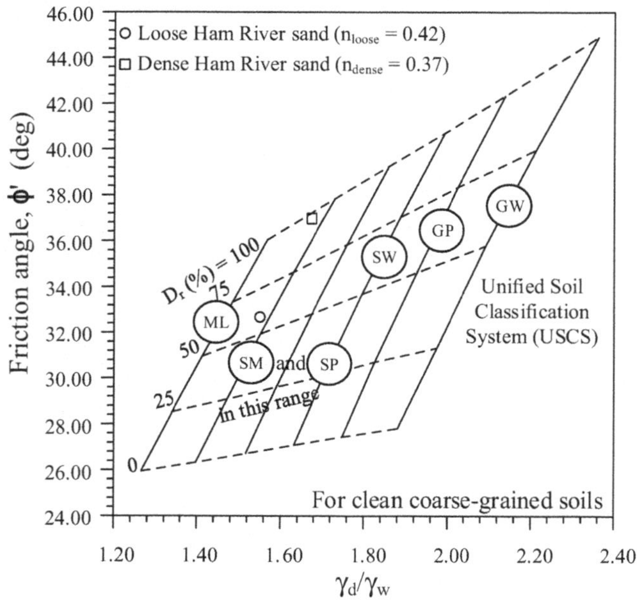

Fig. 5. Friction angle of Ham River sand

$$m_v = \frac{1}{E'} = \frac{1}{2G(1+\upsilon')} \tag{9}$$

where E′ stands for effective Young's modulus, G is the shear modulus, and υ' is the soil effective Poisson's ratio. The Poisson's ratio for this case equals to unloading/reloading Poisson's ratio, υ_{ur} which is equal to 0.20. The G value was predicted by using Seed and Idriss (1970) as shown in Eq. (10).

$$G = k_1\sqrt{p'} \tag{10}$$

wherein the p′ is the mean effective stress in MPa and the k_1 coefficient is a coefficient in a function of void ratio or relative density. The k_1 values in this simulation are equal to 344 and 445 for loose sand and dense sand, respectively.

The cofferdam models involve double-wall case and single wall case. The excavation width, B varies between 7.6 cm and 15.2 cm. Meanwhile, the B value equals to half of the seepage tank width which equivalents to 137.5 cm for the single-wall case. The wall penetration depth, H_p diverse from 2.5 cm to 15.2 cm whereas the wall toe was maintained at 7.6 cm above the impermeable boundary, D. The retained sand level was located at 38.0 cm above the wall toe. The wall structure is a brass plate with 0.125-inch thickness and attached to the brass spacers. The previous wall arrangement yields a relatively rigid retaining system with an unnoticeable wall deformation. Therefore, the wall movement is restrained either vertically or horizontally in the numerical simulation. The restriction of wall displacement in this simulation also leads to a simplification of the seepage problem and consistency against the conventional approaches. Moreover, the effect of the wall properties and the strutting system in this simulation will not affect the failure mechanism in the excavation zone. The details of the model tests and the experimental failure head difference, $\Delta H_{w,f}$ for loose sand and dense sand in accordance to Marsland (1953) are listed in Tables 2 and 3, respectively.

Table 2. Detail of model tests arrangement and experimental failure head of loose sand

Mnemonic name	D (cm)	B (cm)	H_p (cm)	H_e (cm)	$\Delta H_{w,f}$ (cm)
L0076025	7.60	7.60	2.50	33.50	7.75 ± 0.50
L0076051	7.60	7.60	5.10	30.90	11.60 ± 0.40
L0076076	7.60	7.60	7.60	28.40	14.50 ± 0.30
L0076102	7.60	7.60	10.20	25.80	17.00 ± 1.00
L0152025	7.60	15.20	2.50	33.50	9.50 ± 0.40
L0152051	7.60	15.20	5.10	30.90	15.30 ± 0.30
L0152076	7.60	15.20	7.60	28.40	19.10 ± 1.20
L0152102	7.60	15.20	10.20	25.80	20.00 ± 1.30
L1375025	7.60	137.50	2.50	33.50	11.30
L1375051	7.60	137.50	5.10	30.90	20.80
L1375076	7.60	137.50	7.60	28.40	29.00

Table 3. Detail of model tests arrangement and experimental failure head of dense sand

Mnemonic name	D (cm)	B (cm)	H_p (cm)	H_e (cm)	$\Delta H_{w,f}$ (cm)
D0076025	7.60	7.60	2.50	33.50	11.40 ± 0.40
D0076051	7.60	7.60	5.10	30.90	13.80 ± 0.30
D0076076	7.60	7.60	7.60	28.40	17.30 ± 0.30
D0076102	7.60	7.60	10.20	25.80	20.50 ± 0.40
D0076127	7.60	7.60	12.70	23.30	22.50 ± 0.40
D0076152	7.60	7.60	15.20	20.80	25.10 ± 0.20
D0152025	7.60	15.20	2.50	33.50	12.30 ± 0.80
D0152051	7.60	15.20	5.10	30.90	19.00 ± 0.50
D0152076	7.60	15.20	7.60	28.40	24.70 ± 0.50
D1375025	7.60	137.50	2.50	33.50	15.50 ± 0.20
D1375051	7.60	137.50	5.10	30.90	24.70 ± 0.40
D1375076	7.60	137.50	7.60	28.40	36.50 ± 0.50

The experimental failure head difference from Marsland (1953) corresponds to a failure criterion that a general movement of 0.2 mm observed in his experiment.

In addition to the discrepancy of excavation width and wall penetration depth in the numerical simulation, the various width of the prism of soil, S were also observed due to the upheave soil movement in the excavation zone as stated in Terzaghi (1943). In Terzaghi (1943), the excess hydrostatic pressure lifts the body of sand in a distance $H_p/2$, and as soon as it rises and deforms, the water-sand soil mixture starts to erupt. Although Terzaghi (1943) indicates a particular ground movement before failure, the deformation of the soil prism at failure is not the interest in this study. In this parametric study, the height of the prism was kept to the constant which equals to the wall penetration depth, H_p.

The aim of the width of the prism variation is to quantify the Terzaghi (1922) finding in the critical distance of the heave of the prism by examining the tendency of the factor of safety for different S/H_p ratio. There are six variations of the normalize S/H_p value which are 0.10, 0.20, 0.25, 0.33, 0.50, 0.67, and 1.00. Then, the hydraulic gradients inside the prism are arithmetically averaged, and the average value is denoted as i_{prism}. The factor of safety of each prism is estimated by Terzaghi (1922) factor of safety equation as shown in Eqs. (1) or (2). The parametric study allows exposing the most critical section at which the factor of safety exhibits the smallest magnitude of the factor of safety.

The observations were also carried out on the distribution of the effective stresses and the normalized hydraulic gradient (i_c/i) contours in the excavation zone, and also on the tendency of the factor of safety value from several methods. The magnitudes of the factor of safety against piping that was evaluated using various approaches were plotted together to conclude the test condition; failure or not failure. When the effective stress value is zero or negative, and the Terzaghi (1922) or Harza (1935) factor of safety value is less than or equal to one, the soil is defined as the failure state.

4 Results and Discussions

The results of the parametric study on the variation of the width of the prism are illustrated in Fig. 6. It shows the common trend line of the Terzaghi (1922) factor of safety against the width of the prism in both loose sand and dense sand is started by a high magnitude of the factor of safety near the sheet pile; then it continuously decreases until it reaches the inflection point. The inflection point is a point at which the trend of the factor of safety change the slope, and it also indicates the lowermost value of the factor of safety. It can be seen in Fig. 6 that the inflection point in loose sand is located at a distance from the wall equals to $H_p/3$, while in the dense sand it equals to $H_p/4$. The smallest factor of safety for each model falls in the range of $H_p/4$ to $H_p/2$ for loose sand and $H_p/5$ to $H_p/3$ for dense sand. This finding indicates the critical prism that its width is within the range is susceptible to piping failure due to heave, and at the same time the prism has a higher probability to rise or deform. In addition to the determination of the critical width, the width of the critical prism is also sensitive to the of excavation geometry, especially to the excavation width, B. The increase of B value changes the critical width of the soil prism, $S_{critical}$ into a narrower size and becomes closer to the wall. This tendency can be observed in both loose sand and dense sand cases. Consequently, in an attempt to generalize the critical section of the prism, the width of the soil prism equals to $H_p/3$ can be considered adequate to represent the most critical dimension of the prism.

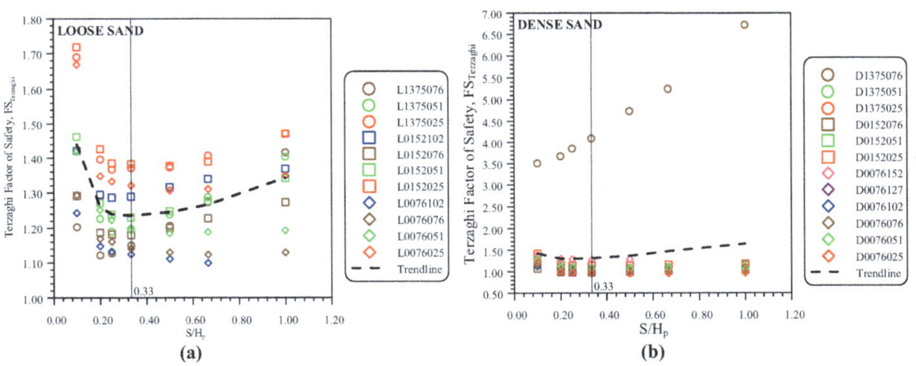

Fig. 6. Terzaghi factor of safety for a various width of the soil prism in (a) loose sand and (b) dense sand

Figure 7 shows the analysis results of the factor of safety magnitude against sand boiling using various methods. The study was carried out by using five approaches; Terzaghi (1922), Harza (1935), Marsland (1953), Harr (1962), and the simplified one-dimensional method of Ou (2006). The Terzaghi (1922) method in this part is assumed that the prism has a width equals to $H_p/2$ as postulated by Terzaghi (1922). Afterward, the hydraulic gradient of the submerged prism within the distance of $H_p/2$ is averaged. Meanwhile, the exit gradient, i_e for the Harza (1935) equation was directly

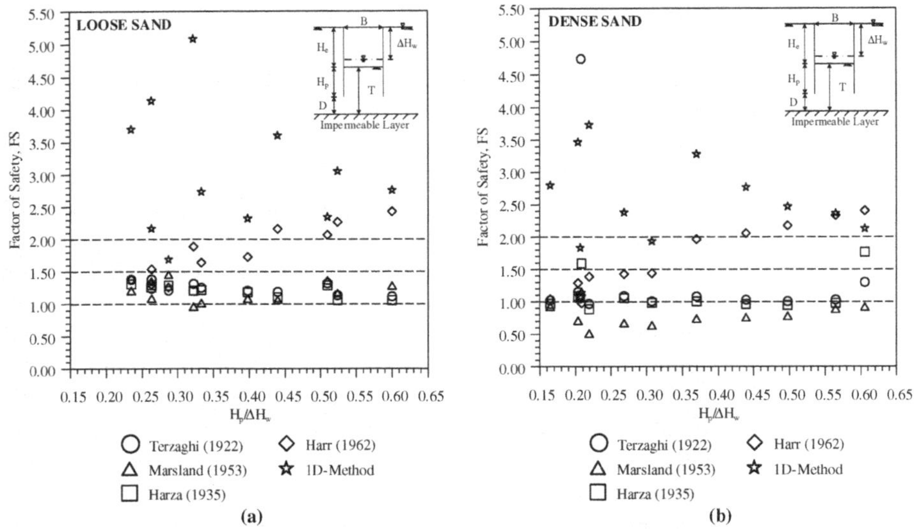

Fig. 7. Factor of safety magnitude in (a) loose sand and (b) dense sand

obtained from SEEP/W program output at the location near the wall. Figures 2a and 4a were used to estimate the factor of safety against sand boiling in accordance to Harr (1962), and Marsland (1953) approaches, respectively. The results exhibit the overall range of the factor of safety value for Marsland (1953) model tests in loose sand ranges from 0.96 to 5.08, whereas the factor safety ranges from 0.51 to 4.74 in dense sand model tests. The common trend in this analysis is the factor of safety values from Terzaghi (1922), Harza (1935), and Marsland (1953) methodologies are close to each other. Specifically, the Terzaghi (1922) and Harza (1935) methods appear to have a close magnitude of the factor of safety in both loose sand and dense sand. The typical value of the factor of safety in loose sand and dense sand based on Terzaghi (1922), Harza (1935), and Marsland (1953) methods are successively 1.21 and 1.10. These results also not only show that the Terzaghi (1922), Harza (1935), and Marsland (1953) techniques correspond to the failure test condition can describe the original circumstances of the test, but also Marsland cases might not be the total failure condition. Although it does not represent the complete failure condition where the flow of soil particles is noticeable, Marsland (1953) did a rigorous observation on the head difference at the onset of the excavation surface deformation.

In contrast to those methods, the simplified one-dimensional (1D-Method) and Harr (1962) methods seem to overestimate the factor of safety magnitude in this case. The average factor of safety from the simplified one-dimensional method is 2.53 for dense sand and 3.05 for loose sand. Whereas, the factor of safety magnitudes are 1.63 and 1.79 for Harr (1962) method in dense sand and loose sand cases, respectively. In addition to the obtained value of the factor of safety, both equations, the Harr (1962) and the simplified one-dimensional methods exhibit a different trend. Harr (1962) has a tendency that the factor of safety increases along with the rise of $H_p/\Delta H_w$ ratio, whereas the trend of 1D-Method goes down in consort with the escalation of $H_p/\Delta H_w$ ratio. That behaviour can be observed in both loose sand and dense sand. Both trends show an unreasoned tendency because the cases that were used in this study are the failure cases. Theoretically, the factor of safety should be close to one or less than one due to the failure condition.

The previous behaviour can be explained by the assumptions that are used in the equation about the geometry of excavation and the soil condition. Both Harr (1962) and 1D-Method assume the excavation width, B and the depth of impermeable layer, T are infinite. In fact, the excavation model has an absolute value of excavation width, B and a certain depth of impermeable layer, T. The change in T and B values will not alter the magnitude of the factor of safety by those methods. On the contrary, those factors affect the extent of the hydraulic gradient in which it also affects the factor of safety. Subsequently, the assumption of the simplified one-dimensional technique that the flow path is measured from the upstream groundwater level to the downstream groundwater level may lead to a relatively high value of the factor of safety because of a relatively long flow path that decreases the magnitude of hydraulic gradient. Although, the effect of the excavation geometry and the depth of impermeable layer are the factors that have to be more reckoned with in the analysis of seepage.

The failure mechanism of an excavation corresponds to groundwater flow is not only determined by the ratio of the hydraulic gradient to critical hydraulic gradient but also can be expressed by the value of effective stress. Figures 8, 9 and 10 show the normalized gradient of i_c/i, effective stress, and total pore water pressure contours for L0076076, D0076076, L0152076, and D0152076 cases successively. The normalized gradient and effective stress contours in Figs. 8 and 9, respectively represent similar patterns in term of the critical zone at which the i_c/i ratio is equal or smaller than one, and the effective stress is equal or smaller than zero. The crucial area for all cases is located near the wall toe. Similar patterns are also observed in the different model case. It explains that the region at which the i_c/i ratio is less than or equal to one, the corresponding effective stress will be equal or smaller than zero. At the same time, the related region will experience significant shear strain due to loss of soil shear strength. Furthermore, the total pore water pressure distributions in Fig. 10 indicates that the highest pore water pressure occurs at the toe of the wall. Nonetheless, the pore water pressure distributions along the excavation width, B are not uniform. The total pore water pressure at the section of B/2 is smaller than that is located near the wall toe.

Figures 8 and 9 illustrate the failure shape always corresponds to a relatively rectangular soil prism. Nevertheless, the width of the prism does not equal to the Terzaghi (1943) suggestion. According to the i_c/i and effective stress contours, the critical zone in the case of dense sand covers a larger region than in the loose sand. The critical area in the loose sand cases usually localized near the wall toe except for a narrow excavation where the wall penetration is larger than or equal to excavation width. Conversely, the effective stress zone that the value equals to zero in the case of dense sand and narrow excavation cases equally distributes near the excavation surface to the wall toe. In the case of narrow excavation, both loose sand and dense sand

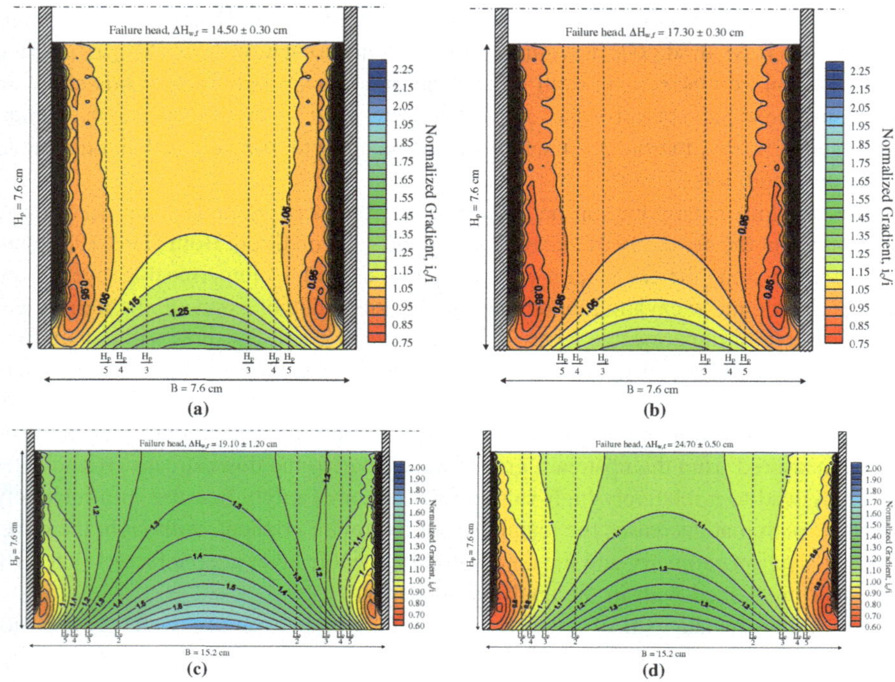

Fig. 8. Normalize gradient, i_c/i contour of (a) L0076076, (b) D0076076, (c) L0152076, and (d) D0152076 cases

exhibit the similar pattern of critical effective stress distribution. The examination of these findings indicates that the most likely possible failure type is the heave of the soil prism for loose sand cases and local failure which is marked the expulsion of the soil-water mixture for dense sand and narrow excavation cases. The concentrated area in the loose sand cases at which the effective stress value is equal or smaller than zero, the pore water pressure acts on the soil body that does not fail. As soon as the pore water pressure exceeds the weight of the soil, the soil will start to deform, and sand boiling occurs. Meanwhile, the zero effective stress in the dense sand and narrow excavation cases that are located near the excavation surface and along the wall penetration depth produces no resistance against the pore water pressure, and it is followed by the expulsion of the groundwater together with transported soil particles to the excavation surface. As a result, the sand boiling occurrence is inevitable.

Referring to Terzaghi (1943) that he explained the general mechanism of piping failure due to heave is caused by the difference of the water head between upstream and downstream that induces water flows near the retaining wall through the soil particles. The groundwater flow in excavation zone contributes to the decrease of the effective stress of the soil. As the effective stress decreases in the excavation area, the soil gradually loses its strength and at a particular condition at which the effective stress is

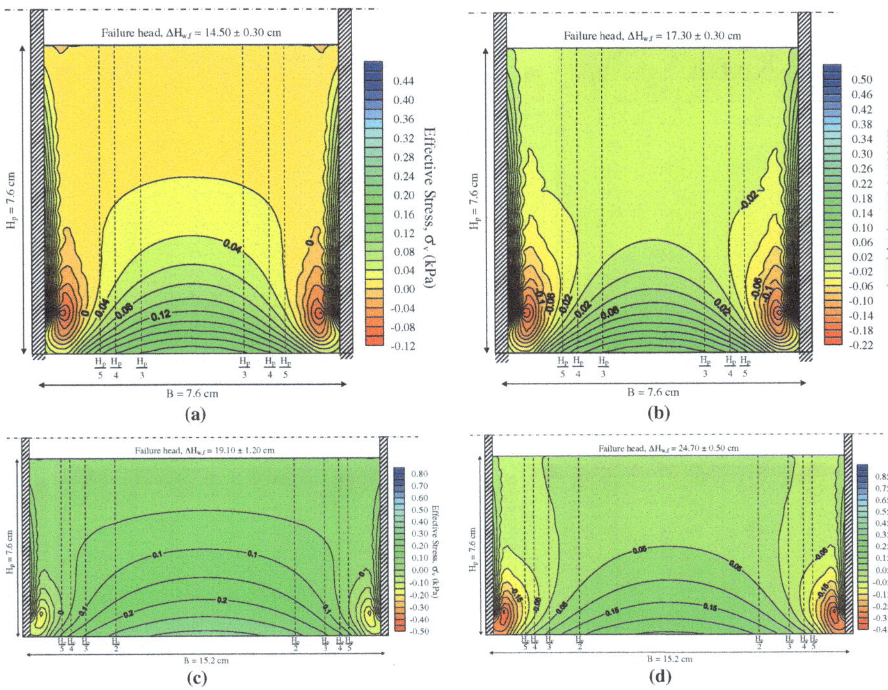

Fig. 9. Effective stress contour of (a) L0076076, (b) D0076076, (c) L0152076, and (d) D0152076 cases

equal to zero the soil cannot sustain any stresses. The soil becomes loose at the downstream side wherein the void ratio, e, porosity, n, and the permeability, k of the soil also escalate. At that condition, the soil prism starts to rise, and sand boiling failure occurrence follows it. A similar mechanism of sand boiling failure in the excavation zone was also observed by Tanaka et al. (2016). Concerning Marsland (1953) experiment results, he explained that the failure in the homogenous loose sand occurs when the amount of excess hydrostatic pressure is sufficient to encounter the submerged weight of the soil prism. Whereas, the failure in the homogeneous dense sand took place when the ratio between i_c and i_e equal to one. Therefore, the presumption using effective stress distribution and a proportion of the hydraulic gradient can be reasonable to describe and predict the failure mechanism against sand boiling failure. Even though the deformation of the excavation surface cannot be captured by this method, observation based on the effective stress distribution in the excavation area is highly suggested to be performed to forecast the critical section and the failure mechanism against sand boiling.

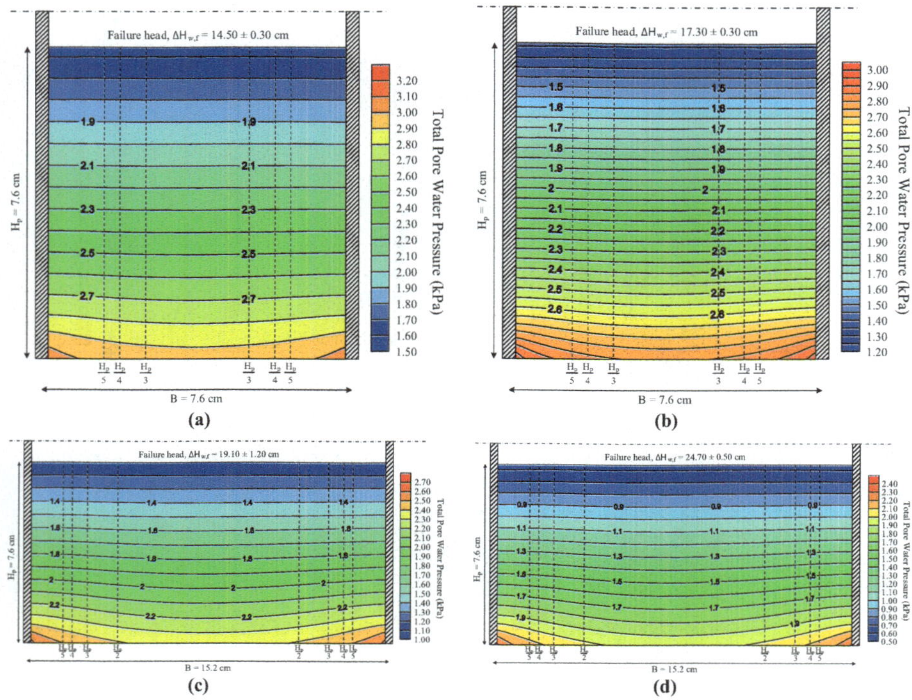

Fig. 10. Total pore water pressure distribution of (a) L0076076, (b) D0076076, (c) L0152076, and (d) D0152076 cases

5 Conclusion

The understanding of the factors affecting sand boiling failure is necessary to estimate the safety of an excavation in cohesionless soil under high groundwater level condition and to predict the failure mechanism. A series of parametric study on the experimental failure case and numerical analysis were carried out by using numerical programs and conventional methods. The observation of the analysis results was focused on the factor of safety magnitude from various methods, hydraulic gradient, and effective stress distribution in the excavation zone. According to the outcomes, the conclusions of this study are shown as follows:

1. By considering the effect of excavation geometry and the existence of impermeable soil layer in the Marsland (1953) case, a distance of $H_p/3$ from the wall with a height equals to H_p can be considered adequate to represent the most critical prism section of which it has the lowermost magnitude of the factor of safety.
2. Terzaghi (1922) and Harza (1935) approaches are consistent with each other and can simulate the closest failure condition of the Marsland (1953) experimental failure model. Nonetheless, Marsland (1953) experimental failure models are not the complete failure model where the discharge of soil-water mixture is observable.

3. In addition to the computation of factor of safety of excavation, observation on the effective stress distribution is highly recommended to be conveyed to predict the critical area and the failure mechanism of sand boiling.
4. According to the previous findings on the i_c/i contours and the effective stress distributions, the most likely possible failure mechanism of loose sand is a piping failure due to heave. Meanwhile, for the dense sand and narrow excavation cases, the failure mechanism is exhibited by the expulsion of the sand-water mixture on the excavation surface.

References

Bishop, A.W., Green, G.E.: The influence of end restraint on the compression strength of a cohesionless soil. Geotechnique **15**(3), 243–266 (1965)

Chapuis, R.P.: Predicting the saturated hydraulic conductivity of sand and gravel using effective diameter and void ratio. Can. Geotech. J. **41**(5), 787–795 (2004)

Harr, M.E.: Groundwater and Seepage, pp. 104–125. McGraw-Hill, New York (1962)

Harza, L.F.: Uplift and seepage under dams on sand. Trans. Am. Soc. Civ. Eng. **100**(1), 1352–1385 (1935)

Kulhawy, F.H., Mayne, P.W.: Manual on Estimating Soil Properties for Foundation Design. EPRI Report EL-6800, Research Project 1493-6, Cornell University, NY (1990)

Marsland, A.: Model experiments to study the influence of seepage on the stability of a sheeted excavation in sand. Geotechnique **3**(6), 223–241 (1953)

Seed, H.B., Idriss, I.M.: Soil Moduli and Damping Factors for Dynamic Response Analyses. Report EERC 70-10, Earthquake Engineering Research Center, University of California, Berkeley (1970)

Tanaka, T., Tachimura, R., Kusumi, S., Nagai, S., Inoue, K.: Experimental findings of 3D seepage failure of soil within a cofferdam. Jap. Geotech. Soc. Spec. Publ. **2**(45), 1608–1613 (2016)

Terzaghi, K.: Der grundbruch on stauwerken und seine verhutung. Die Wasserkraft **17**(24), 445–449 (1922). (in Germany)

Terzaghi, K.: Theoretical Soil Mechanics, pp. 257–262. Wiley, New York (1943)

Ou, C.Y.: Deep Excavation: Theory and Practice, pp. 163–167. Taylor & Francis Group, London (2006)

Case Record of a Strut-free Excavation with Buttress Walls in Soft Soil

Aswin Lim[1,2(✉)] and Chang-Yu Ou[2]

[1] Department of Civil and Construction Engineering, National Taiwan University of Science and Technology, Taipei 10672, Taiwan
aswin_lim@yahoo.co.id

[2] Department of Civil Engineering, Universitas Katolik Parahyangan, Bandung 40141, Indonesia

Abstract. This paper presents a well-documented strut-free excavation case with diaphragm walls, buttress walls, and partial floor slabs as the strut-free retaining system. The final excavation level was 9.2 m, covering an area about 17340 m^2. Because of its large excavation geometry, this project utilized the strut-free retaining system to reduce the construction cost and period. The soil layers above the final excavation level are dominated by the soft to medium clay layer (SPT = 2–7) and the loose silty sand layer (SPT = 2–6). According to the interpretation of field monitoring results, the following significant findings were drawn, such as (1) the measured maximum wall deflection was similar compared with strutted excavation case histories in the Taipei area. The maximum wall deflections to final excavation level ratio (δ_{hmax}/H_e) were between 0.27 and 0.55; (2) at the long-side of diaphragm wall, the pattern of the wall deflections is a cantilever shape with a translational movement at the wall toe and the location of maximum wall deflection was near the top of the wall; (3) at the short-side of diaphragm wall, the pattern of the wall deflections when reaching the final excavation level was a curvature shape and the location of maximum wall deflection was slightly lower than the final excavation level; (4) The maximum ground surface settlements to final excavation level ratio were below $\delta_{vmax}/H_e = 0.3\%$. Although it was quite small, the ground surface settlements extend to the significant distance behind the diaphragm wall; (5) the strut-free retaining system was proven successful to retain soil during excavation.

Keywords: Strut-free system · Buttress walls · Excavation · Soft soil

1 Introduction

In recent decades, buttress walls are widely adopted as an auxiliary measure for the protection of adjacent buildings during excavation in Taiwan (Ou et al. 2008; Hsieh et al. 2015; Hsieh et al. 2016; Hsieh et al. 2017). As an auxiliary measure, the diaphragm wall is mainly supported by steel struts (braced excavation method) and/or concrete floor slabs (top-down excavation method). Buttress walls are concrete walls perpendicular to the diaphragm wall with the limited length. In common practices, buttress walls could be located inside the excavation zone (the inner buttress wall) and/or outside the excavation zone (the outer buttress wall) (Lin and Woo 2007) as

R. Chen et al. (eds.), *Proceedings of the 2nd International Symposium on Asia Urban GeoEngineering*, Springer Series in Geomechanics and Geoengineering,
https://doi.org/10.1007/978-981-10-6632-0_11

illustrated in Fig. 1a. In addition, the shape of buttress walls could be the rectangular shape (R-shape), or the L-shape (Lim et al. 2016) as illustrated in Fig. 1b. Those kinds of buttress walls were used in the case history reported in this paper.

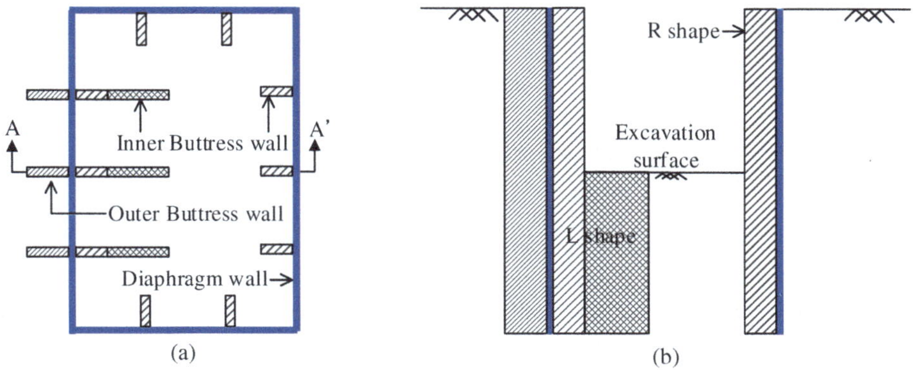

Fig. 1. Schematic diagram of buttress walls: (a) Plan view; (b) Cross-section A-A′ view

Besides as an auxiliary measure, buttress walls are also used as main components in the strut-free retaining wall system. Chuah and Tan (2010) investigated the Tribeca excavation case in Singapore. The excavation geometry was about 75 m 52 m. The Tribeca excavation adopted the counterfort diaphragm wall system to retain 6.6 to 8 m depth of excavation in the Singapore Marine Clay. The retaining system consists of buttress walls combined with embedded diaphragm walls, and counterfort slabs. This system was successfully executed with the maximum wall deflection to final excavation level ratio (δ_{hmax}/H_e)) equal to 1.33%, in which it was quite large compared with the common excavation methods.

The objective of this paper is to report the performance of a case history through field monitoring results. The main features of this case history are (1) the excavation geometry is very large, that is 195.7 m in length and 88.6 m in width, (2) the excavation adopted the strut-free retaining system, and (3) the subsurface conditions are dominated by soft soil.

2 Project Description and Instrumentation

The Xizhi excavation project is located in New Taipei City, Taiwan. The project is considered as a large excavation with the dimension of 195.7 m 88.6 m, and it was excavated to the depth of 9.2 m. It is an industrial building with two floors of basements, and 17 stories above ground. The basement was completed in 3 stages using the periphery top-down method in which floor slabs were casted with a limited width and were located at the excavation circumference. Furthermore, the buttress walls were maintained during the excavation process and demolished after completion of the lower structures. The excavation stages are summarized in Table 1.

Table 1. Construction sequence of excavation project

Stage	Date	Days	Construction activity
Initial	2008/11/20–2009/01/12	0–53	Construct diaphragm walls and buttress walls
	2009/01/12–2009/02/28	53–100	Install monitoring systems
1	2009/03/01–2009/10/28	100–342	Excavate to GL −1.5 m
			Construct partial first-floor slabs (1FL)
2	2009/10/29–2010/02/21	342–458	Excavate to GL −5.0 m
			Construct partial first basement slabs (B1F)
3	2010/02/22–2010/07/28	458–615	Excavate to GL −9.2 m

In addition, the earth retaining system consists of 20.5 m depth of diaphragm walls (0.5 m thick) with several types of buttress walls (0.5 m thick) as depicted in Fig. 2. The spacing between the buttress walls is around 8.5 m and the length of the buttress walls was varied between 4.8 m to 9.7 m. The diaphragm walls, the buttress walls, and the partial floor slabs were casted using concrete with $f_c' = 27.45$ MPa. Moreover, the diaphragm wall and the buttress walls were not founded in the hard-stratum (NSPT > 50). Moreover, deflections of diaphragm wall were measured by eleven in-wall inclinometer casings (SID 01 to SID 11), 25.5 m in depth. Ground surface settlements were measured at five sections perpendicular to the diaphragm wall with settlement markers (SM 01, SM 03, SM 04, SM 05, and SM 07), which was attached to the ground, as shown in Fig. 2.

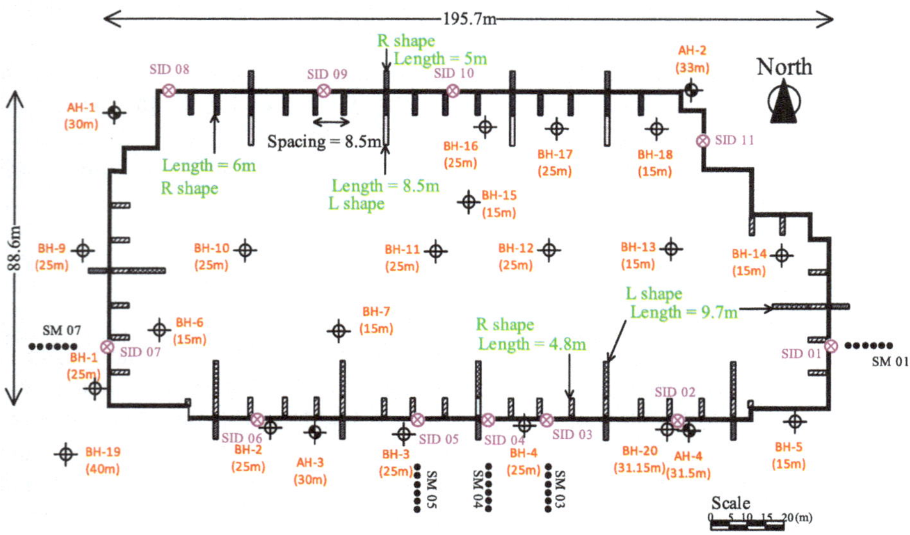

Fig. 2. Excavation geometry and monitoring layout

3 Ground Conditions

The excavation site is located at the terrace deposit of Keelung River which mainly comprises of clay and silty sand layers. In total, twenty initial boreholes (BH-1 to BH-20) and four additional boreholes (AH-1 to AH-4) with various depths were drilled to examine the profile of subsurface soils (Fig. 2). The groundwater level observed during additional boring was located about 10 m below the ground surface. Hence, no dewatering scheme was planned. The ground conditions and soil properties at the site have been investigated by a series of field tests (e.g., standard penetration tests) and laboratory tests (e.g., oedometer tests and triaxial tests). Figure 3 shows the soil profiles and measured soil properties.

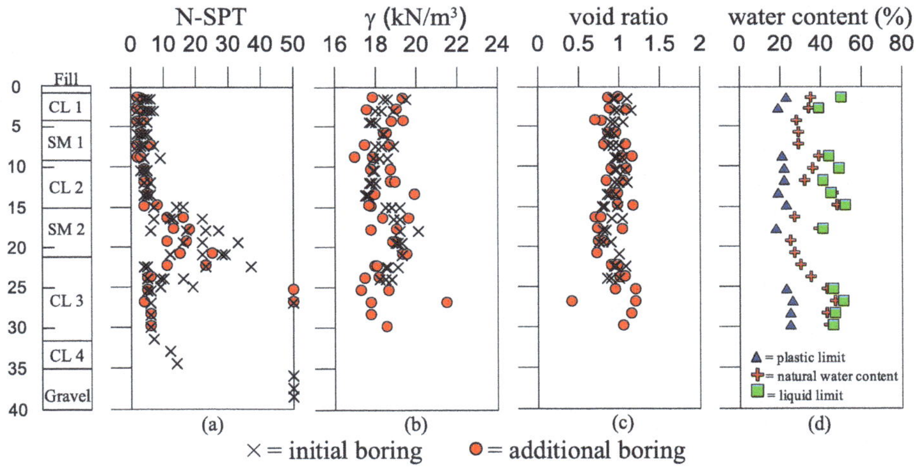

Fig. 3. Profiles of (a) N-SPT values, (b) soil unit weight, (c) void ratio, and (d) Atterberg limits for the project site

The strata of the site could be divided into eight idealized layers. From top to bottom, they are described as follows:

1. The first layer (GL 0–GL −1.0 m) is a layer of fill material which consists of gravels, silty clays, and bricks. The unit weight of this layer is about 19 kN/m^3.
2. The second layer (GL −1 to GL −4.0 m) is a layer of gray silty clay (CL 1), with N_{SPT} values varying from 2 to 7. Its average total unit weight is about 18 kN/m^3, void ratio between 0.8 and 1.1, average natural water content about 34%, liquid limit between 39 and 50, plasticity limit between 19 and 23, compression index (C_c) is 0.3, swelling index (C_s) is 0.03, effective friction angle is about 26° to 28°.
3. The third layer (GL −4 m to GL −9 m) is a silty sand layer (SM 1). The N_{SPT} values varied from 2 to 6 and the average unit weight is 18.5 kN/m^3. Its effective friction angle is about 28° and its natural water content is about 30% to 40%.

4. The fourth layer (GL −9 m to GL −15 m) is a layer of gray silty clay (CL 2), with N_{SPT} values varying from 2 to 9. Its average total unit weight is about 18.2 kN/m^3; average void ratio is 1.04 (GL −9 m to GL −13 m) and 1.08 (GL −13 m to GL −15 m); average natural water content about 40%, liquid limit between 40 and 50, plasticity limit between 20 and 25, compression index (C_c) is 0.3, swelling index (C_s) is 0.03, effective friction angle is about 28°.
5. The fifth layer (GL −15 m to GL −22 m) is a silty sand layer (SM 2). The N_{SPT} values varied from 8 to 38 and the average unit weight is 18.9 kN/m^3. Its effective friction angle is about 32° and its natural water content is about 22% to 30%.
6. The sixth layer (GL −22 m to GL −32 m) is a layer of silty clay (CL 3), with N_{SPT} values varying from 4 to 10. Its average total unit weight is about 17.7 kN/m^3, average void ratio is 1.16, natural water content between 30% to 45%, liquid limit between 45 and 50, plasticity limit between 23 and 25, compression index (C_c) is 0.2, swelling index (C_s) is 0.02, and effective friction angle is about 30°.
7. The seventh layer (GL −32 m to GL −35 m) is a layer of silty clay (CL 4), with N_{SPT} values varying from 12 to 14. Its average total unit weight is about 17.7 kN/m^3, compression index (C_c) is 0.2, swelling index (C_s) is 0.02, and effective friction angle is about 31°.

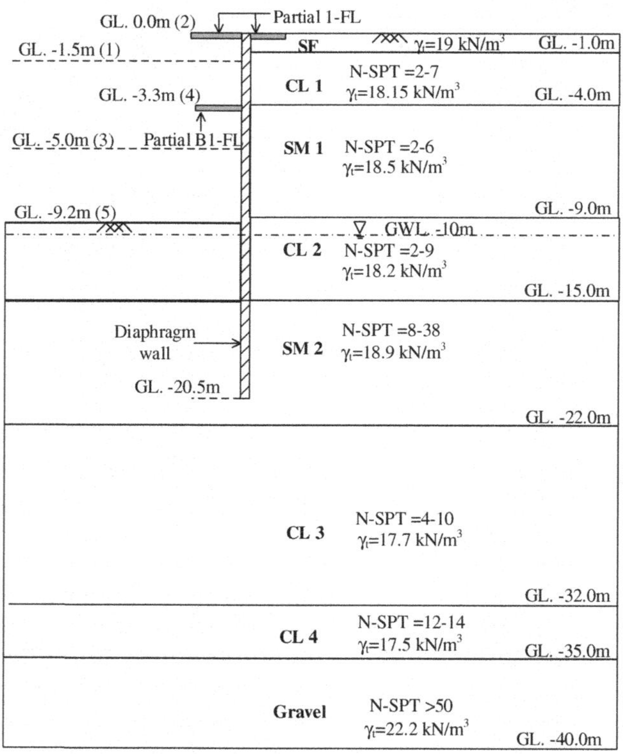

Fig. 4. Excavation and subsurface profile

8. The eighth layer (GL −35 m to −40 m) is a layer of gray gravels (GW). The N_{SPT} values are larger than 50 and the average unit weight is 22.2 kN/m^3. Its effective friction angle is about 38°.

Furthermore, the profile of the subsurface soils and the excavation sequences is illustrated in Fig. 4.

4 Field Monitoring Results and Discussion

The diaphragm wall deflections and its corresponding ground surface settlements at the south side are represented by SID 04, SID 05, SM 04, and SM 05, respectively. In general, the pattern of the wall deflections is a cantilever shape with a translational movement at the wall toe because the toe of the diaphragm wall was not founded in the hard-stratum, as plotted in Figs. 5 and 6. The cantilever shape of wall deflections was also observed by Chua and Tan (2010) in their reported case. It should be noted that the length of the south diaphragm wall is 195.7 m and the location of those inclinometers was close to the center of the diaphragm wall. According to Ou et al. 1996, the plane strain ratio (PSR) was equal to unity. Thus, the measured results could be judged that they were less affected by the corner effect, or more likely in the plane strain condition. In such a condition, it seems that the first partial floor slabs did not function for restraining wall deflections at the top of the diaphragm wall, but the function was more likely similar to a capping beam where it distributed the wall deflection more uniformly at the top of the diaphragm wall. The similar wall deflection shape was also observed at the north side, but with smaller magnitude, as shown in Fig. 7. The smaller magnitude

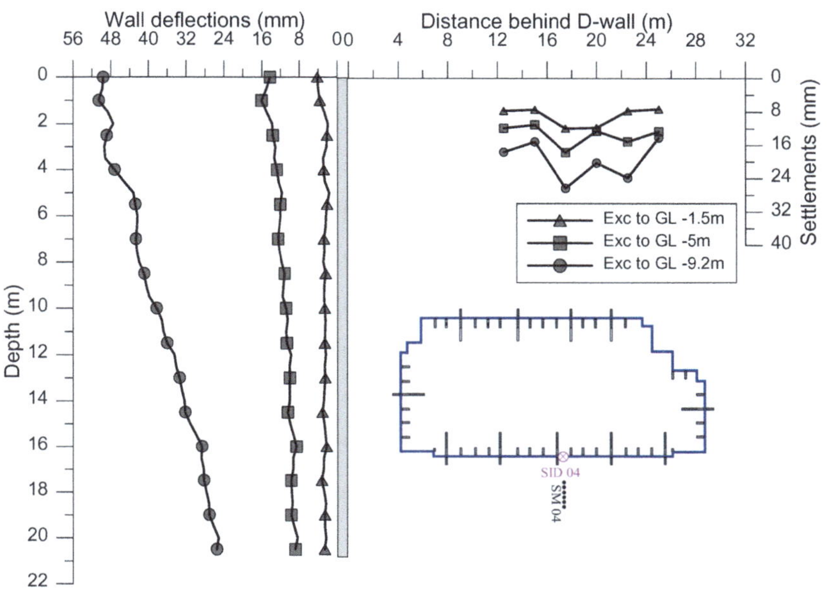

Fig. 5. Measured wall deflections and ground surface settlements of SID 04 and SM 04

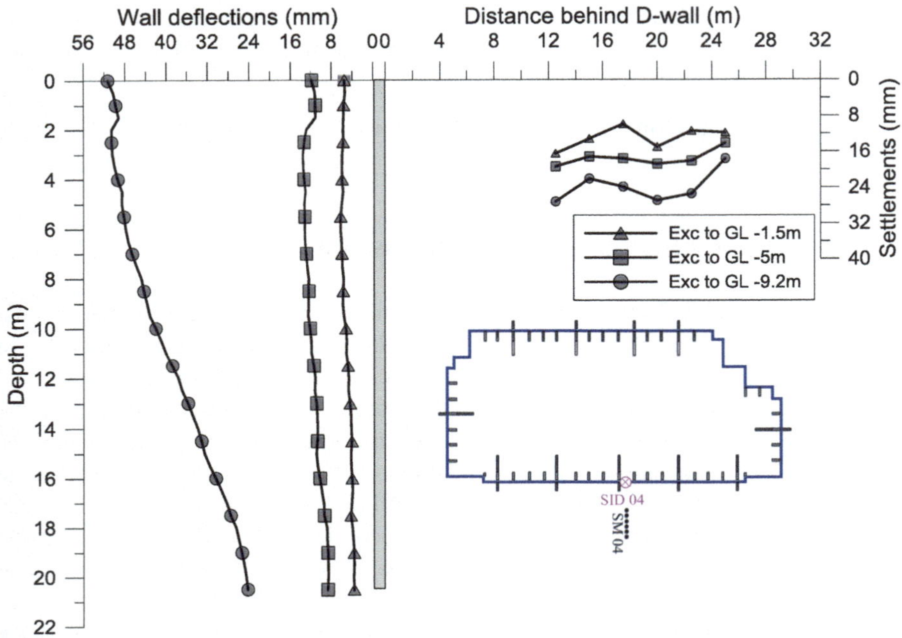

Fig. 6. Measured wall deflections and ground surface settlements of SID 05 and SM 05

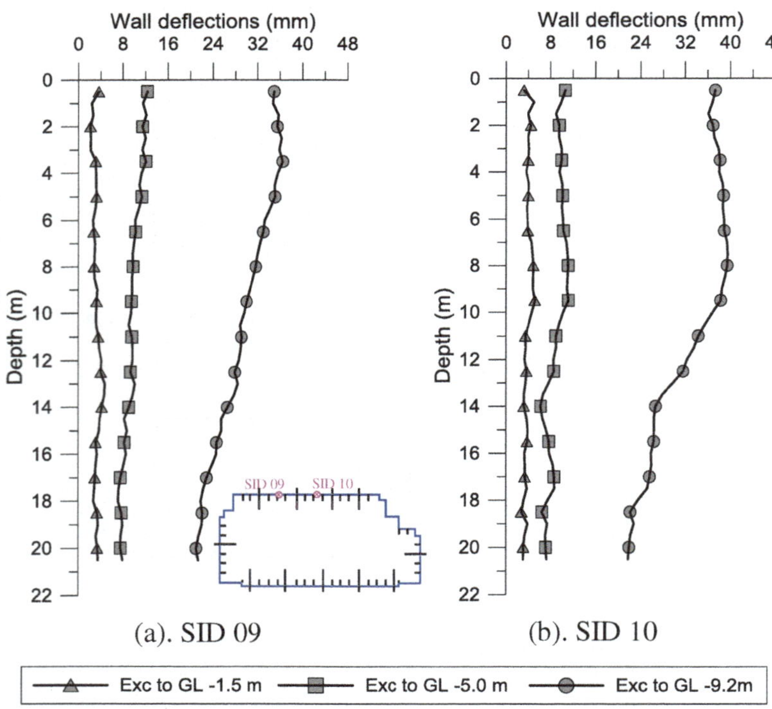

Fig. 7. Measured wall deflections of SID 09 and SID 10

of diaphragm wall deflections was contributed from the subsoil conditions at the north side was stiffer than the south side, especially the CL 1 layer. The N-SPT of CL 1 layer at the north side was 5, meanwhile, the CL 1 layer at the south side was 2. Obviously, the magnitude of wall deflections at the north side was smaller than the magnitude of wall deflections at the south side, and the location of maximum wall deflection was near the top of the wall.

Furthermore, the diaphragm wall deflections and its corresponding ground surface settlements at the west and the east sides are represented by SID 07, SID 01, SM 07, and SM 07, respectively. In general, the pattern of the wall deflections at the final stage is a curvature shape, as depicted in Figs. 8 and 9. It should be noted that the length of the west and east diaphragm wall is 65 m and 50 m, respectively, and the location of those inclinometers was 20 m away from the corner of the diaphragm wall. According to Ou et al. 1996, the PSR of SID 07 and SID 01 were equal to 0.45 and 0.35, respectively. Thus, the measured results could be judged that they were much affected by the corner effect. In such a condition, it seems that the combination of the corner effect and first partial floor slabs could restraint wall deflections at the top of the diaphragm wall. The location of maximum wall deflection was slightly lower than the final excavation level when the shape of the wall deflections was a curvature. Moreover, Table 2 summarizes all of the maximum wall deflection at various excavation stages and locations.

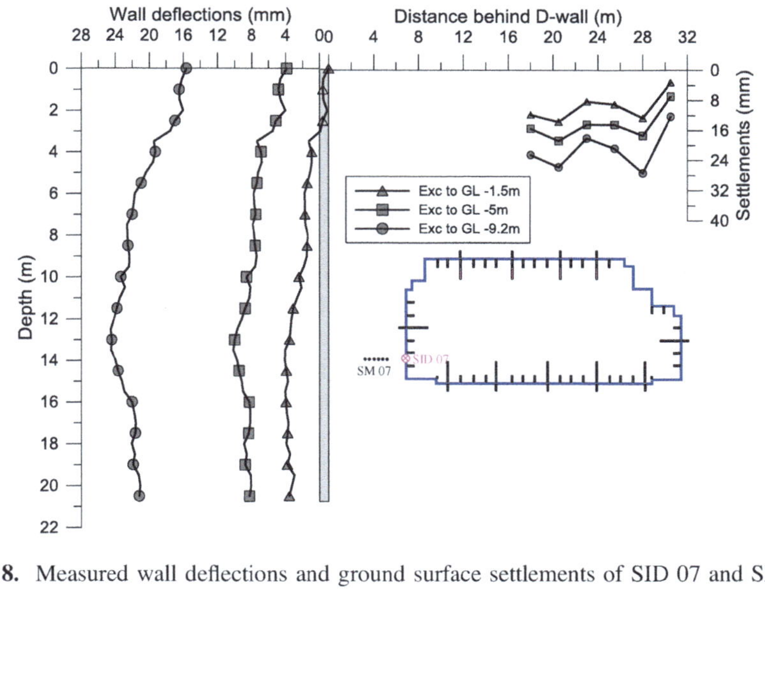

Fig. 8. Measured wall deflections and ground surface settlements of SID 07 and SM 07

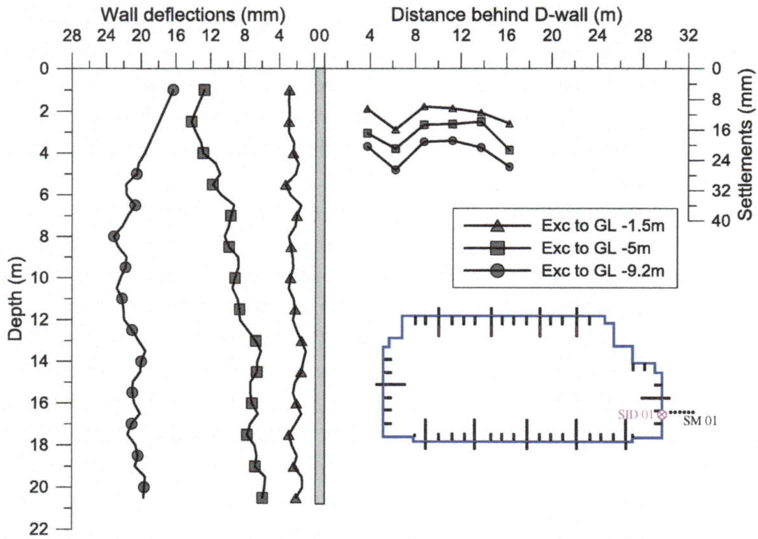

Fig. 9. Measured wall deflections and ground surface settlements of SID 01 and SM 01

Table 2. Summary of maximum wall deflection at various excavation stages and locations

Depth of excavation	Maximum wall deflection (unit: mm)										
	SID 01	SID 02	SID 03	SID 04	SID 05	SID 06	SID 07	SID 08	SID 09	SID 10	SID 11
GL −1.5 m	10.8	7.0	3.8	4.2	6.2	12.4	4.1	3.6	6.5	5.2	5.5
GL −5.0 m	20.9	13.2	9.7	16.1	13.6	18.2	10.2	16.4	15.8	11.2	15.0
GL −9.2 m	25.4	30.3	41.4	50.6	51.3	46.1	24.6	30.8	37.4	39.5	33.6

Figure 10 presents the relationship between measured maximum wall deflection, δ_{hmax}, and final excavation level, H_e. Field data from some published case histories in Taipei Clay (Ou et al. 1993) and Singapore Clay (Chuah and Tan 2010) are also included. It should be noted that the Taipei cases were adopted common excavation methods (i.e. braced excavation or top-down excavation method), meanwhile the Singapore case was adopted the counterfort diaphragm wall system (also a kind of strut-free retaining system). In this project, the maximum wall deflections fell between 0.27 and 0.55% H_e, where the lower and upper values indicate the deflection of diaphragm wall near the corner and the center (PSR = 1.0) of diaphragm wall, respectively. It is quite surprisingly that the maximum wall deflection of the strut-free retaining system was in the range of the maximum wall deflection obtained from common excavation methods (i.e. braced excavation or top-down excavation method). Meanwhile, the Singapore case yielded the maximum wall deflection to final excavation level ratio was 1.33%. For the Singapore case, the groundwater level was near the ground surface level, but the groundwater level was below the final excavation level for this project (Fig. 4). Hence, the possible reason of the relatively small maximum

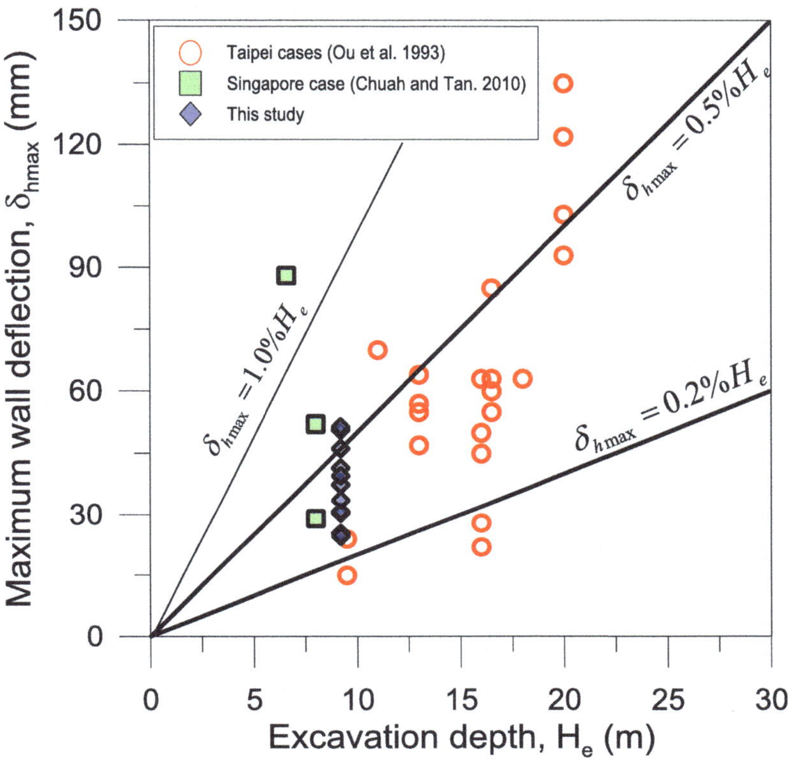

Fig. 10. Relationships between maximum wall deflections and excavation depth

wall deflections measured from this project was the absence of water pressure acting on the strut-free retaining system (from GL −0.0 m to GL −10 m).

Figures 5, 6, 8, and 9 present the ground settlement profiles of SM 04, SM 05, SM 07 and SM 01, respectively. A positive value indicates settlement. Generally, ground surface settlements increased as excavation proceeded downward. The measured data show that the maximum settlement for all the measured data is about 27 mm and that settlements extend as far as 30 m behind the wall, although the data are scarce from 0 to 12 m behind the wall. The ground surface settlements at various excavation stages and locations are listed in Table 3. The measured maximum ground surface settlements were

Table 3. Summary of ground surface settlement at various excavation stages and locations

Depth of excavation	Maximum ground surface settlement (unit: mm)				
	SM 01	SM 03	SM 04	SM 05	SM 07
GL −1.5 m	16	15	12	17	14
GL −5.0 m	21	20	18	19	19
GL −9.2 m	27	26	26	27	27

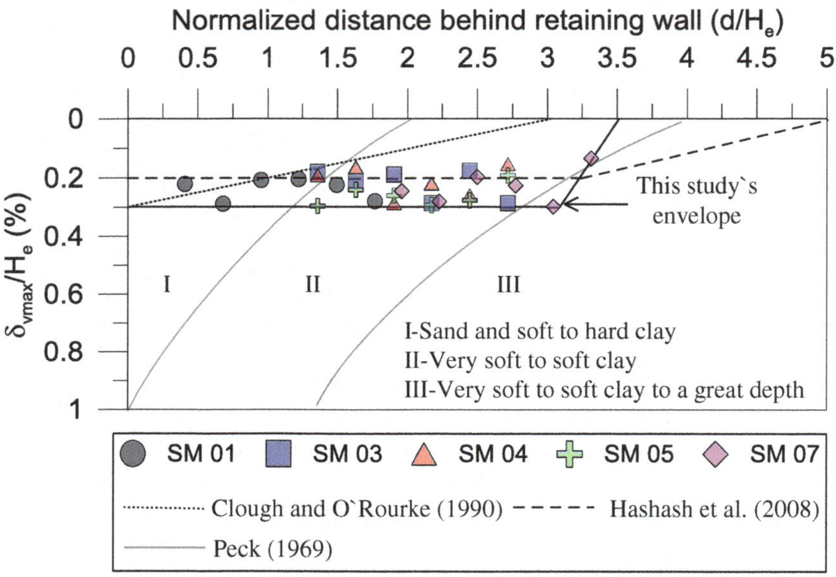

(a)

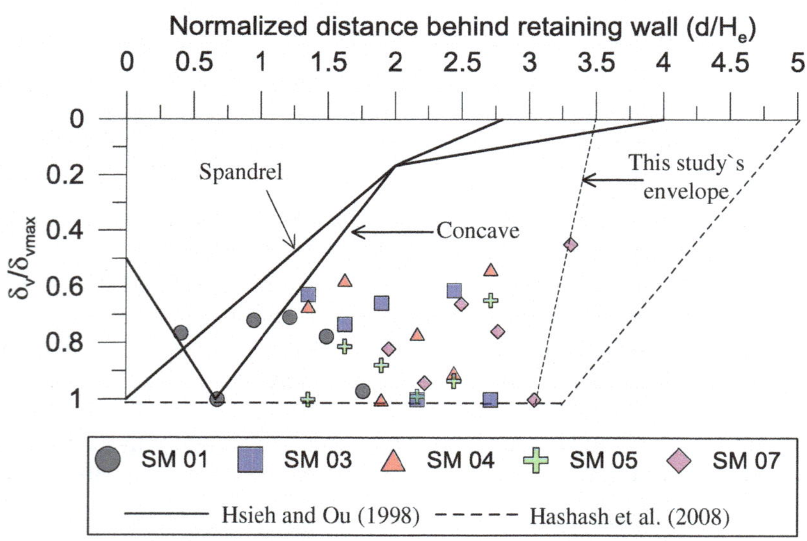

(b)

Fig. 11. Measured ground surface settlement profiles versus those predicted by empirical methods: (a) relationship between normalized ground settlements and normalized distance from wall; (b) relationship between relative ground settlements and normalized distance from wall about 0.3% of final excavation level (H_e), and they were among those concluded by Clough and O'Rourke (1990) for excavation in clay with the boundary 0.15%–0.5% H_e.

Figure 11a depicts the normalized ground surface settlements versus normalized distance behind the excavation. Both quantities are normalized with respect to final excavation level. Most of the measured data were located within the Zone II proposed by Peck (1969) and also slightly below those predicted by Hashash et al. (2008) for medium clay and Clough and O'Rourke (1990) for stiff clay. Although most maximum settlements are less than $\delta_{vmax}/H_e = 0.03\%$, the normalized settlements extend to distances about 3.5 H_e.

Figure 11b shows the relationship between relative ground surface settlement, δ_v/δ_{vmax}, and normalized distance, d/H_e, where δ_v is the ground surface settlement at a certain distance behind diaphragm wall, and δ_{vmax} is the maximum ground settlement along a section. The measured data show that no clear spandrel or concave shapes of the ground surface settlement was observed, rather than it appears that most ground surface settlements fell in the envelope introduced by Hashash et al. (2008) but with smaller normalized distance (d/H_e). The measurements show that even for distances up to 1.5–3H_e, the settlements are on the order of 60% of maximum settlements.

In this study, the envelopes for the clay profiles describe the lower settlement ratios but the wider distribution of settlements behind the wall. For the clay profiles, the envelopes reflect a limit on the lateral extent of settlements to a distance ratio of 3.5. Some of the possible reasons for this settlement characteristic were (1) It seems that the soil behavior is more linear because the relatively small deformations experienced by the soil, (2) Construction activities near the settlement markers might have some effects on measured settlements at distances far away from the excavation.

5 Conclusions

Through a comprehensive instrumentation program, the monitored results of this strut-free large excavation project in soft clay have been interpreted and discussed. Based on the field monitoring results, several conclusions could be drawn:

1. The excavation was successfully constructed with the strut-free retaining system, where buttress walls were integrated with the diaphragm walls and maintained during the excavation process.
2. The measured maximum wall deflection was similar compared with strutted excavation case histories in the Taipei area. The maximum wall deflections to final excavation level ratio (δ_{hmax}/H_e) were between 0.27 and 0.55.
3. At the long-side of diaphragm wall, the pattern of the wall deflections is a cantilever shape with a translational movement at the wall toe and the location of maximum wall deflection was near the top of the wall. Meanwhile, at the short-side of diaphragm wall, the pattern of the wall deflections when reaching the final excavation level was a curvature shape and the location of maximum wall deflection was slightly lower than the final excavation level. It seems that the corner effect played significant roles in determining the magnitude and the shape of wall deflections.

4. The maximum ground surface settlements to final excavation level ratio were below $\delta_{vmax}/H_e = 0.3\%$. Although it was quite small, the ground surface settlements extend to the significant distance behind the diaphragm wall (up to 3.5 × final excavation level).

Acknowledgment. The authors acknowledge the support provided by the Ministry of Science and Technology in Taiwan via grant number MOST103-2221-E-011-070-MY3. The authors would like to thank Sino Geotechnology, Inc. for the provision of detailed geotechnical information on the case study.

References

Chuah, S.S., Tan, S.A.: Numerical study on a new strut-free counterfort embedded wall in Singapore. In: Earth Retention Conference, pp. 740–747 (2010)

Clough, G.W., O'Rourke, T.D.: Construction induced movement of in-situ walls. Geotechnical Special Publication: Design and Performance of Earth Retaining Structures (GSP 25), pp. 439–470. ASCE, Reston (1990)

Hashash, Y., Osouli, A., Marulanda, C.: Central artery/tunnel project excavation induced ground deformations. J. Geotech. Geoenviron. Eng., 1399–1406 (2008). doi:10.1061/(ASCE)1090-0241(2008)134:9(1399)

Hsieh, P.G., Ou, C.Y.: Shape and ground surface settlement profiles caused by excavation. Can. Geotech. J. **35**(6), 1004–1017 (1998)

Hsieh, P.G., Ou, C.Y., Lin, Y.K., Lu, F.C.: Lesson learned in design of an excavation with the installation of buttress walls. J. GeoEngin. **10**(2), 63–73 (2015)

Hsieh, P.G., Ou, C.Y., Hsieh, W.H.: Efficiency of excavations with buttress walls in reducing the deflection of the diaphragm wall. Acta Geotech. **11**(5), 1087–1102 (2016)

Hsieh, H.S., Huang, Y.H., Hsu, W.T., Ge, L.: On the system stiffness of deep excavation in soft clay. J. GeoEngin. **12**(1), 21–34 (2017)

Lim, A., Hsieh, P.G., Ou, C.Y.: Evaluation of buttress wall shapes to limit movements induced by deep excavation. Comput. Geotech. **78**, 155–170 (2016)

Lin, D.G., Woo, S.M.: Three dimensional analyses of deep excavation in Taipei 101 construction project. J. GeoEngin. **2**(1), 29–41 (2007)

Ou, C.Y., Hsieh, P.G., Chiou, D.C.: Characteristics of ground surface settlement during excavation. Can. Geotech. J. **30**, 758–767 (1993)

Ou, C.Y., Chiou, D.C., Wu, T.S.: Three-dimensional finite element analysis of deep excavations. J. Geotech. Eng. **122**(5), 337–345 (1996)

Ou, C.Y., Teng, F.C., Seed, R.B., Wang, I.W.: Using buttress walls to reduce excavation-induced movements. Proc. Inst. Civ. Eng. Geotech. Eng. **161**(GE4), 209–222 (2008)

Peck, R.B.: Deep excavation and tunneling in soft ground: state-of-the-art-report. In: Proceeding of 7th International Conference of Soil Mechanics and Foundation Engineering (ISSMGE), pp. 225–281 (1969)

Characterization of Barrier Effect of Underground Wall in Pumping Tests

Mingfei Zhang[1] and Liyuan Tong[2(✉)]

[1] Institute of Geotechnical Engineering, Southeast University, Nanjing 210096, China
[2] Urban Underground Engineering and Environmental Safety Laboratory of Jiangsu, Nanjing 210096, China
atmu@seu.edu.cn

Abstract. The dewatering of confined aquifer may cause ground settlement, cracking, deformation and tilting of the buildings, and even collapse. It is important and essential to protect environment during the dewatering of the foundation pit project, using the diaphragm wall. So this paper puts emphasis on the barrier effect of diaphragm wall with the data of four pumping tests at Hongzhuang station of Suzhou subway No.4 through the back analysis. The results indicate that (1) the hydraulic barrier corner effect caused by diaphragm wall is existent, and the drawdown of well located in corner of diaphragm wall is 2 times larger than the common; and (2) by calculating the results of comparative analysis, Forchheimer formula is most reasonable, and Kusargent formula is less rational, while Gehalt formula lies in the middle. It is our advice that Forchheimer formula is adaptive in confined water with diaphragm wall. The analyses above show that the hydraulic barrier effect of diaphragm wall has a greater impact on the g drawdown of a well, which can give a clue to similar projects.

Keywords: Barrier effect · Confined aquifer · Dewatering · Forchheimer formula

1 Introduction

Since stepping into the 21st century, more excavations have been reached to the depths of 20–30 m, or even up to 40 m (Luo et al. 2008; Xu et al. 2012). On such conditions, retaining walls, for example diaphragm walls, are all used to retain the surrounding soils and to prevent groundwater into the pits (Pujades et al. 2012a,b; Shen et al. 2008; Wang et al. 2009a; Wu et al. 2015a,b). According to the different position of the bottom of the diaphragm wall in aquifer, three types of seepage around the foundation pit are proposed (Wu 2003): (1) the bottom of the diaphragm wall is located in the impermeable aquifer and the groundwater can be prevented from flowing into the pit; (2) the pit is built in the unconfined aquifer–aquitard–confined aquifer, and the bottom of the underground wall is located in the aquitard in which water can also be prevented effectively; and (3) the underground wall is located in the lower or middle part of the confined aquifer in which the majority of the confined aquifer between inside and

R. Chen et al. (eds.), *Proceedings of the 2nd International Symposium on Asia Urban GeoEngineering*, Springer Series in Geomechanics and Geoengineering,
https://doi.org/10.1007/978-981-10-6632-0_12

outside of the pit is isolated. In the 3rd type, "interaction between underground wall and well" (between concrete walls and well filer tube) can be divided into 4 modes (Wang et al. 2010). More reasonable depth of the continuous concrete wall and the filter tube can be designed using this method (Wang et al. 2009a; Zhou et al. 2010). In addition, the water head variations of confined aquifer is different from aquifer and additional stress generated by precipitation lies funnel-shaped, of which influence scope is determined by Dupuit's formula (Zhang 2012). Retaining walls modify the groundwater flow pattern because the structure reduces totally or partially the aquifer section, which is defined as hydraulic barriers (Pujades et al. 2012b; Wang et al. 2009b). Moreover there is the existence of corner effect in excavation of pit (Liu 1995; Lee et al. 1998).

The object of this paper is to several actual performance of cutoff behavior of underground wall based on field test: (1) The hydraulic barrier corner effect and (2) the optimal equation for the radius of influence.

2 Barrier Effect of Diaphragm Wall

The hydraulic conductivity and radius of influence are reduced by Dupuit's formula,

$$\mathrm{s} = \frac{Q}{2\pi km}\lg\frac{R}{r} \tag{1}$$

$$r = \exp\left(\lg R - \frac{2\pi skm}{Q}\right) \tag{2}$$

$$k = \frac{0.366Q}{m(s_1 - s_2)}\lg\frac{r_2}{r_1} \tag{3}$$

$$\lg R = \frac{s_1 \lg r_2 - s_2 \lg r_1}{s_1 - s_2} \tag{4}$$

where s means drawdown; k is the hydraulic conductivity; R is the radius of influence; Q is the flow of well or wells; m means aquifer thickness; s_1 and s_2 are the drawdown of the first and second well; r_1 and r_2 are the corresponding observation well distance from pumping well or wells center.

The hydraulic conductivity related to Δh can be given by (Li 2004):

$$k = k_0 \exp(-\lambda\gamma_w \Delta h) \tag{5}$$

Where k, k_0 is the hydraulic conductivity after or before dewatering respectively (m/d); λ is test parameter (the value is 0.01 for loam, and 0.03 for Silty sand in that paper); γ_w is the bulk density of water (kN/m^3Kn/m); Δh means the change of groundwater level (m).

As we all know, there are three common formulae, Kusargent formula, Gehalt formula, and Forchheimer formula (based on big well method), which are expressed by respectively:

$$R = 2s_w\sqrt{kH_0} \tag{6}$$

$$R = 10s_w\sqrt{k} \tag{7}$$

$$R = \frac{r_w}{\sin[(\pi/2) \times (s_R/s_w)]} \tag{8}$$

Where s_w is the drawdown of pumping well (m); s_R is opted according to the research needs, such as 0.2, 0.5 or 1 (m); r_w means the radius of 'big well'; H_0 means the height of the water level (m).

3 Data

Suzhou is located at the lower reaches of the Yangtze River in southeastern Jiangsu, China. Hongzhuang station is located under the intersection of new plan of Dongwu Road and Yuehu Road, of which pit length is 471 m. According to the geotechnical investigation report of Hongzhuang station (Fig. 1), the main types of groundwater in the pit are the unconfined water (①$_2$ and ①$_3$) in the shallow clay aquifer, the first confined water (⑤$_2$) in the shallow aquifer of silty clay and the second confined water (⑦$_4$ and ⑨) in the deep aquifer of silt. Table 1 shows the result of the indoor penetration test and their thickness for aquifers. To obtain the hydraulic conductivity of the first confined water, the four field pumping tests (called Test 1, Test 2, Test 3, Test 4) are carried out from January, 2012 to December, 2013 (Table 2). Dewatering well layout plan for Test 2, Test 3 and Test 4 are shown in Fig. 2. The pumping well is single in Test 1 and 2, while multi-well in Test 3 and 4. According to the primary design, the first field pumping test is before excavation and dewatering of the pit, so the depth of the diaphragm wall in test 1 is 0 m, others are all 40 m. There are three wells in Test 1, one pumping well and two and observation wells, and their distances between observation wells and pumping are 6.92 m and 16.86 m. Three drawdown

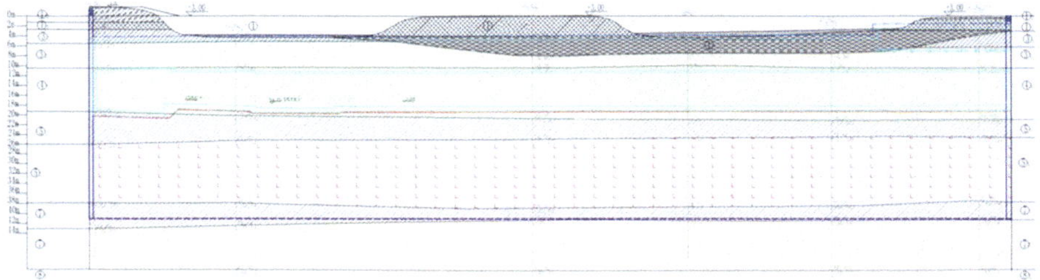

Fig. 1. Geological section of Hongzhuang station of Suzhou subway No. 4

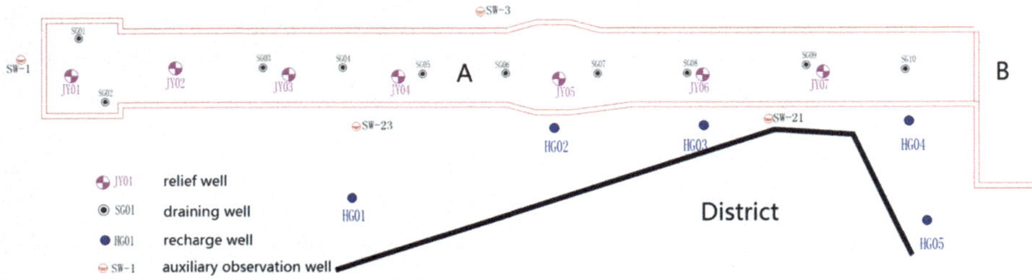

Fig. 2. Dewatering well layout plan for Test 2, Test 3 and Test 4

Table 1. Soil parameters obtained by laboratory penetration test

Layer	Composition of each layer	Thickness (m)	Unit weight (kN/m^3)	The hydraulic conductivity	
				K_V(cm/s)	K_H(cm/s)
①$_2$	Plain fill	0.60–4.40	18.8	4.4E-7	5.4E-7
①$_3$	Fill clipped silt	0.60–40	17.9	7.5E-7	8.4E-7
③$_1$	Clay	0.80–3.80	19.7	5.8E-7	6.2E-7
③$_2$	Silty clay	2.50–7.10	19.4	1.5E-6	2.5E-6
④$_1$	Silty clay	7.40–12.20	18.9	1.2E-5	2.6E-5
⑤$_1$	Silty clay	2.70–6.70	18.6	6.4E-6	7.2E-6
⑤$_2$	Silty sand clipped silty clay	10.70–14.7	19.5	4.7E-4	5.2E-4
⑦$_1$	Silty clay	1.90–6.90	18.8	2.2E-6	3.1E-6
⑦$_3$	Silty clay	1.50–12.90	18.6		
⑦$_4$	Silt	1.70–11.60	19.4		
⑧$_2$	Silty clay	4.00–13.30	19.4		
⑨	Silt	1.00–7.60	19.8		
⑩$_1$	Silty clay	0.80–4.80	19.1		

Table 2. Survey of tests

Type	Time	Well	Pit depth	Wall depth
Test 1	2012.1.18	Single	0	0
Test 2	2013.11.3	Single	20	40
Test 3	2013.11.3	Multi-well	20	40
Test 4	2013.12.11	Multi-well	20	40

order is carried out in Test 1, and the corresponding average flow is 300.8 m^3/d, 195.5 m^3/d, 98.2 m^3/d (Table 3). The depths of three wells are all 40 m.

Pumping test results in Test 1, Test 2 and 3, Test 4, are shown in Tables 3, 4 and 5, respectively. And Fig. 3 tells us the Drawdown-Time Curve of observation well in Test 4. The average flow of the pumping wells in Test 2, 3 and 4 is shown in Table 6, and flow in Test 1 shown in Table 3.

Table 3. Pumping test results in Test 1

Drawdown order	Flow (m^3/d)	Drawdown in the well (m)			The hydraulic conductivity (m/d)	Radius of influence (m)
		Pumping	Observation 1	Observation 2		
First	300.8	6.59	4.31	2.95	2.98	116
Second	195.5	4.42	2.63	1.86	3.42	145
Third	98.2	2.47	1.71	1.18	2.50	122

Table 4. Pumping test results in Test 2 and 3

Type	JY01	JY04	JY06	HG02
Well depth	38	38	38	38
Drawdown in Test 2 (m)	2.67	4.77	4.02	0.7
Distance from pumping well (JY05) in Test 2 (m)	98	32.3	28.9	12.4
Drawdown in Test 3 (m)	9.3	11.57	5.85	1.7
Distance from pumping wells (JY03, JY05)in Test 3 (m)	71	5	56.1	29

Table 5. Pumping test results in Test 4

#	Observation well out of the pit						Observation well in the pit		
	HG05	HG04	HG03	HG02	HG01	SW21*	JY01	JY06	SG01
The initial water depth (m)	7.770	8.279	8.596	8.089	8.865	8.96	8.714	8.759	7.986
The final water depth (m)	10.371	11.033	12.264	11.702	12.027	10.97	21.576	20.391	9.327
Drawdown (m)	2.601	2.754	3.668	3.613	3.162	2.01	12.862	11.62	1.341
Distance from pumping wells (m)	86	78	37.5	14.4	42.3	37.5	91	36	

Table 6. The average flow of the pumping wells in Test 2,3 and 4(m^3/h)

Tests	JY03	JY04	JY05	JY07
Test 2	–	–	17	–
Test 3	14	–	13	–
Test 4	13.0	9.0	7.10	12.81

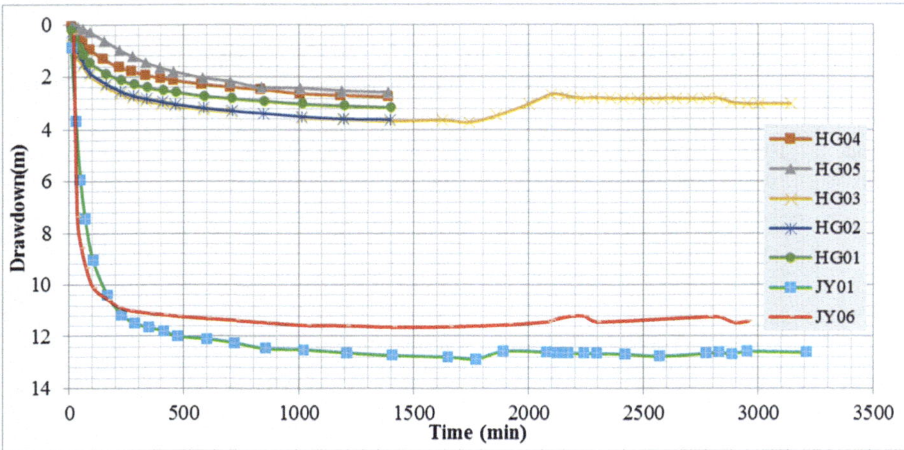

Fig. 3. The drawdown-time curve of observation well in Test 4

4 Analysis and Discussion

The hydraulic barrier effect of diaphragm wall will bring about large difference drawdown of well inside and outside the pit. Well HG02 lies inside the pit, and is selected as a reference object owing to its smaller variation range of drawdown. So Drawdown in the Well in Test 2 and 3 are shown in Table 7. It seems that the drawdown of well outside the pit is 3 ~ 7 times higher than that inside. Since without diaphragm wall, Test 1 has not been discussed.

Table 7. Drawdown in the well in Test 2 and 3 (m)

Tests	JY01	JY04	JY06	HG02
Test 2	2.67	4.77	4.02	0.7
Test 3	9.3	11.57	5.85	1.7

Test 2, 3 and 4 are all later than Test 1, so the hydraulic conductivity should be gradual change. If k_0 in Test 1 adopts 2.96 m/d; and γ_w does 10 kN/m^3; and Δh is 20 m; and λ adopts 0.015, we can obtain k = 0.147 m/d, which should be regarded as the hydraulic conductivity in Test 2, 3 and 4. With formula (3) and (4), only when s_1 divided by two inside the pit, the hydraulic conductivity are all nearly about 0.147 m/d (Table 8). That is to say, it makes 2 times larger than the common, owning to hydraulic barrier effect of diaphragm wall.

Table 8. Hydraulic parameters in Test 2, 3 and 4 with formula (3) and (4)

	Chosen wells	s_1/2(m)	s_2(m)	r_1(m)	r_2(m)	Q(m^3/d)	m(m)	k(m/d)	R(m)
Test 2	JY01,JY04	1.335	4.77	98	32.3	17	5	0.17	151
Test 3	JY01,JY06	5.85	5.85	56.1	71	27	5	0.17	177
Test 4	JY01,JY06	6.431	11.62	91	36	41.91	5	0.24	287

Suppose if without diaphragm wall, i.e., the well outside the pit is supposed as inside. So those parameters inside the pit can be used for calculating the corresponding drawdown and distance from pumping wells outside. There is only one well HG02 outside the pit in Test 2 and 3, so it is regarded as control group (inverse well), while in Test 4, HG04 is selected by reason of its more reasonable hydraulic parameters in Table 9. Hydraulic parameters with back calculation in Test 2, 3 and 4 with formula (2) are shown in Table 10, where r_I and r_{I*} means actual distance and distance back analysis respectively, between inverse well and pumping well or wells center. So r_{I*} - r_I means (about 100 m) the difference value between actual and calculated one, in other words, vertical hydraulic barrier effect seems makes horizontal distance increase about 100 m (5 times of the insertion depth of diaphragm wall).

Table 9. Hydraulic parameters in Test 4 outside the pit with formula (3) and (4)

	Chosen wells	s_1(m)	s_2(m)	r_1(m)	r_2(m)	Q(m^3/d)	*m*(m)	k(m/d)	R(m)
Test 4	HG05,HG04	2.601	2.754	86	78	41.91	5	0.85	452
Test 4	HG05,HG01	2.601	3.162	86	42.3	41.91	5	1.69	2308
Test 4	HG04,HG01	2.754	3.162	78	42.3	41.91	5	2	4852

Table 10. Hydraulic parameters with back calculation in Test 2, 3 and 4 with formula (2)

	Well	Q(m^3/d)	*m*(m)	k(m/d)	R(m)	r_I(m)	r_I^*(m)	r_I^*-r_I(m)
Test 2	HG02	17	5	0.17	151	12.4	120.4	108
Test 3	HG02	27	5	0.17	177	29	127	98
Test 4	HG04	41.91	5	0.24	287	78	176	98

Those calculated radii of influence are shown in Table 11, based on three formulas mentioned above. Because the average radius of influence in Test 1, 2, 3 and 4 is 128 m, 151 m, 177 m, 287 m inside the pit, contrast to Table 11, Forchheimer formula is more rational, and Kusargent formula is less rational, while Gehalt formula lies in the middle. It is our advice that Forchheimer formula are better in confined water with retaining walls.

Note: R_J, R_K, and R_F, are the radius of influence for Kusargent formula Gehalt formula and Forchheimer formula. And s_R means three drawdowns of well inside the pit corresponding to three radia of influence for Forchheimer formula.

Table 11. Radius of influence with Kusargent formula Gehalt formula and Forchheimer formula

	r_w(m)	s_w(m)	k(m/d)	H_0(m)	R_J(m)	R_K(m)	s_R(m)	R_F(m)
Test 2	7	12.25	0.17	20	51.2	45.8	0.2/0.5/1	273/109/55
Test 3	7	13	0.17	20	53.4	47.7	0.2/0.5/1	290/116/58
Test 4	7	13	0.24	20	63.4	56.7	0.2/0.5/1	290/116/58

5 Conclusions

In order to study barrier effect, a case study on Hongzhuang subway station is presented in this paper, with the data of four field pumping tests. The most important findings of our study can be summarized as follows. The hydraulic barrier corner effect is existent. And on the basis of inversion parameters, the drawdown of well located in corner of diaphragm wall is 2 times larger than the common. Vertical hydraulic barrier effect seems makes horizontal distance increase about 100 m (5 times of the insertion depth of diaphragm wall). By calculating the results of comparative analysis, Forchheimer formula is most reasonable, and Kusargent formula is less rational, while Gehalt formula lies in the middle. It is our advice that Forchheimer formula and Gehalt formula are better in confined water with diaphragm wall.

Acknowledgments. Much of the research work described herein was funded by the National Natural Science Foundation of China (NSFC) (Grant No. 4157020433) and Project of the National Twelve-Five Year Research Program of China (Grant No. 2012BAJ01B02). These financial supports are gratefully acknowledged. The authors would like to express their appreciations to the editor and anonymous reviewer for their valuable comments and suggestions.

References

Lee, F.H., Yong, K.Y., Quan, K.C., Chee, K.T.: Effect of corners in strutted excavations: Field monitoring and case histories. J. Geotech. Geoenviron. Eng. **124**(4), 339–349 (1998)

Li, X.: Fully coupling analysis of seepage and stress for deep foundation pit based on displacement back analysis. Chin. J. Rock Mechan. Eng. **23**(8), 1269–1274 (2004). (in Chinese)

Liu, K.X.: Three dimensional analyses of deep excavation in soft clay. Ph.D. thesis, National University of Singapore, Singapore (1995)

Luo, Z.J., Zhang, Y.Y., Wu, Y.X.: Finite element numerical simulation of three-dimensional seepage control for deep foundation pit dewatering. J. Hydrodyn. Ser. B **20**(5), 596–602 (2008)

Pujades, E., Carrera, J., Vázquez-Suñé, E., Jurado, A., Vilarrasa, V., Mascuñano-Salvador, E.: Hydraulic characterization of diaphragm walls for cut and cover tunnelling. Eng. Geol. **125**, 1–10 (2012a)

Pujades, E., López, A., Carrera, J., Vázquez-Suñé, E., Jurado, A.: Barrier effect of underground structures on aquifers. Eng. Geol. **145–146**(6), 41–49 (2012b)

Shen, S.L., Han, J., Du, Y.J.: Deep mixing induced property changes in surrounding sensitive marine clays. J. Geotech. Geoenviron. Eng. **134**(6), 845–854 (2008)

Wang, J., Guo, T., Wu, L., Zhu, Y., Yiqun, T., Yang, P.: Mechanism and application of interaction between underground wall and well in dewatering for deep excavation. Chin. J. Undergr. Space Eng. **06**(3), 564–570 (2010). (in Chinese)

Wang, J., Hu, L., Wu, L., Tang, Y., Zhu, Y., Ping, Y.: Hydraulic barrier function of the diaphragm wall in the pit of subway station and its optimization. Environ. Geol. **57**(2), 447–453 (2009a)

Wang, J.H., Han, X., Zhou, H.L., Zhang, Z.M.: Analytical and semi-analytical methods for the seepage field dammed by underground structures. Hydrogeol. Eng. Geol. **36**(2), 13–17 (2009b)

Wu, L.G.: Design and construction of dewatering engineering and theory of seepage in deep excavation. China Communication Press, China, pp. 24–26 (2003). (in Chinese)
Wu, Y.X., Shen, S.L., Xu, Y.S., Yin, Z.Y.: Characteristics of groundwater seepage with cut-off wall in gravel aquifer. I: Field observations. Can. Geotech. J. **52**(10), 1526–1538 (2015a)
Wu, Y.X., Shen, S.L., Yin, Z.Y., Xu, Y.S.: Characteristics of groundwater seepage with cut-off wall in gravel aquifer. II: Numerical analysis. Can. Geotech. J. **52**(10), 1539–1549 (2015b)
Xu, Y.S., Ma, L., Du, Y.J., Shen, S.L.: Analysis of urbanisation-induced land subsidence in Shanghai. Nat. Hazards **63**(2), 1255–1267 (2012)
Zhang, J.: Study on technology and environmental effect of pressure-relief of confined water in Hangzhou deep pit. Doctoral dissertation, Zhejiang University. Ph.D. thesis, Zhejiang University, Hangzhou, China (2012)
Zhou, N., Vermeer, P.A., Lou, R., Tang, Y., Jiang, S.: Numerical simulation of deep foundation pit dewatering and optimization of controlling land subsidence. Eng. Geol. **114**(3–4), 251–260 (2010)

Comparison of Analytical and Numerical Analysis Results of Deep Excavation for Nhon – Hanoi Urban Railway Project in Hanoi, Vietnam

Nguyen Chau Lan[1(✉)] and Truong Quang Manh[2]

[1] University of Transport and Communications, Hanoi, Vietnam
nguyenlan1981@gmail.com
[2] Asian Institute of Technology, Pathum Thani, Thailand

Abstract. Deep excavation is an indispensable part of constructions on subsoil in Hanoi, Vietnam. There are 8 urban railways which were constructed with many deep foundations in the main terminals in Hanoi. The 8th terminal at the Nhon – Hanoi Station route is more complex than others due to the configuration of the terminal. The terminal is about 8 m deep. Bored piles with 1.2 m in diameter were constructed at the bottom of the pit. In order to ensure stability of the pit, Larsen IV steel sheet piles were constructed with a length of 12 m. In addition, two rows of soil anchors with 14 m in length and 15.2 mm in diameter have also been adopted to strengthen the foundation pit.

Analytical method using the Geo5 program was conducted to design the steel sheet piles at the site. Numerical method using the Plaxis program was also conducted to verify the design of the steel sheet piles. In the analytical analysis, exact solutions are obtained by solving differential equations directly with some assumptions and limitations, meanwhile in the numerical analysis the derivatives in different equations are replaced by some sort of approximation and solutions. Comparison of the deformation of steel sheet piles, internal forces on the piles, and internal forces on the anchors obtained from the analytical and numerical methods were provided in the paper. It was concluded that the predicted values obtained from the analytical method should be more conservative than those obtained from the numerical analysis.

Keywords: Deep excavation · Larsen IV steel sheet pile · Analytical analysis · Numerical analysis · Anchor

1 Introduction

According to planning, there will have 4–5 urban railways in Hanoi. The 3rd route starts from Nhon to Hanoi Station. The project with the full length of 12.5 km contains 9.6 km elevated and 2.9 km underground. Correspondingly, there are 11 elevated terminals and 4 underground terminals. Designed length of the train is 19 to 20 m. The speed is 80 km/h.

The 3rd route has connected stations. In which, the 8th terminal is located in front of University of Transport and Communications, includes main bus station with a large

R. Chen et al. (eds.), *Proceedings of the 2nd International Symposium on Asia Urban GeoEngineering*, Springer Series in Geomechanics and Geoengineering,
https://doi.org/10.1007/978-981-10-6632-0_13

traffic volume. During the construction of the foundation pit for piers of the terminal, Larsen IV steel sheet piles combined with ground anchors were proposed to stabilize the pit.

The 8th terminal at the Nhon – Hanoi Station route is deeper and more complex than others due to the configuration of the terminal that requires appropriate construction methods as well as calculations.

Soil anchors have been applied for deep excavations in many projects all over the world (Briaud and Lim 1999; Weatherby 1998). The anchors are often combined with steel sheet piles to strengthen the excavation (Harris and Bond 2008; Weatherby 1998). They are installed to transfer a tension load in bearing strata that can be soil or rock. The specifications for the soil anchor calculations are sufficient such as BS 8081 (1989), AASHTO LRFD 2007 (SI version) or LRFD 2012.

For transportation projects, the AASHTO LRFD is normally applied. However, projects with the anchors in urban transportation in Vietnam are not mentioned, this paper introduces an example for the soil anchor in the urban railway project.

In the project, Geostructure analysis (Geo5 v19) was used to calculate and audit the steel sheet pile structure with the ground anchors. Using Geostructure Analysis software (Bentley-US) allows the calculation in construction stages, taking into account both internal stability and overall stability of the structure. In which, Sheeting check function is suite for making basic design of the required length of the structure, calculating the internal forces or verifying the cross-sections based on traditional analytical methods.

Plaxis 2D was used to compare to Geo 5 program. Plaxis 2D is one of the most powerful tools for geotechnical engineers. It is a finite element program package specifically intended for the analysis of deformation and stability in geotechnical engineering projects. Although the modeling of the soil itself is an important issue, many projects involving the modeling of structures and the interaction between structures and soils would be analyzed in Plaxis 2D.

2 Designation and Instructions

2.1 General on the Project

The main document is technical design documentation of foundation measures for the project "Pilot Urban Railway in Hanoi: Nhon - Hanoi Station Section" by investor: Hanoi People's Committee - Hanoi Public Transport and Tram Development Management Board.

In order to construct a pillar platform, design consultant uses Larsen piling method around the pit and proceeds to excavate the soil in the foundation pit, finally build the platform. Due to the large depth of excavated soil, the pile wall near the elevated road has stabilized by the soil anchors (Fig. 1).

At the 8th terminal, both the 12 m-long-steel sheet piles and the 12 m-long anchors are installed. Spacing between the anchors is 2 m.

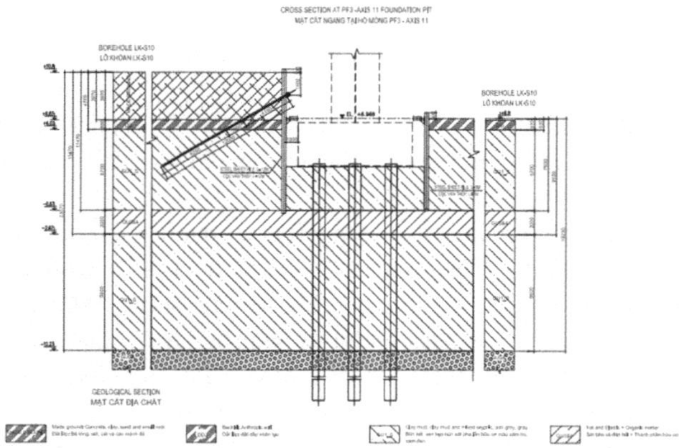

Fig. 1. Cross section of the foundation pit at the 8th terminal

2.2 Input Parameters

2.2.1 Traffic Load

Load due to vehicles acting on one side of Thu Le park is calculated with a designed truck with an axial load of 145 kN.

The designed truck is shown in Fig. 2 below. The traffic range is 8.5 m, each separated by 1.2 m. In addition, construction load such as excavator and crane load are also considered. All construction and traffic load are calculated and shown in Table 1. From the calculations, we select q = 42.65 kN/m^2 for the excavation design (Fig. 3).

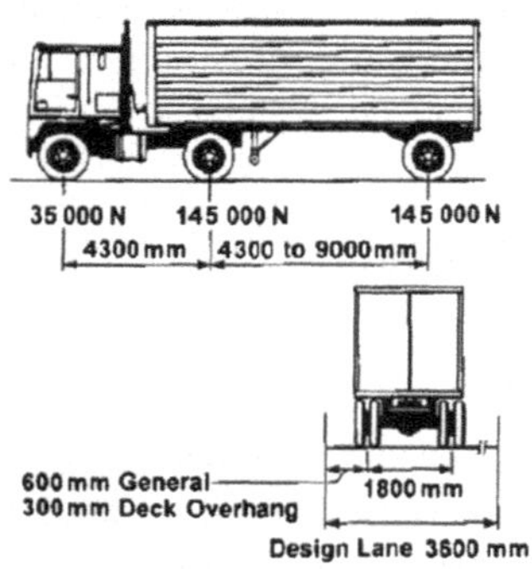

Fig. 2. The designed truck

2.2.2 Geological Conditions

There are 4 layers in the scope of calculation. The top layer is DD layer including concrete, brick, sand and clay, thickness is 3.97 m; Made ground layer: backfill, silty sand, thickness is 0.8 m; GU1_s layer: lean clay with some organic, firm to stiff, thickness is 6.7 m; GU3&4 layer: Fat and elastic (CH-MH) + Organic matter, thickness is 2 m. Properties of the layers are described in Table 2.

Table 1. Traffic load

Load	Load factor	Value (kN/m^2)
Traffic load	1.25	42.65
Excavator	1.3	15.33
Crane	1.3	25.97

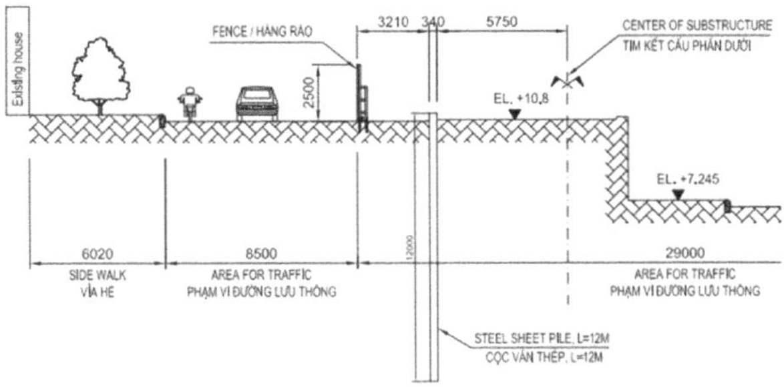

Fig. 3. The truck layout

Table 2. Soil properties

No.	Name	Φ_{ef} (^{o})	C_{ef} (kPa)	γ (kN/m^3)	δ (^{o})	E_{def} (MPa)	$E_{elastic}$ (MPa)
1	DD layer	14.5	0	19.0	14	3.0	15
2	Made ground	14.5	0	19.0	14	3.0	15
3	GU1_s	15.0	42	19.4	11	8.0	30
4	GU3&4	14.0	20	15.9	11	8.5	32

2.3 Materials

2.3.1 Larsen IV

Steel type for Larsen IV is CT3, detail properties are as follow (Table 3):

Table 3. Steel sheet pile properties

Cross section	Dimension			Area		Moment of inertia		Section modulus	
	B	H	t	Per pile	Per unit width	Per pile	Per unit width	Per pile	Per unit width
	mm	mm	mm	cm^2	cm^2/m	cm^4	cm^4/m	cm^3	cm^3/m
SP-IV	400	170	16	96.9	243	4670	38600	362	2270

2.3.2 Anchor

Geometric features of anchor cables: Use 2 cables 15.2 mm (Tables 4 and 5).

Table 4. Anchor ground features

No.	Length (m)	Root (m)	Slope (°)	Spacing (m)	Pre-stressing (kN)
1	12	2	30	2	90

Table 5. Anchor properties

No.	Geometric features of cable	Unit	Value
1	The nominal diameter	mm	15.24
2	The nominal cross section area	mm^2	140
3	Tensile strength	MPa	1860
4	Minimum load of damage	kN	260.4
5	Yield strength	MPa	1670
6	Elastic modulus	MPa	197000
7	The nominal weight	Kg/1 m length	1.102

2.3.3 Scopes of the Calculation

- Checking of the sheet piles L = 12 m in construction stages
- Checking of the ground anchors

3 Analytical Analyses

Calculate and check the sheet piles L = 12 m and the ground anchors in construction stages by using Geo5:

- **Stage 1** Pressing the sheet pile (L = 12 m). Excavate to ground anchor elevation +8.508 (Fig. 4).
- **Stage 2** Install the ground anchors (Fig. 5).

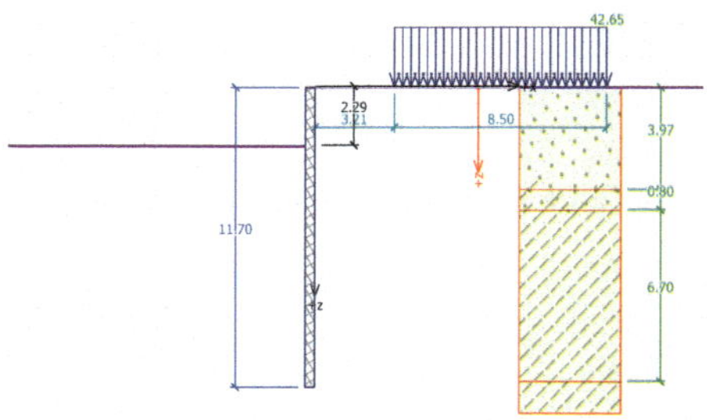

Fig. 4. Model of foundation pit in Geostructure analyses in stage 1

- **Stage 3** Excavate to bored pile elevation +7.245 (Fig. 6).
- **Stage 4** Excavate to wale and struts elevation. Install wale and struts at elevation +6.986, frame fixed frame support and cofferdam by welded (Fig. 7).
- **Stage 5** Excavate to bottom blinding concrete elevation +2.953 (Fig. 8).

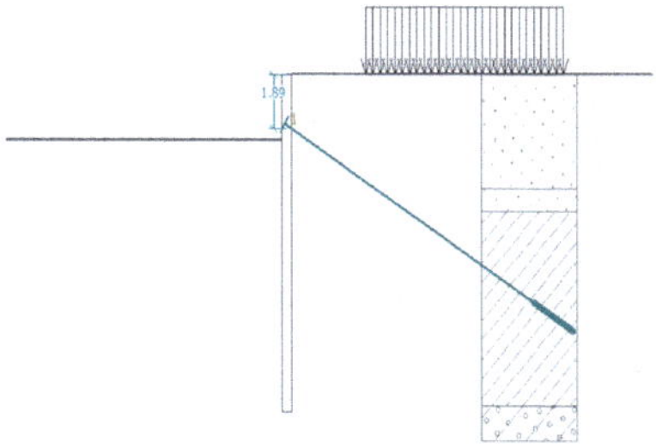

Fig. 5. Model of foundation pit in Geostructure analyses in stage 2

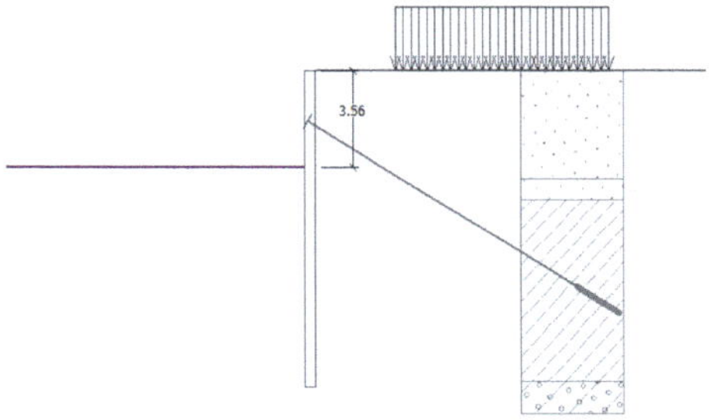

Fig. 6. Model of foundation pit in Geostructure analyses in stage 3

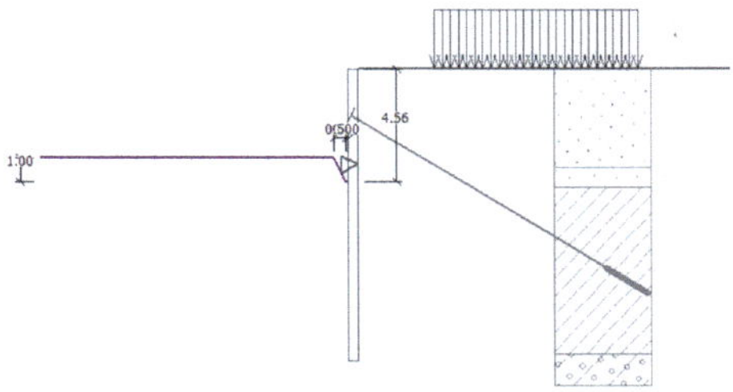

Fig. 7. Model of foundation pit in Geostructure analyses in stage 4

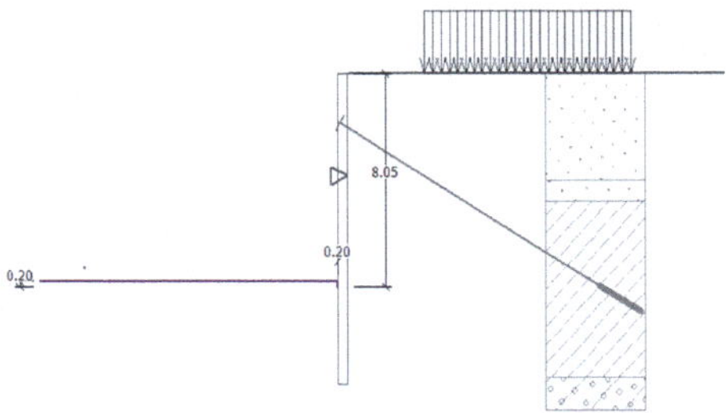

Fig. 8. Model of foundation pit in Geostructure analyses in stage 5

4 Numerical Analyses

Numerical analyses were performed to study the lateral movement and forces of the steel sheet piles. Mohr-Coulomb model is used in the analyses in order to simulate the behavior of the soil layers. Figure 9 shows mesh of the excavation domain in the final stage of the construction with the construction load at one side of the excavation pit.

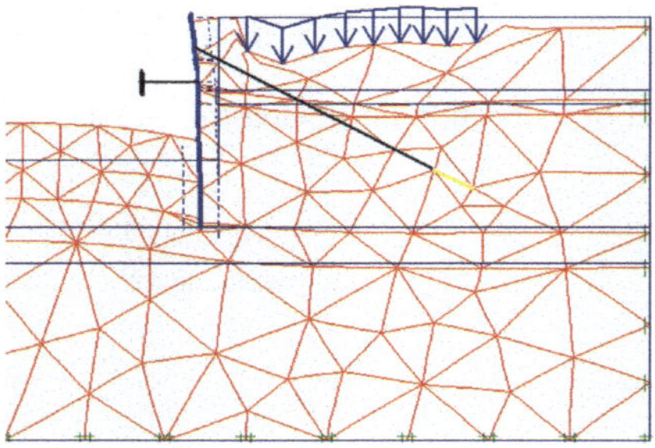

Fig. 9. Deformed mesh at final stage of the construction

5 Results

5.1 Analyses Results

- **Stage 1** Press sheet piles (L = 12 m). Excavate to the ground anchor elevation +8.508 (Fig. 10).

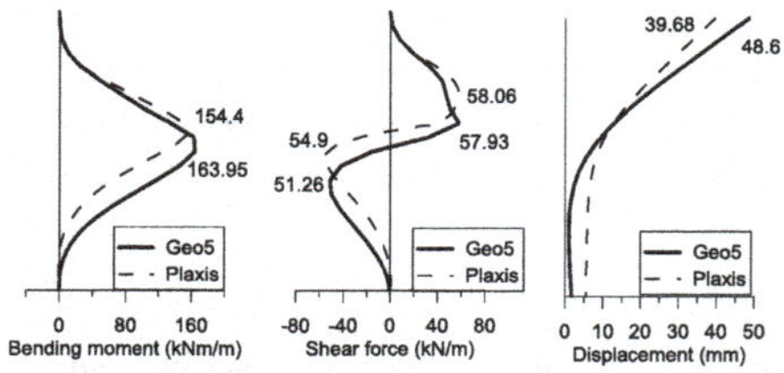

Fig. 10. Diaphragm of bending moment, shear force and displacement at stage 1

- **Stage 2** Install the ground anchors (Fig. 11).

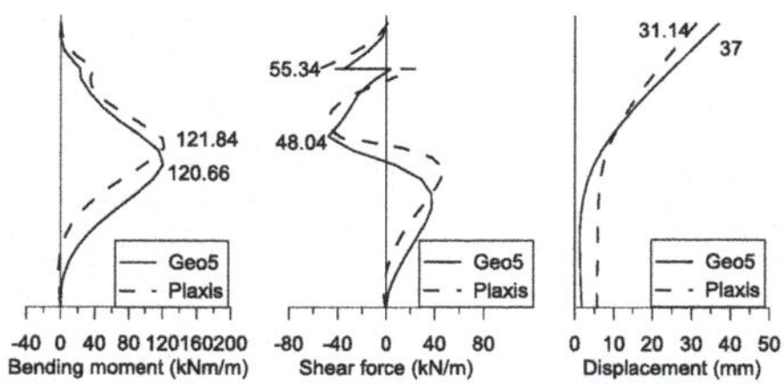

Fig. 11. Diaphragm of bending moment, shear force and displacement at stage 2

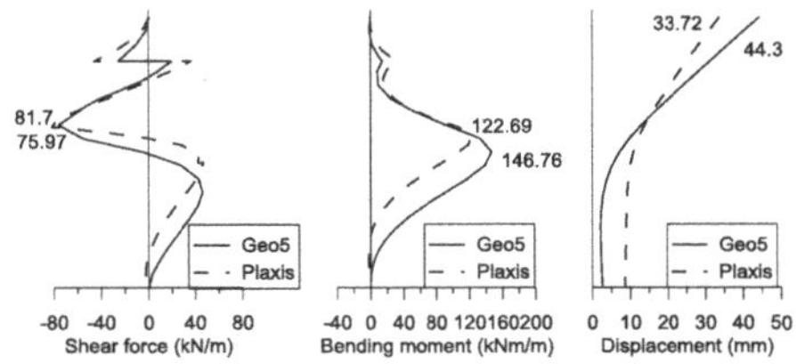

Fig. 12. Diaphragm of bending moment, shear force and displacement at stage 3

- **Stage 3** Excavate to bored pile elevation +7.245 (Fig. 12).
- **Stage 4** Excavate to wale and struts elevation. Install wale and struts at elevation +6.986, frame fixed frame support and cofferdam by welded (Fig. 13).
- **Stage 5** Excavate to bottom blinding concrete elevation +2.953 (Fig. 14).

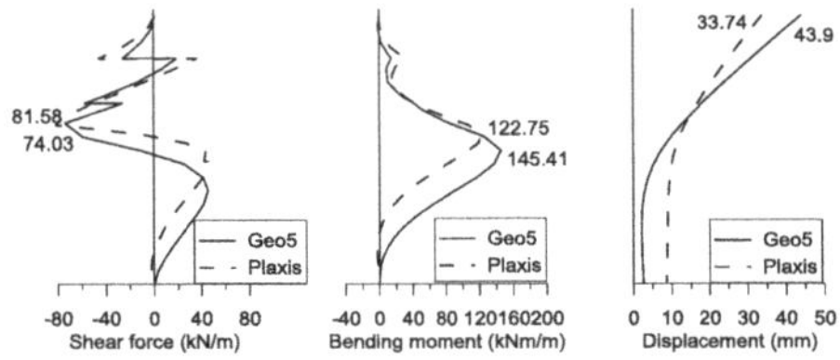

Fig. 13. Diaphragm of bending moment, shear force and displacement at stage 4

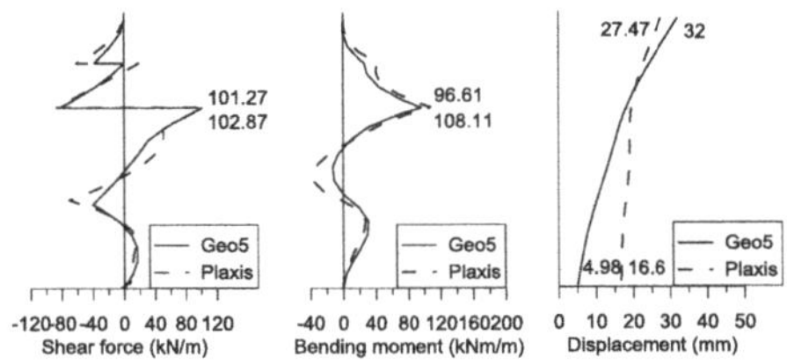

Fig. 14. Diaphragm of bending moment, shear force and displacement at stage 5

5.2 Checking for Steel Sheet Pile

Maximum bending moment and shearing force values will be used to check the cross section of the sheet piles (Table 6).

Table 6. Maximum values in the sheet piles

Result (Maximum)	Geo5	Plaxis	Unit
Bending moment	163.95	154.40	kNm/m
Shear force	101.27	102.87	kN/m
Displacement	48.60	39.68	mm

Cross section of piles ensures the bearing capacity. At the same time, the horizontal displacement of the pile is small, meets the allowable displacement of the anchor pile wall.

5.3 Checking for Ground Anchor

The anchor force values during the construction stages analysed in Geo5 and Plaxis are shown in Table 7.

Table 7. Anchor force

Stage	2	3	4	5
Geo5 (kN/m)	90	104.36	103.9	89.26
Plaxis (kN/m)	90	93.73	93.74	91.22

The length of the ground anchors L = 12 m, meet the requirements of bearing capacity.

6 Discussions

The procedure for deep foundation design using the steel sheet piles in combination with the ground anchors was presented in detail. Based on the results, the following conclusions can be drawn:

- The calculated results are quite compatible in Geo5 and Plaxis.
- The bending moments and shear forces in Plaxis agreed with the calculated results in Geo5 for each construction stage.
- The displacements at the toe of steel sheet pile tend to be larger and at the head of sheet pile are smaller in Plaxis compare to Geo5.
- The internal force values in Geo5 are more conservative than in Plaxis.

The differences between analytical and numerical analyses in the lateral movement of the sheet piles and the ground anchors can be explained by the differences of modulus parameters. In details, the moduli used in Geo5 were deformation moduli that were determined by the in-situ tests. Meanwhile, the moduli used in Plaxis were elastic modulus obtained from laboratory tests.

7 Conclusions

Both analytical and numerical analyses can be used successfully in analyses of steel sheet pile for deep excavation. The results of those analyses show good agreement in the internal forces of the piles for stages of construction.

Analytical analysis with the sheeting check function can be easily used with in-situ tests. Meanwhile, numerical analysis with sophisticated boundary conditions should be used in routine design and construction of steel sheet pile excavation.

This paper summarized the results of calculating internal force and displacement of steel sheet pile in stages and the method of anchored auditing in the soil applied to deep foundation in urban works.

References

AASHTO LRFD: Bridge Design Specifications, SI version (2007)

AASHTO LRFD: Bridge Design Specifications, 6th edn. (US) (2012)

Briaud, J.L., Lim, Y.: Tieback walls in sand: numerical simulation and design implications. J. Geotech. Geoenviron. Eng. **125**(2), 101–111 (1999)

BS 8081: Code of practice for ground anchorages (1989)

Harris, A., Bond, A.: Decoding Eurocode 7. Taylor & Francis, London (2008)

Project Reference: Elevated stations, Package number: HPLMLP/CP-02, December, 2014

Weatherby, D.E.: Design Manual for Permanent Ground Anchor Walls. Federal Highway Administration, Washington, DC (1998)

Oanh, N.C., Tram, D.T.V., Vinh, T.C.T.: Finite element analysis of a braced excavation in marine soft clay. In: Geotechnics for Sustainable Infrastructure Development - Geotec Hanoi 2016, Phung (edt) (2016). ISBN 978-604-82-1821-8

Consideration of the Pile-Soil Friction for Earth Pressure of Limited Soil for Double-Row Piles

A. Yi-jun Zhou[1,2(✉)], B. Ai-jun Yao[1,2], and C. Gang Lei[1,2]

[1] The Key Laboratory of Urban Security and Disaster Engineering, Ministry of Education, Beijing University of Technology, Beijing, China
zyjfcb@126.com
[2] College of Architecture and Urban Planning, Beijing University of Technology, Beijing, China

Abstract. The soil between the front-row pile and the back-row pile is defined as the limited soil in the structure of the double-row pile. However, the Rankine theory is inappropriate for analyzing the properties of the limited soil. Considering the tangential friction of pile-soil for the double-row pile, this paper develops formula of earth pressure of limited soil when it was in limited equilibrium state. The simplified formula of earth pressure intensity of limited soil is also founded. The analysis shows that the result stemmed from the formula equals to that of numerical simulation (FLAC3D). Therefore, the formula of limited soil is reasonable and valuable. The values of earth pressure intensity of limited soil are the same among different width (2d–5d). They are less than that of positive earth pressure. The value of earth pressure intensity at the lower level of the limited soil is far more less than that of positive earth pressure.

Keywords: Double-row pile · Limited soil earth pressure · Pile-soil friction

1 Introduction

At present, the double-row pile supporting structure has been applied in foundation pit engineering and slope engineering (Liu et al. 2014; Jing and Wei 2010; Nie et al. 2008; Cui et al. 2006). Wang et al. (2015a) analyze the mechanical properties of two types of soils arch for double-row pile, acquiring the ultimate bearing capacity of the soil arches. Numerical simulation is used to investigate the mechanical properties of double-row pile structure during deep foundation pit excavation (Sun 2009; Wang et al. 2015b). Liu and Fu (2011) study computational methodology and stress patterns, making stress behavior of double-row pile is in accordance with the practical project. Involving the pile-soil interaction, a new plane linear finite element model is developed (Zheng et al. 2004). Numerous papers mainly focus on earth pressure calculation of soil retaining wall. However, earth pressure between the double-row pile is merely analyzed. Three gaps exit among current researches. (1) Assumptions proposed by several computational models are unreasonable (Ma et al. 2008; Gao 2000; Li and Guo 2008). For instance, some of them suppose that the soil between the double-row pile is

R. Chen et al. (eds.), *Proceedings of the 2nd International Symposium on Asia Urban GeoEngineering*, Springer Series in Geomechanics and Geoengineering,
https://doi.org/10.1007/978-981-10-6632-0_14

rigid-plastic material, neglecting soil-pile interaction. (2) Due to the complicated earth pressure the distribution of earth pressure is imprecisely presented based on current numerical and experimental method. (3) Current researches seldom define the soil between the front-row pile and back-row pile as the limited soil. Consequently, This paper develops a computational model of limited soil of double-row pile, considering the soil-pile interaction. The formula of limited soil under the tangential friction of pile-soil for double-row pile is also founded. The results show that the formula is accuracy compared with numerical simulation's outcome.

2 Calculation Model

The Technical Specification of Retaining and Protecting for Building Foundation Excavation (JGJ120-2012) argues that the distance between the front-row pile and back-row pile must be 2d–5d (d is the diameter of pile) (Liu et al. 2003). Therefore the soil between the front-row pile and back-row pile is not defined as the semi-infinite soil but the limited soil (LS). This paper proposes a calculation model of limited soil pressure, considering the tangential friction of pile-soil, based on the previous research results and the Rankine theory. When the limited soil reaches the positive limited state, the calculation model above the slip surface is shown in Fig. 1.

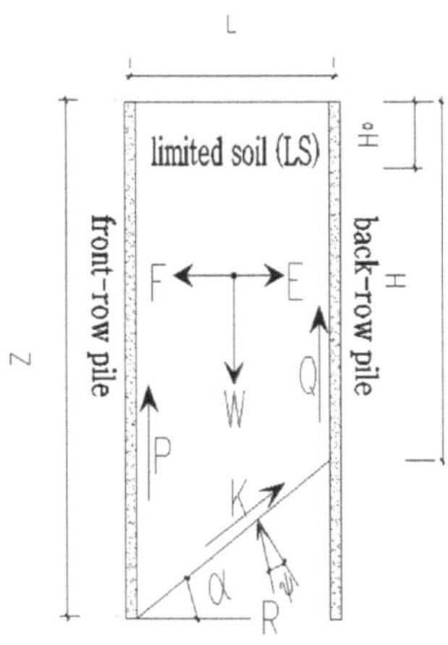

Fig. 1. Computation model of limited soil between double-row pile

The hypothetical condition of the model is that the slip surface is a straight line when the limited soil has reached the positive limited state. The angle between the line and the horizontal direction is α; the space between the front-row pile and the back-row pile is L; the depth of compute point is Z; the gravity of the limited soil W; the supporting force, from the stabled soil block, for the limited soil is R and the friction of the interface is K. Generating from front-row pile and back-row pile, the horizontal forces on the limited soil are E and F as well as the tangential friction are P and Q.

3 Stress Analysis

3.1 Establish the Equation

Assumptions:

(1) The limited soil is the homogeneous;
(2) The limited soil stays in the limit equilibrium state;
(3) The value of stress of limited soil is equal at the same depth.

According to the geometric relationship and the Rankine theory, the gravity W is shown in the Eq. (1) and the relationship between E and F in Eq. (2).

$$W = \gamma LZ - \frac{1}{2}\gamma L^2 \tan \alpha \tag{1}$$

$$\frac{E}{F} = \frac{\frac{1}{2}\gamma(Z - H_0)^2 K_a}{\frac{1}{2}\gamma(H - H_0)^2 K_a} = \left(\frac{Z - H_0}{Z - L \tan \alpha - H_0}\right)^2 \tag{2}$$

K is shown as formula (3), identifying the friction between the LS and the fixed soil below LS.

$$K = \frac{Lc}{\cos \alpha} \tag{3}$$

The LS can be self-stability and does not interact with surroundings according to the Rankine earth pressure theory that the value of earth pressure can be less than 0 among a certain depth. On the contrary, the LS can interact with surroundings beyond the depth. The formula of depth H_0 is shown in Eq. (4) (Wang et al., 2014), when the value of earth pressure is 0. Consequently, H_0 need to be subtracted when calculating the pile-soil interaction.

$$H_0 = \frac{2c}{\gamma\sqrt{K_a}} \tag{4}$$

The tangential friction P and Q of LS is shown in Eq. (5), when the LS is in the limited equilibrium state, according to the Mohr-Coulomb theory

$$P = E \tan \delta + \bar{c}(Z - H_0) \tag{5}$$

$$Q = N \tan \delta + \bar{c}(H - H_0) \tag{6}$$

In the formula: γ-gravity of the soil (kN/m^3); c, φ-cohesion and internal friction angle; $\bar{c}, \delta$-cohesion and friction angle for interface of the pile-soil.

The stress of the soil between front and back-row pile is in limited equilibrium state, shown in formula (7).

$$\begin{cases} P+Q+K\cdot\sin\alpha+R\cdot\sin(\alpha-\varphi)=W \\ E+K\cdot\cos\alpha=F+R\cdot\sin(\alpha-\varphi) \end{cases} \tag{7}$$

Formula (8) springs from the combination between Eq. (7) and equations from (1) to (6).

$$E=\frac{2\bar{c}H_0-Lc[\tan\alpha+\cot(\alpha-\varphi)]+(\gamma L-2\bar{c})(Z-\frac{1}{2}L\tan\alpha)}{\left[1-(\frac{Z-H_0}{Z-L\tan\alpha-H_0})\right]\cot(\alpha-\varphi)+\left[1+(\frac{Z-H_0}{Z-L\tan\alpha-H_0})\right]\tan\delta} \tag{8}$$

3.2 Simplified Analysis

The Rankine earth argues that the angle between the slide plane and principle stress σ_1 acting face is 450 + ψ/2. However, the angle between the α and σ_1 will deflect due to the friction of pile-soil. While the action line of σ_1 near the front-row pile turns clockwise, the angle will be less than 450 + ψ/2. On the contrary, when the action line of σ_1 rotates counterclockwise, the angle will be more than 450 + ψ/2. Accordingly, the angle between the α and σ_1 is 450 + ψ/2 whether the interface of pile-soil is smooth or not. Formula (9) can be obtained by substituting $\tan a=\cot(a-\varphi)=\sqrt{K_P}$ into Eq. (8)

$$E=\frac{2\bar{c}H_0-2Lc\sqrt{K_p}+(\gamma L-2\bar{c})(Z-\frac{1}{2}L\sqrt{K_p})}{\left[1-(\frac{Z-H_0}{Z-L\tan\alpha-H_0})\right]\sqrt{K_p}+\left[1+(\frac{Z-H_0}{Z-L\tan\alpha-H_0})\right]\tan\delta} \tag{9}$$

$\sqrt{K_p}$- passive earth pressure coefficient

The formula (9) is the earth pressure equation for the limited soil between the front-row pile and back-row pile. The formula (9) shows that value of earth pressure reaches maximum when $\alpha = 45^0$ + ψ/2, considering the friction of the pile-soil interface rough.

3.3 Earth Pressure Intensity

According to current researches, the differential expression of earth pressure intensity is $e=\partial E/\partial Z$. However, the formula (9) is too complex to solve. The research adopts the method of finite difference to solve the intensity of earth pressure approximately. The accuracy of calculation result could be higher and higher as the Z_1 and Z_2 getting closer. In this paper, Z_1 and Z_2 meet the accuracy requirements by the validation.

The formula (10) presents earth pressure intensity.

$$e_{Z=\frac{z_1+z_2}{2}}=\frac{\Delta E}{\Delta Z}=\frac{E_{Z=Z_2}-E_{Z=Z_1}}{Z_2-Z_1} \tag{10}$$

4 Numerical Simulation

Numerical simulation is applied to validate formula (9) because it can eliminate irrelevant factors in site monitoring by establishing idealized model. The FLAC3D software best fits limit equilibrium analysis for double-row pile, for simulating the change of earth pressure for double-row pile more accurately than other software in the foundation pit excavation.

4.1 Calculation Parameter Selection

Tables 1, 2 and 3 show the parameter of both the front-row pile and the back-row pile which are in rectangular arrangement.

Table 1. Soil layer of model calculation parameters

Name of soil	Thickness/m	Severe/kN·m^{-3}	Cohesion/kPa	Friction angle ψ/(°)	Bulk modulus *K*/MPa	Shear model *G*/MPa
Homogeneous soil	40	20	10	18	33.1	11

Table 2. Geometrical parameters of supporting structure

Depth of foundation pit/m	Diameter of pile/m	Distance of front-row pile/m	Distance of double-row pile/m	Embedded depth/m	Length of pile/m	Length and wide for section of beam/m
15.2	0.8	1.6	2.4	11.2	25.6	0.8 × 0.8

Table 3. Calculation parameters support structure

Concrete of pile and beam	E (pile and beam)/GPa	μ	E Reinforcement/GPa	Diameter of reinforcement/m	Amount of reinforcement for pile and beam
C30	24	0.2	200	0.028	18/24

4.2 Meshing and Simulation of Construction

The size of numerical model is 52 m × 6.4 m × 40 m. The Mohr- Coulomb model is used establish the double-row pile; the pile-unit, beam-unit and shell-unit are respectively applied for the pile, beam and slope. The model does not consider groundwater impact, because it has been precipitated prior to the foundation pit excavation.

The process of excavation is shown in following:

(1) Establishing the model of stratum, calculating the initial stress field under the action of gravity, and setting the displacement to 0.
(2) Establishing double-row pile model, calculating the stress field after finishing the construction of double-row pile, and setting the displacement to 0.
(3) Excavating the depth about 2.4 m, excavating the rest in every depth about 1.6 m till reaching the depth 15.2 m, then calculating the maximum unbalance stress and the distribution for stress field.
(4) Analyzing of calculation and results, as shown in Figs. 2 and 3.

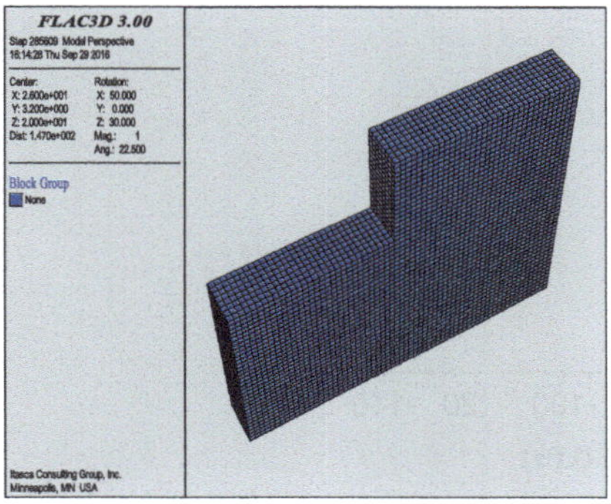

Fig. 2. Model meshing

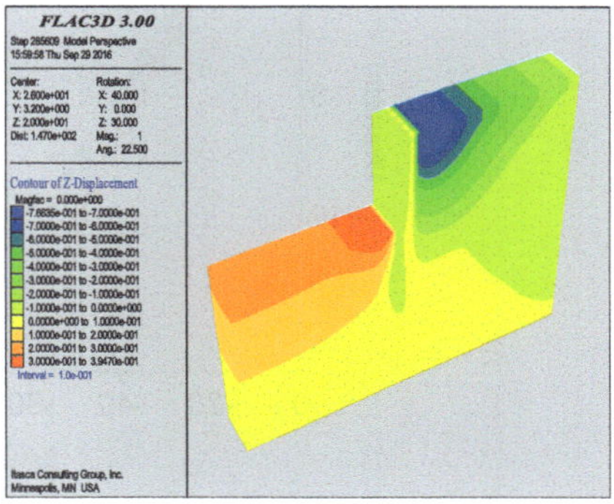

Fig. 3. Excavation of vertical displacement cloud

5 Comparative Analysis

5.1 Comparison Between Theoretical Calculation and Numerical Simulation

The reference Li and Guo (2008) points out that the value of soil-pile boundary strength reduction factors $\bar{c}, \delta$ can be chosen from 25% to 50% of that of fill. Therefore, this paper opts for one third of medium value when calculating.

The values of earth pressure intensity with depth are obtained and shown in Table 4, while the parameters from Tables 1 and 2 are substituted into the formula (9) and (10).

Table 4. Theoretical calculation value of earth pressure intensity

Depth/m	2	3	4	5	6	7	8	9	10	11	12	13	14	15
Earth pressure/kPa	7.41	10.5	17.03	23.57	28.14	31.73	36.08	38.97	42.58	46.1	48.8	50.7	51.6	52.5

The results of numerical simulation, theoretical calculation and Rankine theory are compared and shown in Fig. 4. The value of Rankine positive earth pressure increased linearly with the depth, more than the value of numerical simulation and theoretical calculation. This result shows that Rankine theory is more suitable for semi-infinite soil rather than limited soil. It will lead to larger calculation results when applied to limited soil and result in using redundant materials in construction.

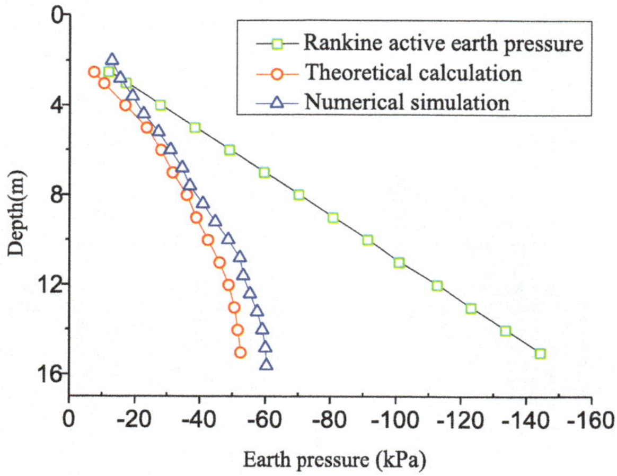

Fig. 4. Calculates the comparison chart

The trend of theoretical calculation for earth pressure is the same as that of numerical simulation. The values of earth pressure intensity both of theoretical calculation and numerical simulation decrease as the depth increases. The earth pressure intensity of the limited soil between front-row pile and back-row pile does not increase linearly with depth, and the slopes of the curves decreases near the bottom of foundation pit.

The value of numerical simulation fits well with that of theoretical calculation, validating the formula (9).

Considering the influence of width of limited soil, this paper takes the width L = 2d, L = 3d, L = 4d, L = 5d (d is diameter of pile) into the formula (9) and (10) separately. The comparison of earth pressure intensity among different width is shown in Fig. 5.

As can be seen from Fig. 5, the trend of earth pressure intensity for limited soil is the same basically in different width. The values of earth pressure intensity in different width are all less than that of Rankine's positive earth pressure. The value of earth pressure intensity for the lower part of the limited soil between front-row pile and back-row pile is far less than that Rankine positive earth pressure.

The intensity of earth pressure at the same depth increases gradually with the increase of width of limited soil, and becomes more close to the value of Rankine active positive earth pressure.

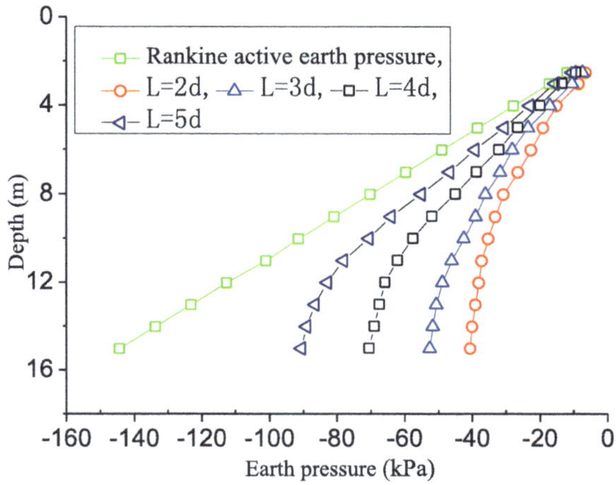

Fig. 5. Comparison of earth pressure intensity in different width

It is use of the FLAC 3D finite difference software for establish the soil of different width to study the influence to the displacement and bending moment. As shown in Figs. 6, 7, 8 and 9.

Knowing from the Figs. 6 and 7, the maximum horizontal displacement of the front and back-row piles are all getting smaller and smaller with increase of width for limited soil. The maximum displacement of the front-row piles are 18.52 mm, 16.77 mm, 15.45 mm and 14.33 mm with increase of width 2d, 3d, 4d and 5d, and the maximum displacement of the back-row piles are 17.6 mm, 17.29 mm, 17.06 mm and 16.93 mm with increase of width 2d, 3d, 4d and 5d. The displacement of the front-row piles is smaller than that of the back-row piles. Specifically the average displacement of front-row piles displacement decreases by 9% and the average displacement of back-row piles decreases by 2%. It is shown that there is a great influence for the displacement of front-row piles with the width of limited soil changes. The front-row piles have play a important role for resisting earth pressure.

The position of maximum displacement for front-row piles is different from that of the back-row piles. the position of maximum displacement for front-row piles is 4 m–4.5 m below the top of pile; the position of maximum displacement for back-row piles is 7 m–8 m below the top of pile. The displacement of front-row piles is larger than that of back-row pile in the same depth.

Knowing from the Figs. 8 and 9, the maximum bending moment of front-row piles is 7 m–8 m below the top of pile, and then the bending moment becomes smaller as the depth increases. The maximum bending moment of front-row pile and back-row piles is increasing gradually with the width of limited soil increases. Specifically the maximum bending moment of the front-row piles are −211.6 kN·m, −242.3 kN·m, −261.9 kN·m and −274.6 kN·m with increase of width 2d, 3d, 4d and 5d, and the maximum bending moment of the back-row piles are −214.1 kN·m, −242.4 kN·m, −253.7 kN·m and −256.5 kN·m with increase of width 2d, 3d, 4d and 5d. The value of maximum

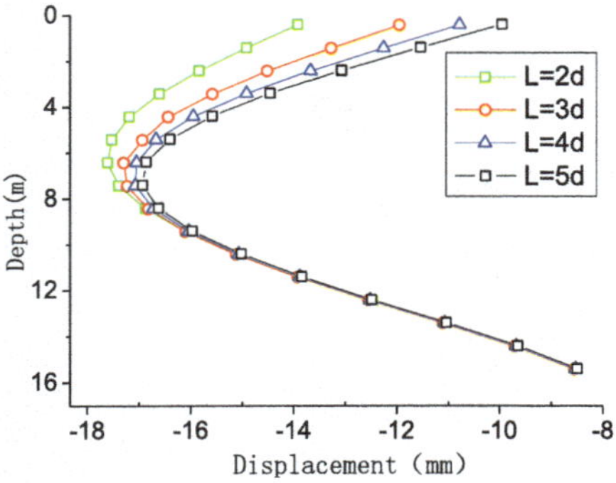

Fig. 6. Displacement diagram of front-row pile

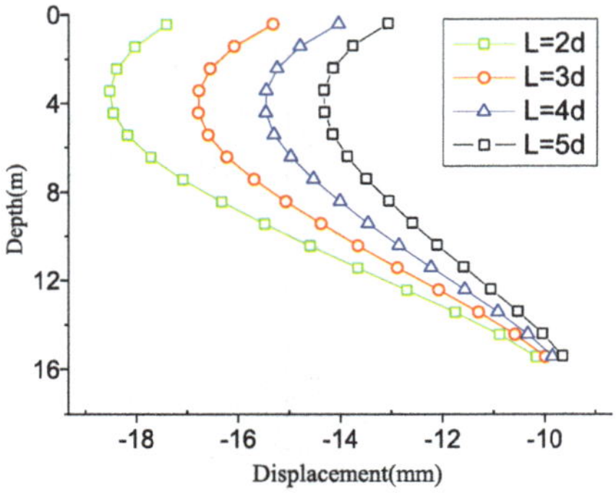

Fig. 7. Displacement diagram of back-row pile

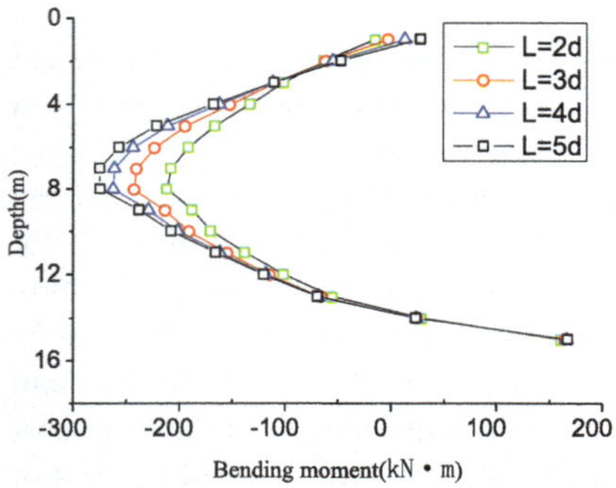

Fig. 8. Bending moment of front-row pile

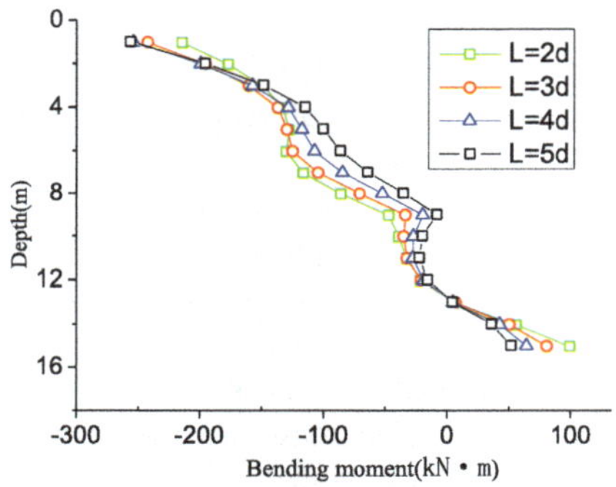

Fig. 9. Bending moment of back-row pile

bending moment for front-row pile is the same as the back-row pile basically, and that is to say the double-row piles can work better together.

6 Conclusion

(1) The research shows the calculation formula of limited soil between front-row pile and back-row pile, considering the tangential friction between pile and soil. The calculation results of the formula are best fit those of numerical simulation, and have a better accuracy. The calculation model of limited soil in this paper has a more practical engineering value.

(2) The value of earth pressure of limited soil between front-row pile and back-row pile is different from that of the Rankine theory which increases linearly with the depth. The slope of pressure curve decreases near the bottom of foundation pit.

The value of earth pressure intensity of the lower part of the limited soil is far less than that of Rankine positive earth pressure.

(3) The intensity of earth pressure at the same depth increases gradually with the increase of width of the limited soil when the width changes from 2d to 5d, and become more close to the value of Rankine positive earth pressure.

References

Liu, B.S., Chen, X.H., Wang, X.Q., Chen, Y.: Development potential of Chinese construction industry in the new century based on regional difference and spatial convergence analysis. KSCE J. Civil Eng. KSCE **18**(1), 11–18 (2014)

Jing, Z.S., Wei, D.: Discussion on application of double row piles files in Wu Han area. Resour. Environ. Eng. **24**(2), 141–143 (2010)

Nie, Q.K., Hu, J.M., Wu, G.: Deformation and earth pressure of a double-row piles retaining structure for deep excavation. Rock Soil Mech. **29**(11), 3089–3094 (2008)

Cui, H.H., Zhang, L.Q., Zhao, G.J.: Numerical simulation of deep foundation pit excavation with double-row piles. Rock Soil Mech. **27**(4), 662–666 (2006)

Wang, Y., Zhao, B.: Multilayer soil arching effect calculation and soil pressure analysis in double-row anti-sliding piles. Beijing Univ. Technol. **41**(8), 1193–1199 (2015a)

Sun, Y.: Research on calculation method of double-row anti-sliding structure under sliding surface. Rock Soil Mech. **30**(8), 2971–2977 (2009)

Wang, X.H., Xie, L.Z., Zhang, M.: Numerical simulation and characteristic of a double-row piles retaining structure for deep excavation. J. Cent. South Univ. (Sci. Technol.) **45**(2), 596–602 (2015b)

Liu, Q.S., Fu, J.J.: Research on model and parameters of double-row piles based on effect of pile-soil contact. Rock Soil Mech. **32**(2), 481–494 (2011)

Zheng, G., Li, X., Liu, C.: Analysis of double-row piles in consideration of the pile-soil interaction. J. Build. Struct. **18**(1), 99–106 (2004)

Ma, P., Qin, S.Q., Qian, H.T.: Calculation of active earth pressure for limited soils. Chin. J. Rock Mech. Eng. **27**(Suppl. 1), 3070–3074 (2008)

Gao, Y.L.: The calculation of finite earth pressrue. Build. Struct. **16**(5), 53–56 (2000)

Li, F., Guo, Y.C.: Analytical study on active soil pressure from finite soil body in construction pit. Build. Struct. **24**(1), 15–19 (2008)

Liu, C.Q., Li, X., Zhang, Y.P.: Shaking table test and analysis of double row pile retaining structure. China Civil Eng. J. **46**(2), 190–195 (2003)

Wang, H.L., Song, E.X., Song, F.Y.: Calculation of active earth pressure for limited soil between existing building and excavation. Eng. Mech. **31**(4), 76–81 (2014)

Different Methods of Excavation for Basement in Jakarta: Design, Reality and Associated Problems

Paulus P. Rahardjo(✉)

Universitas Katolik Parahyangan, Bandung, Indonesia
rahardjo.paulus@gmail.com

Abstract. Jakarta is a wide area covering different soils conditions and different surrounding. As metropolitan city, Jakarta has serious problems due to difficulties in the methods of excavation, geology, water table, and disturbance to neighbors. Of particular interest is there is no fixed standard of design and each designer has used different methods and references.

Many constructions have been successful, and some construction fail due to lack of understanding and knowledge on the soils condition. The soft soils condition is part of the cause of failures. These soft layers spread from south to north of the city owing the fact that old or ancient rivers were there and at present, they could not be identified after covered by sediments. Soil Investigation should be carefully planned to reveal the variety of ground condition. In most cases, the soft soils are still consolidating and hence some failures have been caused due to ignorance on the presence of excess pore pressures. On the other hand, in some areas strong lahar formation also exist and excavation lead the fact that they are stable upon deep excavation over 15 m even without protection or minor treatment. The area could change drastically within a distance of less than 50 m. In central Jakarta surrounding the National Monuments, loose to medium sands are found. The excavation dewatering has caused severe problems mainly settlement of neighborhood that subsequently sued the owner of the buildings. The City of Jakarta now prohibit dewatering by pumping during excavation.

The methods employed by designers and contractors vary depending on their experiences. Lessons learned from success and failures have been documented by the author. This paper discusses case histories of all common methods of excavation for the basement in Jakarta City from design perspectives and facts of recorded failures as well as successful measurements. A discussion on deep excavation for Jakarta MRT is also presented.

Keywords: Excavation · Safety factor · Deformation · Under-consolidation · Lahars

1 Introduction

Problems of deep excavations in urban areas like Jakarta and other big cities in Indonesia is critical due to the need of spaces for parking of high rise buildings. In the last 20 years, the design and construction of basement in Jakarta tend to be deeper say from 4 levels

R. Chen et al. (eds.), *Proceedings of the 2nd International Symposium on Asia Urban GeoEngineering*, Springer Series in Geomechanics and Geoengineering,
https://doi.org/10.1007/978-981-10-6632-0_15

to 6 levels of basement or even more. Other need for excavation is also on demand such as for MRT and underground storage.

The consequence of construction of a deep excavation can induce lateral movement as well as settlement of the ground which can have serious effects on adjacent buildings. The adjacent building may be damaged due to the excessive differential settlement caused by soil movement induced by the wall deflection and heave at the base of the excavation. This potential problem could be even amplified if the deep excavation is conducted in the congested urban area where existing structures are located in close proximity to the excavation.

Papers concerning these issues have been published every where and hundreds or even thousands of research on this topics are available, hence the main purpose of this paper is to present experience in specific location in the Jakarta area, mainly for the purpose of sharing "lessons learned from common practice to specific soil conditions".

A number of the authors experience in handling deep excavation in Jakarta and advice to the local contractors or owners as they face difficulties or even failures will be addressed. Different methods have been proposed depending on soil condition.

2 Geological and Soil Condition

The soil condition in Jakarta varies between those relatively strong at the south part of the city and soft to very soft in the north coastal zone. Hence the main issue to be considered in design and construction of the basement depend primarily on the insitu site condition. Figure 1 shows the geology map of Jakarta metropolitant city and boundary of the soil conditions are clearly shown on the map. Location of Case Studies are also shown on the map.

Fig. 1. Geology map of Jakarta and location of case studies

The south part of Jakarta is generally stronger sedimentary soils and volcanic soil. Lahar formation is generally found at a depth of 10–20 m which consist of lightly to strongly cemented sand. In contrast, the north parts of Jakarta is very soft so soft marine clay 15–17 m thick underlain by thin dilluvium clay with 2–6 m thick, then at a depth of 21–25 m sand lenses normally exist almost every where varying thickness of 2–15 m. Generally medium to stiff clay are located below these sand lenses to a depth of 40–50 m then becoming very stiff to hard to 100 m or even deeper. In some areas, the second sand lenses may be found at 30 m also with varying thickness.

3 Challenge for Design of Braced Excavation in Jakarta City

3.1 Problems of Dewatering

Despite of the challenge of uncertainty, the main issue for design is pro and con between cost and safety. Other important factors to be considered are soil condition, the proximity of the excavation from the neighbour buildings and how reliable is the soil information from the investigation report.

Some interesting problems due to the existence of loose sands and underconsolidating very soft soil layers might be of interest. These two soil conditions have been the cause of failures in excavation issue either due to lateral ground movement of unwanted effect of the excavation. Due to many problems caused by dewatering, it is presently prohibited to lower down water table by dewatering. Hence the designer shall apply other methods (Fig. 2).

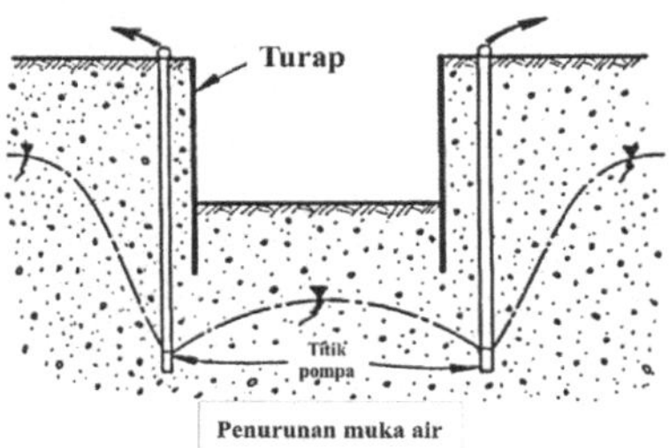

Fig. 2. Dewatering during excavation is not recommended or prohibited in practice for Jakarta City

When soils have high permeability, the practice in Jakarta always request the designer to avoid ground water flow into excavation area by closed system such as by use of cut-off wall or diaphragm wall (Fig. 3). This method is possible if the existence of clay layers are not quite deep from the base of excavation and the cut off system shall be constructed sufficiently deep into the impermeable layer. If impermeable layer is down below the base of excavation, then to reduce tye ground water flow, sufficient thickness of grouting can be an alternative. However, this method does not guaranty full water tightness. *Cut off wall* can also be designed by use of bentonite cement. In practice, most design in Jakarta is by using either secant piles or diaphragm wall.

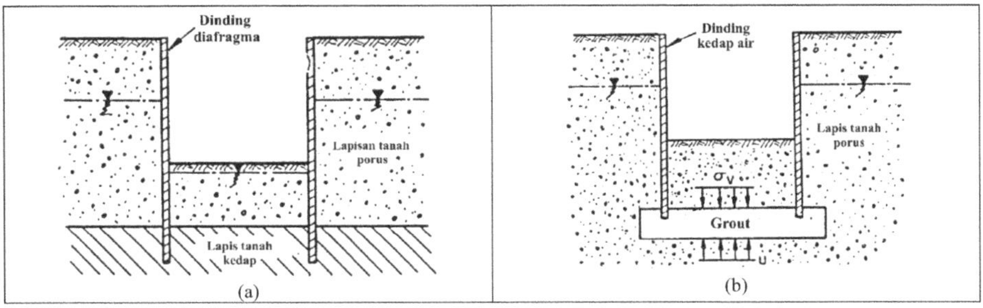

Fig. 3. (a) Reduce dewatering by *cut off wall* penetration through impermeable layer (b) or by grout at below excavation base layer when impermeable layer is too deep

Recently precast diaphragm wall is also used in practice. The excavation has to be conducted sufficiently deep into the impermeable layer, however the precast element can be positioned higher for economy reason. The precast panel method is desirable due to quality, however the present product is limited to 400 mm thick and limited length of 12 m. For larger size the difficulty is on the lifting and transporting the element (Fig. 4).

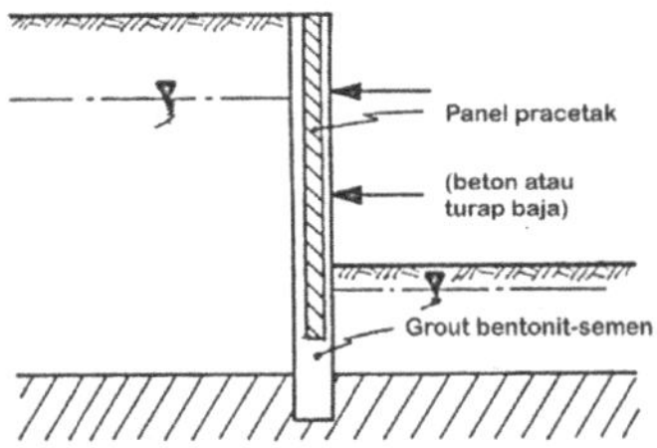

Fig. 4. Use of precast diaphragm wall

When the base of excavation find clay or silty soils relatively impermeable, then dewatering by pumping may not be relevant, then the popular way is by the method of passive dewatering. The water is collected in a pit and then pumping is allowed. This method must be carefully understood by the design engineer, because once the base concrete slab is poured, water will cause uplift and often contractors are not aware of this phenomena.

The best practice that we recommend is by constructing trenches and fill them by drain material or gravels to minimize the buoyancy effect. The unwanted effect of the dewatering was presented as an example in this paper.

3.2 Problems of Excavation in Consolidating Marine Clay Layer in Jakarta

Another challenge that most designers not aware of is the existence of underconsolidating soft marine clay in the north area of Jakarta. During testing in laboratory, the excess pore pressure of these soils disappeared due to stress release as the sample are opened. When designing the protection of excavation, designers only view water pressure based on hydrostatic condition. The existing pore pressure is neglected. Some failures or excessive lateral movement have occured. The author always use Cone Penetration Test to identify and quantify the excess pore pressure (Rahardjo 2015).

Research by Cox has shown that most alluvium in South East Asia are still consolidating. The degree of consolidation was found to depend on its sedimentation rate and type of soils. In most soil investigation report, the soils are described as normally consolidated (Fig. 5).

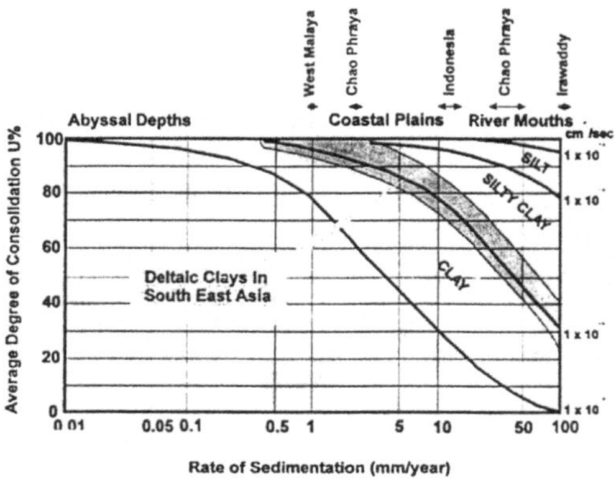

Fig. 5. Degree of consolidation of soft Deltaic Clays in South East Asia (Cox 1970)

4 Excavation at South Jakarta

Deep excavation South of Jakarta is in general in better soil condition where water table is normally low at 6–8 m below ground level. This condition is favorable to allow the excavation with minimum protection.

Favorable soil conditions mean:

- The subsoil has relatively medium to high shear strength, the upper layer are generally medium to stiff clay or cemented sands.
- Low ground water table so that pumping is not necessary or much less.
- There are not many building surrounding the excavation area.

Protection could be with only shotcrete and or soil nailing. Figure 6 show example of excavation with minimum protection in South Jakarta Area before covered by shortcrete. This type of excavation is about 9–10 m deep and the distance to the buildings on the order of 10–15 m. Usually inclinometers and pizometers are installed to monitor the behavior of the excavation. Notice that no water table of wet layers were detected in these excavation (Fig. 7).

Fig. 6. Example of excavation at south of Jakarta with minimum protection

Fig. 7. Another example of 9 m depth of excavation at south of Jakarta

In some cases, the first 3 m consists of soft soil or recent sediments of clay with a range of SPT = 3–5 and undrained shear strength of Su = 20–30 kPa. This layer is normally protected by shotcrete cover and soil nailings. Figure 8 is an example of this case where a berm is provided at the boundary between soft clay layer and lower stiffer layer. The lower layer could be strong and stiff clay or cemented sands (lahar origin) and can be stable with very steep cuts of H:V = 1:5. The problems with the existence of sandy layer is water seepage. To avoid significant drawdown of the ground water, despite of dewatering by pumping, a passive way to dry the site is by horizontal drains as shown on Fig. 9.

Fig. 8. Excavation with combined soil nailing at upper slope and steep cut for excavation at south of Jakarta

Fig. 9. The use of horizontal drains to drain out ground water seepage for stable slope at South Jakarta

5 Use of Inclinometers Inside Soldier Pile to Measure Displacement and Estimate Forces in Piles

A simple method of pile behavior in stabilizing slope is measuring the pile top movement using theodolite. However this measurement is lack of accuracy due to the fact that it can not be used to observe movement below the ground. The use of inclinometers are common and usually they are installed in the soils behind the retaining system. The author has installed the inclinometers to measure the behavior of the soldier piles and to estimate shear forces and moment in the soldier piles. Pile behavior may be better measured using inclinometers embedded into the piles during pile installation. Basically it is assumed that the deflection of the piles will be automatically represented by the deviation of the inclinometers. The measurement can be done from time to time during construction or during the course of the slope movement (Fig. 10).

Fig. 10. Strong layers of soils alternating with the cemented sands are found at South Jakarta

The first step is to read the initial reading of the inclinometers. The initial reading means the position of the inclinometers prior to any changes or movement. The movement of the inclinometers then can be measured using inclinometers' probe inserted into the inclinometer. Based on the deflection of the inclinometers, the shear forces and bending moments can be estimated using simple formula based on finite difference method. Figure 11 shows how deflection is defined in a beam. Similarly this can also be used for piles (Figs. 12 and 13).

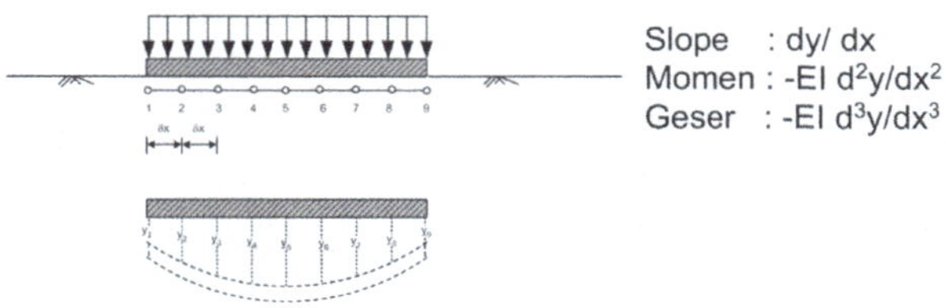

Fig. 11. Deformation of a beam and the magnitude of slope and forces in differential form

Fig. 12. Installation of inclinometers inside bored piles before excavation

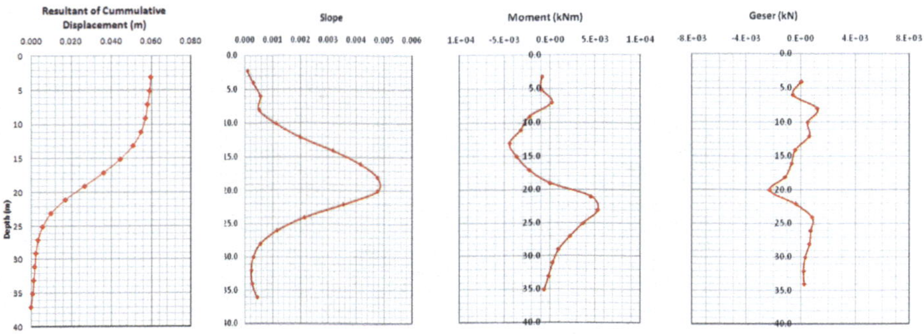

Fig. 13. Example of displacement measurement and calculation of forces in boredpiles

In most cases, the calculated movement is much higher than the measured displacement. This situation may be used to back calculate the soil parameters. Based on the author experience, the soil moduli in excavation (unloading condition) is much higher than the actual moduli although the value has been increased 2–3 times. An example of this experience is shown in Sovereign Tower Project. Free standing soldier piles of 800 mm diameters with c/c spacing of 1.6 m were used for protection of excavation of 12 m deep at south Jakarta (Figs. 14 and 15).

Fig. 14. Protection of excavation by soldier piles at South Jakarta

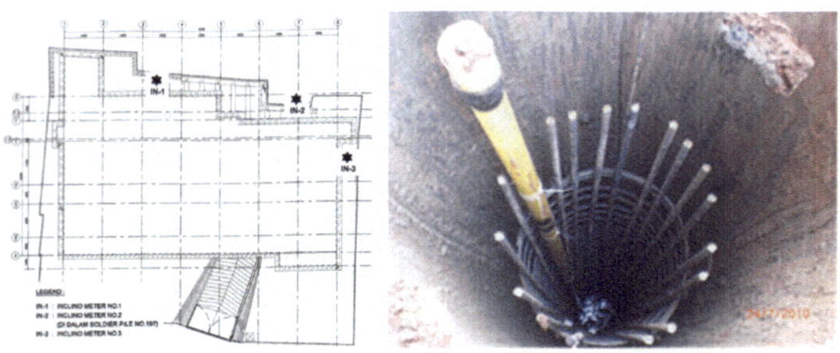

Fig. 15. Inclinometers were installed inside soldier piles at South Jakarta

The calculated movement of the soldier piles analysed by 2DPlaxis with free standing condition was 141 mm and with two level of struts was 70 mm. However until end of excavation, the actual movement of the soldier piles were in the range of 4.0–9.0 mm as shown on Fig. 16.

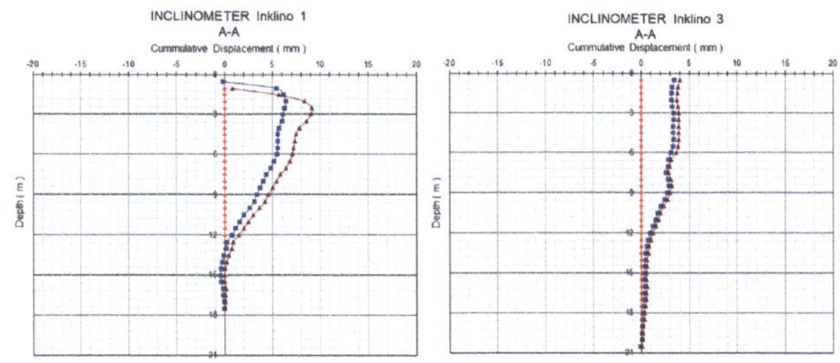

Fig. 16. Displacement measured by Inclinometers inside soldier piles

Comparable soil condition near the area of Sovereign Tower is a project (GKM Tower) located east Jalan Simatupang where similar soil condition and the same dimension of bored piles with the same spacing were used as soldier piles except that soil nailings were installed 4 m below ground level. The soil condition is soft clay with N = 2–4 to 14 m depth but water table is low. Underlying this soft clay are stiff to hard clay (N = 15–38), dense sand or cemented sand. Soil parameters were derived from laboratory tests as well as insitu testing (pressuremeter test). Dilatometer test results and Pressuremeter Test data are shown on the following figures. The data have been used to estimate the insitu stress condition and the moduli for analysis. The results of analysis using the data is better than automatically generated data from the computer (Fig. 17). The soil moduli is normally assumed 3 times the results of insitu tests.

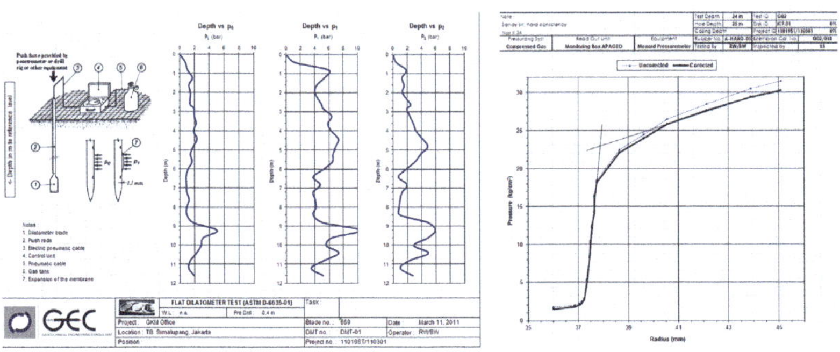

Fig. 17. Insitu test results for better estimate of the design parameter

The results of Pressuremeter Test is shown on Table 1. Po of pressuremeter test represent the insitu total horizontal stress which is more representative for the assumption as the initial pressure. Upon excavation this pressure will automatically drops together with development of the deformation along soldier piles. The moduli also represent in situ moduli under insitu stress condition. However since excavation in under unloading condition, the moduli has been multiplied by 3 for input in the analysis. It is of interest that even though the N-SPT values of the upper soils layer is soft, the dilatometer and the pressuremeter tests confirm the soil is stiffer than expected. Figure 18 shows measured displacement (from inclinometers) and calculated shear force and moment compared to the results of analysis. The original design was protection by soldier piles and installation of struts, however due to difficulties in construction of struts and considering soil condition shows promising strength, the redesign has been changed using soil naillings.

Table 1. Result of Pressuremeter test and Dilatometer test for GKM tower

Borehole	Depth (m)	N_{SPT}	P_o (kg/cm^2)	P_Y (kg/cm^2)	P_L (kg/cm^2)	E_M (kg/cm^2)
BH-01	6.00–7.00	2	0.75	1.74	4.6	66.73
	12.00–13.00	4	1.83	5.69	10.5	181.07
BH-02	18.00–19.00	28	2.13	5.56	9.25	348.68
	24.00–25.00	36	4.25	18.2	39	1686.51

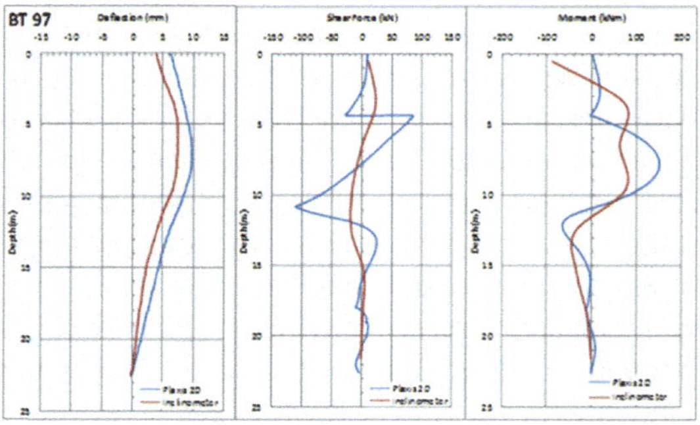

Fig. 18. Results of analysis and measured displacement, shear forces and moments for GKM Tower

The 2 D Plaxis analysis has been used to calculate the forces in the soldier piles, and is plotted on Fig. 18 with measured forces calculated from the inclinometer. The model was analysed with assumption that the position of struts or soil nailings was represented by fixed end anchor. This model results in the high and sharp shear forces, however the re-calculated forces based on the inclinometer shows smooth curves.

6 Excavation at Central Jakarta

In Central Jakarta, the soil condition is mixed depending on its location. Most locations have medium consistency clay but other might have soft layer at significant depth. Water table is high at 2–3 m below ground surface. At certain location in Jakarta specially in the Menteng Area, loose sands are found and these layers have caused difficulties for causing significant effect due to the dewatering problem since the drawdown can reach sufficiently far distance as far as 150–200 m away (Fig. 19).

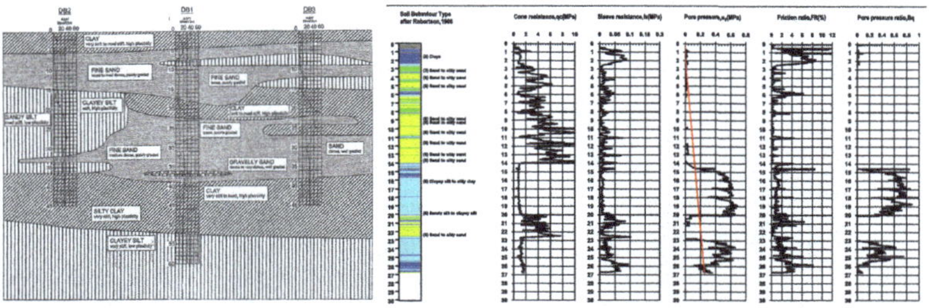

Fig. 19. Typical soil condition in Central Jakarta

Problems of dewatering in sandy soils during excavation are sometimes overlooked. More attention are being focused on stability and deformation problems, particularly of the earth retaining structures. However, the effect of dewatering might be disastrous depending on the reliability of the system and the soils condition. Despite of the necessity for draining an excavated area, method of selected dewatering system may cause serious failure due to lowering of the water table. There are two types of causes of failures, the first one being internal erosion due to piping and the second one due to settlement caused by the increase of effective stress related to ground water lowering. Measurement of water discharge is one factor that can be used to describe the severity of the dewatering problem and the second one is usually deals with the magnitude of settlement and the extent of the subsidence. To overcome the problems, an approach has been adopted.

This condition requires careful design and strict monitoring system to avoid the unwanted effect of dewatering. Rahardjo et al. (2016) reported a drop of water table in Central Jakarta causing damages to buildings surrounding the project site as shown on Figs. 20 and 21.

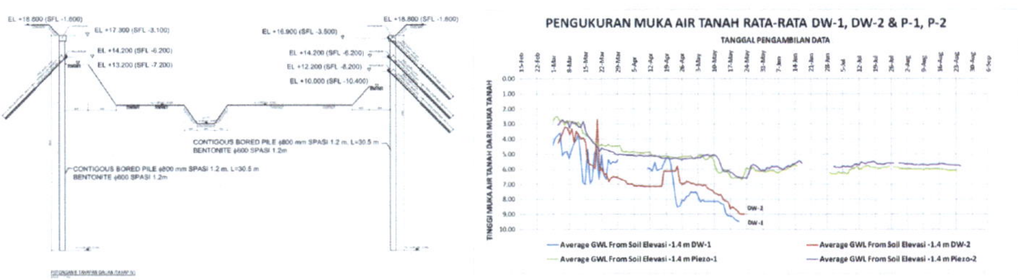

Fig. 20. Excavation system and monitoring of ground water table (Rahardjo 2016)

Fig. 21. Typical damages of buildings due to settlement caused by dewatering (Rahardjo 2016)

Solutions for this problems have been immediately carried out by terminating the pumping and rebackfill the excavation (at this point, the excavation was also terminated at 9 m depth out of the design elevation of 12 m depth) and finally reduction of the basements from 3 level basement to 2 level basement. The continuation of dewatering is by passive dewatering system where water freely flow and collected in sumpit for further disposal.

The excavation system in Central Jakarta can be done by braced excavation, anchored D-wall or anchored Secant Piles or by the top down construction method. To avoid dewatering problem, the drainage are conducted passively with sumpit and pumping on the ground surface. A case history of excavation in this area is represented by top down construction method as shown on Fig. 22. The openings were located in the middle part of the tower. In this method, struts were substituted by the floor slab supported in the periphery and at the location of the column. Care was to be given to watch the movement of column (Fig. 23).

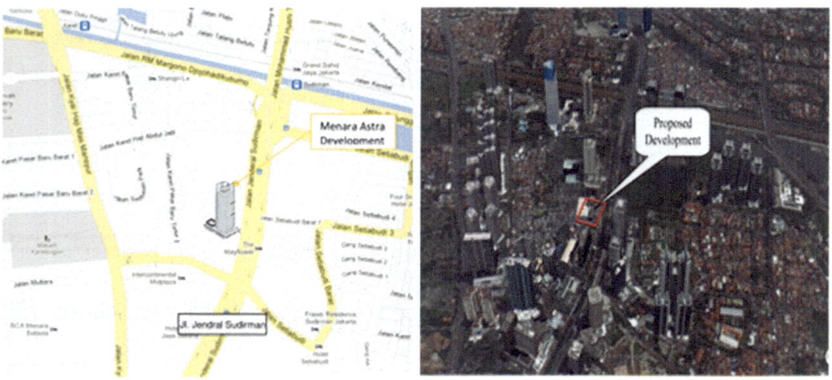

Fig. 22. Case history of excavation in central Jakarta

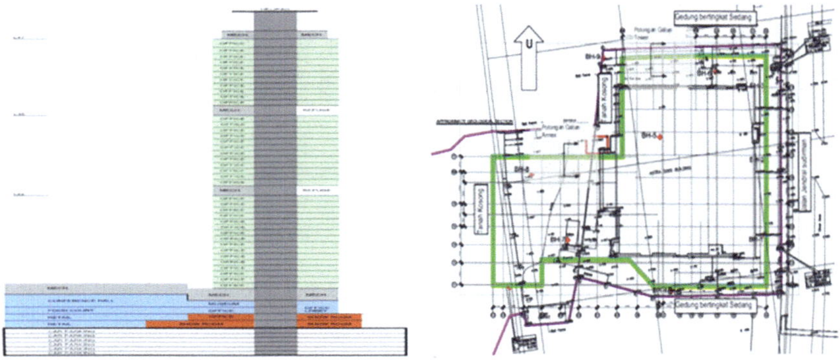

Fig. 23. Cross section and plan view of the project

This project consists of two buildings with estimated area of 7930 m^2 and 215 m in height, namely a 50-floors tower and 5-floors Annex. This building will have a 6-floors basement. Due to proximity of the construction to its neighbour, top down construction method was proposed with first floor and third floor concrete slabs were skipped. The soil condition is soft clay 6–8 m depth underlain by stiff clay. Lenses of dense sands were found continuously at depth of 17 m–22 m depth to 21 m–27 m. The bearing layer is very stiff to hard clay (Fig. 24).

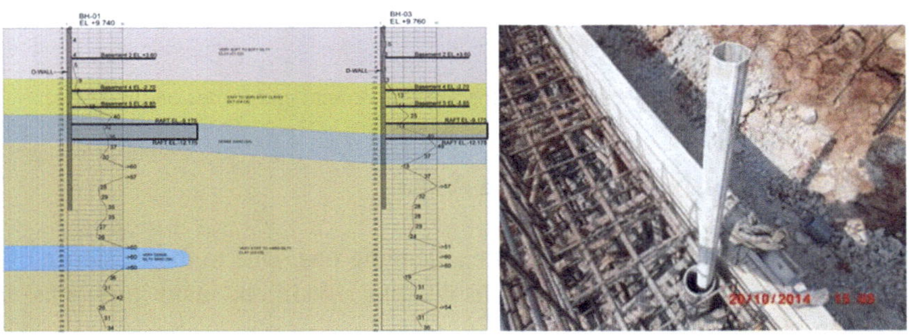

Fig. 24. Soil condition and scheme of excavation

Monitoring of the D-wall movement and water table by pizometers were carried out through out the excavation to warn any excessive displacement of the wall and the fluctuation of water table. The deformations were predicted using 3D Finite Element Method using conservative soil parameters. As high as 150 mm were obtained as shown on Fig. 25. However the actual deformation of the system is only 40 mm (Fig. 26).

Fig. 25. Stage 2 excavation to 4th level Basement and measurement of D-wall movement

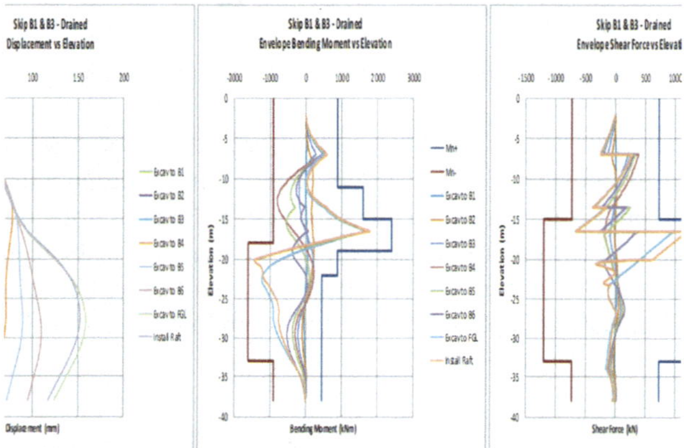

Fig. 26. Results of design analysis for wall movement and forces on the D-wall

7 Excavation at North Jakarta Area

Excavation at North Jakarta is the most challenging one. Due to the existence of soft soils specially because of the underconsolidating condition, many problems arise. Common problems are:

- Stability problems during excavation, because the soils are not only very soft, but also residual excess pore pressure may cause additional load on the retaining system and no passive resistance can be expected.
- Deformation problem may cause substantial movement of the wall and settlement of the neighbour.
- The piles that were driven prior to excavation may be tilted during the excavation period.

Important factors that influence the stability of excavation in soft soils include the low shear strength of the soils and stress level, creep and strain softening, effect of existing pore pressure due to underconsolidation and influence of excavation on nearby building and foundation (Wong 1985; Rahardjo 2005). A study by Rahardjo (2005) has shown that underconsolidation may affect safety factor even for shallow excavation of

3 m protected by sheetpile. Figure 27 shows the influence of the degree of consolidation (representing underconsolidation for OCR < 1) on the safety factor. Hence an excavation of 2.5–3.0 m is very critical in Jakarta depending on whether one can detect the existing pore pressure. Rahardjo et al. (2008), Setionegoro (2013) and Rahardjo (2015) developed a method to determine the Underconsolidation using CPTu which is very useful to in the assessment of soil condition for excavation in soft soils.

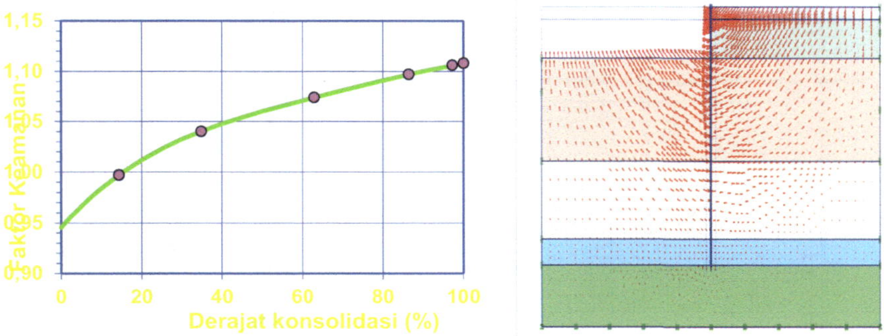

Fig. 27. Influence of the degree of consolidation of the safety factor of excavation in soft soils

Classical example of failures in excavation in North Jakarta area were discussed by Poulos (1991) and Zanusi (1991) where an excavation of unprotected excavation has caused excessive movement of 1 m and failure of building foundation at the project site. The following is chronology of the event:

- 25 August '89: Foundation of Hotel Completed
- 31 Oktober '89: Office Foundation completed
- August '90: Upper Structure of Hotel and Office Completed
- September ~ November '90: Foundation for mall completed
- 20 October '91: Excavation for *ground tank*
- Local failure (slide) during excavation, installation of bamboo piles and sheet piles. Excavation continued
- 30 October '90: Pile head of the mall moved 1 m. Excavation terminated
- 21 November '90: Excavation rebackfilled: No compaction on backfill, Backfill completed in 3 weeks
- 01 and 02 Januari '91: Cracking sound heard by witness (3 times)
- Early morning 02 Januari '90 the corner of office settled 140 mm
- Grouting was conducted but no effect

Zanusi (1991) reported that the reason for this failures were due to no design for safety of excavation, foundation installation induced excess pore water pressure, and failure of piles were due to induced moment exceeding the moment capacity of the pile. Poulos (1991) reported with more detail study on the behavior of the pile based on numerical calculation. Excessive movement of piles (about 1.0 m) caused overloading on piles under the office and hotel. Delay of soil movement (from the start of excavation to the event of soil cracking) may be due eto creep which is time dependent. Failure of beam and base slab caused building tilting. According to Poulos (1991), soil movement caused increasing lateral load and moment on piles. Analysis by computer program ERCAP explain that at pile movement of 20 cm, has caused the pile to reach its ultimate moment capacity (Figs. 28 and 29).

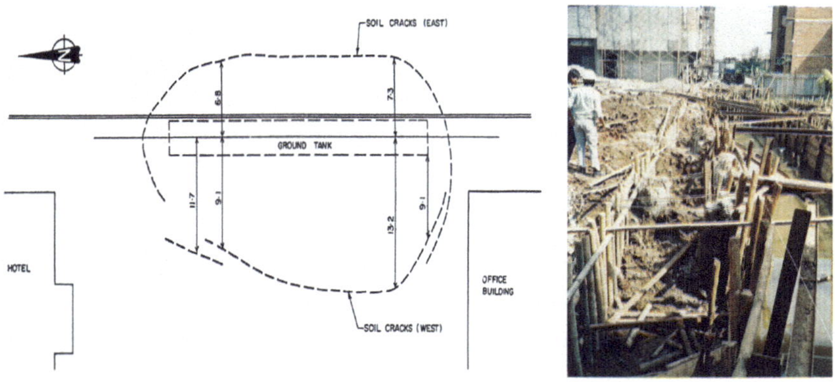

Fig. 28. Movement of boredpiles and cracks showing sliding (Poulos 1992; Zanusi 1992)

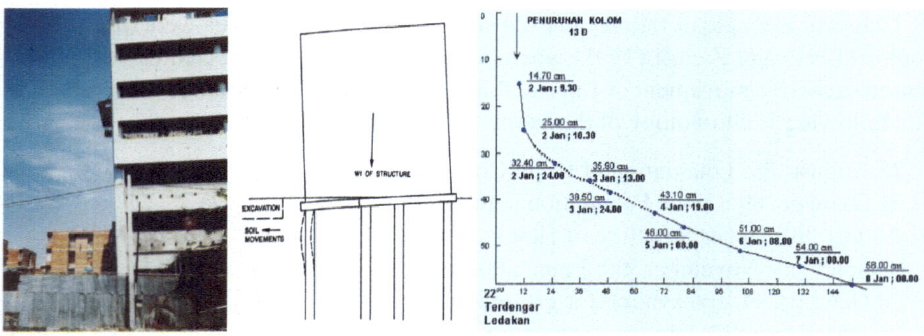

Fig. 29. Tilting of building and settlement of nearest column to excavation (Poulos 1992; Zanusi 1992)

Rahardjo (2005) revisited the problem with geotechnical review including: soil condition, phenomena of under consolidation, effect of excavation of Soft Soil on Pile Foundation, modeling by Finite Element Method and conducted study on the possibility of progressive failures/slides. By use of CPTu and supporting data, Rahardjo (2005) has shown that the soil at depth 5 m–11 m is underconsolidating as shown on Figs. 30 and 31.

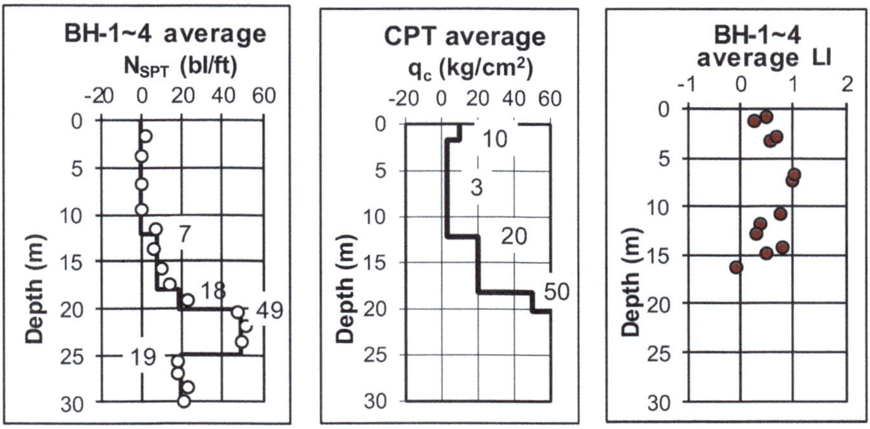

(a)

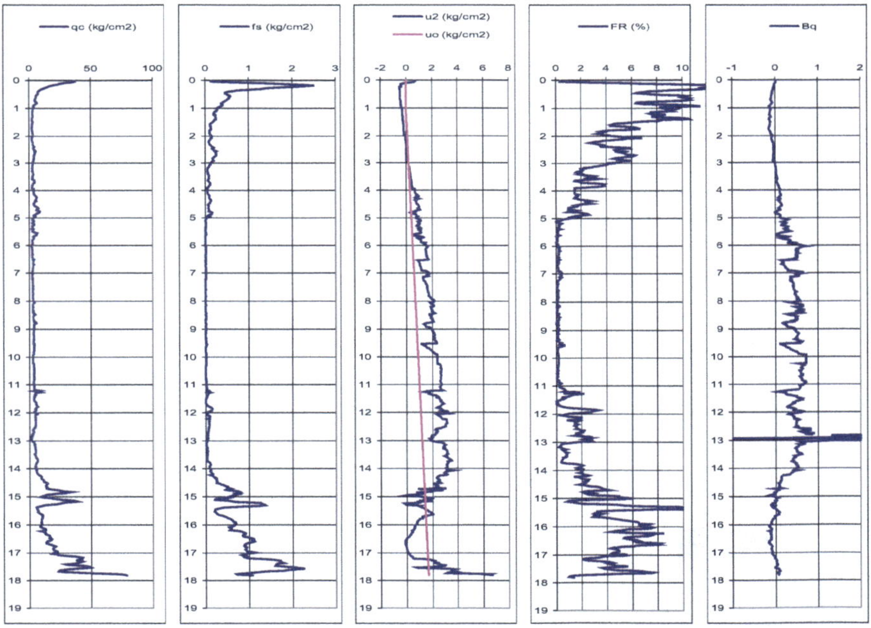

(b)

Fig. 30. Soil condition at North Jakarta area showing underconsolidation (Rahardjo et al. 2015)

Fig. 31. Excavation in soft marine clays North Jakarta

Based on the study, Rahardjo (2005) reported that the soil is consolidating and residual excess pore pressure was found causing Safety Factor of the excavation calculated by Bjerrum and Eide (1956) = 0.26–0.69, hence causing problem of instability, deep sliding was detected at depth of 5–7 m, and due to lateral movement and restraint of pile foundation by the building, the pile failed. The excess pore pressure may be combined by the effect of pile installation.

Further study on excavation at North Jakarta was reflected from the excavation for basement with 12 m depth protected using diaphragm wall of 800 mm thick embedded 25 m and supported by struts or ground anchor at different end. The investigation show that the soft marine clay is consolidating. Movement was measured 120 mm at the end of the excavation or about 1% of the excavation depth.

Further north at the coastal area, an excavation of 5 m was protected by sheetpiles 12 m long, anchored by 600 mm boredpiles at the outer side with 600 mm diameter and 18 m depth. Inclinometers were installed behind the excavation. Large movement occured to more than 150 mm as shown on Fig. 32.

Fig. 32. Movement of retaining system in North Jakarta Excavation

8 Conclusions Summary

1. Excavation in South Jakarta is generally in good soils and may be conducted by open cut with minimum protection (soil nailing, anchor, shotcrete).
2. For excavation in soft soils at North Jakarta there are risks. Excavation in under-consolidating soils need attention on existing pore pressures and very low strength. Uncontrolled excavation of soft soil could result in significant movement of soil mass that could endanger nearby structures.
3. Monitoring of excavation is important part for safety, performance, and warning system (sometimes good for back analysis).
4. Determination of soil parameters are important input.

References

Poulos, H.G.: Investigation of foundation failures in Jakarta. Report (1991, unpublished)

Rahardjo, P.P.: Construction method for deep excavation in soft soils. In: Proceeding Seminar Perkembangan Pondasi di Indonesia. Universitas Katolik Parahyangan (2005)

Rahardjo, P.P.: Case histories of deep excavation in Jakarta and consideration for design of basement construction. Lecture at Indonesian Geotechnical Society, Bandung (2014)

Rahardjo, P.P.: CPTu in consolidating soils. In: Proceeding 5th International Conference on Site Characterization. ISSMGE, Gold Coast, Australia (2016)

Setionegoro, N.: Penentuan Derajat Konsolidasi Menggunakan uji CPTu. disertasi doktor. Universitas Katolik Parahyangan (2013)

Zanusi, F.X.: Kajian Kegagalan Pondasi di Jakarta. Report (1991, unpublished)

Evaluation of Performance of Walls in Deep Excavations Using Wall Deflection Paths

R. Hwang(✉) and L.W. Wong

Moh and Associates, Inc., New Taipei City, Taiwan
hwang.richard168@gmail.com

Abstract. This paper discusses the concept of wall deflection path and reference envelope and illustrates its applications on evaluating the performance of diaphragm walls in deep excavations carried out in thick soft deposits. Numerical analyses were performed for a cut-and-cover section of the Taipei Metro to verify this concept and to quantify the influences of four factors, i.e., the preloading of struts, wall thickness, wall length, and width of excavation on wall deflections. It has been found that the width of excavation is the most influential factor on maximum wall deflections while the thickness and the length of walls have dominating influence on toe stability.

Keywords: Taipei Metro · Underground construction · Deep excavation · Wall deflection path · Reference envelope

1 Introduction

Structures adjacent to deep excavations were frequently damaged to a certain degree as ground moved, or even collapsed in disastrous events. Serious failures of ground may cause traffic congestion or disrupt the function of lifelines, leading to chaos and panics of people, and may lead to injunction of the constructions. Therefore, it is vital to take precautions whenever a deep excavation is carried out in the neighbourhood of sensitive structures, vital utilities, and trunk roads. This is particularly true for metro constructions which are normally carried out in densely populated city areas with heave traffic. On the other hand, the performance of the retaining system of a deep excavation may be affected by the structures in the vicinity. For example, the basements of highrise buildings together with retaining structures left in-place after the completion of construction are very likely to reduce wall deflections. For metro constructions, there are always entrances, ventilation shafts, etc., structurally connected to underground stations and the rigidity of the diaphragm walls is much increased due to the irregularity of the walls, and the wall deflections are much reduced as a result.

Ground settlement, which is one of the primary factors affecting the structures adjacent to excavations, is closely related to wall deflections. Wall deflections thus become the most important subject in evaluating the performance of diaphragm walls in deep excavations and assessing the potential risks of damaging the structures in vicinity. Conventionally numerical analyses are conducted by using two-dimensional models to evaluate the performance of retaining structures, i.e., walls and struts. Since the existing underground structures adjacent to excavations are normally omitted in

R. Chen et al. (eds.), *Proceedings of the 2nd International Symposium on Asia Urban GeoEngineering*, Springer Series in Geomechanics and Geoengineering,
https://doi.org/10.1007/978-981-10-6632-0_16

these analyses, it is unrealistic to compare the results obtained with the observed performance of walls. The conclusions made in such a way may be misleading. For these reasons, it is desirable to have a means to quantify the influence of adjacent structures and many other factors which may affect wall deflections, so the performance of walls can be faithfully evaluated.

Discussed herein is the concept of wall deflection path and reference envelope firstly suggested in Moh and Hwang (2005) and Hwang et al. (2006). Applications of this concept in evaluating the performance of diaphragm walls in the Taipei Basin were later illustrated in Hwang and Moh (2007, 2008) and Hwang et al. (2007a, 2016). Numerical analyses have been subsequently conducted by Hsiung and Hwang (2009); Chao et al. (2010a, b); Hwang et al. (2012) and Hwang and Moh (2017) to substantiate the concept and to study the sensitivity of various factors on wall deflections.

This study is an extension of the above-mentioned studies aiming at generalization of the conclusions reached therein by providing data obtained from additional numerical analyses.

2 Case Studied

Wall deflections are routinely monitored by using inclinometers. The readings obtained are inevitably affected by the movements at the tips which are assumed to be fixed and wall deflections at other depths are calculated accordingly. Moh and Hwang (2005) and Hwang et al. (2007b) recommended to calibrate inclinometer readings to account for the movements at the inclinometer tips by assuming that the joints between the struts at the first level and the diaphragm walls would not move once these struts were preloaded. This

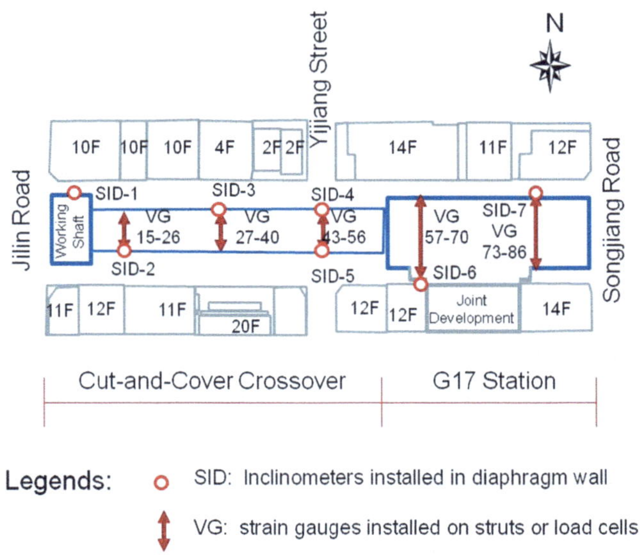

Fig. 1. Layout of G17 Station of the Green Line of Taipei Metro and the adjoining cut-and-cover crossover

recommendation was based on the finding that the changes in the lengths of these struts would be minimal as the load increments and/or decrements in the struts at the 1st level would be very small. To illustrate this point, the readings obtained in the cut-and-cover construction for the crossover, refer to Fig. 1, next to G17 Station (Songjiang Nanjing Station) of Line 3 (the Green Line) of Taipei Metro are discussed herein. The excavation was carried out to depths varying from 17.7 m to 20.2 m below the ground level in 7 stages and the pit was retained by diaphragm walls of 1 m in thickness installed to a depth of 35 m and braced by steel struts at 6 levels as depicted in Fig. 2.

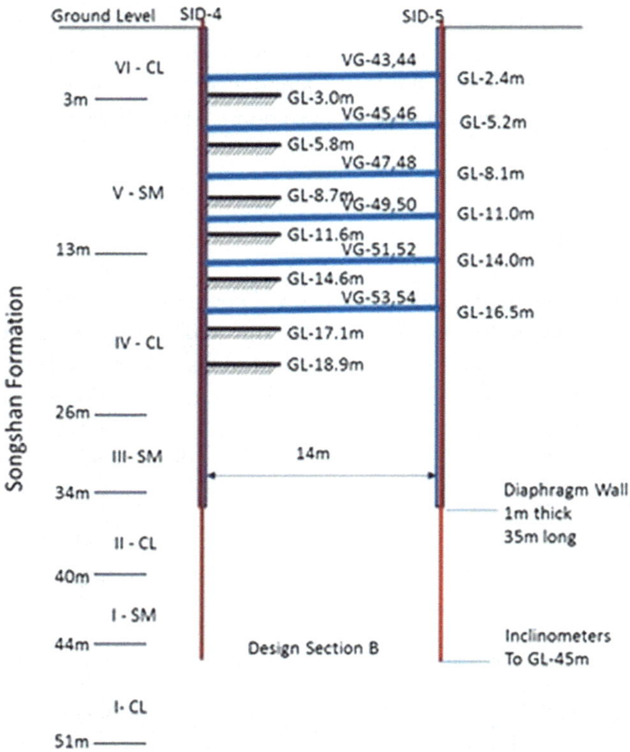

Fig. 2. Excavation scheme for the crossover next to G17 Station of Taipei Metro

3 Correction of Inclinometer Readings

Figure 3 shows the strut loads recorded by two strain gauges, namely, VG-43 and VG-44, installed on the strut at the 1st level between Inclinometers SID-4 and SID-5 which were installed in the diaphragm walls and extended by 10 m below the toes of walls. As can be noted, the strut was preloaded to 71 tones, which is the average of the readings of the two gauges, at the beginning. The load in the strut increased to an average of 87 tons in the second stage of excavation and dropped to a minimum of −3 tons in the subsequent stages. For a strut length of 14 m and a Young's Modulus,

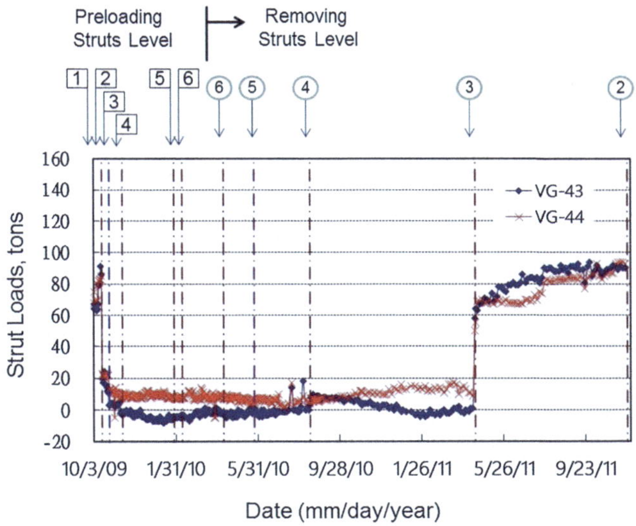

Fig. 3. Loads in the 1st level strut at the locations of Inclinometers SID-4 and SID-5

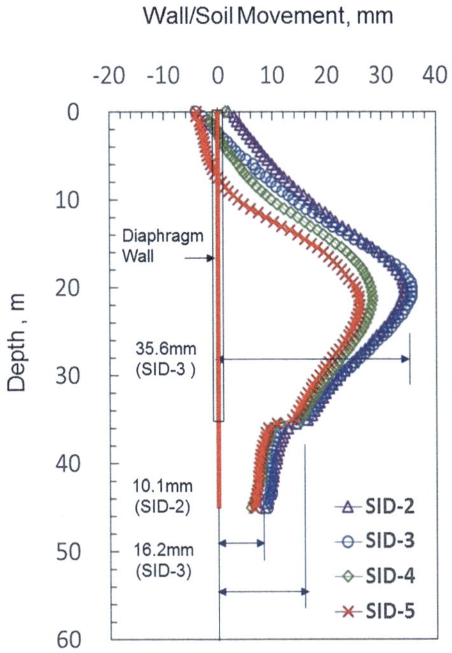

Fig. 4. Inclinometer readings, duly corrected to accounted for the movements at the tips

i.e. the E-value, of 200,000 N/mm^2 for steel, the increment of 16 tones, i.e., from 71 tons to 87 tons, corresponds to a shortening of 0.6 mm of the strut, or an inward movement of 0.3 mm at each end; and the decrement of 74 tons, i.e., from 71 tons to −3 tons, corresponds to a lengthening of 3 mm of the strut, or an outward movement of 1.5 mm at each end. Movements of such magnitudes are negligible for practical purposes and the joints between the struts and the diaphragm walls can indeed be assumed fixed for calibrating the inclinometer readings at other depths.

The readings, duly calibrated accordingly, obtained at the final excavation stage by the 4 inclinometers installed in the diaphragm walls on the two sides of the crossover are shown in Fig. 4. A maximum wall deflection of 35.6 mm was obtained at the location of SID-3. It is interesting to note that the diaphragm wall toes moved by 16.2 mm at the location of SID-3. It has become quite common nowadays to install inclinometers in diaphragm walls and stop at the toe levels of the walls to reduce construction costs. The toe movements of such magnitudes were, nearly, 50% of the maximum wall deflections for the case studied and back analyses would certainly lead to misleading conclusions if inclinometer readings were not corrected. It can also be noted from Fig. 4, the tips of the inclinometers still moved by, as much as, 10.1 mm, or 28% of the maximum wall deflections, even with a 10 m extension below the toes of the diaphragm walls.

4 Numerical Analyses

Numerical analyses were conducted by using the two-dimensional finite element computer program PLAXIS (PLAXIS BV 2011) to compare the performance of the walls with the theoretical predictions. Figure 5 shows the finite element model adopted. The strength parameters of soils suggested by the Detailed Design Consultant of the project are given in Table 1 (MAA 2005). The Mohr-Coulomb Model was adopted to

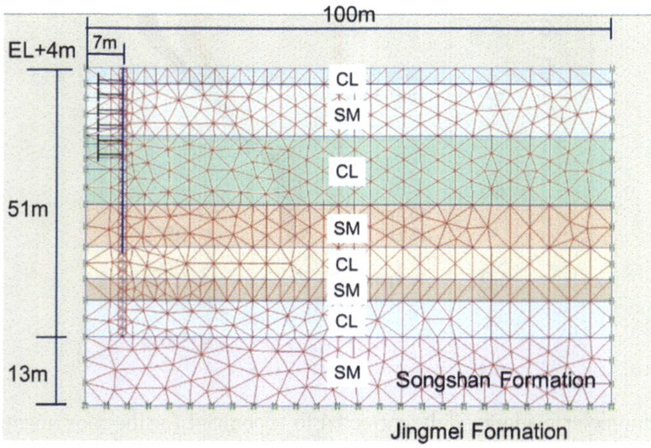

Fig. 5. Finite element model for the crossover next to G17 Station

Table 1. Soil properties and soil parameters adopted in the PLAXIS analyses (MAA 2005)

Depth (m)	Soil type	γ_t (kN/m^3)	N (blows)	S_u (kPa)	c′ (kPa)	Φ′ (deg)	E′ (MN/m^2)	Poisson's ratio, υ′
0–3	CL	18.8	4	25			12.5	0.35
3–13	SM	19.2	8	–	0	32	20.0	0.30
13–26	CL	18.6	6	65			32.5	0.35
26–34	SM	19.4	13	–	0	32	32.5	0.30
34–40	CL	18.9	13	110			55.0	0.35
40–44	SM	19.7	21	–	0	32	52.5	0.30
44–51	CL	19.9	20	145			72.5	0.35

simulate the stress-strain behaviour of soils. The Young's moduli, E′, were correlated to soil strengths by using the following empirical relationships:

$$E' = 500\, Su\, (for\ clayey\ soils) \tag{1}$$

$$E' = 2.5\, N\, (in\ MPa\ for\ sandy\ soils) \tag{2}$$

in which Su = undrained shearing strength, and N = blow counts in standard penetration tests (SPT). Young's moduli are not uniquely related to soil strengths. The coefficients relating these two depend on many things and can only be established empirically by back analyses. The values shown above are applicable at this site in conjunction with the soil strengths given in Table 1.

The deepest borehole reached a maximum depth of 51 m and a remark is noted in the borehole log indicating that the soils below this depth are Type SM (silty sand). A N value (blow counts in standard penetration tests) of 25 was thus assumed for the soils below the depth of 51 m. Based on the local geology, it was estimated that the gravelly Jingmei Formation lies at a depth of, roughly, 64 m. The Jingmei Formation is

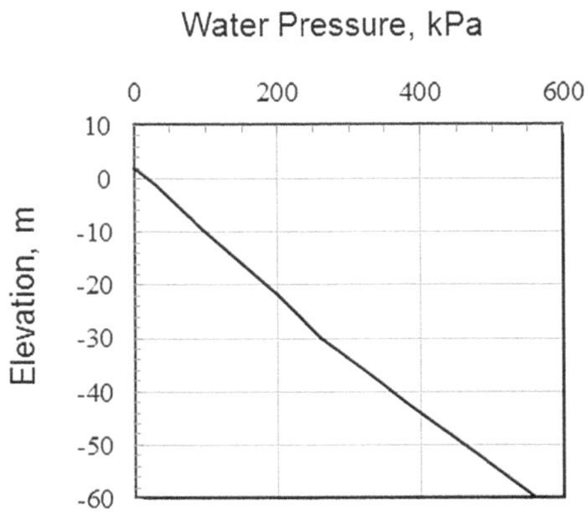

Fig. 6. Groundwater pressures on the outer face of diaphragm wall adopted in the numerical analyses

a hard gravelly strata and can be deemed as a hardpan for engineering purposes, therefore the bottom of the finite element model was assumed at this depth. The water pressures acting on the outer face of the diaphragm wall are shown in Fig. 6. Inside the pit, the water level was assumed to be at a depth of 1 m below the bottom of the excavation in each stage.

The diaphragm walls were simulated by plate elements and an E value of 25,000 MPa was adopted for concrete with an unconfined compressive strength, i.e. f′c value, of 28 MPa. The EI (I = moment of inertia) and EA (A = sectional area) values of the diaphragm walls were reduced by 30%, giving a value of 1,464 MN*m for the former and 17,570 MN/m for the later, following the normal practice to account for the influence of cracking, creep and relaxation during excavation. The structural properties of the struts and the magnitude of preloading adopted in the analyses are shown in Table 2.

Table 2. Stiffness of struts adopted in numerical analyses

Level	Members	Sectional area A (cm^2)	Stiffness, AE/S (MN/m)	Preload kN/m
1	1H350x350x12x19	1 × 173.9	891	178
2, 3	1H400x400x13x21	1 × 218.7	1121	339, 494
4	1H428x407x20x35	1 × 360.7	1849	428
5, 6	1H414x405x18x28	1 × 295.4	1514	458, 538

Note: Spacing between struts S = 4 m

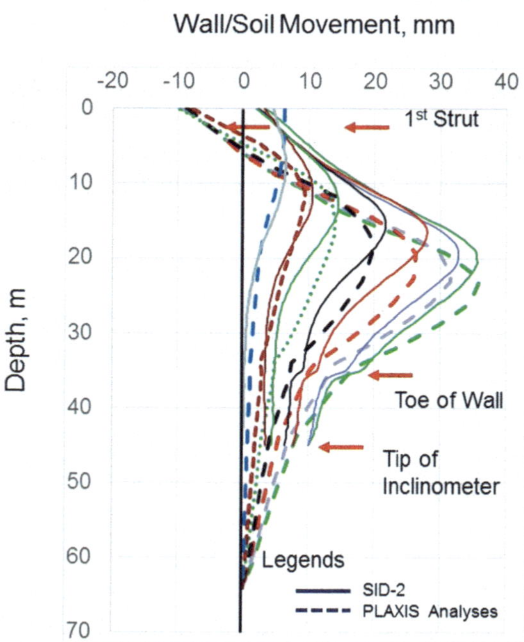

Fig. 7. Comparison of inclinometer readings with computed wall deflections for the case with 100% preloads at all levels

Of the 4 inclinometers installed on the two sides of the crossover, SID-2 and SID-3 gave the largest wall deflections, refer to Fig. 4. The first set of readings of SID-3 was taken after the struts at the first level had been preloaded, therefore, the readings are inappropriate for discussion. The wall deflections obtained from the analyses are compared with the readings obtained by Inclinometer SID-2 in Fig. 7, and as can be noted, the maximum wall deflections in all the stages of excavation are in a fair agreement with the readings of Inclinometer SID-2.

4.1 Effects of Preloading of Struts

The profile of wall deflections obtained from the numerical analyses for the 1st stage of excavation is in a very good agreement with the inclinometer readings. On the other hand, a large outward movement was computed as a result of the preloading of the struts at the first level as depicted in Fig. 8(a). These struts were preloaded to 72.5 tons (the average of 2 readings) on 1 January 2009 and, as depicted in Fig. 8(b), the inclinometer readings taken on 5 January indicated that the wall had hardly moved as compared to those taken on 27 December 2008.

The large outward movements obtained from numerical analysed were presumably due to the use of Mohr-Coulomb Model which much under-estimates the soil moduli in the early stages of excavation. Secondly, the preloads are line load applied to all the struts at the same level simultaneously in two-dimensional numerical analyses; while in reality, struts were preloaded individually, one by one. Each time a strut was preloaded, the load was essentially a point load resisted by the entire wall and the wall movement would be smaller than what would be if all the struts at the same level were preloaded simultaneously. Wall movements due to subsequent preloading of neighbouring struts

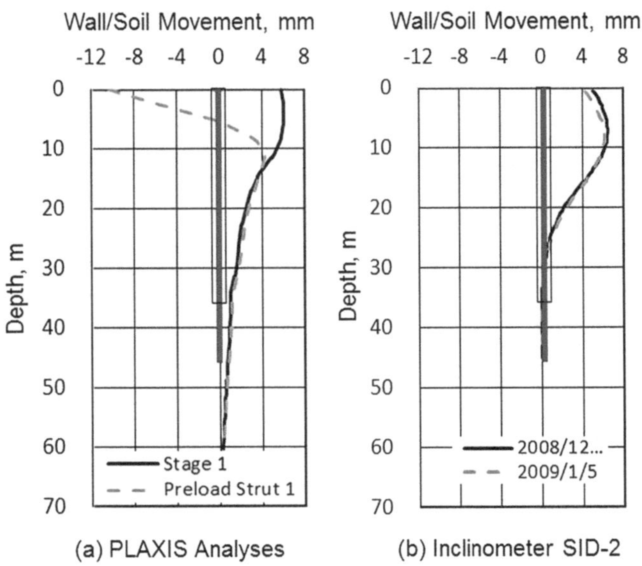

Fig. 8. Influences of preloading the strut at the first level on wall deflections

would be small because the loads in the struts which had already been preloaded were not sustaining.

Figure 9 shows the fact that the strut load dropped to nearly a half in a few days subsequent to preloading presumably due to the preloading of neighbouring struts. This residual load can be considered as an effective load in all the struts at the first level and should be the load to be adopted in numerical analyses. However, this reduction appears to be applicable to the first level only. As depicted in Fig. 10, strut loads at the second level did not drop much after preloading.

The computed wall deflections with the preloads at the first level reduced by 50% are compared with the inclinometer readings in Fig. 11. As can be noted, the outward movements shown in Fig. 7 are much reduced and the agreement between the computed wall deflections and the inclinometer readings is much improved. However, the improvement is limited to shallow depths and the maximum wall deflection in each stage of excavation is unaffected.

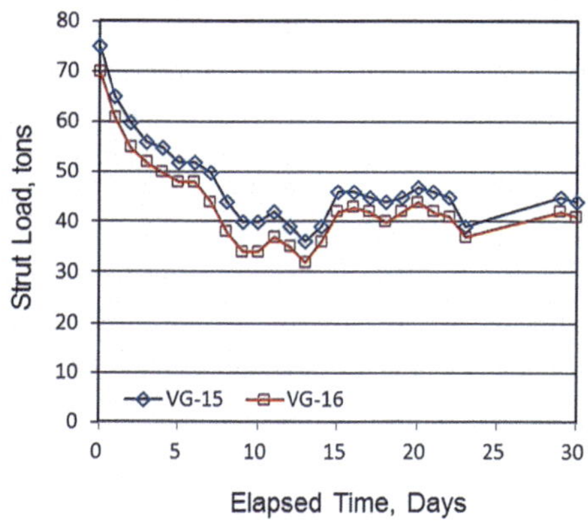

Fig. 9. Strut loads at the first level subsequent to preloading

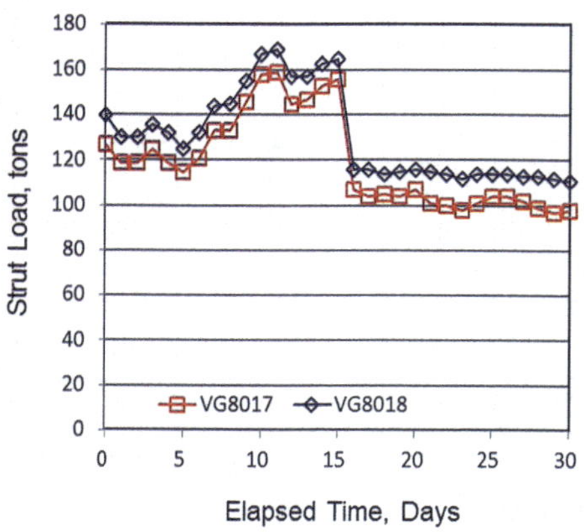

Fig. 10. Strut loads at the second level subsequent to preloading

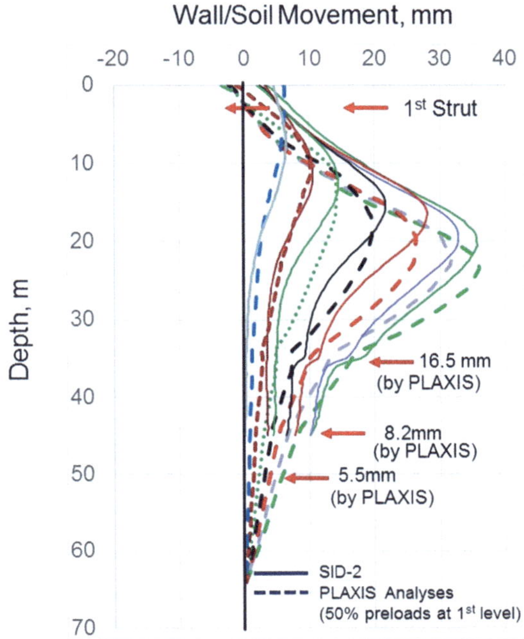

Fig. 11. Comparison of inclinometer readings with the computed wall deflections for the benchmark case with the preloads at the first level reduced by 50%

The case shown in Fig. 11 is considered as the "benchmark case" for parametric studies. For academic interest, analyses have also been carried out for the case with preloads in all stages of excavation ignored and the results are shown in Fig. 12. As can be noted that the shapes of the deflection profiles computed agree very well with those obtained based on the inclinometer readings. However, the maximum wall deflection increases from 36 mm to 44.5 mm at the end of the final excavation.

4.2 Movements at Diaphragm Wall Toes

What is of great significance is the fact that the large movements at the toes of the diaphragm walls and at the tips of inclinometers shown in Fig. 4 were confirmed by the analyses. The computed movements at the first strut level, the toe of diaphragm walls and at the tips of inclinometers are listed in Table 3. As can be noted, the movement at the 1st strut level increased from 1.53 mm after preloading of the strut (inward) to 2.59 mm at the end of Stage 2 excavation, and then dropped to 0.17 mm at the end of the final excavation. These increments and decrements agree well with those deduced from strut loads in Sect. 3. Movements of such magnitudes are indeed negligible in comparison with the large movements at the diaphragm wall toes and at the tips of inclinometers. The procedure of calibrating inclinometer readings by assuming that the diaphragm wall at the first strut level would not move once the struts were preloaded is thus proved to be valid.

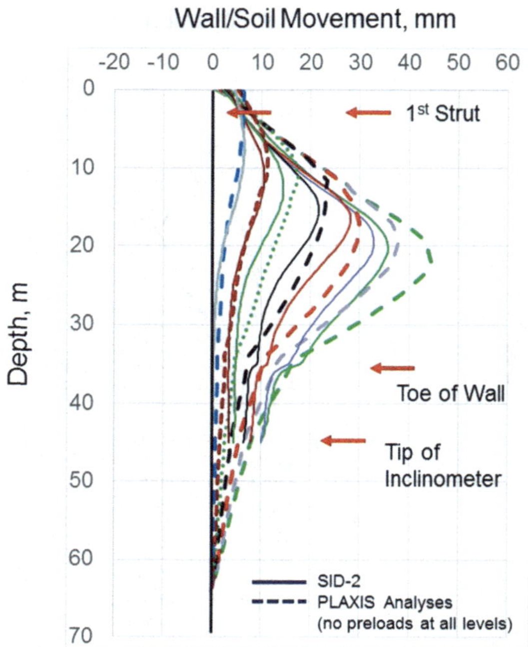

Fig. 12. Comparison of inclinometer readings with the computed wall deflections for the case without preloads at all levels

Table 3. Comparison of inclinometer readings with movements computed for the benchmark case

Location	Lateral movement, mm					
	1st Strut level		Diaphragm wall toe		Inclinometer tip	
Excavation stages and struts preloading	PLAXIS	SID2	PLAXIS	SID2	PLAXIS	SID2
Stage 1	6.26	5.90	1.09	0.21	0.76	0
Preload 1st Strut	1.53	5.26	1.20		0.82	
Stage 2	2.59	5.26	2.80	3.41	1.63	3.49
Preload 2nd Strut	0.94		2.92		1.70	
Stage 3	0.47	5.26	4.54	4.87	2.79	4.35
Preload 3rd Strut	0.62		4.89		2.65	
Stage 4	0.27	5.26	7.39	9.03	3.85	6.34
Preload 4th Strut	0.43		7.10		4.06	
Stage 5	0.25	5.26	10.31	11.16	5.05	7.74
Preload 5th Strut	0.33		10.42		5.09	
Stage 6	0.23	5.26	13.06	15.63	6.89	9.99
Preload 6th Strut	0.23		13.33		6.36	
Stage 7	0.17	5.26	16.50	16.35	8.15	10.11

Notes: Positive values for inward movements and negative values for outward Movements
Readings of Inclinometer SID-2 have been corrected to account for the movements at the tip

5 Concept of Wall Deflection Path and Reference Envelope

The drastic differences between the wall deflection profiles for different inclinometers shown in Fig. 4 fully demonstrate the ambiguity faced in back analyses. Different results will be obtained if a different set of inclinometer readings is selected to compare with the results of numerical analyses. To establish a consistent methodology for back analyses, Moh and Hwang (2005) suggested to plot wall deflections versus depth of excavation, in a log-log scale, and designate such plots as "wall deflection paths". It was further suggested to take the upper envelope of the wall deflection paths obtained at a specific site, or for a specific set of parameters, as "reference envelope" to compare with the results of conventional two dimensional numerical analyses which are usually conducted for excavations in green field without underground structures or utilities in the vicinity. This suggestion was based on the belief that wall deflections are likely to be reduced by the presence of underground structures and facilities as well as many factors and the upper envelope of wall deflection paths will be closer to what will be obtained in numerical analyses carried out for excavations in green field.

For convenience, reference envelopes are defined by the wall deflections for a depth of excavation of 4 m, i.e., Δ4 and the wall deflection projected to a depth of excavation of 100 m, i.e., Δ100. Accordingly, the reference envelope for the case of interest can be expressed as Δ4 = 6 mm and Δ100 = 250 mm as depicted in Fig. 13. The projection of reference envelope to a depth of excavation of 100 m, however, is merely for convenience in defining reference envelopes and by no means implies the validity of the relationship between wall deflection and depth of excavation beyond the final depth of excavation.

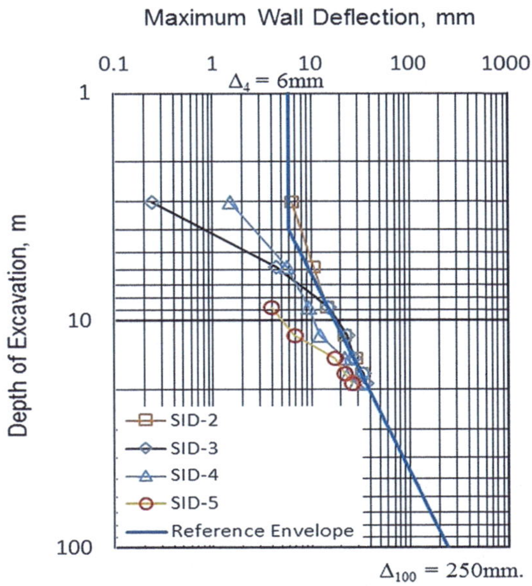

Fig. 13. Observed wall deflection paths and reference envelopes for the crossover

Excavation and preloading of struts at excavation sites are never carried out in the ways specified in designs. They are carried out in a rather unpredictable sequence as site constraints and project progress always prevail. Coordination among subcontractors is often difficult, and as such, over-excavation occurs rather frequently and delay in strutting is quite common. For these reasons, the data for shallow excavation are erratic and should be ignored, and only the data for excavations exceeding 10 m in depth should be considered in establishing reference envelopes as suggested in previous studies (Moh and Hwang 2005; Hwang et al. 2006, 2007a, b). After all, deep excavations are the ones of primary concern.

For numerical analyses, as discussed in Sect. 4.1, wall deflections at shallow depths are likely to be affected by how the preloads at the first and the second levels are applied at the site, therefore the data points for shallow excavations should be ignored. The wall deflection path obtained by PLAXIS analyses for the benchmark case, i.e., with the preload at the first strut level reduced by 50%, is given in Fig. 14 and can be defined by Δ4 = 6 mm and Δ100 = 250 mm. This wall deflection path is identical to the reference envelope given in Fig. 13.

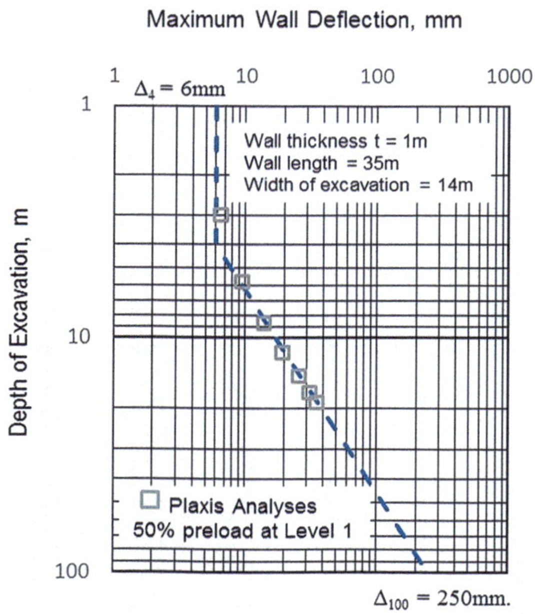

Fig. 14. Wall deflection path obtained by PLAXIS analyses for the benchmark case

6 Factors Affecting Wall Deflections

The concept of wall deflection path is a quite useful tool in evaluating the influences of various factors on wall deflections for excavations in deep soft deposits. For illustration, the influences of four of the primary factors, namely,

(1) Preloading of struts,
(2) Wall thickness,
(3) Wall length,
(4) Width of excavation,

which affect the magnitudes of wall deflections are studied and discussed as follows.

6.1 Influence of Preloading of Struts

As depicted in Fig. 8, the preloads applied at the first strut level affect the wall deflections in the early stages of excavation. However, as can be noted by comparing Fig. 11 with Fig. 7, the reduction of the preloads at the first level affect the maximum wall deflection at each stage of excavation only slightly. On the other hand, Fig. 12 shows the fact that the maximum wall deflection would be much increased if preloads at all the strut levels are omitted. The influences of preloads on wall deflection paths for these three scenarios can easily be visualized from Fig. 15. The reduction of preloads at the first strut level does not affect the Δ100 value while the omission of preloads at all the strut levels increases the Δ100 value from 250 mm to 400 mm. The Δ4 values remain the same as they, in theory, correspond to the first stage excavation before preloading.

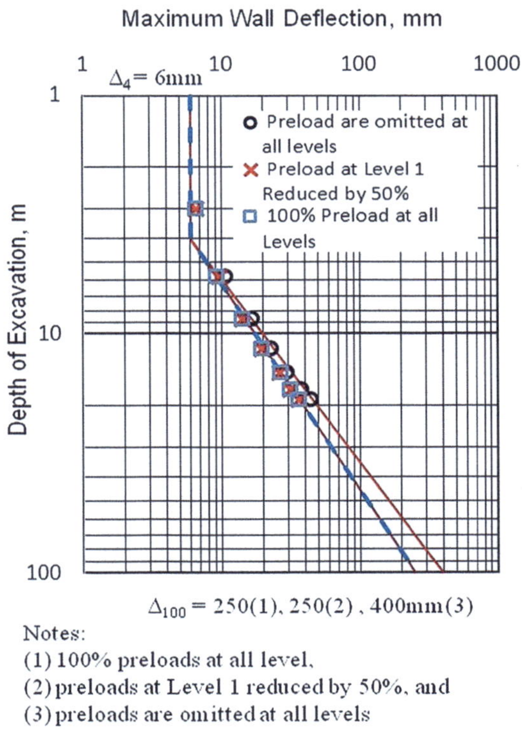

Fig. 15. Influences of preloading of struts on wall deflections obtained by PLAXIS analyses

6.2 Influence of Wall Thickness

The wall deflection paths for walls of 0.6 m and 1.5 m in thickness are depicted in Fig. 16. This figure together with Fig. 14 show that the Δ100 value increases from 180 mm to 250 mm as the thickness of wall decreases from 1.5 m to 1 m and further increases to 320 mm as the thickness of the wall decreases to 0.6 m. The Δ4 values, which correspond to the deflection in Stage 1 excavation before preloading of the struts, are unaffected by the thickness of diaphragm wall. For a depth of excavation of 18.9 m, the maximum wall deflections will be 31 mm, 36 mm and 41 mm for walls with thicknesses of 1.5 m, 1 m and 0.6 m, respectively. It is interesting to note by comparing Fig. 15 with Fig. 16 that the preloads have similar effects as thickening the diaphragm wall, i.e., increasing the rigidity of the retaining system.

For academic interest, analyses were carried out assuming that the excavation were proceeded further below the design depth of 18.9 m using the same excavation scheme just to see how the walls would perform. Excavation from the depths of 18.9 m to 28.0 m are supported with the level 7 to level 10 struts. The stiffness for these struts follows that of the level 6 strut (Table 3). As can be noted from Fig. 16, the deflection paths remain linear till a depth of excavation of 23.8 m is reached. Beyond this depth, the rapid increases in wall deflections indicate instability of the diaphragm wall toes. It should be noted that only the force equilibrium was considered in this exercise with all other factors, such as seepage and blow-in, excluded in the analyses. Furthermore, the wall was assumed to be linear elastic with unlimited structural capacity. Therefore readers are warned not to take this finding for granted in designs.

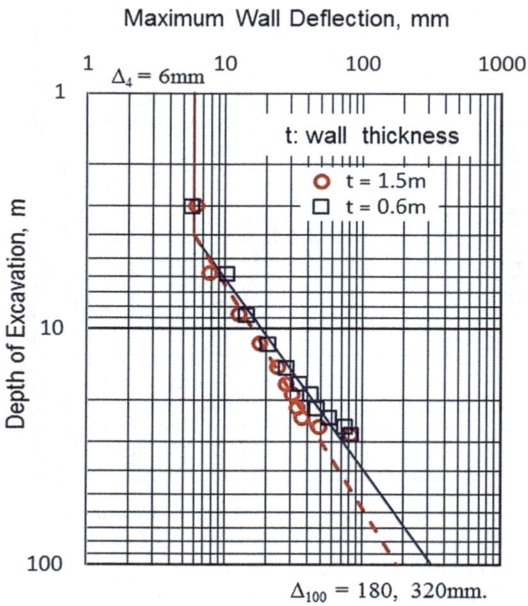

Fig. 16. Influences of thickness of diaphragm wall on wall deflections obtained by PLAXIS analyses

6.3 Influences of Wall Length

Similarly, analyses were carried out to a depth of 28 m to see how wall deflections would be affected by the lengths of walls. Figure 17 shows the wall deflections for walls of 30 m, 35 m and 40 m in length. As can be noted, for excavations to the depth of 18.9 m, the maximum wall deflections are essentially the same in these 3 cases. The wall deflections start to deviate noticeably from a depth of excavation of 18.9 m. For walls of 40 m in length, deflections were slightly below the reference envelope, suggesting that lengthening walls does help in maintaining toe stability.

6.4 Influences of Width of Excavation

Analyses were performed for different widths of excavation and the results are given in Fig. 18. As can be noted that the $\Delta 4$ values are more or less proportional to the width of excavation. This is one of the reasons that only the inclinometers for excavations of similar widths, i.e., SID-2 to SID-5 are included in the comparison and other inclinometers are excluded. It is emphasized again that only the data between depths of excavation in the range of 10 m to 20 m should be considered in constructing wall deflection paths.

What is even more interesting is the fact that the 4 wall deflection paths converge to a same $\Delta 100$ value. This finding is consistent with what was given in Hwang et al. (2012), (Chao et al. 2010a, b) and Hsiung et al. (2016).

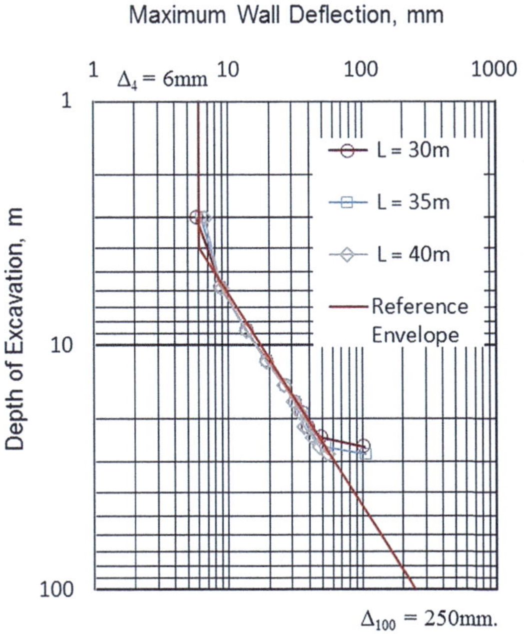

Fig. 17. Influences of wall length on wall deflections obtained by PLAXIS analyses

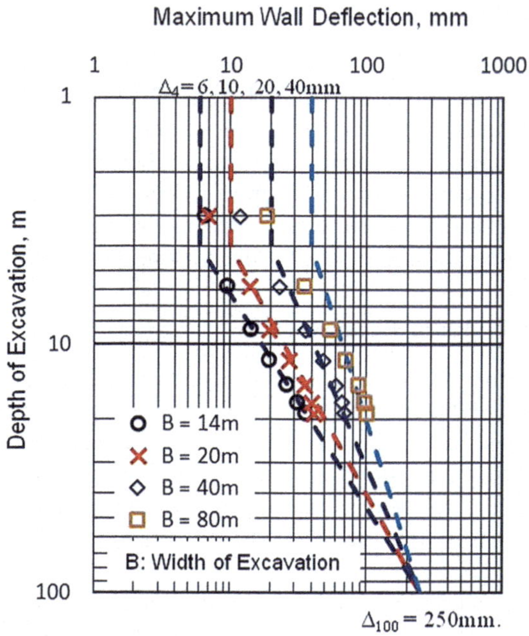

Fig. 18. Influences of width of excavation on wall deflections obtained by PLAXIS analyses

7 Conclusions

Numerical analyses have been conducted for a cut-and-cover section of the Taipei Metro. The results of analyses verify the various input parameters such as the Young's moduli for clay and sand deposits, the stiffness for the struts and the magnitude for strut preloading. The calibrated numerical model validates the wall deflection path concept, which plots wall deflections versus depth of excavation, in a log-log scale. The influences of four factors, i.e., the preloading of struts, wall thickness, wall length, and the width of excavation on wall deflections were quantified by using this concept.

The foregoing discussions lead to the following conclusions:

1. The concept of wall deflection path is a very useful tool to evaluate the performance of diaphragm walls in deep excavations in thick soft sand and clay deposits with high groundwater tables.
2. The influences of various factors can be readily quantified by comparing the wall deflection paths obtained with the reference envelope which is the wall deflection path representative of a specific site or for a specific combination of influencing factors.
3. The width of excavation, the thickness of wall and the magnitude of preloads have pronounced influence on the maximum wall deflections.
4. The preloads have similar effects as thickening the diaphragm wall and, hence, reduce the maximum wall deflections.

5. The influence of the length of wall to wall deflections will be insignificant until a critical depth of excavation is reached. Such critical depth will depend on the bending stiffness of the wall or base heave modes of failure.

References

Chao, H.-C., Change, J.F., Huang, R.H.: Evaluation of performance of diaphragm walls in deep excavation by using deflection path method. In: Proceedings of the Seventeenth Southeast Asian Geotechnical Conference, Taipei, Taiwan, 10–13 May 2010, pp. 390–393 (2010a)

Chao, H.-C., Hwang, R.N., Chin, C.T.: Influence of tip movements on inclinometer readings and performance of diaphragm walls in deep excavation. In: Proceedings of 2010 Earth Retention Conference, 1–4 August Bellevue, Washington, Geotechnical Special Publication No. 208, American Society of Civil Engineers, pp. 326–333 (2010b)

Hsiung, B.C.B., Hwang, R.N.: Evaluating performance of diaphragm walls by wall deflection paths. Special Issue on Excavation and Tunneling in Geotechnical Engineering, Southeast Asian Geotechnical Society, pp. 81–90 (2009)

Hsiung, B.C., Dan, D.S., Lum, C.W.: Evaluation of performance of diaphragm walls by wall deflection paths for deep excavations in Central Ha Noi. Geotech. Eng. J. SEAGS AGSSEA **47**(1) (2016). ISSN 0046-5828

Hwang, R.N., Moh, Z.C.: Deflections paths and reference envelopes for diaphragm walls in the Taipei Basin. J. GeoEng. **2**(1), 1–12 (2007). Taiwan Geotechnical Society, Taipei, Taiwan

Hwang, R.N., Moh, Z.C.: Evaluating effectiveness of buttresses and cross walls by reference envelops. J. GeoEng. **3**(1), 1–11 (2008). Taiwan Geotechnical Society, Taipei, Taiwan

Hwang, R.N., Moh, Z.C.: Deep excavations in Taipei Metro Construction. Geotech. Eng. J. SEAGS AGSSEA **48**(2) (2017). ISSN 0046-5828, Bangkok, Thailand

Hwang, R.N., Moh, Z.C., Kao, C.C.: Design and construction of deep excavations in Taiwan. In: Seminar on the State-of-the-Practice of Geotechnical Engineering in Taiwan and Hong Kong, Hong Kong, 20 January (2006)

Hwang, R.N., Moh, Z.C., Kao, C.C.: Reference envelopes for evaluating performance of diaphragm walls. In: 13th Asian Regional Conference, December, Kolkata, India, pp. 505–508 (2007a)

Hwang, R.N., Moh, Z.C., Wang, C.H.: Toe movements of diaphragm walls and correction of inclinometer readings. J. GeoEng. **2**(2), 61–72 (2007b). Taiwan Geotechnical Society, Taipei, Taiwan

Hwang, R., Lee, T.Y., Chou, C.R., Su, T.C.: Evaluation of performance of diaphragm walls by wall deflection paths. J. GeoEng. **2**(1), 1–12 (2012). Taiwan Geotechnical Society, Taipei, Taiwan

Hwang, R.N., Wang, C.H., Chou, C.R., Wong, L.W.: Deep excavations in Taipei Basin and performance of diaphragm walls. Geotech. Eng. J. SEAGS AGSSEA **47**(2) (2016). ISSN 0046-5828, Bangkok, Thailand

MAA: Design Summary, Construction Contract CG292, Songshan Line, submitted to Department of Rapid Transit Systems, Taipei City Government, by Moh and Associates, Inc., Taipei, Taiwan (2005)

Moh, Z.C., Hwang, R.N.: Geotechnical considerations in the design and construction of subways in urban areas. In: Seminar on Recent Developments on Mitigation of Natural Disasters, Urban Transportation and Construction Industry, 30 November – 2 December, Jakarta Indonesia (2005)
PLAXIS BV: Reference Manual. PLAXIS BV, Amsterdam, The Netherlands (2011)

Influence of Excavation Construction on Adjacent Existing Retaining Structure

Chen Xiaoyu[1], Hu Minyun[1(✉)], Zhang Yong[1], and Yuan Jing[2]

[1] Zhejiang University of Technology, Hangzhou, China
huminyun@zjut.edu.cn
[2] Zhejiang Province Institute of Architectural Design and Research, Hangzhou, China

Abstract. The objective is to examine the influence of excavation construction on an adjacent retained excavation and to determine the safe spacing between them. Firstly, the variation of the earth pressure acting on the retaining walls due to adjacent unloading was studied based on the limit equilibrium theory and the assumption of plane slip. Then, the expression of the safe spacing b between two adjacent excavations is discussed theoretically. Secondly, a parametric study was carried out by finite element method simulations to study the influence of the depth of excavation, the spacing between two excavations, the soil strength, and the surcharge on soil surface between two adjacent excavations. The proposed earth pressure and the recommend expression of the safe spacing by Eq. (6) are evaluated. Suggestions are also given for the sake of safety control for two adjacent excavations.

Keywords: Adjacent retaining structure of excavation · Finite element method · Unloading earth pressure · Deformation response · Safe spacing

1 Introduction

Adjacent excavations are common cases in crowded urban areas in China due to the increasing demand of underground space exploitation. The adjacent excavations can be of different construction purposes, such as basement, subway station, and constructed by different companies with different methods. In this way, the safety of the pre-existing retaining structure can be affected not only by its own excavation process, but also by side unloading due to the adjacent excavation [1]. Therefore, the interaction of two adjacent excavations is now paid more and more attention. The effects of adjacent excavations can be as follows: (1) The deformation of the retaining structure of the pre-excavated pit and the axial forces of the lateral struts, if set up, can be changed due to the unloading of an adjacent excavation; (2) The settlement of the surface soil between the adjacent pits will increase due to the superposition effect; (3) The earth pressure acting on the adjacent retaining structures will decrease due to unloading.

The distance is crucial to the interaction of the two adjacent excavations. When the spacing between the adjacent pits is small, the earth pressure acting on the retaining structures will be influenced because the failure condition of the retained soil mass in

R. Chen et al. (eds.), *Proceedings of the 2nd International Symposium on Asia Urban GeoEngineering*, Springer Series in Geomechanics and Geoengineering,
https://doi.org/10.1007/978-981-10-6632-0_17

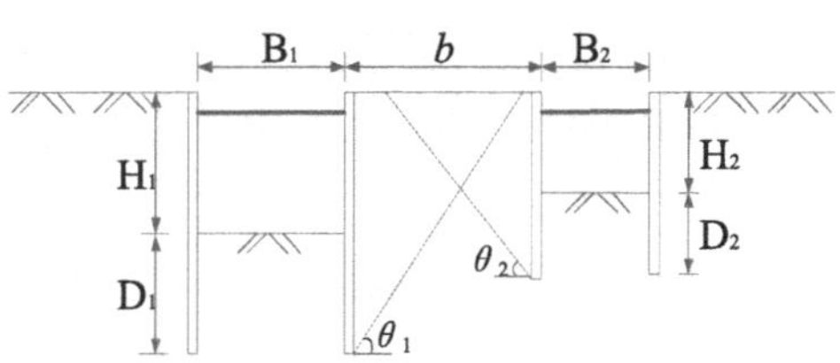

Fig. 1. Interaction of two adjacent excavations

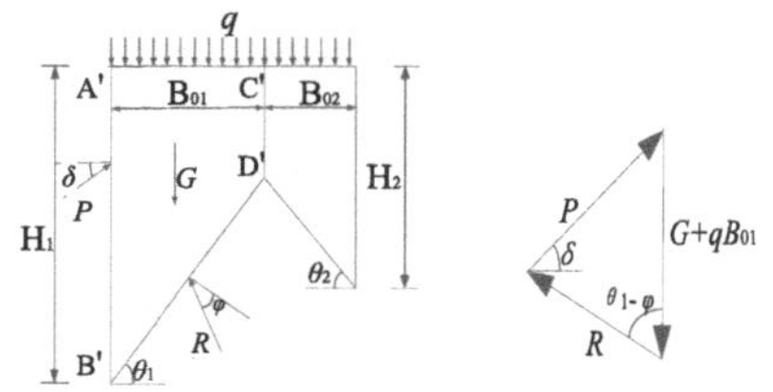

Fig. 2. Finite width of soil between two adjacent excavations

between will be changed. The interaction between two adjacent excavations due to earth pressure change is illustrated in Fig. 1. Thus, the classical earth pressure theory is no longer applicable to the calculation of the earth pressure with a limited soil mass [2–4].

There are a few studies on the earth pressure on a retaining structure from a limited soil mass and on the critical spacing between two excavations. Ma et al. [5] investigated the earth pressure by the limit equilibrium theory, and referred (H is the depth of the excavation) as a critical spacing between two pits. Gao et al. [6] proposed a solution based on upper bound theory, and suggested a critical spacing of $H/\tan\,(45^\circ + 2)$. Li et al. [7] applied limit equilibrium theory and plane slip wedge assumption, and suggested a critical spacing of $0.75H$. Yue et al. [8] took the principle of silo force and considered the friction between the retaining wall and soil, and derived the critical spacing of$(H_1 + H_2)/\tan\,(45^\circ - \varphi/2)$.

However, most of the above researches are for the case in which the foundation pit is adjacent to an existing underground structure. That is different from the case of two adjacent excavations. Some investigations of the interaction of two adjacent excavations have been carried out by site measurement [9–11] and by numerical simulations [12–15], whereas the mechanism of the interaction between two adjacent excavations is not clear, and the influential factors need to be studied. In this study, the earth pressure acting on the adjacent retaining structures is analyzed under limit equilibrium condition. Then the safe spacing b between two adjacent excavations is discussed. Finally the ground and structure behavior of the retained excavation under adjacent unloading were analyzed by numerical simulations.

2 Earth Pressure and Safe Spacing

In order to analyze the earth pressure acting on the retaining structure when the limited soil mass between two excavations reaches the active limit equilibrium state, four assumptions are introduced as: (1) when it is on the point of sliding, the soil between the retaining wall and a trial failure surface is in limit equilibrium; (2) the shear strength parameter c will be taken as zero; (3) a surcharge on soil surface between two adjacent excavations is considered; (4) the failure surface of limit equilibrium can develop on two opposite directions and intersect at C′D′ behind the retaining walls, the forces acting on C′D′ is neglected. The diagrammatic sketch of limited earth pressure analysis is shown in Fig. 2. The analysis is under plain strain condition.

2.1 Earth Pressure with Limited Soil

For the failure condition, the soil wedge is in equilibrium under its own weight (G), the reaction to the force (P) between the soil and the wall, the reaction (R) on the failure plane, and the vertical force due to surcharge ($Q = qB$). The weight of the sliding wedge A′B′C′D′ can be expressed as

$$G = \gamma\left[B_{01}(H_1 + \frac{q}{\gamma}) - \frac{B_{01}^2}{2}\tan\,\theta_1\right] \tag{1}$$

where θ_1 is the angle between the slip surface and the horizontal plane; B_{01} is the calculated width of the limited sliding soil; q is surface surcharge; H_1 is the depth of the first excavation A.

At the limit state, the expression of P can be determined by the geometric relation of force polygon as follows.

$$P = \frac{(G + qB_{01})\sin(\theta_1 - \varphi)\cos\,\delta}{\cos(\theta_1 - \varphi - \delta)} \tag{2}$$

When force P between the soil and the retaining wall reaches the maximum value (active earth pressure state), the corresponding sliding surface can be determined at

$$\frac{\partial P}{\partial \theta} = 0 \tag{3}$$

by using the derivative method. Let $x = \tan\theta$, then Eq. (3) gives

$$\begin{aligned} &- B_{01}^2\cos\,\varphi\,\sin(\varphi + \delta)x^2 - B_{01}^2\cos\,\varphi\,\cos(\varphi + \delta)x \\ &+ (B_{01}/H_1 + B_{01}q/\gamma)\cos\,\delta + B_{01}^2/2\sin\,\varphi\,\cos(\varphi + \delta) = 0 \end{aligned} \tag{4}$$

where δ is the frictional angle between the soil and the retaining wall. From Eq. (4), at active limit state

$$\tan\,\theta_1 = -\frac{1}{\tan(\varphi + \delta)} + \sqrt{\frac{1}{\tan^2(\varphi + \delta)} + \frac{\tan\,\varphi(1 - \tan\,\delta\,\tan\,\varphi) + 2(H_1/B_{01} + q/\gamma B_{01})(1 + \tan^2\,\varphi)}{\tan\,\delta + \tan\,\varphi}} \tag{5}$$

Angle θ_1 between the slip surface and the horizontal plane can be determined by Eq. (5). Thereby the active force acting on the retaining wall can be obtained by Eq. (2), and the active earth pressure can be derived by differentiating P along the depth.

2.2 Safe Spacing Analysis

The spacing b between two adjacent pits is thought to be crucial to the interaction of the two retaining structures during excavation, whereas the depths of the excavations have significant effect on the interaction when b has been determined by the projects plan. With the depths of the adjacent pits increase, the possible sliding soil surfaces behind the retaining wall may touch each other or overlap. As a result, the supposed soil wedges may change from triangles to trapezoids. If we define the Safe Spacing as the spacing between the two adjacent excavated pits which has no effect on the earth pressure acting on either retaining wall, as shown in Fig. 3, the expression of b can be

$$b = B_{01} + B_{02} = \frac{H_1}{\tan\,\theta_1} + \frac{H_2}{\tan\,\theta_2} \tag{6}$$

For excavation 1 in Fig. 3, $\tan\theta_1 = H_1/B_{01}$ is substituted into formula (5) to obtain H_1

$$H_1 = B_{01}\tan\,\varphi + \sqrt{B_{01}^2\tan^2\varphi + B_{01}^2\frac{\tan\,\varphi}{\tan(\varphi+\delta)} + 2\frac{B_{01}q}{\gamma}(\tan\,\varphi + \frac{1}{\tan(\varphi+\delta)})} \tag{7}$$

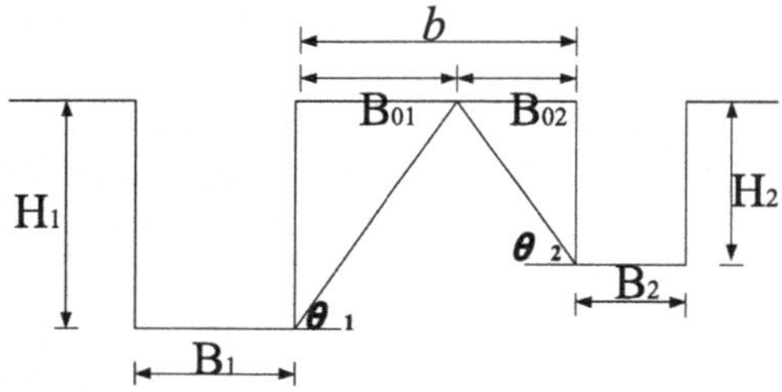

Fig. 3. Safe spacing schematic diagram

If the parameters φ, δ, q, *and* γ are known, then B_{01} can be calculated through Eq. (7). And in the same way, we can get B_{02}. The spacing b is thereby determined.

From Eq. (5), we know that the safe spacing b is related to the depths of both excavations, soil strength, and the surface surcharge. Thereafter, in order to discuss the variation of b with the influential factors, we assume the depths of the two excavations are the same H, giving $\theta_1 = \theta_2$, $b = 2H/\tan\theta = 2B_{01} = 2B_{02}$

Figures 4 and 5 gives the variation of b/H under different q. It can be seen from Fig. 4 that the safe spacing b increases with the depth of excavation and decreases with the increase surcharge q between two adjacent excavations. By comparing Fig. 4 with Fig. 5, it is clear that the safe spacing b decreases with φ increase. For the calculated cases, the variation of b is between $1.0H$ and $1.9H$. The safe spacing b is then suggested to be more than $2H$ (the depth of the excavation). The safe spacing indicates that over this value of b, the interaction between two excavations can be neglected.

From the Fig. 2, we can see that the more the δ is, the less the active earth pressure p. But that is different from safe spacing. Figure 6 gives the variation of b/H under different δ. It can be seen that the safe spacing b increases with the δ.

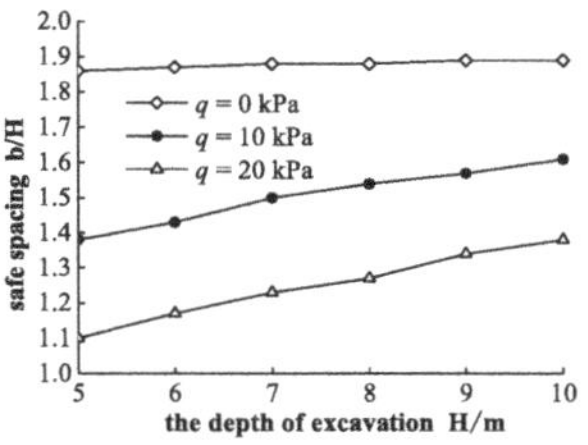

Fig. 4. b/H vs. H under different q ($\delta = 10°$, $\varphi = 15°$)

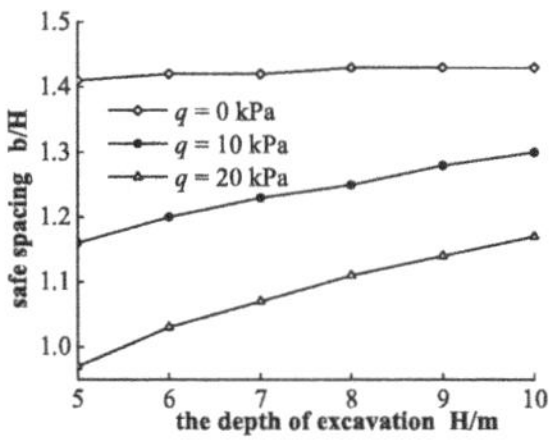

Fig. 5. b/H vs. H under different q ($\delta = 10°$, $\varphi = 25°$)

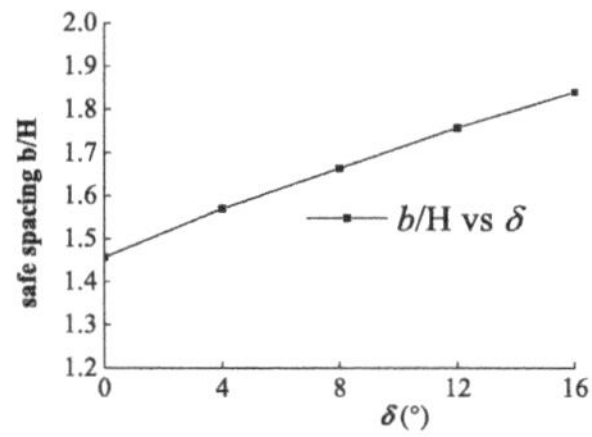

Fig. 6. b/H vs. H under different δ (q = 0 kPa, H = 10 m)

3 Numerical Modeling and Projects

Two-dimensional finite element method simulations are carried out to investigate the interaction between two adjacent excavations under different construction plans. When the lengths of the parallel foundation pits are big enough, two-dimensional numerical simulation is convenient and can give reasonable analysis [16]. The 2D FEM code of Abaqus was adopted for the plane strain analysis for the interaction between two excavations.

An existing excavation called Excavation A is retained by a single row of reinforced bored piles (D = 1.0 m). The retaining wall is treated as a diaphragm wall of thickness of 0.8 m according to the stiffness equivalent. A new adjacent excavation called Excavation B is of the same type of retaining wall and the same thickness. The retaining wall is simulated by beam element, and the embedding ratio is 1:1. The reinforced concrete horizontal props are set up at 2 m below the ground level with the section of 0.5 m × 0.5 m. The elastic modulus of the reinforced concrete of the props is 20GPa and the Poisson's ration is 0.2, and also simulated as beam element. The depth of excavation A is assumed to be 13 m and the width is 8 m. The depth of excavation B is varied to simulate the influence of new excavation process, and the width is 8 m, too. The soil in the numerical model is homogeneous and simulated by CEP4 element. The bulk density of soil is 18.5 kN m^{-3} and other parameters are illustrated in Table 1.

Table 1. Soil mechanical property parameters

Soil	Depth/m	E/MPa	v	c/kPa	φ/(°)
Clay	100	25	0.3	0	20

The distance between the edge of the retaining wall and model boundary is set to be 5 times of the excavation depth. The distance between the excavation bottom and model boundary is set to be 6 times of the excavation depth. The left and right sides of the model are horizontally constrained, and for the bottom of the model, both horizontal and vertical movements are prescribed to be zero. Figure 7 gives the 2D element mesh for this study.

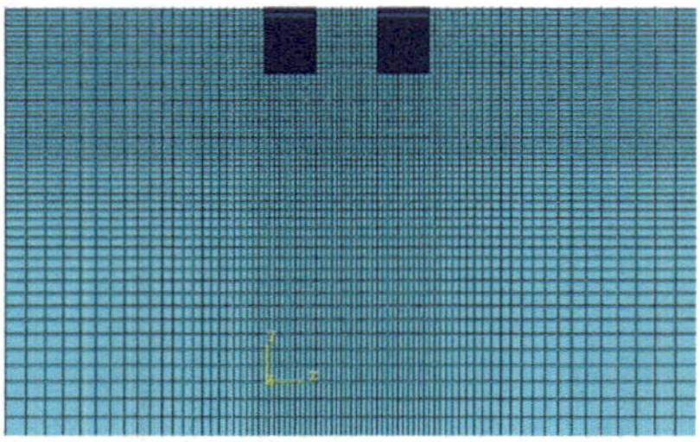

Fig. 7. 2D Finite element mesh model

The Mohr Coulomb model is used to describe the soil's behavior, but this model exhibits less power when dealing with the soil's characteristics of deformation, e.g. base heave can be increased seriously after excavating if the soil's strength parameters keeps constant, which makes the simulation result of base heave seem unreasonable if compared to the actual value. This can be solved by improving the elastic modulus of soil at the bottom of excavation. The elastic modulus of soil at the bottom of excavation is set to be three times of the value above bottom [17].

In this study, the focus is to explore the deformation response of retained excavation A to an adjacent excavation B. The varying parameters are the depth of Excavation B, spacing between two excavations, soil's strength, and the surcharge on soil surface between two adjacent excavations. When a special influencing factor is in study, the other parameters are kept constant.

4 Results and Analysis

4.1 Earth Pressure Distribution

Figures 8 and 9 give the earth pressure distribution along the retaining wall of Excavation A from numerical solution with $1H$ and $2H$ spacing from Excavation B, respectively. The earth pressure distribution curves from different theoretical solutions are also shown in Figs. 7 and 8 for comparison.

It can be seen from Fig. 7 that the earth pressure acting on the retaining wall of Excavation A is reduced after Pit B is excavated. But, with the increase of spacing, as in Fig. 9 where the spacing between two excavations has been increased from 1H to $2H$, the reduction of earth pressure on retaining wall is less than that in Fig. 7. It implies that when the spacing between two adjacent excavations exceeds a certain value - about $2H$ in this case, the adjacent excavation can only have a trivial and negligible influence on the earth pressure acting on the retaining wall of the pre-excavation.

The spacing by comparing the numerical results and the theoretical solutions as shown in Figs. 8 and 9, it can be seen that the solution proposed in this paper is more reasonable than other theoretical methods, although there is a difference between the numerical and theoretical solutions, which may be due to the cohesion of the soil and the displacement of the retaining wall constrained by the props and the embedded wall.

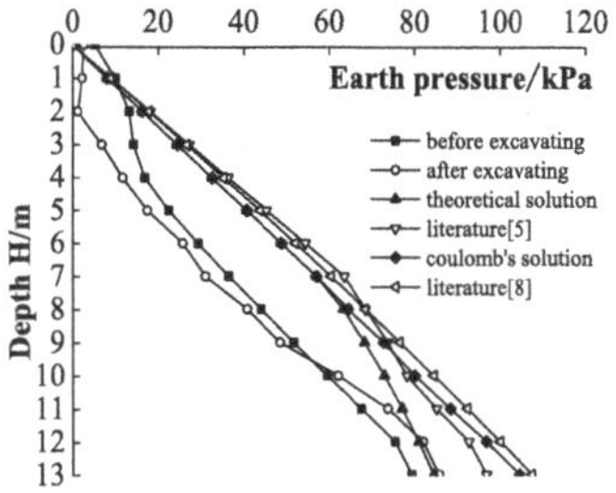

Fig. 8. Earth pressure on retaining wall-2 (Spacing 1*H*)

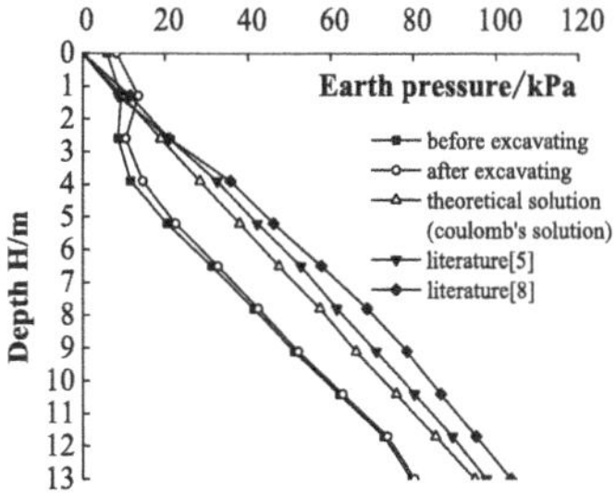

Fig. 9. Earth pressure on retaining wall-2 (Spacing 2*H*)

4.2 Effect of Spacing

Numerical simulations were carried out to investigate the influence of the spacing between two adjacent foundation pits. Three spacing was set up as being 1.0*H*, 2.0*H* and 3.0*H*. In order to identify the deformation of retaining walls, for the first constructed Excavation A, the retaining wall along the adjacent side is called wall-2, and the one along the opposite/further side is called wall-1.

Figure 9 gives the deformation curves of wall-1 and wall-2 before and after Excavation B is constructed with different spacing. It can be seen from Fig. 10 (a)–(c) that the deformation of wall-1 (on the further side) increase laterally when Excavation B is constructed, and the smaller the spacing is, the more increment occurs. Instead, it shows, from Fig. 10 (d)–(f), that the deformation of wall-2 changes in an opposite way, the smaller the spacing is, the more reduction of lateral deformation

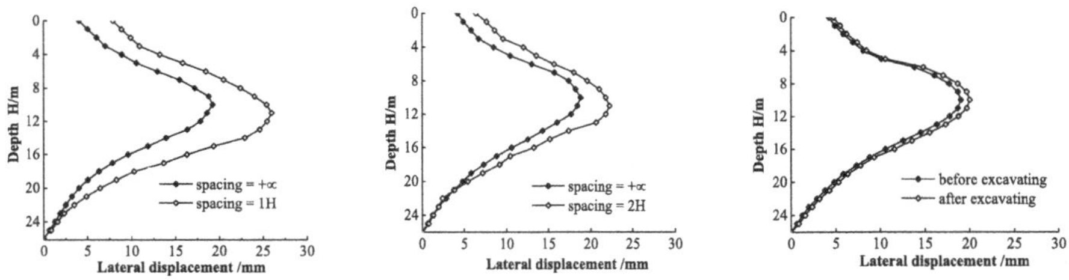

(a) Lateral displacement of wall-1 (b) Lateral displacement of wall-1 (c) Lateral displacement of wall-1

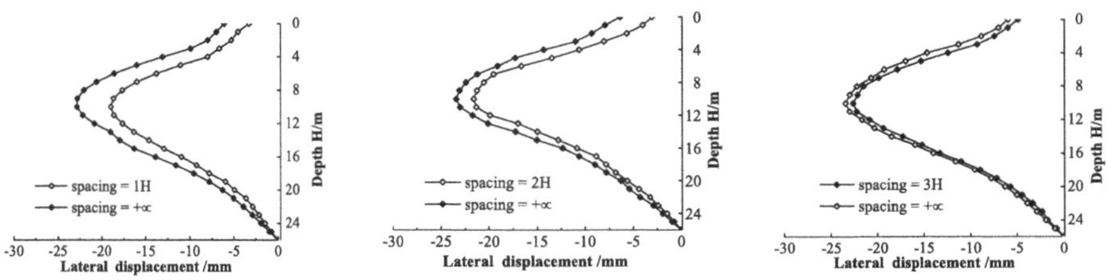

(d) Lateral displacement of wall-2 (e) Lateral displacement of wall-2 (f) Lateral displacement of wall-2

Fig. 10. The lateral displacement of the retaining wall of Excavation A happens. It seems that the whole retained Excavation A tends to tilt back to the adjacent Excavation B.

In Figs. 10(c) and (f), it is clearly shown that, when the spacing between Excavation A and Excavation B is 3*H* (*H* is the depth of both excavations), the lateral displacements of wall-1 and wall-2 change little when Excavation B is constructed, indicating a negligible effect from Excavation B on the retained Excavation A and a reasonable safe spacing of 3*H*.

4.3 Effect of the Pit Depth

The depth of adjacent excavation can have an influence on the safe spacing determination. A series of numerical simulations were carried out with different depths of Excavation B – 0.5*H,* 1.0*H* and 1.5*H*, when the spacing and depth of Excavation were kept constant. The spacing between two adjacent excavations was three times of the depth of Excavation A and the depth of Excavation A was *H*.

Figure 11 shows that the effect of Excavation B on the retained pre -constructed Excavation A is related to the depth of pit B, when the spacing of the two excavations keeps unchanged. It can be seen as well that when the depth of pit B is increased, its effect on Excavation A is also increased, especially when the depth of Pit B exceeds H, the depth of Excavation A. The maximum value of lateral displacement of wall-1 is increased from 19.1 mm to 21.2 mm, with an increment of about 11%, when the depth of Pit B varies from H to 1.5*H*. At the same time, the maximum displacement value of wall-2 is decreased from 23.2 mm to 19.6 mm, of about 16%. From Eq. (7), the depth of excavation within a safe spacing is constrained, therefore if the depth is increased, the interaction between two adjacent excavations is inevitable.

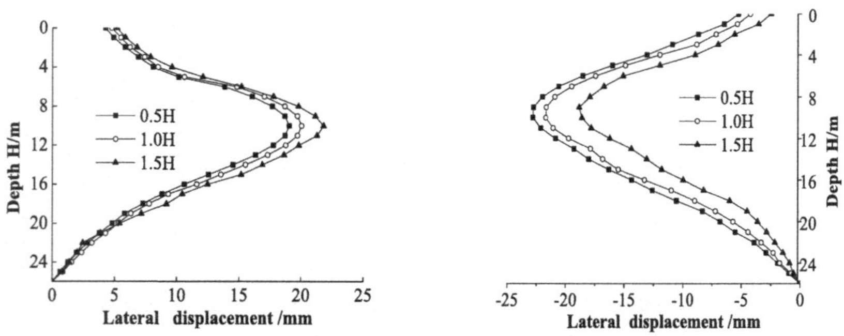

(a) Lateral displacement of wall-1 (b) Lateral displacement of wall-2

Fig. 11. The lateral displacement under different depth of Excavation B (Spacing = 3*H*)

4.4 The Effect from the Strength of Soil

The soil's strength parameters directly affect the angle of the slip surface (Eq. 5) and then on the value of the earth pressure acting on retaining wall. In this section, the

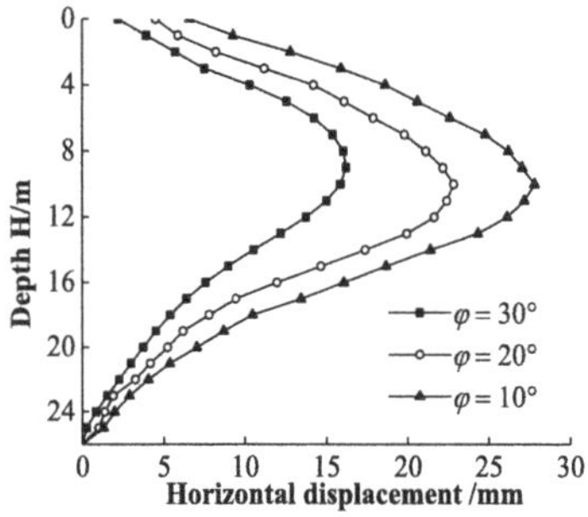

Fig. 12. The lateral displacement of wall-1 under different φ

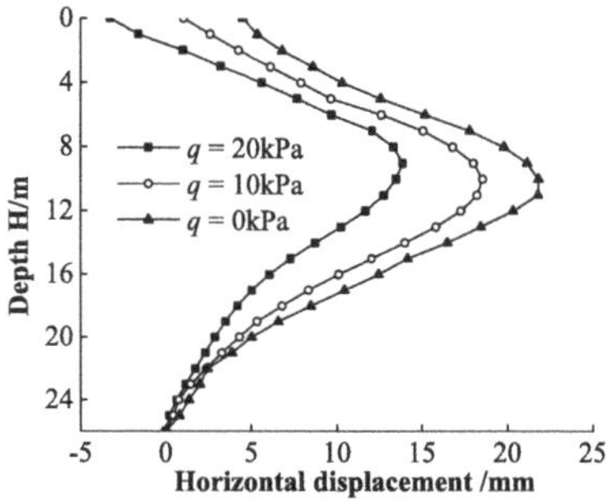

Fig. 13. The lateral displacement of wall-1 under different surcharge q

influence of φ on the interaction of two adjacent excavations is discussed by numerical simulations with $\varphi = 10°$, 20° and 30°. Figure 12 gives the change of lateral displacement of the retaining walls of Excavation A with different friction angles of soil, when the spacing of the two excavations is set to be $2H$.

It can be seen from Fig. 12 that the maximum value of lateral displacement of wall-1 with an adjacent excavation is decreased from 23.8 mm to 18.3 mm, when φ value of soil is changed from 30° to 10°. This indicates that the safe spacing between two adjacent excavations is related to φ, as compared in Figs. 4 and 5, and can be ensured by improving the strength of soil in between.

4.5 The Effect from Surcharge

In the construction of adjacent excavations, the soil between the foundation pits may be surcharged by transportation or by temporary stacked materials. In order to investigate the effect of the surcharge between two adjacent excavations on the deformation of the retaining walls, numerical simulations were carried out with different surcharge magnitudes, when the spacing in between was $2H$.

Figure 13 shows the lateral displacement distribution curve of wall-1 under different surcharge q. It can be seen that with the increase of the surface surcharge loading between the pits, the lateral displacement of Wall-1 is decreased, indicating a make-up to the influence of Excavation B and the safe spacing is ensured. However, the make-up by q is not reliable as the surcharge is usually not uniformly distributed.

5 Conclusion

By setting up the model of two adjacent excavations, and based on the theory of limit equilibrium theory, the active soil pressure on retaining wall due to adjacent excavation is studied by considering the friction between retaining wall and soil, and the surcharge on soil surface between two adjacent excavations. Thereby, the safe spacing b of two adjacent excavations is discussed by theoretical analysis. A parametric study was also carried out by finite element method simulations to explore the deformation response due to interaction. The conclusions are as follows:

(1) The safe spacing b is related to the strength parameter of soil, the depth of excavation, and the surcharge on soil surface between the two adjacent excavations. The interaction between two excavations can be observed by the change of the earth pressure acting on the retaining wall. If the safe spacing is defined as no interaction happens to the retaining walls, then, when the two adjacent excavations are of the same depth H, the suggested safe spacing is about $2H$.
(2) By FEM simulations, the influence of adjacent excavation on the pre-constructed one is investigated under different spacing, different depth of latter excavation, different soil's strength parameter, and different surcharge in between. It is found that, the deeper the adjacent excavations is, the more induced deformation of the retaining wall in the pre-constructed excavation occurs. This results show consistence with the theoretical analysis based on Eq. (6).
(3) For the sake of safety control of the adjacent retained excavations, the depths of the excavations should be concerned, and if necessary, the soil in between should be improved before the excavation and the surcharge on surface in between should be taken in serious if it is not uniformly distributed.

References

1. Huang, M.S., Wang, W.D., Zheng, G.: A review of recent advances in the underground engineering and deep excavations in soft soils. Chin. J. Geotech. Eng. **45**(06), 146–161 (2012)
2. Jin, Y.B., Liu, J.B.: Discussion on calculation method of earth pressure of earth-strip between neighboring excavations. Rock Soil Mech. **30**(12), 3759–3764 (2009)
3. Wang, H.L., Song, E.X., Song, F.Y.: Calculation of active earth pressure for limited soils between existing building and excavation. Eng. Mech. **31**(04), 76–81 (2014)
4. Liu, J.Z., Miao, F.: Computation of earth pressure for limited width soil by upper limit method. Chin. J. Undergr. Space Eng. **12**(S2), 577–581 (2016)
5. Ma, P., Qin, S.Q., Qian, H.T.: Calculation of active earth pressure for limited soils. Chin. J. Rock Mech. Eng. **27**(S1), 3070–3074 (2008)
6. Gao, Y.L.: The calculation of finite earth pressure. Build. Sci. **16**(05), 53–56 (2000)
7. Li, Y.F., Guo, Y.C.: Analytical study on active soil pressure from finite soil body in construction pit. Build. Sci. **24**(01), 15–18 (2008)
8. Yue, S.Q., Zuo, R.Y., Lu, Z.: A method for calculating active earth pressure of soil piece with a finite width between adjacent foundation pits. Rock Soil Mech. **37**(7), 2063–2069 (2016)
9. Ding, Z., Wang, J.Y., Zhou, Y.: Monitoring analysis of the interrelationship of synchronous construction in adjacent foundation pit. Chin. J. Geotech. Eng. **48**(S2), 124–130 (2015)
10. Chen, D.J.: Study on excavation displacement transformation of adjacent pits in shanghai south railway station. Build. Sci. **21**(06), 59–63 (2005)
11. Zhang, R.H.: Analysis of mutual impact of nearby foundation pits construct ion in deep and large foundation pit. Build. Constr. **30**(10), 869–871 (2008)
12. Lin, H.: Characteristic analysis of interaction effect of adjacent excavation. Fuzhou University (2013)

13. Xu, F.C.: The interactional deformation behavior of synchronized excavation of adjacent deep foundation pits. Huaqiao University (2015)
14. Zhang, Z.G., Huang, M.S., Dong, W.W.: Evaluation of deformation response for adjacent tunnels due to soil unloading in excavation engineering. Tunn. Undergr. Space Technol. **38**(9), 244–253 (2013)
15. Do, T.N., Ou, C.Y., Chen, R.P.: A study of failure mechanisms of deep excavations in soft clay using the finite element method. Comput. Geotech. **73**(4), 153–163 (2016)
16. Xu, L.: Finite element analysis of size effect of foundation. Tianjin University (2007)
17. Song, G., Song, E.X.: Selection of soil constitutive models for numerical simulation of foundation pit excavation. Eng. Mech. **31**(05), 86–94 (2014)

In-situ Experimental Investigation on Pullout Performances and Horizontal Bearing Properties of Bored Piles

Linze Shi and Changguang Qi(✉)

Faculty of Architecture, Civil Engineering and Environment,
Ningbo University, Ningbo, China
qichangguang@163.com

Abstract. By combination of practical engineering, the slow maintenance loading method was employed to conduct the field static load test studies of vertical pullout performance and horizontal bearing characteristics for single bored pile. According to the inclinometer pipes embedded in the bored pile during the construction, the variation laws of deep horizontal displacement of pile was investigated in the process of foundation pit excavation. The main conclusions can be summarized as: (1) in the vertical pullout test of single pile, the rates of deformation of pile top under the loading and unloading have some discrete and volatility; (2) in the horizontal static load test of single pile, the relationship between horizontal load and horizontal displacement is nonlinear; (3) the site monitoring shows that the horizontal deformation on the retaining pile top is large, and there is a significant inflection in the position where the steel supporting is installed. In addition, the most dangerous part is not on top of the retaining pile, but at 1/2 $\sim$ 5/6 depth of foundation pit.

Keywords: Bored pile · Pullout test · Horizontal static load test · Field test · Load transfer characteristics

1 Introduction

Excavation of deep foundation pit will cause a series of geotechnical problems and challenges, and especially in recent years, pile foundation bearing problem caused by excavation of deep foundation pit has gradually become a new focus (Sowa 1970; Grigoryan and Khabibullin 1980; Ismael 1989; Dyson and Randolph 2001; Shelke and Patra 2011; Papadopoulou et al. 2014; Bhardwaj and Singh 2015). The Yangtze River, as the first major river in China, has widely distributed floodplains and the geological and hydrological conditions are extremely complicated and special in the floodplain area. At present, there are many researches on the vertical bearing capacity of the bored pile in the floodplain area of Yangtze River; however, there is little research on the vertical uplift bearing capacity of the bored pile caused by the large foundation pit excavation in the floodplain area and the horizontal bearing capacity of single pile (He 2004; Lu et al. 2005; Xu et al. 2009; Wang et al. 2007; Wang et al. 2017). Therefore, the bored pile is subjected to single pile uplift and horizontal static load test and the deformation law of the retaining pile is monitored on site in combination with a

R. Chen et al. (eds.), *Proceedings of the 2nd International Symposium on Asia Urban GeoEngineering*, Springer Series in Geomechanics and Geoengineering,
https://doi.org/10.1007/978-981-10-6632-0_18

cross-river channel connecting line project so as to study the uplift and horizontal bearing properties of borehole bored piles in the floodplain area of Yangtze River.

2 Overview of Support Project and Test Site

2.1 Project Overview

The total length of the cross-river channel is about 2.5 km, of which the length of the under-river tunnel is about 1.2 km. The foundation pit layout is as shown in Fig. 1. The bored pile is located in the B1 area with the safety level of grade one and the environmental protection level of grade two. The support structure of foundation pit is mainly composed of underground continuous wall, bored pile + cement soil mixing pile, SMW pile and gravity retaining wall of cement soil mixing pile. The support adopts horizontal reinforced concrete support combined with steel support, bracing and angle brace.

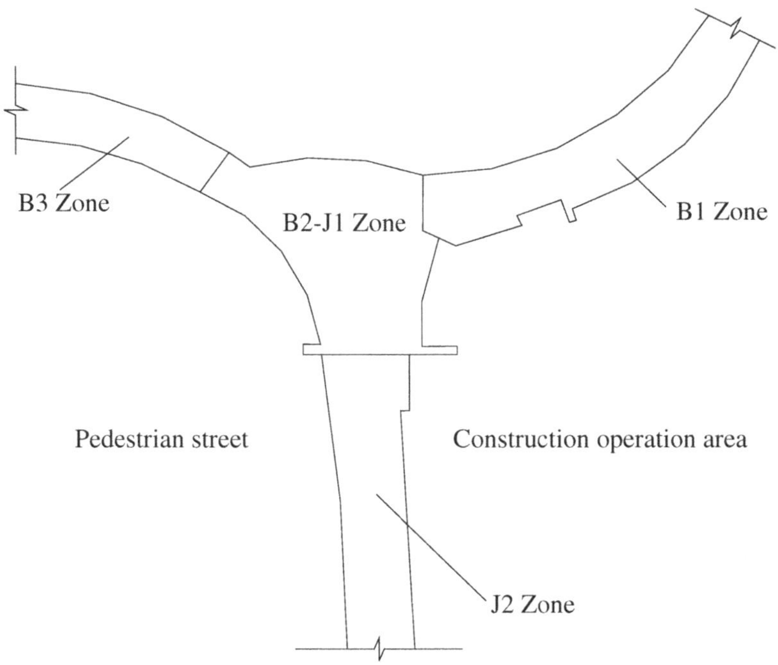

Fig. 1. Foundation pit layout

2.2 Geological Condition of Test Site

This survey reveals that all the upper parts of the stratum are Quaternary unconsolidated sediment, the underlying white subfamily bedrock; the stratum within the survey depth cam be divided into five layers according to the forming courses, burying depth and rock weathering degree of rock-soil layer, and then sub-divided into a number of sub-layers according to the lithological characteristics and physical and mechanical properties of rock-soil layer; the specific physical and mechanical parameters of each soil layer are shown in Table 1.

Table 1. Physical and mechanical parameters of soil layer

Layer no.	Name of soil layer	Thickness of soil layer(m)	w/ %	γ/ (kN/m^3)	e	E_s/ MPa	N/ Hit
①	Filling	3.5	32.6	19.3	0.870	4.6	8
②$_1$	Clay, Silty Clay	1.5	35.0	18.8	0.969	3.6	5
②$_2$	Muddy Silty Clay	9.0	42.0	18.0	1.173	3.0	3
②$_3$	Silty Clay with Silt	3.6	35.8	18.2	1.023	4.0	7
③	Silt	5.8	26.5	19.4	0.748	12.3	12
④	Silty Clay	3.0	33.0	18.7	0.960	4.8	16
⑤$_1$	Silty-fine Sand	4.0	24.7	19.3	0.737	15.7	42
⑤$_2$	Medium-coarse Sand	5.3	17.0	20.0	0.572	18.3	38

Note: w is the moisture content, γ is the soil's unit weight, e is the porosity ratio, E_s is the compression modulus, and N is the standard penetration number.

2.3 Vertical Uplift Static Load Test of Single Pile

The equipment layout of single pile vertical uplift static load test is as shown in Fig. 2. The uplift system is composed of reaction beam, jack, reaction support and anchor; the reaction support is C30 reinforced concrete and the vertical distance from its centre to

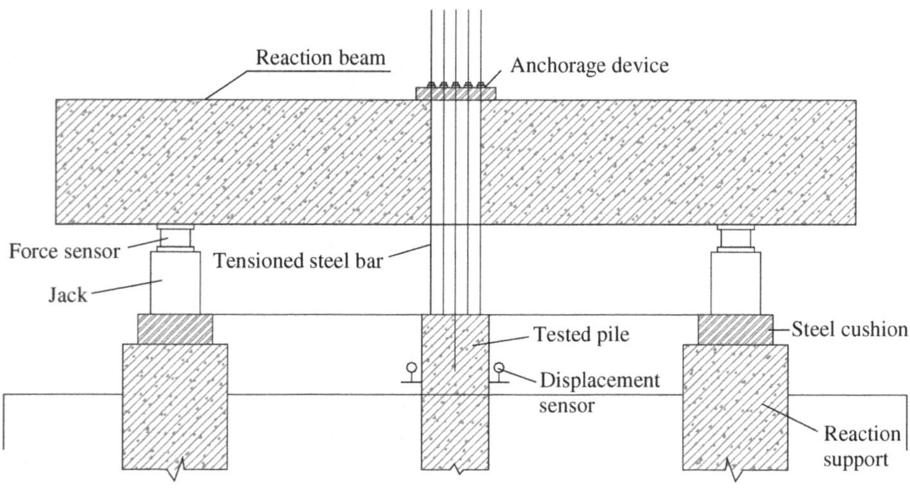

Fig. 2. Schematic diagram of reaction system layout of single pile vertical uplift test

the test pile is 1.8 m. Reaction beam is 4 m in length and the output equipment uses two 1500 kN jacks specific for rock.

The test adopts slow maintain load method, using two reaction supports and setting the jack to add load. The test uses JCQ-503C automatic observation system, which can automatically add pressure, stabilize pressure, do reading, record and control stability standard. KB20 #, KB34 #, KB40 # piles are selected to be carried out load test with the design parameters of piles as shown in Table 2. KB20 # test pile shall be loaded to 351 kN and KB34 # and KB40 # test piles shall be loaded to 1066 kN, which shall be stabilized before being unloaded.

Table 2. Design parameters of pile

Pile No.	Pile Length/m	Pile Diameter/mm	Characteristic Value of Uplift Bearing Capacity of Single Pile/kN
KB20 #	20	1000	350
KB34 #	25	1000	530
KB40 #	25	1000	530

2.4 Horizontal Static Load Test of Single Pile

The layout section of reaction system test equipment of single pile horizontal static load test is as shown in Fig. 3, and C52 #, C57 #, C63 #, D1 #, D10 # and D15 # piles are selected to carry out the horizontal static load test.

In the pile horizontal static load test, jack is used to apply horizontal force to act at about 0.8 m below the pile top and the horizontal force is applied to the pile axis through the spherical hinge; single multi-cycle loading and unloading method is used; the loading force at each stage shall be 1/10 of the vertical limit compression bearing capacity of the single pile of 1/10; the first stage is twice the step load; loading shall continue until the design requirement is reached. After load is applied at each stage, displacement readings shall be observed at 5, 30, 45, 60 min; afterwards, readings shall be observed every 30 min until it reaches relative stability. The unloading value at each stage is 2 times the load value at each stage. After unloading at each stage, the readings of residual displacement shall be observed once every; after reading twice, read after 30 min and then unload the load at the next stage; after the unloading of all loads, the maintenance time shall be 3 h; reading time is the 15, 30, 60 min, and afterwards, the reading shall be observed once every 30 min.

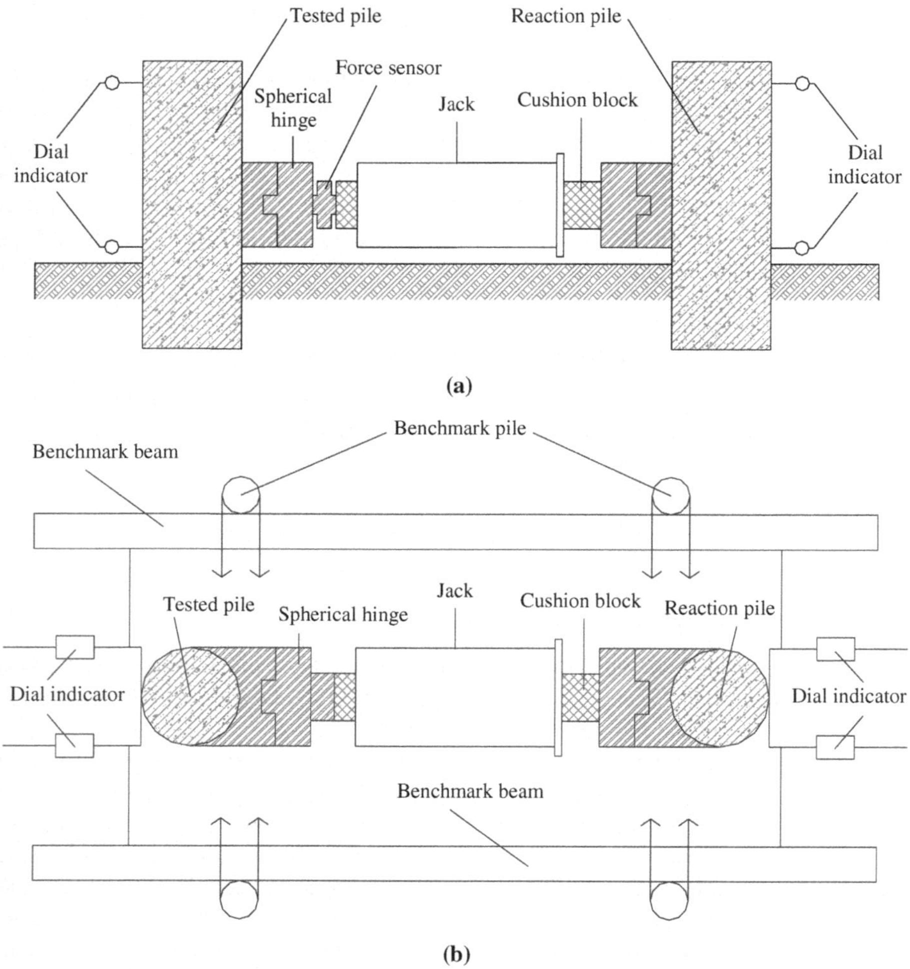

Fig. 3. Schematic diagram of reaction system layout of single pile horizontal static load test: (a) Profile; (b) Top

2.5 On-site Monitoring

Two bored piles in the MK0 + 168 ~ 197.8 section of B1 area of underground traffic project are used as the monitoring object so as to learn the lateral deformation of the different side walls of the foundation pit; the location of the monitoring pile is as shown in Fig. 4. The installation section of retaining pile detection equipment is as shown in Fig. 5; ZL1-2, ZL2-2 and ZL3-2 respectively represent the support in steel pipe.

In the construction of bored pile of foundation pit support, the PVC inclinometer pipe with outer diameter of 70 mm and inner diameter of 59 mm shall be bound at the lateral side of the reinforcement cage in the design position as shown in Fig. 6. The bottom and top of the inclinometer pipe are slightly shorter than the length of the

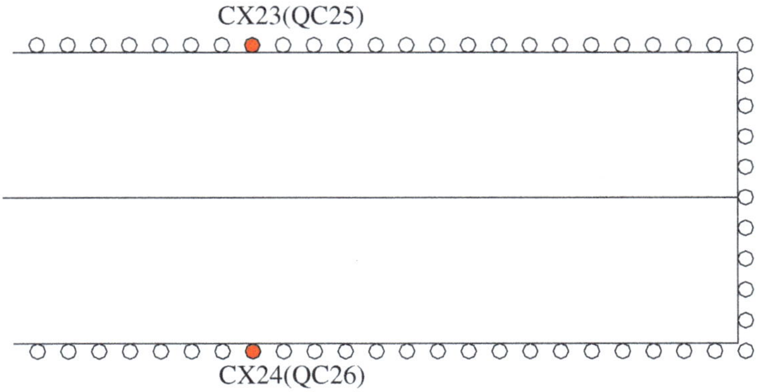

Note: 1) QC26 and QC25 are the number of the pile where the monitoring point is located
2) CX23 and CX24 are the monitoring point number of the piles

Fig. 4. Monitoring point layout

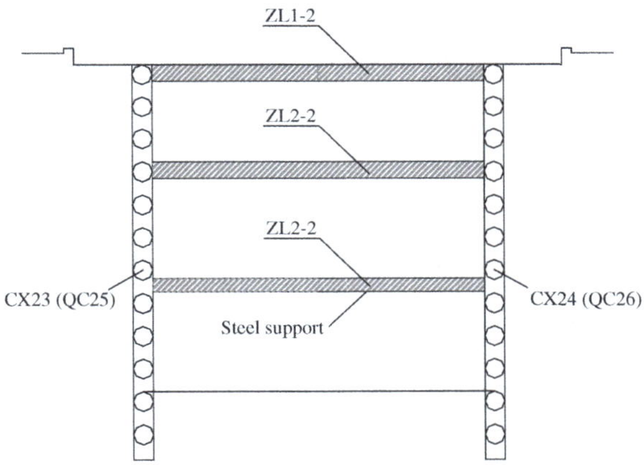

Fig. 5. Installation section of retaining pile detector t

reinforcement cage by about 5 cm. At the same time, it is necessary to ensure that there must be a group of cross guide grooves of the inclinometer pipe perpendicular to the sideline of foundation pit; the outer side of inclinometer pipe at the position 0.5 m above and below the excavation surface of top beam of the support shall be added with Φ109PVC casing protection. A sampling point shall be set every 0.5 m to collect the initial data at each point of the conduit and the data at each point shall be collected according to the construction schedule.

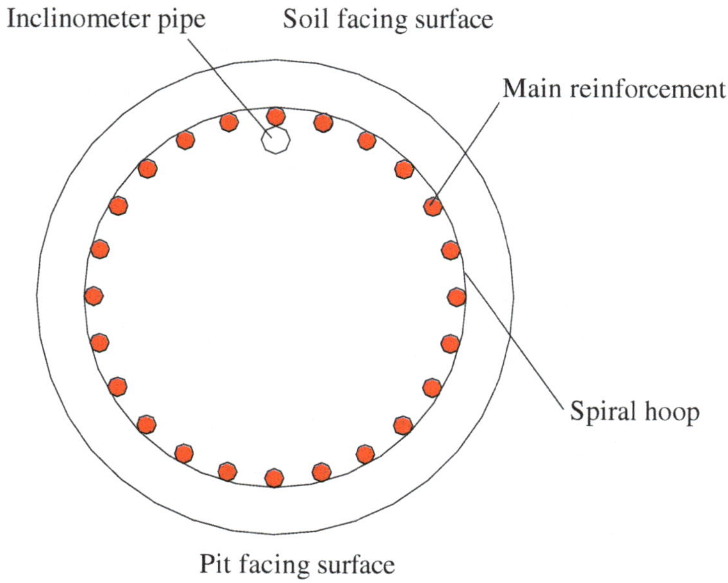

Fig. 6. Embedded diagram of inclinometer hole of bored pile

3 Test Data Analysis

3.1 Analysis of Vertical Uplift Static Load Test Result of Single Pile

Figure 7 is the pull force-uplift amount curve of the KB20 #, KB34 #, KB40 # test piles. It can be seen that the load curve from the first stage load to the fifth stage is relatively gentle and after the fifth stage of load, the curve becomes steep, indicating that with the increase of load, the friction at lower part of the pile plays a significant role with the test results summarized in Table 3; it can be seen that the bearing characteristics of the uplift piles are similar to the compressive piles in the working mechanism. Under the action of the pile top load, elastic deformation occurs to the pile and there's relative displacement between the pile side and the soil around the pile; the friction on the side of pile will be gradually displayed and the axial force will be transferred from top to bottom.

Figure 8 is the time-uplift amount curve of the KB20 #, KB34 #, KB40 # test piles. It can be seen that the change of uplift amount under the action of load at the initial stage is relatively slow, and the change of uplift amount under the action of load at the later stage is relatively quick. When the load reaches the design load unloading process, the resilience value will gradually increase with time, and stabilize in the end.

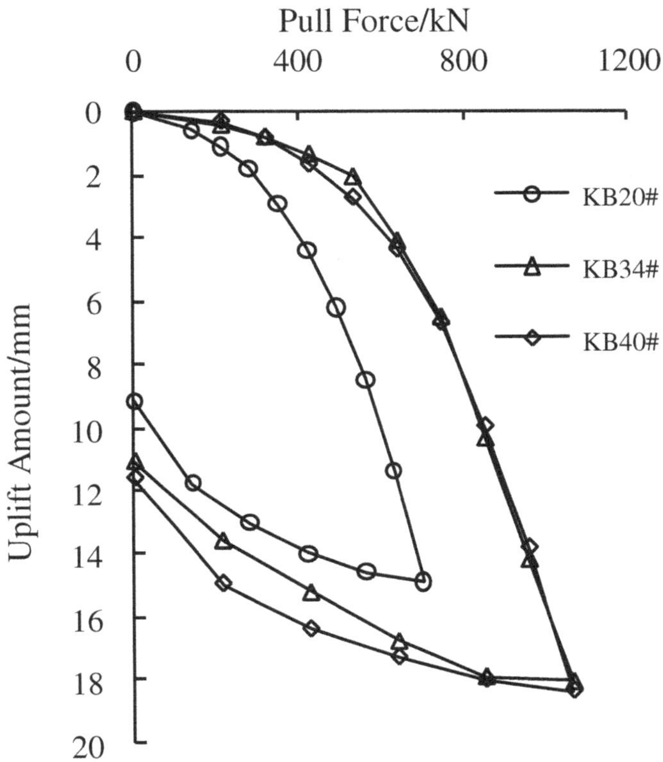

Fig. 7. Single pile load – uplift amount change curve

Table 3. Summary of vertical static load test results

Pile no.	Characteristic value of uplift bearing capacity of single pile/kN	Maximum load/kN	Pile top uplift amount/mm	Resilience value/mm
KB20 #	350	700	15	6
KB34 #	530	1060	18	7
KB40 #	530	1060	18	7

3.2 Analysis of Horizontal Static Load Test Results

The change curves of horizontal displacement with horizontal load of the 6 test piles are as shown in Fig. 9. It can be seen that the horizontal displacement also increases with the increase of the horizontal load; however, due to the large pile diameter, the increase of load with the increase of the horizontal displacement at each stage is not very large. When loaded to the 4th stage load, the horizontal displacement increases with the increase of horizontal load. And the horizontal displacement and the horizontal load are nonlinearly changed throughout the loading process. Under the action of horizontal force, the bored pile will have horizontal displacement, and at the same time,

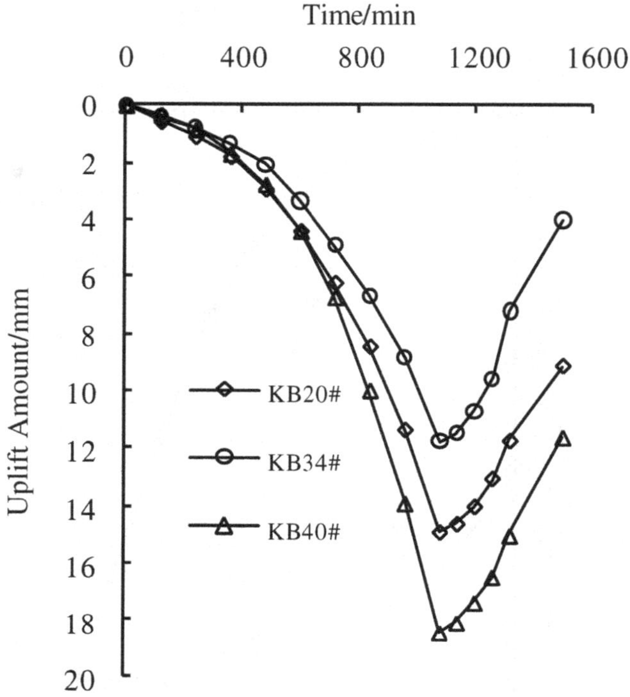

Fig. 8. Single pile uplift amount and time change curve

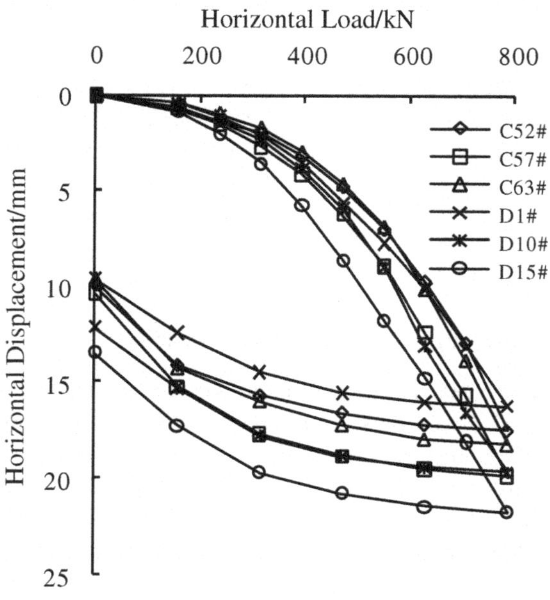

Fig. 9. Single pile horizontal load action curve

Table 4. Summary of test pile horizontal static load test results

Pile no.	Maximum test load/kN		Horizontal displacement/mm	
	Required	Actual	Accumulated	After unloading
C52 #	790	790	18	10
C57 #	790	790	20	10
C63 #	790	790	18	10
D1 #	790	790	16	10
D10 #	790	790	20	12
D15 #	790	790	22	13

small rotation or deflection will occur to the pile. The test results are summarized in Table 4.

Figure 10 shows the time-level displacement curves of C52 #, C57 #, C63 #, D1, D10, D15 test piles. It can be seen that the horizontal displacement change at the initial stage of the load is relatively slow, and the horizontal displacement change at the later stage of the load is relatively fast. In the unloading process after horizontal load reaches the design load, the resilience value of horizontal displacement gradually increases with the change of time, and finally becomes stabilized.

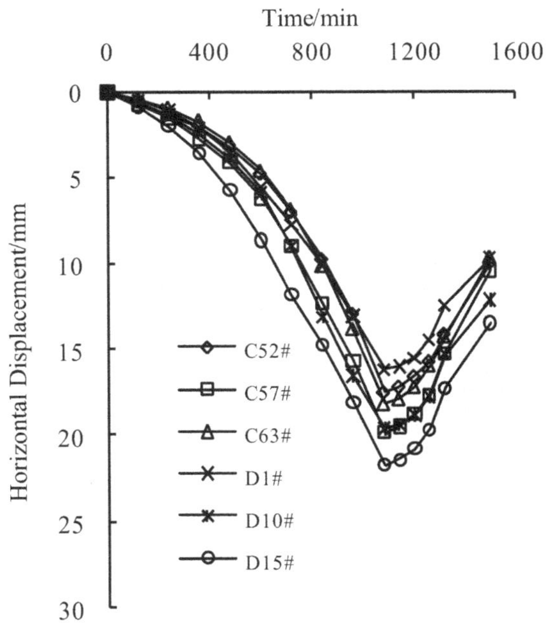

Fig. 10. Time change curve of single pile horizontal displacement

3.3 On-site Monitoring and Analysis of Retaining Piles During Construction Period

The data of the foundation pit shall be analysed selecting the representative location and the data at the measuring points of CX23 and CX24 shall be analysed. The deformation diagram of the key position of the retaining pile of No.CX23 measuring point is as shown in Fig. 11. It can be seen that the retaining pile is deformed and tilted outward at the position of the crown beam with the relatively small amplitude throughout the construction process, tilted towards the foundation pit at the position where the second and third steel supports are installed in the construction process and gradually increased 51 days after the excavation, which, however, tends to be stable with the progress of the project. The bottom of the foundation pit has a large deformation at the beginning of the excavation and is tilted towards the outside of the foundation pit, but is tilted toward the foundation pit 51 days after the excavation and remains stable. The bottom of the retaining pile is tilted toward the foundation pit during the excavation, but remains stable all the time. At the measuring point where the third layer of steel is installed, the tilting value is the largest among all the measuring points, which increases drastically from the 46th day to the 55th day after excavation. According to the construction record, the earthwork proceeds to the position of the third layer of steel support at this time and the foundation pit is excavated to the bottom during this period; due to the unloading of the soil mass, the retaining pile is increased and deformed during this period; due to the fact that the upper part of the soil has formed the stable state with the acting force of the soil mass, in addition, the third-layer steel support has just been installed, therefore, the maximum deformation will occur in this position. In the 115th day after excavation, the deformation of this position reaches the maximum of about 35 mm, considering that the third-layer steel support has just been dismantled at this time, deformation suddenly increases as a result. However, with

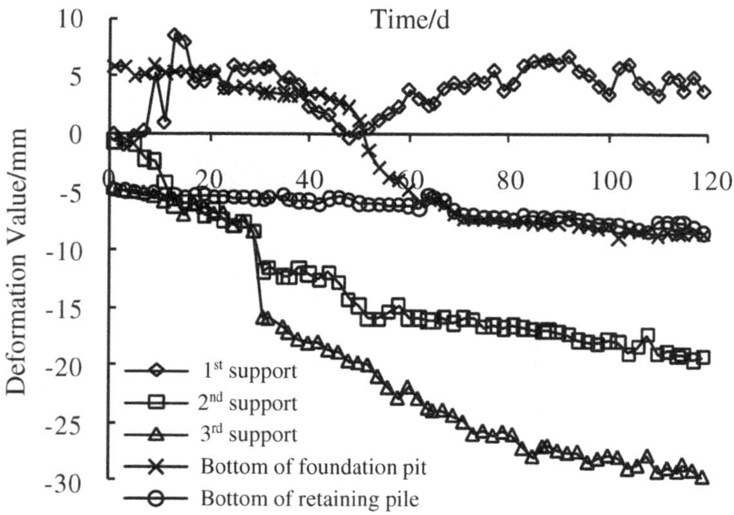

Fig. 11. Variation diagram of the variation of deformation with time at CX23 measuring point

the pouring of the side wall, the displacement will quickly be reduced, and returned to the safe range. The deformation of the retaining structure is closely related to the construction process of the foundation pit. The adoption of different construction process and the varied exposure time of soil mass after excavation will have different effects on the stability of the foundation pit. When such effects are reflected to the supporting structure, the deformation of the supporting structure varies with the excavation process.

The deformation values of the key sections of the retaining pile at the measuring points of CX23 and CX24 are as shown in Figs. 12 and 13. It can be seen that the analysis of the variation of the horizontal displacement of the key position of CX23 and CX24 piles with the construction process shows that under the supporting effect, the observation of horizontal displacement at different depth of the retaining pile can reflect the actual deformation of the retaining pile and its measurement is less affected by the outer circumstances and the data results are more stable, which is the key project in the monitoring of foundation pit construction. Therefore, in the project, the pile top and the horizontal displacement of the pile body shall be comprehensively analyzed and measuring results shall be mutually checked so as to fully grasp the displacement of the entire foundation pit. In addition, the largest and most dangerous part of the deformation of the retaining pile is not at the top of the pile, but at 1/2 ~ 5/6 of the depth of the foundation pit, that is, between 9 m ~ 15 m.

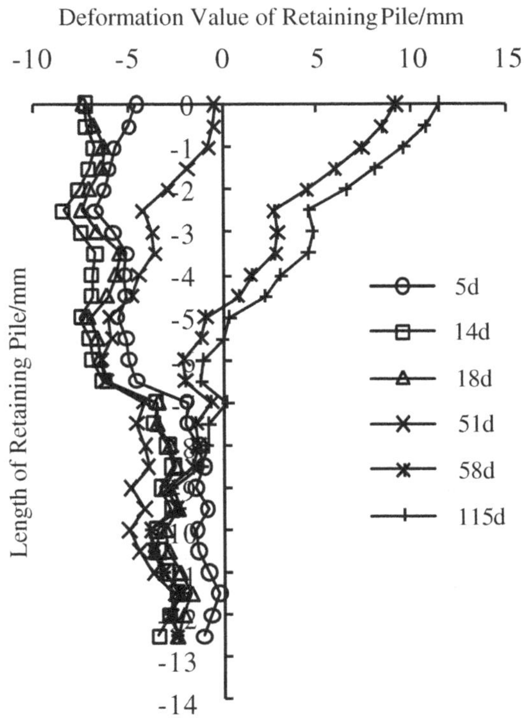

Fig. 12. Deformation curve of CX23 retaining pile

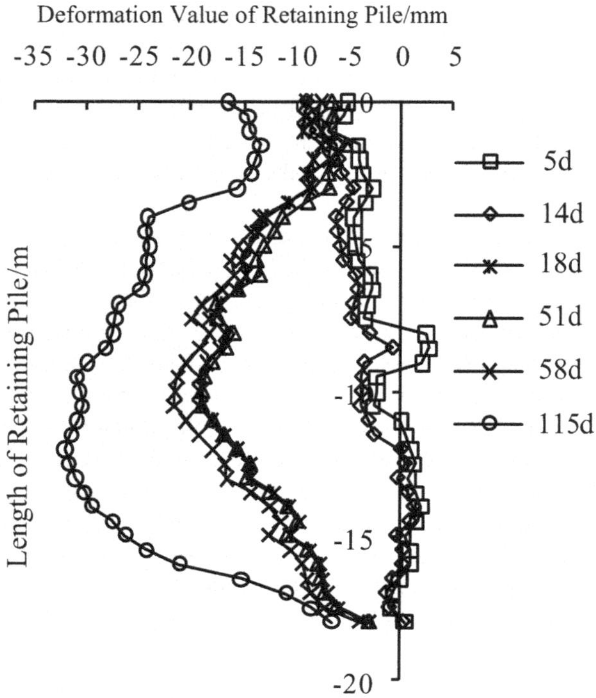

Fig. 13. Deformation curve of CX24 retaining pile

4 Conclusion

(1) Under the action of uplift load, the change of uplift amount is relatively gentle at the initial stage of the load; with the increase of the load, the change of uplift amount is bigger and the friction force of the lower part of the pile gradually plays a significant role; the uplift amount is rebounded about 40% after the unloading of the load.

(2) Under the horizontal force, the horizontal displacement increases with the increase of the horizontal load, but because of the large pile diameter, the load at each stage does not have a very large increase of the horizontal displacement. When a certain load is loaded, with the increase of the horizontal load, the increase of horizontal displacement will be relatively large. And the horizontal displacement and the horizontal load are nonlinearly changed throughout the loading process. Under the action of horizontal force, the bored pile will have a horizontal displacement, and at the same time, the pile body will have a small rotation or deflection.

(3) The horizontal deformation at the top of the retaining pile is large, and there is obvious inflection in the position where the steel support is installed. With the earth excavation, the horizontal deformation of the middle part of the retaining pile gradually increases, the horizontal deformation of the pit bottom is small, and the spatial effect of the foundation pit is obvious. The maximum and most

dangerous part of the retaining pile is not at the top of the pile, but at the 1/2 ~ 5/6 of the depth of the foundation pit, that is, between 9 m and 15 m.

Acknowledgements. The work was supported by Natural Science Foundation of Ningbo City (Plan No.: 2017A610317) and National Natural Science Foundation of China (Project No.: 51508282).

References

Bhardwaj, S., Singh, S.K.: Influence of load obliquity on pullout capacity of micropile in sand. Indian Geotech. J. **45**(2), 1–9 (2015). doi:10.1007/s40098-014-0123-5

Dyson, G.J., Randolph, M.F.: Monotonic lateral loading of piles in calcareous sand. J. Geotech. Geoenvironmental Eng. **127**(4), 346–352 (2001). http://dx.doi.org/10.1061/(ASCE)1090-0241(2001)127:4(346)

Grigoryan, A.A., Khabibullin, I.I.: Experimental investigation of the distribution of stresses in large bored situ-cast piles. Soil Mech. Found. Eng. **17**(3), 102–106 (1980). http://dx.doi.org/10.1007/BF01710396

He, J.: Testing study on anti-draw bearing properties of cast-in-place rock-socked pile. Chin. J. Rock Mech. Eng. **23**(2), 315–319 (2004). (In Chinese). http://dx.doi.org/10.3321/j.issn:1000-6915.2004.02.028

Ismael, N.F.: Field tests on bored piles subject to axial and oblique pull. J. Geotech. Eng. **115** (11), 1588–1598 (1989). http://dx.doi.org/10.1061/(ASCE)0733-9410(1989)115:11(1588)

Lu, F.R., Chen, R.P., Chen, Y.M., Ying, J.G.: Field tests on compression and uplift behavior of micropiles in soft ground. China Civil Eng. J. **38**(3), 99–105 (2005). (In Chinese). http://dx.doi.org/10.3321/j.issn:1000-131X.2005.03.017

Papadopoulou, K., Saroglou, H., Papadopoulos, V.: Finite element analyses and experimental investigation of helical micropiles. Geotech. Geol. Eng. **32**(4), 949–963 (2014). http://dx.doi.org/10.1007/s10706-014-9771-6

Shelke, A., Patra, N.R.: Effect of compressive load on uplift capacity of cast-in-situ bored piles. Geotech. Geol. Eng. **29**(5), 927–934 (2011). http://dx.doi.org/10.1007/s10706-011-9423-z

Sowa, V.A.: Pulling capacity of concrete cast in situ bored piles. Can. Geotech. J. **61**(7), 482–493 (1970). http://dx.doi.org/10.1139/t70-060

Wang, D.Y., Lan, C., He, G.C., Wang, C.Z.: Research on lateral support behavior of large diameter rock-socketed cast-in-place piles at river port by laboratory test. Chin. J. Geotech. Eng. **29**(9), 307–1313 (2007). (In Chinese). http://dx.doi.org/10.3321/j.issn:1000-4548.2007.09.005

Wang, X.Q., Cui, Y.L., Zhang, S.M., Shi, Q.T., Li, B.: Field test investigation on bearing characteristics of diaphragm walls in high pressure water foundation of YangYze River floodplain. Chin. J. Rock Mech. Eng. **36**(3), 773–780 (2017). (In Chinese). http://dx.doi.org/10.13722/j.cnki.jrme.2016.0047

Xu, Z.H., Wang, J.H., Wang, W.D.: Deformation behavior of deep excavations retained by bored pile wall in soft soil. Rock Soil Mech. **30**(5), 1362–1366 (2009). (In Chinese). http://dx.doi.org/10.3969/j.issn.1000-7598.2009.05.031

On Methods of Estimating the Base Stability of Excavations in Clay

T.N. Do(✉)

Thuyloi University, Hanoi, Vietnam
dotuannghia@tlu.edu.vn

Abstract. This study investigated three conventional methods in estimating stability of wide excavations in clay, including the load factor method, the strength factor method, and the slip circle method. Estimates of the factor of safety and required wall embedded depth were compared with those from the finite element method (FEM) using strength reduction technique, which were herein treated as the standard values. Results showed that wall friction had a pronounced effect on the factors of safety by the load factor method and the strength factor method but the slip circle method. For constant s_u/σ_v' (normalized undrained shear strength) case, the H_p/H_e (wall embedded depth to excavation depth) ratios determined by the three hand-calculation methods were smaller than those by the FEM as N_b (the stability number) < 7.0 but larger as $N_b \geq 7.0$. For constant s_u (undrained shear strength) case, as increasing s_u, the H_p/H_e ratios determined by the load factor method were illogical. Estimates by the strength factor method and the slip circle method were larger than those by the FEM as N_b < 4.5 but smaller as $N_b \geq 4.5$.

Keywords: Excavation · Stability analysis · Finite element method

1 Introduction

So far, the load factor and the strength factor methods have been often recommended by CIRIA report 104 (Padfield and Mair 1984) and BS 8002 (British Standard Institute 1994) for estimating stability of an embedded retaining wall. On the other hand, the slip circle method has been widely employed to estimate stability of excavations in clay in most of Asian countries. The method is advocated by TGS (Taiwanese Geotechnical Society 2001) and JSA (Japanese Society Architecture 1988). Very recently, the finite element method (FEM) with reduced shear strength has been significantly improved to analyze the stability of braced excavations. The method has many advantages over the conventional hand-calculation ones, i.e. the failure surface of soil comes out naturally without any assumptions, etc. Do et al. (2016) demonstrated that the method in associated with modelling an elastoplastic support system was capable of predicting failure mechanism of real excavation cases in clay. Also, stability of excavations, which was strongly related to that of walls, could be estimated using the method.

In this study, the load factor method, the strength factor method, and the slip circle method will be examined in estimating stability of wide excavations in clay. Excavations were retained by the wall with multiprop levels. Variations of wall penetration,

R. Chen et al. (eds.), *Proceedings of the 2nd International Symposium on Asia Urban GeoEngineering*, Springer Series in Geomechanics and Geoengineering,
https://doi.org/10.1007/978-981-10-6632-0_19

soil strength (s_u/σ_v' and s_u), and friction between wall and soil were taken into account. Results of the above three methods were compared to those from the FEM with reduced shear strength and consideration of the elastoplastic behavior of the support system. The results from the FEM was treated as the standard method. A large number of failure and successful excavations in clay were also adopted for further validation.

2 Methodology

2.1 Load Factor Method and Strength Factor Method

The methods were originally established for the wall with one prop level. In this study, they would be extended for the multipropped wall.

The load factor method estimates the stability of the wall under effects of lateral earth pressures in the active and passive zones. Considering the short-term behavior of the subsoil (saturated clay), these extreme earth pressures are calculated as follows:

$$\sigma_a = \sigma_v - 2s_u\sqrt{1 + c_w/s_u} \quad (1)$$

$$\sigma_p = \sigma_v + 2s_u\sqrt{1 + c_w/s_u} \quad (2)$$

where σ_a and σ_p = active and passive earth pressures, respectively; σ_v = overburden pressure; s_u = undrained shear strength of soil; c_w = adhesion between wall and soil.

As shown in Fig. 1, this method takes into account the wall part from the lowest strut level to the wall toe in equilibrium analysis. The factor of safety of stability (F_{LF}) is calculated using the following equation:

$$F_{LF} = P_p L_p / P_a L_a \quad (3)$$

where P_p and P_a = resultant forces of the passive and active earth pressures; L_p and L_a = distance from the lowest strut level to the acting point of P_p and P_a;

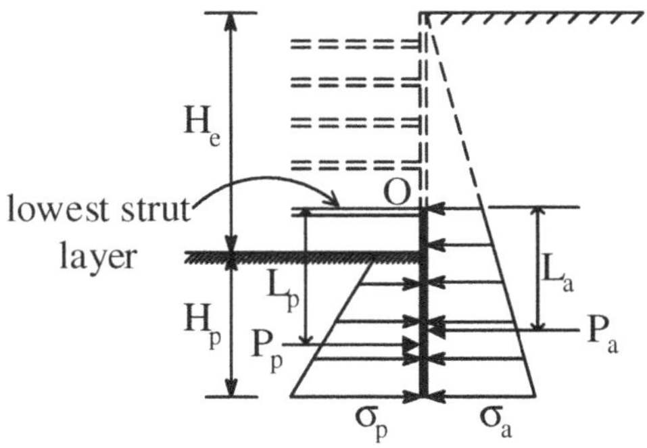

Fig. 1. Load factor and strength factor methods

The strength factor method considers stability of the same wall part but defines the maximum coefficient applied to reduce soil strength at which equilibrium state remains as the factor of safety (F_{SF}):

$$F_{SF} = s_{u,org}/s_{u,lim} \tag{4}$$

where $s_{u,org}$ and $s_{u,lim}$ = undrained shear strength of soil at the beginning and the limiting equilibrium state.

2.2 Slip Circle Method

The slip circle method is performed using the following procedure: (1) assume a trial circular failure surface; (2) calculate the ratio of the resistant moment to the driving moment; (3) repeat steps (1) and (2) until the smallest ratio is obtained. Then, the method will treat the smallest ratio as the factor of safety (F_{SC}) and the corresponding surface as the failure surface of the excavation. Since the failure surface often centers at the lowest strut level, one of the trial surfaces can be observed in Fig. 2. The ratio of the resistant moment to the driving moment is calculated as follows:

$$\frac{M_r}{M_d} = \frac{R\int_0^{\left(\frac{\pi}{2}\right)+\alpha} s_u(Rd\theta)}{W\left(\frac{R}{2}\right)} = \frac{\int_0^{\left(\frac{\pi}{2}\right)+\alpha} s_u(Rd\theta)}{W/2} \tag{5}$$

where R = radius of the trial surface; s_u = undrained shear strength of soil; W = self-weight of the soil body within the DEFG area.

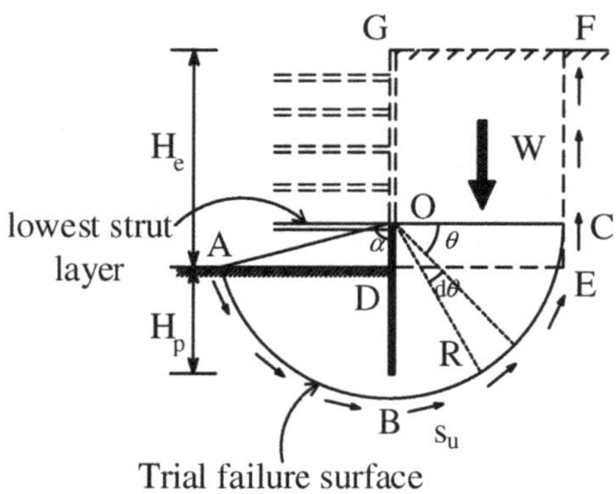

Fig. 2. Slip circle method

2.3 FEM with Reduced Shear Strength

The FEM with reduced shear strength methods makes use of the computer program to estimate the stability of the excavation. It follows that the strength parameters of soil will be reduced constantly until numerical solutions diverge. Divergence of numerical solutions is defined as failure of the excavation and the maximum SR ratio is the factor of safety (FFEM) of the excavation. Details of the method can be seen elsewhere, e.g. Do et al. (2016).

2.4 Excavation Geometry, Construction Sequence, and Soil Profile

Figure 3 plots the excavation geometry used in this study. The final excavation depth was 18 m and reached within six stages. The excavation width was assumed to be very large (i.e., 150 m) in order to avoid the overlapping effect of the failure surfaces of soil below the excavation bottom. The subsoil was a thick deposit of clay so that the influence of the hard stratum was eliminated. The ground water table was located on the ground surface.

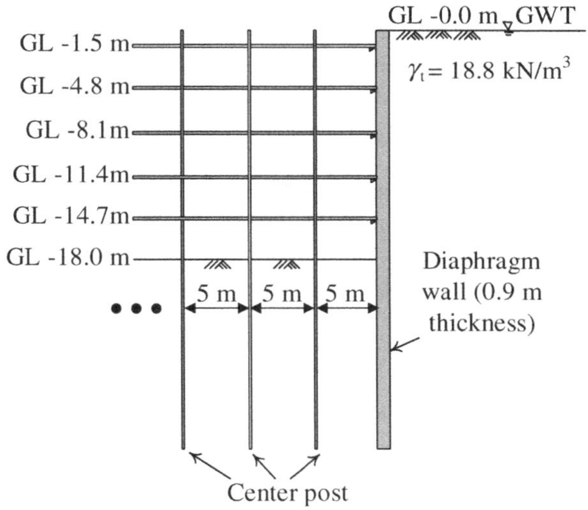

Fig. 3. Excavation geometry

As shown in Fig. 3, the support system was composed of a 0.9-m-thick diaphragm wall and five levels of horizontal struts. It was assumed that the horizontal spacing of struts was 6 m and that of center posts was 5 m. Dimensions of the structural elements used to support the typical excavations was summarized in Table 1. Details of the selection of the structural elements were described in the study by Do (2016).

Effects of wall embedded depth and wall friction were studied by varying H_p/H_e (wall embedded depth/excavation depth) and c_w/s_u, respectively. To examine the effects of the undrained shear strength of soil, s_u/σ_v' and s_u values were varied. The excavations were modelled according to the way described by Do (2016).

Table 1. Dimensions of structural elements

Strut layer	s_u/σ_v' = const			s_u = const (kPa)		
	0.18	0.22	0.30	70	80	90
1	H350x350x12x19	H350x350x12x19	H350x350x12x19	H250x250x9x14	H200x200x8x12	H200x200x8x12
2	4H250x250x9x14	2H350x350x12x19	2H350x350x12x19	H350x350x12x19	H350x350x12x19	H350x350x12x19
3	2H400x400x13x21	4H250x250x9x14	2H350x350x12x19	H350x350x12x19	H350x350x12x19	H350x350x12x19
4	2H400x400x13x21	4H250x250x9x14	2H350x350x12x19	H350x350x12x19	H350x350x12x19	H350x350x12x19
5	2H400x400x13x21	4H250x250x9x14	2H350x350x12x19	H350x350x12x19	H350x350x12x19	H300x300x10x15
Center post	H350x350x12x19	H350x350x12x19	H350x350x12x19	H250x250x9x14	H200x200x8x12	H200x200x8x12

3 Results and Discussion

3.1 Effects of Wall Embedded Depth (H_p/H_e) and Wall Friction (C_w/S_u)

Figure 4 shows the variation of factor of safety (F_{LF}) estimated by the load factor method with H_p/H_e corresponding to c_w/s_u values from 0 to 1. The c_w/s_u values of 0 and 1 indicate the walls with smooth and rough surfaces, respectively. When the s_u/σ_v' is constant or s_u increases linearly with depth, it can be observed from Fig. 4a that F_{LF} gradually increases with H_p/H_e. However, the incremental rate of F_{LF} reduces when H_p/H_e is very large (i.e., $H_p/H_e = 2.5$). As expressed previously in Eqs. 1 and 2, both the active and passive earth pressures are affected by the undrained shear strength of soil. Since the soil strength increases with depth at this case, the passive earth pressure will develop faster than the active one with depth. As a result, F_{LF}, which is the ratio between the resistant moment by the passive earth pressure to the driving moment by the active earth pressure, increases significantly.

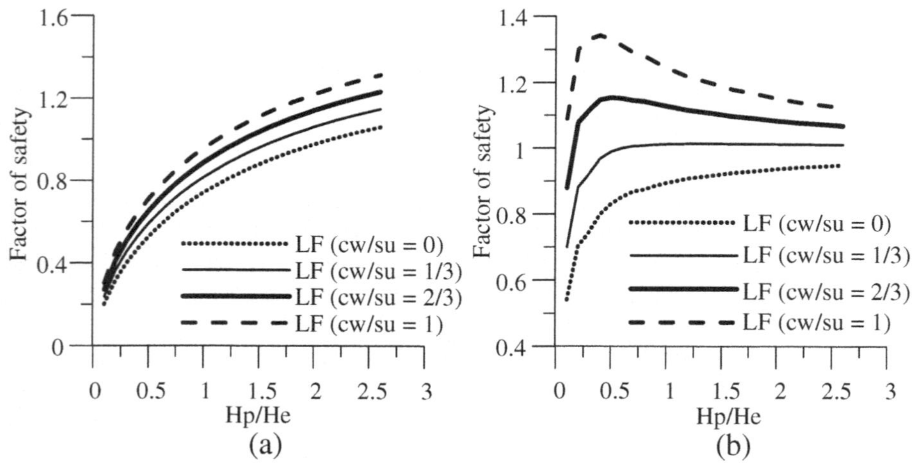

Fig. 4. Factor of safety of the load factor method vs. H_p/H_e with different c_w/s_u values as: (a) s_u/σ_v' = const and (b) s_u = const

On the other hand, when s_u is constant with depth (Fig. 4b), as $c_w/s_u = 0$ and 1/3, F_{LF} initially grows with H_p/H_e and then becomes constant. As $c_w/s_u = 2/3$ and 1, F_{LF} also firstly rises with H_p/H_e but drops to a constant value as the H_p/H_e ratio is greater than 0.5. A clear peak of the relationship between F_{LF} and H_p/H_e can be observed as $c_w/s_u = 1$. Consequently, a specific F_{LF} value (e.g., $F_{LF} = 1.2$) corresponds to two H_p/H_e ratios. Since s_u is constant with depth, based on the Eqs. 1 and 2, the soil strength does not strongly affect the passive and active earth pressures. Mathematically, the resisting moment provided by the passive earth pressure firstly increases faster than the driving moment by the active earth pressure for small H_p/H_e values. However, when H_p/H_e is very large, the first moment rises slowly and approaches the latter. Therefore, F_{LF} does not always increases with the wall embedded depth, which is illogical.

The variations between factor of safety (F_{SF}) estimated by the strength factor method with H_p/H_e and c_w/s_u are shown in Figs. 5a and b. As shown in Fig. 5a, when s_u/σ_v' is constant or s_u increases with depth, F_{SF} rises linearly with H_p/H_e. In case that s_u is constant (Fig. 5b), F_{SF} firstly increases with H_p/H_e and then becomes constant as the H_p/H_e ratio is large (e.g., $H_p/H_e \geq 0.5$). The development of F_{SF} with c_w/s_u is very clear. There is the fact that the strength factor method considers the stability of the wall part (from the lowest strut layer to the wall toe) under the passive and active earth pressures, which is similar to the load factor method. Therefore, the variations of F_{SF} with H_p/H_e and c_w/s_u can be explained using the same reasons as discussed previously.

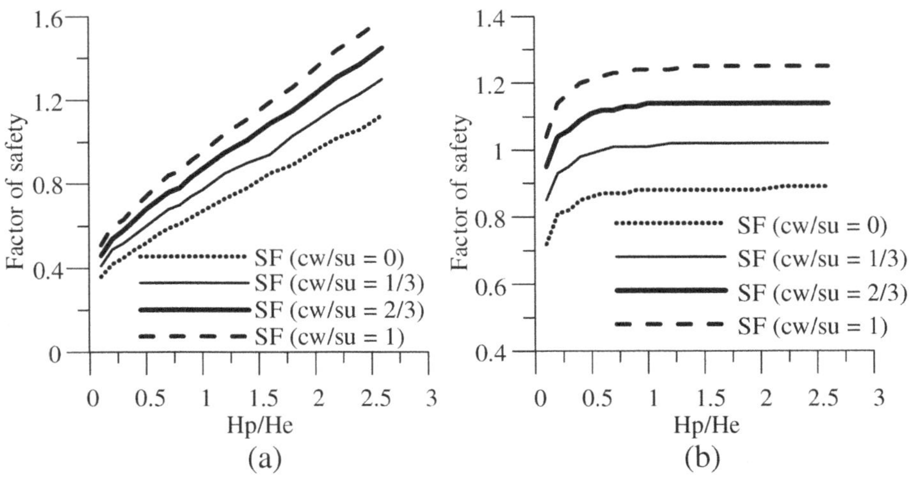

Fig. 5. Factor of safety of the strength factor (SF) method vs. H_p/H_e with different cw/c values as: (a) $s_u/\sigma_v' =$ const and (b) $s_u =$ const

As shown in Fig. 6, when s_u/σ_v' is constant or s_u increases with depth, the factor of safety (F_{SC}) of the slip circle method rises linearly with H_p/H_e. As discussed in the Methodology section, the failure surface has a center at the lowest strut layer and passes through the wall toe is directly adopted in estimation of F_{SC}. Therefore, the surface size will develop simultaneously with H_p/H_e. Together with the growth of s_u

with depth, the resisting moment provided by the soil shear strength along the surface increases significantly, which leads to the development of F_{SC}. On the other hand, when s_u is constant with depth, the resistant moment develops slowly with the failure surface size as compared with the driving moment by the soil weight behind the wall. As a result, F_{SC} gradually rises at the beginning and then reaches a constant value for very large H_p/H_e ratios (i.e., $H_p/H_e = 2.5$). Note that since the method treats the wall part (from the lowest strut layer to the wall toe) and the soil as a united body experiencing excavation-induced movement, the c_w/s_u ratio is implicitly assumed to be 1.0. Hence, the effect of c_w/s_u is not investigated.

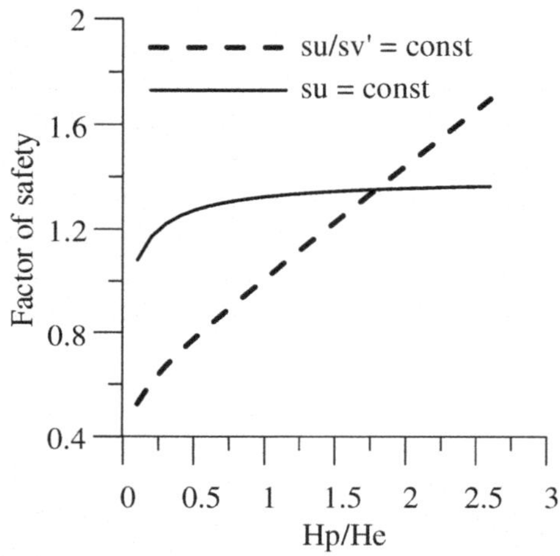

Fig. 6. Factor of safety of the slip circle (SC) method vs. H_p/H_e as: (a) su = const and (b) su/sv′ = const

3.2 Effects of Undrained Shear Strength of Soil

$s_u/s_v' = const$

Figure 7 shows the comparison of H_p/H_e ratios determined by the load factor method ($H_p/H_{e,\ LF}$), the strength factor method ($H_p/H_{e,\ SF}$), the slip circle method ($H_p/H_{e,\ SC}$), and the FEM ($H_p/H_{e,\ FEM}$) for different s_u/σ_v' values. For a given s_u/σ_v' value, H_p/H_e is separately determined by these methods in order to have the corresponding factor of safety equal to 1.2. For example, when $s_u/\sigma_v' = 0.22$ in order to have $F_{SC} = 1.2$ by the slip circle method, the required $H_p/H_{e,\ SC}$ ratio should be 1.4. The factor of safety of 1.2 was selected based on the fact that this number has been broadly adopted by practicing engineers for stability of excavations, especially in Japan and Taiwan. In addition, the stability number, N_b, is employed to roughly indicate stability of excavation. N_b is calculated as follows: $N_b = \gamma H_e/s_u$ where γ = unit weight of soil; H_e = excavation depth; s_u = undrained shear strength of soil between the excavation bottom and the influenced depth of excavation. As shown in the figure, when s_u/σ_v' is increased from 0.16 to 0.36 (N_b = 4.2–9.5), H_p/H_e calculated by all of the methods

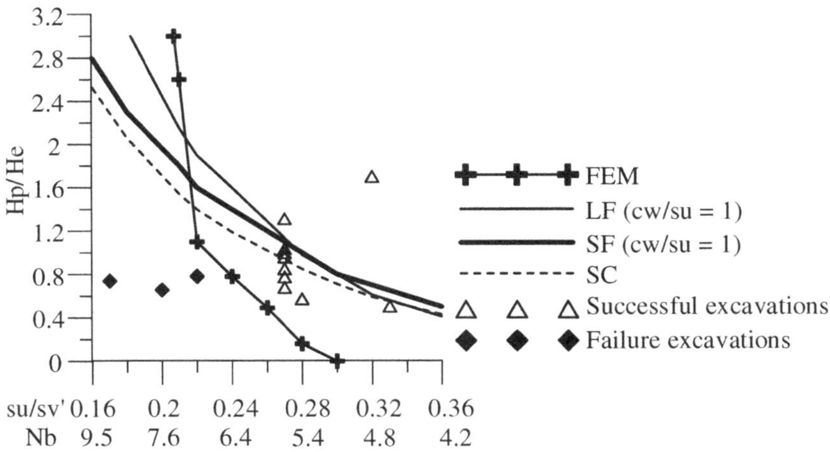

Fig. 7. H_p/H_e estimated by different methods at factor of safety of 1.2 and $s_u/\sigma_v' = \text{const}$

generally decreases. It is observed that $H_p/H_{e,\,SF}$ and $H_p/H_{e,\,SC}$ are very close to each other. $H_p/H_{e,\,LF}$ is greater than $H_p/H_{e,\,SF}$ and $H_p/H_{e,\,SC}$ as $s_u/\sigma_v' \leq 0.3$ ($N_b \geq 5.1$) but close to them as $s_u/\sigma_v' > 0.3$ ($N_b < 5.1$). The relationships between H_p/H_e by the three methods and s_u/σ_v' are generally smooth curves. Also, $H_p/H_{e,\,FEM}$ and s_u/σ_v' have the curved relationship. However, at the beginning, when s_u/σ_v' is small (i.e., $s_u/\sigma_v' = 0.22$), $H_p/H_{e,\,FEM}$ is highly sensitive to the change of s_u/σ_v' as indicated by the nearly vertical part of the curve in the figure. As compared with $H_p/H_{e,\,FEM}$, which is considered as a standard value, the three hand-calculation methods would underestimate the required wall embedded depth (H_p/H_e) as $s_u/\sigma_v' \leq 0.22$ ($N_b \geq 7.0$) and overestimate the depth as $s_u/\sigma_v' > 0.22$ ($N_b < 7.0$).

For validation, H_p/H_e ratios of case histories, including failure and successful excavations, are added into Fig. 7. As shown in the figure, the predicting curves of the four methods locate above the black points of the failure cases so that the required H_p/H_e ratios are greater than those adopted in the field. Therefore, failure of the cases, in view of the four methods, is reasonable. At the successful cases, the predicting curve of the FEM stays below the white points of the cases while those of the load factor method, the strength factor method, and the slip circle method lie among these points. Hence, these methods require the smaller or equivalent H_p/H_e ratios as compared with the real ones and also indicate the success in excavation of the cases.

$s_u = const$

Figure 8 plots the variations of H_p/H_e ratios determined by the load factor method ($H_p/H_{e,\,LF}$), the strength factor method ($H_p/H_{e,\,SF}$), the slip circle method ($H_p/H_{e,\,SC}$), and the FEM ($H_p/H_{e,\,FEM}$) for different s_u values. For a given s_u value, the methods determine the H_p/H_e ratio to obtain the corresponding factor of safety at 1.2 (i.e., as s_u = 70 kPa, $H_p/H_{e,\,SC} = 0.59$ by the slip circle method at which $F_{SC} = 1.2$). It is observed that when s_u is increased from 68 kPa to 92 kPa (N_b = 3.7–5.0), the relationship between $H_p/H_{e,\,SC}$ and s_u is a smooth curve. $H_p/H_{e,\,SC}$ reduces significantly when s_u is

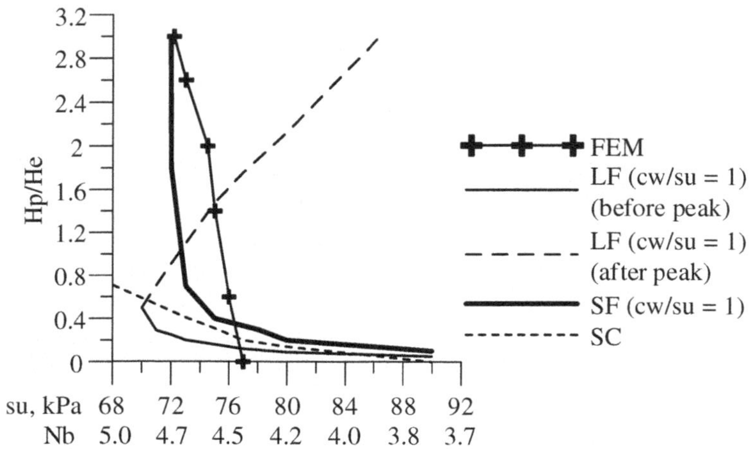

Fig. 8. H_p/H_e estimated by different methods at factor of safety of 1.2 and s_u = const

approximate 76 kPa ($N_b = 4.5$) but slowly when $s_u \geq 76$ kPa ($N_b \leq 4.5$). $H_p/H_{e,\ SC}$ is smaller than $H_p/H_{e,\ SF}$ as $s_u \leq 76$ kPa ($N_b \geq 4.5$) but close to $H_p/H_{e,\ SF}$ as $s_u > 76$ kPa ($N_b < 4.5$). On the other hand, results by the load factor method are not reasonable because this method generates two H_p/H_e ratios corresponding to each given s_u value. When $s_u \leq 76$ kPa ($N_b \geq 4.5$), $H_p/H_{e,\ FEM}$ is highly sensitive to s_u, as indicated by the steep curve in the figure. However, when $s_u > 76$ kPa ($N_b < 4.5$), it is not required to penetrate the wall into the subsoil or $H_p/H_{e,\ FE} = 0$ because the factor of safety estimated by the FEM is greater than 1.2 corresponding to these s_u values. As compared with results by the FEM, it can be seen that the strength factor method and the slip circle method underestimate the required wall embedded depth (H_p/H_e) as $s_u \leq 76$ kPa ($N_b \geq 4.5$) and overestimate the depth as $s_u > 76$ kPa ($N_b < 4.5$).

4 Conclusion

In this study, the load factor method and the strength factor method (recommended by CIRIA report 104 and BS 8002) as well as the slip circle method (advocated by TGS (2001) and JSA (1988)) have been investigated in estimating stability of wide excavations. The excavations are performed in a thick deposit of clay and supported by the wall with multiprop levels. Results of these methods are compared with those of the FEM, which is treated as the standard method (Do et al. 2016). On the basis of this study, some conclusions can be drawn as follows:

1. The development of the friction between wall and soil can significantly raise factors of safety estimated by the load factor method and the strength factor method because it increases the passive earth pressure and reduces the active earth pressure acting on the wall. However, result of the slip circle method is not affected by wall friction since the method does not consider the friction between wall and soil in calculation.

2. When the H_p/H_e (wall embedded depth to excavation depth) ratio increases, factors of safety by the conventional methods (the load factor method, the strength factor method, and the slip circle method) will rise linearly if the s_u/σ_v' ratio of subsoil is constant with depth. It is due to the fact that the wall can prevent surrounding soil from sliding toward excavation, hence, the wall embedded depth has the pronounced effect on stability of the excavations. However, if the s_u of subsoil is constant with depth, factors of safety estimated by these methods do not significantly improve with H_p/H_e. The reason is the surrounding soil can slide deeply below the wall toe toward the excavations. The failure surfaces of soil cover the whole excavations. It is the bearing capacity of soil against the plastic heave at the excavation bottom that dominates stability of the excavations instead of the wall embedded depth.
3. When the effect of the soil strength is investigated, the H_p/H_e ratio is estimated by the methods in order to have the factor of safety at 1.2, which is widely adopted by practicing engineers. It is observed that when the s_u/σ_v' ratio of subsoil is constant with depth, the strength factor method and the slip circle method always give similar H_p/H_e ratios for different s_u/σ_v' values. The required H_p/H_e ratios of the load factor method are greater than those of the above methods as $s_u/\sigma_v' \leq 0.3$ ($N_b \geq 5.1$) but close to them as $s_u/\sigma_v' > 0.3$ ($N_b < 5.1$). If compared with the FEM, the three hand-calculation methods will underestimate the required wall embedded depth as $s_u/\sigma_v' \leq 0.22$ ($N_b \geq 7.0$) and overestimate the depth as $s_u/\sigma_v' > 0.22$ ($N_b < 7.0$). Also, these four methods can give the reasonable H_p/H_e ratios as compared with those adopted in the real failure and successful excavations.
4. When the s_u of subsoil is constant with depth, the required H_p/H_e ratios estimated by the slip circle method are smaller than those by the strength factor method as $s_u \leq 76$ kPa ($N_b \geq 4.5$) but close to them as $s_u > 76$ kPa ($N_b < 4.5$). The load factor method generates illogical results since it requires two different H_p/H_e ratios corresponding to each of s_u values. If compared with the FEM, the required H_p/H_e ratios of the strength factor method and the slip circle method would be smaller than those of the FEM as $s_u \leq 76$ kPa ($N_b \geq 4.5$) and greater as $s_u > 76$ kPa ($N_b < 4.5$).

References

BS 8002 (British Standard Institute): Code of practice for earth retaining structures, London (1994)

Do, T.N., Ou, C.Y., Chen, R.P.: A study of failure mechanisms of deep excavations in soft clay using the finite element method. Comput. Geotech. **73**, 153–163 (2016)

Do, T.N.: A study of stability of deep excavations in clay with consideration of a full elastoplastic support system. Ph.D. Dissertation, National Taiwan University of Science & Technology, Taipei, Taiwan, R.O.C (2016)

JSA (Japan Society of Architecture): Guidelines of design and construction of deep excavations, Tokyo (1988)

Padfield, C.J., Mair, R.J.: Design of retaining walls embedded in stiff clays. CRIR Report No. 104, England, pp. 83–84 (1984)

TGS (Taiwanese Geotechnical Society): Design specifications for the foundation of the building, Taipei, Taiwan, R.O.C (2001)

Railway Underground Adjacent Construction and Design Issues in Soft Soil of Taipei Urban Area

Chung-Ren Huang[1,2,3(✉)], Jimmy Chen[1,2,3], Wen-Jen Chang[1,2,3], Tai-Shen Pan[1], Chen-Yuan Liao[2], and Chung-Yi Tsai[3]

[1] Sinotech Engineering Consultant, Ltd., Taipei, Republic of China
churen47@mail.sinotech.com.tw
[2] Office of Rapid Transit Systems, Department of Transportation, Taoyuan City Government, Taoyuan City, Republic of China
[3] Railway Reconstruction Bureau, Ministry of Transportation Communication, Taipei, Republic of China

Abstract. Low soil strength and high ground water level are the geotechnical engineering characteristics of the soil deposit in Taipei urban area. The railway underground project of Taiwan Railway Administration (hereinafter called the "TRA")/Taiwan High Speed Railway (hereinafter called the "THSR") is majorly constructed in congested metropolitan area with high density building and limited space. The key issue is to ensure the security of existing railway operation and to protect the buildings close to the construction site with soft soil condition. This article makes a summary about the design and construction feature for the Nangang extension of the railway underground project in Taipei city. Design features include pile foundation, underpinning slab and temporary supporting posts to withstand the loading of temporary tracks and platforms, ground improvement to increase the strength of soil to ensure the stability of 20 m deep excavation and bearing capacity of cut & cover tunnel with the condition of soil strength Su = 2 t/m^2 and groundwater level about 1 m below ground level. Construction features include the removal schedule of existing tunnel and station, new tunnel and station construction process, risk management, safety monitoring system of deep excavation, and control of construction vibration and noise. Prevention design concept that reduce risk and ensure continuous operations of the TRA in the construction process of the project can be adopted as a reference in similar case for future engineers.

Keywords: Adjacent construction · Underpinning slab · Risk management · Monitoring system · Cut & cover tunnel

1 Introduction

Taiwan railway construction enter its underground stage in 1983. The existing railway must maintain its operation function during the underground construction period. The railway underground project of TRA/THSR is constructed in congested Taipei metropolitan area with high density building and limited space. Some key issues of

R. Chen et al. (eds.), *Proceedings of the 2nd International Symposium on Asia Urban GeoEngineering*, Springer Series in Geomechanics and Geoengineering,
https://doi.org/10.1007/978-981-10-6632-0_20

design and construction should be addressed to ensure the security of existing railway operation and to protect the buildings close to the construction site with low soil strength and high ground water level, the geotechnical engineering characteristics of the soil deposits in Taipei urban area.

2 Project Description

This article summaries the design and construction feature for part of Nangang extension of the Taipei city railway underground project. The project includes Songshan station underground section (CL308 lot) and Hulin Street tunnel section (CL309 lot). CL308 lot includes TRA and THSR tunnel in the Songshan station, with tunnel length about 612 m and 574 m respectively (see Fig. 1). Major items are as follows,

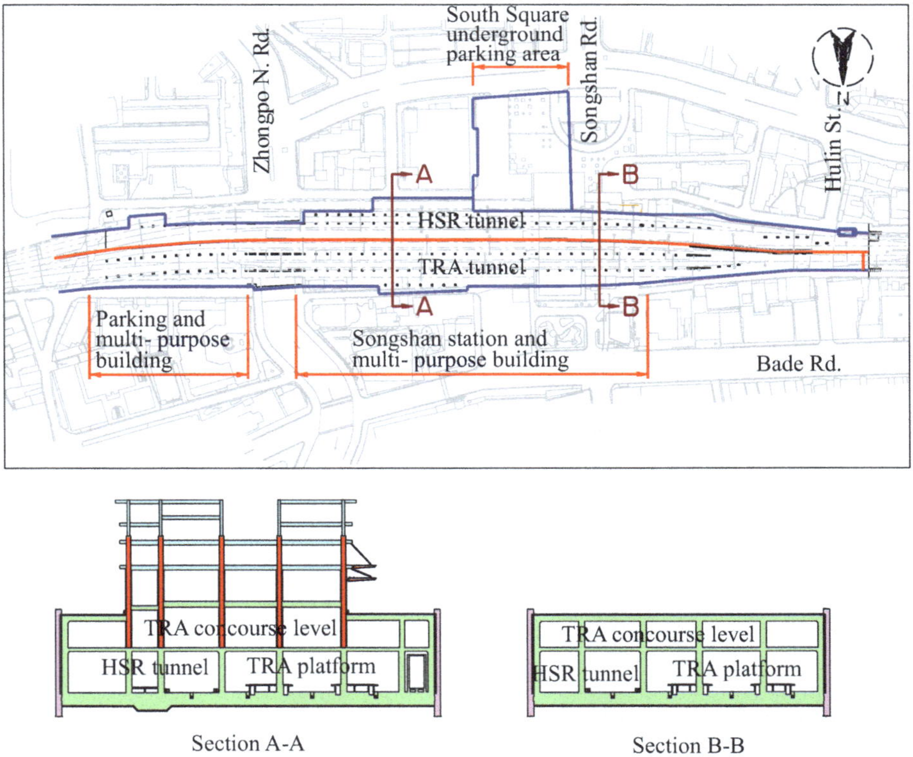

Fig. 1. Songshan station underground section (CL308 lot)

(1) Songshan station and multi-purpose building, with 2-story underground and 11-story superstructure above the ground (above the 3rd belong BOT), the length is about 253.75 m, the width is about 55 to 74 m.
(2) Parking and multi-purpose building, with 3-story underground and 14-story superstructure above the ground (above the 2nd belong BOT), the length is about 105 m, the width is about 46 to 55 m.
(3) South Square underground parking building, with 4-story underground column slab box structure.

CL309 lot is coordinate with the CL308 lot to replace the existing ground track and tunnel by underground track and new lower longitudinal slope tunnel. CL309 lot tunnel has total length of about 946 m, from east to west is divided into U-type approach section, ╙┘-type approach section, double-tube tunnel Sect. 1 and double-tube tunnel Sect. 2 as shown in Fig. 2.

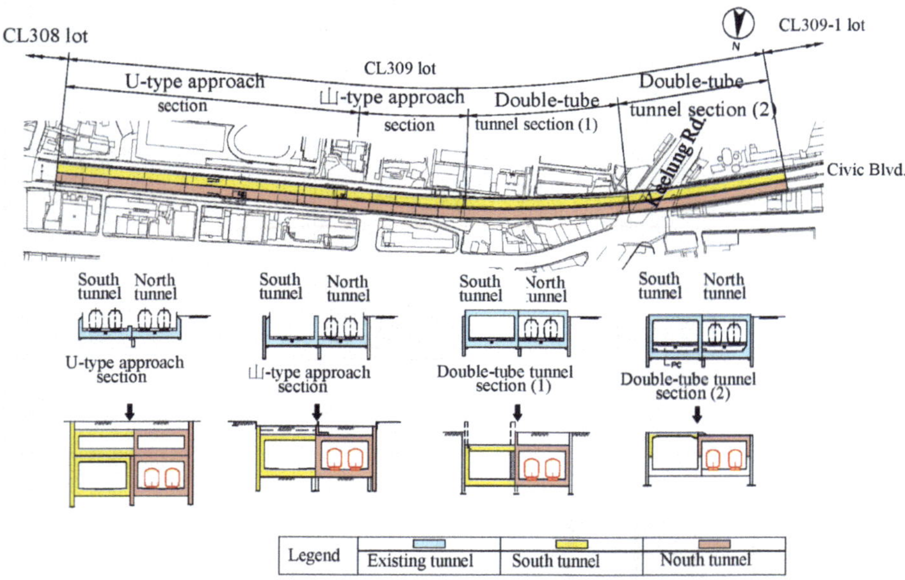

Fig. 2. Hulin Street tunnel section (CL309 lot)

3 Geological and Groundwater Condition

This project is located on the east side of the Taipei Basin. The bottom of the basin formed mainly by the Tertiary sedimentary rocks, and the rock was covered by Quaternary deposit. The thickness of the Songshan Formation on top of Quaternary in this project is about 43 m. All structure of the station and tunnel in this project is located within the soil of the Songshan Formation. But the pile foundation is set deep within

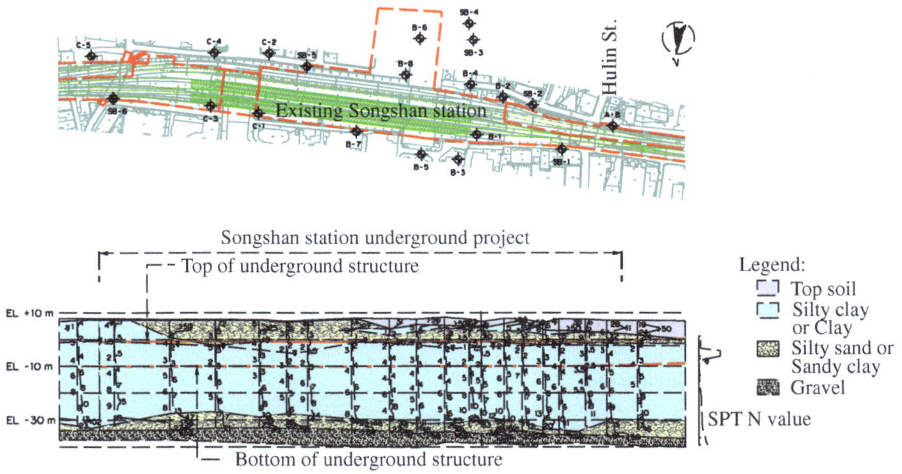

Fig. 3. Geological investigation results

Chingmei gravel layer. Project site is adjacent to the Keelung River, soil is mainly composed of silty clay and silty sand according to the geological investigation results as shown in Fig. 3.

According to the recording results of piezometers, the groundwater level in the Songshan Formation is about 1 m below the ground surface. Groundwater pressure distribution above the depth of 15 m appears to be in a static water pressure state. Below the depth of 15 m, the water pressure distribution is lower than the hydrostatic pressure, groundwater pressure at depth of 40 m is about 11.5 t/m^2 lower than the static water pressure. On the other hand, the groundwater pressure in the gravel layer showed a rising trend each year.

Table 1. Simplified soil parameters

Depth (m)	Soil classification	Thickness (m)	N value	γt (t/m^3)	φ (degree)	δ (degree)	C (t/m^2)	Su (t/m^2)
–	CL	4	3	1.91	30	10	0	2.5
7	SM	3	6	1.9	31	10.33	0	-
18	CL	11	3	1.81	29	9.67	0	2.5
27	CL	9	5	1.86	30	10	0	3
7	CL	10	9	1.9	32	10.67	0	6.5
41.5	SM	4.5	25	1.98	34	11.33	0	-
>41.5	GM	-	>100	2.2	38	12.67	0	-

The excavation depth of the Songshan station is about 19 m. The excavation depth of the South Square underground parking building is about 23.5 m. Bottom of excavation is located in the clay layer of Songshan Formation, which natural water content is about 29.5–41.9%, liquid limit is about 29.9–41.0%. Table 1 summarizes the simplified soil parameters.

In some soil, the natural water content is slightly greater than or close to the liquid limit, due to this reason, the soil has the characteristics of high fluidity and low self-support. If the depth of diaphragm wall penetration is insufficient, there is a risk of squeeze out, heave or blow-in problem. The sensitivity of soil parameters affects the analysis results and the protection measures for adjacent buildings should be taken into account to ensure the overall stability of the excavation.

4 Design Considerations and Construction Planning

4.1 General Consideration

The main considerations for design and construction planning of this project are as follows:

(1) It is necessary to maintain normal operation of the railway and station during construction period.
(2) Reduce the levy of land and demolition of existing buildings, make full use of existing railway land to avoid secondary levy land caused by construction.
(3) Ensure the quality of the new cut & cover tunnel.
(4) Construction feasibility (including safety, difficulty, etc.).
(5) Project cost control.
(6) Construction schedule and cooperate with the temporary railway track switch work.
(7) Road traffic maintenance during construction period.
(8) Protection of adjacent buildings during construction.

4.2 Design of Songshan Underground Station

Existing Songshan station has four platforms, two island platforms and two side platforms, 8 track lines (No. 1 and No. 4 to No. 8 track line for landing, No. 2 and No. 3 track line for temporary stay). The area of No. 1 to No. 5 track lines and No. 1, No. 2 platforms were utilized for build two island platforms and one side platform, five temporary track lines. It was named the south temporary station. The south temporary station is consisted of south diaphragm wall, pile foundation, steel support post pile and rail track underpinning slab to maintain the operation of TRA. South temporary station was completed in sequence as illustrated in Fig. 4. Construction photo of rail track underpinning slab, post and station structure are shown in Fig. 5.

Loading of temporary station, train, track and platform were beared by rail track underpinning slab. Excavation construction was held below the rail track underpinning slab. The volume of rail track underpinning slab and live load were quite large, resulting in the great axial force of the steel support post pile to withstand the large loading. The length of steel support post pile is more than 16 m to withhold enough space for excavating to the bottom of the station. The steel support post pile of the

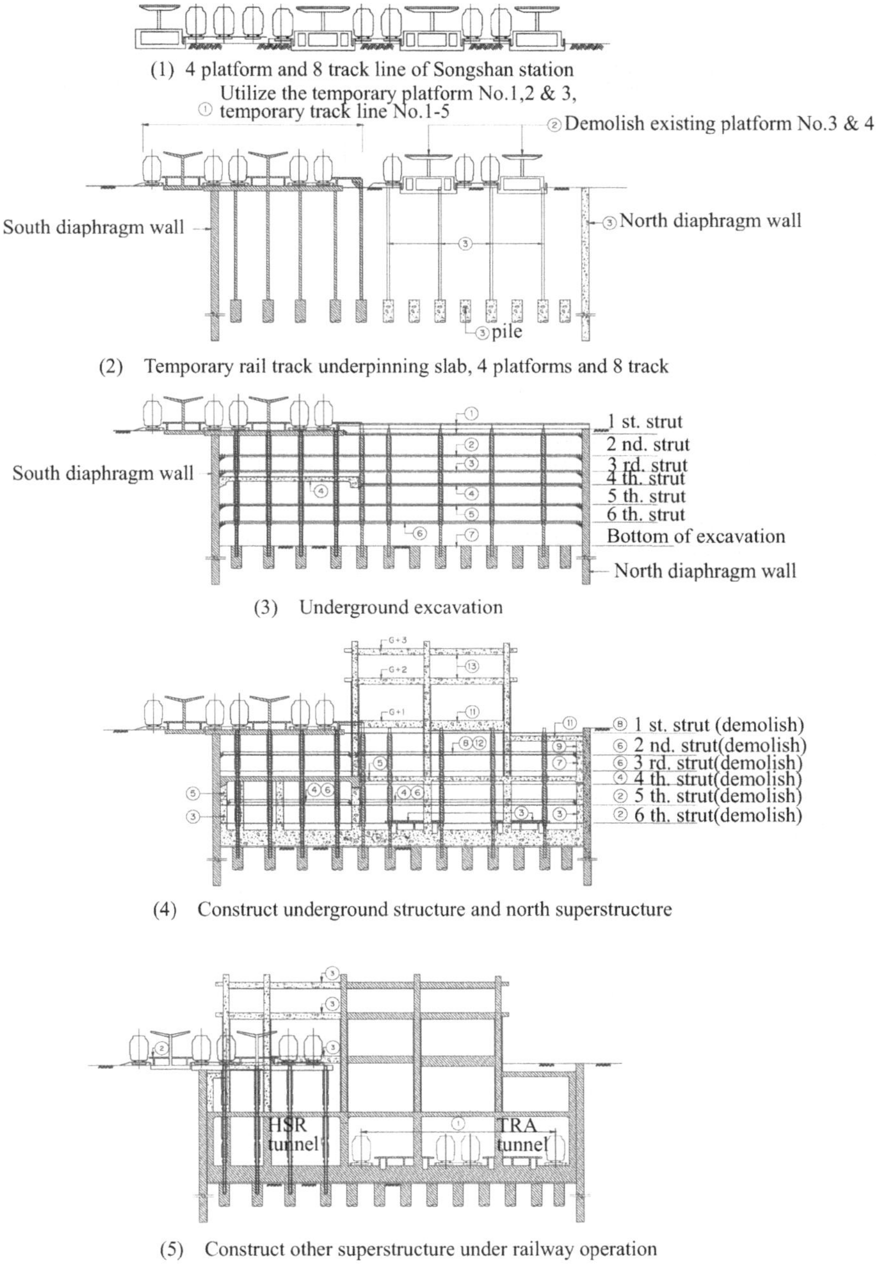

Fig. 4. Construction sequence of South temporary station and Songshan station

Fig. 5. Photo of rail track underpinning slab, post and station structure

excavation support system was inserted into pile foundation which is deep into Chingmei gravel.

The location of steel support post pile is arranged with the location of columns of the Songshan station. Spacing in east-west direction is 4.375 m, and 5.5 m in north-south direction. Max axial force of the steel support post pile is more than 200t. Besides the axial force, 1/200 construction vertical error and derived bending moment are also considered in the steel support post pile design. Due to the uncertainty of the construction vertical error direction and difference of the strong or weak axis of steel support post pile, composite ⊞ type steel were used in this project.

Pile foundation design can be divided into three conditions according to different stages.

(1) After the completion of the underground station, superstructure building body above the ground is not fully completed, piles provided function to withstand uplift water pressure force. To prevent the underground structure from floating due to groundwater uplift pressure, piles are used to withstand vertical uplift force.
(2) After the completion of superstructure building body above the ground, the load of the superstructure is increased, the pile foundation is used to withstand the vertical downward force.
(3) During underground station excavation stage, pile foundation is used as support system to withstand the vertical force or temporary load from rail track underpinning slab.

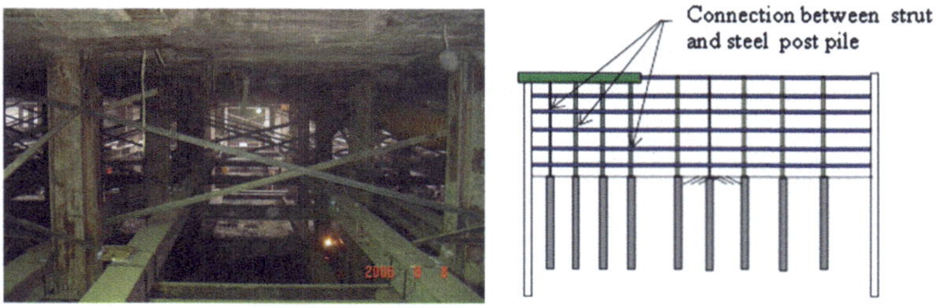

Fig. 6. Supplement with welded angle steel between strut and steel post pile

Uplift force of pile foundation design is to be about 150 tons, the vertical bearing capacity is 350 tons, the pile diameter is 1.6 m, the depth is about 43.5 m, the upper 18.85 m in the area of underground structure is empty bored.

As design load of the steel post pile is about 200 tons, general U-shaped steel bar lock method does not meet the requirement. Additional steel angles or high strength bolts have to be welded on the steel post piles and struts to reduce the vibration caused by train running through the temporary rail track underpinning slab. The additional steel angles or tension rods are welded in appropriate locations to minimize the movement of support system as shown in Fig. 6.

4.3 Design of Hulin Street Cut and Cover Tunnel

Double-tube tunnel Sect. 1 of Hulin street cut and cover tunnel is different from general excavation works, both the existing slurry wall and the site improvement are utilized for construction.

Excavation depth of the new tunnel is about 12 m, distance from the bottom of the existing tunnel structure is about 2.0 m–6.0 m. Due to the existing slurry walls can't meet the stability requirements of excavation. In order to avoid the use of private land, high pressure jet grouting (JSG) site improvement were adopted below the bottom of excavation, it can improve the soil strength of excavation area to meet the overall stability of the excavation work. The procedure of construction is shown in Fig. 7. Photos of existing tunnel and demolition are shown in Fig. 8.

5 Excavation Support System and Adjacent Building Protection Measure

5.1 Analysis of Hulin Street Cut and Cover Tunnel Excavation

A software known as RIDO was used for analysis of excavation on Hulin street cut and cover tunnel, in accordance with the soil conditions and construction procedures to simulate the overall excavation work. The analysis procedure is as follows:

(1) Simulate the existing tunnel excavation construction procedures.

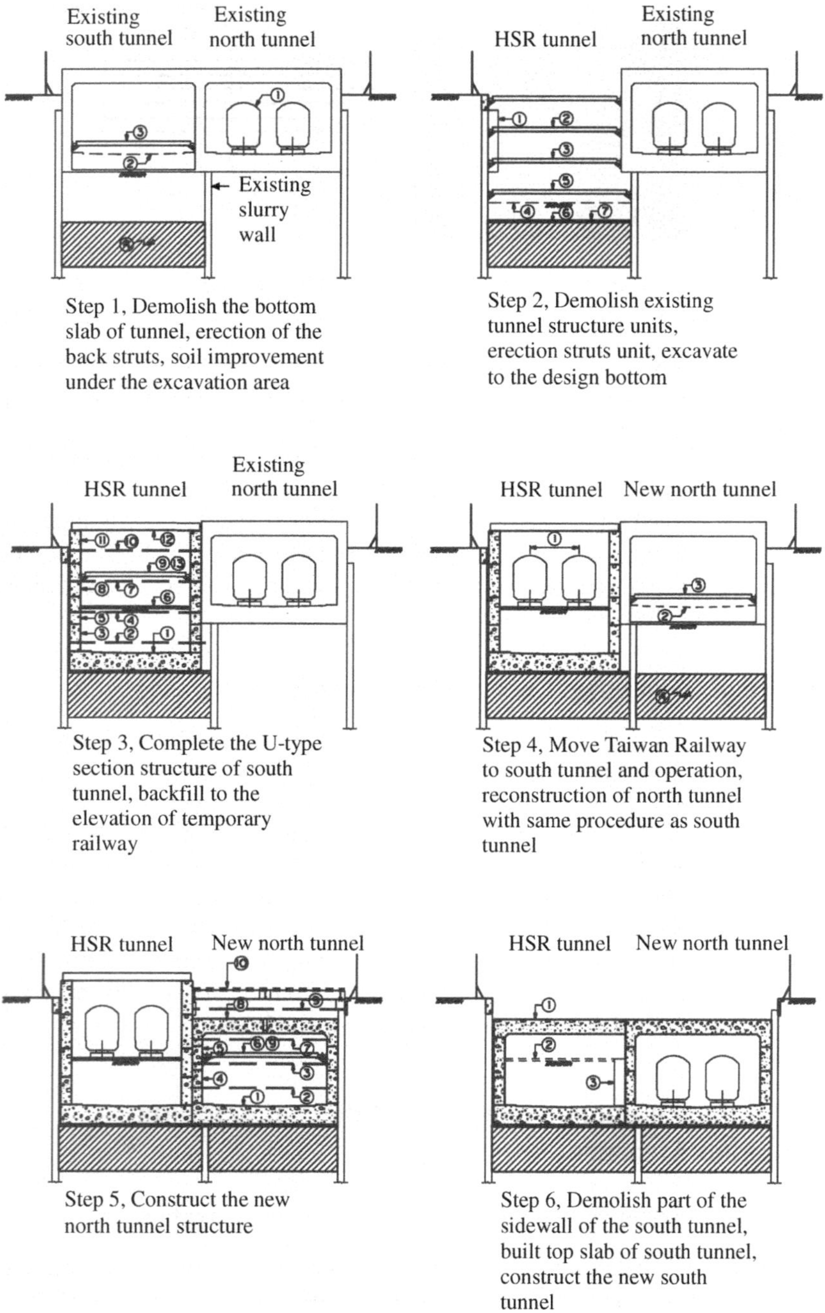

Fig. 7. Construction procedure of Hulin street cut and cover tunnel

Fig. 8. Photos of existing tunnel and demolition

(2) Consider the different demolition unit, remove the existing tunnel structure bottom plate and erection of the back struts.
(3) Ground improvement under the excavation.
(4) Remove the top and side walls of existing tunnel structure, remove the back struts.
(5) Set up the required support system and excavate.
(6) Completion of tunnel structure construction and demolition of support system.
(7) Check the deflection and stress of existing slurry wall during the new tunnel construction.

The ground improvement design under the existing tunnel is high pressure jet grouting pile diameter of 1.0 m with qu strength reaching 10 kg/cm^2 or more, due to demand of water proof. The effect of ground improvement was considered in RIDO analysis with the existing tunnel unit-by-unit removal sequence. Each removed unit is 8 m–12 m long and effects on adjacent existing unit support are also considered in the analysis. According to the results of the design and analysis, the maximum displacement of existing 0.6 m thick slurry wall occurred at the construction stage of the south side of the tunnel, at 8.65 m below the ground surface, resulting in 68.6 mm displacement, the maximum bending moment of the wall occurred 7.5 m below the ground surface, its value is 76.2 t-m, see Fig. 9. The reinforced bar of existing slurry wall met the stress demand after review.

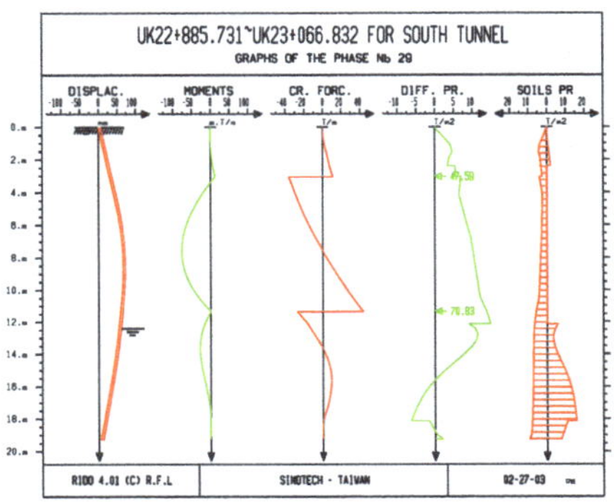

Fig. 9. Analysis result of Hulin street cut and cover tunnel excavation

5.2 Analysis of Excavation for Songshan Underground Station Excavation

To ensure the safety of adjacent buildings and reduce the risk of railways operation during Songshan station underground excavation process, investigation of existing buildings (total of 60 buildings near the influenced area of excavation) and excavation subsidence analysis was carried out before excavation. Subsidence caused by the deep excavation may occur during 3-part stages: (1) during the period of construction of diaphragm wall trench, (2) after the completion of the diaphragm wall, due to excavation of underground and movement of support system, (3) Long term settlement after the completion of the project. The 3-part results of analysis are superimposed on the quadratic curve to simulate the subsidence curve. As an example, a building with foundation depth of 3.75 m, with 5 m distance from diaphragm wall (40 m in length), the settlement value of subsidence curve is shown in Fig. 10 and Table 2.

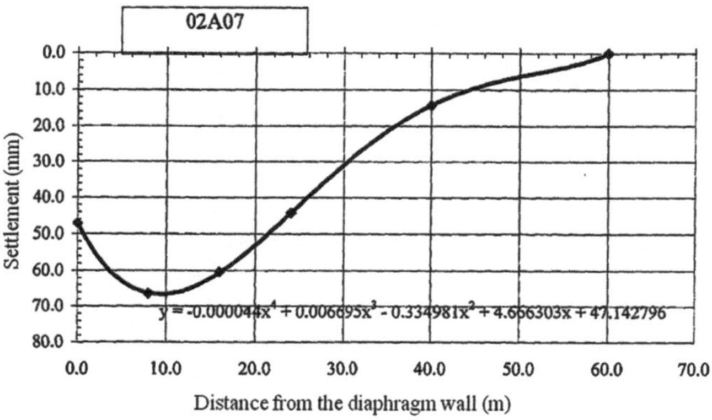

Fig. 10. Example of subsidence curve

Table 2. Settlement value of subsidence curve

Distance from the diaphragm wall (m)	0	8	16	24	40	60
Due construction of diaphragm wall (mm)	9.1	9.1	6.8	4.5	0	0
Due excavation (mm)	23.1	46.1	46.1	33.8	9.2	0
Due consolidation (mm)	15	11.3	7.5	6	5	0
Total settlement (mm)	47.1	66.4	60.4	44.3	14.2	0

5.3 Additional Ground Improvement During Construction

Ground improvement was adopted to protect adjacent buildings during construction process for the important cases as follows.

Using all casing piles to replace diaphragm wall in area connected with sewage pipe (diameter 150 cm). In order to ensure sealing effect, 60 cm diameter CCP grouting improvement was adopted in the retaining side of the excavation as shown in Fig. 11.

The water catchment pit with 22.5 m long, 11.5 m wide and 1.5 m deep was set up at the bottom of excavation, grouting improvement pile was adopted to increase the soil stiffness and strength. The diameter of grouting pile was 60 cm, with spacing 50 cm form 1 m wide and 5 m long, being acting as buttress for diaphragm wall shown in Fig. 12. This improvement can reduce the deflection of diaphragm wall and ensure the safety of excavation.

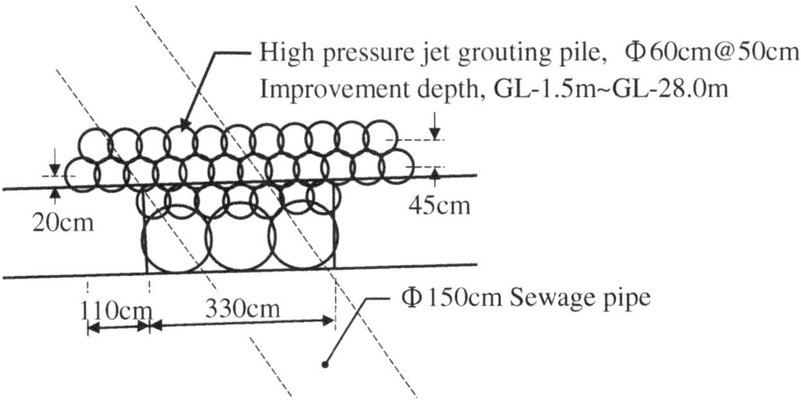

Fig. 11. Grouting improvement in area connect with sewage pipe

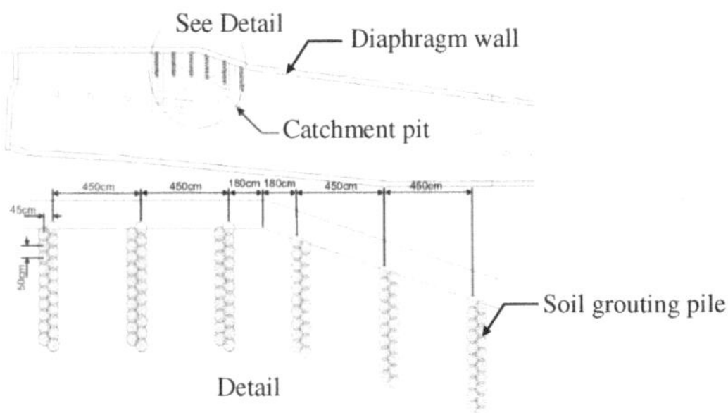

Fig. 12. Location of improvement being as buttress

6 Reading Data of Monitoring System

The principle of safety monitoring system installation for Songshan station excavation is shown in Fig. 13. The layout of monitoring instrument for Songshan station excavation and surrounding area are show in Fig. 14. The alarm value, action value and recommend measures of different instruments as summarize in Table 3, the allowed value range of building tiltmeters are shown in Table 4. Also the scale and extent of vibration generated by the construction process is constantly monitored and reviewed

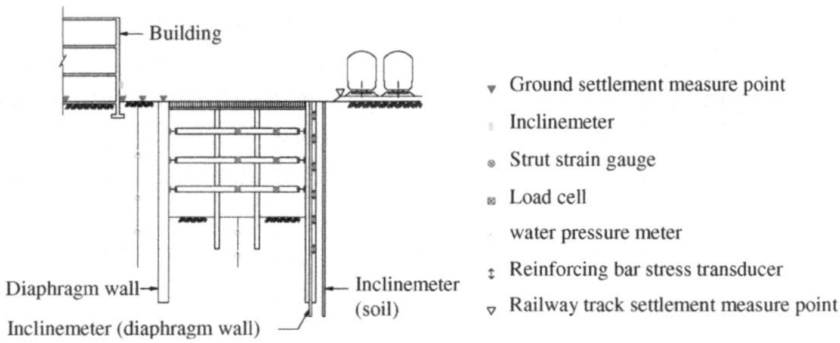

Fig. 13. The principle of safety monitoring installation

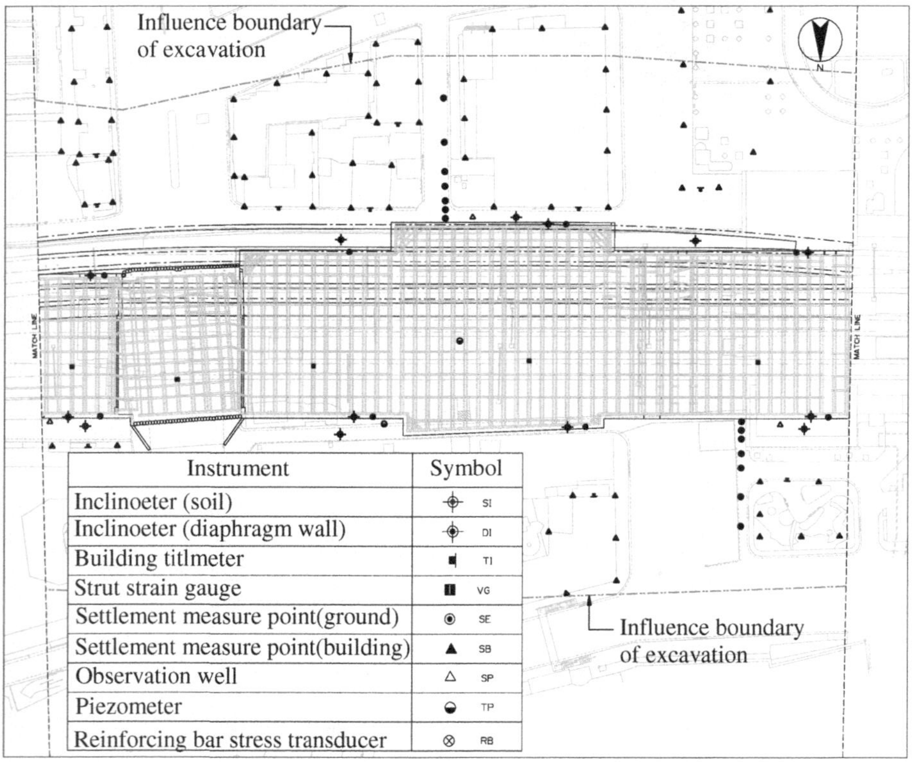

Fig. 14. Monitoring instrument layout of Songshan station excavation and the surrounding area

as to whether it affects the precision equipment in the adjacent buildings. Alarms are issued as the vibration exceeds the allowed values according to specification, for effective early warning and control. Table 5 express the allowed value of vibration.

The settlement of the rail track underpinning slab during the construction period is about 8–14 mm, settlement of railway track is about 3–7 mm, there is no unusual behaviour. Displacement reading of inclinometer in soil is about −19–90 mm.

Table 3. Safety monitoring instruments of excavation

Instrument	Alarm value	Action value	Recommend measures
Inclinometer(soil)	47–65 mm	58–81 mm	Review and check building movement, building protection
Piezometer	Different elevation according depth of excavation	2 m above alert water level	Install point well or deep well, grouting improvement
Inclinometer (diaphragm wall)	52–72 mm	65–90 mm	Review and check building movement, building protection
Settlement measure point of railway track	7 mm	13 mm	Review and check building movement, building protection
Settlement measure point	20 mm	25 mm	
Strut strain gauge	90% of design loading	125% of design loading	Add strut, grouting improvement
Reinforcing bar stress transducer	2500 kg/cm^2	300 kg/cm^2	

Table 4. Allowable values of building tiltmeter

Displacement type	Building type	Alarm value	Action value	Recommend measures
Max settlement (mm)	Brick	20	25	Review and check building movement, Building protection
Differential settlement		1/3125	1/2500	
Max settlement (mm)	RC independent foundation	32	40	
Differential settlement		1/1500	1/1200	
Max settlement (mm)	RC mat foundation	48	60	
Differential settlement		1/1250	1/1000	

Table 5. Allowable values of vibration

Type	Alarm value	Action value
Sinusoidal vibration (gal)	140	175
Random vibration (gal2/Hz)	50	75
livelihood vibration (Db)	62	82

Although some reading exceeds management value, there is no tendency to increase after the end of the excavation. In the case of pumping water during excavation of catchment pit or over excavation, some reading of strut strain gauge exceeds management value.

Because the construction sequence of the Hulin Street tunnel is different from general excavation support method, no relevant case can been as reference. The effectiveness of site improvement will directly affect the safety of the excavation.

In order to increase the safety of construction, safety monitoring system was installed along tunnel alignment to control the construction risk by observation from monitoring instruments.

The results of the monitoring system showed that the maximum displacement of the southward inclination in soil occurs at 2.5 m below ground surface during the tunnel reconstruction, and deformed 3.90 cm toward the direction of excavation, the maximum ground surface settlement is 1.23 cm. The maximum displacement of the northward inclination in soil occurs at 3.5 m below the surface during the tunnel reconstruction, and deformed 3.39 cm toward the direction of excavation, the maximum ground surface settlement is 1.93 cm.

7 Conclusion

In the past, in cut & cover underground tunnel construction project, the temporary support system of excavation including the post pile, strut, wale and deck are only used to withstand road traffic load. Rail track underpinning slab in this project is the first case to withstand the railway operating load, including the platform, trains, electric pole, load quantity is about 3 times more than general road load. Through the high capacity of steel support post pile, rail track underpinning slab and pile foundation, excavation of this project had successed to avoid the settlement, to reduce the risk of the railway operation, and protected adjacent building. Some changes of construction method and sequence, additional ground improvement, rigorous monitoring and fast measure during abnormal condition are the necessary process to ensure the safety of the excavation in the soft soil strata of Taipei urban area.

Overall, during the construction period, no obvious displacement and settlement happened in the surrounding environment of project site, it shows the project process was successfully and completely managed in the design and construction stages.

Reliability-Based Design of Basal Heave Stability for Braced Excavation Using Three Different Methods

Wan-Huan Zhou[1(✉)], Yu Mu[1], Xiao-Hui Qi[1], and Shuaidong Yang[2]

[1] Department of Civil and Environmental Engineering, University of Macau, Zhuhai, Macau SAR, China
hannahzhou@umac.mo

[2] Pearl River Water Conservancy Science Research Institute, Pearl River Water Resources Commission of the Ministry of Water Resources, Guangzhou, China

Abstract. Terzaghi's (T), Bjerrum-Eide's (BE) and the slip circle (SC) methods are widely used to evaluate the basal heave stability in reliability-based design of braced excavations. The three methods produce different reliabilities for the same excavation problem. It is still unclear which of these methods is most conservative or most economical. Hence, this study compares the reliabilities of basal heave stability for the three methods for both wide and deep excavations. The probability of failure (P_f) for a given factor of safety (FS) and coefficient of variation (COV) for soil parameters is calculated using the first-order reliability method. The P_f − FS design charts for different methods are compared. The results demonstrate that the SC method is more conservative than the other two methods. For wide excavations, the same level of the factor of safety results in a much smaller P_f for the SC method than for the T method. For deep excavations, the P_f for the SC method is also smaller than for BE method. In addition, the P_f − FS design curves for T are more sensitive to the penetration depth of the wall than the other two methods. These results serve as a guideline for the method selection of basal heave stability evaluation in geotechnical practice.

Keywords: Reliability-based design · Basal heave stability · Braced excavation · Terzaghi's method · Bjerrum-Eide's method · Slip circle method

1 Introduction

Basal heave stability is required for the design of braced excavations in clay. Traditionally, the factor of safety (FS) is used to evaluate the safety level of geotechnical structures. Usually, the FS index cannot consider the variability of loads and resistances. Nowadays, reliability-based design (RBD) methods are increasingly being used in geotechnical practice. One of the advantages of RBD approaches is that they promote and improve decision-making (Phoon 2008). Although the focus in RBD has generally been on the format and calibration of design-checking equations, RBD presumes that the goal is to achieve an acceptable level of reliability for the design.

R. Chen et al. (eds.), *Proceedings of the 2nd International Symposium on Asia Urban GeoEngineering*, Springer Series in Geomechanics and Geoengineering,
https://doi.org/10.1007/978-981-10-6632-0_21

Taking this as a premise, designers and engineers can capitalize on a range of possible alternatives to achieve the desired reliability. Today, RBD methods are incorporated into the design code of many areas, such as the load and resistance factor design codes adopted by AASHTO (AASHTO 1994); Eurocode 7 developed in European (Eurocode 7 1993); and Geo-Code 21, established in Japan (Honjo and Kusakabe 2002). However, some design codes still lack rigorous calibration. There are limited studies including uncertainty in geotechnical engineering (Qi et al. 2016).

The limit equilibrium methods are used to calculate stability. The major factors affecting the safety of the braced excavation projects are soil properties, loadings, size of the braced excavation (width, depth and length) and penetration depth of the wall. Besides, the effect of soil density, soil suction, initial water content on the volumetric water content (Zhou et al. 2016) and the underground water seepage system (Su et al. 2017) could be considered. It is fulfilling to reduce the uncertainty in the braced excavation system for more rational RBD (Qi and Zhou 2017), and they can be accounted for using a probabilistic approach. One popular method, which is applied in this paper is the reliability index (Hasofer and Lind 1974). This approach represents the failure probability through the mean value and standard deviations of the values for these factors (Low 2005). It has been adopted in the design of many geotechnical structures.

Terzaghi's (T) method (Terzaghi 1943), Bjerrum-Eide's (BE) method (Bjerrum and Eide 1956) and the slip circle (SC) method (JSA 1988; TGS 2001) are widely used to evaluate the basal heave stability of the reliability-based design of braced excavations. The three methods can produce different reliabilities for the same excavation problem. It is still unclear which method is most economical. Hence, this study compares the reliabilities of basal heave stability for the three methods in both wide and deep excavations. Moreover, the relationship between P_f and FS is discussed with different coefficient of variations (COVs) for soil parameters.

2 Three Design Methods and Reliability Index for Evaluating Basal Heave Stability

The schematics of the three applied calculation methods are presented in Table 1. Based on the bearing capacity theory, the T and BE methods assume homogeneous properties of soil, whereas SC method does not. Moreover, the SC method is semi-empirical and is commonly employed in Asia (JSA 1988; TGS 2001).

The required FS value in codes is 1.2 for the BE and SC methods (JSA 1988; TGS 2001) and 1.5 for the T method (JSA 1988). For wide excavations that had an excavation depth (He) smaller than the excavation width (B), the T method presented the design equations shown in Table 1. Compared to the T method, the BE method is a more appropriate choice for deep excavations in which He > B (Cai et al. 2002). Moreover the SC method can be used for both wide and deep excavations.

Traditionally, the FS is determined by the relation between the load (S) and resistance (R) of the system in geotechnics. Numerous researchers have proven that this relationship may include model uncertainties. It considers only a limited description of the real domain of safety because of the effect of the uncertainties on the design

Table 1. Summary of the three design methods based on limit equilibrium of excavation

Methods	Calculation equation	Schematic
Slip circle (JSA, 1988; TGS, 2001)	$FS = \frac{M_r}{M_s} = \frac{s_{ub} dt^2 [\pi/2 + \cos^{-1}(H_S/dt)]}{\gamma H_e (dt^2/2) + q(dt^2/2)}$ (1) M_r: resisting moment M_s: driving moment H_e: excavation depth H_p: penetration depth H_s: the distance from the excavation base to the last strut dt: radius of the failure, which is the minimum between [B/$\sqrt{2}$, D] Su_b: undrained shear strength along the failure arc q: surcharge	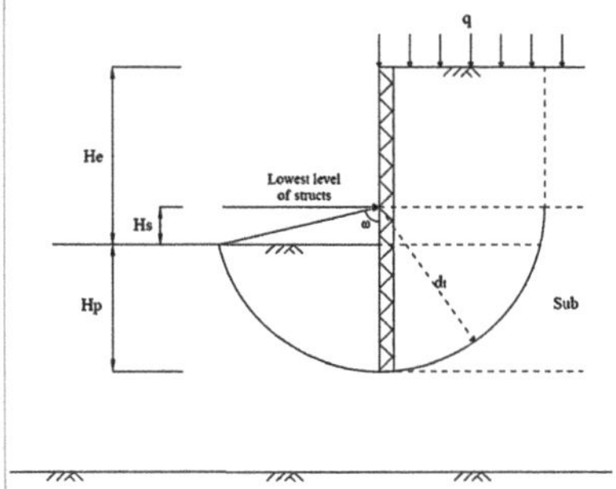
Terzaghi (Terzaghi, 1943)	For $H_p \leq$ dt $FS = \frac{5.7 s_{ub} dt + s_{uh} H_e}{(\gamma H_e + q) dt}$ (2) For Hp > dt $FS = \frac{5.7 s_{ub} dt + s_{uh} H_e + s_{ud} \cdot H_P - dt)}{(\gamma H_e + q) dt}$ (3) H_e: excavation depth H_p: penetration depth B: excavation width D: the distance between the excavation base and the hard stratum dt: radius of the failure, which is the minimum between [B/$\sqrt{2}$, D] Su_h, Su_d and Su_b: undrained shear strength along the failure arc q: surcharge	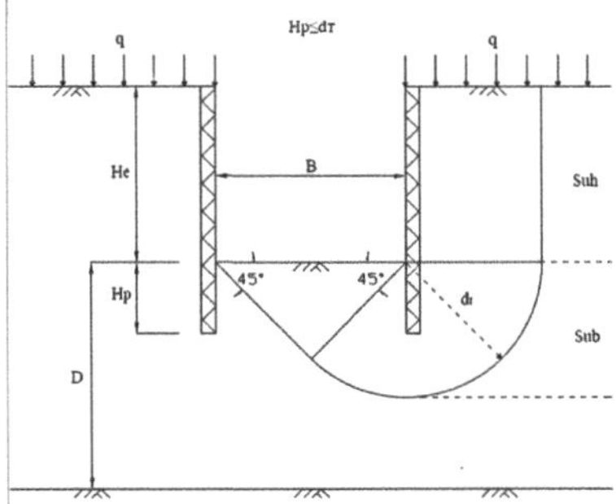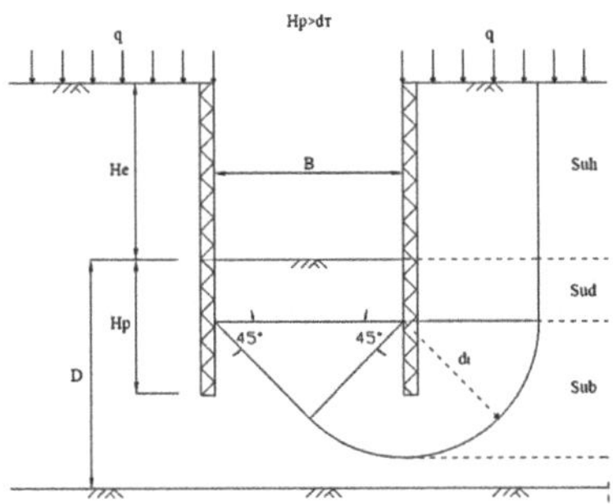

(*continued*)

Table 1. (*continued*)

Methods	Calculation equation	Schematic
Bjerrum-Eide's (Bjerrum and Eide, 1956)	$FS = \frac{N_c S_{ub}}{\gamma H_e + q}$ (4) For $H_p \leq$ dt: $N_c = 5(1 + 0.2B/D)[1 + 0.2H_e/B]$ For Hp > dt: $N_c = 5(1 + 0.2B/D)[1 + 0.2H_{ec}/B]$ N_c: Skempton's(Skempton, 1951) bearing capacity factor H_e: excavation depth H_p: penetration depth H_{ec}: the distance from the ground surface to the center of the failure arc B: excavation width D: the distance between the excavation base and the hard stratum dt: dt is the minimum between $[B/\sqrt{2}, D]$ Su_b: undrained shear strength along the failure arc q: surcharge	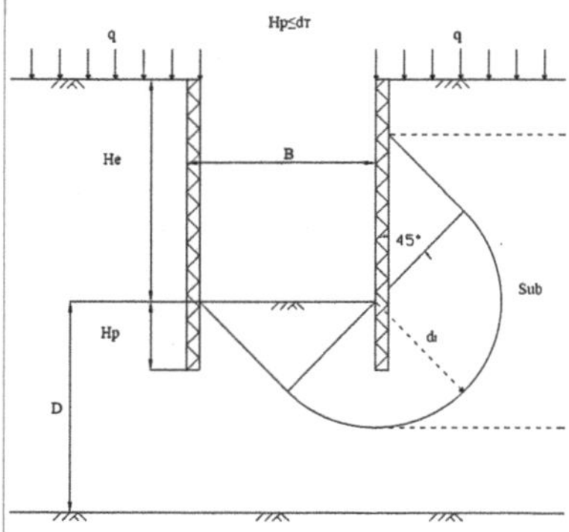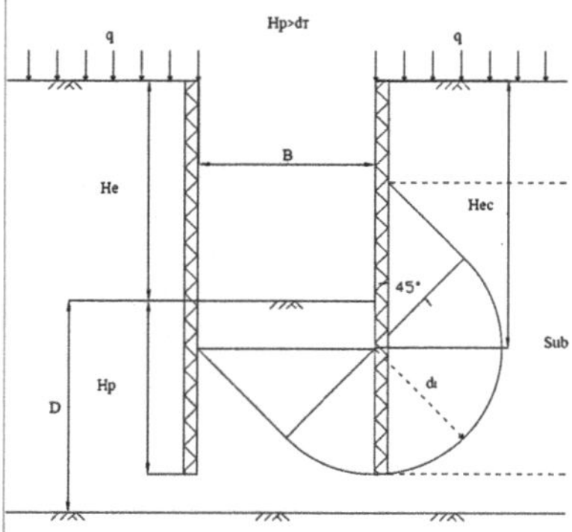

parameters of P_f. The reliability index approach is an advisable alternative approach to RBD.

The depth and geometry of the sliding surface is the fundamental basis for the mechanism evaluating (Xu et al. 2016). The failure performance can be proposed by a limit-state surface function (G(**x**)):

$$G(x) = R - S = 0 \quad (5)$$

Mathematically, G(**x**) < 0 is denoted as the failure domain where **x** is a vector denoting the random variables.

In general, P_f can be expressed by reliability index β Theoretically, β can be estimated by the mean value (m) and standard deviation (σ) of R and S as $\beta = (m_R - m_S)/\sqrt{\sigma_R^2 + \sigma_S^2}$ for the whole system.

When R and S are in accordance with standard Gaussian space, the limit-state surface is computed as

$$G(x) = a_0 + a_1X_1 + a_2X_2 + La_nX_n \quad (6)$$

where a_i is constants (i = 0,1,2,...n) and the X_i terms stand for uncertain variables. The reliability index can be expressed as (Li et al. 1993; Nowak 2000)

$$\beta = \frac{a_0 + \sum_{i=0}^{n} a_i m_{X_i}}{\sqrt{\sum_{i=1}^{n} (a_i \sigma_{X_i})^2}} \quad (7)$$

where m_{X_i} is the mean value and σ_{X_i} is standard deviation. P_f can be derived from the cumulative density function of β :

$$P_f = 1 - \Phi(\beta) \quad (8)$$

where, for example, for a wide braced excavation with B = 20 m and He = 10 m, based on the T method with the related random variables: $s_{ub}(m_{sub} = 30$ kPa, $\sigma_{sub} = 6$ kPa), $s_{uh}(m_{suh} = 15$ kPa, $\sigma_{suh} = 3$ kPa), $q(m_q = 8$ kPa/m, $\sigma_q = 1.6$ kPa/m), and $\gamma(m_\gamma = 15\text{kN/m}^3$, $\sigma_\gamma = 3\text{kN/m}^3)$, dt = 14.14, the FS expression shown in Eq. (2) is FS ≈ 1.03. G(**x**) can be determined as

$$G = 5.7s_{ub}dt + s_{uh}H_e - \gamma H_e dt - qdt \quad (9)$$

and Eq. (7) can used to express the reliability index *b*, and $P_f = 1-\Phi(\beta) = 0.3022$.

3 Comparison and Discussion

Design parameters such as excavation depth and penetration depth of the wall are important for RBD in geotechnical structures. The effect of different values of penetration is discussed using three methods. Furthermore, the effect of excavation depth is considered for both wide and deep excavations. The following figures present a series of results about the relationships between FS and P_f as the required FSs increase from 1.1 to 2.0. For all the analyses, the width of the excavation (B) is 20 m, and the depth below the base of the excavation to the hard stratum (D) is 45 m. Other unchanged parameters are the surcharge q ($m_q = 8$ kPa/m, $\sigma_q = 0.8$ kPa/m), and the unit weight of the clay γ ($m_\gamma = 15\text{kN/m}^3$, $\sigma_\gamma = 1.5\text{kN/m}^3$). H_s, which is the distance from the excavation base to the last strut, is 2 m for the SC method. The mean values of s_{ud} and s_{uh} are assumed to be the same ($m_{sud} = m_{suh} = 20$ kPa), and the COVs of undrained shear strength (s_{ud}, s_{uh} and s_{ub}) ranges from 0.1 to 0.6.

3.1 The Effect of Penetration Depth of the Wall

Figures 1, 2 and 3 show the relationships between the probability of failure and factor of safety using three different methods that have different values of penetration depth $H_{p.}$

As the schematic shows in Table 1, the H_p value should not be smaller than the radius of the failure surface dt (14.14 m). Thus, three H_p values (14.14 m, 20 m, 40 m) are adopted. For a required FS, the relationship between P_f and FS shows that a large FS don't mean a small level of failure. P_f changes as the COV_{su} changes. For example, Fig. 1 shows that when FS = 1.5, P_f increase from 0.0158 to 0.1253 as the COV_{su} increase from 0.2 to 0.4. Figures 2 and 3 also show this trend. Across the different penetration depths of the wall, the T method is more susceptible than the SC and BE methods, as shown in Figs. 1, 2 and 3. This proves that the curves for the T method are more sensitive to the penetration depth of the wall than the other two methods.

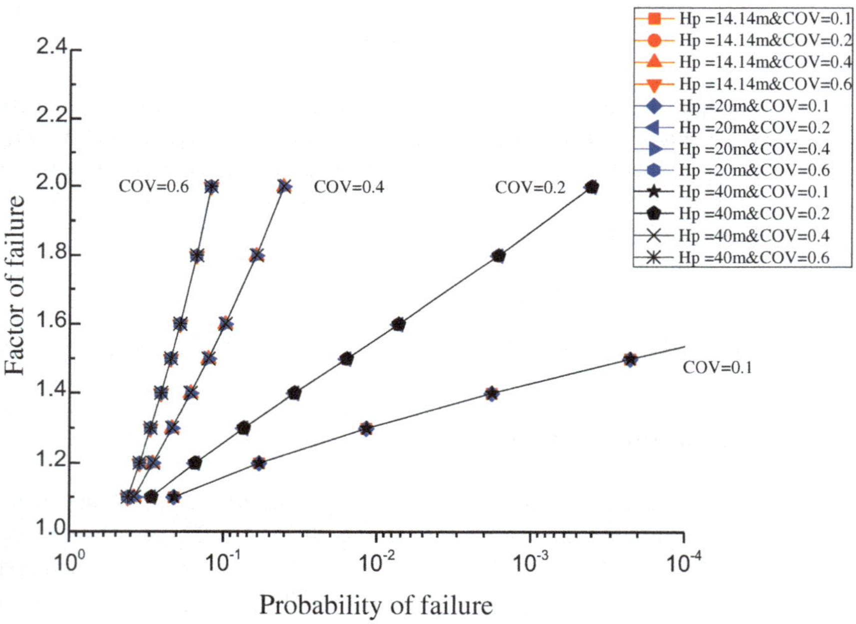

Fig. 1. Relationships between P_f and FS by slip circle method with different penetration depths

3.2 Comparison of Three Methods

The S and T methods are applicable for wide excavations ($H_e \leq B$). Figures 4 and 5 show the relationships between the probability of failure and factor of safety by comparing the SC method with the T method at different values of excavation depth $H_{e.}$ Two H_e/B ratios ($H_e/B = 0.5$ and $H_e/B = 1$) are adopted. For wide excavations, the

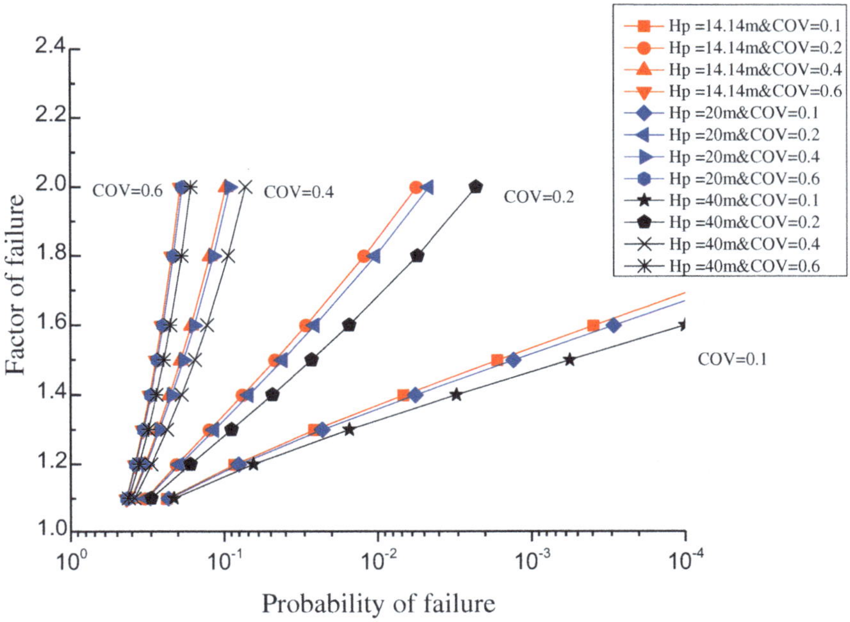

Fig. 2. Relationships between P_f and FS by Terzaghi's method with different penetration depths

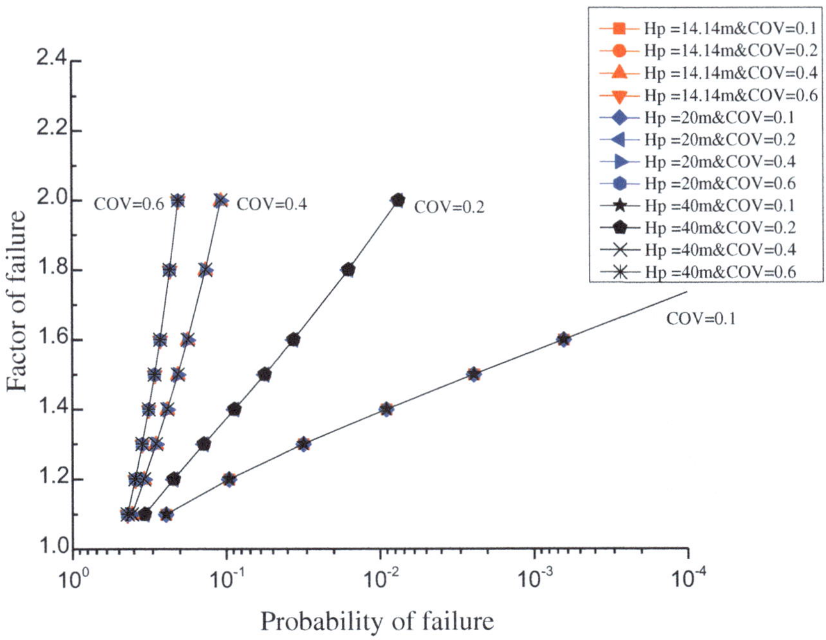

Fig. 3. Relationships between P_f and FS by Bjerrum-Eide's method with different penetration depths

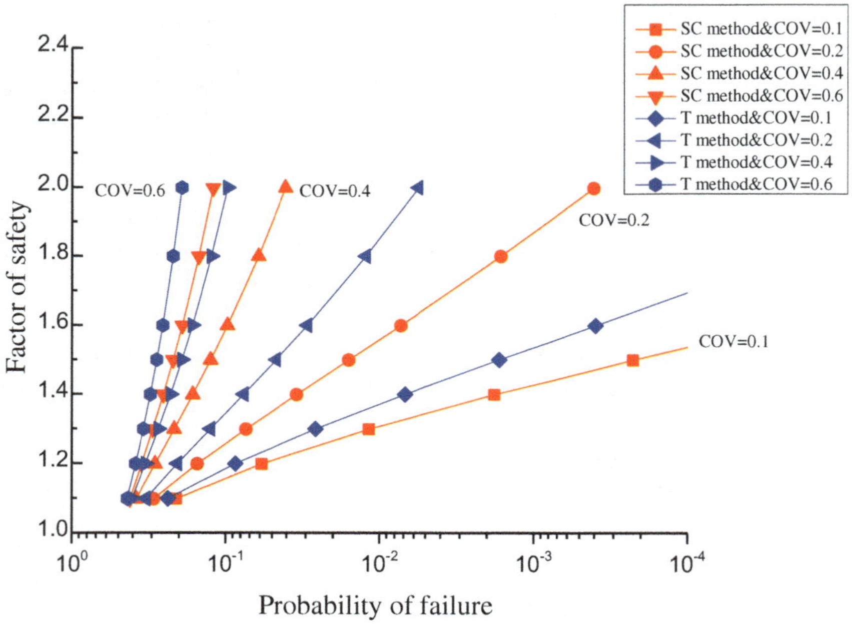

Fig. 4. Comparison of the slip circle method and Terzaghi's method with $H_e/B = 0.5$

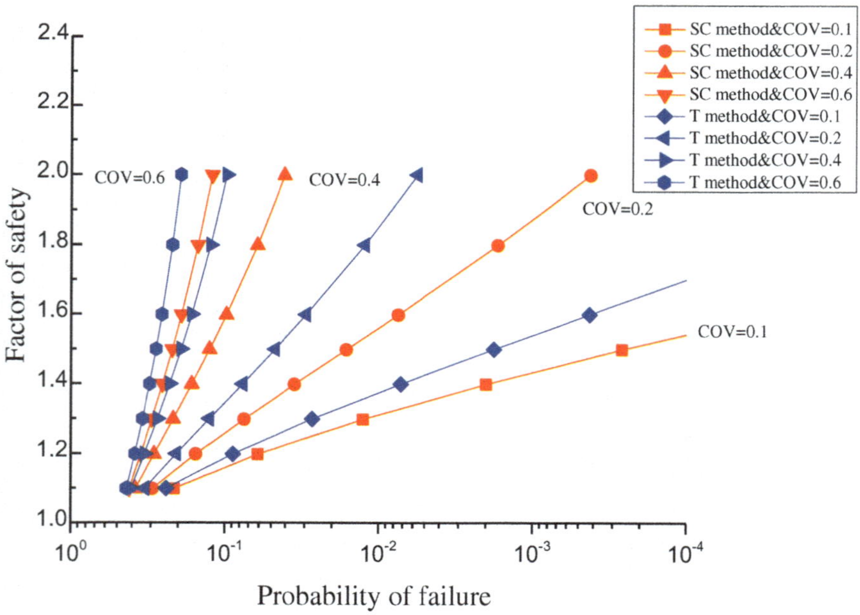

Fig. 5. Comparison of the slip circle method and Terzaghi's method with $H_e/B = 1$

same level of factor of safety results in a smaller P_f for the SC method than for the T method. For example, in Fig. 4, when FS = 1.3 and H_e = 10 m, the P_f of SC method is 0.0117, which is about half of that for T method (0.0258). Moreover, a required FS don't only present one certain value of P_f. The P_f values increase as the COV_{su} values increase.

Figures 6 and 7 present the results on deep excavation design (H_e>B). The P_f and FS values are calculated based on S and BE methods. Figures 6 and 7 show the relationships between the probability of failure and factor of safety by comparing the SC method with the T method using two different H_e/B ratios (H_e/B = 1.5 and H_e/B = 2). For deep excavations, the same level of factor of safety results in a smaller P_f when the SC method is used than when the BE method is used. For instance, in Fig. 7, when FS = 1.3 and H_e = 40 m, the P_f of the SC method is 0.0131, which is about one third of that for BE method (0.0330). The value of H_e/B increases from 1.5 to 2, and the relationship between the probability of failure and factor of safety is one of positive correlation. The P_f values increase as the COV_{su} increases.

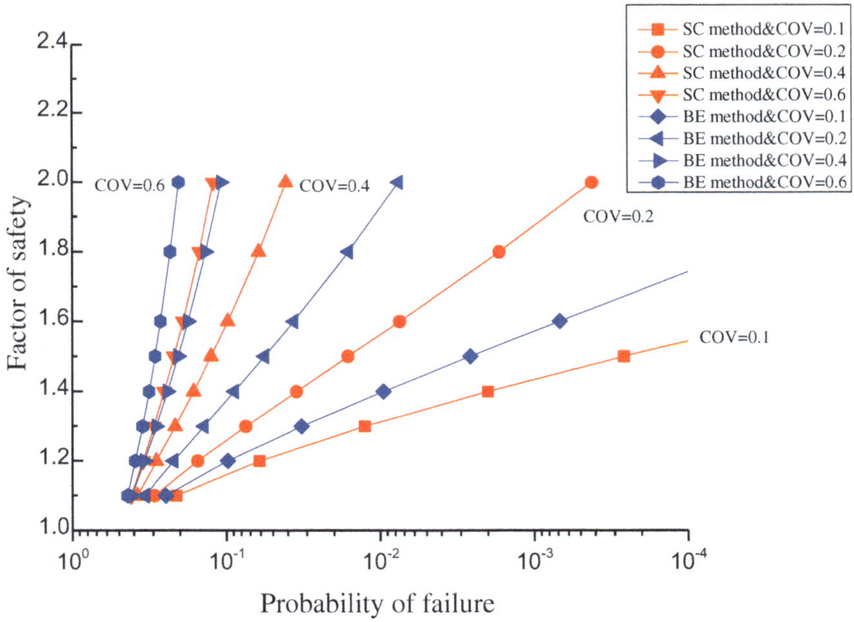

Fig. 6. Comparison of the slip circle method and Bjerrum-Eide's method with H_e/B = 1.5

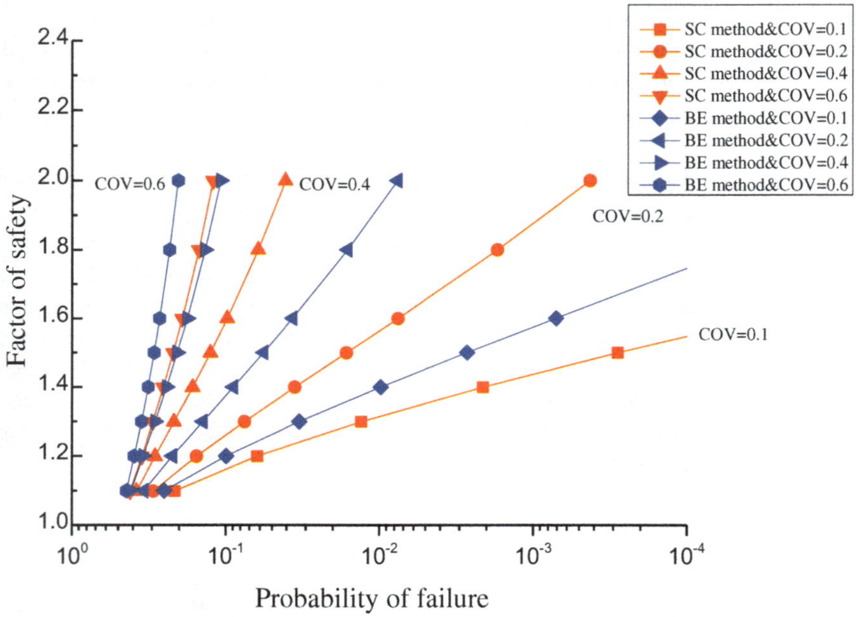

Fig. 7. Comparison of the slip circle method and Bjerrum-Eide's method with $H_c/B = 2$

4 Conclusion

This study employs Terzaghi's, Bjerrum-Eide's and the slip circle methods to consider the influence of basal heave stability on reliability-based design of braced excavations. For this purpose, the P_f − FS design charts are presented and the relationship between P_f and FS shows that a large FS don't mean a small level of failure. The comparison demonstrates that the SC method is more economical than the other two methods for the same target P_f. In other words, for a required range of P_f values, such as when the value should be smaller than 0.001 in RBD, the minimum required FS value for the SC, T, and BE method is 1.41, 1.50 and 1.50, respectively. In addition, the T method is more sensitive to the penetration depth of the wall than the other two methods. In conclusion, these results provide a guideline for the method selection of basal heave stability evaluation in geotechnical practice.

Acknowledgements. The authors gratefully acknowledge financial support from the University of Macau Research Fund (MYRG2015-00112-FST) and the Macau Science and Technology Development Fund (FDCT): 125/2014/A3.

References

American Association of State Highway and Transportation Officials: AASHTO LRFD Bridge specifications (SI unit), First Ed. AASHTO, Washington, DC (1994)

Bjerrum, L., Eide, O.: Stability of strutted excavations in clay. Geotechnique **6**(1), 32–47 (1956)

Eurocode 7: Geotechnical design, general rules. European committee for standardization. Danish Geotechnical Institute, Copenhagen (1993)

Hasofer, A.M., Lind, N.C.: Exact and invariant second-moment code format. J. Eng. Mech. Div. **100**(1), 111–121 (1974)

Honjo, Y., Kusakabe, O.: Proposal of a comprehensive foundation design code: Geo-code 21 ver. 2. In: International Workshop on Foundation Design Codes and Soil Investigation in view of International Harmonization and Performance Based Design, pp. 95–101 (2002)

JSA: Guidelines of Design and Construction of Deep Excavation. Japanese Society of Architecture, Tokyo, Japan (1988)

Li, K.S., Lee, I.K., Lo, S.C.R.: Limit state design in geotechnics. In: Proceedings of the Conference on Probabilistic Methods in Geotechnical Engineering, pp. 29–42. Balkema, Rotterdam (1993)

Low, B.K.: Reliability-based design applied to retaining walls. Geotechnique **55**(1), 63–75 (2005)

Nowak, A.S.: Reliability of Structures. McGraw-Hill, New York (2000)

Phoon, K.K. (ed.): Reliability-Based Design in Geotechnical Engineering: Computations and Applications. CRC Press, London (2008)

Qi, X.H., Zhou, W.H.: An efficient probabilistic back-analysis method for braced excavations using wall deflection data at multiple points. Comput. Geotech. **85**, 186–198 (2017)

Qi, X.H., Zhou, W.H., Li, D.Q.: Uncertainty analysis of slope stability considering geological heterogeneity. In: Proceedings of 12th International Symposium on Landslides and Engineered Slopes – Experience, Theory and Practice, pp. 1685–1693 (2016). doi:10.1201/b21520-209

Skempton, A.W.: The bearing capacity of clays. British Building Research Congress, Institute of Civil Engineers, pp. 180–189 (1951)

Su, L.J., Xu, X.Q., Geng, X.Y., Liang, S.Q.: An integrated geophysical approach for investigating hydro-geological characteristics of a debris landslide in the Wenchuan earthquake area. Eng. Geol. **219**, 52–63 (2017)

TGS: Design Specifications for the Foundation of Buildings. Taiwan Geotechnical Society, Taipei, Taiwan (2001)

Terzaghi, K.: Theoretical Soil Mechanics. Wiley, New York (1943)

Xu, X.Q., Su, L.J., Zhu, H.H.: Analysis on shear wave velocity structure of a gravel landslide based on dual-source surface wave method. Landslides **14**, 1127–1137 (2016). doi:10.1007/s10346-016-0780-9

Zhou, W.H., Garg, A., Garg, A.: Study of the volumetric water content based on density, suction and initial water content. Measurement **94**, 531–537 (2016). doi:10.1016/j.measuremen-t.2016.08.034

Risk Assessment of Slope Accounting for the Cross-Correlation of Soil Properties

Hong-zhan Cheng(✉) and Jian Chen

Key Laboratory of Geomechanics and Geotechnical Engineering,
Institute of Rock and Soil Mechanics, Wuhan 430071, China
cheng2217@126.com

Abstract. The spatial variability of soil properties is inherent and apparent, and the cross-correlation of them is also exhibited. Research on risk of slope failure considering the cross-correlation of soil properties is not sufficient now. Accounting for this disadvantage, the cross-correlation random fields describing the spatial variability of soil properties are generated by using the covariance matrix decomposition method, and the procedure of the method is briefly introduced. Failure probability and slip mass are employed to quantify the risk of failure. A simple method programming method with the let-in Fish code in $FLAC^{3D}$ is proposed to determine slip mass. Taking a frictional/cohesive slope for example, a series of sensitivity analyses are performed through Monte Carlo simulation, and effects of the cross-correlation of soil shear parameters on probability, mechanism and risk of failure are investigated. The results indicate that the cross-correlation has a significant influence on the stability of slope. Ignoring the negative cross-correlation of shear strength parameters, both failure probability and failure risk will be overestimated. With the cross-correlation coefficient increasing from negative to positive, both failure probability and failure risk increase, and the critical slip surface becomes more scattered. Slip mass is more sensitive to the variation of cross-correlation coefficient than safety of factor. The paper also suggests that the influence of failure mechanism should be considered in risk analysis of slope. These conclusions can improve our understanding about the effect of the cross-correlation of soil properties on the slope stability, and provide reference for risk assessment of slope failure.

Keywords: Slope · Cross-Correlation · Random field · Risk of failure

1 Introduction

Due to the complex formation process and various internal and external geological forces, the physical and mechanical properties of natural rock and soil are characterized by spatial variability [1, 2]. It can be described that properties of soil at different locations are not consistent. The nearer the location, the closer the properties, and indicates the stronger the correlation of the two. The correlation decreases along with the increase of the distance between the locations, and the difference of their properties be larger. The correlation is called auto-correlation for single parameter, and cross-correlation for two

R. Chen et al. (eds.), *Proceedings of the 2nd International Symposium on Asia Urban GeoEngineering*, Springer Series in Geomechanics and Geoengineering,
https://doi.org/10.1007/978-981-10-6632-0_22

or more parameters. As far, random field proposed by Vanmarcke [3] is widely used to describe the spatial variability of soil properties associating with high accuracy.

In random field model, an additional parameter, namely, scale of fluctuation was proposed and introduced in comparison with random variable model. In addition, the correlation function is used to establish the relation between the 'point' and 'local averaging', which takes the structure of parameters in space into full consideration. At present, statistical analysis on the correlation distance of formation parameters in different areas has been performed by numerous researchers, and a lot of useful results have been obtained. Meanwhile, it also provides valuable references for parameter value in the following projects [4–7]. On the other hand, Fan [8] counted the results of 26 groups of shear tests, and found that there is a negative correlation between cohesion c and internal friction angle φ, and the correlation coefficients are mostly between −0.3 and −0.9. Tang et al. [9] listed the correlation coefficients between c and φ of rock and soil mass in different areas, and all the correlation coefficients are less than 0, i.e., the negative correlation is also observed. Although research on the cross-correlation of rock and soil parameters is not deep enough due to the limitation of the test conditions and the sample quantity, but it can be determined that the cross-correlation between parameters is truly existed.

In terms of slope stability, several reliability analysis methods were proposed considering the spatial variability of soil properties. Then, effects of probability distribution form and auto-correlation distance of soil parameters on failure probability and failure mechanism were explored [10–13]. The results indicate that ignoring the spatial variability induces a significantly overestimation of the reliability of slope, and provide a useful reference for the following studies. However, only the variability of one or several independent parameters is considered in their study, and few attempts have been made to study the influence of the cross-correlation of them. Considering the cross-correlation of parameters, Cho [12], Tang et al. [9] analyzed the influence of cross-correlation coefficients of c and φ on slope failure probability. It should be pointed out that the results of them mainly concentrate on the statistical analysis of slope failure probability, but seldom involve the consequence assessment and risk evaluation of slope failure. Hicks [14], Huang [15], Xiao [16], Ali [17] and Li [18, 19] adopted random field model to calculate safety of factor and the volume of slip mass considering the spatial variability of single soil parameter, and established a method for quantifying the damage caused by slope failure. In general, study on the risk of slope failure considering the spatial variability of parameters needs to be further studied, particularly involving the cross-correlation of soil properties.

In the present paper, a risk assessment method for slope failure was proposed. Covariance matrix decomposition method was employed to the simulation of the cross-correlation of shear strength parameters. Combined random field theory with finite difference software, Random finite difference method (RFDM) for slope reliability and failure risk analysis was developed. Taking a two-dimensional friction/cohesive soil slope for example, the influence of the cross-correlation of shear strength parameters on the risk of slope failure was studied based on the Monte Carlo method.

2 Simulation of Auto-Correlation and Cross-Correlation Random Field

At present, there are several simulation methods for random field in common used, including covariance matrix decomposition method, local subdivision average method and fast Fourier transform method (FFT) [20]. Among which, covariance matrix decomposition method has the advantages of simple calculation, high accuracy and strong applicability to random field simulation of anisotropic or irregular grids. Therefore, in this paper, this method is applied to simulate random field of soil parameters. A brief introduction is given below to the simulation process of auto-correlation random field and cross-correlation random field, respectively.

Firstly, the domain studied is represented by the discrete points. Afterwards, correlation coefficients between any two discrete points x_i, x_j $(i, j = 1,2,\ldots,n)$ are calculated by using the correlation function based on the coordinate of them. Then, covariance matrix $C_{n\times n}$ is established. Since the correlation function is only related to the relative distance between any two points, therefore the matrix $C_{n\times n}$ is a positive definite one, and it can be decomposed by Cholesky decomposition:

$$C = LU = LL^T \tag{1}$$

where: L and U are lower triangular and upper triangular matrix, respectively. L^T is the transpose of L.

Let Y indicates the column vector composed of n independent normal variables, then the normal Gaussian auto-correlation random field Z can be represented as:

$$Z = LY \tag{2}$$

According to the distribution for uncertain parameters, simple mathematical transfer based on the Gaussian random field Z can realize the simulation of non- Gaussian or non-stationary auto-correlation random field.

In terms of the simulation of the cross-correlation random field, both the cross-correlation and auto-correlation should be considered [21, 22]. Covariance matrix decomposition method can be also used to simulate the cross-correlation random field model with a little more complicated procedure. Take the simulation of cross-correlation random field of two parameters A_1 and A_2 for example here. The cross-correlation coefficient between them is ρ_{12}. $\rho_{12} = \pm 1$ represents a liner correlation between the two, $\rho_{12} = 0$ indicates that the two are independent of each other.

Auto-correlation covariance matrix C_{11} and C_{22} are constructed respectively as the simulation of the auto-correlation random field. The cross-correlation covariance matrix C_{12} is established according to the cross-correlation function supposed. In general, A_1 and A_2 have the same auto-correlation function, and the cross-correlation function of them can be seen as the produce of ρ_{12} and the auto-correlation function. Suppose the two realizations of cross-correlation random field R_1 and R_2 have the following expression:

$$\left.\begin{array}{l} R_1 = L_{11}U_1 \\ R_2 = L_{12}U_1 + L_{22}U_2 \end{array}\right\} \tag{3}$$

where: U_1 and U_2 are vectors composed of n independent normal variables, and the two are independent of each other. n dimensions lower triangular matrix L_{11}, L_{12} and L_{22} are to be solved.

To ensure that both R_1 and R_2 meet the auto-correlation and cross-correlation, R_1 and R_2 should satisfy the following expressions, and matrix L_{11}, L_{22} and L_{12} can be determined by solving the expressions.

$$\left.\begin{aligned} Cov(R_1, R_1) &= C_{11} = L_{11}L_{11}^T \\ Cov(R_1, R_2) &= C_{12} = L_{11}L_{12}^T \\ Cov(R_2, R_2) &= C_{22} = L_{12}L_{12}^T + L_{22}L_{22}^T \end{aligned}\right\} \tag{4}$$

Take slope model in Cho [12] for example, considering the influence of spatial variability of c and φ. Suppose both of them follow lognormal distribution, the mean and coefficient of variation are (10 kPa, 0.3) and (30°, 0.2), respectively. Considering the spatial variability of soil parameters in vertical direction is obviously stronger than that in horizontal direction, horizontal auto-correlation distance θ_h and vertical auto-correlation distance θ_v are assumed as 20 m and 2 m. The auto-correlation and cross-correlation of c and φ are characterized by exponential correlation functions (Eqs. 5 and 6). Figure 1 shows the two-dimensional slope random fields of three cross-correlation coefficients ($\rho_{c,\varphi} = -0.5$,0 and 0.5) based on the assumption that c and φ have the same auto-correlation.

$$\rho_{c,c(\phi,\phi)}(\tau_h, \tau_v) = \exp(-\frac{|\tau_h|}{\theta_h} - \frac{|\tau_v|}{\theta_v}) \tag{5}$$

$$\rho_{c,\phi}(\tau_h, \tau_v) = \rho_{c,\phi}\exp(-\frac{|\tau_h|}{\theta_h} - \frac{|\tau_v|}{\theta_v}) \tag{6}$$

In Fig. 1, the left and the right figures show random field of c and φ, respectively. As can be seen from Fig. 1, considering the spatial variability of shear strength parameters, c and φ of each element of them are different, but strength parameters between adjacent elements are relatively close. The variation degree of shear strength parameters in horizontal direction is obviously smaller than that in vertical direction, which is due to the large horizontal auto-correlation distance. In Fig. 1a, when $\rho_{c,\varphi}$ is -0.5, namely, c and φ show a negative correlation, the value of c and φ has the opposite trend in random fields. In the same location, when the value of c is larger, the value of φ is smaller. In Fig. 1c, when c and φ show a positive correlation, the value of c and φ has the same trend. Therefore, obvious weak or strong zones are formed in the random fields, which will have an important impact on the stability of the slope. When $\rho_{c,\varphi}$ is 0, namely, c and φ are independent of each other, there is no obvious correlation between the two. Considering the cross-correlation between parameters, random fields would be constrained to some extent, and the dispersion of stochastic analysis results is changed. This conclusion will be illustrated in detail below.

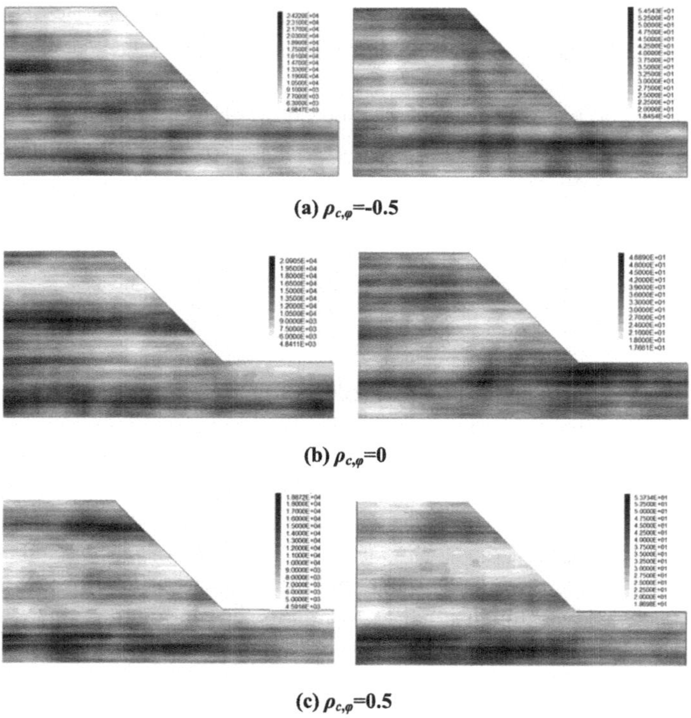

Fig. 1. Cross-correlation random fields

3 Risk Assessment Method for Slope Failure

In traditional slope risk analysis, the method for calculating factor of safety is usually based on the assumed critical slip surface, but it is not suitable for statistical analysis considering the spatial variability of soil parameters. In general, the larger the scale of slip mass, the more serious the consequence. Huang [15] put forward the method for quantify risk of slope failure by volume of slip mass, and established the formula with the following expression:

$$R = C \times P_f \tag{7}$$

where R indicates the slope risk, C represents the quantitative consequence of slope failure, P_f is probability of failure. It is necessary to state that this paper focuses on the study of effects of the cross-correlation on slope failure risk. Volume of slip mass is also employed as an approximate quantitative index of consequence of slope failure. Note that the critical slip surface of each stochastic calculation is not constant, which induces an uncertain slip mass. Therefore, Huang [15] made further improvements to the Eq. (7):

$$R = \sum_{i=1}^{n_f} P_{fi} \times C_i \tag{8}$$

where: n_f is the number of stochastic calculation; P_{fi} is probability of failure corresponding to the consequence of C_i obtained from ith random field.

The method for searching the critical slip surface based on the calculated largest shear strain increment is proposed and programed using the Fish language in FLAC^{3D}. Imaging a point at the top of slope, multiple rays are scattered from the point to the slope. Select the element of the maximum shear strain increment as the target one on each ray, connect all the target elements, and the critical slip surface is formed. The point imagined above should avoid the situation that the ray is tangent to the slip surface as much as possible. Otherwise, the accuracy of the target elements determined will be reduced.

In Fig. 2, two typical slope cases are selected to validate the proposed method. The frictional/cohesive slope in Case 1 is selected from references [13], slope height is 5 m with a ratio 1:2. The frictional/cohesive slope in Case 2 is selected for references [12, 15], slope height is 10 m with a ratio 1:1. As can be seen from Fig. 2, the critical slip surface determined agrees very well with that obtained by limit equilibrium method (Bishop method), and the corresponding factor of safety calculated by strength reduction method in FLAC^{3D} is also basically coincident. Therefore, the method used to search the slip surface and calculate the volume of slip mass is proved.

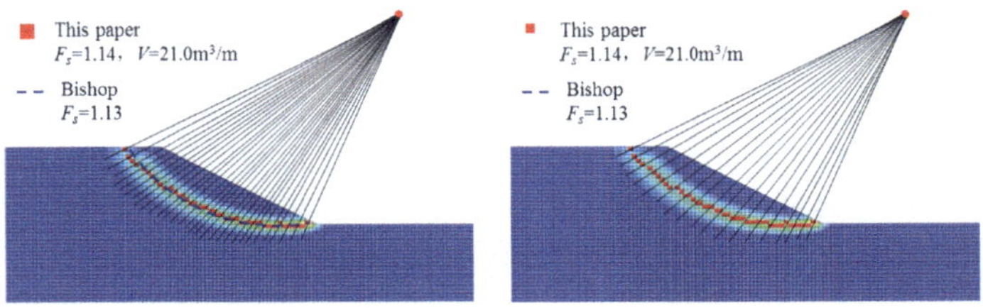

(a) Case 1 **(b) Case 2**

Fig. 2. Illustration of the research of slope slip surface

Figure 3 shows the procedure of the risk analysis on slope failure due to the cross-correlation between soil parameters by random finite difference method (RFDM), including the following 5 steps.

(1) Determine the geometry of slope, select the constitutive model of soil, and determine statistics of uncertain parameters (mean, standard deviation, distribution, auto-correlation distance, cross-correlation coefficient and correlation function).
(2) Discrete the slope into points and simulate the auto-correlation and cross-correlation random fields by using the covariance matrix decomposition method.
(3) Substitute random fields of soil parameters into numerical models. Then, factor of safety is calculated by using the strength reduction method, and slip mass is determined by using the method mentioned above.
(4) Within the frame of Monte-Carlo simulation, step 2 and 3 are repeated until the number of stochastic analysis N is greater than the assumed one N^*.
(5) Calculate the risk R by using Eq. (8).

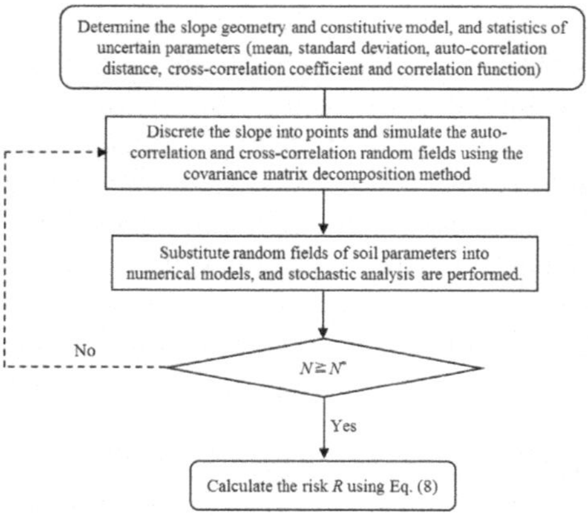

Fig. 3. Implementation procedure of the risk analysis of slope failure using RFDM

4 Case Study

4.1 Format Illustration

In this part, a typical slope model is adopted to study the influence of spatial auto-correlation and cross-correlation of c and φ on the risk of slope failure. The slope selected is the same as the one shown in Fig. 2b, which is convenient to be compared with the results of Cho [12]. Considering that the parameters E and ν have little effect on the results and the degree of spatial variability of γ is low, therefore soil parameters other than c and φ are considered as determined values, and the specific parameters are selected as shown in the Table 1.

Table 1. Physico-mechanical parameters in the model

Soil parameters	Means	Coefficients of variation	Notes
Weight, γ/(kN/m^3)	20	/	Constants
Poisson's ratio, ν	0.3	/	
Elastic modulus, E/MPa	100	/	
Friction angle, φ/(°)	30	0.2	Lognormal
Cohesion, c/kPa	10	0.3	Exponential correlation functions, $\theta_h = 20$ m; $\theta_v = 2$ m

4.2 Probability of Failure

In stochastic analysis, factor of safety F_s is obtained by each stochastic calculation. In this paper, 1000 stochastic calculations are performed for each cross-correlation coefficient $\rho_{c,\varphi}$ between c and φ, and statistics of F_s (mean and coefficient of variation) for various $\rho_{c,\varphi}$ are analyzed, as shown in Fig. 4. The mean of F_s decreases linearly with the increase of $\rho_{c,\varphi}$. Obviously, when the correlation is negative, the slope stability is higher, and the slope stability is relatively poor when the correlation is positive, which is consistent with the conclusion of references [12, 23]. Meanwhile, compared with the deterministic solution (Fig. 2b), the means of F_s are smaller, and the minimum mean of F_s is 1.15 with a decrease of 5%. As can be seen from Fig. 4b, coefficient of variation of F_s increases linearly with the increase of $\rho_{c,\varphi}$. It means that the larger the $\rho_{c,\varphi}$, the higher the scatter of F_s in stochastic analysis. At the same time, it is shown that the results obtained by using RFDM and random limit equilibrium method of Cho [12] agree well with each other. Because factor of safety depends on the average shear strength of all cross elements of slip surface, coefficient of variation of F_s is smaller than that of input parameters c and φ.

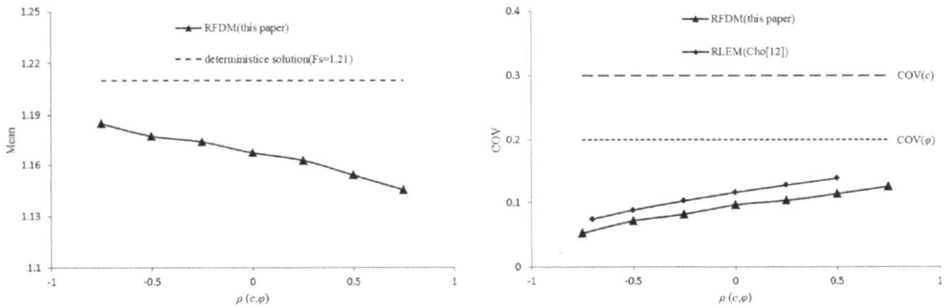

(a) Mean **(b) Coefficient of variation**

Fig. 4. Effects of the cross-correlation of c, φ on the mean and coefficient of variation of F_s

The cumulative probability curve of F_s is given in Fig. 5. Clearly, the smaller the $\rho_{c,\varphi}$, the more concentrated the curves. Figure 6 shows the relationship of P_f and $\rho_{c,\varphi}$. As expected, P_f increases with the increasing $\rho_{c,\varphi}$, and the increase is obvious, which is basically consistent with the conclusion of references [12], and the importance of cross-correlation of c and φ in reliability analysis of slope is also indicated.

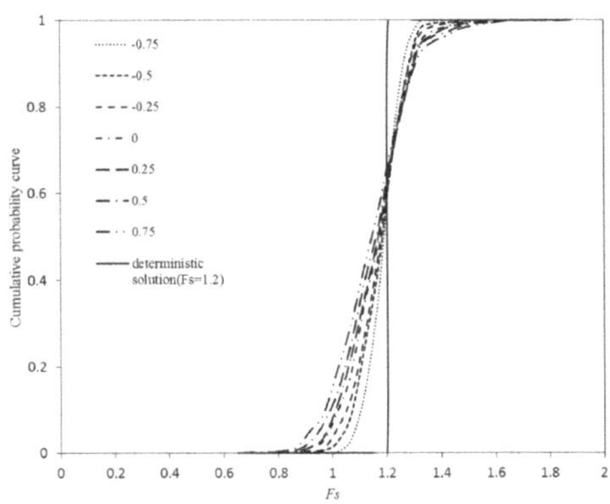

Fig. 5. Effects of the cross-correlation of c, φ on the cumulative probability of F_s

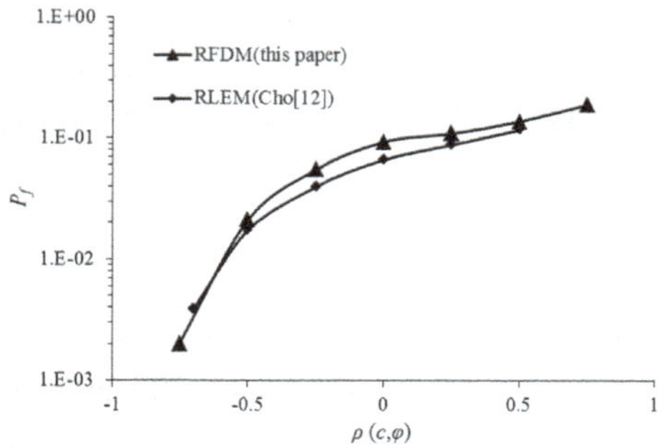

Fig. 6. Effects of the cross-correlation of c, φ on the failure probability of F_s

4.3 Slip Mass

Possible failure mechanisms in stochastic analysis are studied before the statistical analysis on the volume of slip mass. Three typical slope failure mechanisms are selected in Fig. 7. The grey scale shows the distribution of random field c and φ. The left is the distribution of random field c, and the right is the distribution of random field φ. The deeper the color, the higher of the shear strength. The line indicates the critical slip

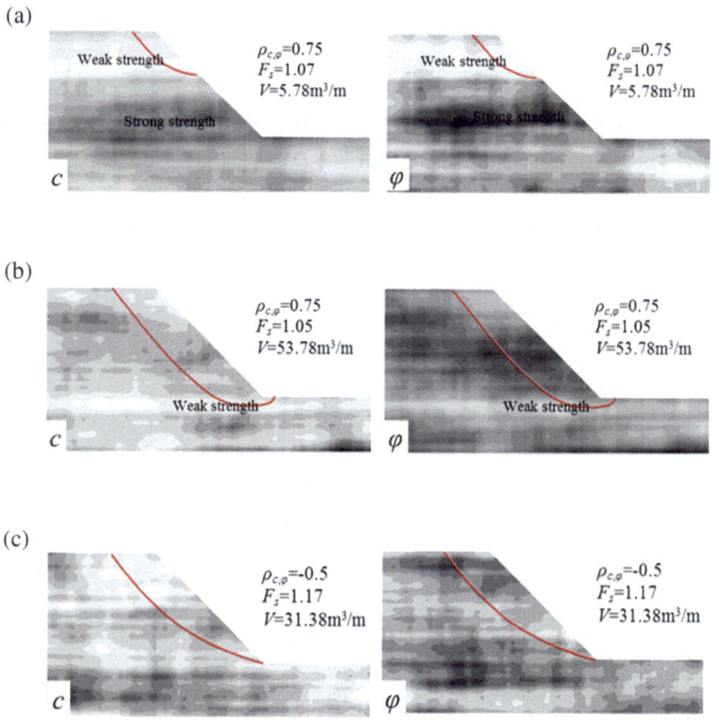

Fig. 7. Three typical slope failure mechanisms and the corresponding random fields

surface determined by the maximum shear strain increment. Obviously, considering the spatial variability of c and φ, the critical slip surface is no longer constant. Figure 7a shows a typical face failure for a positive correlation ($\rho_{c,\varphi} = 0.75$). At the top of slope, c and φ are small and shear strength is low, therefore slope failure occurs first at the top. Figure 7b shows a toe failure, because there is a weak zone below the toe, so the failure depth is slightly deeper than deterministic results. The failure mechanism shown in Fig. 7c is close to the deterministic result. When $\rho_{c,\varphi}$ is −0.5, it's difficult to form the weak zone of smaller c and φ or the strong zone of larger c and φ in random fields, which makes the critical slip surface in stochastic analysis is close to the deterministic solution. Predictably, when $\rho_{c,\varphi}$ is less than 0, the distribution of the critical slip surface will be more concentrated.

Figure 8 illustrates effects of $\rho_{c,\varphi}$ on coefficient of variation of V. It can be seen that coefficient of variation of V is obviously larger in comparison with that of F_s, which indicates the effect of $\rho_{c,\varphi}$ on failure mechanism is more obvious. The cumulative probability curve of V under different $\rho_{c,\varphi}$ is given in Fig. 9. As can be seen, the distribution of V becomes more scattered when $\rho_{c,\varphi}$ is increased from −0.75 to 0.75, which is consistent with the change of F_s. When there is a negative correlation between c and φ, it's difficult to form a relatively concentrated weak or strong zone in random fields, so the location of the critical slip surface fluctuates little.

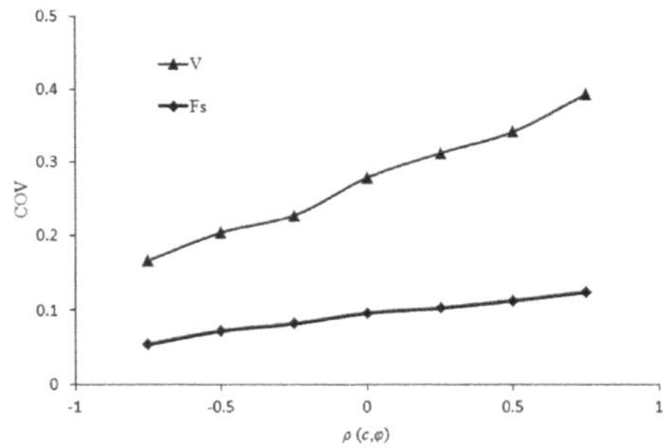

Fig. 8. Effects of the cross-correlation of c, φ on the coefficient of variation of V

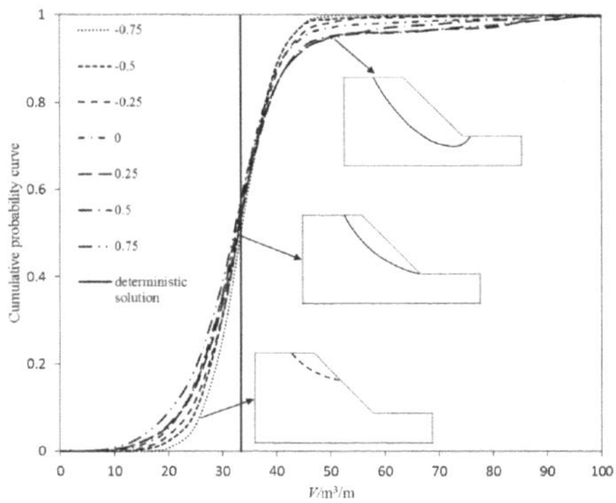

Fig. 9. Effects of the cross-correlation of c, φ on the cumulative probability of slide volume

When there is a positive correlation between c and φ, and there will be an obvious weak zone and strong zone in random fields (Fig. 7), which will affect the location of the critical slip surface to some extent, and lead to the scatter of V. Taking $\rho_{c,\varphi} = -0.5$ and $\rho_{c,\varphi} = 0.5$ for example, Fig. 10 shows the distribution of V in the form of a scatter plot. The line shows the cumulative probability distribution of F_s. Scatterplot shows the corresponding distribution of V. Obviously, when $\rho_{c,\varphi}$ is 0.5, the distribution of the scatter plot is more scattered, and the probability of large-scale slope sliding is also higher.

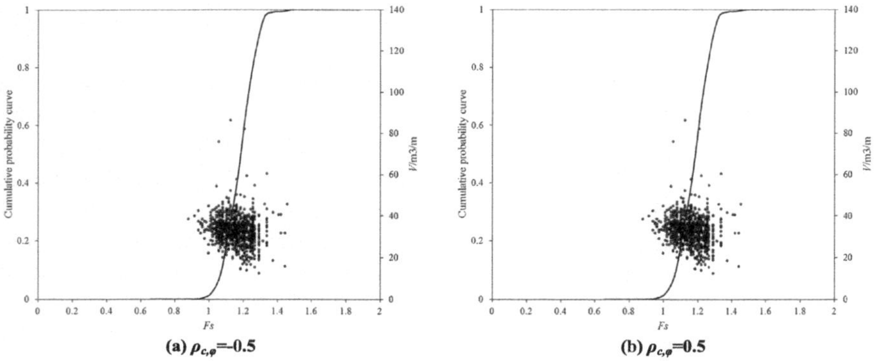

Fig. 10. Relationship of factor of safety and slide volume for $\rho_{c,\varphi} = -0.5$ and $\rho_{c,\varphi} = 0.5$

4.4 Risk Assessment

Figure 11 shows the relationship between the risk and the cross-correlation coefficient $\rho_{c,\varphi}$, and it is compared with the failure probability curve P_f. According to Eq. (8), the risk $R1$ is obtained by the accumulated volumes of slip mass when factor of safety is less than 1 in stochastic analysis. The risk $R2$ is the product of failure probability and volume of slip mass (33.4 m^3/m) in deterministic analysis without considering the variety of slope failure mechanisms. As can be seen, both $R1$ and $R2$ increase with the increasing $\rho_{c,\varphi}$, which is consistent with the change of P_f. Ignoring the influence of failure

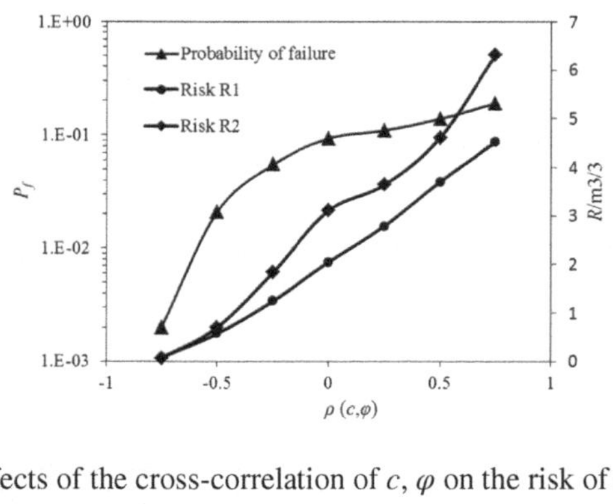

Fig. 11. Effects of the cross-correlation of c, φ on the risk of slope failure

mechanism leads to an overestimation of risk of slope failure by comparing the curve of $R1$ with $R2$, and the importance of failure mechanism on risk assessment is also illustrated.

5 Conclusions

Considering the spatial variability of shear strength parameters, the present paper investigated the effect of the cross-correlation of them on the slope stability. The following conclusions are obtained.

(1) The cross-correlation of shear strength parameters shows a significant effect on probability of failure. The increasing cross-correlation coefficient leads to a larger probability of failure. In other words, ignoring the negative cross-correlation reduces an underestimation of slope stability.
(2) The cross-correlation of shear strength parameters can affect the distribution of weak and strong zones in random fields. Then, multi failure mechanisms are obtained in stochastic analysis, which is different from deterministic analysis.
(3) The comparison between coefficient of variation of factor of safety and slip mass shows that the cross-correlation has more significant effect on the latter.
(4) The calculation risk increases with the cross-correlation coefficient increasing, which is similar as the change of failure probability.

It should be noted that this paper merely investigated the slope stability in homogenous soil. Further studies should be made to study the reliability of multilayered soil slope stability. In addition, ignoring the change of failure mechanism reduces an overestimate of slope risk, which is only used for the slope case in the paper. More attempts can be performed in terms of risk assessment on various slopes.

References

1. Zhang, Z., Liu, S.: The variability and evaluation of parameters of rock and soil. China Civ. Eng. J. **28**(6), 43–51 (1995)
2. Zhao, H., Feng, X.T., Zhang, D.X., et al.: Spatial variability of geomechanical parameter estimation via ensemble Kalman filter. Rock Soil Mech. **28**(10), 2219–2223 (2007)
3. Vanmarcke, E.: Random Fields: Analysis and Synthesis. MIT Press, Cambridge (1983)
4. Li, X., Xie, K., Yu, Y.: Research of the characteristics of correlation distance on soil properties indexes. China Civ. Eng. J. **36**(8), 91–95 (2003)
5. Phoon, K.K., Kulhawy, F.H.: Characterization of geotechnical variability. Can. Geotech. J. **36**(4), 612–624 (1999)
6. Liu, C.N., Chen, C.: Spatial correlation structures of CPT data in a liquefaction site. Eng. Geol. **111**(1), 43–50 (2010)
7. Lloret-Cabot, M., Fenton, G.A., Hicks, M.A.: On the estimation of scale of fluctuation in geostatistics. Georisk Assess. Manage. Risk Eng. Syst. Geohazards **8**(2), 129–140 (2014)
8. Fan, M.: Probabilistic properties of shear strength parameters φ, c of refilled clay. Chin. J. Geotech. Eng. **19**(4), 100–104 (2000). (in Chinese)

9. Tang, X.S., Li, D.Q., Rong, G., et al.: Impact of copula selection on geotechnical reliability under incomplete probability information. Comput. Geotech. **49**(4), 264–278 (2013)
10. Hicks, M.A., Samy, K.: Influence of heterogeneity on undrained clay slope stability. Q. J. Eng. Geol. Hydrogeol. **35**(1), 41–49 (2002)
11. Griffiths, D.V., Huang, J.S., Fenton, G.A.: Influence of spatial variability on slope reliability using 2-D random fields. J. Geotech. Geoenvironmental Eng. **135**(10), 1367–1378 (2009)
12. Cho, S.E.: Probabilistic assessment of slope stability that considers the spatial variability of soil properties. J. Geotech. Geoenvironmental Eng. **136**(7), 975–984 (2010)
13. Qi, X.-H., Li, D.-Q., Zhou, C.-B., et al.: Stochastic analysis method of critical slip surfaces in soil slopes considering spatial variability. Chin. J. Geotech. Eng. **35**(4), 745–753 (2013)
14. Hicks, M.A., Spencer, W.A.: Influence of heterogeneity on the reliability and failure of a long 3D slope. Comput. Geotech. **37**(7), 948–955 (2010)
15. Huang, J., Lyamin, A.V., Griffiths, D.V., et al.: Quantitative risk assessment of landslide by limit analysis and random fields. Comput. Geotech. **53**(3), 60–67 (2013)
16. Xiao, T., Li, D.Q., Cao, Z.J., et al.: Three-dimensional slope reliability and risk assessment using auxiliary random finite element method. Comput. Geotech. **79**, 146–158 (2016)
17. Ali, A., Huang, J., Lyamin, A.V., et al.: Simplified quantitative risk assessment of rainfall-induced landslides modelled by infinite slopes. Eng. Geol. **179**(10), 102–116 (2014)
18. Li, D.Q., Xiao, T., Cao, Z.J., et al.: Enhancement of random finite element method in reliability analysis and risk assessment of soil slopes using subset simulation. Landslides **13**(2), 1–11 (2016)
19. Li, D.Q., Xiao, T., Cao, Z., et al.: Slope risk assessment using efficient random finite element method. Rock Soil Mech. **37**(7), 1994–2003 (2016)
20. Fenton, G.A.: Simulation and analysis of random fields. J. Publ. Health **29**(1), 40–47 (1990)
21. Griffiths, D.V., Fenton, G.A.: Bearing-capacity prediction of spatially random c-ϕ, soils. Can. Geotech. J. **40**(1), 54–65 (2003)
22. Chen, J., Wang, Z., Rong, H.: 'Two sides of one' method for inversion of correlated parameters random fields. Rock Soil Mech. **37**(6), 1773–1780 (2016). (in Chinese)
23. Li, D.Q., Qi, X., Zhou, C., et al.: Reliability analysis of infinite soil slopes considering spatial variability of soil parameters. Chin. J. Geotech. Eng. **35**(10), 1799–1806 (2013)

Tunnel and Underground Engineering

A Study on the Influence of Water Layer on the Seismic Response of Immersed Tunnel

X. Cheng[1(✉)], L. Jing[1], J. Cui[2], Y. Li[1], R. Dong[1], and G. Wang[2]

[1] Key Laboratory of Earthquake Engineering and Engineering Vibration of China Earthquake Administration, Institute of Engineering Mechanics, China Earthquake Administration, Harbin, China
chengxinjuniem@163.com
[2] School of Civil Engineering, Guangzhou University, Guangzhou, China

Abstract. Immersed tunnel is widely adopted in underwater transportation system. According to literature reviews and current studies, no experiments for the influence of water layer on the dynamic behavior of immersed tunnel under seismic loading have been reported. In order to solve this problem, a series of shaking table tests were conducted in this study. Two groups of tests are designed. One is the immersed tunnel model in saturated sand, the other one is the immersed tunnel in saturated sand under a water layer. This study highlights the experimental results composed of the excess pore pressure of soil layer, the acceleration of tunnel model, and the peak dynamic strain of tunnel model. These results indicate that for vertical earthquake loading, the water layer influences little on the seismic behavior of soil. Three is no apparent difference between pore pressure of saturated sand and saturated sand under a water layer. However, the water layer significantly affects the dynamic response of the tunnel model. The dynamic strain of the tunnel model located in saturated ground under the water layer is larger than that in saturated ground.

Keywords: Immersed tunnel · Shaking table tests · Seismic vibration direction · Water layer · Seismic response

1 Introduction

Immersed tunnels (Grantz 2011; Rasmussen 1995; Jensen et al. 2006; Cheng et al. 2008) usually act as connecting parts between two adjacent islands. Compared to the mountain tunnels and underground metros, immersed tunnels are often built in underwater condition. When an immersed tunnel subjected to vertical earthquake ground motions, the earthquake-resistance of the immersed tunnel is the main concern for security design under dynamic water pressure. Therefore, the affects of a water layer on the immersed tunnel system under vertical earthquake should be take into study.

A water layer, which normally generates dynamic water pressure due to the ground motion. The way of the dynamic water pressure influences on the ground soil and the tunnel itself is of importance to a comprehensive understanding of the dynamic behavior

R. Chen et al. (eds.), *Proceedings of the 2nd International Symposium on Asia Urban GeoEngineering*, Springer Series in Geomechanics and Geoengineering,
https://doi.org/10.1007/978-981-10-6632-0_23

of an immersed tunnel. Based on theoretical method (Zhao et al. 2011; Li 2013; John and Zahrah 1987; Okamoto and Tamura 1973; Nobuyuki et al. 1990), there are some theoretical studies on the dynamic water pressure on immersed tunnels. Peng et al. (2008) constructed a structure-soil-water coupling system to study the behavior of immersed tunnel under dynamic water pressure, and found that under horizontal earthquake loading, the dynamic water pressure influences little on an immersed tunnel. Based on finite element method, Chen et al. (2006) investigated the seismic responses of an immersed tunnel under different earthquake waves and found that the response of immersed tunnel seems more intensive when a water-soil coupling effect was taken into consideration.

However, only few experimental results on the immersed tunnel system under dynamic water pressure induced by earthquake. Kasper et al. (2008) carried out 2D wave flume tests of Busan-Geoje immersed tunnel (Jeong and Kim 2012; Janssen et al. 2006; Odgaard et al. 2006), through the tests he found that the water wave could cause the uplift force of the immersed tunnel. During the tests, the water wave were generated from the wave flume. Cheng et al. (2017) studied the seismic performance of immersed tunnel at three different sites by shaking table tests under horizontal earthquakes. So the experimental verification of effects of water layer on immersed tunnel under seismic shaking is needed.

To figure out the influence of a water layer on an immersed tunnel under vertical ground motion, an experimental investigation on the dynamic water pressure on the immersed tunnel is introduced in the paper. There are two groups of shaking table tests in the investigation. First one is an immersed tunnel located in saturated ground (test 1), and another one is an immersed tunnel located in saturated ground under water (test 2). A reduced scale 1:30 is used to the design tunnel model, and vertical shaking loading is factually applied to a experimental model. After a series of shaking table tests, the pore water pressure, the structural acceleration, and the dynamic structure strain are obtained, a comparative study on the experimental results then conducted.

2 Background

The immersed tunnel model presented in the paper is based upon the Hongkong-Zhuhai-Macau (H-Z-M) immersed tunnel (Yin 2014; Hussain et al. 2011; Zhang et al. 2015) which is a super-large submarine engineering. The tunnel part is approximately 6 km in length (shown in Fig. 1), which consists of 33 tunnel elements connected by flexible tunnel joints (Liu et al. 2013). As the Fig. 2 shows, the length of one tunnel element is 180 m, and the cross-sectional size of the tunnel element is 37.95 m × 11.4 m. The flexible tunnel joint contains a GINA seal, an OMEGA seal, and multi-groups of shear keys.

Fig. 1. Sketch plan of the H-Z-M project

Fig. 2. Information of the H-Z-M element

3 Experiment Design

Figure 3 shows the immersed tunnel model adopted in the tests, the geometry size of a transverse section of the tunnel model element is shown in Fig. 4. In these tests, three tunnel elements are connected by two flexible rubber joints which are simplified of real tunnel joints. The dimension of the tunnel elements is 1.26 m in width, 0.38 m in height. Compared to the H-Z-M immersed tunnel, the testing model here is rectangular as the small contribution of tunnel transverse shape to the effect researched in the paper. A multifunctional laminar shear container (Sun et al. 2012) is used to set the immersed-tunnel system.

Fig. 3. Experimental model

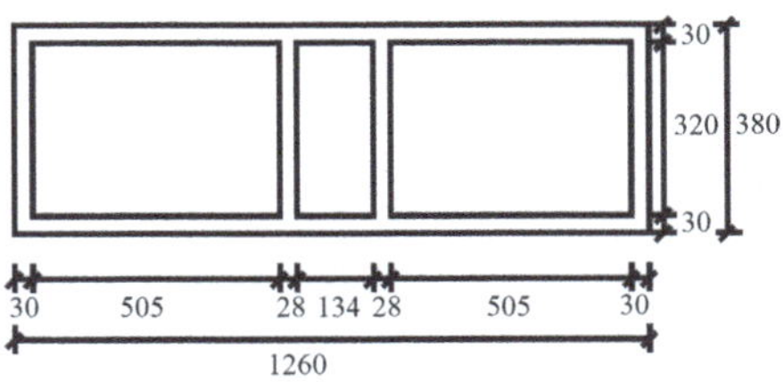

Fig. 4. Dimension of tunnel element

3.1 Site Conditions

As for test 1, the ground soil is made of saturated sand with 1.2 m in height. As for test 2, the ground soil is also made of saturated sand with 1.2 m in height, furthermore, there is a water layer beyond the soil surface with 0.2 m in height. During the soil manufacturing process of these two tests, the saturated density of ground soil keeps the same (2015 kg/m^3).

3.2 Data Collection Setup

To obtain the necessary data, a number of transducers are adopted in the tests. They are accelerometers, pore water pressure sensors, and strain gauges. The details of setup of transducers are presented in Fig. 5. It should be noticed that accelerometers A4 and 14 are installed on the elements 2 and 3.

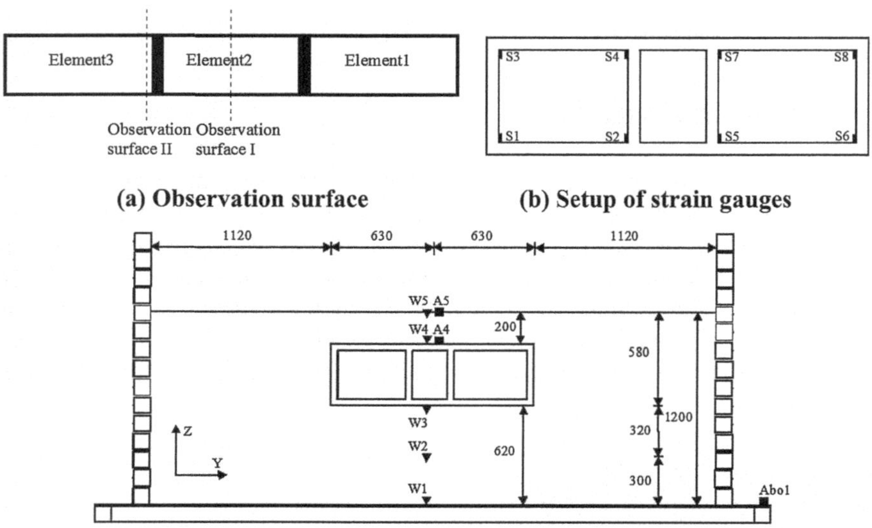

(a) Observation surface

(b) Setup of strain gauges

(c) Setup of accelerometers and pore water pressure sensors in Observation surface I

Fig. 5. Setup of data sensors (here W - pore pressure sensor, A - accelerometer, and S - strain gauge.)

3.3 Data Collection Setup

The tunnel model system is placed in the multifunctional laminar shear container. In the shaking tests, the loading pattern is vertical seismic shaking (Z direction in Fig. 5(c)). The loading cases of two groups of tests are the same. Since the shaking table system has been taken into use for a long time, there exists a discrepancy between input and output recordings. It should be stated that the input excitations analyzed here are the acceleration records reading from the accelerometer set on the shaking table. Here, El-Centro and Kobe waves are selected as loading waves. The loading seismic wave amplitudes ranged from 0.1 g to 0.4 g. Figure 6 shows the output waves from loading cases 1 and 2, and the loading cases of the tests are summarized in Table 1.

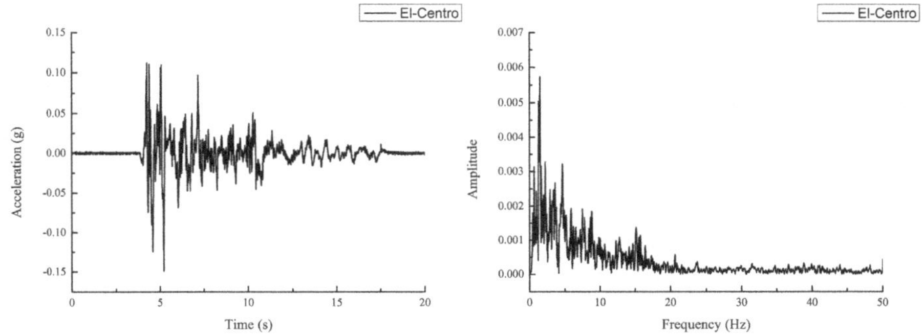

(a) Acceleration time history of El-Centro wave (b) Fourier spectrum of El-Centro wave

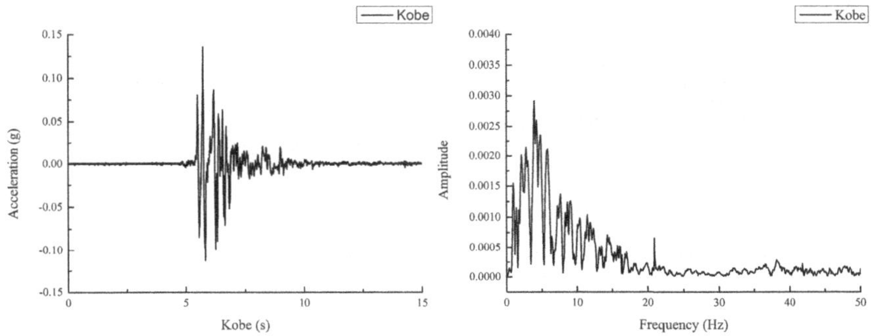

(c) Acceleration time history of Kobe wave (d) Fourier spectrum of Kobe wave

Fig. 6. Output recordings from loading cases 1 and 2

Table 1. Loading cases, where PGA means the peak ground acceleration (unit: g).

Item	Loading case	Loading direction	PGA (g)
1	El-Centro-1 (EL-1)	Z	0.1
2	Kobe-1 (KB-1)	Z	0.1
3	El-Centro-2 (EL-2)	Z	0.2
4	Kobe-2 (KB-2)	Z	0.2
5	El-Centro-3 (EL-3)	Z	0.4
6	Kobe-3 (KB-3)	Z	0.4

4 Experiment Results and Analysis

4.1 Excess Pore Pressure Ratio

Earthquake loading can cause pore pressure of ground soil. In the basis of recording data from W1~W5, the excess pore pressure ratio (PR) then can be calculated by formula (1):

$$PR = \frac{U}{\sigma} \tag{1}$$

where U is short for excess pore pressure, and σ is short for initial effective stress.

Figure 7 shows the development trends of PR in sand deposits of test1. Apparently, under different input excitations, there are little differences of PRs of W1~W5. The PRs induced by El-Centro-3 larger than those by Kobe-3. As compared to the Kobe wave, the El-Centro with low predominant frequency makes more contribution to the soil. The same phenomenon also can be drawn in Fig. 8. In addition, the growth trends of pore pressure of test 2 match well with those of test 1.

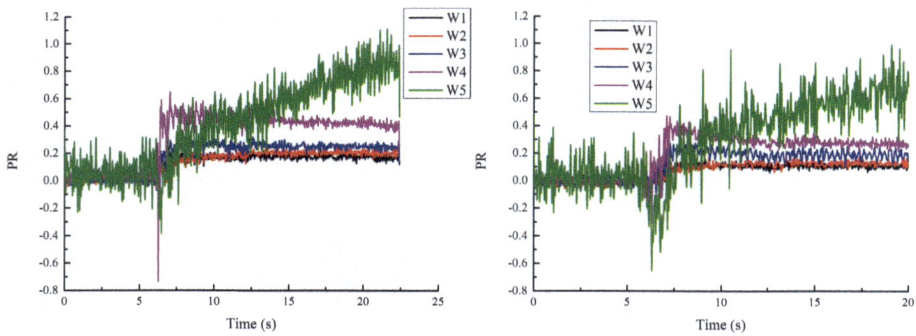

(a) EL-3 **(b) KB-3**

Fig. 7. PR results of test 1

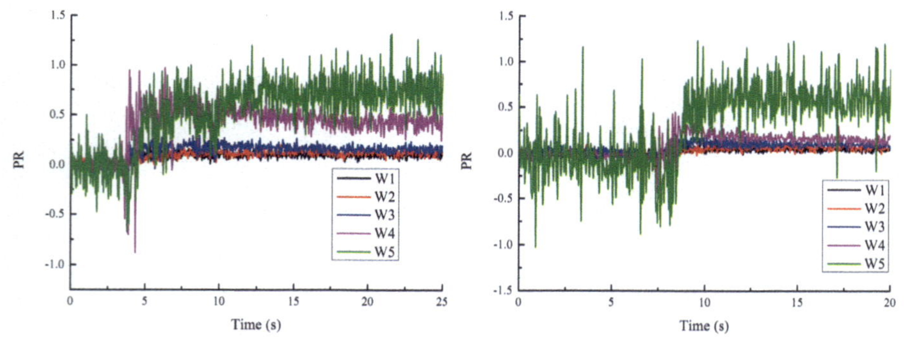

(a) EL-3 **(b) KB-3**

Fig. 8. PR results of test 2

4.2 Acceleration of Tunnel Elements

Tunnel elements acceleration results can be obtained from A 4 and A14 as mentioned before. As Figs. 9 and 10 shown, there is a difference between the acceleration of tunnel elements 2 and 3. Compared to element 3, the acceleration of element 2 has reduced

14%, 0.2%, 12%, and 4% corresponding to EL-3 of test1, KB-3 of test1, EL-3 of test2, and KB-3 of test2 respectively. That is due to the shock absorption function of the flexible joint. The relative movement trend between two tunnel elements in test 1 agrees well with that of test 2.

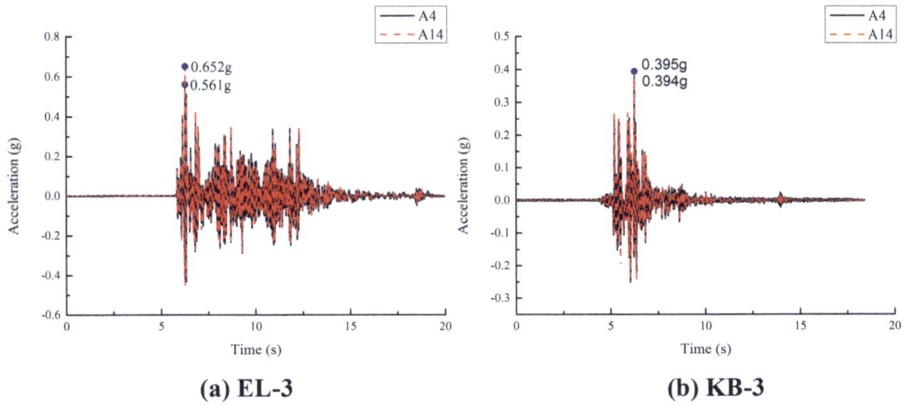

Fig. 9. Structural acceleration results of test 1

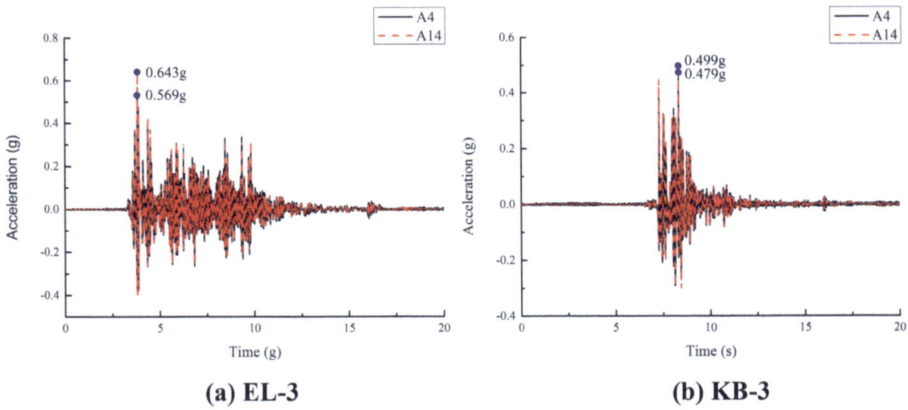

Fig. 10. Structural acceleration results of test 2

4.3 Dynamic Structural Strain

Through the strain gauges S2, S3, S4, S5, S7, and S8, the peak dynamic strain of the tunnel model of tests 1 and 2 can be calculated as shown in Table 2. Larger the amplitude of input excitation is, larger the structural dynamic strain response is. As a whole, the dynamic strain of the tunnel model in test 1 is smaller than that of test 2. The water above the soil layer in test 2 increases the load of the tunnel model. Moreover, the dynamic water pressure can be generated during the shaking process, this also make the tunnel model to bear more pressure. For the tunnel model in both two tests, the largest dynamic strain value occurs at S8.

Table 2. Peak dynamic strain of the tunnel model

Item	Test 1						Test 2					
	EL-1	EL-2	EL-3	KB-1	KB-2	KB-3	EL-1	EL-2	EL-3	KB-1	KB-2	KB-3
2	8	11	28	6	9	13	9	12	33	7	10	18
3	10	12	14	5	8	12	10	14	16	6	8	14
4	7	14	23	7	9	11	9	12	27	10	13	19
5	13	80	84	14	21	25	17	84	89	16	24	43
7	6	11	11	4	7	6	6	12	14	4	5	7
8	84	103	206	58	101	183	100	124	211	76	113	200

To study the growth rate of structural dynamic strain of tests 1 and 2, SR (dynamic strain growth rate) is recommended to be analyzed, which can be drawn from formula (2):

$$SR = \frac{\left(R_j - R_i\right)}{R_i} \tag{2}$$

where R_i represents the tunnel element peak dynamic strain under EL-1 and KB-1, R_j represents the tunnel element peak dynamic strain under EL-2, EL-3, KB-2, and KB-3.

Based on the data listed in Table 2, dynamic strain development curves under EL-2, EL-3, KB-2, and KB-3 of two tests are drawn in Fig. 11. The SR value of the tunnel model under EL-3 and KB-3 is larger than that under EL-2 and KB-2. In general, there is no significant difference in SR development trend between tests 1 and 2. Under EL-2 and EL-3 the fastest dynamic strain growth points of tests 1 and 2 are both S5. The lowest dynamic strain growth points of tests 1 and 2 are S3, S8 under EL-2, while the lowest dynamic strain growth points of tests 1 and 2 are both S3 under EL-3. As for test1, the largest SR value occurs at S7 and S8, the smallest SR value occurs at S4 and S7 corresponding to KB-2 and KB-3. Under KB-2 and KB-3, the largest SR value of tests 1and 2 both occurs at S5 and the smallest SR value both occurs at S7.

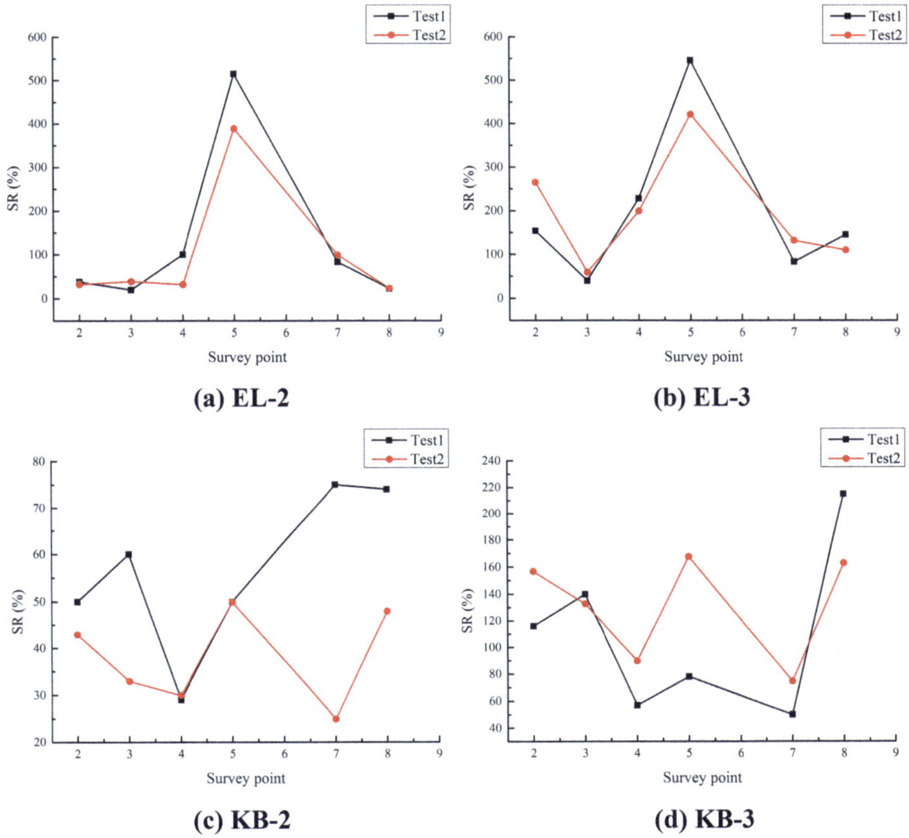

Fig. 11. SR of the tunnel model in tests 1 and 2

5 Conclusions

The experimental results of tests 1and 2 are presented in the paper to study the influence of water layer on an immersed tunnel system. Based on this goal, a series of shaking table tests were conducted. Immersed tunnel model located in both saturated ground and saturated ground under a water layer have been considered. Through the analysis on testing results, the conclusions are given below:

1. The water layer contributes little to the pore water pressure of saturated sand, and makes no obvious influence on the sand liquefaction. Earthquake wave riches in low frequency contributes more to the soil dynamic response.
2. The water layer does no influence on the acceleration of immersed tunnel model, the flexible joint can absorb shock energy in a certain extent.
3. The comparison between the dynamic strain of immersed tunnel model in tests 1and 2 is carried out. The dynamic strain of immersed tunnel model in test1 is a bit larger

than that of test 2. This proves that the water layer affects a lot on the dynamic strain of the immersed tunnel structure.

References

Grantz, W.C.: Immersed tunnel settlements. Tunn. Undergr. Space Technol. **16**, 203–210 (2011)

Rasmussen, N.S.: Immersed tunnels. Struct. Eng. Int. **5**(4), 213 (1995)

Jensen, O.P., Olsen, T.H., Kim, C.W., Heo, J.W., Yang, B.H.: Construction of immersed tunnel in off-shore wave conditions Busan-Geoje project south korea. Tunn. Undergr. Space Technol. **21**(3), 333 (2006)

Cheng, L., Liu, X., Gu, C.: Research on development of immersed tunnel engineering at home and abroad. Water Resour. Power **17**(3), 226–231 (2008). (in Chinese)

Zhao, M., Du, X., Liu, J., Liu, H.: Explicit finite element artificial boundary scheme for transient scalar waves in two-dimensional unbounded wave guide. Int. J. Numer. Meth. Eng. **87**(11), 1074–1104 (2011)

Li, P.: Numerical analysis for longitudinal seismic response of tunnels in saturated soils. DR Thesis, Tsinghua University, Beijing, China (2013). (in Chinese)

John, C.M.S., Zahrah, T.F.: Aseismic design of underground structures. Tunn. Undergr. Space Technol. **2**(2), 165–197 (1987)

Okamoto, S., Tamura, C.: Behavior of subaqueous tunnels during earthquakes. Earthquake Eng. Struct. Dynam. **1**(3), 253–266 (1973)

Nobuyuki, M., Tadashi, O., Hiroyoshi, K., Atsushi, S.: Application of tunnel seismic prediction method under special condition. J. Appl. Spectrosc. **52**(3), 267–271 (1990)

Peng, H.K., Meng, G., Ding, Q., Xu, F.: The Influence of hydrodynamic pressure on immersed tunnel under earthquake excitation. J. Shanghai Jiaotong Univ. **42**(6), 1027–1031 (2008). (in Chinese)

Cheng, G., Gao, B., Zhao, Y.: Earthquake-resistant research of immersed tunnels. J. Seismol. Res. **29**(1), 60–64 (2006). (in Chinese)

Kasper, T., Steenfelt, J.S., Pedersen, L.M., Jackson, P.G., Heijmans, R.W.M.G.: Stability of an immersed tunnel in offshore conditions under deep water wave impact. Coast. Eng. **55**(9), 753–760 (2008)

Jeong, S., Kim, J.: The immersed tunnel and bridges of Busan-Geoje fixed link. Struct. Eng. Int. **22**(1), 20–25 (2012)

Janssen, W., Haas, P.D., Yoon, Y.H.: Busan-Geoje link: immersed tunnel opening new horizons. Tunn. Undergr. Space Technol. **21**(3), 332 (2006)

Odgaard, S.S., Jensen, O.P., Kasper, T., Yoon, Y.H., Chang, Y., Park, R.Y.: Design of long immersed tunnel for highway in offshore conditions Busan-Geoje fixed link. Tunn. Undergr. Space Technol. **21**(3), 334 (2006)

Cheng, X.J., Jing, L.P., Cui, J., Li, Y.Q., Dong, R.: Shaking-table tests for immersed tunnels at different sites. Shock Vib. **2017**, 1–11 (2017). Article ID 2546318

Yin, H.: Key technologies applied in design and construction of artificial islands and immersed tunnel of Hong Kong-Zhuhai-Macau bridge (HZMB) project. Tunn. Constr. **34**(1), 60–66 (2014). (in Chinese)

Hussain, N., Wong, C., Carter, M., Kwan, S., Mak, T.W.: Hongkong Zhuhai Macao link. Procedia Eng. **14**, 1485–1492 (2011)

Zhang, Q.M., Li, Q.W., Li, K.F., Fan, Z.H.: Quality control for concrete durability when subjected to chloride diffusion in Hongkong-Zhuhai-Macau sea link project. Eng. Mech. **32**(3), 175–182 (2015). (in Chinese)

Liu, P., Ding, W., Yang, B.: Joints and gasket seismic response of deep-water long immersed tube tunnels. J. Tongji Univ. **41**(7), 984–988 (2013). (in Chinese)

Sun, H.F., Jing, L.P., Meng, X.C., Wang, N.W.: A three-dimensional laminar shear soil container for shaking table test. J. Vib. Shock **31**(17), 26–32 (2012). (in Chinese)

An Empirical Estimation of Settlement-Time Relationship Due to Shield Tunneling in Gravelly Soils

Yung-Show Fang[1(✉)], Lin-Yao Kuo[2], Chuo-Ming Lin[3], and Ying-Chu Shih[1]

[1] Department of Civil Engineering, National Chiao Tung University, Taibei 30010, Taiwan, China
ysfang@mail.nctu.edu.tw
[2] Bureau of High Speed Rail, MOTC, Banciao District, Taibei 22041, Taiwan, China
[3] Trinity Foundation Engineering Consultants Co., Ltd., Taibei 10685, Taiwan, China

Abstract. This paper reported the surface settlement-time relationship due to shield tunneling in gravelly soils. Field data monitored during the construction of Taoyuan International Airport MRT, Hukou power-cable tunnels, and Beijing Metro Line 1 tunnels were collected and analyzed. Based on the field data, the hyperbolic model was proposed to simulate the settlement-time relationship above the centerline of the tunnel. Based on the field measurements, it was found that the settlement-time relationship due to shield tunnelling could be property approximated with the hyperbolic model. The hyperbolic constant 1/a represented the surface settlement that occurred during the first day after the face of shield passed the monitoring sensor. For shield tunnelling in gravelly soils, the mean of initial settlement speed was 2.5 mm/day. The constant 1/b represented the long-term ultimate settlement due to shield tunnelling. For tunnelling in gravelly soils, the measured maximum settlement was only 4.70 to 8.06 mm. These values were much lower than the long-term settlements due to shield tunnelling in clay, and clay-and-sand grounds. The reason for these extremely small settlements was most probably due to the high stiffness and strength of the gravelly soils.

Keywords: Gravelly soil · Hyperbolic model · Settlement-time relationship · Shield tunnelling · Ultimate surface settlement

1 Introduction

In recent years, due to the rapid development of urban areas, many public facilities such as the mass rapid transit (MRT) tunnels, underground sewer tunnels, and high-voltage power-cable tunnels have been constructed. Because of the disruptive effects of the cut-and-cover method, it is becoming more popular to employ the underground tunneling to pass beneath heavily trafficked commercial areas.

R. Chen et al. (eds.), *Proceedings of the 2nd International Symposium on Asia Urban GeoEngineering*, Springer Series in Geomechanics and Geoengineering,
https://doi.org/10.1007/978-981-10-6632-0_24

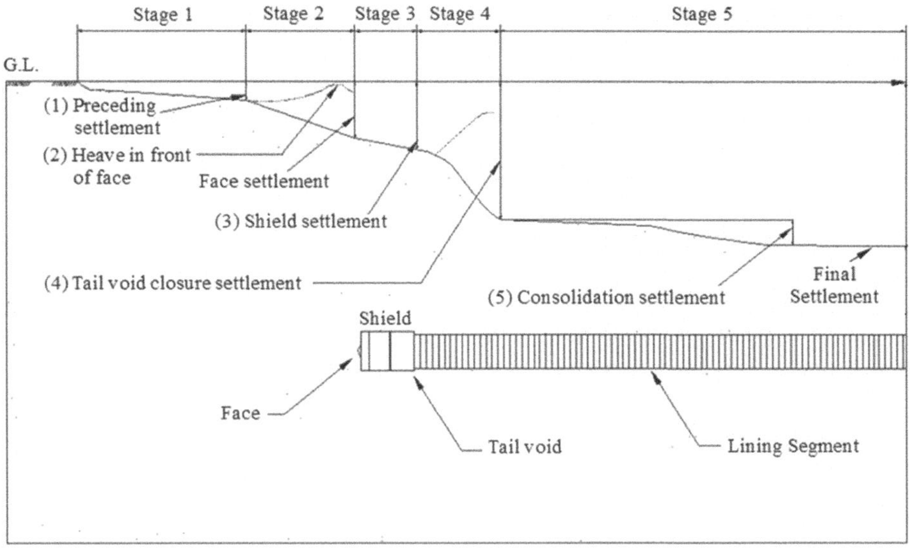

Fig. 1. Longitudinal settlement profile due to shield tunnelling (redrawn after Japan Society of Civil Engineers 1996)

Ground settlement usually happens with the progress of shield tunnelling. Possible reasons for ground movement to occur are associated with the loss of soil due to excavation. Japan Society of Civil Engineers (1996) reported the following types of ground movements that might happen due to shield excavation: (1) preceding settlement; (2) heave and settlement in front of face; (3) shield settlement; (4) tail-void closure settlement; and (5) consolidation settlement, as illustrated in Fig. 1. In this paper, the shield tunnelling for the construction of Taoyuan International Airport (TIA) MRT was introduced. The gravelly geological condition of the tunnelling project was briefly described. The Earth-Pressure-Balanced (EPB) shield machine and tunnelling operation was summarized.

This paper studied the rate of ground settlement as a result of shield tunnelling by analyzing the time-settlement data monitored in the field. Hwang et al. (1995) suggested that the consolidation settlement from 10 to 100 days after the passage of the face and log t could be represented by a straight line. Fang et al. (1993) suggested that the time-settlement relationship due to tunnelling with an EPB shield could be modelled by the hyperbolic model. Based on 10 sets of field monitored data obtained at Taoyuan Airport MRT, Hukou power-cable tunnels, and Beijing Metro Line 1 tunnels, the hyperbolic constants a and b for tunnelling in gravelly soils were investigated. The time-settlement curves estimated with the hyperbolic model were compared with field monitored data.

2 Tunnelling for Taoyuan International Airport MRT

The objective of Taoyuan International Airport MRT is to provide airport passengers with a safe, convenient, comfortable and high quality transit service. TIA MRT system between Taipei and Jhongli linked the Taoyuan International Airport to the Taiwan High Speed Rail (HSR), Taiwan railway systems, and Taipei MRT systems. The total budget for this project was approximately US$ 3.1 billion.

As illustrated in Fig. 2, the project starts from Taipei Main Station (A1 Station), passes through Sanchong, Sinjhuang, Linkou, Taoyuan International Airport (A12-A14a Stations), and terminates at Jhongli Train Station (A21 Station). The route intersects Taiwan High Speed Rail at HSR Taoyuan Station at A18 Station. The total route length is 51.03 km, consisting of 10.92 km of underground section and 40.11 km of elevated section. In Fig. 2, the TIA MRT system has 22 stations, consisting of 7 underground and 15 elevated stations. The commercial service of the Airport MRT began on March 2, 2017.

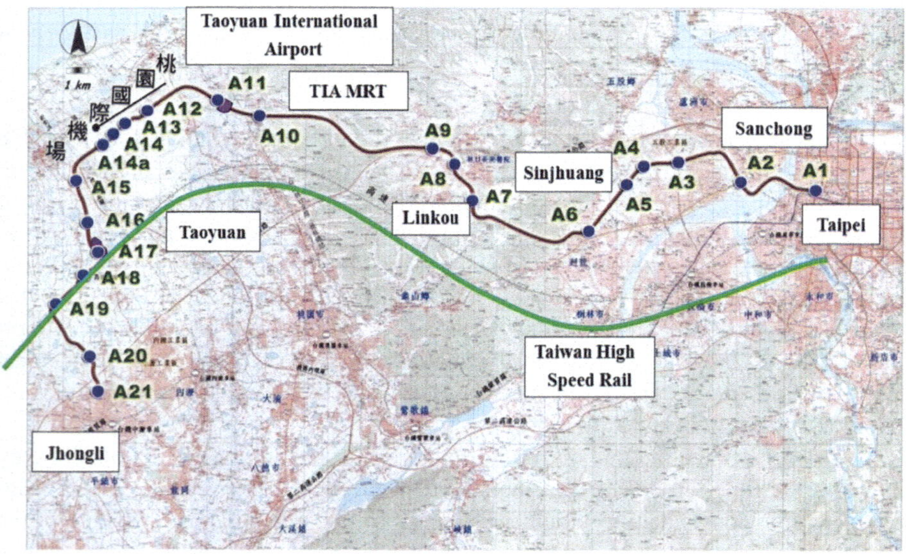

Fig. 2. Taoyuan International Airport MRT system

Figure 3 shows the express train entered the Chang-Gung Memorial Hospital (A8) Station. On an express train, it takes 36 min to travel from the Taoyuan Airport Terminal 1 Station (A12 Station) to Taipei Main Station (A1 Station). The electric train set has a maximum operational speed of 100 km/hr.

Fig. 3. Express train entering Chang-Gung Memorial Hospital Station

2.1 Subsoil Condition

Figure 4 shows the geological profile of Lot CU02A of Taoyuan Airport MRT. Shield tunnelling stated from shaft A11-P1, driven under through shaft A11-P2, station A12 under Terminal 1, station A13 under Terminal 2, station A14 under Terminal 3, station 14a under Airport Hotel, and ended at shaft A14-P2. From Tunnel I to V, tunnelling was divided into five sections, with a total length of 3,622 m. The subsoils excavated were mainly cobble and gravelly soil of Taoyuan plateau. The tunnelling of Lot CU02A started on April of 2008, and was completed on January of 2010.

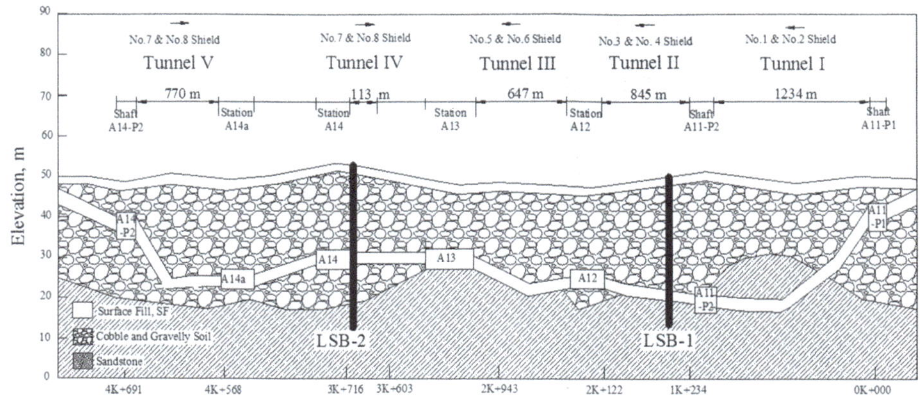

Fig. 4. Geological profile of Taoyuan International Airport

On descending order, properties of the three sublayers shown in Fig. 4 were as follows:

(1) Surface fill (SF): surface laterite layer, at 0–3.5 m below ground level, SPT N-value of 8, average density 1.85 ton/m^3.
(2) Cobble and gravelly soil: 14.7–34.3 m thick, SPT N-value greater than 50, average density 2.0 ton/m^3.
(3) Weathered sandstone: 20–50 m thick, SPT N-value greater than 50, average density 2.1 ton/m^3.

Ground water table was at 3 to 5 m below ground level. It was clear in Fig. 4 that most of shield tunnelling was conducted in cobble and gravelly soils. Based on large-diameter boring with steel casing at LSB-1 and LSB-2, it was found that the cobble and gravelly matrix included 48–55% cobble, 30–32% gravel, 11–16% sand, and 4% of fines. The average cobble and gravel content was 70%–80%, and the largest cobble size was 600 mm.

2.2 Shield Tunnelling in Gravelly Soils

For the 8 EPB shield machines used for this project, 6 were made by Okumura Machinery Corporation (shown in Fig. 5) and 2 were manufactured by Hitachi Zosen Corporation. The outer diameter of the shield was 6,240 mm. The outer and inner diameter of the lining segments were 6,100 and 5,600 mm, respectively. The reinforced concrete segments were 250 mm-thick and 1,000 mm-long. The tail void clearance was 70 mm. Backfill grouting with a mixture of fluid A (cement + bentonite + stabilizer) and fluid B (water glass) was carried out to fill the tail void and reduce ground settlement.

Fig. 5. No. 1 EPB shield machine used for tunnelling of Lot CU02A

2.3 Monitoring Sections

For Lot CU02A, tunnelling alignment passed near several sensitive structures of the Airport, such as the terminal building, tower, and taxiway. For safety reasons, it was necessary to set up monitoring sensors along the tunnelling route. For this project, monitoring sections were established every 80 to 100 m along the centreline of tunnels. Surface settlement indicators (SSI) were arranged on the sections to observe the variation of ground settlement troughs, and settlement-time relationship due to shield tunnelling.

3 Hyperbolic Settlement-Time Relationship

Based on field data, Fang et al. (1993) suggested that the time-settlement relationship due to shield tunnelling could be estimated with the hyperbolic model as follows:

$$S(t) = \frac{t}{a + bt} \quad (1)$$

where S(t) = Surface settlement above the tunnel centreline in mm; t = time in days after the face passed the sensor; a and b are hyperbolic constants. In this study, it was intended to find out whether or not the hyperbolic model shown in Fig. 6 could be used to describe the settlement-time curve due to shield tunnelling in cobble and gravelly soils.

3.1 Determination of Constants a and b

To determine the constants a and b, Eq. (1) could be rewritten as:

$$\frac{t}{S(t)} = a + bt \quad (2)$$

As the data measured with SSI-1021 at the Taoyuan Airport were plotted on a graph with t/S(t) as vertical coordinate, and t as horizontal coordinate, a straight line was obtained by linear regression as shown in Fig. 7. The tunnel diameter was 6.10 m, and the tunnel centreline was located at the depth of 16.43 m. In Fig. 7, constant b was the slope of the straight line, and constant a was the vertical coordinate intercept.

3.2 Physical Meaning of Constants a and b

3.2.1 Hyperbolic Constant b

The physical meaning of constants a and b was illustrated in Fig. 6. For the hyperbolic model expressed in Eq. (1), as time approached infinity, the L'Hopital's rule was used to determine the ultimate settlement S_{max} = 1/b. This indicated that the reciprocal of b stood for the long-term settlement above the tunnel centreline due to tunnelling.

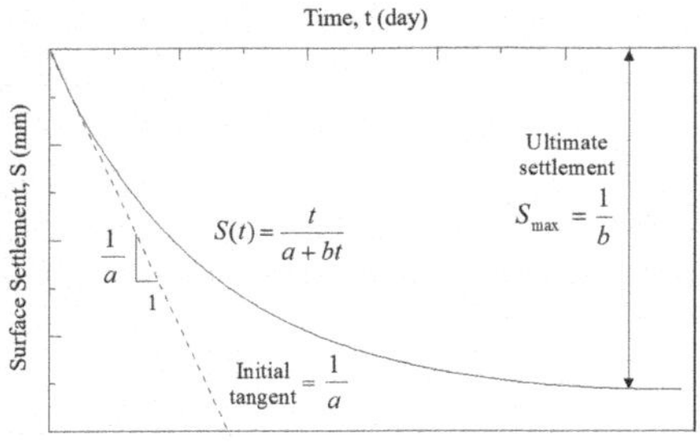

Fig. 6. Hyperbolic model with constants a and b

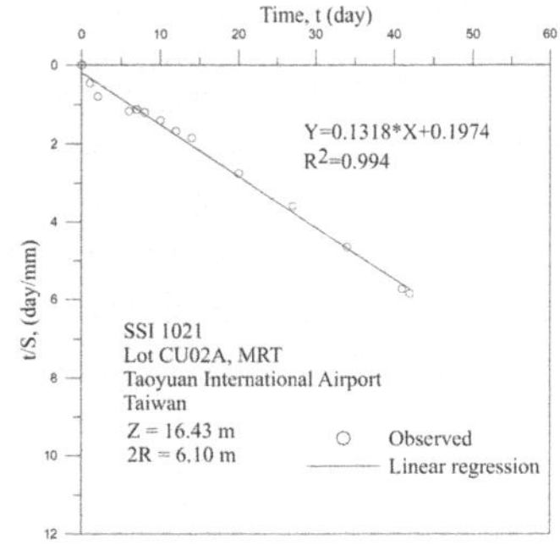

Fig. 7. Determination of constants a and b

3.2.2 Hyperbolic Constant a

In Eq. (1), if the settlement S(t) was differentiated with respect to time t, the hyperbolic constant a was the reciprocal of the slope of the settlement curve at t = 0. The constant 1/a represented the subsidence speed during the initial moment. A small a value indicated that the settlement due to tunnelling would be completed in a relatively short period.

4 Comparison of Monitored and Estimated Settlement-Time Data

To justify the applicability of the proposed method, the settlement-time relationship measured at the Taoyuan Airport MRT, Hukou power-cable tunnels, and Beijing Metro Line 1 tunnels were compared with hyperbolic estimation.

4.1 Taoyuan Airport MRT Tunnels

The estimated and monitored settlement-time relationships reported by CECI (2009) were illustrated in Fig. 8. The outer diameter of the shield was 6.10 m, and the depth of tunnel center below SSI 1021 was 16.34 m. Based on Fig. 7, the hyperbolic constant a determined was 0.20 day/mm, and the constant b was 0.13 1/mm. Figure 8(a) showed the measured settlement data was in fairly good agreement with the hyperbolic estimation. Figure 8(b), (c), and (d) illustrated the surface settlement data measured at SSI 1057, SSI 1081 and SSI 1092, and their hyperbolic simulation. In these figures, the monitored data and hyperbolic estimation were in fairly good agreement. The constants a and b for these relationships were summarized in Table 1.

4.2 Hukou 161 KV Power-Cable Tunnels

Figure 9(a) showed the settlement-time relationship measured at SSI 06 above the Tai-Power 161 kV power-cable tunnel centerline, located at Hukou, Taiwan

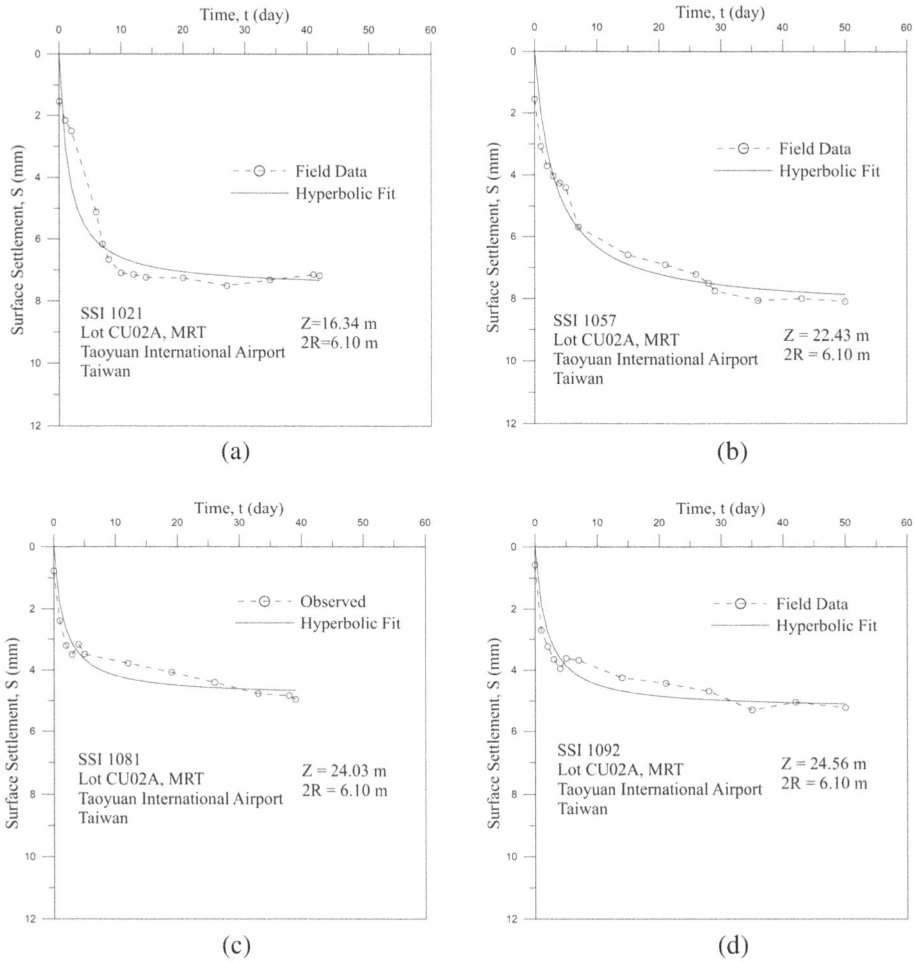

Fig. 8. Field measured and estimated settlement-time relationship for Taoyuan Airport MRT tunnels

(Chieh-Hsing/Shimizu 2009). The outer diameter of the shield was 6.55 m, and the depth of tunnel centerline was 17.48 m. The excavated ground consisted of 10–25% cobble, 40–55% gravel, 10–15% sand, and 3–7% of silt and clay. As illustrated in Table 1, the hyperbolic constants a and b were 0.22 day/mm and 0.17 1/mm, respectively. At this monitor point SSI 06, the measured settlement data were in fairly good agreement with the estimated curve. Comparison of monitored data with estimated relationship for SSI 17, SSI 28, and SSI 39 were displayed in Fig. 9(b), (c) and (d), respectively. It may be observed from these figures that the measured settlement data matched fairly well with the estimated hyperbolic curves.

Table 1. Cases of shield tunnelling in gravelly soils

Case	Soil type	Monitored sensor	Tunnel diameter D (m)	Tunnel depth Z (m)	Maximum surface Settlement S_{max} (mm)	a (day/mm)	1/a (mm/day)	b (1/mm)	1/b (mm)	Reference
Taoyuan International Airport MRT Taiwan	Cobble and Gravelly soil Cobble = 48–55% Gravel = 30–32% Sand = 11–15% Silt & Clay = 4–6%	SSI 1021	6.10	16.34	7.59	0.20	5.06	0.13	7.70	CECI (2009)
		SSI 1057	6.10	22.43	8.06	0.38	2.61	0.12	8.33	
		SSI 1081	6.10	24.03	4.86	0.33	2.98	0.20	5.00	
		SSI 1092	6.10	24.56	5.05	0.34	2.89	0.20	5.00	
Hukou161 kV Power-cable Tunnel	Cobble and Gravelly soil Cobble = 10–25% Gravel = 40–55% Sand = 10–15% Silt & Clay = 3–7%	SSI 06	6.55	17.48	5.92	0.22	4.45	0.17	5.88	Chieh-Hsing/Shimizu (2009)
		SSI 17	6.55	17.55	6.70	0.79	1.26	0.14	7.14	
		SSI 28	6.55	17.48	5.44	0.47	2.09	0.18	5.56	
		SSI 39	6.10	15.27	5.88	0.89	1.16	0.17	5.88	
Beijing Metro Line 1	Gravel with Sand	SSI 01	5.4	18.0	5.50	0.63	1.58	0.16	6.25	Zhou and Wang (2007)
		SSI 04	5.4	18.0	4.70	1.05	0.95	0.18	5.56	

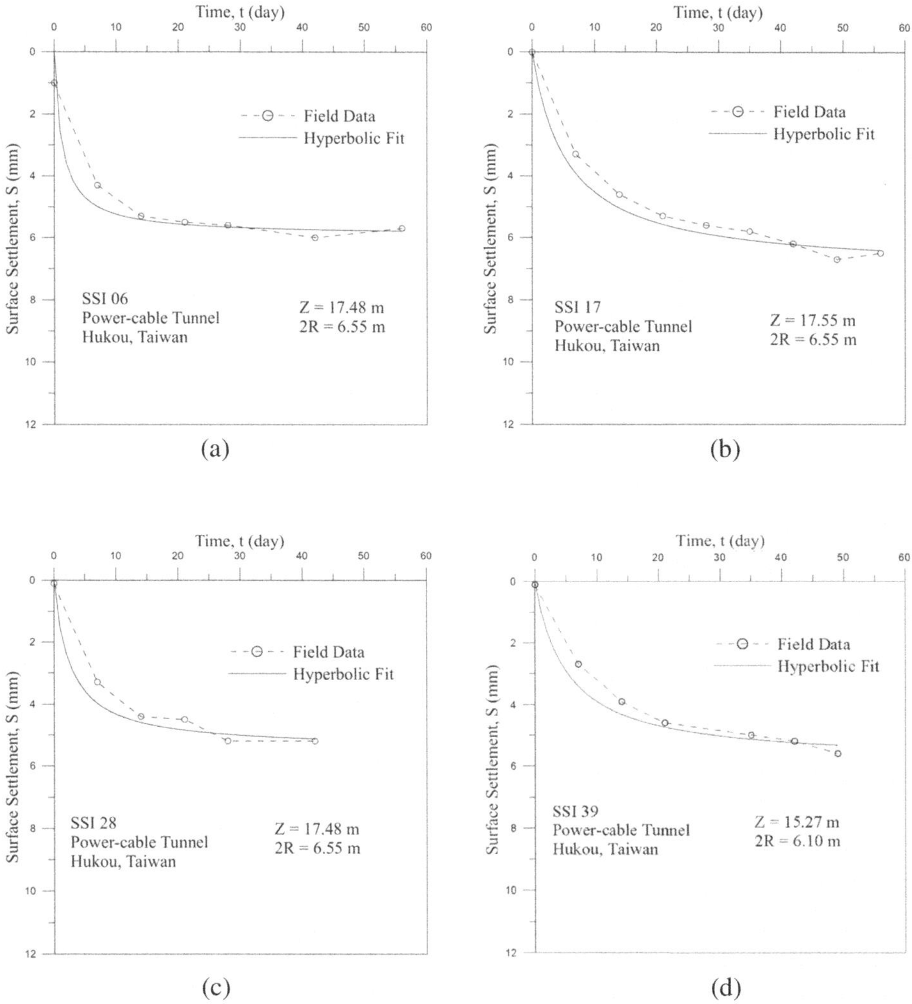

Fig. 9. Measured and estimated settlement-time relationship for Hukou power-cable tunnels

4.3 Beijing Metro Line 1 Tunnels

Zhou and Wang (2007) reported the settlement-time relationship monitored for Beijing Metro Line 1 tunnel construction. For this project, the outer diameter of the shield was 5.4 m, and the depth of tunnel centerline was 18.0 m. The excavated soils included mostly gravel and sand. In Table 1, for the data measured at SSI 01, the calculated constant a = 0.63 day/mm, and b = 0.16 1/mm. Figure 10(a) showed the measured and estimated settlement-time curves were in good agreement. A similar settlement-time trend for SSI 04 could be observed in Fig. 10(b).

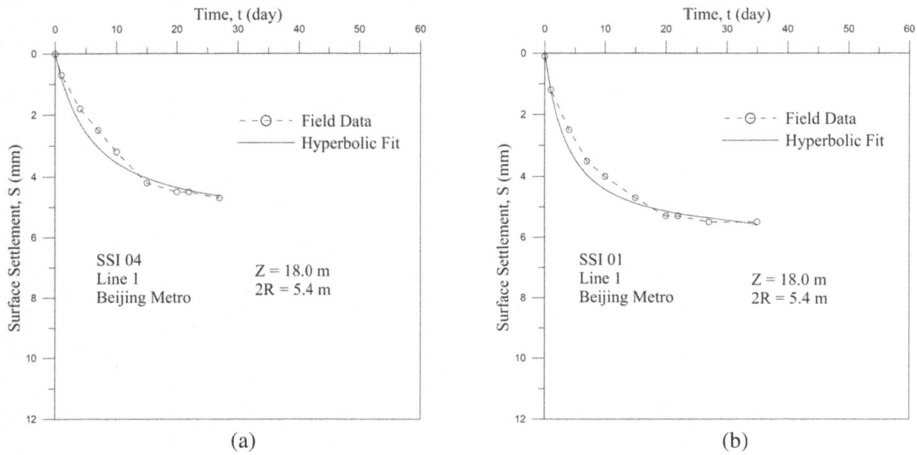

Fig. 10. Measured and estimated settlement-time relationship for Beijing Metro Line 1 tunnel

4.4 Discussion of Initial Settlement Speed

As introduced in Sect. 3.2.2, the physical meaning of constant 1/a approximated the surface settlement occurred during the first day after the face of shield passed the monitoring sensor. With a unit of mm/day, 1/a was the initial surface settlement speed due to shield tunnelling. Based on the 10 sets field settlement-time data collected in the study, the hyperbolic constants a and b were summarized in Table 1. For shield tunneling in gravelly soils, the mean of initial settlement speed (constant 1/a) was about 2.5 mm/day.

4.5 Discussion of Ultimate Settlement

In Sect. 3.2.1, it was stated that the constant 1/b actually represented the long-term ultimate settlement due to shield tunnelling. Based on the 10 sets of settlement data summarized in Table 1, the mean ultimate settlement was only 6.23 mm.

Based on data obtained from EPB shield construction in Japan, Fujita (1982) reported that the maximum surface settlement induced in clay was 60 ± 25 mm. The maximum settlement due to tunnelling in clay-and-sand ground was 20 ± 10 mm. For shield tunnelling in gravelly soils, in Table 1 the estimated ultimate settlement was only 5.00 to 8.33 mm, and the measured maximum settlement was 4.70 to 8.06 mm. These values were much lower than the long-term settlements due to shield tunnelling in clay, and in clay-and-sand grounds. The reason for these extremely small settlements was most probably due to the high stiffness and strength of the gravelly soils.

4.6 Limitations of the Method

The main limitations of the proposed hyperbolic settlement-time relationship are discussed in this section. When the hyperbolic model was used, at the time of 0, the estimated settlement would be zero, and the settlement-time curve would pass the origin point.

However, as illustrated in Fig. 1, before the face of shield reached the monitor point, the preceding settlement, and heave or subsidence in front of the face had already occurred, which could not be covered by the hyperbolic method.

In general, the tunnel construction must be continued and completed within the period specified in the contract. For the Taoyuan Airport MRT, the speed of tunneling was about 3 to 5 m/day. For the Hukou 161 kV power-cable tunnel, the average tunneling speed was 2 to 4 m/day. For the Beijing Metro Line 1, the speed of tunneling was about 3 m/day. However, in case some special difficulties were encountered and the construction was halted, such as removing of unexpected underground obstacles, the estimation of settlement-time relationship with the hyperbolic model might not be suitable.

Besides, the proposed method could be used for tunnelling in a single soil layer. This method could be used to estimate the surface settlement-time relationship only. The prediction of subsurface settlement, horizontal displacement, and pore pressure variation were not possible.

5 Conclusions

Based on the study regarding the time-settlement relationship above the tunnel centerline due to shield tunnelling in gravelly soils, the following conclusions were drawn:

1. Based on the field measurements for the construction of Taoyuan Airport MRT tunnels, Hukou power-cable tunnels, and Beijing Metro Line 1 tunnel, it was found that the settlement-time relationship due to shield tunnelling could be properly approximated with the hyperbolic model.
2. The hyperbolic constant 1/a represented the surface settlement that occurred during the first day after the face of shield passed the monitoring sensor. For shield tunnelling in gravelly soils, the mean of initial settlement speed was approximately 2.50 mm/day.
3. The hyperbolic constant 1/b represented the long-term ultimate settlement due to shield tunnelling. For shield tunnelling in gravelly soils, the measured maximum settlement was only 4.70 to 8.06 mm. These values were much lower than the long-term settlements due to shield tunnelling in clay, and clay-and-sand grounds. The reason for these extremely small settlements was most probably due to the high stiffness and strength of the gravelly soils.

Acknowledgements. The authors wish to acknowledge the Bureau of High Speed Rail, CECI Engineering consultants, Inc., and RSEA Engineering Corporation/Okumura Corporation Joint Venture for providing valuable information and support. The financial assistance of the Ministry of Science and Technology (Grant No.: MOST 105-2221-E-009-021-MY3) is gratefully acknowledged.

References

CECI Engineering Consultants Incorporation: Taiwan Taoyuan International Airport access MRT System-Construction Proposal for Lot CUO2A, Taipei, Taiwan (2009). (In Chinese)

Chieh-Hsing Construction Co. and Shimizu Corporation Co.: Chu-Kung Extra High Voltage Substation Outlet kV Power-cable Turnkey Project: Shield Tunnel Construction Proposal, Taipei, Taiwan (2009). (In Chinese)

Fang, Y.S., Lin, S.J., Lin, J.S.: Time and settlement in EPB shield tunnelling. Tunn. Tunn. **25**, 26–27 (1993)

Fujita, K.: Prediction of settlements by shield tunnelling. In: Proceedings, International Conference on Soil Mechanics, vol. 1, pp. 239–246 (1982)

Hwang, R.N., Ju, D.H., Tsai, M.S., Fang, Y.S.: Soft ground tunnelling in Taiwan. In: Proceedings, US/Taiwan Geotechnical Engineering Collaboration Workshop, Taipei, pp. 77–90 (1995)

Japan Society of Civil Engineers: Japan Standard for Shield Tunnelling, 3 edn. (1996). (in Japanese)

Zhou, S., Wang, J.A.: The surface settlement analysis of curved tunnel shield. Chin. J. Undergr. Space Eng. **3**(5), 909–913 (2007). (In Chinese)

Application of Horizontal MJS Piles in Tunneling Beneath Existing Twin Tunnels

Pin Zhang[1,2,3], Yuan Liu[1,2,3], Xin Kang[1,2,3], Ke Zhong[4], and Renpeng Chen[1,2,3(✉)]

[1] MOE Key Laboratory of Building Safety and Energy Efficiency, Hunan University, Changsha 410082, China
chenrp@hnu.edu.cn

[2] National Joint Research Center for Building Safety and Environment, Hunan University, Changsha 410082, China

[3] College of Civil Engineering, Hunan University, Changsha 410082, China

[4] Changsha Metro Group Co., Ltd., Changsha, China

Abstract. This paper presents a case of a single tunnel excavated beneath closely spaced existing twin tunnels with reinforced MJS piles in Changsha, China. Due to soil disturbance and stress re-distribution induced by tunneling, the safety of overlying tunnels during excavating becomes a serious issue. To solve this problem, the Metro Jet System (MJS) method was used. The response of stratum and overlying tunnels are studied during the installation of horizontal MJS piles and shield tunneling, by monitoring pore water pressure of surrounding stratum, additional stress and vertical displacement of tunnel in the water-rich sand layer. It indicates that the pore water pressure increased rapidly during the period of jetting-grout and the maximum excess pore water pressure reached 35 kPa. The distribution of additional stress at monitoring cross-section is studied in different construction stages. The vertical displacement of the two existing tunnels presents distinct characters due to temporary road load on the tunnels. A satisfactory match between the proposed fitting curves and the measured settlement data of north tunnel is presented. Overall, MJS horizontal pile can effectively reduce the settlement of existing tunnel.

Keywords: MJS · Shield tunnel · Field monitoring

1 Introduction

With the rapid development of subway and underground engineering, the cases of shield crossing existing tunnels has been growing increasingly. Generally, the underground has complex geological conditions. The new tunnels excavated underneath are very close to the existing tunnels. The structural safety and normal operation becomes a serious issue in shield tunnel design.

Many studies about the interaction that between closely spaced tunnels were given by field measurements, model tests, theoretical analysis and numerical simulation. Kim et al. (1998) and Byun et al. (2006) conducted a model test to study the influence of tunnel excavation on buried pipes in homogeneous soils. The two-stage analysis method

R. Chen et al. (eds.), *Proceedings of the 2nd International Symposium on Asia Urban GeoEngineering*, Springer Series in Geomechanics and Geoengineering,
https://doi.org/10.1007/978-981-10-6632-0_25

was used to simplify the problem by using Winkler foundation model and elastic homogeneous half-space foundation model. Zhang et al. (2009) established a continuous elastic method with elastic layered half space soil model to study this problem for the first time. But they only considered the influence of soil loss. Addenbrooke and Potts (2001) and Chehade and Shahrour (2008) used FEM to analyze the impact of tunnel excavation on adjacent tunnels. Fang et al. (2015) presented a case of closely spaced twin tunnels constructed beneath two other existing closely spaced tunnels. Their study showed the settlement profile of the existing tunnels displayed a double trough "W" shape. However, in terms of construction methods, very few studies were available on the special ways to solve this problem and assess effectiveness.

Metro line 4 in Changsha was excavated beneath two other existing closely spaced tunnels (Metro line 2) by using MJS piles to improve the soils between Metro lines 2 and 4. Based on field measurements, the response of the surrounding stratum and overlying tunnels during the construction of MJS piles was studied as well as the influence of excavation on the overlying tunnels.

2 Project Outline

Metro Line 2 in Changsha was currently operating, running west-east. The two tunnels of Metro line 2 are referred to as the north and the south tunnel. The two tunnels were constructed by earth pressure balanced shield. The clearance between the two tunnels is 15 m. The Metro line 4, running north-south, was constructed by the same method. The two tunnels of Metro line 4 are referred to as the east tunnel and the west tunnel. The angle between the two lines is 60° at Yingwan station. The tunnel interaction is located 28.8 m and 38.7 m of line 4 from the Yingwan station, respectively. A temporary road is above the north tunnel. Two shafts with the size of 8 m long and 6 m wide, were set on the line 4 trajectory. The horizontal MJS piles were constructed in the shaft. The shortest distance between the line 2 and the shaft is 5.9 m, as shown in Fig. 1.

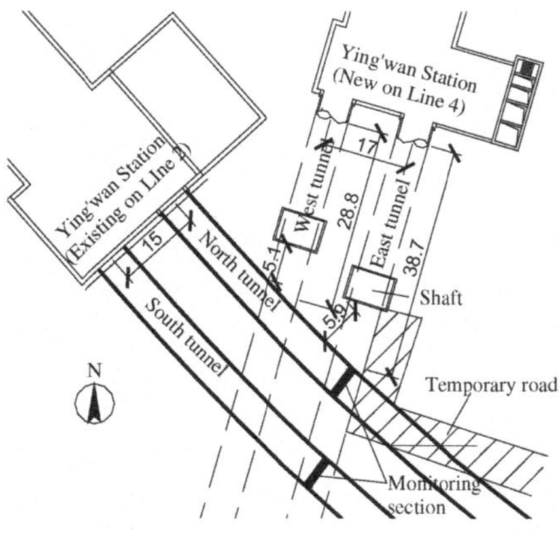

Fig. 1. Plan layout of metro line 4 crossing line 2 (unit: m)

The buried depths of existing and new tunnel crown were about 8.8 m, 17.8 m respectively. The vertical clearance between the two lines was only 3 m. In order to protect the existing tunnels, 13 MJS piles were constructed on the top of each new tunnel of Metro line 4. These piles had a target diameter of 2.0 m and length of 42 m. The longitudinal profile and the layout of MJS piles are shown in Figs. 2 and 3.

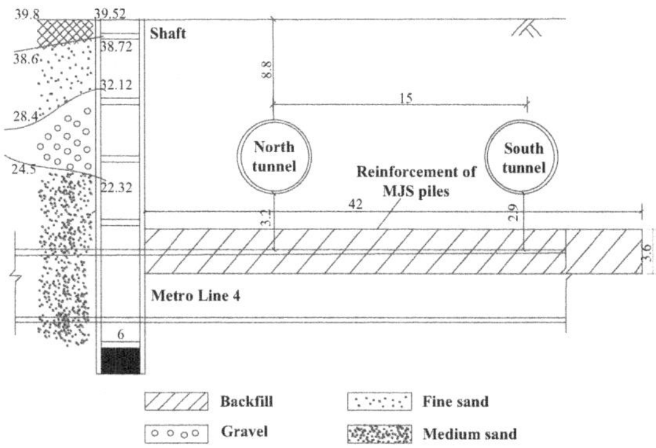

Fig. 2. Longitudinal profile of the tunnel with soil layers (unit: m)

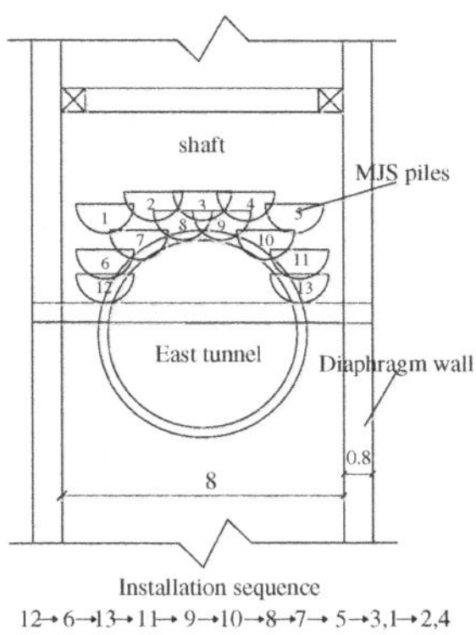

Fig. 3. Schematic diagram of MJS piles (unit: m)

Figure 2 plots the soil profile. The stratum comprises a total of four layers. The top layer to a depth of 1.2 m is backfill ①$_1$. The fine sand ④$_3$ is located at a depth of 1.2–11.4 m. The gravel ④$_5$ is about 3.9 m thick. The medium sand ④$_4$ is more than 10 m thick. The groundwater table is found to be 3 m below the ground surface. Table 1 gives an overview of the properties for each layer. The new tunnels were excavated within the medium sand ④$_4$ layer and MJS piles were constructed at the same layer. The particle gradation experiment shows that particle size greater than 0.25 mm is 55% of the total weight, and those greater than 0.5 mm accounts for 23%.

Table 1. Overview of soil properties

Soil layer	γ (kN/m^3)	γ_{sat} (kN/m^3)	E_{oed} (MPa)	υ	c' (kPa)	φ' (°)
①$_1$ backfill	19	19.2	6.4	0.3	46	18.5
④$_3$ Fine sand	19.6	19.8	32	0.3	2	32
④$_5$ Gravel	20	20	34.9	0.27	0	37.4
④$_4$ Medium sand	19.5	19.5	37.8	0.3	2	34

3 MJS Method and Monitoring

The construction steps are illustrated in Fig. 4. An antipriming pipe, 142 mm in diameter, was drilled to the soil by the drilling system (Fig. 4a). Figure 4b shows the process of horizontal jet-grouting and withdrawing. The grout shrouded by pressurized air was injected vertically at the tip of the antipriming pipe to erode surrounding soils and pipe was slowly withdrawn. Meanwhile the discharged spoil was drained to the spoil tank through the center of antipriming pipe. Figure 4c shows the antipriming pipes were removed from the borehole when the MJS pile was completed. Finally, the boreholes were immediately sealed with a block to prevent the mixture of grout and water from flowing out.

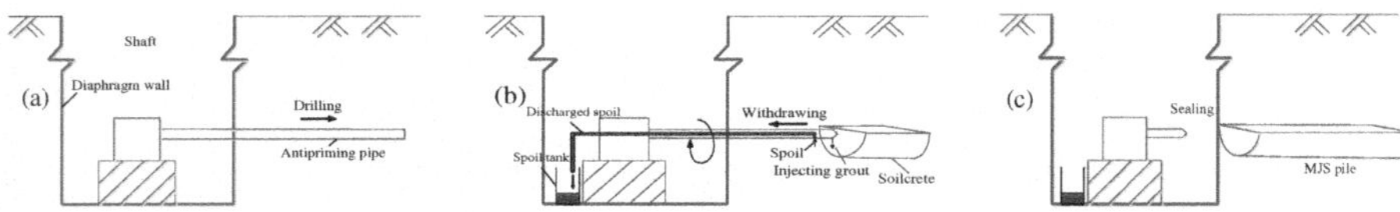

Fig. 4. Construction steps of MJS horizontal pile: (a) Drilling with antipriming pipe, (b) Jet grouting and withdrawing, (c) Completing and sealing the borehole.

The site monitoring program is summarized as follows:

(1) Pore water pressure of surrounding stratum. On the east side of the east tunnel, the pore water pressure test holes 1, 2 were located near the MJS pile. Two monitoring holes were 9 m and 6 m away from the center of the east tunnel, and were 5.8 m and 2.8 m away from the center of the MJS piles, respectively. Each test hole was installed with 3 piezometers (P1-P3, P4-P6). The vertical spacing of the piezometer was 2 m (Fig. 5).

(2) Additional stress of overlying tunnel. To monitoring the additional stress condition of cross-section (Fig. 1) of the line 2, the strain gauges were installed inside the segments to measure the circumferential strain of the segment. The locations of the strain gauges were shown in Fig. 6, in which the numbers in the brackets indicate the sensors assigned to the north tunnel.

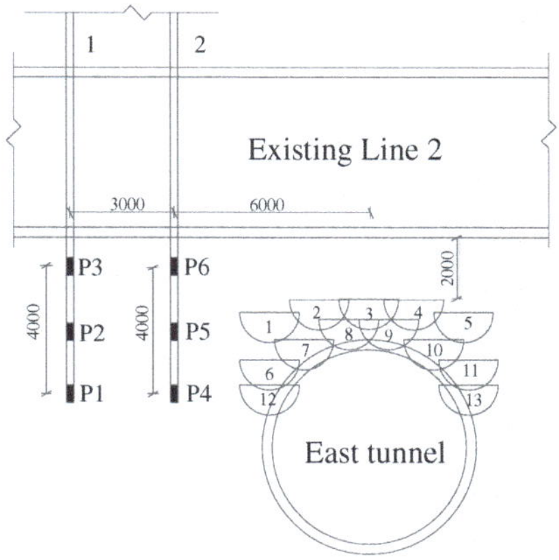

Fig. 5. Layout of piezometers in soils (unit: mm)

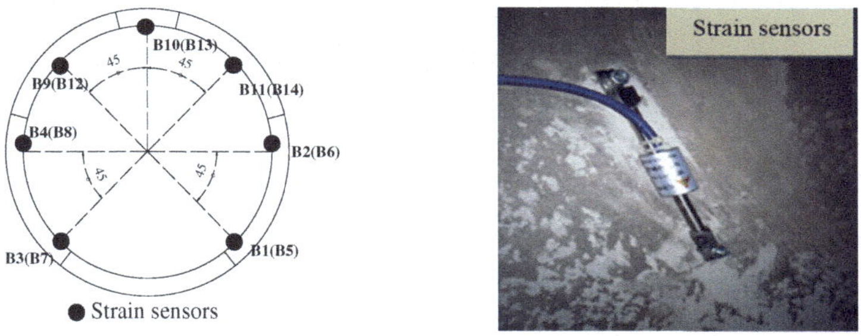

(a) Layout scheme **(b) Site layout**

Fig. 6. Layout of strain sensors on tunnel surface

(3) Vertical displacement of overlying tunnel. The displacement monitoring points along the ballast bed of Metro line 2 were installed at regular interval of 5 m. The displacement monitored by a hydrostatic level system. Hydrostatic level cells were fastened (Fig. 7).

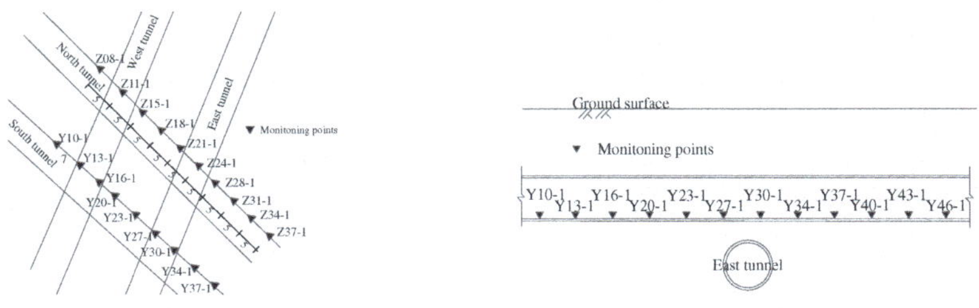

(a) Plan layout of monitoring points (unit: m) **(b) Cross-section view of monitoring points**

Fig. 7. Layout of monitoring points of tunnel settlement

4 Analysis of Monitoring Data

4.1 Pore Water Pressure

No. 12 pile (denoted as C12) was the first constructional MJS pile. The maximum increase of excess pore water pressure, Δμmax induced by the installation of C12 is plotted against r/R_c (see Fig. 8), where r is the distance between the column centerline and the piezometer, and R_c is the radius of MJS pile. Figure 8 indicates that the excess pore-water pressure decreased with increasing distance away from the jet grouting position. The excess pore water pressure becomes negligible when r/R_c is greater than 15.

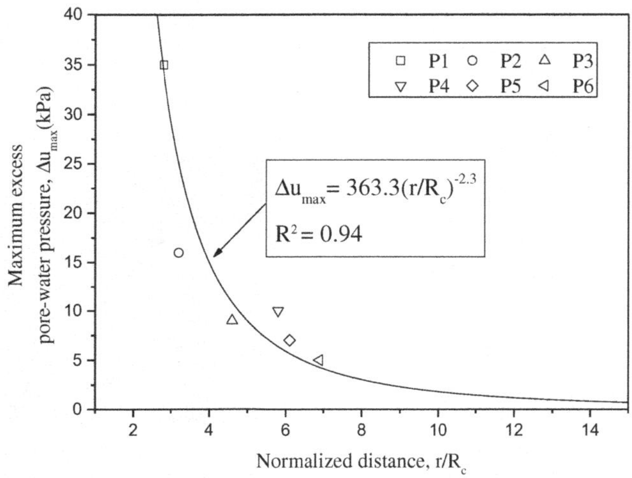

Fig. 8. Relationship between and maximum excess pore water pressure and distance from column centerline

4.2 Stress in Tunnel

The tensile stress of tunnel is specified positive and the compress stress of tunnel is negative. After the construction of 13 MJS piles were completed, the stress in right track position B1, the right arch position B2, the left track position B3, and the left arch position B4, were negative in Fig. 9a. The maximum additional compress stress was 1.64 MPa. However, the additional tensile stress was 0.6 MPa in B9. Due to the skew of the line 2 and the line 4, the additional stress distribution in the inner side of the monitoring section was nearly oval. Consequently, the distribution of grout pressure beneath the existing tunnels was asymmetry.

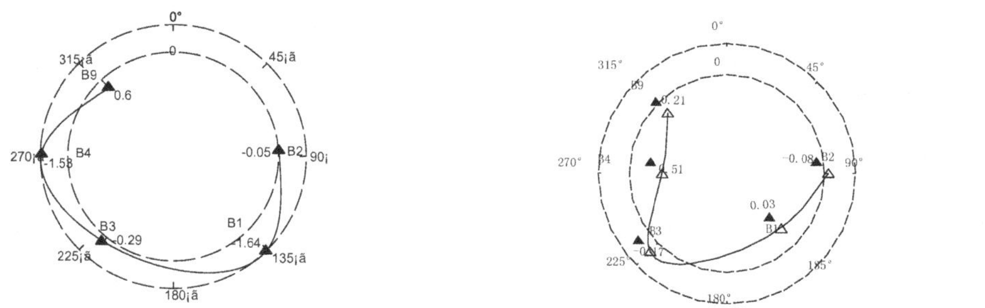

(a) MJS piles finishing **(b) Excavating away from overlying tunnels**

Fig. 9. Additional stress of monitoring section

Figure 9b shows the final additional stress profile of the cross-section when the shield was away from the south tunnel. The earth pressure on both sides of the tunnels remains essentially unchanged while unloading in the vertical direction of line 2, because the soil beneath line 2 was excavated. It was equivalent to the passive extrusion process in triaxial experiment. The additional stress of the tunnel is shown in Fig. 9b.

4.3 Settlement Profile of Existing Tunnels

The vertical soil settlement, S_v, due to tunneling, can be described by a Gaussian curve (Peck 1969):

$$S_v = S_{\max} \exp\left[-\frac{x^2}{2i^2}\right] \tag{1}$$

Where S_{max} = maximum settlement; and i = trough width parameter corresponding to the distance measured from the centerline of the tunnel to the inflection points of the curve. Unfortunately, in many cases the Gaussian curve is not satisfactory to accurately describe the soil settlement. To obtain a better fit to the observed soil settlement, Vorster et al. (2005) suggested a modified Gaussian curve of the following form:

$$S_v = \frac{n}{(n-1) + \exp\left[\alpha (x/i)^2\right]} S_{\max}$$

$$n = e^{\alpha} \frac{2\alpha - 1}{2\alpha + 1} + 1 \tag{2}$$

Where n = shape function parameter controlling the width of the profile; and α = parameter to ensure that i remains the distance to the inflection point and therefore has the same definition as in Eq. (1). Note, Eq. (2) is the Gaussian curve [Eq. (1)] when $n = 1$. Although Eqs. (1) and (2) are used to calculate the settlement of surface. The deformation pattern of surface and existing tunnel is distinct. However, in this case, the settlement of the overlying tunnels also can be predicted using Eq. (2) by adjusting the value of α.

After shield was away from south tunnel, in order to describe the settlement form of the overlying tunnel, the settlement trough curve shown in Fig. 10 is plotted according to the site monitoring data and the theoretical value calculated by Eq. (2). The data of south tunnel are appropriately fitted by Eq. (2) and the settlement profile displays a single-trough "U" shape. About 20 m away from the centerline of the east tunnel, the tunnel settlement is almost 0. Conversely, the points of north tunnel are inappropriately fitted by Eq. (2) due to the temporary road load (Fig. 1). A summary of the data obtained by fitting and calculation is shown in Table 2.

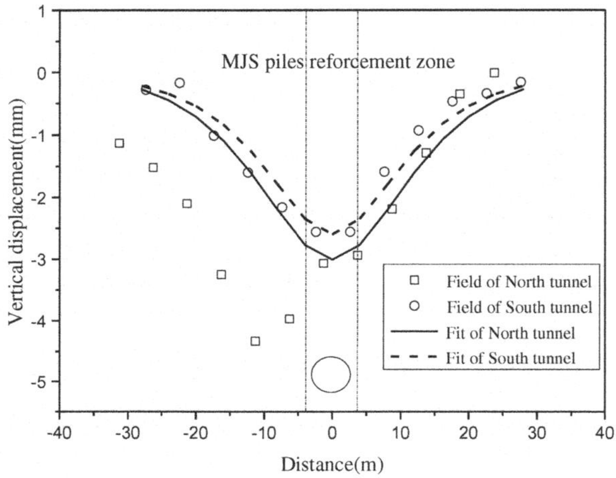

Fig. 10. Settlement trough of existing tunnel

Table 2. Settlement parameters of overlying tunnels

Monitoring points	Monitored maximum settlement (mm)	Fitted maximum settlement (mm)	Trough width (m)	α
North tunnel	4.33	3.01	8.70	0.15
South tunnel	2.56	2.60	7.60	0.10

Combined with the excavation parameters on the east tunnel and the response of the north tunnel, the chamber pressure and thrust increased, so the ultimate settlement of the south tunnel was smaller than the north one. Overall, the disturbance of excavation on the overlying tunnels was obviously reduced by strengthening the soil layer between the existing tunnels and the new one.

5 Conclusion

This paper presents a case of a single tunnel constructed beneath two other existing closely spaced tunnels with reinforcement of MJS piles. Based on the field measurement, some are summarized as follows:

(1) During the period of MJS pile grouting, the pore water pressure was found to increase in the surrounding stratum and the maximum excess pore water pressure decreased rapidly with increasing distance from the location of jetting. At a distance of 15 m from the jet-grouting, the generated excess pore-water pressure was considered to be negligible.
(2) Due to the skew of the line 2 and line 4, the induced stress distribution in the facing monitoring section shows oblique oval shape when the construction of 13 piles were completed. The overlying tunnel was squeezed in the horizontal direction during the period of tunneling. The maximum additional stress was 1.58 MPa during construction process.
(3) The maximum settlement of overlying tunnel was only 4.33 mm with the MJS horizontal piles. The data of south tunnel were appropriately fitted by modified Gaussian curve, while the settlement distribution of north tunnel was asymmetric due to random traffic load.

Acknowledgements. The present work was carried out with the support of the National Key Research and Development Program of China (2016YFC0800207), National Natural Science Foundation of China (41472244), the Provincial Key Research and Development Program of Hunan (0105679005), the Industrial Technology and Development Program of Zhongjian Tunnel Construction Co., Ltd. (17430102000417) and the Research Program of Construction Engineering of Shenzhen (20151118003B).

References

Kim, S.H., Burd, H.J., Milligan, G.W.E.: Model testing of closely spaced tunnels in clay. Geotechnique **48**(3), 375–388 (1998)

Byun, G.W., Kim, D.G., Lee, S.D.: Behavior of the ground in rectangularly crossed area due to tunnel excavation under the existing tunnel. Tunn. Undergr. Space Technol. **21**(3/4), 361–367 (2006)

Zhang, Z.G., Huang, M.S., Wang, W.D.: Response analysis of existing tunnels due to adjacent tunnelling in multi-layered soils. Chinese J. Geotech. Eng. **31**(4), 600–608 (2009)

Addenbrooke, T.I., Potts, D.M.: Twin tunnel interaction: surface and subsurface effects. Int. J. Geomech. **1**(2), 249–271 (2001)

Chehade, F.H., Shahrour, I.: Numerical analysis of the interaction between twin-tunnels: influence of the relative position and construction procedure. Tunn. Undergr. Space Technol. **23**(2), 210–214 (2008)

Fang, Q., Zhang, D.L., Li, Q.Q., et al.: Effects of twin tunnels construction beneath existing shield-driven twin tunnels. Tunn. Undergr. Space Technol. **45**, 128–137 (2015)

Peck, R.B.: Deep excavations and tunneling in soft ground. In: Proceedings of the 7th International Conference Soil Mechanics and Foundation Engineering. Mexico City, pp. 225–290 (1969)

Vorster, T.E.B., Klar, A., Soga, K., et al.: Estimating the effects of tunneling on existing pipelines. J. Geotech. Geoenviron. Eng. **131**(11), 1399–1410 (2005)

Evaluate the Impact on Adjacent MRT Station from Shield Tunneling by Numerical Analysis and Risk Management – Taking Taipei MRT Circular Line for Example

Chen-Wei Hung(✉), Chien-Ming Lai, and Fu-Lai Su

Sinotech Engineering Consultants, Ltd., Taipei, Taiwan
hcw@mail.sinotech.com.tw

Abstract. The first and second phase of Taipei Mass Rapid Transit (MRT) networks were operated now, and the additional routes were approved under design and construction. Although the developed MRT network provides great convenience to passengers, the planned new routes and stations usually overlap the operated ones. Therefore, the complexity and difficulty of design and analysis are gradually increased, and the geotechnical analysis tools should also be improved. In this project, risk identification and rating are carried out at first to realize the risk which might happen at each construction schedule. The strategies in accordance with each risk item are figured out at the design stage to mitigate the risk as much as possible. Hence, the numerical analysis (1-D, 2-D, 3-D) follows the strategy suggestions to simulate whether the effect of these risk strategies is enough or not. The feedback database of previous projects from the in-situ monitoring results could provide some basis for numerical analysis to assume the suitable parameter. Therefore, the analysis results could simulate the actual condition closer, and the analysis results should satisfy the related standard demands. For monitoring the safety of the operation station at all times, automatic monitoring system is set up to monitor the deformation immediately during the whole life cycle.

The major goal is that "Prevent in design, Manage in construction, Monitor after completing", and the passengers' safety could be ensured under constructing the new routes. The evaluation and analysis of the impact on the existing stations from the nearby constructions were discussed in this study, and the conclusions could provide some suggestions for the design and risk management.

Keywords: Risk management · Automatic monitoring · Feedback database · Numerical analysis

1 Introduction

This project is located at the intersection of Beixin Road and Minquan Road in Taipei, Taiwan. Dapinglin station the existing operation station is constructed below Beixin Road, and the new station of Circular Line (Y6 station) is built below Minquan Road. The Y6 station passes through and is adjacent to Dapinglin station; therefore, the

R. Chen et al. (eds.), *Proceedings of the 2nd International Symposium on Asia Urban GeoEngineering*, Springer Series in Geomechanics and Geoengineering,
https://doi.org/10.1007/978-981-10-6632-0_26

transfer between Y6 station and Dapinglin station could be linked. In addition, the shield tunnel of Circular Line should pass through the bottom of Dapinglin station. This project is the first case that passes through the existing operation station in Taiwan. The relationship between Y6 station and Dapinglin station is shown in Fig. 1.

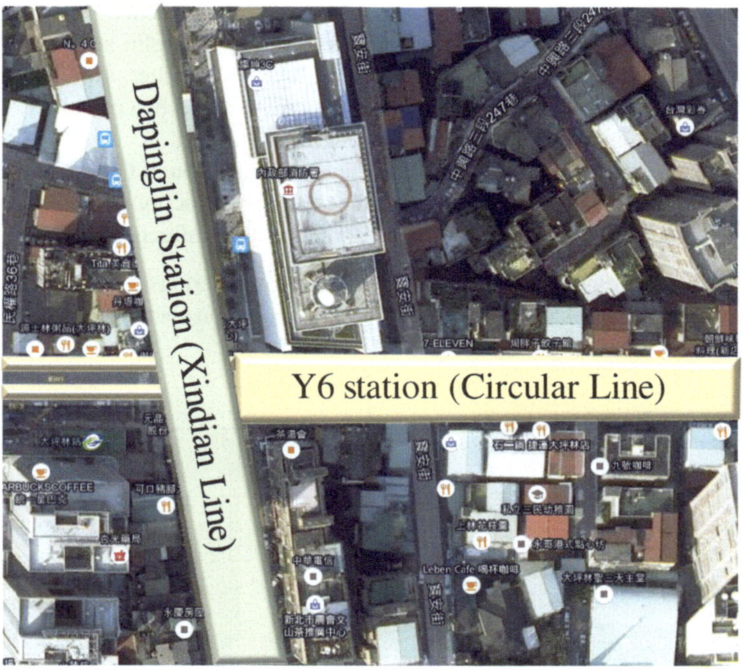

Fig. 1. Relationship between Y6 station and Dapinglin Station

The soil distribution profile is shown in Fig. 2, and the major soil type is Chingmei gravel Formation. The design ground water table is located below ground surface about 8 m, and belongs to static state.

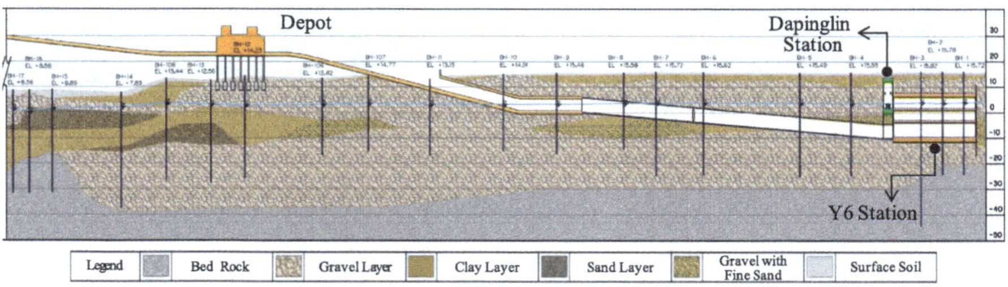

Fig. 2. Soil distribution profile

The challenges and risks which this project might face are listed as follows.

1. The deep excavation and diaphragm wall construction of Y6 station is adjacent to Dapinglin station. The existing diaphragm wall of Dapinglin station is not smooth because of the excessive dredge; therefore, the new diaphragm wall construction might be influenced. Besides, the loss of water, hole collapse and deformation of new diaphragm wall construction in gravel layer might influence Dapinglin station.
2. The new route of Circular Line should pass through Dapinglin station the existing station belonging to Xindian Line. Therefore, the loss of water, sand and excessive settlement due to the Y6 shield tunnel construction should be controlled rigorously to make sure the safety of operating Dapinglin station.
3. This project is the first case in Taiwan which passing through the operating station. The key point in this project is to decrease the construction risk for the nearby station, and the customers' comfortable should not be influenced.

Due to the importance of this project, the value engineering is used to program the best plan in the design stage. There are three construction methods being chosen to

Table 1. Value engineering

Risk matrix			Extent of impact (I)						
			Level	I1	I2	I3	I4	I5	I6
			Description	Insignificant	Marginal	Significant	Critical	Catastrophic	Disastrous
Workers' safety			Death					<5	≥5
			Serious injury				<5	≥5	
	Light injury		≥3 days sick leave			<5	≥5		
			< 3 days sick leave		<5	≥5			
Construction schedule			Schedule delay	≤3 days	3~14 days	14 days~2 months	2~6 months	6 months~2 years	≥2years
Property loss			Loss cost	≤10,000	10,000~ 100 thousand	100 thousand~ 1 million	1 million~ 10 million	10 million~ 100 million	>100 million
Users' safety			Death					<5	5~50
			Serious injury				<5	5~50	51~500
			Light injury			<5	5~50	51~500	>500
Operation service			System break off			<20 mins	Few hours	1 day	1 week
			Route break off		20~60 mins	Few hours	1 day	1 week	1 month
			Station closed	<20 mins	Few hours	1 day	1 week	1 month	Few months
Occurrence probability (P)	Level	Description	Occurrence probability (times/year)	I1	I2	I3	I4	I5	I6
				Insignificant	Marginal	Significant	Critical	Catastrophic	Disastrous
	P6	Frequent	>100	R3	R1	R1	R1	R1	R1
	P5	Probable	1~100	R4	R2	R1	R1	R1	R1
	P4	Occasional		R4	R3	R2	R2	R1	R1
	P3	Remote		R4	R4	R3	R3	R2	R2
	P2	Improbable		R4	R4	R4	R4	R3	R3
	P1	Incredible		R4	R4	R4	R4	R4	R4

solve the passing problem: pipe-roofing method, ground freezing method and shield tunneling method. Shield tunneling method is chosen eventually because of the lower construction risk and the limits of the construction situation (e.g. width of road). Value engineering also used in each construction steps to decrease the security risk as much as possible, and the rating about value engineering is discussed in Table 1.

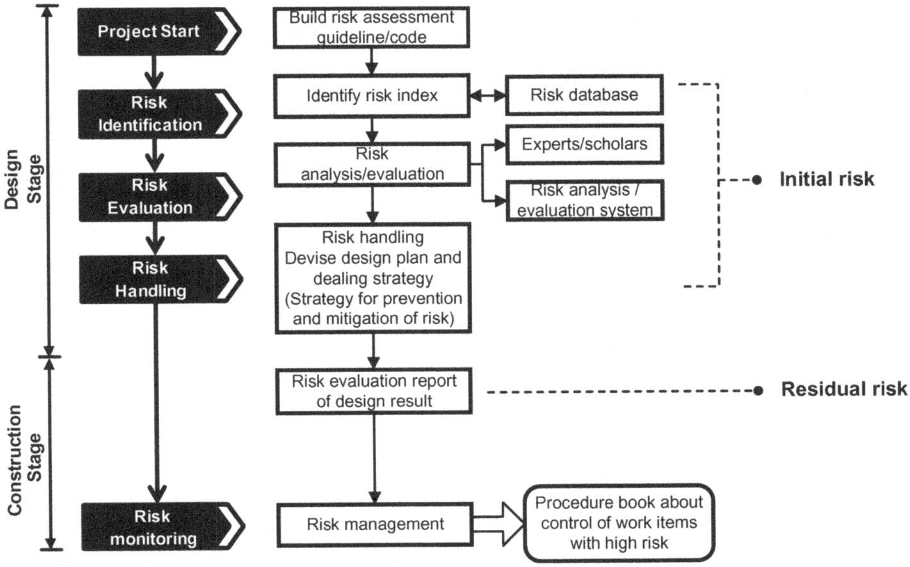

Fig. 3. Flow chart of risk evaluation and inspection

The flow chart of risk evaluation and inspection is illustrated in Fig. 3. In this flow chart, the risk assessment guideline during the whole project life cycle is discussed in both design stage and construction stage. In addition, the definition of each risk assessment stage explains as follows. The monitoring results would feedback to risk identification and risk assessment could implement again to decrease the risk carefully.

1. Risk identification: Identify the events that might endanger or influence the project.
2. Risk evaluation: Evaluate the occurrence probability of risk events and the extent of impact.
3. Risk handling: Propose the strategy for prevention and mitigation of risk.
4. Risk monitoring: Check and trace the implement condition of strategies.

2 Construction Process and Value Engineering

Through comparing each construction method with one another, shield tunneling method is chosen as the proper method. The construction process of shield tunneling method programmed for this case is illustrated in Fig. 4, and the details of each steps is

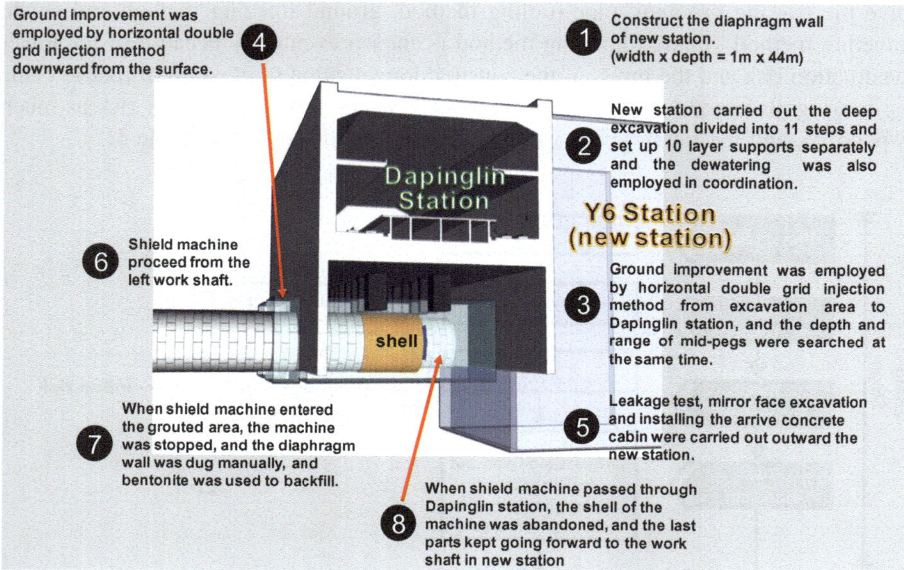

Fig. 4. Construction process of shield tunneling method

explained as follows with the corresponding risk identification, evaluation, handling and monitoring in both design and construction stage.

2.1 Constructional Steps

2.1.1 Step 1. Build New Diaphragm Wall for New Station

Risk identification	Strategy for prevention and mitigation of risk	
	Design stage	Construction stage
1. The construction risk of new diaphragm due to the outward deformation of existing diaphragm wall belonging to Dapinglin station 2. The impact on Dapinglin station due to leakage and collapse of diaphragm wall	1. Use full-casting secant pile to cut the extra concrete to make sure the workability and verticality of new diaphragm wall 2. Investigate geology in detail. Make sure the range and distribution depth of gravel layer and the size of gravel 3. Consider the monitoring plan to observe and prevent the leakage condition	1. The energy of grapple should strong enough to grab bed rocks and the gravels 2. The mix ratio of stabilizing solution should possess the function of leakage prevention 3. Implement automatic monitoring plan of the station indeed

2.1.2 Step 2. Deep Excavations and Dewatering for New Station

Risk identification	Strategy for prevention and mitigation of risk	
	Design stage	Construction stage
1. The deformation of operation station due to adjacent excavation 2. The deformation during deep excavation because of excessive dredge or delay setting up of supports	1. Estimate soil parameters carefully 2. Inspect the penetration depth of diaphragm wall and dewatering of ground water according to the design code, and the results should be included in design results 3. According to the rules of prohibition/limit building, 2D analysis software is available, and make sure the analysis results could satisfied the rules	1. Demand construction contractor to implement the works of diaphragm wall construction and dewatering by the rules in contract 2. Demand the quality control of diaphragm wall construction to avoid boiling 3. Automatic monitoring in the station operates all the time to monitor the deformation 4. Excavate and set up the supports step by step on schedule

2.1.3 Step 3. Ground Improvement and Search Obstacles Under Dapinglin Station

Risk identification	Strategy for prevention and mitigation of risk	
	Design stage	Construction stage
1. The influence of heave and settlement on Dapinglin station during grouting	1. Analyze whether the settlement satisfied the rules of prohibition/limit building or not during crossing the existing station 2. The overall ground improvement is selected, and it is good for reducing settlement, controlling shield machine, and preventing the leakage of mirror face excavation	1. Search the depth and the location of obstacles under Dapinglin station while carrying out the horizontal ground improvement 2. Automatic monitoring in the station is implemented in coordination with horizontal grouting

2.1.4 Step 4. Ground Improvement from the Other Direction

Risk identification	Strategy for prevention and mitigation of risk	
	Design stage	Construction stage
1. The influence on existing operation station during vertical low pressure grouting	1. Select the suitable grouting method 2. The grouting pressure and the monitoring system during construction are prescribed in the rules of design code	1. Automatic monitoring in the station is implemented in coordination with vertical grouting

2.1.5 Step 5. Leakage Test, Mirror Face Excavation and Install Arriving Cabin

Risk identification	Strategy for prevention and mitigation of risk	
	Design stage	Construction stage
1. The risk caused by the mirror face excavation of new/old diaphragm wall 2. The leakage risk occurs when mirror face excavate is carried out under Dapinglin station	1. The leakage grouting between new and old diaphragm wall before mirror face excavation should be noted in the rules 2. In order to reduce vibration influencing on station and new/old diaphragm wall, outer ring mirror face excavation could be handled by flushing manually 3. Add the rules about leakage test. Since the diaphragm walls on both sides have been grouted, the length of leakage test could be lengthened 4. Specify to install the full section arriving cabin for reducing the arriving risk	1. If there exists leakage when implement leakage test between new/old diaphragm walls, the grouting improvement should be carried out immediately 2. Flushing or chain-saw is used to excavate mirror face 3. Install arriving cabin indeed in accordance with the rules 4. Shell of shield machine was abandoned usually to reduce settlement and construction risk

2.1.6 Step 6. Shield Machine Proceed

Risk identification	Strategy for prevention and mitigation of risk	
	Design stage	Construction stage
1. Choose unsuitable type of shield machine inducing much lower construction speed or much larger ground settlement 2. Teeth wear of shield machine during proceeding leads to stoppage when boring in the gravel layers	1. Realize in-situ grain size distribution condition of gravel layers from boring investigation reports 2. Construction contractor should consider the location and timing to replace the teeth of shield machine 3. Type of shield machine should be chosen in accordance with grain size of gravel layers, and this demand should be listed in the rules	1. Submit evaluation report about the choice of type of shield machine based on boring investigation information 2. Enhance the control of management values of shield driving force and backfill grouting, etc

2.1.7 Step 7. Dig Diaphragm Wall Manually and Backfill with Bentonite

Risk identification	Strategy for prevention and mitigation of risk	
	Design stage	Construction stage
1. The risk happens when workers open the cabin and cause the leakage of sand and water 2. The influence on Dapinglin station when digging diaphragm wall	1. Finish ground improvement from the ground surface at first 2. The following additional remarks should be listed in the rules: (1) Leakage test should be carried out before opening the cabin. If leakage happens, grouting should be implemented by shield machine (2) Leakage test would be implemented outside diaphragm wall (under Dapinglin station) after workers walking out the cabin. Make sure the condition is safe, then dig diaphragm wall manually (3) The diaphragm wall should be broken step by step, and also backfill stepwise	1. The demands which additional remarks list in the rules in design stage should be carried out actually in construction stage 2. Automatic monitoring apparatus should be set up in order to control the deformation in coordination with construction steps

2.1.8 Step 8. Shield Machine Keep Going After Abandoning the Shell

Risk identification	Strategy for prevention and mitigation of risk	
	Design stage	Construction stage
1. The settlement occurs while shield machine passing through Dapinglin station 2. The risk happens when shield machine reach the arriving cabin and open the cabin	1. The overall ground improvement is selected to reduce settlement 2. The full section arriving cabin is selected to install at the arriving section to decrease leakage risk 3. Both up and down-track arrive at the arriving cabins by abandoning the shell to reduce the settlement and leakage risk	1. The precision of shield machine should be controlled carefully 2. Leakage test should be carried out before reaching the arriving cabin. If there is leakage condition, grouting should be implemented by shield machine again 3. Use flushing and man power to break diaphragm wall step by step, and also backfill stepwise until reaching mirror face 4. Install full section cabin 5. After the head of shield machine inlays in diaphragm wall and the cabin, the abandoned shell and the flexible connect steel ring are welded together to prevent all the leakage possible routes. Therefore, the arriving works are finished 6. Automatic monitoring apparatus should be set up to control the deformation in coordination with each construction steps

3 Numerical Analysis

In this project, one-dimensional (RIDO), two-dimensional (PLAXIS) and three-dimensional (MIDAS GTS) analysis software was used separately to simulate the true construction conditions with strategies mitigating construction risk. There are two major problems in this project: one is the deep excavation of new station (Y6 station) adjacent to Dapinglin station, and another is the shield tunnel crossing under Dapinglin station. The influence of deep excavation was simulated by both RIDO and PLAXIS. On the other hand, the influence of passing through Dapinglin station was analyzed by PLAXIS and MIDAS GTS.

Different analysis software has different functions and different sensitive to soil parameter; therefore, the feedback database from the in-situ monitoring results provides the important basis to verify whether this software is suitable to that condition or not. In addition, the feedback database also could provide some guideline to adjustment the parameters to simulate much closer to real condition. RIDO has been used in Taipei MRT excavation for more than ten years. The analysis results from RIDO has been verified by the in-situ feedback database, and the maximum diaphragm wall deformation of in-situ monitoring results are probably 70–80% of excavation analysis results. Although RIDO is a simple and high reliable analysis tool, the deformation of station due to excavation could not be simulated. Therefore, PLAXIS was used to realize the deformation of station due to deep excavation, and the results from RIDO and PLAXIS could compare with each other to make sure the correction. According to compare the analysis results from RIDO with those from PLAXIS, the results from PLAXIS are probably 83–85% of those from RIDO. In other words, the analysis results from PLAXIS are much closer to the in-situ monitoring results (monitoring results = 70–80% RIDO). The analysis results about the right diaphragm wall of Y6 station by RIDO and PLAXIS are listed in Table 2 and Fig. 5. Besides, the excavation analysis results from PLAXIS show in Table 3, and are all smaller than standard demand.

Table 2. Analysis results about the right diaphragm wall of the new station

Item	RIDO	PLAXIS
Max. wall deformation (cm)	3.16 cm	2.69 cm
Max. moment (t-m/m)	219 t-m/m	183 t-m/m

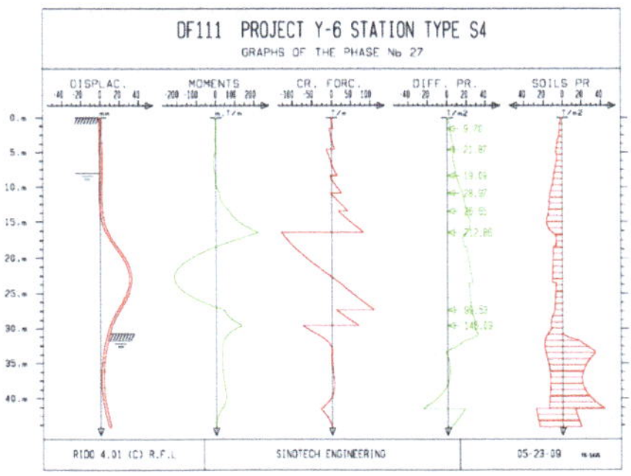

RIDO result

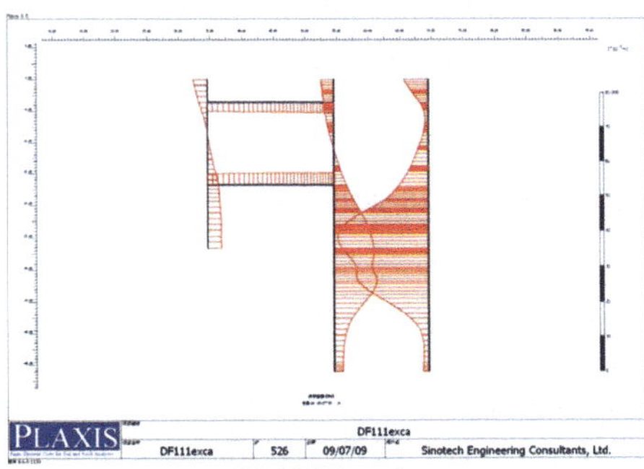

PLAXIS result

Fig. 5. Analysis results about the right diaphragm wall of the new station

Table 3. Compare safety standard with analysis results form PLAXIS (excavation)

Item	Safety standard	PLAXIS
Vertical deformation	10 mm	0.3 mm (heave)
Horizontal deformation	10 mm	3.7 mm
Vertical incline (East-West)	3/5000	0.072/5000 (heave)
Horizontal distortion	3/5000	0.881/5000

For realizing the impact on Dapinglin station when the shield tunnel passing through, both PLAXIS and MIDAS GTS are used to analyze the danger to the operation station. The feedback information system about the ground settlement during shield tunnel boring is built up with lots of monitoring sections and its settlement data. Through feedback information system, choose the projects with similar geologic condition to this project, and then the gap parameter (g) could be get in basis of these real monitoring feedback data. Therefore, the ground loss ratio could be determined with Eq. (1), and one of the function of PLAXIS, contraction of tunnel lining (contraction ratio), could be used to simulate the ground settlement caused by shield tail ground loss condition, as Eq. (2) shows. In this project, gap parameter is about to 21.08 mm; therefore, contraction ratio probably equals to 0.7%.

$$\text{Ground loss ratio} = \frac{\pi(R+g/2)^2 - \pi R^2}{\pi R^2} = \frac{4gR+g^2}{4R^2} \tag{1}$$

$$\text{Contraction ratio} = \frac{(6.1+g)^2 - 6.1^2}{6.1^2} \tag{2}$$

where R = radius of the runnel; g = gap parameter get by feedback information system

Table 4. Compare safety standard with analysis results (passing)

Item		Safety standard	PLAXIS	MIDAS GTS
Vertical deformation		10 mm	8.71 mm (settle)	7.3 mm (settle)
Horizontal deformation		10 mm	–	–
Vertical incline	North-South	3/5000	2.75/5000	2.86/5000
	East-West		–	1/5000 (settle)
Horizontal distortion		3/5000	–	–

The analysis results from PLAXIS about the deformation of operation station are listed in Table 4. Both max. vertical deformation and max. vertical incline are smaller than standard demand. Although there are many verified cases for PLAXIS about excavation and tunnel boring, the influences of the panel thrust of shield tunnel machine, thrust of jack lift, pressure of immediate backfill grouting and shield tail ground settlement to the operation station and surrounding structures still should be simulated and analyzed by 3D numerical analysis software. In this project, MIDAS GTS was used to evaluate the danger to the operation station during the shield tunnel passing. The related settings as analysis are listed as follows.

(1) The panel thrust of shield tunnel machine = soil pressure at rest + 0.2 kg/cm^2 = $k_0 \times \sigma_V$ + 0.2 kg/cm^2.
(2) The thrust of jack lift = panel thrust of shield tunnel machine + friction of shield machine casing.

Table 5. Compare safety standard with total analysis deformation results

Item		Safety standard	Total analysis deformation result
Vertical deformation		10 mm	8.71 − 0.3 = 8.41 mm (PLAXIS)
			7.3 − 0.3 = 7 mm (MIDAS GTS)
Horizontal deformation		10 mm	3.7 mm
Vertical incline	North-South	3/5000	2.75/5000 (PLAXIS) 2.86/5000 (MIDAS GTS)
	East-West		0.982/5000[a]
Horizontal distortion		3/5000	0.881/5000

[a] (1/5000) − (0.072/5000) = 0.982/5000

The analysis results are listed in Table 4, and compare with results from PLAXIS. From Table 4, the comparing results of PLAXIS with MIDAS GTS are similar, that is to say, the analysis results are reliable.

However, the actual deformation of the operation station should take both the influence of deep excavation (analysis results from RIDO and PLAXIS) and shield tunnel passing (analysis results from PLAXIS and MIDAS GTS) into considering. The total deformation conditions are listed in Table 5. All the analysis results are satisfied the safety standard; therefore, the results prove that the risk identified and strategies mitigating risk before construction are workable.

4 Automatic Monitoring System

In order to monitor the deformation of the operation station and the track system all the time during the whole life cycle of construction and operation, the automatic monitoring system is set up in the operation station as Fig. 6 showing. According to the experience from Taiwan high speed rail and MRT, the monitoring apparatus used in this project are total station, EL Beam and tiltmeter, and the major function of each apparatus is listed in Table 6. Total station scans the whole section automatically at the certain timing, and the frequency could reach 10–15 min. EL beam could measure the vertical deformation of the structure. With the experience of other projects, the manual measurement points are needed to measure regularly to modify the error of EL beam. Tiltmeter could measure the tile and deformation of structure, but the actual deformation of the track could not be get directly; therefore, tiltmeter is usually treated as assist tool.

Besides, the automatic monitoring system should be collocated with automatic and wireless transmission system to access and monitor the safety of the station at any time. The automatic monitoring system should have the warning function, and the warning message should be notified to the staff immediately to make the emergency response at once. Each apparatus has different precision, advance and shortcoming; hence the judgment should be made by taking all information into considering.

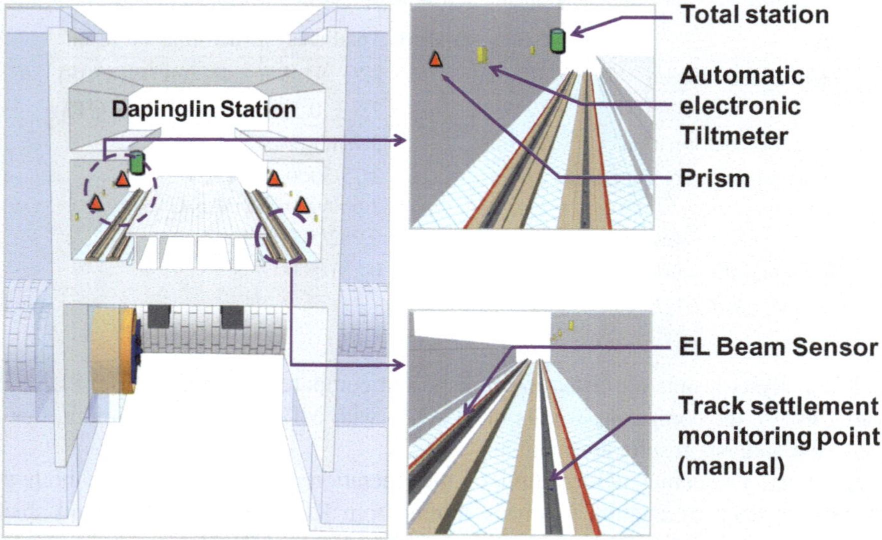

Fig. 6. Automatic monitoring system

Table 6. Function of automatic monitoring apparatus

Automatic monitoring apparatus	Measurement object
Total station	3D deformation of track plate
EL beam	Vertical settlement of structure
Tiltmeter	Amount of structure incline

5 Conclusion and Suggestion

There are more and more cases about the overlapping between new and old MRT routes; hence, how to control the safety of the existing operation station as earlier as possible is important. The conclusion and suggestion are listed as follows.

1. In this project, the risk management has been carried out at first in design stage to identified each possible risk at every construction schedule, and the related mitigating strategies has also been figured out to reduce the risk.
2. One-dimensional (RIDO), two-dimensional (PLAXIS) and three-dimensional (MIDAS GTS) analysis software was used separately to simulate the true construction conditions with strategies mitigating construction risk. The feedback database could provide some guideline to adjustment the parameters to simulate much closer to real condition. The analysis results in this project all satisfy the standard demand.
3. In order to monitor the deformation of the operation station and the track system all the time during the whole life cycle of construction and operation, the automatic monitoring system is set up in the operation station. Besides, the automatic

monitoring system should be collocated with automatic and wireless transmission system to access and monitor the safety of the station at any time. Each apparatus has different precision, advance and shortcoming; hence the judgment should be made by taking all information into considering.

Shield Tunneling in Gravels – Difficulties and Resolutions

J.F. Chang[1(✉)], C.R. Chou[1], Y.F. Lai[1], and Y.H. Hsieh[2]

[1] MAA Group Consulting Engineers, Taipei, Taiwan
jungfeng.chang@maaconsultants.com
[2] Department of Rapid Transit Systems, Taipei City Government, Taipei, Taiwan

Abstract. This paper presents a case study in a Taipei MRT project where the major stratification that shield tunneling passed is gravels. In view of ground condition and tight schedule, the contractor proposed an "embrace" idea for the design of the cutter head. Such an idea allowed the tunnel boring machine (TBM) to "swallow" rather than break the cobbles or boulders during excavation, thus reducing frequency and time in exchanging cutter bits that were worn out. It was however unfortunate that the TBMs encountered drift woods and unexpectedly big boulders. While the chamber was open for investigation, most cutters were found so seriously damaged that the structure of the cutting wheel was also affected. Over a series of process in evaluation and reinforcement, the tunnel section was finally completed without further difficulties. The paper first introduces the background of the project and the recordings of the incident. Its origin, evaluation, and resolutions are then illustrated. Lessons learnt from the study case are provided at the end of the paper.

Keywords: Shield tunneling · Gravels · Taipei MRT

1 Introduction

Gravels are principally distributed around the Central Range of Taiwan that are transported downstream by debris flow, floods, or turbidity currents and deposited in the river valleys, alluvial fans, floodplains, beaches, and even deep sea beds (e.g., Teng 1996). In recent years, networks of transportation (e.g., rapid transits) and lifeline (e.g., water, electricity, and wastewater pipelines) have been expanded from urban areas where soft grounds are always encountered to the outskirts where gravels become the major characteristics of the ground condition. Tunneling in gravels using either shield or pipe jacking methods is implemented through these projects and experience has been accumulated (e.g., Lan et al. 2008; Jhuo et al. 2009) for the engineering behavior and construction difficulties.

This paper introduces a project in the Taipei MRT system where tunnel boring machine (TBM) was stuck in gravels. The background of the incident is first presented followed by cause investigation and the corresponding resolutions. The study could provide a reference for similar cases in the future.

R. Chen et al. (eds.), *Proceedings of the 2nd International Symposium on Asia Urban GeoEngineering*, Springer Series in Geomechanics and Geoengineering,
https://doi.org/10.1007/978-981-10-6632-0_27

2 Background

The project ranges from an operating terminal station extending southwards to a new underground station (with a crossover) in front of a technical park. The total length of the project is about 1,950 m where the tunnel section is about 1,529 m long with 4 cross passages (i.e., CP1 to CP4) connecting the two single tunnels for the up- and down-tracks. The overburden of the tunnels is between 13 and 24 m and major ground condition that tunneling would encounter includes gravels, sandstones, silty clays, and silty sands. Figure 1 shows the long and cross profile of the project with street map and ground condition, respectively.

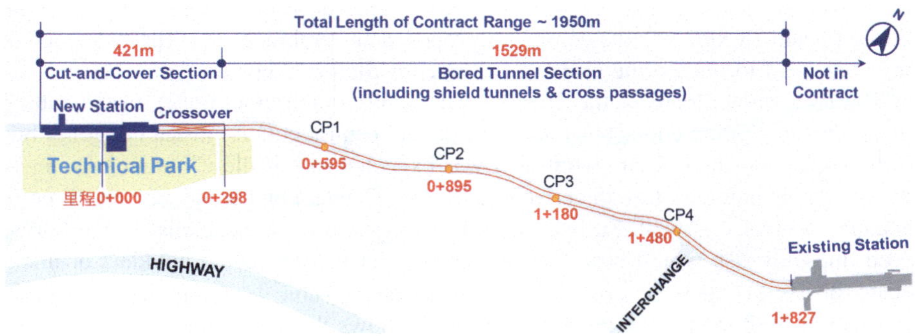

(a) Long Profile

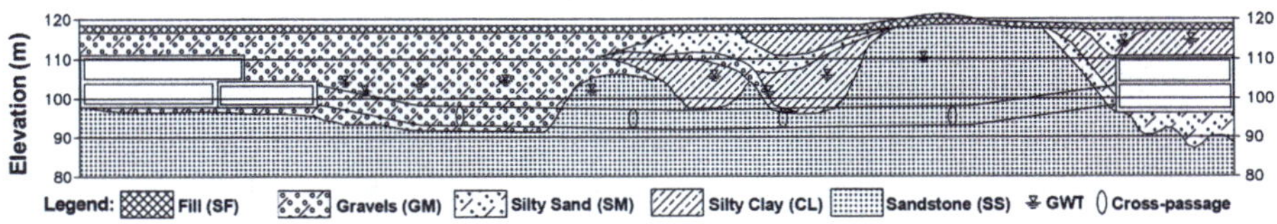

(b) Cross Profile

Fig. 1. Scope of the project

It is noted that since the project is the extension from an existing operating line tunneling was planned to start from the shaft at the new station. After TBM arrived at the existing station, the shields would be abandoned on site before the connection between tunnels and the existing station commenced.

As illustrated by Fig. 1, tunneling would encounter gravels from the departure end (at mileage about 0 + 298 m) to about 45 m away from CP2 (at mileage about 0 + 850 m) with a total length of about 552 m. After that, tunneling would mainly encounter argillaceous sandstone or sandstone interbedded with shale for the next 900 m (at mileage about between 0 + 850 m and 1 + 750 m). Mixed ground of alluvium soils and rocks could however be expected within this section. Tunneling would then meet alluvium soils (such as silty sands and silty clays) for the final 77 m (from mileage about 1 + 750 m to the existing station) before arriving at the existing operating station.

In view of complex ground condition for the tunnel section, the TBM design shall consider not only the dimension and strength of cobbles or boulders that shall be tackled in the first section but also the characteristics of rocks and soils confronted in the latter section. In addition, the groundwater table exhibited a high level (about 3 to 4 m below ground surface) in the alluvium soils around existing station and relative low level (about 11.3 to 12.5 m below ground surface) for the rest of the project. A hydrostatic pressure distribution with depths was observed for the entire project.

Since more than one-third length of tunnel section would be gravels (with average permeability of $1.5*10^{-2}$ cm/s and uniaxial compression strength between 941 and 1,773 kg/cm^2) and nearly 60% of total length would be poorly cemented sandstone (with average permeability of $4.6*10^{-5}$ cm/s and uniaxial compression strength between 1 and 38 kg/cm^2), the contractor proposed an "embrace" idea for the design of the cutter head to release the pressure in the construction schedule.

The idea was to increase the opening ratio as possible (about 66%) on the cutting wheel, thus equipping cutter bits (such as side and central cutters) instead of roller bits on the spokes or outer circumference. It was expected to "swallow" rather than break the cobbles or boulders to reduce wear and tear of cutters and frequency and time in exchanging cutters as well. Such an idea was supposed to be beneficial to the boring speed; the only risk would however be when bigger-than-expected boulders or unexpected obstacles, such as drift woods, are present. Table 1 summarizes the basic information of TBM for the project, and Fig. 2 shows the associated photos.

Table 1. Basic information of tunnel boring machine (TBM) for the project

Item	Description
Closed-Face TBM	Modified earth pressure balance (EPB) type
Cutter head	Spoke (opening ratio 66%)
Cutting wheel	Flat type
Cutting wheel bearing	Central bearing
Cutters	Cutter bits, reinforced and central cutter bits, fishtail cutting edge
Screw conveyer	Shaftless ribbon type
Maximum boulder size passing screw conveyer	690

(a) TBM Factory Inspection (b) Shaftless Ribbon Type Screw Conveyer

Fig. 2. TBM adopted for the project

3 Event Recordings

3.1 Up-Track Tunnel Construction

The construction of up-track tunnel commenced from the breakthrough of excavation face at the departure shaft for the TBM #1 on May 13 of 2011. Following the initial excavation, equipment transfer, and formal excavation, the advancing rate significantly dropped at the 89th ring (i.e., R0089) on July 02 of the same year. The excavation was finally suspended at the R0091 on July 09 as a result of extreme boring resistance.

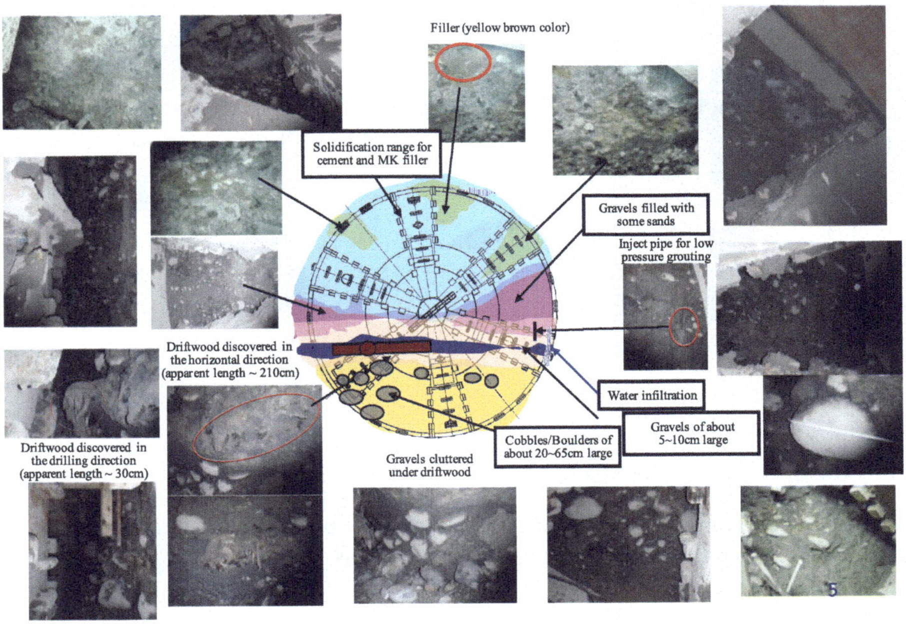

Fig. 3. Excavation face condition on up-track TBM (source: contractor's presentation materials)

It was until ground improvement (i.e., low-pressure grouting) had been implemented and the chamber was opened that a bunch of drift woods was discovered at the middle layer of the excavation face falling within a range of approximately 2.1 m long and 0.3 m wide. In addition, cobbles and boulders of about 0.2 to 0.6 m in diameter were piled up at the bottom half of the excavation face (Fig. 3). Most of cutter bits on the TBM #1 were suffered extraordinary damage and even the cutting wheel and the spokes were serious affected, as shown in Fig. 4. The excavation was resumed on February 05 of 2012 after the obstacles were manually removed and the cutter bits and cutting wheel structure were reinforced.

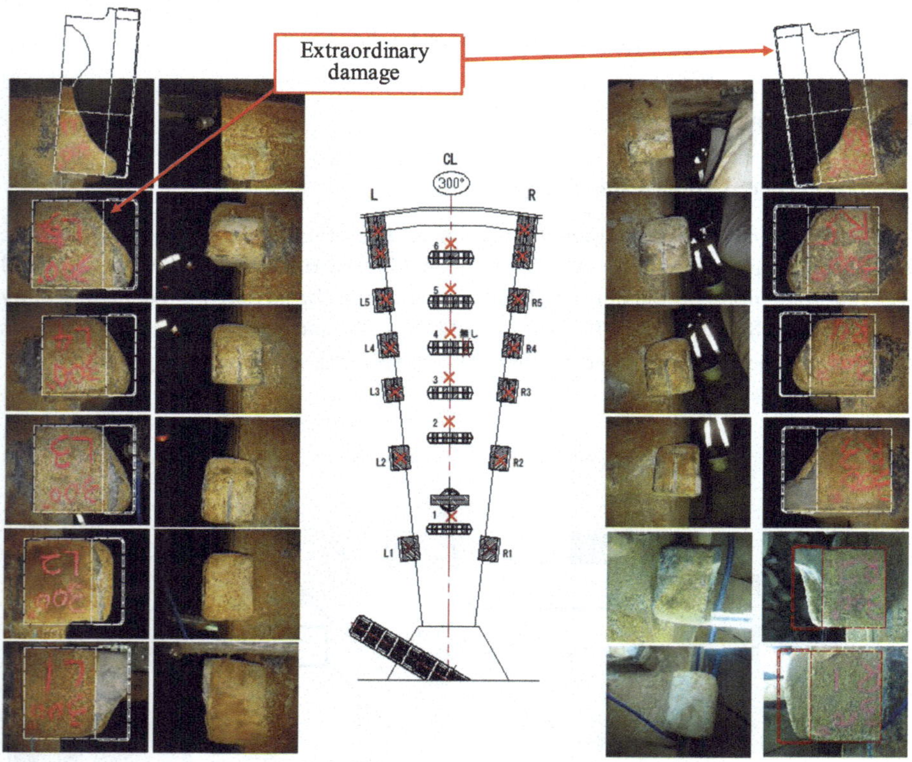

Fig. 4. Damage of cutter bits on the 300-degree spoke of up-track TBM (source: Contractor's Presentation Materials)

3.2 Down-Track Tunnel Construction

The construction of down-track tunnel began from the breakthrough of excavation face for the TBM #2 on July 2, 2011, nearly two months after the TBM #1 had broken through the departure shaft. Same as TBM #1, the TBM #2 encountered extreme resistance in tunnel boring and excavation was forced to set back at R0124 on August 23. When low-pressure grouting was again implemented around the tunnel face and chamber was opened, drift woods were found to gather at the middle layer of the face with a range of about 2.4 m long and 0.6 m wide. Boulders of size ranging from 0.45 m to 0.57 m were gathered at the bottom half of the face, as shown in Fig. 5. The cutter bits and cutting wheel of TBM #2 were as observed on the TBM #1 suffered extraordinary damage, as shown in Fig. 6. Following manual removal of the obstacles and reinforcement of the cutter bits and spoke structure, tunnel boring recommenced on February 25, 2012, about 20 days after TBM #1 resumed boring.

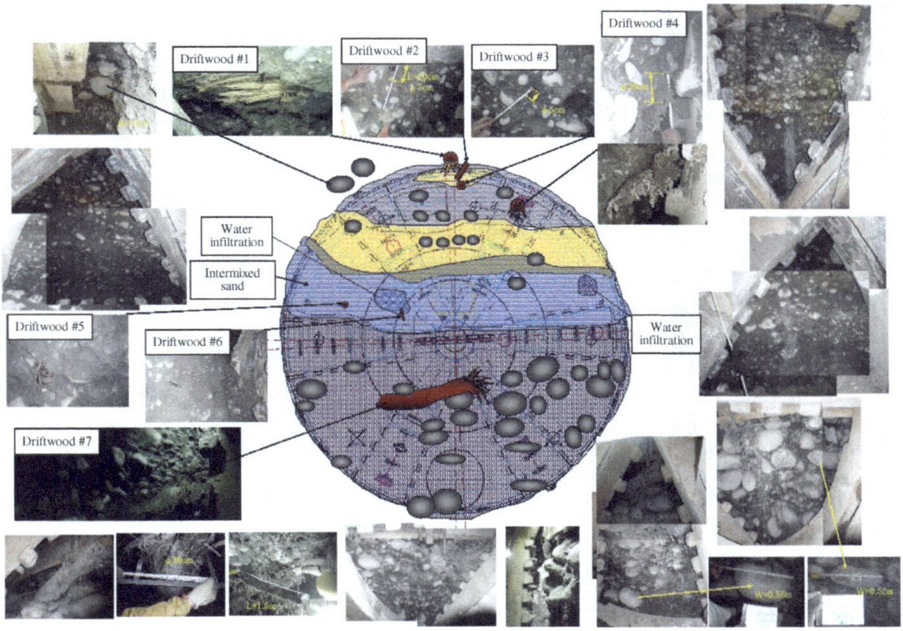

Fig. 5. Excavation face condition on down-track TBM (source: Contractor's Presentation Materials)

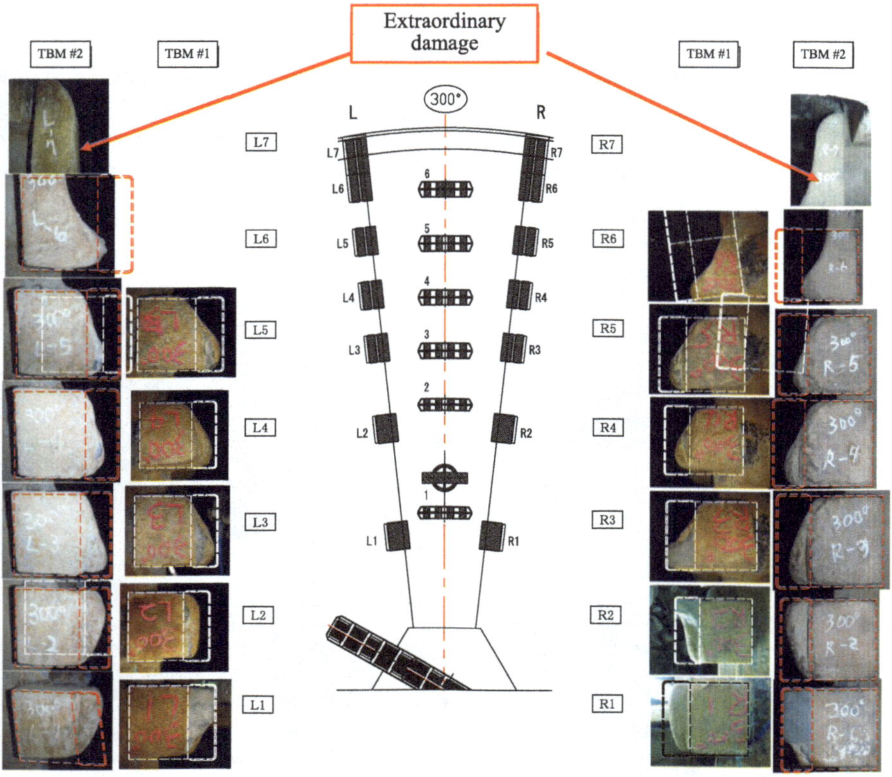

Fig. 6. Damage of cutter bits on the 300-degree spoke of down-track TBM (source: Contractor's Presentation Materials)

4 Cause Investigation

Summarizing the construction records of shield boring, preliminary conclusion for the causes of excavation difficulties on the two TBMs may include use of roller bits, control of additives, presence of drift woods, and wear and tear of cutter bits. These factors are investigated and described in the following sections.

4.1 Use of Roller Bits

According to past boring experience in gravels, though the cutter head and bits configuration of TBMs may vary from different manufactories those who tend to use spoke type to increase opening ratio and "embrace" cobbles or boulders would still install roller bits to break unexpectedly big boulders. The contractor of the project, however, did not configure roller bits to support the "embrace" idea on the cutter head design. Before obstacles were encountered, the advancing rate of about 5 to 6 rings a day was no difference from general excavation. While the chamber was opened and boulders of size greater than 0.5 m was observed to pile up on the bottom half of the face, it was inferred that the "embrace" idea without support of roller bits tends to have difficulties in handling bigger boulders.

4.2 Control of Additives

To remain a stable excavation face for the modified earth pressure balanced (EPB) shield tunneling, a key issue is the management of plastic muck flow aiming to increase fluidity of soil paste in the chamber and the excavated materials can be expelled smoothly. Based on the construction records before the obstacles were encountered (Fig. 7), the recordings illustrated that the variation of earth pressure, torque of cutting wheel, total thrust, and discharge volume increased gradually with TBM advance. Further, the discharge volume derived from screw conveyor and muck cars revealed poor relevance. It was implied that the selection and adjustment of additives were yet in the best condition, leading to poor liquidity of the excavated materials and thus unsteady discharge and possible clog in the chamber.

4.3 Drift Woods

As illustrated by Fig. 7, readings for torque of cutting wheel and total thrust showed an extraordinarily increase trend after TBM confronted obstacles, implying the possible presence of drift woods that blocked excavated material from entering the chamber. Since TBM was not able to break the woods, cobbles and boulders were forced to accumulate outside the cutting wheel that accelerated wear and tear of cutter bits and the drilling process was halted at last.

4.4 Wear and Tear of Cutter Bits

According to the investigation report, the wearing coefficient for the cutter bits equipped on the sides of spokes, with a total of 76 bits for each TBM, ranged from 0.206 to 0.694 mm/km for the TBM #1 and from 0.194 to 0.524 mm/km for the TBM #2, all of which highly exceeded the design value of 0.025 mm/km.

For the reinforced cutters equipped along the centerline of the spokes, with a total of 35 bits for each TBM, the wearing coefficient fell within a range between 0.033 and 0.415 mm/km for the TBM #1 and 0 and 0.392 mm/km for the TBM #2, most of which surpassed the design value of 0.050 mm/km. Through visual inspection, however, some cutters were damaged by direct impact of gravels not truly by wear and tear. It was thus indicated that the allocation and wearing capacity of the cutter bits did not match up very much. Furthermore, the wearing detector that was equipped on the spokes to provide warnings for the wearing bits was totally worn out during excavation and lost function entirely for the incident.

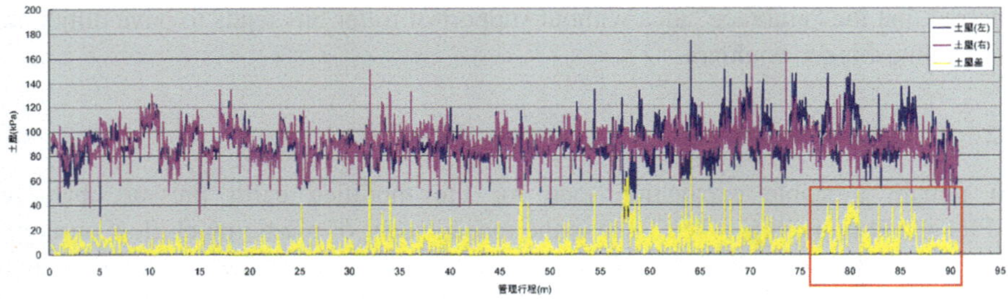

(a) Earth Pressure

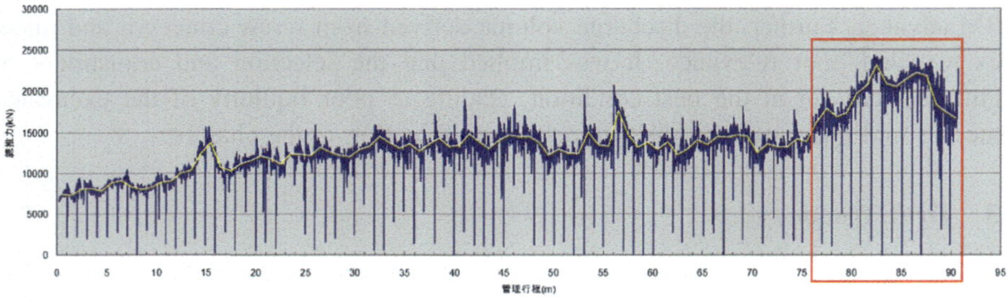

(b) Total Thrust

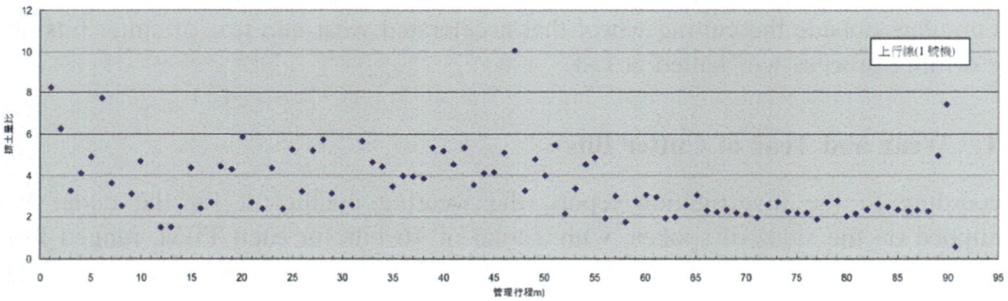

(c) Ratio of Calculated Muck Car to Screwed Spoils

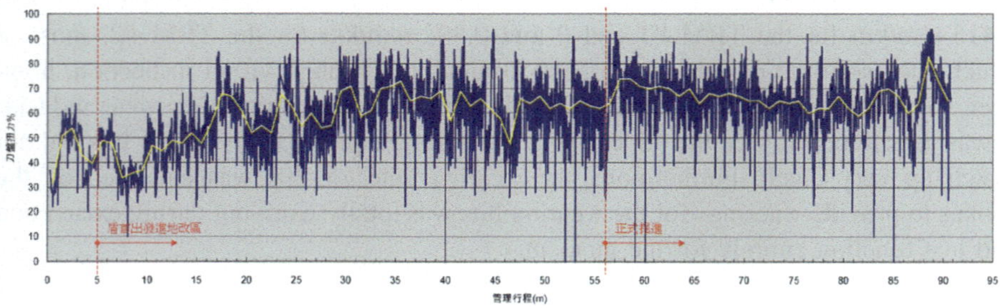

(d) Torque on Cutting Wheel

Fig. 7. Construction records of TBM excavation (Up-track line)

5 Resolutions and Strategies

5.1 Cutting Wheel Remodeling

Since the cutting wheel and bits were seriously damaged and the TBM was not able to be pulled out of ground the remodeling work, including replacement and reinforcement of the cutter head, could only be conducted within constrained space. The ideas of the remodeling were briefed in the following paragraphs and the layouts of the cutter head before and after remodeling are as sketched in Fig. 8.

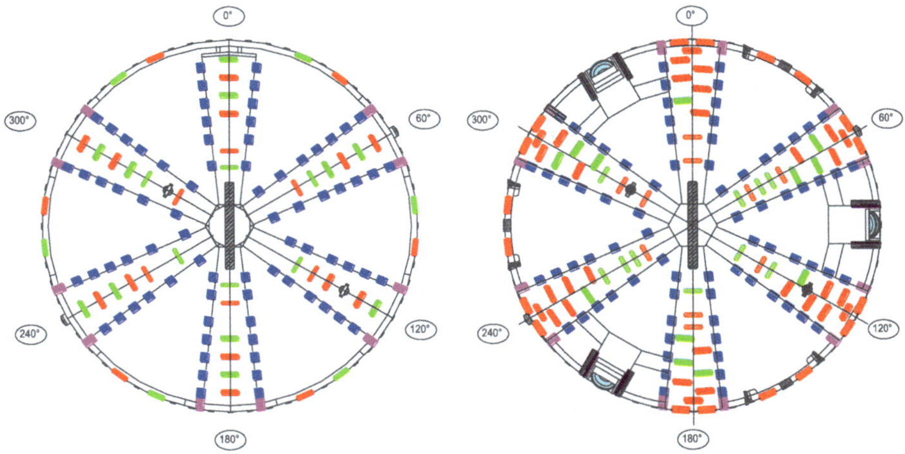

Fig. 8. Cutting bits allocation before and after TBM remodeling

(1) Enhance cutting capability for drift woods.
The original spacing for the reinforced cutters within the inner circle was 300 mm. The spacing was shortened to 150 to 200 mm and number of cutter bits was therefore increased from 35 to 62 sets to enhance cutting capability for drift woods.

(2) Reinforce breaking and removal capability for gravels accumulated on the outer circle and increase the wearing capability
   (a) Outer reinforced cutter bits – 12 sets for original design and increased to 18 sets.
   (b) Cutter bits – 10 sets for original design and increased to 12 sets.
   (c) 3 sets of outer protection bits, 3 sets of roller bits, 6 sets of protection bits for roller bits, and 6 sets of scraper bits were additionally equipped.

(3) Enhance wearing capability on the rings of cutter head.
Anti-wearing steel plates of 100 and 30 mm wide were equipped on the outer circumference of the ring.

(4) Reduce exceeding breakage of gravels that leads to damage of hard insert unit and extraordinary wear and tear of cutter bits.
The way of inserting the hard unit of the cutter bits was changed from welding attachment to embedment to prevent fall out while TBM is boring through gravels and driftwoods.
(5) Cut off damaged spokes and replace new one by welding.
In view that adding roller bits and more cutter bits on the cutting wheel could induce more friction and torque during TBM boring, thus leading to temperature increase in oil pressure, sealing ring, bearing, and clutch, the cooling device for operating oil (air cooling), return pipe for oil (water cooling), and compartment (water cooling) and a good ventilation environment shall be properly maintained during TBM boring.

5.2 Excavation Management

In addition to remodeling of cutting wheel, the earth pressure, discharge and additive during tunnel boring shall be properly managed. The advance rate is adjusted following the torque variation on the cutting wheel. To maintain the earth pressure on the excavation face, additives shall be added for a stable face, fluidity of discharge, and prevention of blow out.

Since the wearing detector did not work throughout the event, a daily inspection was then carried out to control the wearing condition of the bits.

6 Conclusions and Suggestions

While shield tunneling is conducted in gravels, the purpose of applying "embrace" or "swallow" idea for cutter head design is to enhance advance rate and prevent cutter bits from exceeding wear and tear. The construction risk is however escalated accordingly. This is because tunneling would be stuck when drift woods or unexpectedly large boulders are encountered. It most of time could be resolved by implementing grouting and then manual removal of obstacles; the schedule and budget are both affected anyhow.

Therefore, though increasing opening ratio is still a right direction equipping roller bits is a way to enhance TBM's resilience against unexpected condition and thus reduce construction risks. As to the exceeding wear and tear on cutter bits, it is suggested to have exchange and inspection plans in advance that could facilitate tunneling in gravels very much.

Acknowledgements. Permission for publishing this paper from MAA Group Consulting Engineers and Department of Taipei Rapid Transit Systems (DORTS) of Taipei City Government is gratefully appreciated. Special thanks to Dr. R.N. Hwang who helped prepare the manuscript.

References

Jhuo, Y.B., Huang, M.L., Liao, M.Y., Liu, W.N.: Technical discussion on shield tunneling in gravels. In: The 8th Cross-Strait Conference on Tunnels and Underground Construction, pp.C3-1–C3-12 (2009)

Lan, C.J., Yen, C.H., Li, Y.C., Tang, C.W.: A case study of long span pipe jacking technique in Taichung gravel formation. Sino-Geotechnics **115**, 83–90 (2008)

Teng, L.S.: Geological background of the gravel formations of Taiwan. Sino-Geotechnics **55**, 5–24 (1996)

Three Dimension Seepage Flow Numerical Modeling of Slurry Shield with the Permeable Filter Cake

Xinsheng Yin[2], Fanyan Meng[2], and Renpeng Chen[1,2](✉)

[1] College of Civil Engineering, Hunan University, Mountain Yuelu, Changsha 410082, China
chenrp@hnu.edu.cn

[2] MOE Key Laboratory of Soft Soils and Geoenvironmental Engineering, Department of Civil Engineering, Zhejiang University, 866 Yuhangtang Road, Hangzhou, China

Abstract. The filter cake on the tunnel face has a significant influence on project of the slurry-driven shield. During the excavation, only permeable filter cake is allowed to form because the rotating speed of cutter wheel is fast. Through the filter cake, the water of slurry will infiltrate into the soil. The water enhances the hydraulic head and the risk the failure of tunnel face. In order to determine the extent of hydraulic head, the analysis of seepage flow with the considering of the permeable filter should be conduct. A three-dimensional numerical model with the permeable filter cake was established for the analysis of seepage flow. Based on the numerical results, the hydraulic head field was obtained under different conditions. The hydraulic head field shows that the existence of filter cake seriously affects the hydraulic head field near the tunnel face. The hydraulic head tends to hydrostatic head, at the points where is $2D$ in front of tunnel face, horizontal $1.5D$ to the lining of tunnel or soil surface. When the hydraulic conductivity of soil increases, the hydraulic loss in filter cake increases from 0.16 to 0.5 the difference between slurry and hydrostatic pressure.

Keywords: Bentonite slurry · Hydraulic conductivity · Hydraulic head · Seepage flow · Excess pressure

1 Introduction

The slurry shield has been widely used to construct the tunnels, especially for the tunnels under the river. The pressurized bentonite slurry is frequently used to stabilize the tunnel face in sandy soil. Under the pressure, the water of slurry infiltrates into the ground. Some of the small particles of slurry plug the pores on the tunnel face. And the large particles of slurry bridge the pores on the tunnel face. These particles form the filter cake, which becomes thicker with the increasing filtration time (Tien 2012). Some researchers (Li et al. 2009; Kim and Tonon 2010) presumed that filter cake is impermeable, because the hydraulic conductivity of filter cake is far less than the hydraulic conductivity of soil (Yeo et al. 2005; Chung and Daniel 2008). Based on the impermeable filter cake, a principle instructs the slurry pressure must be equal to the sum of three components: the hydrostatic pressure at center of the tunnel face, lateral effective stress

R. Chen et al. (eds.), *Proceedings of the 2nd International Symposium on Asia Urban GeoEngineering*, Springer Series in Geomechanics and Geoengineering,
https://doi.org/10.1007/978-981-10-6632-0_28

at the tunnel axis and addition pressure (20 kPa) (Zhang et al. 2004). The density of the slurry is 1200 kg/m^3, the slurry pressure increases by 12 kPa when the height decreases by one meter. And the gradient of the earth pressure (including the hydrostatic pressure and lateral effective stress) is about 14–16 kPa/m. According to the principle above, the slurry pressure may be lower than the earth pressure. The probability of the tunnel face failure exists when the filter cake is presumed impermeable.

However, the filter cake is permeable, because rotating cutter wheel removed the filter cake at a certain time interval (about 12 s) (Broere 2001). And only the permeable filter cake is formed in 12 s (Yin et al. 2016). The permeable filter cake results the rising hydraulic head in soil. If the filter cake with a high hydraulic conductivity, the failure of tunnel face will occurs. For instance, during the advance of a sewage gallery of 2.8 m diameter in Zurich, the bentonite slurry established only a low-quality filter cake, which led to various surface failures (Fritz et al. 2002). The failure will be prevented base on a better understanding of the seepage flow with the consideration of filter cake. Hence, the seepage flow analysis for slurry shields continues to be a major research topic.

Analytical and numerical methods were applied to analyze the seepage flow near the tunnel face of slurry shield. Anagnostou and Kovári (1994) assumed that the scope of seepage force is equal to the slurry penetration distance. And the seepage flow was considered based on the empirical equation. Broere and von Tol (2000) ignored the filter cake and presumed that the penetration distance had an influence on the seepage flow. An analytical solution was suggested to calculate the pore pressure on the tunnel axis. However, this solution cannot deduce the three-dimension hydraulic head field. The omitting of the permeable filter cake may lead an insufficient result. Kim and Tonon (2010) and Li et al. (2009) used the numerical method to study the problem on the slurry shield. In the existed numerical model, the filter cake is supposed to be impermeable. The hydraulic head field from numerical result does not match the field measurement. A three-dimension numerical model with the permeable filter cake should be established for calculating the hydraulic head field near tunnel face.

In this paper, the permeable filter cake on the tunnel face under transient-state groundwater flow conditions is considered and applied to the three-dimension numerical model. The analysis of seepage flow is performed based on the numerical results.

2 Numerical Modeling

In order to study the distribution of the hydraulic head around the tunnel face, the three-dimensional numerical modeling was performed by finite element program COMSOL. Figure 1 shows the typical model for a tunnel with the diameter D of 15 m located at C/D ratio equaling to 1, C is the overburden thickness. The model is 150 m long, 60 m high and 60 m wide. The size of the model is large enough to eliminate the boundary effect. The model is filled with soil except the tunnel. A horizontal tunnel with a length of 45 m is located in the model and 30 m above the bottom. One end of the tunnel is connected to the model side wall. A rectangle tunnel cross-Sect. (15 m × 5.89 m) is adopted for simplicity (Perazzelli et al. 2014). Due to the vertical symmetry plane, the computational domain consists of one-half of the system. The mesh was automatically generated by computer and the configuration is physics-controlled mesh with a normal element size. The size of automatically generated soil element is 1 to 3 m.

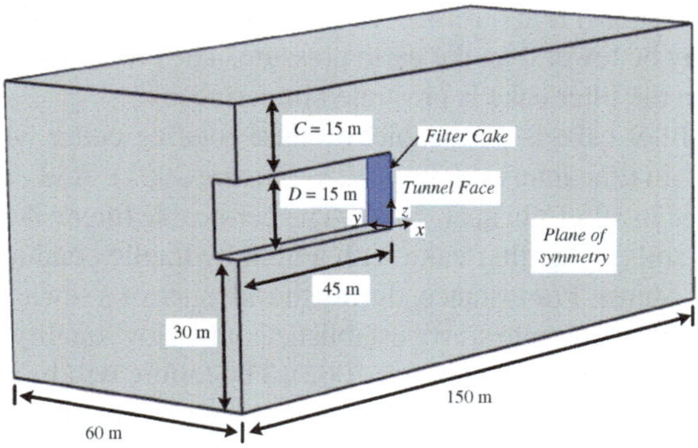

Fig. 1. Schematic view of the numerical model

In the model, a thin block attached to the tunnel face was created (Fig. 1). This block was used to simulate the filter cake of slurry shield. The width of the block is 5.89 m and the height of the block is 15 m. Since the shape of the block must be in consistency with the tunnel face. The thickness of the block is different with the actual thickness of the filter cake. When filtration time is 60 min, a filter cake is formed in constant pressure (71–120 kPa) dead-end filtration, whose thickness is ranged from 3 to 5 mm (Yin et al. 2016). If the thickness of the block is equal to the thickness of filter cake immediately, the FEM solutions will be not convergent. Since the maximum size of filter cake element is smaller than 5 mm, which is only 0.5 percent of the minimum size of the soil element. The FEM convergence is sensitive to the element size; the size difference between two adjacent elements must be controlled. After conducting a series of FEM analysis with different thicknesses of the block, the numerical result was reasonable when the thickness of the block was equal to 35 mm. Thus the 35-mm thin block was adopted to simulate the filter cake. The mesh of the filter cake was physics-controlled mesh with a normal element size. It resulted that tetrahedrons of filter cake element and soil element share the same side at the boundary.

The no-flow boundary was prescribed to the tunnel lining, the vertical surfaces and bottom of the model. The boundary condition of hydraulic pressure was applied on the tunnel face to simulate the slurry pressure. The slurry pressure increases linearly with depth with a gradient of 12 kPa/m, because the unit weight of the slurry is 1.2 kN/m^3. The excess pressure s' is equal to the difference of the slurry pressure and the hydrostatic pressure at the center line of the tunnel. h_F is the hydraulic head at the center of tunnel face.

At the far-field boundary, the hydraulic head h_0 has no influence on the pore pressure when s' keeps constant. The seepage field only depends on the s' at tunnel face. The water table was taken equal to the initial elevation of the top surface of the ground. Assuming a sufficient groundwater discharge from the surface, the inflow effect of the tunnel does not cause the rise of the water level. It was taken into consideration by prescribing the height of the model to the water table.

3 Model Parameters

The filter cake is formed on the tunnel face, when the pressurized slurry is applied. The filter cake is made from the slurry particles filtrated by the soil. Under slurry pressure, the porosity of the filter cake decreases. The hydraulic conductivity of filter cake is lower than the soil. Since the porosity of filter cake is small and the slurry particles are bentonite, which is a kind of clay. The low hydraulic conductivity of filter cake leads to a hydraulic loss in the filter cake. This hydraulic loss has a relationship with the seepage force on the tunnel face. Thus the hydraulic conductivity of filter cake is a significant parameter for seepage flows analysis.

On the tunnel face, hydraulic resistance R, the ratio between the thickness and hydraulic conductivity of filter cake, is varying dynamically when the cutter wheel is rotating. At a certain point on tunnel face, the filter cake grows with the time because the slurry particles accumulate during cake filtration. But the filter cake will be removed by cutter when spoke of the cutter wheel passes that point. The spoke passes that point at a certain time interval when the cutter wheel is rotating. The filter cake grows until being removed and this process occurs repeatedly. The time for filter cake growing t can be calculated with Eq. (1).

$$t = \frac{\alpha}{w} \tag{1}$$

α is the angle between the past spoke and the line between the certain point and center of the tunnel face; w is the angular velocity of the cutter wheel. The maximum time for filter cake growing depends on the w and the angle between two adjacent spokes. For example, a cutter wheel with five spokes rotating at 1 rpm, it means the angle between adjacent spokes is 72° and the time interval of each spoke passing is 12 s. Thus the maximum time for filter cake growing is 12 s. along the reverse direction of the rotation, the time for filter cake growing increases with the increasing α. For instance, $w = 60°/\text{s}$, when α increases one degree, the time for filter cake growing increases 0.167 s at this point.

In order to determine the hydraulic conductivity of filter cake during excavation, the relationship between R and time must be clarified. This relationship was investigated by a series of constant pressure (71–120 kPa) dead-end filtrations (Yin et al. 2016). The result shows that R increases quickly from 175 m/s to about 3.5×10^4 m/s with the increasing time, in the first 5 s. Then R tends to a value between 1.8×10^6 m/s and 4.0×10^6 m/s in the following 295 s. If the thickness of filter cake is presumed to be constant and equal to 35 mm. When the slurry infiltration starts, the hydraulic conductivity of filter cake is decreases quickly from 2×10^{-4} m/s to about 1×10^{-6} m/s within first 5 s, then slowly decreases to a stable value between 8.8×10^{-9} m/s and 2×10^{-8} m/s in the following 295 s. An attention should be paid because the hydraulic conductivity of filter cake is an equivalent value when the thickness of filter cake is 35 mm. This value is not the actual hydraulic conductivity of filter cake.

The relationship between hydraulic conductivity and time is applied to analyze the hydraulic conductivity of filter cake on the tunnel face during excavation. For a cutter wheel with five spokes rotating at 1 rpm, the angle between adjacent spokes is 72°. Five spokes divide tunnel face into five sectors. In each sector, the hydraulic conductivity is ranged from 2×10^{-4} m/s to about 1×10^{-6} m/s. On the tunnel face, the hydraulic conductivity at a certain point changes dynamically, because the cutter wheel is rotating. It is assumed that the hydraulic conductivity of filter cake on the tunnel face has a mean value k_{fm}. k_{fm} is 1×10^{-6} m/s, because the average hydraulic conductivity of filter cake keeps about 1×10^{-6} m/s in the first 12 s of filtration.

It was assumed that no particles in the slurry penetrate into the soil, the hydraulic conductivity of the ground k_s kept constant. Different k_s (5×10^{-5} m/s, 2×10^{-4} m/s and 5×10^{-4} m/s) were considered to study the influence of k_s on the hydraulic head field. The other characteristic parameters are listed in Table 1.

Table 1. Soil parameters used in the finite element analysis

Parameter	Value
Young's modulus, E (kPa)	2.972×10^4
Poisson's ratio, v	0.35
Porosity, n	0.39
Density of grain, r (kg/m^3)	2650
Saturated density of soil, r_s (kg/m^3)	2000
effective unit weight of soil, γ' (kN/m^3)	10

4 Process of Analysis

The analysis started by setting the boundary condition on the tunnel face as a no-flow boundary. Then the steady-state calculation was conducted for initial hydrostatic pressure. The data resulted from this step was inherit for the next calculation. The boundary condition of hydraulic pressure on the tunnel face was enabled. After that, the transient-state calculation was conducted. The pore pressure increases with time during excavation. When the shield stops to assemble linings, the time is enough to form an impervious filter cake. The pore pressure dissipates during assembling linings because the inflow effect of the tunnel is disabled. It is assumed that all the excess pore pressure dissipates during this time. Thus the pore pressure reaches the peak when the shield halts. The time for transient-state calculation was equal to excavation time t. For an excavation speed of 5 cm/min and a lining length of 1.5 m, t is 30 min.

5 Numerical Result and Discussion

5.1 Typical Hydraulic Head Field

A typical hydraulic head field is analyzed in this section. In this case the diameter of tunnel D is 15 m, C/D is 1, k_{fm} is 1×10^{-6} m/s, k_s is 2×10^{-4} m/s and t is 30 min. The typical contour lines of hydraulic head h based on the numerical result are shown in Fig. 2. h in the filter cake is higher than the initial hydraulic head h_0 (60 m), because the pressurized slurry on the tunnel face raises the pore pressure. At the center of tunnel face, h is equal to 65 m. A dramatic change of h occurs on the permeable filter cake. The difference of h between both sides of filter cake is about 1.5 m. It leads to a strong seepage force on the tunnel face and its direction is along the tunnel axis. h in the soil around the tunnel face is increased. For example, h is 63 m at the tunnel axis and 1/3 D ahead of the tunnel face. This position has a lower h than the center of tunnel face. The loss of h causes a seepage force in the soil. The direction of the seepage force is from the tunnel face to the front soil surface, which can be judged from the contour lines of h. The seepage force in the soil and filter cake will improve the tunnel stability.

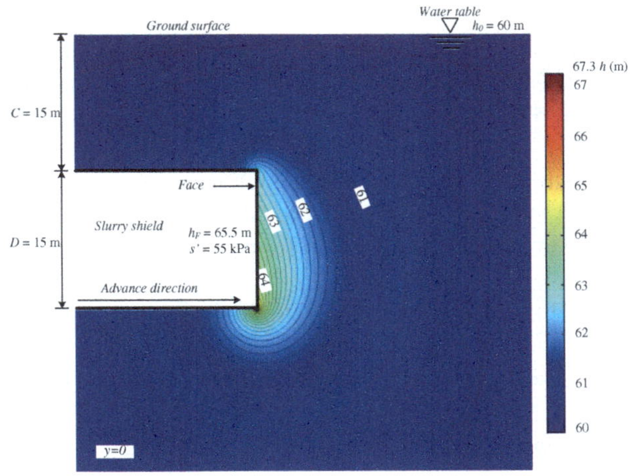

Fig. 2. Typical contour lines of the hydraulic head (numerical results for s' = 55 kPa, C/D = 1, $k_s = 2 \times 10^{-4}$ m/s, $k_{fm} = 1 \times 10^{-6}$ m/s; $y = 0$).

5.2 Hydraulic Loss in Filter Cake and Soil

The normalized hydraulic head $(h\text{-}h_F)/(h_0\text{-}h_F)$ is adopted for the quantitative analysis of the hydraulic loss. $(h\text{-}h_F)/(h_0\text{-}h_F)$ is benefit for comparing the hydraulic loss under different numerical cases. The influence of k_s and s' on the $(h\text{-}h_F)/(h_0\text{-}h_F)$ are investigated based on the numerical computation. $(h\text{-}h_F)/(h_0\text{-}h_F)$ under varying k_s (5×10^{-5}, 2×10^{-4} and 5×10^{-4} m/s) are calculated in the first group. s' is constantly equal to 55 kPa. The rest configuration of numerical model is as following: D is 15 m, C/D is 1 and k_{fm} is

1×10^{-6} m/s. $(h\text{-}h_F)/(h_0\text{-}h_F)$ at tunnel axis is obtained for analyzing the hydraulic loss in filter cake and soil, as line AB in Fig. 3.

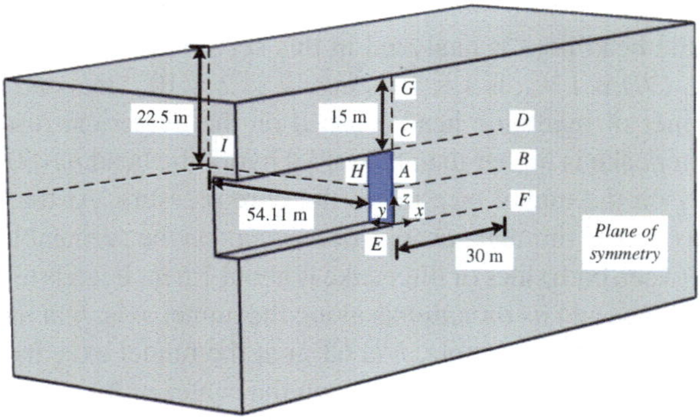

Fig. 3. Characteristic location in the numercial model

Figure 4 shows the numerically computed distribution of the normalized hydraulic head $(h\text{-}h_F)/(h_0\text{-}h_F)$ along line AB $(h\text{-}h_F)/(h_0\text{-}h_F)$ increases substantially in the filter cake ($0 < x < 35$ mm) and tends to 1 at x = 2*D*. The hydraulic loss in the filter cake depends on k_s: the smaller k_s, the less hydraulic loss will be. For example, at $x = 35$ mm, $(h\text{-}h_F)/(h_0\text{-}h_F)$ is 0.49 for $k_s = 5 \times 10^{-4}$ m/s, $(h\text{-}h_F)/(h_0\text{-}h_F)$ is 0.36 for $k_s = 2 \times 10^{-4}$ m/s and $(h\text{-}h_F)/(h_0\text{-}h_F)$ is 0.16 for $k_s = 5 \times 10^{-5}$ m/s. On the contrary, the hydraulic loss in the soil increases with the decreasing k_s. For example, hydraulic loss in soil is 0.51 $(h_F\text{-}h_0)$ for $k_s = 5 \times 10^{-4}$ m/s and hydraulic loss is equal to 0.84 $(h_F\text{-}h_0)$ for $k_s = 5 \times 10^{-5}$ m/s. For the hydraulic loss in the filter cake and soil in front of the tunnel face, k_s is a key factor.

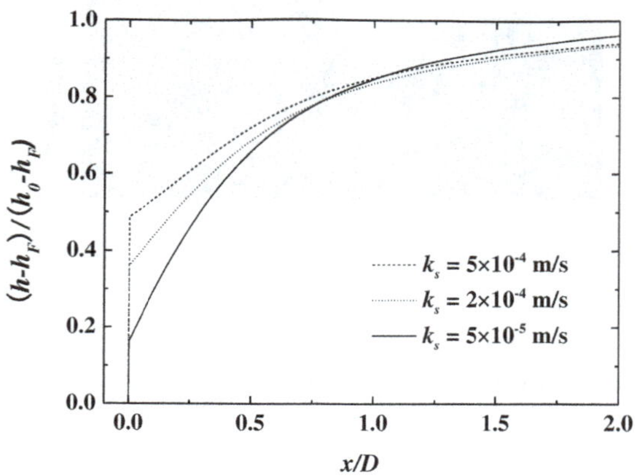

Fig. 4. Distribution of normalized hydraulic head $(h\text{-}h_F)/(h_0\text{-}h_F)$ under different hydraulic conductivity k_s ($C/D = 1$, $s' = 55$ kPa and $k_{fm} = 1 \times 10^{-6}$ m/s, Line AB).

6 Conclusions

In this study, the hydraulic head field near the tunnel face of slurry shield was studied using the numerical analysis. A finite element analysis program, COMSOL, was used to establish the three-dimensional numerical model. In the model, A block, its thickness is 35 mm and hydraulic conductivity is 1×10^{-6} m/s, was adopted to simulate the filter cake on tunnel face. Based on the numerical results, the seepage flow analysis was conducted. To verify the calculated results of the numerical analysis, a comparison with the field measurement was performed. The results of the numerical study on the hydraulic head field with the consideration of filter cake on tunnel face leads to the following conclusions.

The existence of filter cake seriously affects the hydraulic head field near the tunnel face. In the front of tunnel face, the hydraulic head decreases dramatically in the filter cake. At the tunnel axis, the hydraulic head decreases by 0.39 $(h_F\text{-}h_0)$ from tunnel face to the outer side of filter cake. In the soil, the hydraulic head decreases with the increasing distance to the tunnel face. The hydraulic head tends to hydrostatic head, at the points where is $2D$ in front of tunnel face, horizontal $1.5D$ to the lining of tunnel or soil surface.

The hydraulic loss in filter cake increases with the increasing hydraulic conductivity of soil. On the contrary, the hydraulic loss in soil decreases when hydraulic conductivity of soil increases. For s' is 55 kPa, $C/D = 1$, hydraulic conductivity of filter cake is 1×10^{-6} m/s, hydraulic loss increases from $0.16(h_0\text{-}h_F)$ to $0.5(h_0\text{-}h_F)$ when hydraulic conductivity of soil increases from 5×10^{-5} to 2×10^{-4} m/s, at the tunnel axis.

References

Anagnostou, G., Kovári, K.: The face stability of slurry-shield-driven tunnels. Tunn. Undergr. Space Technol. **9**(2), 165–174 (1994). doi:10.1016/0886-7798(94)90028-0

Broere, W., van Tol, A.F.: Influence of infiltration and groundwater flow on tunnel face stability. In: Kusakabe, O., Fujita, K., Miyazaki, Y. (eds.) Geotechnical Aspects of Underground Construction in Soft Ground, pp. 339–344. CRC Press, Tokyo (2000)

Broere, W.: Tunnel face stability & new CPT application. Ph.D. thesis, Delft University of Technology, Delft, Netherlands (2001)

Chung, J., Daniel, D.: Modified fluid loss test as an improved measure of hydraulic conductivity for bentonite. Geotech. Test. J. **31**(3), 1–10 (2008). doi:10.1520/GTJ100005

Fritz, P., Stengele, R., Heinz, A.: Modified bentonite slurries for slurry shields in highly permeable soils. In: Fourth International Symposium Geotechnical Aspects of Underground Construction in Soft Ground, Toulouse, France (2002)

Kim, S., Tonon, F.: Face stability and required support pressure for TBM driven tunnels with ideal face membrane–drained case. Tunn. Undergr. Space Technol. **25**(5), 526–542 (2010). doi: 10.1016/j.tust.2010.03.002

Li, Y., Emeriault, F., Kastner, R., Zhang, Z.: Stability analysis of large slurry shield-driven tunnel in soft clay. Tunn. Undergr. Space Technol. **24**(4), 472–481 (2009). doi:10.1016/j.tust. 2008.10.007

Perazzelli, P., Leone, T., Anagnostou, G.: Tunnel face stability under seepage flow conditions. Tunn. Undergr. Space Technol. **43**, 459–469 (2014). doi:10.1016/j.tust.2014.03.001

Tien, C.: Principles of Filtration. Elsevier, Amsterdam (2012)

Yeo, S., Shackelford, C., Evans, J.: Consolidation and hydraulic conductivity of nine model soil-bentonite backfills. J. Geotech. Geoenviron. **131**(10), 1189–1198 (2005). doi:10.1061/(ASCE)1090-0241(2005)131:10(1189)
Yin, X., Chen, R., Li, Y., Qi, S.: A column system for modeling bentonite slurry infiltration in sands. J. Zhejiang Univ. Sci. **17**(10), 818–827 (2016). doi:10.1631/jzus.A1500290
Zhang, F., Zhu, H., Fu, D.: Shield Tunnelling Method. China Communication Press, Beijing, pp. 192–194 (2004). (In Chinese)

Foundation and Foundation Treatment

A Load Transfer Approach for Studying the Load-Deformation Response of Vertically Loaded Single Pile

Qingyuan Lu[1,2(✉)] and Qiang Luo[1,2]

[1] School of Civil Engineering, Southwest Jiaotong University, Chengdu, China
luqingyuan@my.swjtu.edu.cn
[2] MOE Key Laboratory of High-Speed Railway Engineering, Chengdu, China

Abstract. A proper load-transfer (or *t-z*) model describing the pile and surrounding soil interaction is important for accurately predicting the behavior of piles. This paper presents a nonlinear prediction approach of load-displacement response for vertically loaded single pile based on a load transfer method by introducing a proposed *t-z* model for pile-soil interface. The proposed *t-z* model comprises a hardening/softening mechanism which is able to capture the hardening and softening behavior of pile-soil interactions. Combined with a linear load-displacement relationship of pile tip, a highly effective computer program has been developed for nonlinear analysis of the load-displacement behavior of a vertically loaded single pile. And also, predictions of *t-z* curves and load-settlement curves from the proposed approach have been compared with the measured values from two well-documented field experimental pile loading case studies. The results of the comparisons indicate that the predictions show good agreement with the measured values.

Keywords: Vertically loaded single pile · Load transfer · Pile-soil interface · *t-z* model · Load-settlement curve

1 Introduction

Pile foundation is commonly used for support loading from super structure of vertical forces because of its high capacity, small settlement, good suitability and wide applicability. The analysis of load-deformation response of single piles is an essential basis for pile foundation design. The most reliable method to determine the response of the piles is based on the results obtained from pile load tests, but it can be expensive and time-consuming. Consequently, the theoretical method for analyzing the load-settlement behavior of single piles is essential for engineering design.

A number of theoretical methods are used to predict load-settlement for single pile, including the load-transfer method (Kraft et al. 1981), the shear displacement method (Randolph and Wroth 1978; Guo and Randolph 1999), the finite-element method (Said et al. 2009), and other simplified analytical methods (Zhang and Zhang 2012). The load-transfer method has widely used for prediction of the load-deformation relationship for vertically loaded single pile due to its simplicity and capability of incorporating nonlinear soil behavior. The load-transfer (or *t-z*) curve is the graphical representation

R. Chen et al. (eds.), *Proceedings of the 2nd International Symposium on Asia Urban GeoEngineering*, Springer Series in Geomechanics and Geoengineering,
https://doi.org/10.1007/978-981-10-6632-0_29

of the unit skin friction attributable to the vertical displacement of the pile relative to the surrounding soil in each geological stratum, and varies with depth, soil-pile interface friction angle, and type of pile installation. Numerous *t-z* models or functions have been proposed for the prediction of the *t-z* curves, including empirical functions (Colye and Reese 1966; Vijayvergiya 1969; Reese and O'Neill 1989) and theoretical functions (Randolph and Wroth 1979; Kraft et al. 1981). While the empirical functions are very site-specific and may produce error up to 100%, for example, Coyle and Reese (1966) reported that the error between the computed and actual bearing capacities of piles varies from −52% to +61% using their empirical *t-z* curve. The theoretical functions are more attractive because of the flexibility with which site specific strength, stress-strain behavior and stress history of soils, pile-soil interaction as well as pile installation can be readily incorporated.

To account for the nonlinearity in the stress-strain hardening response of soil, the theoretical *t-z* curve is usually taken to be hyperbolic function to simulate the relationship between unit skin friction and pile soil relative displacement developed along the pile-soil interface (Kondner 1963; Duncan and Chang 1970). However, laboratory and field observations suggest that skin friction softening phenomenon occurs at pile-soil interface in clay and cemented soils (Randolph and Gourvenec 2011; Zhang and Zhang 2012). The skin friction softening behavior is generally attributable to the decrease in the interface friction angle along the residual shear surface developed in the pile-soil interface (Nanda and Patra 2014). Although the empirical *t-z* curves included skin friction softening behavior reported by Coyle and Reese (1966) and the theoretical *t-z* curves by Vijayvergiya (1969) and Zhang and Zhang (2012), a few theoretical *t-z* models account for the skin friction softening behavior in detail.

This paper presents a nonlinear prediction approach of load-displacement response for vertically loaded single pile based on the load transfer method by introducing a proposed theoretical *t-z* model for pile-soil interface. The proposed theoretical *t-z* model comprises a hardening/softening mechanism which is able to capture the hardening and softening behavior of pile-soil interactions. Combined with a linear load-displacement relationship for pile tip, a highly effective computer program has been developed for nonlinear analysis of the load-displacement behavior of a vertically loaded single pile. The proposed *t-z* model can be used for all soils (clay and sand), and the proposed method can be used for homogeneous soil and inhomogeneous layered soil system. Two case studies with different soil strata are considered for validation.

2 Proposed *t-z* Hardening/Softening Model for Pile-Soil Interface

2.1 Pile-Soil Interaction

The mechanical behavior of pile-soil interaction within the interface is rather complex, which affect by pile-soil interface roughness, composition of soil, relative density, grain-size distribution, shape of soil particles, moisture content of soil, magnitude stress level, and rate of shearing (Evgin and Fakharian 1997). The relationship between shear stress and shear displacement for the pile-soil interface at a same constant normal stress

(σ_{ni}) may be either hardening or softening, which depends on the void ratio in granular soil and on the Atterberg limits in fine soils. For example, Cao et al. (2014, 2015) conducted large-scale shear tests on sand-concrete interface shearing behavior under constant normal stress and found that shear stress mobilized rapidly in the initial shear displacement and thereafter the rate of shear stress mobilization decreases considerably with the shear displacement for relative denseness $D_r = 21.6\%$ as shown in Fig. 1, Type 1. For $D_r = 72.5\%$, after an initial increase in unit skin friction, softening behavior occurred (Fig. 1, Type 2).

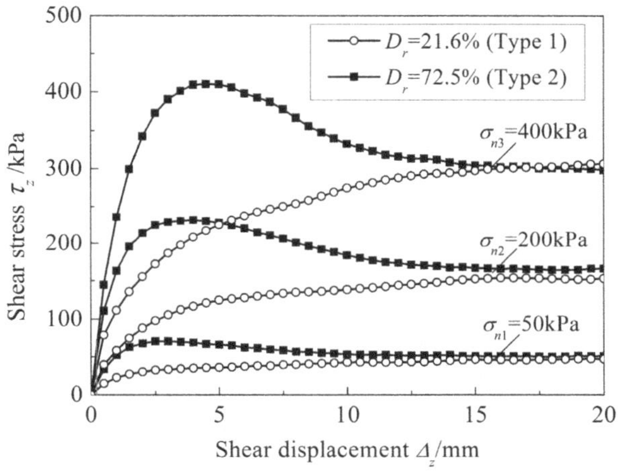

Fig. 1. The variation of shear stress with shear displacement (after CAO et al. 2014, 2015)

2.2 Modified Hyperbolic *t-z* Hardening Model

Based on the schematic diagram shown in Fig. 1, for type 1 in the hardening regime the *t-z* function is assumed of hyperbolic type, it can be written as follows (Duncan and Chang 1970):

$$\tau_z = \frac{\Delta}{a + b\Delta_z} \tag{1}$$

Where τ_z and Δ_z are the unit skin friction and pile-soil relative displacement at give depth, z, respectively; and a, b are empirical coefficients, which can be obtained in the following form:

$$a = \frac{1}{G_0} \tag{2}$$

$$b = \frac{1}{\tau_{zult}} \tag{3}$$

Where, G_0 is initial shear stiffness (tangent of the hyperbolic curve at the origin); τ_{zult} is limiting unit skin friction.

The initial shear stiffness, G_0, is an important parameter related to accurately predictions of the *t-z* curves. Several investigators have reported that the limiting initial shear stiffness value of very small strain varies from 10^{-6} for low plasticity to 10^{-5} for highly plastic clays (Likitlersuang et al. 2013). However, the conventional tests, for example, special triaxial test, obtaining the maximum initial shear stiffness only have 10^{-4} strain. Therefore, it clearly follows that the accurately estimation of the initial shear stiffness is really challenging by using conventional tests. Furthermore, the limitation of traditional hyperbolic *t-z* model is that it does not peak, yet it is obvious that a finite pile-soil relative displacement is required to mobilize maximum unit skin friction. Therefore, using these constants in computation ultimately produces no complete convergence to the experimentally observed relationship between unit skin friction and pile-soil relative displacement at failure. Consequently, an appropriate *t-z* function, which defines maximum unit skin friction with respect to pile-soil relative movement at failure, can be more attractive. Hence

$$\tau_{zult} = \frac{\Delta_{zult}}{a + b\Delta_{zult}} \tag{4}$$

Where, Δ_{zult} is limiting pile-soil relative displacement at failure, in an unconfined compression test of soil sample from near the pile shaft stress level.

Assuming the relationship between initial shear stiffness and tangential shear stiffness at failure can be expressed as the following form:

$$G_0 = nG = n\frac{\tau_{zult}}{\Delta_{zult}} \tag{5}$$

Where, G is tangential shear stiffness at failure; n is constant.

Substitute Eqs. (2), (4) and (5) into Eq. (1), the modified hyperbolic *t-z* hardening model can be obtained in the following form:

$$\frac{\tau_z}{\tau_{zult}} = \frac{n\Delta_z}{\Delta_{zult} + (n-1)\Delta_z} \quad (0 \leq \Delta_z \leq \Delta_{zult}) \tag{6}$$

Figure 2 shows the performance of the model by comparing it with the actual test results, the computed results from Eq. (6). The experimental databases used in Fig. 2 were taken from the results reported by Wang et al. (2012). It is observed that the proposed *t-z* model shows good agreement with the measured value. In addition, the advantage of the proposed *t-z* model is that the degree of nonlinearity of pile-soil interaction behavior can be described by the parameter n. The proposed *t-z* model would be degenerate to the traditional theoretical *t-z* model, for example perfectly plastic, liner elastic-perfectly plastic models, and the empirical models, as shown in Fig. 3.

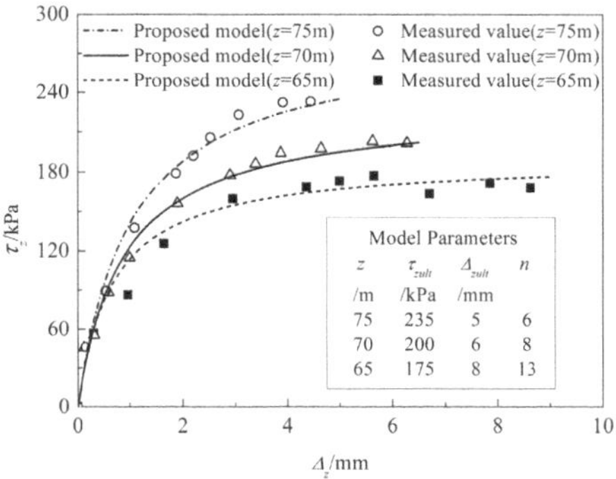

Fig. 2. Computed and experimental *t-z* relationships

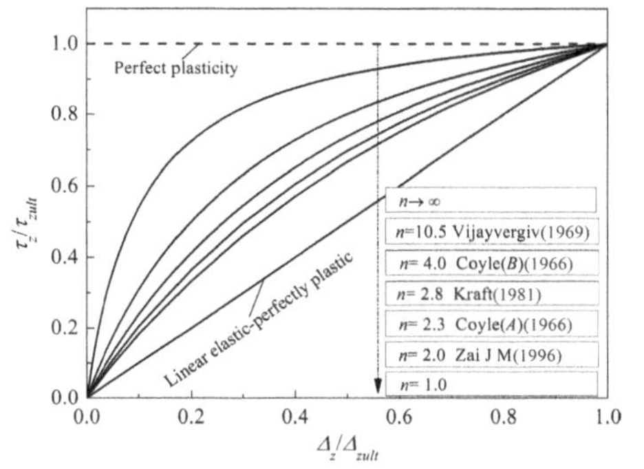

Fig. 3. Normalized load transfer hardening models

2.3 Proposed *t-z* Hardening/Softening Model

As already observed in Fig. 1 in the case of type 2, the initial hardening phase with increasing value of shear displacement (Δ_z), after a peak of maximum mobilized shear stress at failure ($\Delta_z = \Delta_{zmax}$, $\tau_z = \tau_{zmax}$), evolves into shear stress softening until a final plateau corresponding to the ultimate state ($\Delta_z = \Delta_{zult}$, $\tau_z = \tau_{zult}$). Thus, the *t*-*z* model in the softening regime can be governed by the following equation:

$$\tau_z = \frac{n\Delta_z}{\Delta_{zult} + (n-1)\Delta_z}\tau_{zult} + f\frac{\Delta_z}{\Delta_{z\,max}}e^{\left(1-\frac{\Delta_z}{\Delta_{z\,max}}\right)}\tau_{z\,max} \tag{7}$$

Where, *f* indicates the amplitude of shaft shear stress hardening/softening. It can be observed from Eq. (7) that, when the parameter $f = 0$, it reflects the *t-z* hardening model, whereas it is the *t-z* softening model for $f > 0$. The peak of unit skin friction may occur only in high strength of soil structure, for example in overconsolidated soil or in dense sand. A parametric study of the influence of parameter *f* on the evolution of *t-z* curves is given in Fig. 4. The laboratory data used in Fig. 4 were taken from the results reported by Cao et al. (2015). It shows that the bigger the value of *f*, the greater the peak of parabolic curve is obtained, which corresponds eventually to the denser soil. Figure 4 also shows that for a given relation of unit skin friction and pile-soil relative displacement softening behavior, there is only a determined value of *f*.

Figure 5 shows *t-z* curves plotted from the present model is generally consistent with the measured values given by Wang et al. (2012) and the predicted results using *t-z* traditional softening models (Exponential model and Parabola model) proposed by Zou et al. (2011) and Vijayvergiya (1969). It is clear that the proposed *t-z* model shows a well good agreement between the calculated value and the measured value than the traditional *t-z* softening models, particularly for the pile-soil relative displacement greater than the maximum value.

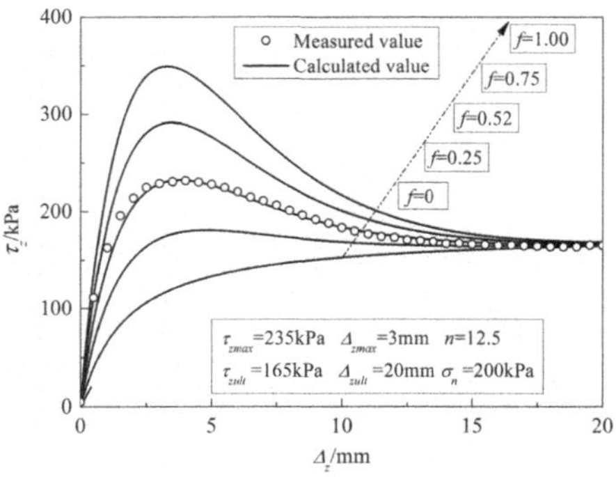

Fig. 4. Parametric study of f

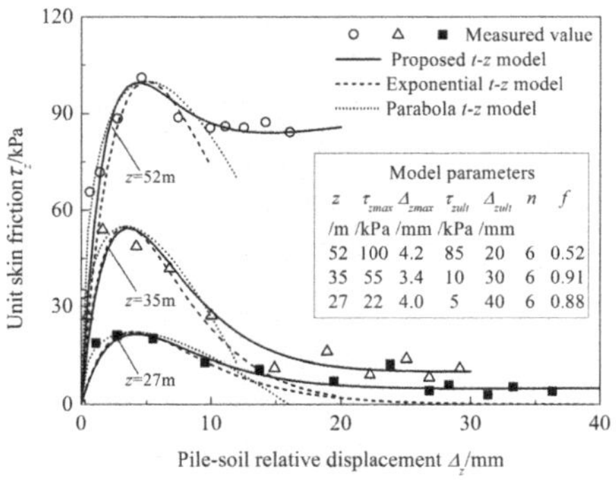

Fig. 5. Comparison of the calculated value and situ test value

3 Load Transfer Curve for Pile Tip

The accuracy of predicting the load-deformation behavior of vertically loaded single pile is not only dependent on the pile-surrounding soil interaction but also dependent on the interaction between pile tip and the bearing stratum soil.

In practical engineering, the ratio of pile length respect to its diameter of greater 20, the mobilized tip resistance at the design load may represent a small percentage of the limiting tip resistance, with the corresponding degree of mobilization typically less than 20% at twice the design load (Zhu and Chang 2002). Besides, it is difficulty in complete clearing of the bearing stratum soil of a borehole in bored pile construction, which will lead to a greater pile tip displacement when the tip resistance is fully mobilized unless the pile is short and rests directly on rocks. In addition, an excess pile tip settlement implies an excess pile head settlement that would not be permitted in practice. Therefore, a linear relationship between pile tip resistance and tip displacement is proposed, as shown in Eq. (8).

$$Q_b = \frac{E_{sb} D_b}{1 - v_{sb}^2} \Delta_b \tag{8}$$

Where, Q_b and $\varDelta_b$ are pile tip resistance and displacement, respectively; E_{sb} and v_{sb} are compressive modulus and Poisson's ratio of bearing stratum soil, respectively; and D_b is pile diameter at the tip.

4 Model Parameters Study

The proposed *t-z* hardening/softening model requires six parameters for computation: degree of nonlinearity of pile-soil interaction (n), limiting and maximum pile-soil relative displacement ($\varDelta_{zult}$, $\varDelta_{zmax}$), unit skin friction at limit and maximum (τ_{zult}, τ_{zmax}), and the strength of soil structure (f). These parameters have physical meaning soils can easily be determined from laboratory or field experiments. The properties of the soil sample for a laboratory test should be close to the behavior of the soil around the pile shaft in the field.

4.1 Degree of Nonlinearity (*N*)

Based on the field single pile load test databases in different soils presented by the authors (Rollins et al. 2005; Zhang and Zhang 2012) are used to determine the degree of nonlinearity of pile-soil interaction (n), as shown in Fig. 6. In Fig. 6, the skin friction, τ_z, is normalized by the limiting unit skin friction, τ_{zult}, and the pile-soil relative displacement, Δ_z, is normalized by the pile shaft diameter, D. It can be observed from Fig. 6 that, the value of n varies from 2 to 13, and the trend line of n is 5 of the value of coefficient of determination term, $r^2 = 0.977$.

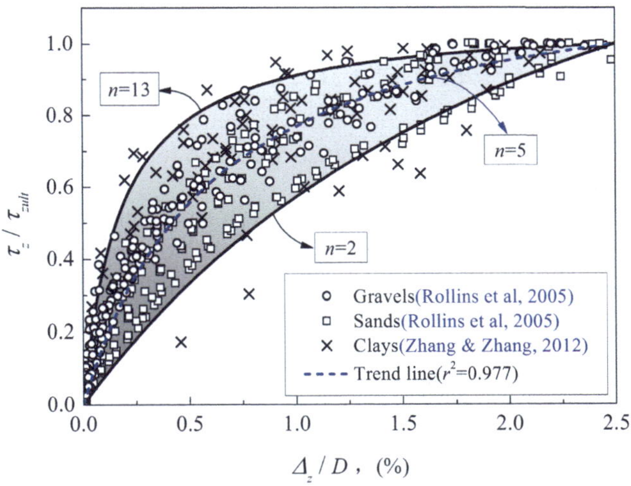

Fig. 6. Effect of n of normalized t-z curves

4.2 Maximum/Limiting Unit Skin Friction and Pile-Soil Relative Displacement (τ_{Zmax}, Δ_{Zmax}, τ_{Zult}, and Δ_{Zult})

Based on the results of the load tests on 18 instrumented piles (Zou et al. 2011; Wang et al. 2012; Zhang and Zhang 2012) in different soils for skin friction softening behavior, the relationship between τ_{zult}/τ_{zmax} and $\Delta_{zult}/\Delta_{zmax}$ is obtained, as shown in Fig. 7. It can be noted that the ratio of the limiting to the maximum unit skin friction, τ_{zult}/τ_{zmax}, is generally from 0.75 to 1.00 and the ratio of $\Delta_{zult}/\Delta_{zmax}$ ranges from 1.5 to 3.0.

It is well known that the interface behavior of an actual pile at a site is affected by many complicated factors, such as construction methods, pile types, soil types, stratigraphy, and loading procedure, which may influence the magnitude of parameter Δ_{zult}. The value of Δ_{zult} can be determined experimentally. Generally, the values of

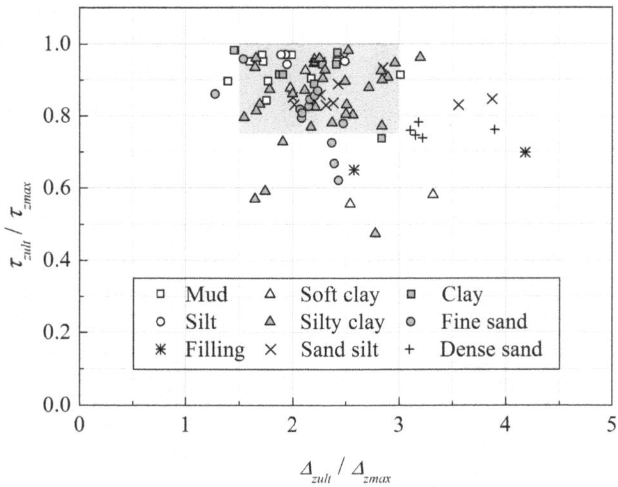

Fig. 7. The ratio of τ_{zult}/τ_{zmax} and $\Delta_{zult}/\Delta_{zmax}$ developed from load test databases

parameter have been found to be in the range 5–25 mm for 0.7–1.1 m pile diameter in different soils (Zhang and Zhang 2012).

From above discussion, the parameter Δ_{zmax} can be capture from the ratio of $\Delta_{zult}/\Delta_{zmax}$ in Fig. 7, and the value of τ_{zult} can be computed from the ratio of τ_{zult}/τ_{zmax} (see Fig. 7). Therefore, the maximum unit skin friction, τ_{zmax}, must be determined firstly.

Numerous skin friction design approaches have been proposed for the load-transfer analysis. These design approaches generally fall into two main categories: (1) total stress design methods using undrained analysis (Skempton 1959), and (2) effective stress design approach using drained analysis (Burland 1973). Despite a number of significant contributions being made to the literature on the earlier category, many of these approaches have a number of inherent drawbacks, for example soil behavior is governed by effective stress and complex stress-strain changes occur during the pile installation which cannot be completely describes using the initial undrained strength profile. In an attempt to overcome many of the drawbacks associated with total stress design approaches, Burland (1973) advocated an effective stress design approach using the drained strength of soils. In Burland's method, the evaluation of the maximum unit skin friction, τ_{zmax}, at a given depth, z, along a vertical pile, which is controlled by the horizontal effective stress at failure $\sigma_h'(z)$, and the pile-soil interface friction angle δ', is linked to the vertical effective stress $\sigma_v'(z)$ through an empirical parameter beta, β.

$$\tau_{z\max} = \sigma_h'(z) \tan \delta' = \beta \sigma_v'(z) \tag{9}$$

A particularly attractive feature of the β method is that assuming the lateral earth pressure coefficient, K_f, is equal to the lateral earth pressure coefficient at rest, K_0, and the pile-soil interface friction angel, δ', is equal to the effective internal friction angle φ' (see Eq. (10)). The range of β is from about 0.15 to 0.30 in the practical range φ' (range of 10° to 45°). That is, almost any reasonable estimate for φ' gives the same computed skin resistance. However, it still remains to be seen from a load test whether it is correct.

$$\beta = K_f \tan \delta' \approx K_0 \tan \varphi' = (1 - \sin \varphi') \tan \varphi' \tag{10}$$

Figure 8 is a data plot from Reese and O'Neill (1988), Rollins et al. (2005) and Park et al. (2010) that was obtained from back computing more than one hundred field load tests in gravel, gravelly sand, sand and clay soil. It can be observed that very high values of β, up to 7.2, can occur at shallow depth. The values of β decrease with an increase in depth and asymptotically approach the normally consolidated (NC) range beyond a depth of approximately 20 m or so. These findings are reasonable because, in most natural soil profiles, the overconsolidation ratio (OCR) generally decreases with increasing depth. It also can be seen form the Fig. 8 that at a given depth larger values for β are expected passing from clay to gravel.

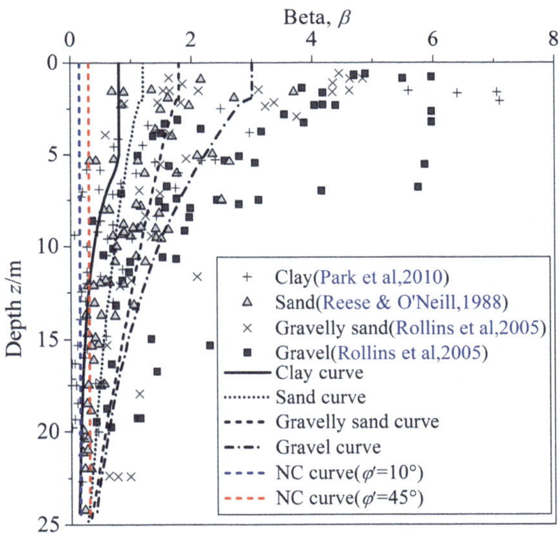

Fig. 8. Experimental β values and its design curves for different soil types

Consideration of soil property variation in depth can make a substantial difference, the suggested different depth depending β curves in different soil types are developed by the authors (Reese and O'Neill 1988; Rollins et al. 2005; and Park et al. 2010), as shown in Fig. 8. It is clear that making a load-displacement prediction that compares favorably with a load test is more a coincidence than the result of using the Eq. (10), particularly for short piles.

4.3 Strength of Soil Structure (*f*)

The parameter *f* can be determined by back-analysis of field load test results. A series of field test (Zou et al. 2011; Wang et al. 2012; Zhang and Zhang 2012) show that the values of parameter f have been found to be in the range 0.45-0.55 for different soils (e.g., mud, clay, sandy silt, silty clay).

5 Algorithm for Load-Settlement Analysis of a Vertically Loaded Single Pile

The steps involved in developing load-settlement curves are similar to the procedure discussed by Coyle and Reese (1966). Some of the steps proposed by Coyle and Reese (1966) have been modified to accommodate computer coding and faster computation. The detailed steps are given as follows:

(1) Divide a single pile into m segments from the pile end to the pile head, as shown in Fig. 9.
(2) Assume a small pile tip displacement, Δ_b, and determine the pile tip resistance, Q_b, from Eq. (8).
(3) Assume the vertical moment at middle height of first pile segment, Δ_{c1}, equal to the movement of the pile tip, Δ_b. Based on the proposed t-z function as given in Eq. (7), the unit skin friction of pile segment 1, τ_{z1}, developed in the interface between pile segment 1 and soil can be obtained using the assumed value of Δ_{c1}.

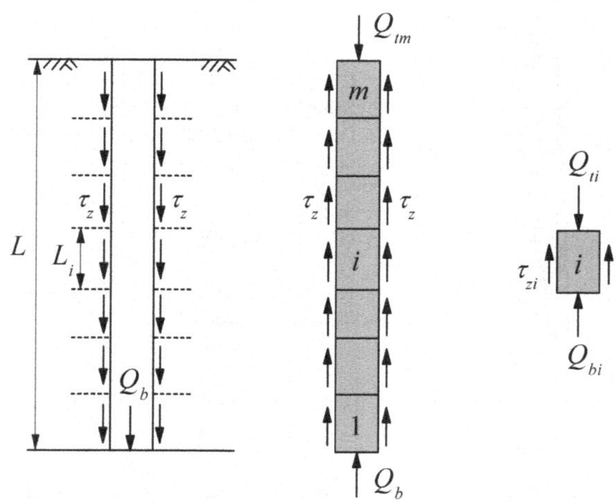

(a) Stresses on soil (b) Stresses on pile (c) Pile segment

Fig. 9. Stresses acting on adjacent soil, pile and pile segment

(4) Compute load at top of pile segment 1, $Q_{ti}|_{i=1}$, by adding load found in step (3) to the known load acting at the bottom of the segment, Q_b, according to the force equilibriums Eq. (11) of a pile segment (Fig. 9(c)).

$$Q_{ti} = Q_{bi} + \pi D L_i \tau_{zi} \tag{11}$$

(5) Assuming a linear variation of load in the pile segment, the elastic deformation at the midpoint of pile segment 1, $\Delta_{ci}|_{i=1}$, can be computed by:

$$\Delta_{ci} = \frac{(Q_{ti} + 3Q_{bi})L_i}{8E_pA_p} \tag{12}$$

(6) The new movement of the midpoint of pile segment 1, $\Delta_{ci}'|_{i=1}$, can be written as:

$$\Delta_{ci}' = \Delta_{bi} + \Delta_{ci} \tag{13}$$

(7) Compute the difference between new midpoint movement $\Delta_{c1}'$with assumed value of Δ_{c1} from step (3). Check if this difference is less than an acceptable specified value, e.g., 1×10^{-6}m.
(8) Repeat steps (3) through (7) until the requirement for acceptably specified value mentioned in step (10) is met.
(9) Calculate the movement and load at the top of pile segment 1, $\Delta_{ti}|_{i=1}$ and $Q_{ti}|_{i=1}$, respectively, using the following equations:

$$\Delta_{ti} = \Delta_{bi} + \Delta_{ci}' \tag{14}$$

$$Q_{ti} = Q_{bi} + \pi D L_i \tau_{zi}' \tag{15}$$

Where, τ_{zi}' is derived from Eq. (7) and a new midpoint movement, Δ_{ci}'.
(10) Set load at the tip of the next segment above equal to load found in step (4).
(11) Consider the next segment above and assume the movement at the midpoint of this segment equal to the movement of the top of the segment next below. Repeat steps (3) through (9) for this segment.
(12) Repeat step (11) until the top segment of the pile has been considered and load and settlement on the top of pile for the tip movement assumed in step (2) have been determined.
(13) Repeat steps (2) through (12) until all specified pile tip movements have been considered.

6 Verification of Proposed Approach

In order to examine the validity of the proposed approach and procedure outlined in the preceding sections for the analysis of the load-settlement response of vertically loaded single pile, two instrumented pile load tests are analyzed and their results reviewed.

6.1 Case Study 1

Nabil (2001) presented the results of axial compression load tests on two concrete bored piles installed in homogeneous cemented sands on the South Surra, Kuwait. The bored cast-in-site pile of 2.25 m length and 0.1 m diameter, having an elastic modulus of 20 GPa. The cemented sand was reported to have a thickness of 4.5 m, a water content of 3.6%, a bulk density of 18.17 kN/m^3, a compression modulus of 25 MPa, and a limiting pile-soil relative displacement of 25 mm. The peak strength parameter c' and φ' are 20 kPa and 35°, respectively, and the residual strength parameters are 0 and 34°. The input parameters for proposed *t-z* model can be determined by the results of drained triaxial compression tests using undisturbed samples taken from the cemented sand deposit for testing reported by Nabil (2001). The degree of nonlinearity $n = 10$, the ratio of limiting to maximum pile-soil relative displacement $\Delta_{zult} / \Delta_{zmax} = 4.38$, the ratio of unit skin friction at limit to at maximum $\tau_{zult} / \tau_{zmax} = 0.71$, and the strength of soil structure $f = 0.70$. To account for the value of β variation in depth, the value of β located at upper depth of 0.5 pile length is 4.0, and the below depth of 0.5 pile length the $\beta = 3.6$. In the analysis of the load-settlement response of a single pile, the pile is divided into 13 segments with each pile segment of 0.2 m in length, except of the pile head and end segment where the pile segment length is assumed to be 0.025 m.

Figure 10(a) shows that the load-settlement curves at the pile head plotted from the present method is generally consistent with the measured results given by Nabil (2001) and the predicted results using *t-z* softening models proposed by Vijayvergiya (1969) and Zou et al. (2011) at all loading levels. It can be observed that the proposed approach is generally in good agreement with field-observed behavior than the calculated results using traditional *t-z* softening model. In addition, the predicted limiting bearing capacity, Q_u, using parabola and exponential *t-z* softening model is significant less than the measured value. This finding is reasonable because, when the pile-soil relative displacement greater than the maximum value, the unit skin friction quickly decreases with increasing pile-soil relative displacement, which lead to the limiting unit skin friction less than the measured value at limiting pile-soil relative displacement (see Fig. 5).

Figure 10(b) compares the variation in the axial load along the pile shaft obtained from the traditional *t-z* model and the present *t-z* model. It is clear that the axial load values obtained in the present *t-z* model are less than that from the traditional *t-z* model at the same depth. This can be also explained the reason of predicted limiting bearing capacity using parabola and exponential *t-z* softening model is significant less than the measured value.

6.2 Case Study 2

The response of concrete bored piles were installed in Bangkok clay located at three sites has been investigated by Promboon and Brenner (1981). The test results of the pile situated at Site A (Bank of Thailand, Tha Kasem Road) is used for examine the validity of the proposed approach and procedure outlined in the preceding sections for the analysis of the load-settlement response of vertically loaded single pile. The bored cast-in-site pile of 25.8 m length and 1.0 m diameter, and had an elastic modulus of

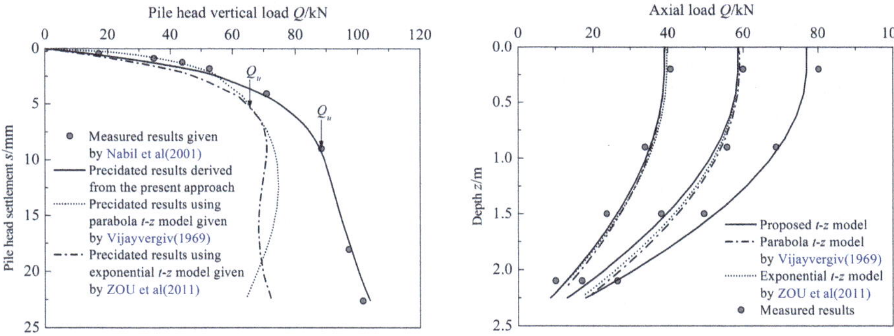

Fig. 10. Comparison of predicted and measured results: (a) load-settlement curves; (b) axial load distributions along the pile shaft

22GPa. The soil strata consist of soft clay, medium clay, stiff sandy clay, and sand at various locations. The soil properties and engineering characteristics have been obtained from the observations reported by Promboon & Brenner (1981). The soil strata are lightly overconsolidated, having an OCR of 2–3.5. The liquid limit, plasticity index, bulk density, and internal friction angle of soil strata vary from 70% to 50%, 40 to 30, 15.5 to 17.0 kN/m^3, and 24° to 25°, respectively.

Promboon and Brenner (1981) reported skin friction softening behavior in the *t-z* curve at depths of 4.5–10.5 and 19.5 m–23.5 m. For other soil strata, skin friction is accord with *t-z* hardening curve. Based on the measured *t-z* curve at different depth, the details about the input parameters for proposed *t-z* model used for prediction are given in Table 1.

The predicted and measured load-settlement curves at the head of the single pile are shown in Fig. 11(a). The computed and measured axial load distribution along pile shaft is shown in Fig. 11(b). The evaluation as reported in Figs. 11 shows good agreement between the computed and measured values. Indirectly the validity of the proposed *t-z* model in Eq. (7) is verified.

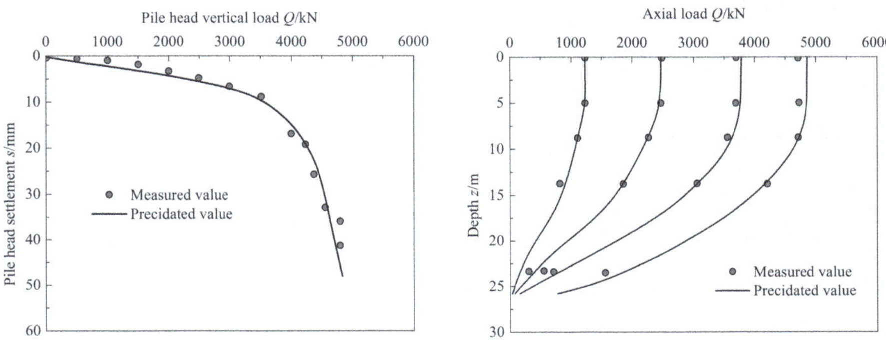

Fig. 11. Comparison of predicted and measured results: (a) load-settlement curves; (b) axial load distributions along the pile shaft

Table 1. Load transfer parameters along pile shaft

i	L_i/m	β	τ_{zmax}/kPa	Δ_{zult}/mm	$\frac{\tau_{zult}}{\tau_{z\,max}}$	$\frac{\Delta_{zult}}{\Delta_{z\,max}}$	n	f
1	0.4	1.6	9.92	20	1.0	1.0	3	0
2	1.0	1.3	28.21	20	1.0	1.0	3	0
3	1.0	1.2	44.64	20	1.0	1.0	3	0
4	1.0	1.0	52.70	20	1.0	1.0	3	0
5	1.0	0.90	61.38	20	0.86	3.5	5	0.62
6	1.0	0.76	70.23	20	0.86	3.5	5	0.62
7	1.0	0.66	72.07	20	0.86	3.5	5	0.62
8	1.0	0.57	71.82	20	0.85	3.5	5	0.65
9	1.0	0.48	68.55	20	0.85	3.5	5	0.65
10	1.0	0.40	63.84	20	0.82	3.5	5	0.67
11	1.0	0.36	63.52	20	0.82	3.5	5	0.67
12	1.0	0.33	61.86	20	1.0	1.0	4	0
13	1.0	0.29	59.09	20	1.0	1.0	4	0
14	1.0	0.27	59.41	20	1.0	1.0	4	0
15	1.0	0.24	56.72	20	1.0	1.0	4	0
16	1.0	0.22	55.58	20	1.0	1.0	4	0
17	1.0	0.20	53.79	20	1.0	1.0	4	0
18	1.0	0.19	54.20	20	1.0	1.0	4	0
19	1.0	0.18	54.28	20	1.0	1.0	4	0
20	1.0	0.17	55.69	20	0.80	3.8	6	0.56
21	1.0	0.17	58.55	20	0.80	3.8	6	0.56
22	1.0	0.16	57.80	20	0.80	3.8	6	0.56
23	1.0	0.15	56.70	20	0.80	3.8	6	0.56
24	1.0	0.14	55.27	20	0.80	3.8	6	0.56
25	1.0	0.13	51.92	20	1.0	1.0	4	0
26	1.0	0.13	54.03	20	1.0	1.0	4	0
27	0.4	0.12	50.46	20	1.0	1.0	4	0

7 Conclusion

An analytical approach for load-displacement response of vertically loaded single pile has been developed based on a load transfer method by introducing a proposed *t-z* model for pile-soil interface considering the hardening/softening behavior of skin friction. There are some meaningful conclusions in the current study which are presented as follows:

1. A development of a theoretical *t-z* function for the pile-soil interface is being established. The function is based on a series of results and hypotheses of the previous field and laboratory investigations. The proposed function is able to describe soil nonlinear stress-strain behavior and skin friction hardening/softening mechanism and the degradation phenomenon for strong structural characteristic of soils such as dense sand and cemented soil. The six input parameters required for

the proposed *t-z* function have physical meanings and can be easily determined from field or laboratory experiments. Case studies of bored piles in different soil strata indicate that the proposed *t-z* function can be describe the load transfer behavior of pile-soil interface with reasonable accuracy.

2. A highly effective computer program for nonlinear analysis of the load-displacement response of a vertically loaded single pile has been developed by using the load transfer method with the proposed *t-z* model. The proposed approach has been validated by two cases studies using very different soil strata. The response of a single pile resulting from two traditional *t-z* softening models including parabola and exponential functions analyses are also compared in order to study the degradation phenomenon of skin friction. The results of the comparisons indicate that the proposed approach is generally in good agreement with field-observed behavior than the calculated results derived from traditional *t-z* softening models.

References

Burland, J.B.: Shaft friction of piles in clay - a simple fundamental approach. Ground Eng. **6**(3), 30–42 (1973)

Coyle, H.M., Reese, L.C.: Load transfer for axially loaded piles in clay. J. Soil Mech. Found. Div. **92**(2), 1–26 (1966)

Cao, W.P., Lu, Q.Y., Liang, P.: Experimental study on shear mechanical properties on dense sand-concrete interface under constant and varying normal stress. Build. Struct. **45**(1), 82–86 (2015). (in Chinese)

Duncan, J.M., Chang, C.Y.: Nonlinear analysis of stress and strain in soils. Soil Mech. Found. Div. J. **96**(5), 1629–1653 (1970)

Evgin, E., Fakharian, K.: Effect of stress paths on the behavior of sand steel interfaces. Can. Geotech. J. **33**(6), 853–865 (1997)

Guo, W.D., Randolph, M.F.: An efficient approach for settlement prediction of pile groups. Géotechnique **49**(2), 161–179 (1999)

Kraft, L.M., Ray, P.R., Kagawa, T.: Theoretical t-z curves. J. Geotech. Eng. Div. **107**(11), 1543–1561 (1981)

Kondner, R.L.: Hyperbolic stress-strain response: cohesive soils. J. Soil Mech. Found. Div. **89**(1), 115–143 (1963)

Likitlersuang, S., Teachavorasinskun, S., Surarak, C., et al.: Small strain stiffness and stiffness degradation curve of Bangkok Clays. Soils Found. **53**(4), 498–509 (2013)

Liang, P., Cao, W.P., Lu, Q.Y.: Experimental research on shear mechanical properties on loose sand-concrete interface under constant/varying normal stress. Build. Struct. **44**(8), 61–65 (2014). (in Chinese)

Nabil, F.I.: Axial load tests on bored piles and pile groups in cemented sands. J. Geotech. Geoenviron. Eng. **127**(9), 766–773 (2001)

Nanda, S., Patra, N.R.: Theoretical load-transfer curves along piles considering soil nonlinearity. J. Geotech. Geoenviron. Eng. **140**(1), 91–101 (2014)

Promboon, S., Brenner, R,P.: Large diameter bored piles in Bangkok clay. In: 10th International Conference on Soil Mechanics and Foundation Engineering, pp. 815–818 (1981)

Park, S., Roberts, L.A., Misra, A.: Characterization of "t-z" Parameters and their variability for auger pressure grouted piles using field load test data. In: GeoFlorida 2010: Advances in Analysis, Modeling & Design, pp. 1757–1766 (2010)

Reese, L.C.: Drilled shafts: construction procedures and design methods. Tunn. Undergr. Space Technol. **5**(1), 156–157 (1989)

Randolph, M.F., Wroth, C.P.: Analysis of deformation of vertically loaded piles. J. Geotech. Eng. Div. **104**(12), 1465–1488 (1978)

Randolph, M., Gourvenec, S.: Offshore Geotechnical Engineering. Spon, New York (2011)

Rollins, K.M., Clayton, R.J., Mikesell, R.C., et al.: Drilled shaft side friction in gravelly soils. J. Geotech. Geoenviron. Eng. **131**(8), 987–1003 (2005)

Skempton, A.W.: Cast in-situ bored piles in London clay. Géotechnique **9**(4), 153–173 (1959)

Said, I., De Gennaro, V.D., Frank, R.: Axisymmetric finite element analysis of pile loading tests. Comput. Geotech. **36**(1), 6–19 (2009)

Vijayvergiya, V.N.: Load movement characteristic of piles. In: 4th Annual Symposium of the Waterway, pp. 269–284 (1969)

Wang, W.D., Li, Y.H., Wu, J.B.: Pile-soil interface shear model of super long bored pile and its FEM simulation. Rock Mech. **33**(12), 3818–3824 (2012). (in Chinese)

Zhu, H., Chang, M.F.: Load transfer curves along bored piles considering modulus degradation. J. Geotech. Geoenviron. Eng. **128**(9), 764–774 (2002)

Zou, J., Zhang, Z.M., Liu, J.W., et al.: A load transfer model considering strain softening of soils. Chin. J. Geotech. Eng. **32**(7), 27–31 (2011). (in Chinese)

Zhang, Q.Q., Zhang, Z.M.: A simplified nonlinear approach for single pile settlement analysis. Can. Geotech. J. **49**(11), 1256–1266 (2012)

A Nonlinear Calculation Approach of Long-Term Settlement for Pile Groups in Layered Soft Soils

Shengyang Feng[1], Suhua Zhou[2](✉), Hongquan Li[3], and Limin Wei[3]

[1] School of Environmental Protection and Safety Engineering, University of South China, Hengyang 421001, China
[2] College of Civil Engineering, Hunan University, Changsha 410082, China
zhousuhua@hnu.edu.cn
[3] School of Civil Engineering, Central South University, Changsha 410075, China

Abstract. The pile settlement in soft soils can sustain for several decades, owing to the creep behavior of soft soils around the pile. This paper put forward a nonlinear method to analyze the long-term settlement of pile groups in layered soft clay soils. This approach uses the load transfer (*t-z*) model to describe the pile-soil interaction by the Burgers model distributed along the pile shaft and the pile base. The advantage is that it can calculate the long-term settlement of pile groups with rigid or flexible high cap. A program named CTSP (Calculating Time-dependent Settlement of Pile foundation) has been developed to analyze the long-term settlement of two pile groups in layered soft clay soils. Comparisons of the calculations and the measurements were presented to validate the effectiveness and accuracy of the proposed approach.

Keywords: Pile groups · Long-term settlement · Soft soil · Viscoelastic

1 Introduction

Because of the creep behavior of soft soils surrounding the pile, the pile settlement in soft soils will continue for a very long time (Booker et al. 1976). Through a lot of long-term pile tests, many researchers (Murayama et al. 1960; Sharman 1961; Yamagata 1963; Bromham et al. 1973) indicated that the pile settlement in soft soils inclines to increase for several decades after the consolidation of soft soils. So when designing pile foundation in soft soils, it is essential to research the pile long-term settlement. But nowadays, most methods focus on the total pile settlement, neglecting the long-term settlement of pile groups in layered soft soil.

Recently, some researches have been focused on this key issue. Edil et al. (1988) used the parabolic equation to describe the shearing creep of soil around the pile and the creep of the pile-soil interface. Guo (2000) proposed a calculation method for the long-term settlement of vertical loaded piles, based on non-linear viscoelastic shaft and base load transfer (*t-z*) models. Wu et al. (2012) used the virtual soil-pile model and the viscoelasticity of soils around pile to establish a method for calculating the single pile

R. Chen et al. (eds.), *Proceedings of the 2nd International Symposium on Asia Urban GeoEngineering*, Springer Series in Geomechanics and Geoengineering,
https://doi.org/10.1007/978-981-10-6632-0_30

time-effect settlement. Feng et al. (2014) established a simple and practical method for the pile long-term settlement in soft soils. In this method, the Mesri creep model is used to describe the behavior of soil under pile tip, and the Mindlin-Geddes method is used to calculate the vertical additional stress of pile bottom. Yang et al. (2014) established a nonlinear analysis method of long-term settlement for bridge pile groups. This method can simulate the actual loading conditions of pile foundation and the consolidation behavior of soil layers below the pile.

In this work, we will establish a non-linear approach of the long-term settlement for the pile groups in layered soft soils. In this approach, the Burgers model will be adopted to describe creep characteristic of soft soils surrounding the pile based on the load transfer (*t-z*) model.

2 Calculation Methods

2.1 Burgers Model of Soft Soils

As we know, soft soils around the pile possess obviously creep characteristic (Poulos and Davis 1968, 1980; Ahmad and Davies 1985). Because the Burgers model can reflect the instantaneous elastic strain, the main creep and the secondary creep of soft soils, this work used the Burgers model to present the long-term deformation of soft soils around the pile.

The Burgers model is made up of a Maxwell model and a Kelvin model in a series, and it can be illustrated in Fig. 1. The relationship of stress $\sim$ strain of soils around pile can be written as (Komamura et al. 1974)

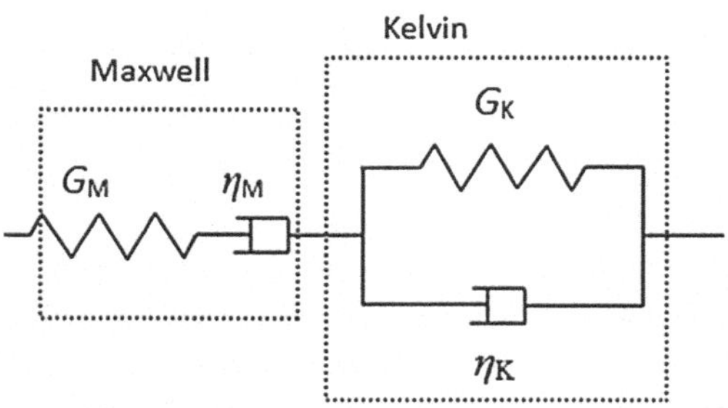

Fig. 1. Burgers Model

$$\tau+\left(\frac{\eta_M}{G_M}+\frac{\eta_K+\eta_M}{G_K}\right)\dot{\tau}+\frac{\eta_M\eta_K}{G_M G_K}\ddot{\tau}=\frac{\eta_M\eta_K}{G_K}\ddot{\gamma}+\eta_M\dot{\gamma} \tag{1}$$

where G_M,G_M are the elastic shear modulus and the viscosity coefficient of the Maxwell model, respectively, $G_M = E_M/[2(1+v)]$, v is the Poisson ratio; G_K, η_K are the elastic shear modulus and the viscosity coefficient of the Kelvin model, respectively, $G_K = E_K/[2(1+v)]$.

Assuming that load on pile top is continuous and stable, and soft soils of pile is in creep stage, so $\tau =$ constant, $\ddot{\tau} = \dot{\tau} = 0$, therefore creep equation of the Burgers model can be written as

$$\gamma = \left[\left(\frac{1}{G_M} + \frac{t}{\eta_M} \right) + \frac{1}{G_K} \left(1 - e^{-\frac{G_K}{\eta_K}t} \right) \right] \tau \quad (2)$$

2.2 Non-Linear Viscoelastic Load Transfer Model

2.2.1 Governing Equation of Single Pile Based on Burgers Model

The pile bearing capacity of a single pile is made up of the lateral friction and the pile tip bearing capacity. For the cylindrical pile, assuming that the pile-soil interface has no slip, the linear governing differential equation of single pile could be expressed as

$$\frac{\partial^2 w}{\partial z^2} = \frac{U}{A_p E_p} \tau \quad (3)$$

where w is the pile's displacement; τ is the shearing strength of the pile-soil interface; U is the perimeter of the pile; A_p is the sectional area; E_p is the elasticity modulus of the pile.

If the Burgers model can describe the stress ~ strain ~ time relationship of soft soils around pile, the pile-soil interaction could be shown in Fig. 2.

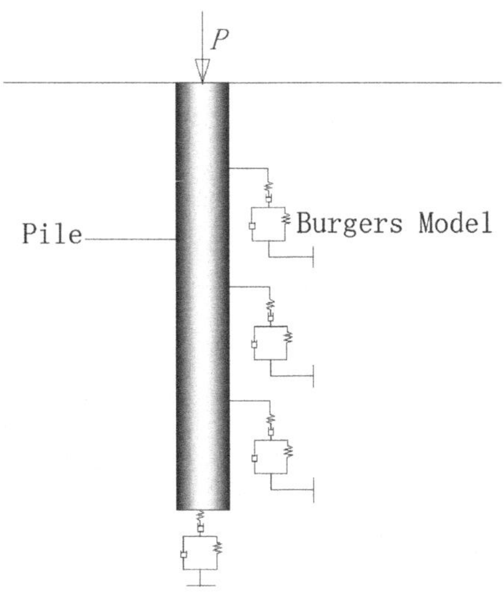

Fig. 2. Calculation model of pile-soil

When Eq. (2) was substituted into Eq. (3), the displacement equation of single pile can be written as

$$\frac{\partial^2 w}{\partial z^2} - \alpha(t)^2 w = 0 \tag{4}$$

where

$$\alpha(t) = \frac{2}{E_p r_0 \left[\left(\frac{1}{G_M} + \frac{t}{\eta_M} \right) + \frac{1}{G_K} \left(1 - e^{-\frac{G_K}{\eta_K} t} \right) \right]}$$

where r_0 is the pile's radius.

The general solution of Eq. (4) is

$$w(z) = c_1 e^{\alpha(t)z} + c_2 e^{-\alpha(t)z} \tag{5}$$

where c_1 and c_2 are the constants.

The equilibrium equation can be expressed as

$$\frac{\partial P(z)}{\partial z} = -2\pi r_0 \tau \tag{6}$$

where $P(z)$ is the pile's axial force.

According to Eqs. (2), (5) and (6), it could be obtained as

$$p(z) = -\alpha(t) A E_p \left(c_1 e^{\alpha(t)z} - c_2 e^{-\alpha(t)z} \right) \tag{7}$$

Combining Eqs. (5) and (7), the axial force and displacement of single pile at z can be written as

$$\begin{Bmatrix} w(z) \\ p(z) \end{Bmatrix} = \mathbf{B}(z) \begin{Bmatrix} c_1 \\ c_2 \end{Bmatrix} \tag{8}$$

where

$$\mathbf{B}(z) = \begin{bmatrix} e^{\alpha(t)z} & e^{-\alpha(t)z} \\ -AE_p \alpha(t) e^{\alpha(t)z} & AE_p \alpha(t) e^{-\alpha(t)z} \end{bmatrix}$$

2.2.2 Recurrence Relation of the Axial Force and Displacement of Single Pile

The top and bottom displacement of No. i pile section can be respectively written as

$$\begin{Bmatrix} w_{ti} \\ p_{ti} \end{Bmatrix} = \mathbf{B}(z_t)\begin{Bmatrix} c_{ti1} \\ c_{ti2} \end{Bmatrix} \tag{9}$$

$$\begin{Bmatrix} \omega_{bi} \\ p_{bi} \end{Bmatrix} = \mathbf{B}(z_b)\begin{Bmatrix} c_{bi1} \\ c_{bi2} \end{Bmatrix} \; (z_b = z_t + l_i) \tag{10}$$

where w_{ti}, w_{bi} are the top and bottom displacement of pile section, respectively; p_{ti}, p_{bi} are the top and bottom axial force of pile section, respectively; z_t, z_b are the top and bottom depth of No. i pile section, respectively; c_{ti1}, c_{ti2} are the constants of top pile section, respectively; c_{bi1}, c_{bi2} are the constants of bottom pile section, respectively.

Boundary conditions of the top and bottom at a pile section are the same, so $c_{ti1} = c_{bi1}$, $c_{ti2} = c_{bi2}$. c_1, c_2 could be eliminated by Eqs. (9) and (10), so

$$\begin{Bmatrix} w_{ti} \\ p_{ti} \end{Bmatrix} = \mathbf{B}(z_t) \cdot \mathbf{B}(z_b)^{-1} \cdot \begin{Bmatrix} w_{bi} \\ p_{bi} \end{Bmatrix} \tag{11}$$

When $\mathbf{K}_i = \mathbf{B}(z_t) \cdot \mathbf{B}(z_b)^{-1}$, so

$$\begin{Bmatrix} w_{ti} \\ p_{ti} \end{Bmatrix} = \mathbf{K}_i \cdot \begin{Bmatrix} w_{bi} \\ p_{bi} \end{Bmatrix} \tag{12}$$

where

$$\mathbf{K}_i = \begin{bmatrix} \cosh[\alpha(t)l_i] & \frac{1}{AE_p\alpha(\mathrm{t})}\sinh[\alpha(\mathrm{t})l_i] \\ AE_p\alpha(t)\sinh[\alpha(\mathrm{t})l_i] & \cosh[\alpha(\mathrm{t})l_i] \end{bmatrix}$$

If a single pile can be divided into n section, the relationship of the top displacement ω_t, the axial force p_t and the bottom displacement ω_b, the axial force p_b can be expressed as

$$\begin{Bmatrix} w_t \\ p_t \end{Bmatrix} = \mathbf{K}_1\mathbf{K}_2\cdots\mathbf{K}_i\cdots\mathbf{K}_n\begin{Bmatrix} w_b \\ p_b \end{Bmatrix} = \mathbf{K}\begin{Bmatrix} w_b \\ p_b \end{Bmatrix} \tag{13}$$

where $\mathbf{K} = \prod_{i=1}^{n} \mathbf{K}_i$.

2.2.3 Flexibility of Single Pile

When soils beneath the pile is in elastic state, the displacement of soils beneath the pile ω_b can be written as (Randolph and Wroth 1979)

$$w_{\mathrm{b}} = \frac{1 - v^2}{2r_0 E} p_{\mathrm{b}} \tag{14}$$

where E is the elasticity modulus of soils beneath the pile.

Assuming soils beneath the pile can be represented by the Burgers model, the viscoelasticity expression of ω_{b} can be obtained as (Wang and Sun 2011)

$$w_{\mathrm{b}} = \frac{1 - v^2}{2r_0 C_{\mathrm{sb}}(t)} p_{\mathrm{b}} \tag{15}$$

where

$$C_{\mathrm{sb}}(t) = \frac{1}{\left[\left(\frac{1}{E_{\mathrm{M}}} + \frac{t}{\eta_{\mathrm{M}}}\right) + \frac{1}{E_{\mathrm{K}}}\left(1 - e^{-\frac{E_{\mathrm{K}}}{\eta_{\mathrm{K}}}t}\right)\right]} \tag{16}$$

where E_{M}, η_{M} are the elastic modulus and the viscosity coefficient of the Maxwell model of soils beneath the pile, respectively; E_{K}, η_{K} are the elastic modulus and the viscosity coefficient of the Kelvin model of soils beneath the pile, respectively.

When unit load is added to the pile bottom, that is, $P_{\mathrm{b}} = 1$, according to Eqs. (13) and (15), the flexibility of single pile can be expressed as

$$f = \frac{w_{\mathrm{t}}}{P_{\mathrm{t}}} \tag{17}$$

2.3 Calculation Approach of Long-Term Settlement for Pile Groups

2.3.1 Governing Equation Considering Interaction Effects with Neighboring Piles

Figure 3 is the sketch map of two piles interaction. It can be seen that, considering interaction effects with neighboring piles, the displacement of the No.1 (or No.2) pile section is made up of 3 parts:

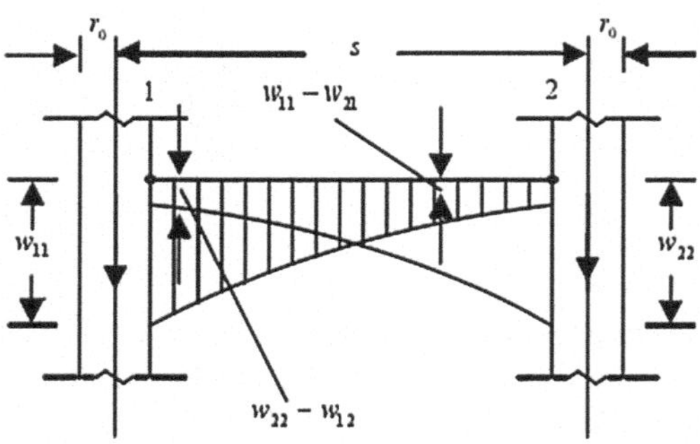

Fig. 3. Interaction between two piles

(1) the displacement of the No.1 (or No.2) pile section under its gravity, $w_{11}(w_{22})$;
(2) the displacement of the No.1 (or No.2) pile section caused by the gravity of the No.2 (or No.1) pile section, $w_{22}(w_{11})$;
(3) the displacement reduction of the No.1 (or No.2) pile section caused by the No.2 (or No.1) pile section under no gravity, $w_{12}(w_{21})$.

Considering effects of interaction with neighboring piles, the displacement of the No.1 (or No.2) pile section w_1, w_2 can be expressed as

$$\begin{cases} w_1 = [w_{11} + (w_{22} - w_{12})] \\ w_2 = [(w_{11} - w_{21}) + w_{22}] \end{cases} \tag{18}$$

Referring to Eq. (17), the matrix form of Eq. (18) can be written as

$$\begin{Bmatrix} w_1 \\ w_2 \end{Bmatrix} = \begin{bmatrix} f_{11} & f_{22} - f_{12} \\ f_{11} - f_{21} & f_{22} \end{bmatrix} \begin{Bmatrix} p_1 \\ p_2 \end{Bmatrix} \tag{19}$$

According to Eq. (19), the load of the No.1 (or No.2) pile section p_1, p_2 can be expressed as

$$\begin{Bmatrix} p_1 \\ p_2 \end{Bmatrix} = \frac{1}{f_{11}f_{12} + f_{22}f_{11} - f_{12}f_{21}} \begin{bmatrix} f_{22} & f_{12} - f_{22} \\ f_{21} - f_{11} & f_{11} \end{bmatrix} \begin{Bmatrix} w_1 \\ w_2 \end{Bmatrix} \tag{20}$$

where f_{11}, f_{22} are the flexibility of the No.1 (or No.2) pile section under its gravity, respectively; f_{12}, f_{21} are the flexibility reduction of the No.2 (or No.1) pile section under no gravity.

Assuming two piles are the same, and under the same load, so

$$\begin{cases} f_{11} = f_{22} = f \\ f_{12} = f_{21} = \Delta f \end{cases}, \begin{cases} w_1 = w_2 = w \\ p_1 = p_2 = p \end{cases}$$

Therefore, Eq. (20) can be expressed as

$$p = \frac{1}{2f - \Delta f} w = k^* w \tag{21}$$

where k^* is the lateral soil stiffness of some pile section under two piles interacting.

The relationship of load and displacement of lateral soil can be written as (Randolph and Wroth 1979)

$$\begin{cases} w(r) = \frac{\tau(z) r_0}{G} \ln(\frac{r_m}{r}), r_0 \leq r \leq r_m \\ w(r) = 0, r > r_m \end{cases} \tag{22}$$

where $w(r)$ is the horizontal displacement of pile lateral soil at r away from the pile axes; G is the shear elasticity of pile lateral soil; r_m is the maximal radius of soil displacement which can be ignored,

$$r_m = 2.5L\rho_m(1 - \upsilon) \tag{23}$$

of which,

$$\rho_m = \frac{1}{G_m L}\sum_{i=1}^{n} G_i l_i \sqrt{\frac{G_m}{G_b}}\left[1 - e^{(1-h/L)}\right] \tag{24}$$

where G_m is the maximal shear elasticity of soil; l_i is the thickness of the ith soil layer; G_b is the shear elasticity of the pile bottom soil; h is the distance from the pile bottom to the rigid soil.

According to Eq. (22), w_{11} can be written as

$$w_{11} = \frac{\tau_1 r_0}{G}\ln\left(\frac{r_m}{r_0}\right) \tag{25}$$

w_{12} can be written as

$$w_{12} = \frac{\tau_2 r_0}{G}\left[\ln\left(\frac{r_m}{r_0}\right) - \ln\left(\frac{r_m}{s}\right)\right] \tag{26}$$

where τ_2 is the shear stress of the No. 2 pile section caused by the load of the No. 1 pile section; s is the distance between two piles.

According to Eq. (25), so

$$f = \frac{w_{11}}{p_1} = \frac{\frac{\tau_1 r_0}{G}\ln\left(\frac{r_m}{r_0}\right)}{2\pi r_0 \tau_1} = \frac{1}{2\pi G}\ln\left(\frac{r_m}{r_0}\right) \tag{27}$$

According to Eq. (26), so

$$\Delta f = \frac{\frac{\tau_2 r_0}{G}\left[\ln\left(\frac{r_m}{r_0}\right) - \ln\left(\frac{r_m}{s}\right)\right]}{2\pi r_0 \tau_2} = \frac{1}{2\pi G}\ln\left(\frac{s}{r_0}\right) \tag{28}$$

If pile groups have n piles, divided into m sections, and the length of the gth pile section of the pile i and pile j $(i, j = 1, 2, \cdots, n\ ,\ i \neq j)$ is l_g, according to Eq. (21), the shear stress of pile lateral τ can be written as

$$\tau = \frac{p}{U l_g} = \frac{k^* w}{U l_g} \tag{29}$$

When Eq. (29) was substituted into Eq. (3), governing equation considering effects of interaction with neighboring piles can be written as

$$\frac{\partial^2 w(z)}{\partial z^2} - \frac{k^*}{AE_p l_g} w(z) = 0 \tag{30}$$

When $\beta^2 = \frac{k^*}{AE_p l_g}$, the general solution of Eq. (30) is

$$w(z) = c_1 e^{\beta z} + c_2 e^{-\beta z} \tag{31}$$

2.3.2 Recurrence Relation of the Axial Force and Displacement Considering Interaction Effects with Neighboring Piles

Referring to Eqs. (6)–(13), considering interaction effects with neighboring piles between the pile i and pile j, the relationship of the top displacement ω_{tij}, the axial force p_{tij} and the bottom displacement ω_{bij}, the axial force p_{bij} can be expressed as

$$\begin{Bmatrix} w_{\text{tij}} \\ p_{\text{tij}} \end{Bmatrix} = \mathbf{B}_{\text{ij1}} \mathbf{B}_{\text{ij2}} \cdots \mathbf{B}_{\text{ijg}} \cdots \mathbf{B}_{\text{ijm}} \begin{Bmatrix} w_{\text{bij}} \\ p_{\text{bij}} \end{Bmatrix} = \mathbf{B}_{\text{ij}} \begin{Bmatrix} w_{\text{bij}} \\ p_{\text{bij}} \end{Bmatrix} \tag{32}$$

of which $\mathbf{B}_{\text{ij}} = \prod_{g=1}^{m} \mathbf{B}_{\text{ijg}}$,

$$\mathbf{B}_{\text{ijg}} = \begin{bmatrix} \cosh[\beta_{\text{ij}} l_g] & \frac{1}{AE_p \beta_{\text{ij}}} \sinh[\beta_{\text{ij}} l_g] \\ AE_p \beta_{\text{ij}} \sinh[\beta_{\text{ij}} l_g] & \cosh[\beta_{\text{ij}} l_g] \end{bmatrix}$$

2.3.3 Flexibility of Pile Groups Considering Interaction Effects with Neighboring Piles

When unit load is added to the pile bottom, $p_{\text{bij}} = 1$, the displacement can be expressed as (Lee 1993)

$$w_{\text{bij}} = \frac{1-\upsilon}{4G_b r_0}(1+\zeta_{\text{bij}}), \zeta_{\text{bij}} = 2r_0/\pi s_{\text{ij}} \tag{33}$$

where s_{ij} is the distance between the pile i and j.

When Eq. (33) was substituted into Eq. (32), w_{tij}, p_{tij} could be obtained. So the integrated flexibility can be written as

$$f_{\text{ij}}^* = \frac{w_{\text{tij}}}{p_{\text{tij}}} \quad (i,j = 1,2,\cdots,n,\ i \neq j) \tag{34}$$

The flexibility of the pile i f_{ii} can be calculated by Eq. (17), so the flexibility of pile groups considering effects of interaction with neighboring piles can be written as

$$f_{\text{ij}} = f_{\text{ij}}^* - f_{\text{ii}}(i,j = 1,2,\cdots,n,\ i \neq j) \tag{35}$$

2.3.4 Calculation Methods

For n pile groups, the displacement of the pile i can be written as

$$w_{\rm i} = \sum_{j=1}^{n} f_{\rm ij} p_{\rm i} \tag{36}$$

where $p_{\rm i}$ is the load on the pile i; when $i = j$, $f_{\rm ij}$ could be calculated by Eq. (17); when $i \neq j$, $f_{\rm ij}$ could be calculated by Eq. (35).

The matrix form of Eq. (36) can be expressed as

$$\begin{Bmatrix} w_1 \\ w_2 \\ \vdots \\ w_{\rm n} \end{Bmatrix} = \begin{bmatrix} f_{11} & f_{12} & \cdots & f_{1\rm n} \\ f_{21} & f_{22} & \cdots & f_{2\rm n} \\ \vdots & \vdots & \vdots & \vdots \\ f_{\rm n1} & f_{\rm n2} & \cdots & f_{\rm nn} \end{bmatrix} \begin{Bmatrix} p_1 \\ p_2 \\ \vdots \\ p_{\rm n} \end{Bmatrix} \tag{37}$$

Assuming that all of pile groups are the same, so $f_{11} = f_{22} = \cdots = f_{\rm nn}$, $f_{\rm ij} = f_{\rm ji}$, that is, the flexibility matrix of Eq. (38) is symmetric matrix.

(1) For the n pile groups with rigid high cap, when the load on pile groups at time t is P(t), so

$$\begin{cases} w_1 = w_2 = \cdots = w_{\rm n} = w \\ p_1 + p_2 + \cdots p_{\rm n} = P(t) \end{cases} \tag{38}$$

(2) For the n pile groups with flexible high cap, because the load on each pile is the same, the load on the pile i is

$$p_{\rm i} = P(t)/n \tag{39}$$

Combining Eqs. (37) and (38), the long-term settlement of the pile groups with rigid high cap can be calculated. Combining Eqs. (37) and (39), the long-term settlement of the pile groups with flexible high cap can be calculated as well. On the basis of the method above, a program CTSP (Calculating Time-dependent Settlement of Pile foundation) compiled by Matlab was carried out to calculate the long-term settlement of pile groups in layered soft clay soil. Program flow chart of CTSP was shown in Fig. 4.

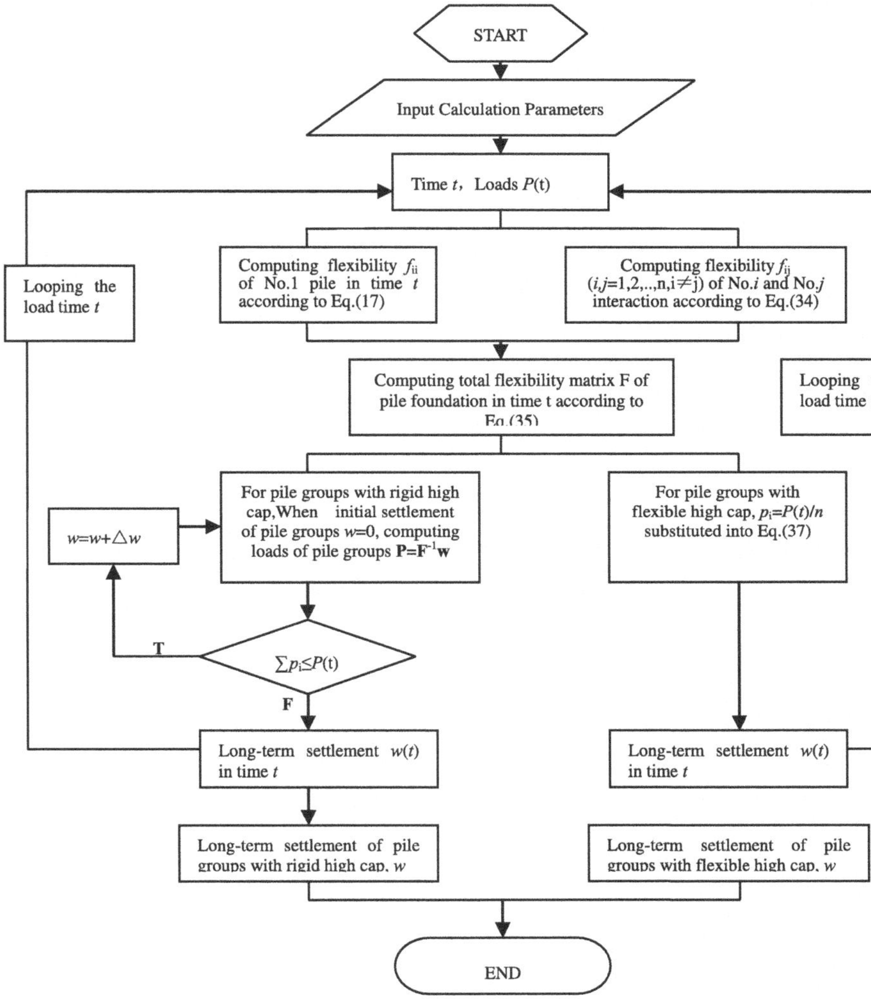

Fig. 4. Program flow chart of CTSP

3 Example and Analysis

3.1 Case One

A high-speed railway bridge pile groups can be considered as the pile groups with rigid high cap, and it was shown in Fig. 5. The pile groups have 9 concrete piles with 1 m diameter, of which the elastic modulus is 32.5 G, and the length is 62 m. The layout of foundation can be seen in Fig. 4 as well, and the calculation parameters of foundation can be listed in Table 1.

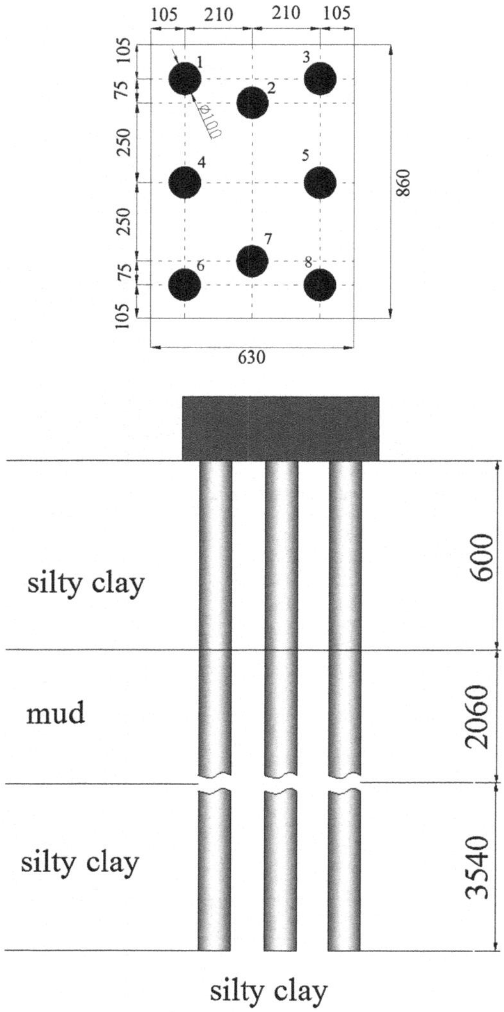

Fig. 5. Pile groups size and layout (Unit:cm)

Table 1. Calculation parameters of foundation

Soils	Layout thickness H/m	Poisson ratio v	E_M/ MPa	E_K/ MPa	η_M/ MPa.d	η_K/ MPa.d
silty caly	6	0.35	15.1	18	1780	5.1
mud	20.6	0.4	7.9	6.5	1068	2.9
silty caly	35.4	0.32	15.7	20.7	1908	4.5
silty caly	pile bottom	0.3	22.5	19.4	2872	7.6

According to pile length, the pile could be divided into 620 sections, that is, every section is 0.1 m. The measured results of the pile groups were compared in Fig. 6 with that theoretically calculated using the CTSP program. Figure 6 shows that the long-term settlement of the pile groups calculated by the CTSP is close to that of the measured. The tendency of the calculated results approximately coincides with the measured. The maximum relative error is no more than 18.8% and the minimum is just 4.6%.

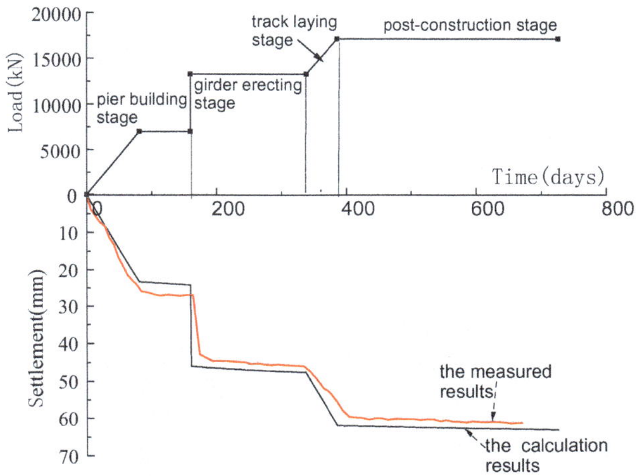

Fig. 6. Comparisons of the calculated and measured long-term settlement of the pile groups

3.2 Case Two

A pile groups with flexible high cap was illustrated in Fig. 7. The pile groups have 322 (14 × 23) concrete piles with 0.8 m diameter, 2.7 m pile distance, 50 m length. The elastic modulus of pile is 32.5 GPa. The foundation soil is silty soft soil. The elastic modulus of the soil is 3.7 MPa, and the poisson ratio is 0.38. The Burgers model parameters could be fitted by the results of long-term soil compression test (Wang et al. 2006; Yan 2007), and E_M = 8.3 MPa, E_K = 5.3 MPa, η_M = 1195 MPa.d, η_K = 3.6 MPa.d. Load on the pile groups is 999570 kN lasting for 10000 days. The pile had been divided into 500 sections, then using the CTSP to calculate the long-term settlement of this pile groups.

In this work, we also used the ABAQUS to calculate the long-term settlement of this pile groups. The Drucker Prager Creep model has been used for the foundation soft soil, with the elastic modulus of 3.7 MPa, the poisson ratio of 0.38, and the parameters of Drucker Prager yield criterion of $\alpha = 0.053$, $k = 21.54$ kPa. Specially, Creep rule of the model was used the "time hardening model", in which the creep model parameters $A = 9.3 \times 10^{-9}$, $m = -0.905$, $n = 1.306$. The linear elastic model was used for the pile and cap with the elastic modulus of 32.5 GPa, and the poisson ratio of 0.17. The pile-soil interface used the coulomb friction model, with the friction coefficient of 0.15. The finite element analysis model of the pile groups was shown in Fig. 8.

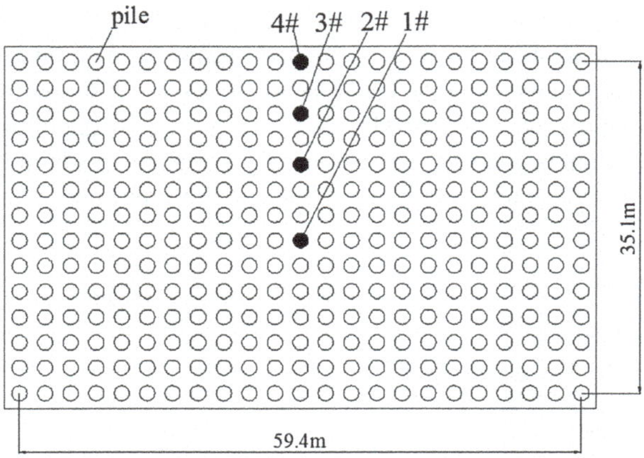

Fig. 7. Pile groups sketch map

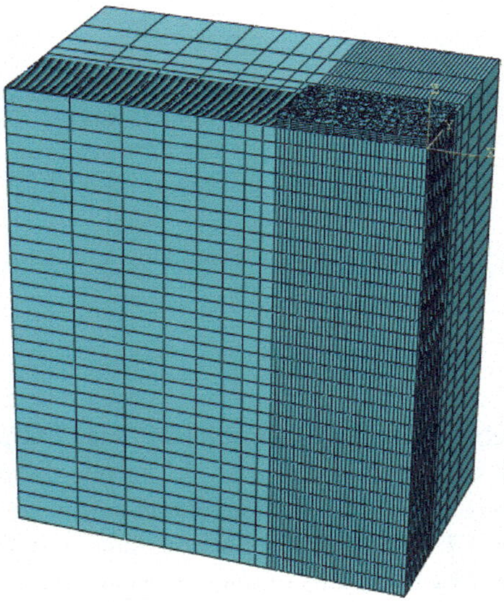

Fig. 8. Pile groups FEM

The theoretical results of the pile 1# ~ 4# in Fig. 7 calculated by the CTSP were compared in Fig. 9 with the numerical results by the ABAQUS. From Fig. 9, the long-term settlement of the pile groups calculated by the CTSP is very close to that of the ABAQUS. The maximum relative error of the pile 1# is no more than 4.5%, the pile 2# 6.7%, the pile 3# 5.7%, and the pile 4# 5.8%. In addition, the long-term settlement of the pile 1# is obviously greater than the other 3 piles, and the settlement of pile groups gradually reduces from the central pile to the outer.

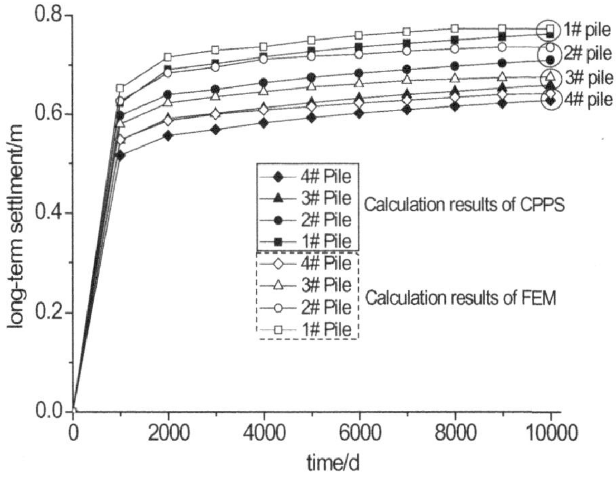

Fig. 9. Comparisons of the theoretical and numerical results

4 Conclusions

In this paper, we established a non-linear approach to investigate the pile long-term settlement in layered soft clay soils. The approach used the load transfer model to describe the interaction between pile and soil by the Burgers model, which is distributed in the shaft and base of the pile. What makes this method special is the way in which it is flexible to describe visco-elastic, non-linear and non-homogeneous characteristic of soft soils surrounding the pile, and also considering interaction effects with neighboring piles. Based on the proposed approach, we developed a computer program to analyze the pile long-term settlement in layered soft soils. Comparisons of the long-term settlement for two piles in layered soft soils were presented to validate its effectiveness and accuracy.

Acknowledgments. This work was financed by the state key laboratory of Safety and Health for Metal Mines (No. 2016-JSKSSYS-08), the scientific research project of Hunan Province Education Department (No. 15C1191), the PHD initial foundation program of the University of South China (No. 2014XQD11), and the science and technology development program plan of Hengyang City (No. 2015KS1706).

References

Ahmad, S., Davies, T.G., Manolis, G.D.: Viscoelastic analysis of piles and pile groups. Int. J. Numer. Anal. Meth. Geomech. **9**(3), 237–252 (1985)

Booker, J.R., Poulos, H.G.: Analysis of creep settlement of pile foundations. J. Geotech. Eng. **102**(1), 1–14 (1976). ASCE

Bromham, S.B., Styles, J.R.: 1973 An analysis of pile loading tests in a stiff clay. In: Proceedings of the First Australian-New Zealand Conference On Geomechanics, Melbourne, Australia, pp. 256–253 (1973)

Edil, T.B., Mochtar, I.B.: Creep response of model pile in clay. J. Geotech. Eng. **114**(11), 1245–1260 (1988)
Feng, S.Y., Wei, L.M., He, C.Y., et al.: A computational method for post-construction settlement of high-speed railway bridge pile foundation considering soil creep effect. J. Cent. South Univ. **21**(7), 2921–2927 (2014)
Guo, W.D.: Visco-elastic load transfer models for axially loaded piles. Int. J. Numer. Anal. Method Geomech. **24**, 135–163 (2000)
Komamura, F., Huang, R.J.: New rheological model for soil behaviour. J. Geotech. Geoenvironmental Eng. **100**, 807–824 (1974)
Lee, C.Y.: Settlement of pile groups-practical approach. J. Geotech. Eng. **119**(9), 1449–1461 (1993). ASCE
Murayama, S., Shibata, T.: The bearing capacity of a pile daiven into soil and its new measuring method. Soil Found. **1**(2), 2–11 (1960)
Poulos, H.G., Davis, E.H.: The settlement behavior of single axially loaded incompressible piles and piers. Geotechnique **18**(3), 351–371 (1968)
Poulos, H.G., Davis, E.H.: Pile Foundation Analysis and Design. Wiley, New York (1980)
Randolph, M.F., Wroth, C.P.: An analysis of vertical deformation of pile groups. Geotechnique **29**, 423–439 (1979)
Sharman, F.A.: The anticipated and observed penetration resistance of some friction piles entirely in clay. In: Proceedings of the 5th International Conference on Soil Mechanics and Foundation Engineering, Paris, France, pp. 135–141 (1961)
Wang, H.G., Wei, L.M., He, X.G.: Determination of three-dimensional rheologic parameters based on long term oedometer test. Chin. J. Geotech. Eng. **28**(5), 669–673 (2006)
Wang, D.D., Sun, J.: Long-term settlement of pile foundation of bridge based on generalized shear displacement method. Chin. J. Geotech. Eng. **33**(sup2), 47–53 (2011)
Wu, W.B., Wand, K.H., Zhang, Z.Q., et al.: A new approach for time effect analysis of settlement for single pile based on virtual soil pile model. J. Cent. South Univ. **19**(9), 2656–2662 (2012)
Yamagata, K.: The yield-bearing-capacity of bearing piles. In: Proceedings of the International Conference on Soil Mechanics and Foundation Engineering, Budapest, Hungary, pp. 325–342 (1963)
Yan, J.Y.: The influence of pile foundation in lifecycle on adjacent existing tunnel. Ph.D. Thesis, Tongji University, Shanghai, China (2007)
Yang, Q., Leng, W.M., Zhang, S., et al.: Long-term settlement prediction of high-speed railway bridge pile foundation. J. Cent. South Univ. **21**(6), 2415–2424 (2014)

Calibrating Resistance Factor and Factor of Safety of Driven Piles Considering Set-up Effects

X.Y. Bian[1(✉)], X.Y. Chen[1], H.L. Lu[1], and J.J. Zheng[2]

[1] School of Resource and Civil Engineering, Wuhan Institute of Technology, Wuhan 430073, China
bianxy@hust.edu.cn
[2] Institute of Geotechnical and Underground Engineering, Huazhong University of Science and Technology, Wuhan 430074, China

Abstract. The ultimate bearing capacity of the driven piles is mainly composed of the ultimate shaft resistance and the ultimate base resistance. The frequent appearance of the set-up effects of the driven piles will be conductive to the improvement of the ultimate shaft resistance after pile driving. The fundamental purpose of this paper is to study the influence of the set-up effects on the design resistance factor and the factor of safety (FS) of the driven piles based on the load and resistance factor design (LRFD) criteria. The set-up effects factor (α) and the ratio (ε) of the initial ultimate shaft resistance to the initial ultimate bearing capacity are used to calibrate the formulas for estimating the resistance factor and FS of the driven piles. Meanwhile, the influence of the factor α and the ratio ε on the design resistance factor and FS of the driven piles are studied. Analysis indicates that the smaller design resistance factor will be determined for the driven piles with the stronger set-up effects, and so it is for the larger ε. And the smaller FS will be estimated for the driven piles with the stronger set-up effects, and so it is for the larger ε. However, analysis also demonstrates that the design resistance factor and the estimation FS are nearly not influenced by the factor α and the ratio ε when they are larger than the certain values.

Keywords: Driven piles · Set-up effects · Resistance factor · Factor of safety · LRFD criteria

1 Introduction

Pile set-up effects refer to the increase of the bearing capacity for the driven piles with increasing time after the completion of driving. The first well-documented case of the pile set-up effects in noncohesive soils was by Tavenas and Audy (1972). The rate and magnitude of the set-up effects are function of many number of factors (Samson and Authier 1986), such as pile type, soil type, pile size, etc. The mechanisms of the pile set-up effects are mainly involved in the stress relaxation (horizontal creep) and the soil aging (caused mainly by rearrangements and interlocking of soil particles) (Chow et al. 1998). In practice, the set-up effects is predominately demonstrated by the increase of the pile shaft resistance (Samson and Authier 1986; Fellenius and Altaee 2002; Tan et al. 2004; Bullock

R. Chen et al. (eds.), *Proceedings of the 2nd International Symposium on Asia Urban GeoEngineering*, Springer Series in Geomechanics and Geoengineering,
https://doi.org/10.1007/978-981-10-6632-0_31

et al. 2005). When accurately accounting for the driven pile set-up effects during the design, the integration of the pile set-up effects can lead to the more cost-effective design as it will reduce the number of piles and/or pile lengths (Lee et al. 2010). Unfortunately, experimental data required for the detailed pile set-up studies rarely exist.

Static or dynamic tests can be performed to evaluate the pile set-up effects; however, it is not feasible in practice to perform these tests over a period of time as acknowledged in the interim report by the American Association of State Highway and Transportation Officials (AASHTO 2007). A number of empirical methods to estimate the pile set-up effects have been proposed by several engineers and researchers, such as Huang (1988), Skov and Denver (1988), Svinkin et al. (1994), and others. Amongst those proposed formulas, the logarithmic relationship proposed by Skov and Denver (1988) has been commonly adopted for the prediction of the set-up behaviors.

Set-up effects are conductive to improving the ultimate bearing capacity of the driving piles. Bullock et al. (2005) presented a method to predict the total pile capacity although this method was conservative. Komurka et al. (2005) applied the factor of safety (FS) to estimate the set-up effects in the driven pile capacity. Liu et al. (2011) studied the effect of the set-up effects on the bearing capacity of the jacked concrete pipe piles in mixed soils, which have been utilized in the load and resistance factor design (LRFD) of the driven piles. Yang and Liang (2006) proposed the first-order reliability method based approach to incorporate the set-up effects into the LRFD design of the driven piles in clay. Yang and Liang (2009) also calibrate the separate resistance factors for various target reliability levels considering the set-up effects of the driven piles in sandy.

Different from the studies by Yang and Liang (2006, 2009), this paper will present a method to calibrate the resistance factor for LRFD of the driven piles considering the set-up effects. In this method, the factors α and ε are used to respectively express the increasing degree of the ultimate shaft resistance with the time, and the ratio of the initial ultimate shaft resistance to bearing capacity. The factor α and ratio ε are further accepted to calibrate the formulas of the resistance factor and FS. Finally, the impacts of the factor α and ratio ε on the resistance factor and FS will also be analysed.

2 Basic Equations

For a driven pile, the ultimate bearing capacity generally includes the ultimate shaft resistance (or shaft friction) and the ultimate base resistance (or end resistance) (e.g., Poulos and Davis 1980; Bian et al. 2016). So, at a time (called as the initial time) after the construction completion of the driven pile considered, the initial ultimate bearing capacity can be expressed as the summation of the initial ultimate shaft resistance and the ultimate base resistance, as follows (Sugai and Matsuo 1993):

$$R_0 = R_s + R_p \tag{1}$$

where R_0 is the initial ultimate bearing capacity; R_s is the initial ultimate shaft resistance; R_p is the ultimate base resistance.

Due to the set-up effects for the driven piles, the ultimate shaft resistance will increase with the time. And the relationship of the ultimate shaft resistance with the initial ultimate shaft resistance can be measured by a factor, α, which makes the ultimate bearing capacity, R, have the new form as follows (Sugai and Matsuo 1993):

$$R = \alpha R_{\mathrm{s}} + R_{\mathrm{p}} \tag{2}$$

The ratio ε of the initial ultimate shaft resistance R_{s} to the initial ultimate bearing capacity R_0 is used to show the effects of the set-up on the ultimate shaft resistance, which is denoted as (Sugai and Matsuo 1993):

$$\varepsilon = \frac{R_{\mathrm{s}}}{R_0} \tag{3}$$

Combined Eqs. (1)–(3), the following formula for estimating the ultimate bearing capacity can be obtained (Sugai and Matsuo 1993):

$$R = [\alpha\varepsilon + (1 - \varepsilon)]R_0 \tag{4}$$

When considering the set-up effects in the driven piles reliability analysis and design, the limit state function can be established as follows:

$$g = [\alpha\varepsilon + (1 - \varepsilon)]R_0 - Q = 0 \tag{5}$$

If the load Q is only involved in dead load Q_{D} and live load Q_{L}, the limit state function expressed as Eq. (5) will becomes into Eq. (6):

$$g = [\alpha\varepsilon + (1 - \varepsilon)]R_0 - Q_{\mathrm{D}} - Q_{\mathrm{L}} = 0 \tag{6}$$

3 Characteristics of α and ε

3.1 Probability Characteristics of α

The factor α can be used to measure the set-up effects of the driven piles on the initial ultimate bearing capacity (accurately speaking, that is the initial ultimate shaft resistance), and much research on it has been conducted during the past half century. Seed and Reese (1955) carried out the driven model pile experiments in silty clay and found that the increase of the ultimate bearing capacity one month after pile driving can be six times as much as the initial ultimate bearing capacity. Eide et al. (1961) found that the increase of the ultimate bearing capacity one month after pile driving can be four times as much as the initial ultimate bearing capacity. Based on other's analysis results, Sugai and Matsuo (1993) reported that α usually ran to three or more for the commonly accepted soil parameters of the soft clay, high possibility running to two or three.

According to the literature which was published by Sugai and Matsuo (1993), if the factor of the highly variable soils could be considered when conducting the study on the pile set-up effects, the gamma distribution, expressed as Eq. (7), is suitable to reflect the mechanical and probabilistic behavior of the important factor α.

$$f_{\alpha}(\alpha) = \frac{\upsilon^{\lambda}}{\Gamma(\lambda)}(\alpha - 1)^{\lambda-1}e^{-\upsilon(\alpha-1)}, (\alpha \geq 1.0) \tag{7}$$

where $\lambda/\upsilon = 1.0$.

3.2 Probability Characteristics of ε

We know, by Eq. (3), the parameter ε is referred as the ratio of the initial ultimate shaft resistance R_s to the initial ultimate bearing capacity R_0. This value depends on the condition of the ground where the pile is driven and upon the type of the pile. With the special ground condition and referring from the characteristics of the ultimate bearing capacity of the driven piles, for example, the driven piles can be divided into two broad groups, friction piles and end-bearing piles. The driven piles from the former group are further subdivided into the pure friction piles and the end-bearing friction piles. The driven piles from the latter group are further subdivided into the pure end-bearing piles and the friction end-bearing piles. It is known that the above analysis results in the driven piles to be sorted into four pile types.

For the friction piles, the ultimate base resistance R_p is general very small, even can be ignored with respect to the ultimate shaft resistance R_s, which means the ultimate bearing capacity R_0 is nearly equal to the ultimate shaft resistance R_s, resulting in the ratio ε to be nearly equal to one; for the end-bearing piles, the results is opposite, and the ultimate shaft resistance R_s is almost zero, which means the ultimate bearing capacity R_0 is nearly equal to the ultimate base resistance R_p, resulting in the ratio ε to be nearly equal to zero; for the end-bearing friction piles and the friction end-bearing piles, the comparisons between the ultimate shaft resistance R_s and the ultimate base resistance R_p are complex, and both R_s and R_p are not zero, by which it is apparent that the value of the ratio ε is between zero and one. From the above discussion, we know that, the former two cases are the extreme conditions and the values of ε obtained are the high level and the low level, respectively. However, the latter two cases are the common conditions for the geotechnical engineering practice.

It is quite certain that the value of ε is distributed between zero and one. Meanwhile, the ratio ε can be assumed to be a random variable due to the uncertainties of soil types, pile material types, test technology and methods, and so on. By Sugai and Matsuo (1993), it is proposed, in the view of probability study, that the random variable ε follows a uniform distribution from zero to one, and the probability density function of ε should be expressed as follows:

$$f_{\beta}(\beta) = \frac{1}{1-0} = 1, \qquad (0.0 \leq \beta \leq 1.0) \tag{8}$$

In this paper, it is impossible to accept the agreement with the points about the random variable ε from the literature reported by Sugai and Matsuo (1993). There are several reasons for it as follows:

(1) $\varepsilon = 0$ represents the pure end-bearing piles, and $\varepsilon = 1$ represents the pure friction piles. They are two extreme cases for the driven piles. In practice, both the pure end-bearing piles and the pure friction piles are nonexistent in pile engineering. Then the values of ε do not commonly appear in the neighborhood of 0 and 1, and not uniformly distribute in these two regions.
(2) $0 < \varepsilon < 1$ represents the friction end-bearing piles or the end-bearing friction piles. This is the common case for the driven piles in pile engineering, and the value of ε frequently distributes between 0 and 1.

The above discussions tell us that the random variable ε will hardly get around the regions nearby 0 and 1, and commonly get around the regions remained from zero to one.

As an example, a database of the ultimate bearing capacity of the driven piles, compiled by Luo (2004), is mentioned again, from which this paper studies on the ultimate shaft resistance and the ultimate bearing capacity of these driven piles. At the same time, the ratio of the ultimate shaft resistance to the ultimate bearing capacity of these driven piles can be computed, namely the random variable ε. The probability characteristics of the random variable ε could be obtained by the statistical analysis, as described in the Fig. 1, including the histogram graph and the cumulative probability distribution. The observations from Fig. 1 seem to conform that the random variable ε mainly distributes from 0.5 to 0.9 with the ratio of more than 90 percent.

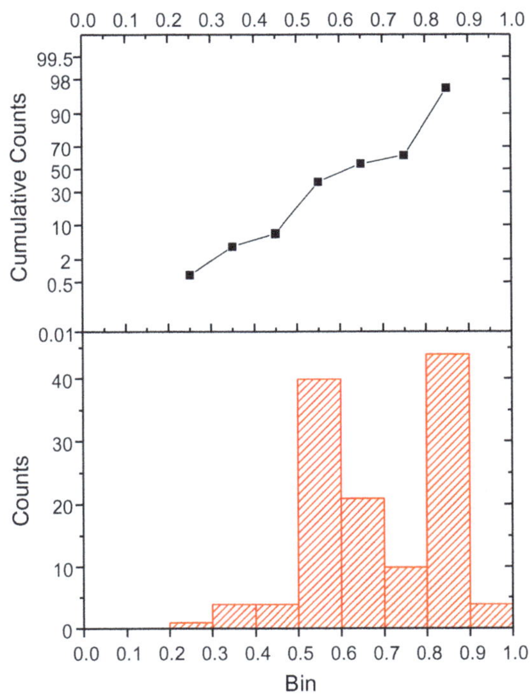

Fig. 1. Histogram graph and cumulative probability distribution of ratio ε

4 Calibration of Load and Resistance Factors

Calibration of load and resistance factors is an important task for geotechnical engineering, which can be conducive to maintain the same levels of load factors for all loads at the different conditions. As proposed in the AASHTO LRFD Specification (AASHTO 2007), the set-up effects have a role influence on the reliability design and calibration of load and resistance factors of the driven piles. When incorporating the set-up effects into the design of the driven piles, the LRFD criteria can be expressed as:

$$\varphi[\alpha\varepsilon + (1-\varepsilon)]R_0 \geq \gamma_D Q_D + \gamma_L Q_L \tag{9}$$

where φ is the resistance factor of the driven piles considering the set-up effects; γ_D and γ_L are the load factors for dead and live loads, respectively.

According to Federal Highway Administration (2001), the initial ultimate bearing capacity R_0 can be estimated in the traditional allowable stress design (ASD):

$$R_0 = R_n = \mathrm{FS}(Q_D + Q_L) \tag{10}$$

where R_n is the nominal (ultimate) resistance of the driven piles.

Substituting Eqs. (4) and (10) into Eq. (9), the resistance factor will be obtained:

$$\varphi = \frac{\gamma_D Q_D + \gamma_L Q_L}{[\alpha\varepsilon + (1-\varepsilon)]\mathrm{FS}(Q_D + Q_L)} = \frac{\gamma_D \rho + \gamma_L}{[\alpha\varepsilon + (1-\varepsilon)]\mathrm{FS}(\rho + 1)} \tag{11}$$

where ρ is the ratio of Q_D/Q_L; FS is the factor of safety in the traditional ASD.

According to the design specifications AASHTO (2007), the load factors γ_D and γ_L are taken as 1.25 and 1.75, respectively, which will be used to determine the resistance factor φ of the driven piles considering the set-up effects in this research. Here FS is taken as 2.33 for computing the resistance factor φ.

Hansell and Viest (1971) reported that the ratio of the dead to live loads for the bridge structures changed with the bridge span lengths. Meanwhile, Hansell and Viest (1971) accepted the following empirical formula to express the relationship of the ratio of the dead to live loads and the bridge span length:

$$\frac{Q_D}{Q_L} = (1 + I)(0.0132l) \tag{12}$$

where I is the dynamic load factor, for example, taken as 0.33 for the LRFD loads; l is the bridge span length in feet (1 feet approximately equal to 0.3048 m).

Many of researchers and engineers usually recommend it reasonable that the values of Q_D/Q_L between 0.5 and 4.0 were accepted for research. Paikowsky (2004) used a Q_D/Q_L ratio ranging from 2.0 to 2.5, while Allen (2005) used a conservative ratio of 3.0. Due to the frequent use of the short span bridges in the State of Iowa, the Iowa DOT used a Q_D/Q_L ratio of 1.5. However, Paikowsky et al. (2004) concluded that the resistance factors were insensitive to the choice of a Q_D/Q_L ratio. To strike a balance between two extremes (0.52 for 9 m and 3.53 for 60 m bridge spans), an average Q_D/Q_L ratio of

2.0 was also selected for study (Ng 2011). Recently, along with the appearance of the super large-span bridge, this paper employs the ratio of Q_D/Q_L expanding to the higher value 5.0.

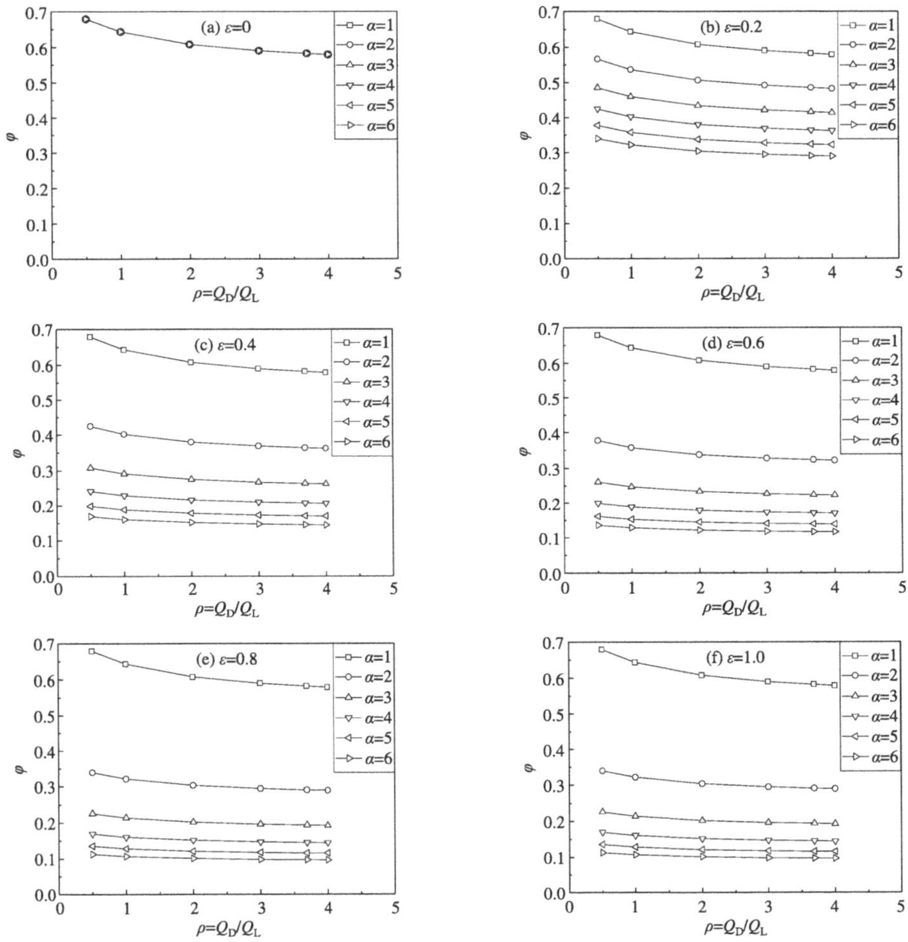

Fig. 2. Variation of resistance factor φ versus ratio of dead to live load considering set-up effects and ratio ε

In order to estimate the resistance factor φ, the computation is respectively conducted with six values of the set-up effects factor α: 1, 2, 3, 4, 5, and 6. Meanwhile, six values (namely ε = 0, 0.2, 0.4, 0.6, 0.8, and 1.0) for the ratio $\varepsilon = R_s/R_0$ are applied. Figure 2 shows the variations of the resistance factor φ with ρ.

According to Fig. 2:

(1) For each ε and α, φ slightly decreases as ρ increases from 0.5 to 5.0 in each curve of φ versus ρ. In particular, when α is not smaller than 2, the curves of φ versus ρ

are all almost horizontal. This indicates that the ratio ρ has slightly influence on the resistance factor φ.

(2) For each figure from Fig. 2(a) to (f), the curves of φ versus ρ corresponding to six levels of the set-up effects (namely $\alpha = 1, 2, 3, 4, 5$, and 6, respectively) go down with an increase of the factor α while the rate gradually diminishes. This demonstrates that the stronger set-up effects of the driven piles will give rise to the smaller design resistance factor. When the set-up effects of the driven piles enhance to a certain extent, the variation of the design resistance factor is very small.

(3) By comparing the figures from Fig. 2(a) to (f), the curves of φ versus ρ with the same set-up effects level go down. This phenomenon is mainly due to the increase of the factor ε from 0 to 1. This indicates that the larger factor ε will make the smaller design resistance factor if the factor α is given. However, for the larger factor ε, the influence of the larger factor ε on the resistance factor φ will reduces. For example, the curves of φ versus ρ with the same set-up effects level in Fig. 2(e) and (f) are nearly identical, in which the factor ε are 0.8 and 1.0, respectively.

FS in the traditional ASD can be calculated as follows (Federal Highway Administration 2001):

$$\mathrm{FS} = \frac{\gamma_{\mathrm{D}} Q_{\mathrm{D}} + \gamma_{\mathrm{L}} Q_{\mathrm{L}}}{[\alpha\varepsilon + (1-\varepsilon)]\varphi\left(Q_{\mathrm{D}} + Q_{\mathrm{L}}\right)} = \frac{\gamma_{\mathrm{D}}\rho + \gamma_{\mathrm{L}}}{[\alpha\varepsilon + (1-\varepsilon)]\varphi(\rho+1)} \tag{13}$$

Equation (13) indicates that FS is function of the set-up effects factor α, the resistance factor φ, and the ratio ε. Here ρ is taken as 3.69, corresponding to the length of 75 m bridge spans, for computing FS. In order to estimate FS, the computation is respectively performed with six values of the set-up effects factor α: 1, 2, 3, 4, 5, and 6. Meanwhile, six values (namely $\varphi = 0.3, 0.4, 0.5, 0.6, 0.7$, and 0.8) for the resistance factor φ are applied. Figure 3 shows the variations of FS with α, φ, and ε.

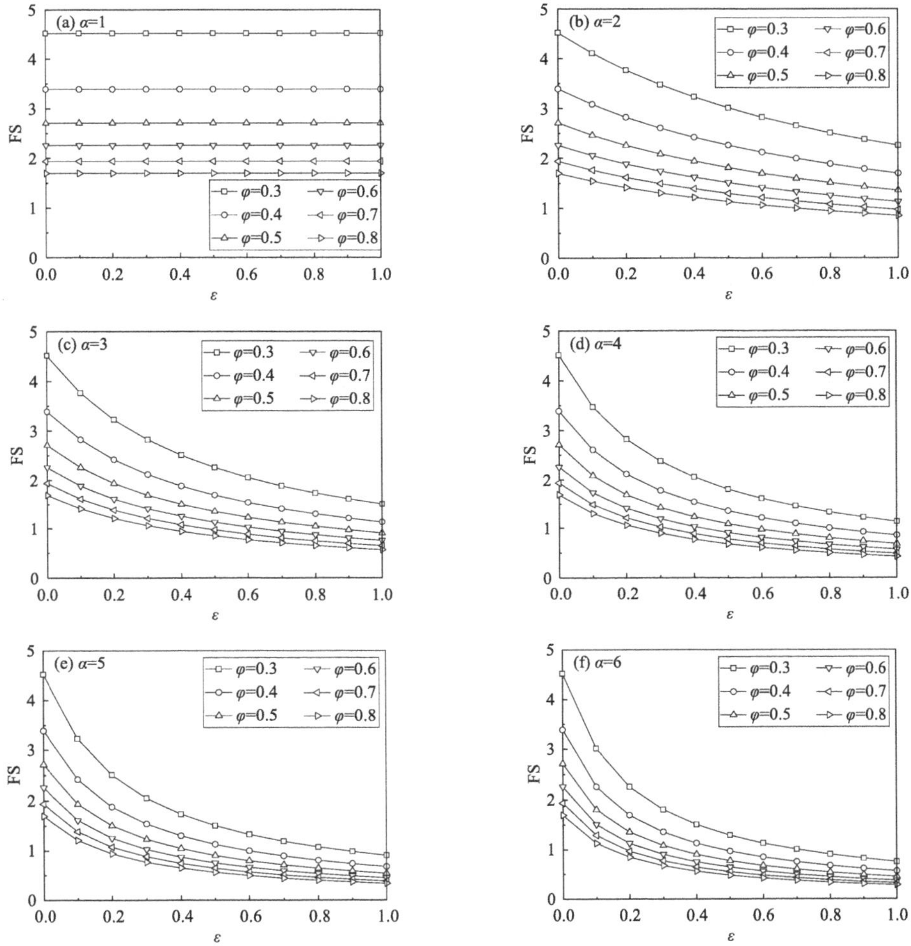

Fig. 3. Variation of factor of safety FS versus ratio ε considering set-up effects and resistance factor

According to Fig. 3:

(1) For each α (except $\alpha = 1$) and φ, FS decreases as ε increases from 0 to 1.0 in each curve of FS versus ε, although the decrease rate gradually diminishes. If α is set as 1, all the curves of FS versus ε are almost horizontal for each resistance factor φ. This indicates that the ratio ε has no influence on FS with $\alpha = 1$.

(2) For each figure from Fig. 3(a) to (f), the curves of FS versus ε corresponding to six levels of resistance factor (namely $\varphi = 0.3$, 0.4, 0.5, 0.6, 0.7, and 0.8, respectively) go down with an increase of the resistance factor φ while the rate gradually diminishes. This demonstrates that the larger design resistance factor of the driven piles will give rise to the smaller FS. When the design resistance factor of the driven piles is determined as the larger values, the variation of FS is very small.

(3) By comparing the figures from Fig. 3(a) to (f), the curves of FS versus ε with the same design resistance factor go down. This phenomenon is mainly due to the enhancement of the set-up effects factor α from 1 to 6. This indicates that the larger factor α will make the smaller FS if the ratio ε is given. However, for the larger factor α, the influence of the factor ε on FS will reduces. For example, the curves of FS versus ε with the same design resistance factor in Fig. 3(e) and (f) are nearly identical, in which the factor α are 5 and 6, respectively.

5 Conclusions

This paper incorporates the set-up effects into the calibration of the resistance factor and FS for the driven piles. The following findings are summarized.

(1) The load ratio $\rho = Q_D/Q_L$ has the slightly influence on the design resistance factor φ. The design resistance factor φ decreases with the increase of the factor α and the ratio ε although the decrease rates gradually diminish.
(2) The FS also decreases with the increase of the factor α and the factor ε, and the decrease rate gradually also diminishes.
(3) When the lager factor α and larger ratio ε are obtained for LRFD, the variation of the factor α and the ratio ε are not the controllable factors for the design resistance factor.

Acknowledgments. The authors would like to express their gratitude to the National Natural Science Foundation of China (51408444, 51378404) and the National Key Research and Development Program of China (2016YFC0800208).

References

AASHTO: AASHTO LRFD Bridge Design Specifications, 4th edn. American Association of State Highway and Transportation Officials, Washington, D.C. (2007)
Allen, T.M.: Development of the WSDOT pile driving formula and its LRFD calibration. Federal Highway Administration (FHWA) Report, Washington State DOT, WA (2005)
Bian, X.Y., Zheng, J.J., Xu, Z.J., Zhang, R.J.: A method to calculate resistance factors for LRFD of driven piles. In: Proceedings of the 6th Asian-Pacific Symposium on Structural Reliability and its Applications, pp. 157–163 (2016)
Bullock, P.J., Schmertmann, J.H., McVay, M.C., Townsend, F.C.: Side shear setup. II: Results from Florida test piles. J. Geotech. Geoenviron. Eng. **131**(3), 301–310 (2005)
Chow, F.C., Jardine, R.J., Brucy, F., Nauroy, J.F.: Effects of time on capacity of pipe piles in dense marine sand. J. Geotech. Geoenviron. Eng. **124**(3), 254–264 (1998)
Eide, O., Hutchinson, J.N., Landva, A.: Short and long-term test loading of a friction pile in clay. In: Proceeding of the 5th ICSMFE, pp. 45–54 (1961)
Federal Highway Administration: Load and resistance factor design (LRFD) for highway bridge substructures. Department of Transportation, Washington (2001)
Fellenius, B.H., Altaee, A.: Pile dynamics in geotechnical practice—six case histories. In: Deep Foundations, vol. 116, pp. 619–631. Geotechnical Special Publication (2002)

Hansell, W.C., Viest, I.M.: Load factor design for steel highway bridges. AISC Eng. J. **8**(4), 113–123 (1971). American Institute of Steel Construction

Huang, S.: Application of dynamic measurement on long H-pile driven into soft ground in Shanghai. In: Proceedings of the 3rd International Conference on the Application of Stress-Wave Theory to Piles, pp. 635–643 (1988)

Komurka, V.E., Winter, C.J., Maxwell, S.G.: Applying separate safety factors to end-of-drive and set-up components of driven pile capacity. In: Geotechnical Applications for Transportation Infrastructure: Proceedings of the 13th Great Lakes Geotechnical and Geoenvironmental Conference, pp. 65–80 (2005)

Lee, W., Kim, D., Salgado, R., Zaheer, M.: Setup of driven piles in layered soil. Soils Found. **50**(5), 585–598 (2010)

Liu, J.W., Zhang, Z.M., Yu, F., Lin, C.G.: Effect of soil set-up on the capacity of jacked concrete pipe piles in mixed soils. J. Zhejiang Univ. Sci. A (Appl. Phys. Eng.) **12**(8), 637–644 (2011)

Luo, S.X.: Research and engineering application on probability limit states design method of pile foundation. Ph.D. thesis, Southwest Jiaotong University, Chengdu, China (2004). (in Chinese)

Ng, K.W.: Pile Setup, Dynamic Construction Control, and Load and Resistance Factor Design of Vertically-Loaded Steel H-piles. Iowa State University, Ames (2011)

Paikowsky, S.G.: Load and resistance factor design (LRFD) for deep foundations. National Cooperative Highway Research Program Report 507, Transportation Research Board, National Research Council, Washington D.C. (2004)

Poulos, H.G., Davis, E.H.: Pile Foundation Analysis and Design. Wiley, New York (1980)

Samson, L., Authier, J.: Change in pile capacity with time: case histories. Can. Geotech. J. **23**(2), 174–180 (1986)

Seed, H.B., Reese, L.C.: The section of soft clay along friction piles. ASCE **81**, 1–28 (1955)

Skov, R., Denver, H.: Time-dependence of bearing capacity of piles. In: Proceedings of the 3rd International Conference on the Application of Stress-Wave Theory to Piles, pp. 879–888 (1988)

Sugai, M., Matsuo, M.: The control of steel pipe pile driving based on reliability analysis of bearing capacity estimation. Soils Found. **33**(3), 40–53 (1993)

Svinkin, M.R., Morgano, C.M., Morvant, M.: Pile capacity as a function of time in clayey and sandy soils. In: Proceedings of the 5th International Conference and Exhibition on Piling and Deep Foundations, pp. 1.11.1–1.11.8 (1994)

Tan, S.L., Cuthbertson, J., Kimmerling, R.E.: Prediction of pile set-up in non-cohesive soils. In: Current Practices and Future Trends in Deep Foundations, vol. 125, pp. 50–65. Geotechnical Special Publication (2004)

Tavenas, F., Audy, R.: Limitations of the driving formulas for predicting the bearing capacities of piles in sand. Can. Geotech. J. **9**(1), 47–62 (1972)

Yang, L., Liang, R.: Incorporating set-up into reliability-based design of driven piles in clay. Can. Geotech. J. **43**(9), 946–955 (2006)

Yang, L., Liang, R.: Incorporating setup into load and resistance factor design of driven piles in sand. Can. Geotech. J. **46**(3), 296–305 (2009)

Deformation Behavior of Geocell Reinforced Granular Cushion Over Soft Soil with Stone Columns

Ling Zhang(✉), Qiang Ou, and Minghua Zhao

College of Civil Engineering, Hunan University, Changsha 410082, China
zhanglhd@163.com, 980071290@qq.com, mhzhaohd@21cn.com

Abstract. This paper developed a simplified model for an embankment resting on geocell reinforced granular cushion over soft soil with stone columns. The geocell reinforced granular cushion was idealized as an Euler-Bernoulli beam by taking beam's geometric nonlinearity into account, while the stone columns were idealized as tensionless and elastic-perfectly plastic springs, and the soft soil was idealized by Kelvin-Voight model to represent its consolidation behavior. The Euler-Bernoulli theory was used to analyze the foundation beam with consideration of the interfacial resistances at the top and bottom of the beam. The influences of soil consolidation and column stiffness on the embankment settlement were also discussed.

Keywords: Geocell reinforced granular cushion · Stone column · Euler-Bernoulli beam · Embankment settlement

1 Introduction

In the last ten years, stone columns in conjunction with geosynthetic-reinforced granular fills have been used more and more extensively, when rapid construction and/or strict deformation control of the road embankment is required. This geosynthetic reinforced granular fill (0.3 m or more) installed above the columns and beneath the embankment serves as a tensioned platform to reduce embankment settlement further.

The analysis of such foundations is relatively more complex and challenging. Most of the works reported in the literature are developed for foundations either reinforced by stone columns (Ambily and Gandhi 2007; Black et al. 2007; Zahmatkesh and Choobbasti 2010; Deb and Dhar. 2013, Zhang et al. 2013a) or geosynthetic layers (Shukla and Chandra 1994; Dash et al. 2003, 2007; Latha et al. 2006; Zhang et al. 2009, 2010a, 2010b, 2012a, 2013b). However, limited studies have been done on the combined use of geosynthetic reinforcement and stone columns. Considering lumped parameter modeling approach mechanical models have been developed to study the settlement response of geosynthetic-reinforced granular fill-soft soil system with stone columns for single layer (Deb et al. 2007) and multilayer (Deb et al. 2008) geosynthetic reinforcements. Maheshwari and Khatri (2011, 2013) developed a simplified model for an infinite beam resting on geosynthetic-reinforced granular bed over soft soil improved with stone columns. In the above mentioned models, the two-dimensional

R. Chen et al. (eds.), *Proceedings of the 2nd International Symposium on Asia Urban GeoEngineering*, Springer Series in Geomechanics and Geoengineering,
https://doi.org/10.1007/978-981-10-6632-0_32

geosynthetic reinforcement, such as geogride was used. However, geocells, the three-dimensional geosynthetic reinforcements, which are quite different with the two-dimensional geosynthetic reinforcement, can be used along with stone columns in the field. Thus, it is necessary to study the geocell-reinforced granular fill resting over soft soil improved with stone columns. The purpose of this paper is to propose a mechanical foundation model for a geocell reinforced granular over soft soil with stone columns, which incorporates the consolidation effect of the soft soil.

2 Model Establishment and the Analysis

To simply the problem, a 2-dimensional section is used to represent a 3-dimensional section with length b in the longitudinal direction of the embankment. Figure 1 shows proposed foundation model for geocell-reinforced granular fill-soft soil system with stone columns where a geocell-reinforced granular fill is placed over the stone column-improved soft soil. In this model, the granular bed is idealized by homogeneous Euler-Bernoulli beam with height h, width b, length l and equivalent elastic modulus E subjected to symmetrical external loads including the concentrated load $P(x)$ and the distributed load $q(x)$.

The stone columns are idealized by tensionless and elastic-perfectly plastic springs, and the normal stress p_{zc} acting on the stone columns can be written as

$$\begin{cases} p_{zc} = k_c w_c & 0 \leq w_c \leq s_{uc} \\ p_{zc} = p_{uc} & w_c > s_{uc} \\ p_{zc} = 0 & w_c < 0 \end{cases} \tag{1}$$

where k_c is the modulus of vertical subgrade reaction of stone columns; s_{uc} is the yield displacement of stone columns; p_{uc} is the ultimate bearing capacity of stone columns; and w_c is the displacement of stone columns.

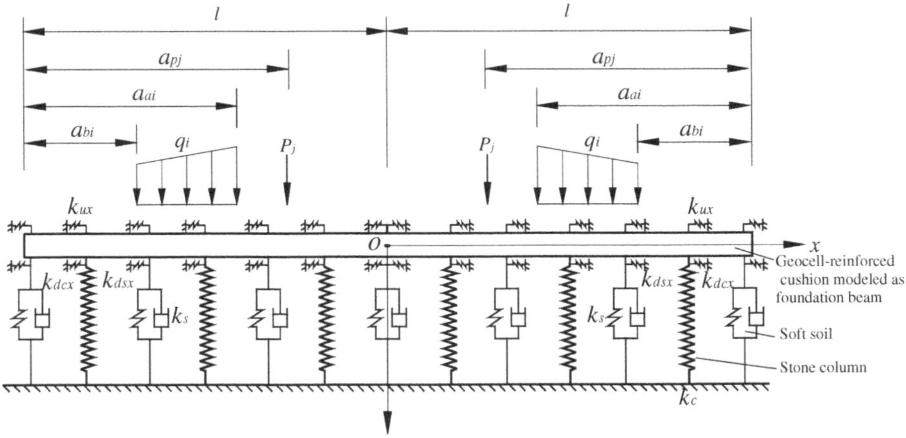

Fig. 1. Foundation model establishment

The consolidation effect due to the inclusion of stone columns has been included in the model. To incorporate the consolidation effect of the soft soil in the present model, the expression for the normal stress p_{zs} acting on the soft soil at time $t > 0$ can be presented by using Terzaghi's effective stress principle as

$$p_{zs} = \sigma' + u_e \tag{2}$$

where σ' and u_e are average effective stress and average excess pore water pressure at time, t. The average excess pore water pressure u_e can be expressed as:

$$u_e = u_0(1 - U) = p_{zs}(1 - U) \tag{3}$$

where u_0 is the initial pore water pressure and U is the average degree of consolidation at any time t, which can be obtained by combining the degree of consolidation in the radial as well as vertical direction in case of stone columns-reinforced soil.

When a bilinear elastic-plastic model is employed to express the effective stress σ', the normal stress p_{zs} acting on the soft soil can be rewritten as:

$$p_{zs} = \frac{\sigma'}{U} = \begin{cases} k_s w_s / U & 0 \le w_s \le s_{us} \\ p_{us}/U & w_s > s_{us} \\ 0 & w_s < 0 \end{cases} \tag{4}$$

where k_s is the modulus of vertical subgrade reaction of soft soil; w_s is the displacement of the soil; s_{us} and p_{us} are the yield displacement and the ultimate bearing capacity of the soil, respectively.

The right half of the model beam is divided into n segments with length l_i (i = 1, …, n), and each segment represents single supporting condition or loading condition. A local coordinate system x_i–z with its origin o' at the left end of the axis line of ith beam segment are set up.

A free-body diagram of one beam segment (T_ith segment) is chosen for force analysis, and the following equations can be obtained (Zhang et al. 2012a):

$$\frac{dT_i}{d\xi_i} = b\left(p_{ux,i} + p_{dx,i}\right) \tag{5}$$

$$\frac{dM_i}{d\xi_i} = Q_i - b\left(p_{ux,i} - p_{dx,i}\right)\frac{h}{2} - T_i w_i' \tag{6}$$

$$\frac{dQ_i}{d\xi_i} = b\left[p_{z,i} - q_i + \left(p_{ux,i} + p_{dx,i}\right)\theta_i\right] \tag{7}$$

where T, Q, M and θ are the tension force, shear force, bending moment and rotation angle within the beam, respectively; TTTpBB$_{uxBB}$ and p_{dxBB} are the interface resistances on the top and bottom faces of the beam; p_z is the vertical subgrade reaction; q is the distributed load acting on the beam.

Based on the basic hypothesis of the Euler-Bernoulli theory, the rotation angle θ_i, bending moment M_i, and the tensile force T_i, can be expressed by:

$$\theta_i = w_i' \tag{8}$$

$$M_i = -EIw_i'' \tag{9}$$

$$T_i = EA \cdot (v_i^0)' + \frac{\chi}{2} EA \cdot (w_i')^2 \tag{10}$$

where v^0 is the horizontal displacement on the neutral axis of the beam; w are the vertical displacement of the beam; "/" and "//"denote respectively the first-order and the second-order partial derivatives with respect to ξ_i; and χ is a nonlinear coefficient. If geometric nonlinearity of the beam is taken into account, $\chi = 1$, otherwise, $\chi = 0$. The value of equivalent elastic modulus E depends on the geocell's type and size, the mattress composition and the unit weight of infilling materials.

The resistances $p\text{BB}_{ux\text{BB}}$ and $p\text{BB}_{dx\text{BB}}$ are assumed to be linear with the relative displacements, as shown below:

$$\begin{cases} p_{ux,i} = k_{ux,i} v_i^u \\ p_{dx,i} = k_{dx,i} v_i^d \end{cases} \tag{11}$$

where $v_{i\text{PBPB}}^{u}$ and $v_{i\text{PBPB}}\text{P}^{d\text{PP}}\text{P}$ are the horizontal displacements at the soil-foundation beam interface on the beam's top and bottom faces, respectively; and the horizontal displacements v_i^u and v_i^d can be calculated by:

$$\begin{cases} v_i^u = v_i^0 - \frac{h}{2} w_i' \\ v_i^d = v_i^0 + \frac{h}{2} w_i' \end{cases} \tag{12}$$

$k\text{BB}_{ux,i\text{BB}}$ and $k\text{BB}_{dx,i\text{BB}}$ are the corresponding moduli of horizontal subgrade reactions, in which $k\text{BB}_{dx,i\text{BB}} = k\text{BB}_{dsx,i\text{BB}}$ for the segments supported by soil alone; $k_{dx,i\text{BB}} = k_{dpx,i\text{BB}}$ for the segments supported by piles; $k\text{BB}_{dsx,i\text{BB}}$ is the modulus of horizontal subgrade reaction on the beam's bottom face for the segments supported by soil; and $k\text{BB}_{dpx,i\text{BB}}$ is the modulus of horizontal subgrade reaction for the segments supported by stone columns.

In order to simplify the derivation process below, the expression of $p_{z,i}$ in Eqs. (1) and (4) is written in a unified form as follows:

$$p_{z,i} = k_i \cdot w_i + g_i \tag{13}$$

where k_i and g_i are coefficients, for pile segments:

$$\begin{cases} k_i = k_c \text{ and } g_i = 0, & 0 \le w_c \le s_{uc} \\ k_i = 0 \text{ and } g_i = p_{uc}, & w_c > s_{uc} \\ k_i = 0 \text{ and } g_i = 0, & w_c < 0 \end{cases}$$

; and for soil segments:

$$\begin{cases} k_i = k_s/U \text{ and } g_i = 0, & 0 \leq w_s \leq s_{us} \\ k_i = 0 \text{ and } g_i = p_{us}/U, & w_s > s_{us} \\ k_i = 0 \text{ and } g_i = 0, & w_s < 0 \end{cases}.$$

Only a linear distributed load is considered in this study, namely $q(\xi BB_{iBB})$ can be expressed as:

$$q_i(\xi_i) = B_i\xi_i + D_i \tag{14}$$

in which B_{iBB} and DBB_{iBB} are constants.

Substitute Eqs. (10)–(12) into Eq. (5), and substitute Eqs. (7)–(9), (11)–(14) into Eq. (6), then the governing equations of beam displacements w and v^0 can be obtained as below:

$$EA \cdot (v_i^0)'' - b(k_{ux} + k_{dx})v_i^0 = \frac{bh}{2}(k_{ux} - k_{dx})w_i' - \chi \cdot EA \cdot w_i'w_i'' \tag{15}$$

$$\begin{aligned} EI \cdot w_i^{(4)} - \left[T_i + \frac{b(k_{ux} + k_{dx})h^2}{4}\right] \cdot w_i'' + b(k_i w_i + g_i) \\ - b(B_i\xi_i + D_i) - \frac{bh}{2}(k_{ux} - k_{dx}) \cdot (v_i^0)' = 0 \end{aligned} \tag{16}$$

Obviously, the above two governing equations with items of w and v^0 are too complicated to solve directly. A decoupled iteration method (Zhang et al. 2012a) is employed to obtain the solutions.

Firstly, neglecting the effect of interface resistances $TTTp_{uxBB}$ and p_{dxBB}, the beam is simply idealized as a classic Euler-Bernoulli beam supported only by vertical elastic-perfectly plastic springs to obtain $w_i^{k=1}$, $\theta_i^{k=1}$, $M_i^{k=1}$ and $Q_i^{k=1}$.

Secondly, regarding $(w_i^k)'$ and $(w_i^k)''$ obtained from the first step as constants in Eq. (15), then the solution of $(v_i^0)^k$ is:

$$(v_i^0)^k = c_{i,1}^k e^{a\cdot\xi_i} + c_{i,2}^k e^{-a\cdot\xi_i} + \varpi_i^k \tag{17}$$

where $a = \sqrt{\frac{b(k_{ux} + k_{dx})}{EA}}$; $\varpi_i^k = -\frac{bh(k_{ux} - k_{dx})\cdot(w_i^k)' - 2\chi\cdot EA\cdot(w_i^k)'(w_i^k)''}{2b(k_{ux} + k_{dx})}$; $c_{i,1}^k$ and $c_{i,2}^k$ are two unknown coefficients. $c_{i,1}^k$, $c_{i,2}^k$, A_i^k, B_i^k, C_i^k and D_i^k

Then, the tensile force T_i^k is:

$$T_i^k = EA\left((u_i^0)^k\right)' + \frac{\chi}{2}EA \cdot ((w_i^k)')^2 = EA \cdot a\left(c_{i,1}^k e^{a\cdot\xi_i} - c_{i,2}^k e^{-a\cdot\xi_i}\right) + \frac{\chi}{2}EA \cdot ((w_i^k)')^2 \tag{18}$$

Thirdly, treating $\left((u_i^0)^k\right)'$ and T_i^k obtained from the second step as constants in Eq. (16), the solution of w_i^{k+1} can be expressed as follows:

$$w_i^{k+1} = A_i^{k+1} f_{A,i}^{k+1} + B_i^{k+1} f_{B,i}^{k+1} + C_i^{k+1} f_{C,i}^{k+1} + D_i^{k+1} f_{D,i}^{k+1} + \frac{Y_i^k}{bk_i} \tag{19}$$

where A_i^{k+1}, B_i^{k+1}, C_i^{k+1} and D_i^{k+1} are four unknown coefficients;

$$f_{A,i}^{k+1} = \begin{cases} e^{\alpha_i^{k+1}\xi_i}\cos(\beta_i^{k+1}\xi_i), & \Delta_i^k < 0 \\ e^{r_{A,i}^{k+1}\xi_i}, & \Delta_i^k \geq 0,\ X_i^k \geq 0; \\ \cos(\varphi_{1,i}^{k+1}\xi_i), & \Delta_i^k \geq 0,\ X_i^k < 0 \end{cases}$$

$$f_{B,i}^{k+1} = \begin{cases} e^{\alpha_i^{k+1}\xi_i}\sin(\beta_i^{k+1}\xi_i), & \Delta_i^k < 0 \\ e^{r_{B,i}^{k+1}\xi_i}, & \Delta_i^k \geq 0,\ X_i^k \geq 0 \cdot; \\ \sin(\varphi_{1,i}^{k+1}\xi_i), & \Delta_i^k \geq 0,\ X_i^k < 0 \end{cases}$$

$$f_{C,i}^{k+1} = \begin{cases} e^{-\alpha_i^{k+1}\xi_i}\cos(\beta_i^{k+1}\xi_i), & \Delta_i^k < 0 \\ e^{r_{C,i}^{k+1}\xi_i}, & \Delta_i^k \geq 0,\ X_i^k \geq 0; \\ \cos(\varphi_{2,i}^{k+1}\xi_i), & \Delta_i^k \geq 0,\ X_i^k < 0 \end{cases}$$

$$f_{D,i}^{k+1} = \begin{cases} e^{-\alpha_i^{k+1}\xi_i}\sin(\beta_i^{k+1}\xi_i), & \Delta_i^k < 0 \\ e^{r_{D,i}^{k+1}\xi_i}, & \Delta_i^k \geq 0,\ X_i^k \geq 0; \\ \sin(\varphi_{2,i}^{k+1}\xi_i), & \Delta_i^k \geq 0,\ X_i^k < 0 \end{cases}$$

$$\alpha_i^{k+1} = \frac{\sqrt{\sqrt{\frac{4bk_i}{EI}} + \frac{X_i^k}{EI}}}{2};\ \beta_i^{k+1} = \frac{\sqrt{\sqrt{\frac{4bk_i}{EI}} - \frac{X_i^k}{EI}}}{2};$$

$$r_{A,i}^{k+1} = \sqrt{\frac{\frac{X_i^k}{EI} + \sqrt{\Delta_i^k}}{2}};\ r_{B,i}^{k+1} = -\sqrt{\frac{\frac{X_i^k}{EI} + \sqrt{\Delta_i^k}}{2}};\ r_{C,i}^{k+1} = \sqrt{\frac{\frac{X_i^k}{EI} - \sqrt{\Delta_i^k}}{2}};$$

$$r_{D,i}^{k+1} = -\sqrt{\frac{\frac{X_i^k}{EI} - \sqrt{\Delta_i^k}}{2}};\ \varphi_{1,i}^{k+1} = \sqrt{\frac{-\frac{X_i^k}{EI} - \sqrt{\Delta_i^k}}{2}};\ \varphi_{2,i}^{k+1} = \sqrt{\frac{-\frac{X_i^k}{EI} + \sqrt{\Delta_i^k}}{2}};\ \Delta_i^k = \left(\frac{X_i^k}{EI}\right)^2 - \frac{4bk_i}{EI};$$

$X_i^k = \left[T_i^k + \frac{bh^2}{4}(k_{ux} + k_{dx})\right]$; and $Y_i^k = -bg_i + bq_i + \frac{bh}{2}(k_{ux} - k_{dx}) \cdot \left((u_i^0)^k\right)'$.

By the way, Eq. (12) also suitable for the vertical displacement of the beam without consideration of the horizontal internal resistances in the first step, only need to set $X_i^k = 0$ and $Y_i^k = -bg_i + bq_i$.

Then, the rotational angle θ_i^{k+1}, bending moment M_i^{k+1}, and shear force Q_i^{k+1} can be derived from the vertical displacement w_i^{k+1}. Then regarding $(w_i^{k+1})'$ and $(w_i^{k+1})''$ obtained from the third step as constants in Eq. (15), the solution of $(v_i^0)^{k+1}$ is similar to that shown in Eq. (17), only need to substituted $(w_i^k)'$ and $(w_i^k)''$ by $(w_i^{k+1})'$ and $(w_i^{k+1})''$. The rest of the step may be deduced by analogy until the following inequalities

$$\begin{cases} \left|\frac{w_i^{k+1} - w_i^k}{w_i^{k+1}}\right| < \zeta \\ \left|\frac{(v_i^0)^{k+1} - (v_i^0)^k}{(v_i^0)^{k+1}}\right| < \zeta \end{cases} \tag{20}$$

are satisfied for all i, where ζ is an acceptable convergence with a tolerance of 10^{-6} in this study.

There are 6 unknown coefficients ($c_{i,1}^k$, $c_{i,2}^k$, A_i^k, B_i^k, C_i^k and D_i^k) for every i segment, and 6n unknown coefficients for the whole beam. The following 6 boundary conditions and 6(n-1) continuity conditions are used to determining these 6n unknown coefficients.

Boundary conditions

$$\begin{cases} u^0|_{x=0} = 0 \\ T|_{x=l} = 0 \end{cases}, \begin{cases} \theta_0 = 0 \\ Q_0 = -P_0/2 \end{cases} \text{ and } \begin{cases} M|_{x=l} = 0 \\ Q|_{x=l} = 0 \end{cases}$$

Continuity conditions

$$\begin{cases} (u_{i,0}^0)^k = (u_{i-1,l_{i-1}}^0)^k \\ T_{i,0}^k = T_{i-1,l_{i-1}}^k \end{cases} \text{ and } \begin{cases} w_{i,0}^k = w_{i-1,l_{i-1}}^k \\ \theta_{i,0}^k = \theta_{i-1,l_{i-1}}^k \\ M_{i,0}^k = M_{i-1,l_{i-1}}^k \\ Q_{i,0}^k = Q_{i-1,l_{i-1}}^k - P_{i,i-1} \end{cases}$$

3 Results and Discussion

3.1 Validation

To validate the present analytical procedure, a hypothetical soil-foundation problem as shown in Fig. 2 was set up to analyze.

The beam with l = 10.0 m, b = 1.0 m, h = 0.4 m and E = 40 MPa was subjected to a symmetrical concentrated load P = 200 kN at points x = −1.5 m and x = 1.5 m and a uniform load q = 60 kN/m in the range of $-4.0\text{ m} \leq x \leq 4.0$ m. The coefficient of soil reactions are k_s = 2,500 kN/m^3, k_{ux} = 4,000 kN/m^3, and k_{dsx} = 2,500 kN/m^3. The diameter of stone column (d) is 0.5 m, and its spacing from center to center (S) is 1.5 m. The coefficient of stone column reactions are k_c = 10 k_s, and k_{dpx} = 10 k_{dsx}. Yield displacement of the soil is 15 mm and that of the stone column is 20 mm.

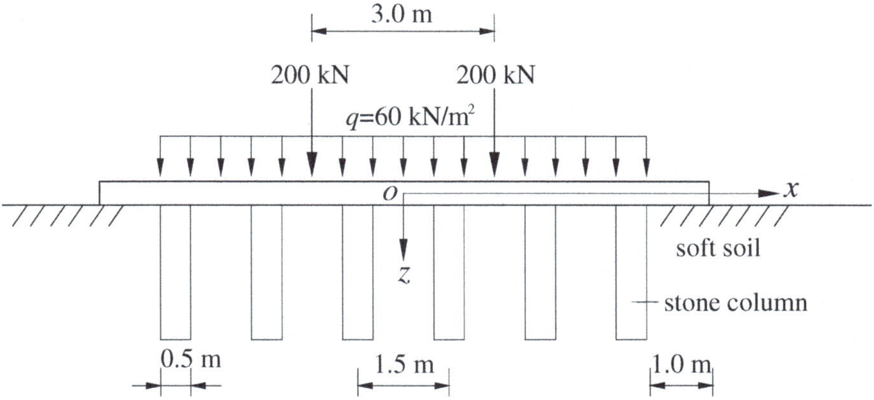

Fig. 2. Model beam used for numerical validation analysis

In the absence of stone columns and without consideration of soil consolidation, a foundation model under consideration reduces to a model same as that presented by Zhang et al. (2012a). Same model beam presented in Zhang et al. (2012a) has been analyzed by substituting the degree of soil consolidation U as 100%, and stiffness of stone columns as that of soil in the proposed foundation model.

The response of foundation beam respectively on elastic and elastic-plastic models has been compared and presented in Fig. 3 in terms of vertical displacement of

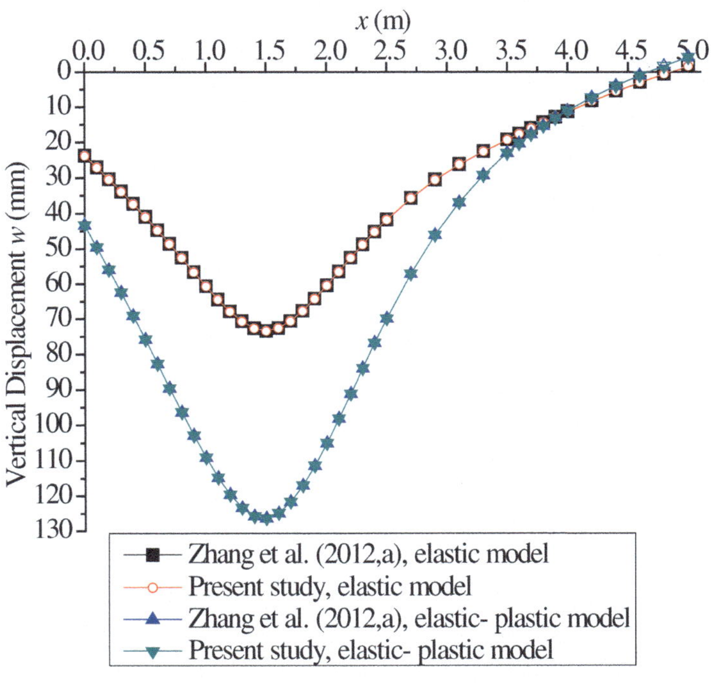

Fig. 3. Comparison of vertical displacements with Zhang et al. (2012a)

foundation beam for typical values of input parameters as mentioned above. It can be observed that, in this case, the foundation beam on elastic-plastic model yields into higher deflection almost along the length of the beam.

Moreover, as shown in Fig. 3, the results from this study have been found to be identical with those obtained from Zhang et al. (2012a). This validated the proposed model and developed computer program.

Further, the proposed model has also been validated by comparing the results from Zhang et al. (2012b) to that from a degenerated case of present study. Zhang et al. (2012b) did not consider both consolidation of soil and geometric nonlinearity of foundation beam and stone column was considered to be linear. In view of this, in the degenerated case of present study, the degree of consolidation of soil has been considered to be 100%, and the coefficient χ related with geometric nonlinearity of beam has been considered to be zero. Further, stone columns have been considered to be linear. Same parameters mentioned above were used for comparison purposes.

From Fig. 4, the results from the present study with elastic foundation model have been found to be identical with those obtained from Zhang et al. (2012b). This validated the proposed model and developed computer program once again. It also can be found again that the foundation beam on elastic-plastic model yields into higher maximum deflection than the beam on elastic model.

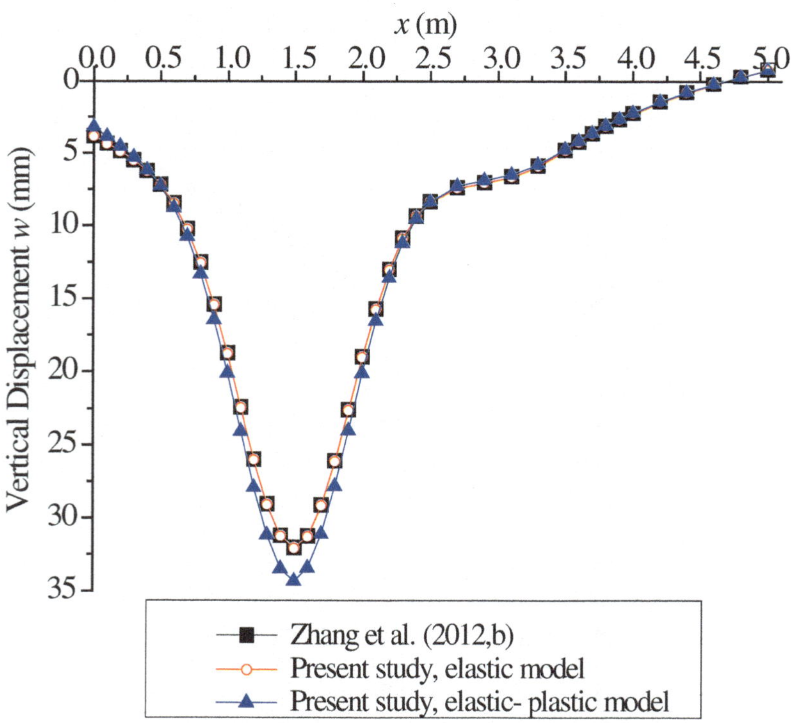

Fig. 4. Comparison of vertical displacements with Zhang et al. (2012b)

3.2 Parametric Study

3.2.1 Influence of Average Degree of Consolidation (*U*)

The influence of average degree of consolidation on the deflection profiles of footing (idealized as beam) has been depicted in Fig. 5. The average degree of consolidation *U* has been varied from 40% to 100%, and the other input parameters keep same as the validation study.

As degree of consolidation increases, the maximum vertical displacement in the footing also increases. As shown in Fig. 5, the maximum settlement has been found to increase significantly from 13.1 mm to 36.1 mm corresponding to the increase in degree of consolidation from 40% to 100%.

Figure 6 shows the variation of maximum settlement with degree of consolidation for different loading conditions. It has been observed once again that the maximum displacement in footing is increased significantly with the increase of the degree of consolidation. Moreover, as the concentrated load increases from 100 kN to 300 kN, the rate of increase of maximum settlement with degree of consolidation becomes high.

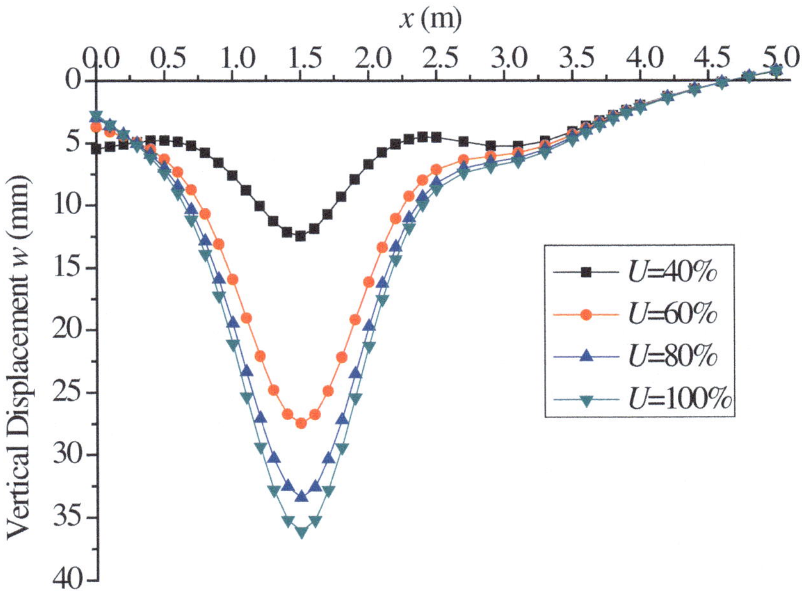

Fig. 5. The influence of average degree of consolidation on displacement of foundation beam

3.2.2 Influence of Stiffness of Stone Columns (K_c)

Comparison of the response of soil-foundation system with and without stone columns has been made by varying the column stiffness k_c from k_s to $10k_s$. The other input parameters keep same as the validation study except $U = 80\%$.

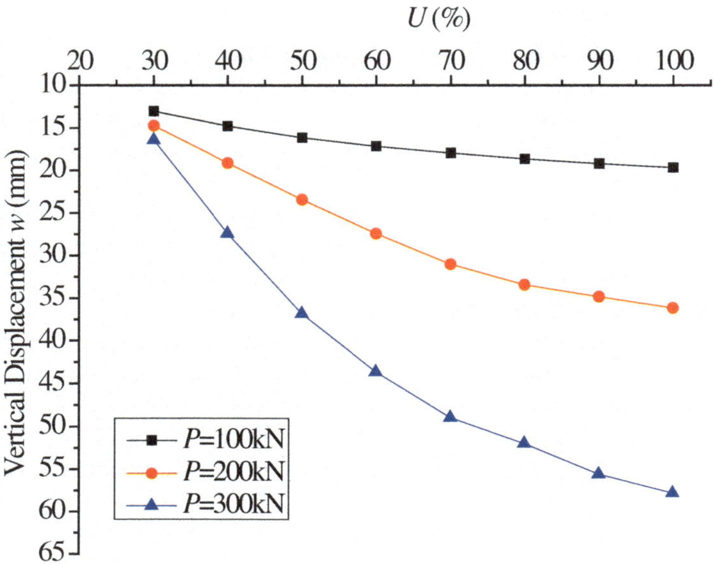

Fig. 6. Variation of the maximum settlement with degree of consolidation for various loading conditions

As expected, the displacement of foundation beam reduces with inclusion of stone columns in the soft soil. From Fig. 7, the maximum settlement at the edge of foundation beam has been found to reduce by 66.2% (from 88.2 mm to 29.8 mm).

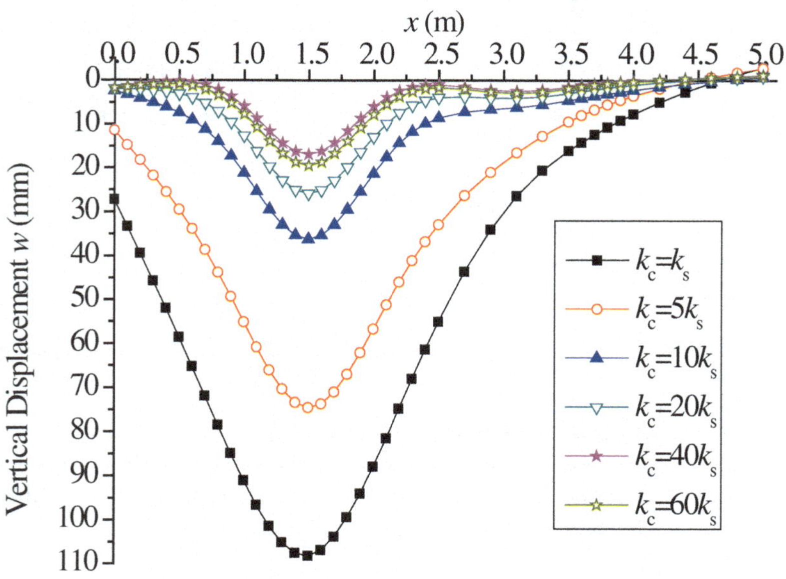

Fig. 7. Influence of stiffness of stone columns on displacement of foundation beam

A study was also conducted to compare the maximum settlement of foundation with different column stiffness. The column stiffness has been varied from $5k_s$ to $60k_s$. It was observed from Fig. 7 that the maximum settlement has been found to reduced significantly from 74.4 mm to 19.4 mm corresponding to the increase in stiffness of pile k_p from $5k_s$ to $40k_s$, reduced by 73.9%. However, as column stiffness further enhances from $40k_s$ to $60k_s$, the maximum settlement has been found to be reduced by 13.4% (from 19.4 mm to 16.8 mm).

4 Conclusion

The deformation behavior of geocell reinforced granular cushion over soft soil with stone columns under symmetrical external loads was investigated in this study. To simply the problem, a 2-dimensional section was used to represent a 3-dimensional section with length b in the longitudinal direction of the embankment, the geocell reinforced granular cushion was idealized as a foundation beam with geometrical nonlinearity, while the stone columns were idealized as elastic-perfectly plastic springs in compression, and the soft soil was idealized by Kelvin-Voight model to represent its consolidation behavior. The Euler-Bernoulli theory was employed to solve the geocell and stone column problem studied in this paper, with consideration of the lateral interfacial shearing resistance at the top and bottom of the beam. This resistance was assumed to be linearly proportional to the relative lateral displacement of the beam. Semi-analytical solutions were proposed to quantify the effects of average consolidation degree and stone columns stiffness.

The results indicate that: The maximum settlement of the foundation increases significantly with the increment of consolidation degree, and its increment rate develops with higher concentrated load acting on the beam. Stone columns are very much effective to reduce the maximum settlement, but the settlement reduction rate is slow down with increasing of the column stiffness.

Acknowledgements. This research was funded through the National Natural Science Foundation of China (NSFC No. 51208191), and the Basal Research Fund Support by Hunan University.

References

Ambily, A.P., Gandhi, S.R.: Behavior of stone columns based on experimental and FEM analysis. J. Geotech. Geoenviron. Eng. **133**(4), 405–415 (2007)

Black, J.A., Sivakumar, V., Madhav, M.R., Hamill, G.A.: Reinforced stone columns in weak deposits: Laboratory model study. J. Geotech. Geoenviron. Eng. **133**(9), 1154–1161 (2007)

Dash, S.K., Rajagopal, K., Krishnaswamy, N.R.: Behavior of geocell reinforced sand beds under strip loading. Can. Geotech. J. **44**(7), 905–916 (2007)

Dash, S.K., Sireesh, S., Sitharam, T.G.: Model studies on circular footing supported on geocell reinforced sand underlain by soft clay. Geotext. Geomembr. **21**(4), 197–219 (2003)

Deb, K., Dhar, A.: Parameter estimation for a system of beam resting on stone column-reinforced soft soil. Int. J. Geomech. **13**(3), 222–233 (2013)

Deb, K., Basudhar, P.K., Chandra, S.: Generalized model for geosynthetic-reinforced granular fill-soft soil with stone columns. Int. J. Geomech. **7**(4), 266–276 (2007)

Deb, K., Chandra, S., Basudhar, P.K.: Response of multilayer geosynthetic reinforced bed resting on soft soil with stone columns. Comput. Geotech. **35**(3), 323–330 (2008)

Maheshwari, P., Khatri, S.: A nonlinear model for footings on granular bed–stone column reinforced earth beds. Appl. Mathe. Model. **35**(6), 2790–2804 (2011)

Maheshwari, P., Khatri, S.: Response of infinite beams on geosynthetic-reinforced granular bed over soft soil with stone columns under moving loads. Int. J. Geomech. **13**(6), 713–728 (2013)

Latha, G.M., Rajagopal, K., Krishnaswamy, N.R.: Experimental and theoretical investigations on geocell-supported embankments. Int. J. Geomech. **6**(1), 30–35 (2006)

Shukla, S.K., Chandra, S.: A study of settlement response of a geosynthetic-reinforced compressible granular fill-soft soil system. Geotext. Geomembr. **13**(9), 627–639 (1994)

Zahmatkesh, A., Choobbasti, A.J.: Settlement evaluation of soft clay reinforced with stone columns using the equivalent secant modulus. Arab. J. Geosci. **5**(1), 103–109 (2010)

Zhang, L., Zhao, M.H., Shi, C.J., Zhao, H.: Bearing capacity of geocell reinforcement in embankment engineering. Geotext. Geomembr. **28**(5), 475–482 (2010a)

Zhang, L., Zhao, M.H., Shi, C.J., Zhao, H.: Nonlinear analysis of a geocell mattress on an elastic–plastic foundation. Comput. Geotech. **42**(5), 204–211 (2012a)

Zhang, L., Zhao, M.H., Shi, C.J., Zhao, H.: Settlement calculation of composite foundation reinforced with stone columns. Int. J. Geomech. **13**(3), 248–256 (2013a)

Zhang, L., Zhao, M.H., Zhao, H.: Deformation analysis of a geocell mattress using a decoupled iterative method. Struct. Eng. Mech. Int. J. **46**(6), 775–790 (2013b)

Zhang, L., Zhao, M.H., Zou, X.W., Zhao, H.: Deformation analysis of geocell reinforcement using Winkler model. Comput. Geotech. **36**(6), 977–983 (2009)

Zhang, L., Zhao, M.H., Zou, X.W., Zhao, H.: Analysis of geocell-reinforced mattress with consideration of horizontal–vertical coupling. Comput. Geotech. **37**(9), 748–756 (2010b)

Zhang, L., Zhao, M.H., Hu, Y.X., Zhao, H., Chen, B.C.: Semi-analytical solutions for geosynthetic-reinforced and pile-supported embankment. Comput. Geotech. **44**(6), 167–175 (2012b)

Investigation of Mechanical Behavior of Foundations Over Unsaturated Sands Reinforced by Stone Columns Using Artificial Neural Networks

S. Hashemi and A. Fotovvat(✉)

Department of Civil and Environmental Engineering, Amirkabir University of Technology (Tehran Polytechnic), Tehran, Iran
Amirreza.fotovvat@gmail.com

Abstract. Installing stone columns has been widely used in recent decades and is considered to be a promising method to improve bearing capacity and reduce settlements. Despite multitudinous efforts to estimate the bearing capacity of foundations over stone columns, there have not been comprehensive studies of stone columns behavior in unsaturated soils. Unsaturated soils encompass characteristics that engender complexities in such analyses. This study deals with artificial neural networks to evaluate mechanical behavior of stone columns in unsaturated granular media. This method proposes a linear regression model with less computational cost in comparison to other methods. The results present that the proposed method has a high compatibility with the experimentally measured data and is able to estimate the bearing capacity of footings over unsaturated soils reinforced by stone columns.

Keywords: Stone columns · Unsaturated soils · Artificial Neural Network · Finite element analysis

1 Introduction

Drilling holes at specified points and filling them with granular material is one way ground improvement prior to construction of foundation. These cylindrical elements, which gain their strength and stiffness from the confinement provided by the surrounding soil, are called Stone Columns. Therefore, the radial strain developed under loading and the interaction between soil and column mobilize the strength of the system (Mohammadizadeh and Asadi 2015). The estimation of this bearing capacity and corresponding settlement is of utmost importance in engineering problems (Sedano and Vanapalli 2011). The soil and Stone materials produce a composite system and the majority of the investigations are focused in fully saturated soil; the unsaturated state of soils is and state in which the degree of saturation of the soil is between two limits, saturated and dry. Therefore many complexities arise when discussing the settlement and bearing capacity of soils in real situations. One of the tools to understand the behavior of unsaturated media is the soil water characteristic curve (Barbour 1998). Recently, computational methods have been widely used to investigate the properties of geomaterials.

R. Chen et al. (eds.), *Proceedings of the 2nd International Symposium on Asia Urban GeoEngineering*, Springer Series in Geomechanics and Geoengineering,
https://doi.org/10.1007/978-981-10-6632-0_33

Artificial Neural Networks has been used in this study to investigate bearing capacity and settlement of a single column installed in granular media. The data required for training and verifying the proposed model are collected from numerical outputs from an axisymmetric model using geostudio modules. The model was verified by experimental results of a single stone-column tested In amirkabir university of technology soil dynamics laboratory (Vaseghi Magvan et al. 2016).

2 Bearing Capacity of Stone Columns

In the last few decades, there have been studies to estimate the bearing capacity of a single and a group stone columns. Brauns (1978) presented that the ultimate bearing capacity of a single stone column (Brauns 1978) that can be estimated as:

$$q_{ult} = \left(q + \frac{2s_u}{\sin 2\delta}\right)\left(1 + \frac{\tan\delta_s}{\tan\delta}\right)\tan^2\delta_p \quad (1)$$

Where s_u is undrained shear strength of the native soil; δ_p and δ_s are failure angle in aggregate material of the stone column and the surrounding respectively; δ is the conical angle of shearing in the native soil. Similarly Mitchell (1981) proposed an empirical method to estimate the ultimate capacity of a single stone column (Mitchell 1981):

$$q_{ult} = s_u.N_p \quad (2)$$

Where N_p is bearing capacity factor estimated to be between 19 and 22 (Barksdale and Bachus 1983).

3 ANN and Intelligent Models

The backward propagation of errors, or backpropagation, is a common method of training artificial neural networks and its structure is shown in Fig. 1. When an input vector is given to the network, it is propagated forward through the network, until it reaches the output layer. The output of the network is then compared to the desired output, and an error value is calculated for each of the neurons in the output layer.

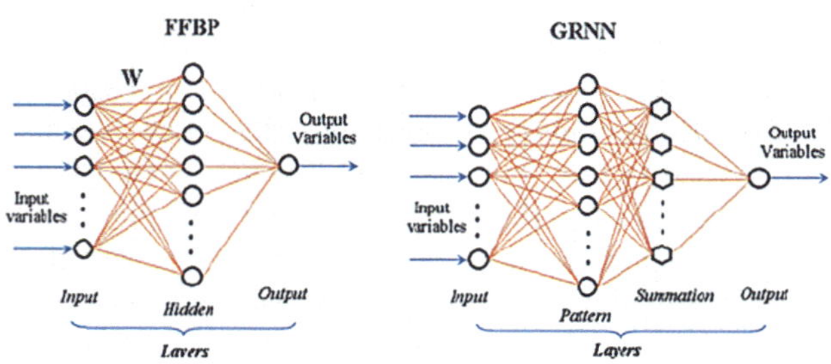

Fig. 1.

Backpropagation algorithm includes an executive index, which is the average of errors squared. The neuron learns from training examples, which in this case consists of a set of tuples where p the input to the network and t is the correct output.

$$\{(P^1, t^1)(P^2, t^2), \ldots, (P^Q, t^Q)\} \tag{3}$$

After inserting the input P(k) to the network, the error signal in the neuron j of the output layer at the moment k is:

$$e_j(k) = t_j(k) - a_j(k) \tag{4}$$

In addition to the mentioned method, a generalized regression neural network (GRNN) is used for function approximation. It has a radial basis layer and a special linear layer. The idea is that every training sample will represents a mean to a radial basis neuron. It uses single-pass learning so no backpropagation is required. Moreover, it can handle noises in the inputs and it has high accuracy in the estimation since it uses Gaussian functions.

Equation (5) shows the mathematical representation of this method. Where $Y_{(x)}$ is the prediction value of input x and d_k is the square of the distance between training samples and input.

$$Y(x) = \frac{\sum_{k=1}^{N} yk^{e^{d_k/2\sigma}}}{\sum_{k=1}^{N} e^{d_{k/2\sigma}}} \tag{5}$$

4 Collecting Data

One of the most important steps in developing a valuable neural model is the collection of data. Experimental tests were carried out (Vaseghi Magvan et al. 2016). Then a 2D axisymmetric FEM analysis model was developed, as for verification (Fig. 2). There was a reasonable trend observed between the results.

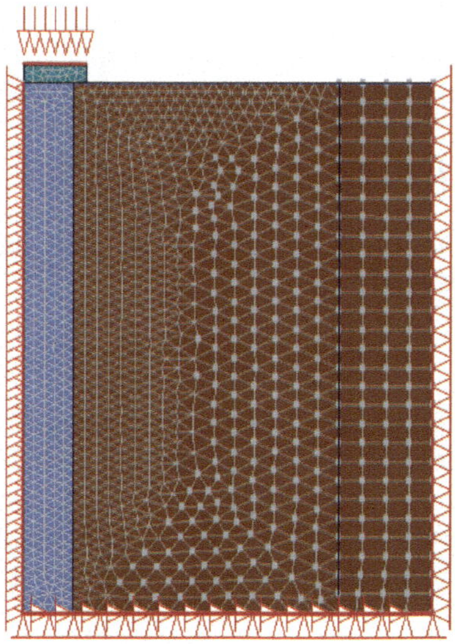

Fig. 2. The finite element model

The reference experimental case consists of a single centered stone column installed in unsaturated firoozkooh sand. The loading system includes a circular rigid loading with a uniform vertical displacement of 1 mm/min. Table 1 shows the properties of the materials obtained by simple shear tests.

Table 1. Mechanical properties of the soil and the column material

	Stone column	Soil
Effective cohesion (kPa)	0	3.5
Internal friction angle	33.9	31.5

The soil and the column are assumed to have hyperbolic and elasti-perfectly plastic behavior respectively, therefore Mohr-coulomb constitutive modeling was applied for the column. By experimental filter paper method, the SWCC of the soil was obtained (Fig. 3). Table 2 shows the SWCC determining characteristics.

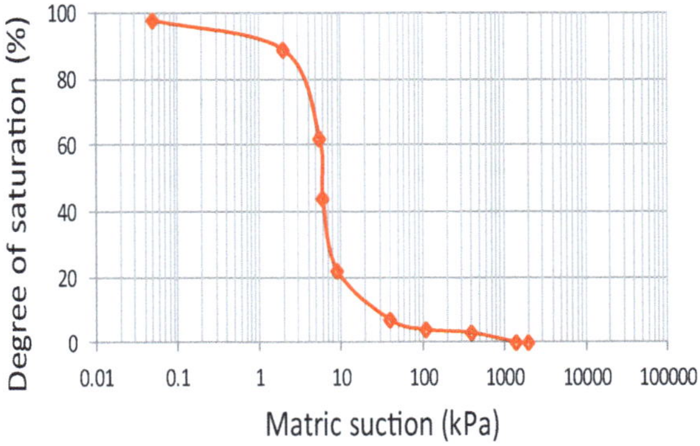

Fig. 3. Soil Water Characteristic Curve of the soil

Table 2. Soil Water Characteristic Curve important values

Air-entry value (kPa)	3.9
Residual suction (kPa)	9.1
Residual degree of saturation (%)	6
Air- entry degree of saturation (%)	92

4.1 Unsaturated Modifications

The cohesion of unsaturated soils can be modified by the following equation (Vanapalli et al. 1996):

$$C = C' + (u_a - u_w)\tan\phi' \frac{(\theta - \theta_r)}{(\theta_s - \theta_r)} \tag{6}$$

Where C is unsaturated cohesion, C' is effective cohesion, ϕ' is effective friction angle, θ_s is the water content at the state of saturation and θ_r is the residual water content.

Negative pore water pressures based on the SWCC of the soil are also applied to consider the matric suctions for the unsaturated soil. Figures 4, 5 and 6 show the compatibility of the numerical and experimental results in degrees of saturations of 4%, 16% and 60% respectively. Which leads to the conclusion the model is verified.

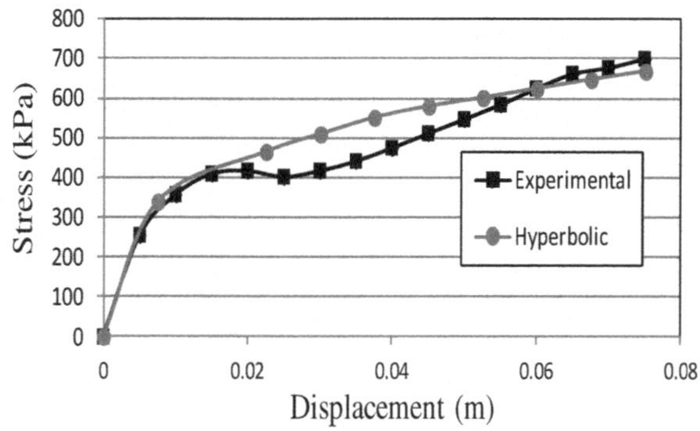

Fig. 4. Stress - displacement for the soil with the degree of saturation of 4%

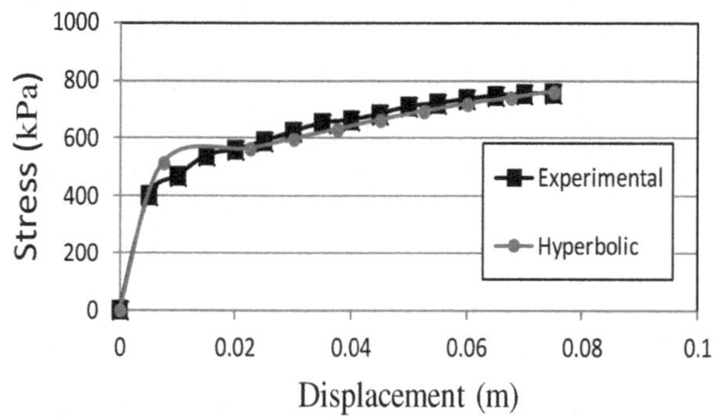

Fig. 5. Stress - displacement for the soil with the degree of saturation of 16%

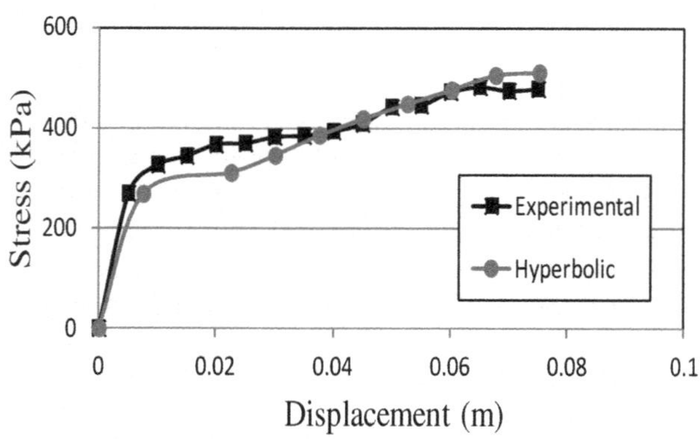

Fig. 6. Stress - displacement for the soil with the degree of saturation of 60%

Then, 187 data were extracted from the numerical model with following input sets:

- The soil degree of saturation, S_r. The degrees of saturation of 90%, 60%, 30%, 16% and 4% were considered as the first set of inputs; which are the saturation degrees verified by the existing data from laboratory.
- The column length, L.
- The column diameter (d_p)
- The footing diameter (B)
- The footing settlement. Settlements of 2, 5, 10, 17, 25, 35 and 50 mm were considered as the last input set.

The outputs were considered to be the bearing pressures of the column corresponding to each set of inputs.

5 Evaluating the Bearing Pressure

In this section three neural network models, generalized regression neural network (GRNN), feed-forward back propagation (FNN) and cascade feed-forward back propagation (CFNN) are utilized to evaluate bearing pressures corresponding to specific settlements of spread footing on composite ground reinforced with stone columns. The inputs represented were selected as input parameters for intelligent models. Predictions of models are illustrated versus experimental bearing pressure in Figs. 7, 8 and 9 for FNN, CFNN and GRNN respectively. Figures 10, 11 and 12 represent the validation regression for the methods respectively. According to these figures, all of the methods precisely capture the value of bearing pressures. Although The GRNN predicts well but there is no optimal method to improve it and it can be computationally expensive. The average error of the models GRNN, FNN and CFNN are 7, 13.5 and 12.6% respectively. Though the results show good prediction, but it has to be taken into account that training too much can result in over-training and can affect the results of the recall-data predictions.

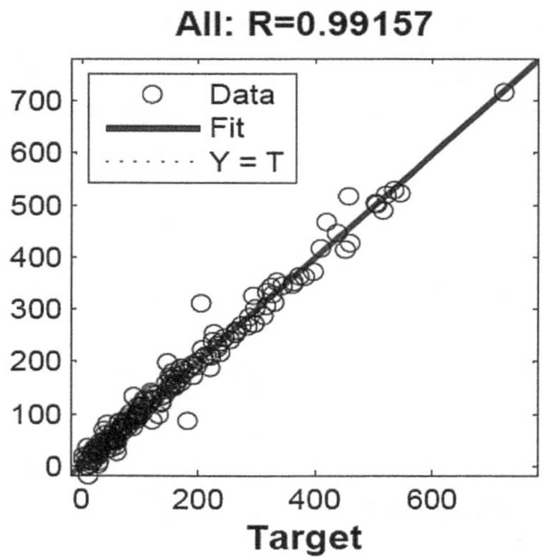

Fig. 7. Predictions of FNN versus experimental bearing pressure

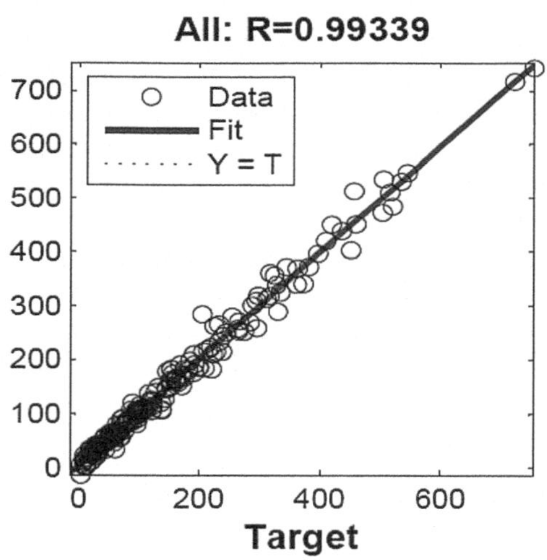

Fig. 8. Predictions of CFNN versus experimental bearing pressure

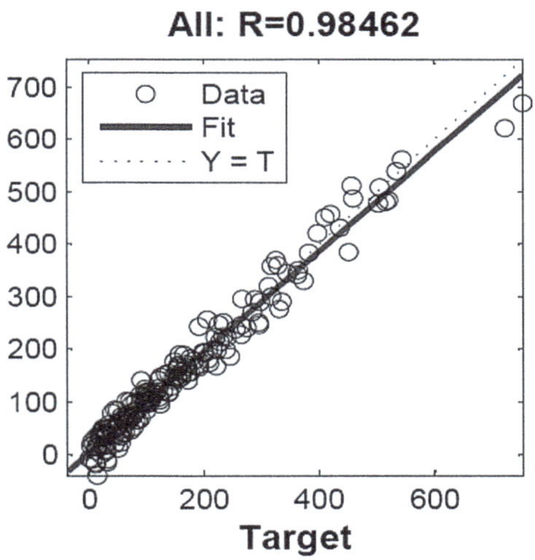

Fig. 9. Predictions of GRNN versus experimental bearing pressure

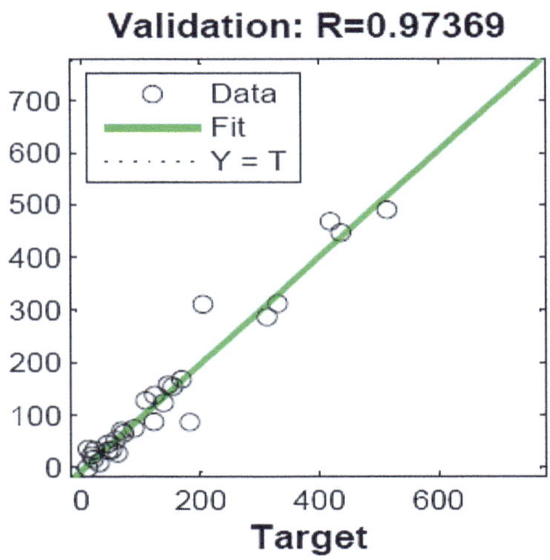

Fig. 10. Validation regression for FNN

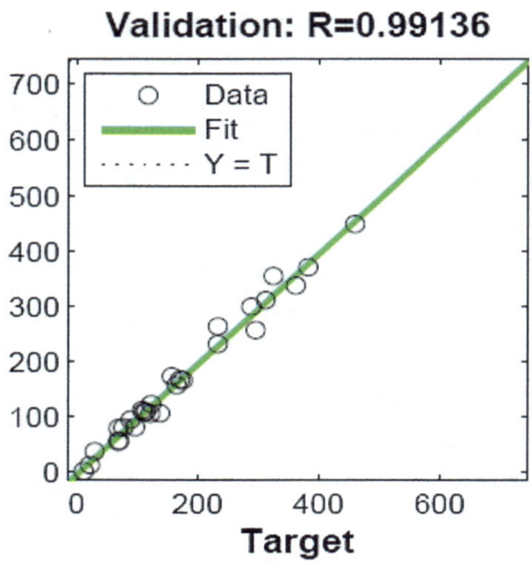

Fig. 11. Validation regression for CFNN

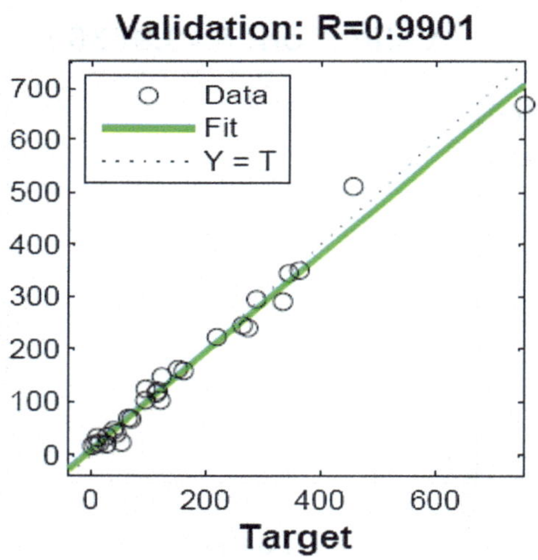

Fig. 12. Validation regression for GRNN

6 Conclusion

In current study, three neural network models were utilized to evaluate bearing pressures of spread footing on granular soil reinforced by a single stone column. The data required for training and verification of the models were collected from a developed numerical model verified with high quality results of experimental tests on a single column published in the literature. The results depict that these models can estimate the bearing pressures under certain settlements of stone columns in unsaturated media with higher accuracy than other existing methods.

References

Barbour, S.L.: Nineteenth Canadian geotechnical colloquium: the soil-water characteristic curve: a historical perspective. Can. Geotech. J. **35**(5), 873–894 (1998)

Barksdale, R.D., Bachus, R.C.: Design and Construction of Stone Columns Volume II, Appendixes. Federal Highway Administration, Washington, D.C. (1983)

Brauns, J.: Initial bearing capacity of stone column and sand piles. In: Proceedings of the Symposium on Soil Reinforcing and Stabilizing Techniques in Engineering Practise, Sydney (1978)

Mitchell, J.K.: Soil improvement-state of the art report. In: Proceedings of the 11th International Conference on SMFE (1981)

Mohammadizadeh, M., Asadi, M.: Estimation of bearing capacity and settlement of spread footings over stone-column-reinforced clay using fuzzy models and artificial neural networks. Indian J. Fundam. Appl. Life Sci. **5**(2), 3038–3050 (2015)

Sedano, J., Vanapalli, S.: Experimental investigation of the relationship between the critical state shear strength of unsaturated soils and the soil-water characteristic curve. Int. J. Geotech. Eng. **5**(1), 1–8 (2011)

Vanapalli, S.K., Fredlund, D.G., Pufahl, D.E., Clifton, A.W.: Model for the prediction of shear strength with respect to soil suction. Can. Geotech. J. **33**(3), 379–392 (1996)

Vaseghi Magvan, S., Imam, S.M., Takallou, M.A., Soroori, M.M.: Experimental study of load-settlement behaviour of stone columns in unsaturated soils. In: 9th National Congress on Civil Engineering, Islamic Republic of Iran, 11 May 2016

One Dimensional Consolidation of Composite Foundation in Multilayered System

J. Jun[1,2(✉)], Y. Minghui[3], and Z. Xinjun[3]

[1] Department of Construction Management, School of Management, Changshu Institute of Technology, Changshu 215500, China
jujun@cslg.edu.cn
[2] Institute of Geotechnical Engineering, School of Transportation, Southeast University, Nanjing 210096, China
[3] Institute of Geotechnical Engineering, Hunan University, Changsha 410082, China

Abstract. An analytical solution, which is easy to implement, for the consolidation of composite foundation in multilayered system is proposed by taking into account the pile column of impervious, and the effect of time-dependent external loads. A relevant computer program has been developed. Then numerical example is presented for multilayer clay profiles under double drainage conditions that validate the predicted excess pore pressure. 15 cases are computed for investigation of the consolidation behaviour of a composite foundation in layered system. Results show that the consolidation behaviour will be greatly effected by the pile modulus, area of replacement ratio, the coefficient of the soil permeability, pile length and construction time. An increase in the values of these parameters significantly accelerate the consolidation rate of a composite foundation. However, the soil modulus plays a less important role on the rate of consolidation.

Keywords: One dimensional solution · Composite foundation · Consolidation · Multilayered system

1 Introduction

The composite foundations reinforced by pile columns (e.g. deep mixing, concrete pile) which have great potential benefits not only to improve the bearing capacity and reduce total and differential settlements of superstructure, but also to accelerate the rate of consolidation of soft soils, have been used worldwide for highway, airport and so on. Consequently, studies on prediction of engineering properties of composite foundation still been attractive nowadays to both practitioners and researchers in Asia.

The consolidation theories of composite foundation are developed on the basis of that for sand-drained well foundations (Barron 1948). Since Yoshikuni (1974) proposed the concept of 'stress concentration' for granular column reinforced foundations, this concept has been taken into account by most of the consolidation theories of composite foundation. In the past decades, various analytical solutions have been developed for this kind of problem (Xie et al. 2001; Taiebat and Carter 2001; Cai and Xu 2002;

R. Chen et al. (eds.), *Proceedings of the 2nd International Symposium on Asia Urban GeoEngineering*, Springer Series in Geomechanics and Geoengineering,
https://doi.org/10.1007/978-981-10-6632-0_34

Wang et al. 2002; Han and Ye 2001; Han and Ye 2002; Lorenzo and Bergado 2003; Zhang et al. 2006; Lei and Liu 2008; Xie et al. 2009a,b; Lu et al. 2010). These theories have made a great contribution to the design and assessment. However, almost all the previous consolidation theories assumed that the pile column is pervious which means that the water within the surrounding soil can flow into the pile. This assumption was not reasonable for cement-mixed columns as it is known that the permeability of cement-mixed columns is much lower than surrounding soil that they can be treated as impervious (Hou and Gong 2000; Cai and Xu 2002). Obviously, the piles are impervious if they are made by concrete. On the other hand, there seems to be no analytical solution for consolidation taking the layered characteristics of soil into consideration as due to mathematical difficulty. Thus, available expressions are generally limited to simple cases. To remedy the disadvantages in previous solutions, in this paper a more rigorous theoretical solution is developed to compute the degree of consolidation of composite foundations by considering the pile column of impervious and embedded in a layered system, and the effect of time-dependent external loads. Then, the consolidation behaviour of the composite foundation is investigated.

2 Governing Equation

The unit cell of one-dimensional consolidation of composite foundation in layered system is idealized as shown in Fig. 1 which considered a pile embedded in a soil deposit that has n layers. Each layer extends to the influence distance of single pile at horizontal direction. The vertical depth to the base of any intermediate layer i is z_i, which means that the thickness of the ith layer (h_i) is $z_i - z_{i-1}$ with $z_0 = 0$ and $z_n = H$. The pile head is at the surface while the base can be embedded in any layer. Define the original point of z coordinate on the ground surface at the centre of the pile head and downwards positively. The ground surface is subjected to an external load ($q(t)$) having the initial value of q_0 and variation with construction time (t_c) which finally reach the limited value (q_u) as shown in Fig. 2. In order to obtain a simplified analytical solution of the excess pore pressure and the consolidation degree of composite foundation, the main assumptions in this analysis are summarized as follows:

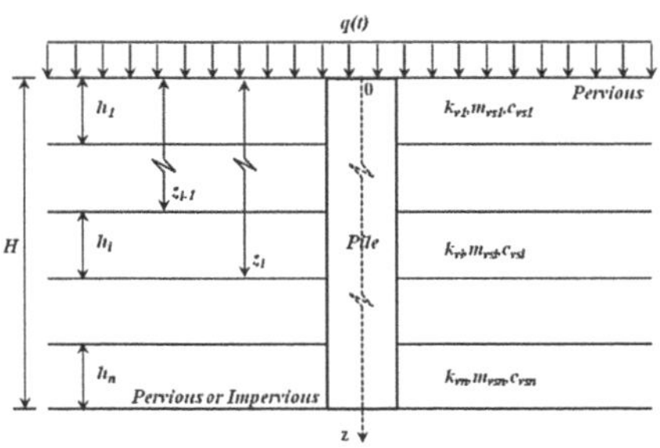

Fig. 1. Schematic diagram for the consolidation of composite foundation in multi-layered deposit

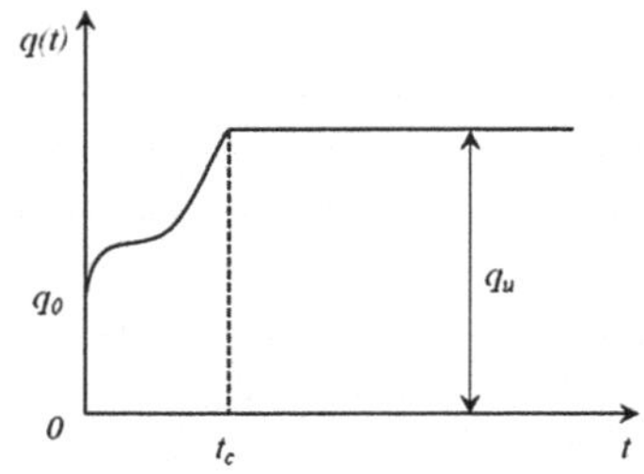

Fig. 2. External load

(1) The equal vertical strain assumption is adopted, that is, the pile and the surrounding soil only deform vertically and have an equal strain at any depth.
(2) The pile is impervious.
(3) Lateral movement, soil arch effect, and disturbance of the soil induced by construction are not considered.
(4) The total average stress in a composite foundation is assumed to be constant.
(5) At the moment of the load being applied, all the load is carried by the excess pore pressure within the surrounding soil. The initial excess pore pressure (u_{i0}) is equal to initial load (q_0).
(6) The effective stress within the soil is constant with depth.
(7) Other assumptions are the same as those used in Terzaghi's one dimensional consolidation theory.

Based on Terzaghi's one dimensional consolidation theory, the surrounding soil satisfies the following equation

$$-\frac{k_{vi}}{\gamma_w}\frac{\partial^2 u_i}{\partial z^2} = \frac{\partial \varepsilon_{vi}}{\partial t}, i = 1, 2, \cdots\cdots, n \tag{1}$$

Where k_{vi}, the vertical permeability of layer i; u_i, the excess pore pressure derived by external load within ith layer; γ_w, the unit weight of water and ε_{vi}, the vertical volumetric strain of ith layer.

At any time, both the pile column and the surrounding soil share the applied load in a composite foundation, *i.e.*

$$M_i\sigma_{pi} + (1 - M_i)\sigma_{si} = q(t) \tag{2}$$

Where σ_{pi} and σ_{si} are total stress in the pile and within the surrounding soil, respectively. And M_i is the area replacement ratio (the ratio of the pile cross-sectional area to the unit cell area).

The equal strain assumption between the pile column and the surrounding soil can be expressed as

$$\partial \varepsilon_{vi} = m_{vsi} \partial \sigma'_{si} = m_{vp} \partial \sigma_{pi} \tag{3}$$

Where σ'_{si} is the effect stress within the soil ($\sigma'_{si} = \sigma_{si} - u_i$). m_{vsi} and m_{vp} are the coefficients of vertical volumetric compressibility of soil and pile, respectively. $m_{vsi} = 1/E_{si}$ and $m_{vp} = 1/E_p$. E_{si} and E_p are the modulus of soil and pile, usually assumed to be constant.

Combining Eqs. (2) and (3), σ'_{si} can be written as

$$\sigma'_{si} = \frac{q(t) - (1 - M_i)u_i}{1 + M_i(N_i - 1)} \tag{4}$$

Where

$$N_i = m_{vsi}/m_{vp} \tag{5}$$

Differentiation of Eq. (4) gives

$$\frac{\partial \sigma'_{si}}{\partial t} = \frac{1}{1 + M_i(N_i - 1)} \left[\frac{\partial q(t)}{\partial t} - (1 - M_i) \frac{\partial u_i}{\partial t} \right] \tag{6}$$

The rate of the soil vertical volumetric strain variation with time, that is

$$\frac{\partial \varepsilon_{vsi}}{\partial t} = m_{vsi} \frac{\partial \sigma'_{si}}{\partial t} \tag{7}$$

Substitutions of Eqs. (6) and (7) into Eq. (1), the following relationship can be obtained:

$$-\frac{k_{vi}}{\gamma_w} \frac{\partial^2 u_i}{\partial z^2} = \frac{m_{vsi}}{1 + M(N_i - 1)} \left[\frac{\partial q(t)}{\partial t} - (1 - M_i) \frac{\partial u_i}{\partial t} \right] \tag{8}$$

Making use of relation

$$c_{vi} = [1 + M_i(N_i - 1)] \frac{k_{vi}}{\gamma_w m_{vsi}} = [1 + M_i(N_i - 1)] c_{vsi} \tag{9}$$

Equation (8) can be further simplified as

$$(1 - M_i) \frac{\partial u_i}{\partial t} = c_{vi} \frac{\partial^2 u_i}{\partial z^2} + \frac{\partial q(t)}{\partial t} \tag{10}$$

Here, define c_{vi}, the vertical consolidation coefficient of the composite foundation. c_{vsi} is the vertical consolidation coefficient of the ith layer.

The boundary conditions are

$$z = 0 : u_1 = 0 \tag{11}$$

$$z = H : \frac{\partial u_n}{\partial z} = 0 \ (for\, pervious\, base) \tag{12a}$$

$$u_n = 0 \ (for\, impervious\, base) \tag{12b}$$

The interface flow continuity conditions are

$$z = z_i : u_{i-1} = u_i, \tag{13a}$$

$$\left(1 - M_{(i-1)}\right)k_{v(i-1)}\frac{\partial u_{(i-1)}}{\partial z} = (1 - M_i)k_{vi}\frac{\partial u_i}{\partial z} \tag{13b}$$

The initial conditions at $t = 0$, assumed in this study to be uniform with depth, is given by

$$t = 0 : u_i = q_0 \tag{14}$$

3 Derivation of Analytical Solution

3.1 The Excess Pore Pressure

Define:

$$a_i = \frac{k_{vi}}{k_{v1}}, b_i = \frac{[1 + M_1(N_1 - 1)]m_{vsi}}{[1 + M_i(N_i - 1)]m_{vs1}}, \rho_i = \frac{h_i}{H}, \mu_i = \sqrt{\frac{(1 - M_i)b_i}{a_i}}, c_i = \frac{h_i}{h_1} \tag{15}$$

Referring to Xie et al. (2002), the solution of Eq. (10) can be written in the form of an infinite series as

$$u_i = \sum\nolimits_{m=1}^{\infty} g_{mi}(z)e^{-\beta_m t}[B_m + C_m T_m(t)] \tag{16}$$

Where

$$g_{mi}(z) = A_{mi}\sin\left(u_i\lambda_m\frac{z}{H}\right) + B_{mi}\cos\left(u_i\lambda_m\frac{z}{H}\right) \tag{17a}$$

$$\beta_m = \frac{\lambda_m^2 c_{v1}}{H^2} \tag{17b}$$

$$T_m(t) = \int_0^t e^{\beta_m \tau}\frac{\partial q(\tau)}{\partial \tau}d\tau \tag{17c}$$

β_m, B_m, C_m, A_{mi}, B_{mi} *and* λ_m are unknown coefficients to be found out; $T_m(t)$ is a function of t, derived by the external load, $q(t)$, in Eq. (6).

Submitting Eqs. (16) and (17a), (17b), (17c) into Eqs. (11) and (13a), (13b) results in

$$A_{m1} \cdot 0 + B_{m1} \cdot 1 = 0 \tag{18}$$

$$A_{mi} \sin\left(\mu_i \lambda_m \frac{z_{i-1}}{H}\right) + B_{mi} \cos\left(\mu_i \lambda_m \frac{z_{i-1}}{H}\right) = A_{m(i-1)} \sin\left(\mu_{i-1} \lambda_m \frac{z_{i-1}}{H}\right) + B_{m(i-1)} \cos\left(\mu_{i-1} \lambda_m \frac{z_{i-1}}{H}\right) \tag{19}$$

$$\begin{aligned}&(1-M_i)k_{vi}\frac{\mu_i\lambda_m}{H}\left[A_{mi}\cos\left(\mu_i\lambda_m\tfrac{z_{i-1}}{H}\right) - B_{mi}\sin\left(\mu_i\lambda_m\frac{z_{i-1}}{H}\right)\right]\\ &= (1-M_{i-1})k_{v(i-1)}\frac{\mu_{i-1}\lambda_m}{H}\left[A_{m(i-1)}\cos\left(\mu_{i-1}\lambda_m\tfrac{z_{i-1}}{H}\right) - B_{m(i-1)}sin\left(\mu_{i-1}\lambda_m\tfrac{z_{i-1}}{H}\right)\right]\end{aligned} \tag{20}$$

From Eq. (18), it can be assumed that

$$\begin{bmatrix} A_{m1} \\ B_{m1} \end{bmatrix} = \begin{bmatrix} 1 \\ 0 \end{bmatrix} \tag{21}$$

Rearranging the Eqs. (19) and (20) leads to the matrix form

$$\begin{bmatrix} \gamma_i & \theta_i \\ \theta_i & -\gamma_i \end{bmatrix} \begin{bmatrix} A_{mi} \\ B_{mi} \end{bmatrix} = \begin{bmatrix} \alpha_i & \eta_i \\ d_i\eta_i & -d_i\alpha_i \end{bmatrix} \begin{bmatrix} A_{m(i-1)} \\ B_{m(i-1)} \end{bmatrix} \tag{22}$$

Where

$$d_i = \frac{(1-M_{i-1})k_{v(i-1)}\mu_{i-1}}{(1-M_i)k_{vi}\mu_i} = \frac{(1-M_{i-1})^{\frac{3}{2}}\sqrt{a_{i-1}b_{i-1}}}{(1-M_i)^{\frac{3}{2}}\sqrt{a_i b_i}} \tag{23a}$$

$$\alpha_i = sin(\mu_{i-1}\lambda_m \frac{z_{i-1}}{H} \tag{23b}$$

$$\eta_i = cos\left(\mu_{i-1}\lambda_m \frac{z_{i-1}}{H}\right) \tag{23c}$$

$$\gamma_i = sin\left(\mu_i\lambda_m \frac{z_{i-1}}{H}\right) \tag{23d}$$

$$\theta_i = cos\left(\mu_i\lambda_m \frac{z_{i-1}}{H}\right) \tag{23e}$$

Define

$$S_i = \begin{bmatrix} \gamma_i & \theta_i \\ \theta_i & -\gamma_i \end{bmatrix}^{-1} \begin{bmatrix} \alpha_i & \eta_i \\ d_i\eta_i & -d_i\alpha_i \end{bmatrix} = \begin{bmatrix} \alpha_i\gamma_i + d_i\eta_i\theta_i & \eta_i\gamma_i\_d_i\alpha_i\theta_i \\ \alpha_i\theta_i - d_i\eta_i\gamma_i & \eta_i\theta_i + d_i\alpha_i\gamma_i \end{bmatrix} \tag{24}$$

Therefore, the coefficients A_{mi} and B_{mi} can be given by the following recurrence expression

$$\begin{bmatrix} A_{mi} \\ B_{mi} \end{bmatrix} = S_i \begin{bmatrix} A_{m(i-1)} \\ B_{m(i-1)} \end{bmatrix}, i = 2, 3, \cdots\cdots, n \tag{25}$$

Considering the boundary condition represented by Eq. (12a), (12b) gives

$$\begin{cases} A_{mn} soc(\mu_n \lambda_m) - B_{mn} \sin(\mu_n \lambda_m) = 0 & \textit{for pervious base} \\ A_{mn} sin(\mu_n \lambda_m) + B_{mn} \cos(\mu_n \lambda_m) = 0 & \textit{for impervious base} \end{cases} \tag{26}$$

Thus, S_{n+1} can be derived by Eq. (27)

$$S_{n+1} = \begin{cases} [\cos(\mu_n \lambda_m) - sin(\mu_n \lambda_m)] & \textit{for pervious base} \\ [\sin(\mu_n \lambda_m) cos(\mu_n \lambda_m)] & \textit{for impervious base} \end{cases} \tag{27}$$

Finally, the eigenvalues, λ_m, are given by solving the following characteristic equation

$$S_{n+1} \cdot S_n \cdot S_{n-1} \cdots\cdots\cdots S_2 \cdot S_1 = 0 \tag{28}$$

In which: $S_1 = \begin{bmatrix} 1 \\ 0 \end{bmatrix}$.

Therefore, the coefficient C_m can be expressed in a detailed form using the method mentioned by Schiffman and Stein (1970):

$$C_m = \frac{\sum_{i=1}^{n} (1 - M_i) m_{vi} \int_{z_{i-1}}^{z_i} g_{mi}(z) dz}{\sum_{i=1}^{n} (1 - M_i) m_{vi} \int_{z_{i-1}}^{z_i} g_{mi}^2(z) dz}$$
$$= \frac{2 \sum_{i=1}^{n} \sqrt{(1 - M_i) a_i b_i} \left[-A_{mi} \left(\eta_{i+1} - \theta_i + B_{mi} (\alpha_{i+1} - \gamma_i) \right) \right]}{\sum_{i=1}^{n} \sqrt{(1 - M_i) a_i b_i} \left[u_i \rho_i \lambda_m (A_{mi}^2 + B_{mi}^2) + (B_{mi}^2 - A_{mi}^2) \left(\alpha_{i+1} \eta_{i+1} - \gamma_i \theta_i \right) - 2 A_{mi} B_{mi} \left(\eta_{i+1}^2 - \theta_i^2 \right) \right]} \tag{29}$$

Where,

$$m_{vi} = [1 + M_i (N_i - 1)] m_{vsi} \tag{30}$$

Then, the coefficient B_m can be derived by using the boundary condition of Eq. (13a), (13b)

$$B_m = q_0 C_m \tag{31}$$

3.2 Average Degree of Consolidation

The average degree of consolidation for the composite foundation in the layered system can be expressed in terms of either excess pore pressure or settlement.

If the maximum drainage path is given by D, the average degree of consolidation defined by excess pore pressure is expressed as

$$U_p = 1 - \frac{1}{D}\int_0^D \frac{u}{u_0} dz \tag{32}$$

And defined by settlement

$$U_s = \frac{S_t}{S_\infty} \tag{33}$$

Where s_∞ is the ultimate settlement and s_t is the settlement at time t.

For layered system, Eqs. (32) and (33) become

$$U_p = \frac{\sum_{i=1}^{n} \int_{z_{i-1}}^{z_i} [q(t) - u_i] dz}{\sum_{i=1}^{n} \int_{z_{i-1}}^{z_i} q(t) dz} \tag{34}$$

$$U_s = \frac{S_{it}}{S_{i\infty}} = \frac{\sum_{i=1}^{n} \int_{z_{i-1}}^{z_i} m_{vi}[q(t) - u_i] dz}{\sum_{i=1}^{n} \int_{z_{i-1}}^{z_i} m_{vi} q(t) dz} \tag{35}$$

3.3 Solution for the Cases of Constant Loading and Ramp Loading

Based on the above theories, the solutions for two particular loading conditions: constant loading and ramp loading can be given in explicit forms.

For the case of constant loading (Fig. 3, line a),

$$q(t) = q_0 = q_u, \frac{\partial q(t)}{\partial t} = 0, T_m(t) = 0 \tag{36}$$

Then the following solutions can be obtained

$$u_i = \sum_{m=1}^{\infty} B_m g_{mi}(z) e^{-\beta_m t} \tag{37}$$

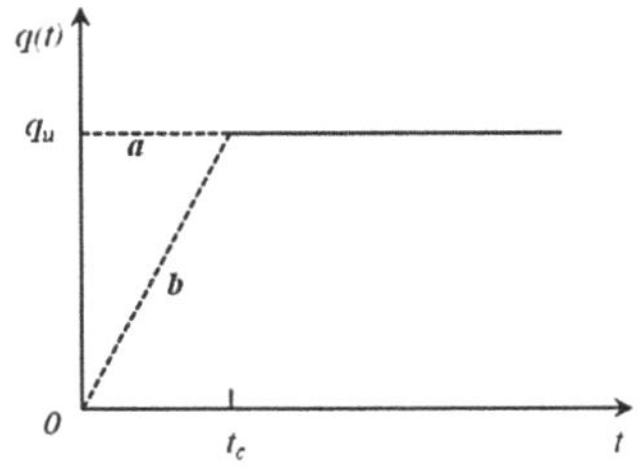

Fig. 3. Loading curve for special cases

$$U_s = 1 - \frac{\sum_{i=1}^{n} \frac{b_i c_i}{\mu_i \rho_i} D_{mi}}{q_u \sum_{i=1}^{n} b_i c_i} \tag{38}$$

$$U_p = 1 - \frac{\sum_{i=1}^{n} \frac{c_i}{\mu_i \rho_i} D_{mi}}{q_u \sum_{i=1}^{n} c_i} \tag{39}$$

Coefficient B_m *and* D_{mi} can be given by

$$B_m = \frac{2q_u \sum_{i=1}^{n} \sqrt{(1-M_i)a_i b_i}\left[-A_{mi}(\eta_{i+1}-\theta_i)+B_{mi}(\alpha_{i+1}-\gamma_i)\right]}{\sum_{i=1}^{n} \sqrt{(1-M_i)a_i b_i}\left[\mu_i\rho_i\lambda_m(A_{mi}^2+B_{mi}^2)+(B_{mi}^2-A_{mi}^2)(\alpha_{i+1}\eta_{i+1}-\gamma_i\theta_i)-2A_{mi}B_{mi}(\eta_{i+1}^2-\theta_i^2)\right]} \tag{40a}$$

$$D_{mi} = \sum_{m=1}^{\infty} \frac{B_m}{\lambda_m} e^{-\beta_m t}\left[-A_{mi}(\eta_{i+1}-\theta_i)+B_{mi}(\alpha_{i+1}-\gamma_i)\right] \tag{40b}$$

And for the case of ramp loading (Fig. 3, line b) that the external load increases linearly with time and becomes constant at time t_c. Consequently, *q(t)* can be written as

$$q(t) = \begin{cases} \frac{t}{t_c} q_u & 0 \le t \le t_c \\ q_u & t > t_c \end{cases} \tag{41}$$

Therefore,

$$\frac{\partial q(t)}{\partial t} = \begin{cases} \frac{q_u}{t_c} & 0 \le t \le t_c \\ 0 & t > t_c \end{cases} \tag{42a}$$

$$B_m = 0 \tag{42b}$$

$$T_m(t) = \int_0^t e^{\beta_m \tau} \frac{\partial q(\tau)}{\partial \tau} d\tau = \begin{cases} \frac{q_u}{\lambda_m^2 T_{vc}}\left(e^{\lambda_m^2 T_v} - 1\right) & 0 \le t \le t_c \\ \frac{q_u}{\lambda_m^2 T_{vc}}\left(e^{\lambda_m^2 T_{vc}} - 1\right) & t > t_c \end{cases} \tag{42c}$$

Thus, the solutions for this condition can be derived as

$$u_i = \begin{cases} q_u \sum_{m=1}^{\infty} \frac{C_m}{\lambda_m^2 T_{vc}} g_{mi}(z)\left(1 - e^{-\lambda_m^2 T_v}\right) & 0 \le t \le t_c \\ q_u \sum_{m=1}^{\infty} \frac{C_m}{\lambda_m^2 T_{vc}} g_{mi}(z) e^{-\lambda_m^2 T_v}\left(e^{-\lambda_m^2 T_{vc}} - 1\right) & t \ge t_c \end{cases} \tag{43}$$

$$U_s = \begin{cases} \frac{T_v}{T_{vc}} - \frac{\sum_{i=1}^{n} \frac{b_i c_i}{h_i} \sum_{m=1}^{\infty} \frac{C_m E_{mi}}{\lambda_m^2 T_{vc}}\left(1 - e^{-\lambda_m^2 T_v}\right)}{\sum_{i=1}^{n} b_i c_i} & 0 \le t \le t_c \\ 1 - \frac{\sum_{i=1}^{n} \frac{b_i c_i}{h_i} \sum_{m=1}^{\infty} \frac{C_m E_{mi}}{\lambda_m^2 T_{vc}} e^{-\lambda_m^2 T_v}\left(e^{-\lambda_m^2 T_{vc}} - 1\right)}{\sum_{i=1}^{n} b_i c_i} & t > t_c \end{cases} \tag{44}$$

$$U_p = \begin{cases} \frac{T_v}{T_{vc}} - \frac{\sum_{i=1}^{n} \frac{c_i}{h_i} \sum_{m=1}^{\infty} \frac{C_m E_{mi}}{\lambda_m^2 T_{vc}} \left(1 - e^{-\lambda_m^2 T_v}\right)}{\sum_{i=1}^{n} b_i c_i} & 0 \le t \le t_c \\ 1 - \frac{\sum_{i=1}^{n} \frac{c_i}{h_i} \sum_{m=1}^{\infty} \frac{C_m E_{mi}}{\lambda_m^2 T_{vc}} e^{-\lambda_m^2 T_v} \left(e^{-\lambda_m^2 T_{vc}} - 1\right)}{\sum_{i=1}^{n} b_i c_i} & t > t_c \end{cases} \quad (45)$$

Where

$$C_m = \frac{2\sum_{i=1}^{n} \sqrt{(1-M_i)a_i b_i}\left[-A_{mi}\left(\eta_{i+1} - \theta_i + B_{mi}(\alpha_{i+1} - \gamma_i)\right)\right]}{\sum_{i=1}^{n} \sqrt{(1-M_i)a_i b_i}\left[u_i \rho_i \lambda_m (A_{mi}^2 + B_{mi}^2) + (B_{mi}^2 - A_{mi}^2)(\alpha_{i+1}\eta_{i+1} - \gamma_i \theta_i) - 2A_{mi}B_{mi}(\eta_{i+1}^2 - \theta_i^2)\right]} \quad (46a)$$

$$E_{mi} = \int_{z_{i-1}}^{z_i} g_{mi}(z)dz = \frac{H}{u_i \lambda_m}\left[-A_{mi}\left(\eta_{i+1} - \theta_i\right) + B_{mi}(\alpha_{i+1} - \gamma_i)\right] \quad (46b)$$

$$T_v = \frac{c_{v1} t}{H^2}, T_{vc} = \frac{c_{v1} t_c}{H^2} \quad (46c)$$

4 Computation Procedure

A computer program has been developed for one-dimensional consolidation analysis of composite foundation in layered system based on above solutions. The main steps in this program are as follows

(a) Choose input parameters n, k_{vi}, E_{si}, γ_w, h_i, M_i, E_{vp}, $q(t)$ and t
(b) Calculate derived parameters N_i, m_{vi}, c_{vi}, H, a_i, b_i, ρ_i, μ_i, c_i and d_i from Eqs. (5), (9), (15), (23a) and (30).
(c) Evaluate sufficient terms in the infinite series in Eq. (28) to obtain an accuracy value of λ_m, only a few terms will usually be sufficient.
(d) Calculate the coefficients α_i, η_i, γ_i, θ_i, A_{mi} and B_{mi} from Eqs. (23a), (23b), (23c), (23d), (23e) and (25).
(e) Calculate the coefficient C_m from Eq. (31)
(f) For constant loading, namely $T_m(t) = 0$, coefficients B_m and D_{mi} can be derived from Eq. (40a), (40b). Then the excess pore pressure and the average degree of consolidation for a composite foundation can be estimated according to Eqs. (37) to (39).
(g) For ramp loading, namely $B_m = 0$, evaluate coefficients T_v, T_{vc}, $T_m(t)$, C_m, E_{mi}, from Eqs. (42a), (42b), (42c) and (46a), (46b), (46c). Finally, the excess pore pressure and the average degree of consolidation for a composite foundation can be calculated by Eqs. (43) to (45).

5 Numerical Verification

Let $N_i = 1$ and $M_i = 0$ ($i = 1,2,\ldots\ldots, n$), which means there is no pile in the foundation. Then the composite foundation in layered system becomes the problem only considered the consolidation of the multilayered soil and the solutions should be equivalent to those given by Lee [18]. To further validate the present solutions, a four-layered soil is investigated whose related parameters are listed in Table 1. Schiffman and Stein (1970) gave the analytical results to this four-layered soil with free drainage at both top and bottom boundaries, in which the applied load is constant with time and of a unit magnitude. The first 10 eigenvalues computed from characteristic Eq. (28) are listed in Table 2. It can be seen that the results shown in Fig. 4 are the same between the Schiffman and Stein's solution and present solution which proved the validity of the method proposed in this study.

Table 1. Parameters of four-layered system

Layer No.	1	2	3	4
h_i(m)	3.05	6.1	9.14	6.1
k_{vi} (10^{-11}m/s)	2.78	8.25	1.17	2.94
E_{si}(MPa)	15.6	24.51	49.02	24.51

Table 2. The first 10 eigenvalues

m	λ_m	m	λ_m
1	5.0375	6	24.3747
2	7.4728	7	29.0247
3	11.5142	8	32.4759
4	17.7089	9	37.2490
5	20.8461	10	41.5802

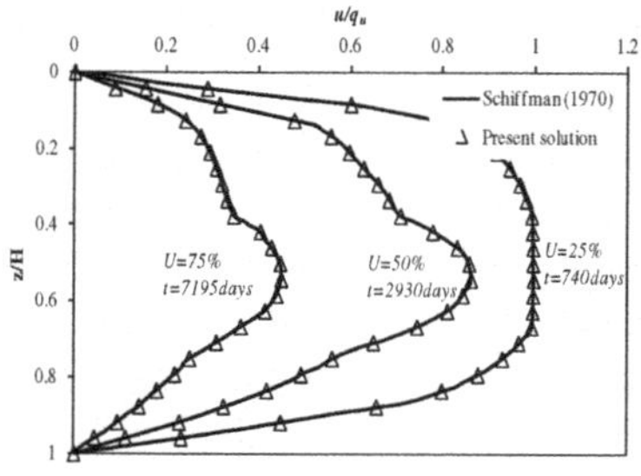

Fig. 4. Excess pore pressure isochrone computed by Schiffman (1970) and present solution

6 Consolidation Behavior of Composite Foundation in Layered System

The above solutions indicate that the parameters E_p, M_i, k_{vi}, L_p/H and t_c govern the consolidation behavior of composite foundation in layered system. 15 cases have been computed using the program developed to investigate how these parameters effect the consolidation behavior of the composite foundation. The ultimate load for all the cases is the same and have a value of 150 kPa. In this study only the average degree of consolidation defined in terms of settlement (U_s) is calculated. Table 3 shows the basic parameters used in this analysis. It is assumed that the homogeneous isotropic soil deposits have perfectly pervious surface at the top and bottom ends of the clay. Figures 5, 6, 7, 8 and 9 are under the condition of constant loading while Fig. 10 is corresponding to the ramp loading situation. The results from the consolidation of natural soil deposits by setting $N_i = 1$ and $M_i = 0$ (i = 1, 2, 3, 4) which referred as case 0, is also plotted in these figures.

Table 3. Geotechnical parameters for case study

Layer no.	1	2	3	4
h_i (m)	3	6	9	6
k_{vi} (10^{-9}m/s)	2	5	1	8
E_{si} (MPa)	8	5	3	15
E_p (Mpa)	200			
q_u (kPa)	150			
M	0.15			

Figure 5 shows the influences of the pile modulus on consolidation behavior of a composite foundation with four layered system, corresponding to E_p = 50, 100, 200, 400, respectively. As can be expected, the U_s–t curve moves generally to the left as the pile modulus increases indicating that higher pile modulus results in higher rate of consolidation and can significantly increase the consolidation rate for the whole system.

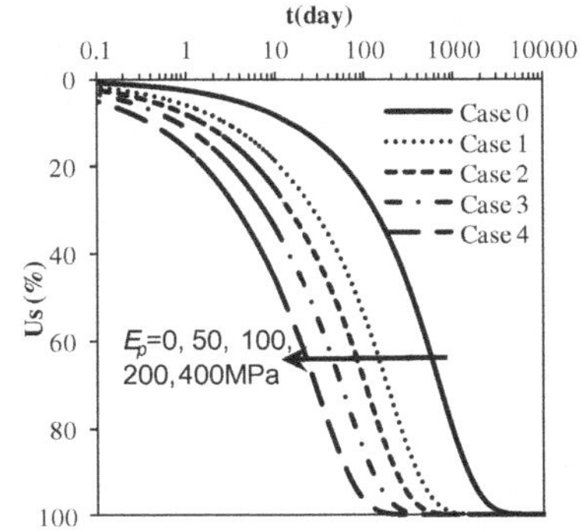

Fig. 5. Influence of pile modulus on the rate of consolidation

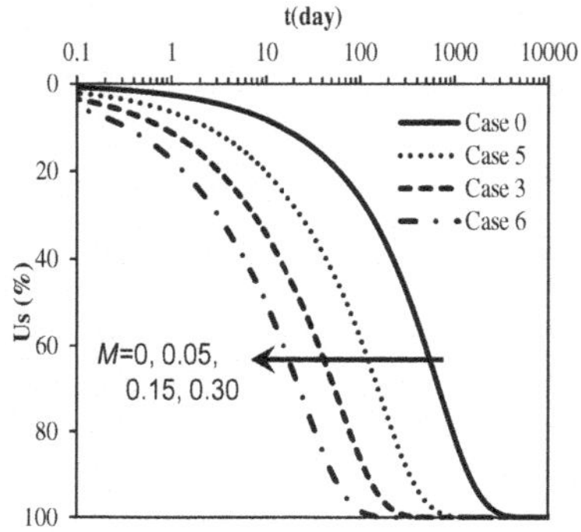

Fig. 6. Influence of replacement ratio on the rate of consolidation

The value of M reflect the area percentage of the soil replaced by the pile in unit cell. Figure 6 shows the variation of this parameter will significantly effect the consolidation rate for the whole composite foundation. It can be obtained that the larger the value of M, the more rapidly the consolidation for the composite foundation. This is because, in such condition, more external load will be transferred to the pile body according to Eq. (2). Thus, less load will be carried by the surrounding soil which increases the dissipation of the excess pore pressure within the soil.

Table 4. Values of the coefficient of permeability for case 7 and case 8

Case no.	Layer no.	1	2	3	4
Case 7	k_{vi} (10^{-10}m/s)	10	30	50	80
Case 8	k_{vi} (10^{-10}m/s)	10	8	5	0.1

The coefficient of the permeability of soil decreases downwards layer by layer in case 7, while in case 8, increases downwards layer by layer (Table 4). In case 9 and 11 both the coefficient of the permeability of the soil and the soil modulus are downwards layer by layer while the difference between the two cases is the pile modulus (Table 5). Figure 7 indicates that the coefficient of the permeability of soil layer will greatly effect the consolidation behavior of the composite foundation in layered system. An increase in the value of this parameter for each layer enhances the consolidation rate. Compare to Fig. 7, it can be seen from Fig. 8 that the average consolidation degree for the whole composite foundation derived from case 9 is similar to that of case 8 as the soil modulus for each layer decreases although the coefficient of permeability of the soil layer is decreased to the same magnitude. With the same situation but reduce the pile modulus to a value of 50 MPa (case 10 and 11), as shown in Fig. 8, case 11 only slightly reduce the consolidation rate comparison with case 10. It therefore implies that the coefficient of soil permeability plays a more important role than the soil modulus on the rate of consolidation for composite foundation with layered system. This trend is more obvious for larger E_p value.

Table 5. Parameters used in cases 8 to case 11

Case no.	Layer no.	1	2	3	4
Case 8	k_{vi} (10^{-10}m/s)	10	8	5	1
	E_{si} (MPa)	8	5	3	15
	E_p (MPa)	200			
Case 9	k_{vi} (10^{-10}m/s)	10	8	5	1
	E_{si} (MPa)	6	4	2	1
	E_p (MPa)	200			
Case 10	k_{vi} (10^{-10}m/s)	10	8	5	1
	E_{si} (MPa)	8	5	3	15
	E_p (MPa)	50			
Case 11	k_{vi} (10^{-10}m/s)	10	8	5	1
	E_{si} (MPa)	6	4	2	1
	E_p (MPa)	50			

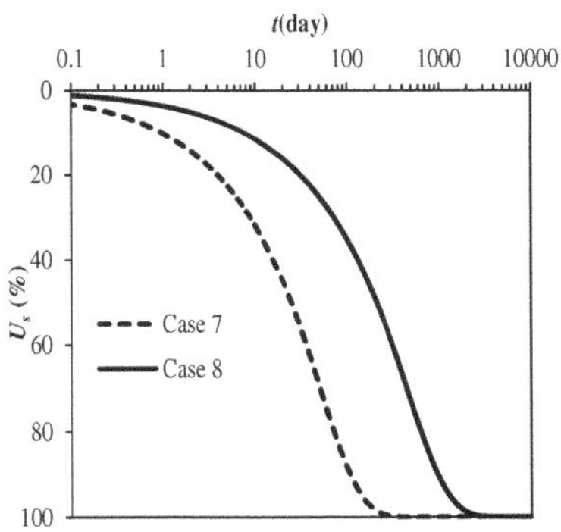

Fig. 7. Influence of the vertical permeability of the soil on the rate of consolidation

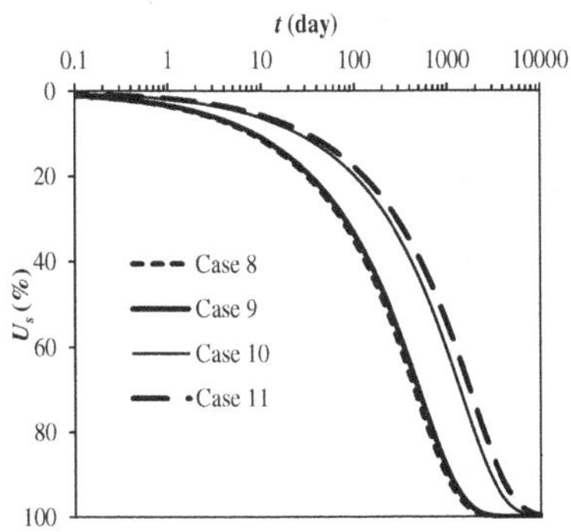

Fig. 8. Influence of the vertical permeability of the soil and soil modulus on the rate of consolidation

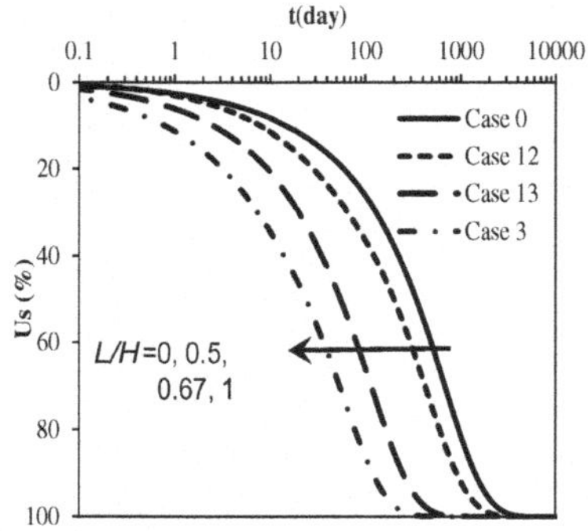

Fig. 9. Influence of pile length on the rate of consolidation

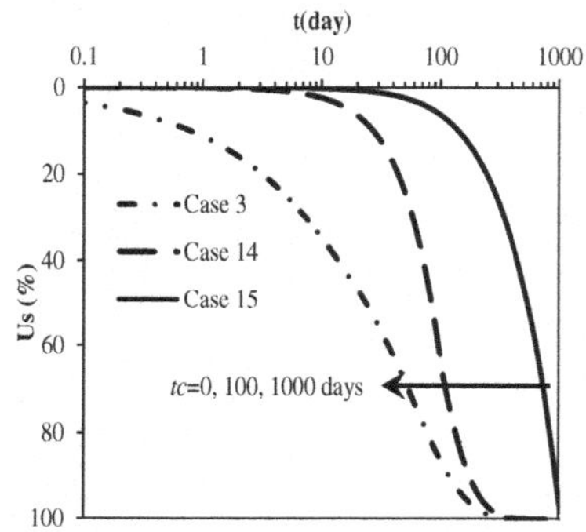

Fig. 10. Influence of the construction time on the rate of consolidation

The influence of pile length (L_p) is plotted in Fig. 9, from which the following conclusions are drawn: the value of L_p/H will significantly effect the consolidation rate of a composite foundation. It can be seen that the consolidation is greatly accelerated with an increase in the value of L_p/H. In the case of $L_p/H = 0$, which means no pile in the foundation, the consolidation rate is at its minimum. While in the case of $L_p/H = 1$, which indicates the pile is fully penetrated into the layered soil, the consolidation rate is at its maximum.

In Fig. 10, $t_c = 0$ indicates that the external load is constant load which means the ultimate load applied instantaneously, the consolidation rate is at its maximum in this case. Compare to the condition of $t_c = 0$, the consolidation ratio is much more slowly when $t_c = 1000\ d$ for the situation of ramp loading.

7 Conclusion

In this study, an efficient general solution and elegant computer program have been developed for the composite foundation consolidation by taking into account multi-layered soil and construction time effects. Comparison to the existing solutions verified that the present solution for this kind of problem is reasonable. The consolidation behavior of a composite foundation in layered system can be effected by the pile modulus, area of replacement ratio, the coefficient of the soil permeability, soil modulus, pile length and construction time. Results show that an increase in the values of E_p, M, k_{vi}, L_p/H and t_c significantly accelerate the consolidation rate of a composite foundation. However, the soil modulus plays a less important role on the rate of consolidation. This trend is more obvious for larger E_p value.

Acknowledgement. Majority of the work presented in this paper was funded by the Natural Science Foundation of China (Grant No. 51608059).

References

Barron, R.A.: Consolidation of fine-grained soils by sand drain wells. Trans. ASCE **113**, 718–742 (1948)

Cai, Y.Q., Xu, C.J.: Consolidation behaviour of cement- and lime-cement-mixed column foundations. J. Zhejiang Univ. Sci. A **36**(5), 507–512 (2002)

Han, J., Ye, S.L.: Simplified method for consolidation rate of stone column reinforced foundations. J. Geotech. Geoenvironmental Eng. **127**(7), 597–603 (2001)

Han, J., Ye, S.L.: A theoretical solution for consolidation rates of stone column reinforced foundations accounting for smear and well resistance effects. Int. J. Geomech. **2**(2), 135–151 (2002)

Hou, Y.F., Gong, X.N.: The permeability of cement-mixed material. J. Zhejiang Univ. **34**(2), 189–193 (2000). (in Chinese)

Lei, C., Liu, S.Y.: Consolidation calculation of soft ground improved by T-shape deep mixing. Geosustainability and Geohazard Mitig. **178**, 620–627 (2008). ASCE, GSP

Lorenzo, G.A., Bergado, D.T.: New consolidation equation for soil-cement pile improved ground. Can. Geotech. J. **40**(2), 265–275 (2003)

Lu, M.M., Xie, K.H., Guo, B.: Consolidation theory for a composite foundation considering radial and vertical flows within the column and the variation of soil permeability within the disturbed soil zone. Can. Geotech. J. **47**(2), 207–215 (2010)

Schiffman, R.L., Stein, J.R.: One-dimensional consolidation of layered systems. J. Soil Mech. Found. Div. **96**, 1499–1504 (1970). ASCE

Taiebat, H.A., Carter, J.P.: A semi-analytical finite element method for three-dimensional consolidation analysis. Comput. Geotech. **28**(1), 55–78 (2001)

Wang, R.C., Xie, K.H., Guan, S.H.: Analytical solutions for consolidation of composite ground with granular columns under time dependent loading. J. Zhejiang Univ. Sci. A **36**(1), 12–17 (2002)

Xie K.H., Gao P., Xie X.Y.: Consolidation theory for soft clays reinforced by cement of granular columns. In: Proceedings of the 10th International Conference of the Association for Computer Methods and Advances in Geomechanics, America, Tucson, pp. 1263–1270 (2001)

Xie, K.H., Xie, X.Y., Jiang, W.: A study on one-dimensional nonlinear consolidation of double-layered soil. Comput. Geotech. **29**(2), 151–168 (2002)

Xie, K.H., Lu, M.M., Hu, A.F., Chen, G.H.: A general theoretical solution for the consolidation of a composite foundation. Comput. Geotech. **36**(1–2), 24–30 (2009a)

Xie, K.H., Lu, M.M., Liu, G.B.: Equal strain consolidation for stone columns reinforced foundation. Int. J. Numer. Anal. Methods Geomech. **33**(15), 1721–1735 (2009b)

Yoshikuni, H., Nakanodo, H.: Consolidation of soils by vertical drain wells with finite permeability. Soils Found. **14**(2), 36–42 (1974)

Zhang, Y.G., Xie, K.H., Wang, Z.: Consolidation analysis of composite ground improved by granular columns considering variation of permeability coefficient of soil. Ground Modif. Seismic Mitig. **152**, 135–142 (2006). ASCE, GSP

Performance of Raft Foundation for a High-Rise Building

Kheng-Boon Chin[1(✉)], Yoshikatsu Ogawa[1], and Peng-Boon Ng[2,3]

[1] Overseas Department, Kiso-Jiban Consultants Co., Ltd., Singapore city, Singapore
CHIN_kb@kiso.com.sg
[2] Kiso-Jiban Singapore Pte. Ltd., Singapore city, Singapore
[3] KCL Consultants Pte. Ltd., Singapore city, Singapore

Abstract. In Singapore, high-rise buildings are predominantly supported by piles or piled raft foundations. Only very limited high-rise-buildings are supported by un-piled raft foundations. Amongst the scarce amount of un-piled raft foundations, almost all of them have a deep embedment depth with at least several basement levels. This paper aims to present the performance of an un-piled raft foundation with shallow embedment depth for a 30-storey high-rise building. Close monitoring of building settlement has been carried out during and post construction. The recorded maximum settlement of the 30-storey Tower One was on the order of 20 mm. This value was well within the design expectation and found to be very satisfactory. Back analysis has been carried out to determine the elastic modulus using finite element method program, known as Plaxis 3D. The back-analysed elastic modulus of the competent residual soils was found to be on the order of 500 MPa and compared with the elastic modulus derived from extensive in situ field tests (plate load and pressuremeter) and laboratory tests (oedometer and triaxial). The results showed that the second loading cycle of field tests generally give the more reasonable estimate as compared to the laboratory tests. Using limited data, an empirical relationship between elastic modulus and standard penetration testing blow counts has been proposed under small-strain foundation loading. Finally, raft foundation is recommended for future high-rise buildings to be constructed on competent residual soils.

Keywords: Raft foundation · High-rise building · Finite element · Plaxis 3D · Residual soils · Site investigation

1 Introduction

In Singapore, high-rise buildings are predominantly supported by piles or piled raft foundations. Only very limited high-rise-buildings are supported by un-piled raft foundations. These are probably due to complex heavy loadings, unfavourable ground conditions and authority's stringent requirements. Amongst the scarce amount of un-piled raft foundations, almost all of them have a deep embedment depth with at least several basement levels. This paper aims to present the performance of an un-piled raft foundation with shallow embedment depth for a 30-storey high-rise building.

R. Chen et al. (eds.), *Proceedings of the 2nd International Symposium on Asia Urban GeoEngineering*, Springer Series in Geomechanics and Geoengineering,
https://doi.org/10.1007/978-981-10-6632-0_35

The high-rise building known as Tower One, is located at the fringe of Southern Singapore. The construction took slightly more than 2 years to be completed. Close monitoring of building settlement has been carried out up to six months post construction. The observed settlement was back-analysed and the deformation parameter was compared with the elastic modulus derived from field and laboratory tests.

2 High-Rise Building

The commercial building is a 30-storey high tower block (known as Tower One) of reinforced concrete design, with no basement. A typical frame of Tower One is presented in Fig. 1. The Tower One is about 55 m by 77 m in plan and 150 m in height. The building loads to be transferred to the foundations comprised 623 MN in corewall and 1347 MN in columns. The raft foundation has a thickness of 3.4 m, with full embedment in the ground.

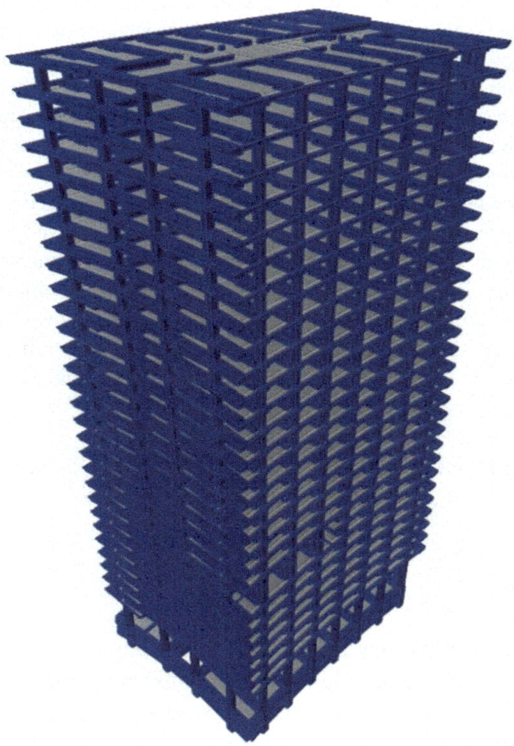

Fig. 1. Typical frame of 30-storey Tower One (Courtesy of KK Lim & Associates)

3 Ground Condition

During detailed design, a total of eleven exploration boreholes has been drilled nearby and within the footprint of the 30-storey Tower One to assess the ground condition. Some of the borehole drilling depths were up to 83 m, conforming to the Eurocode, where a depth of 1.5B (i.e., B is the width of raft) is required. The field exploration tests revealed that the subsurface soil of the site consisted of two layers, namely a thin layer of fill top soil and underlain by residual soils, originated from sedimentary rock (locally known as Jurong Formation rocks). The underlying rocks encountered were of sandstone, mudstone, siltstone and tuff. A typical soil profile within the Tower One's footprint is presented in Fig. 2. Based on drilling record, the groundwater was approximately 1 m below ground surface.

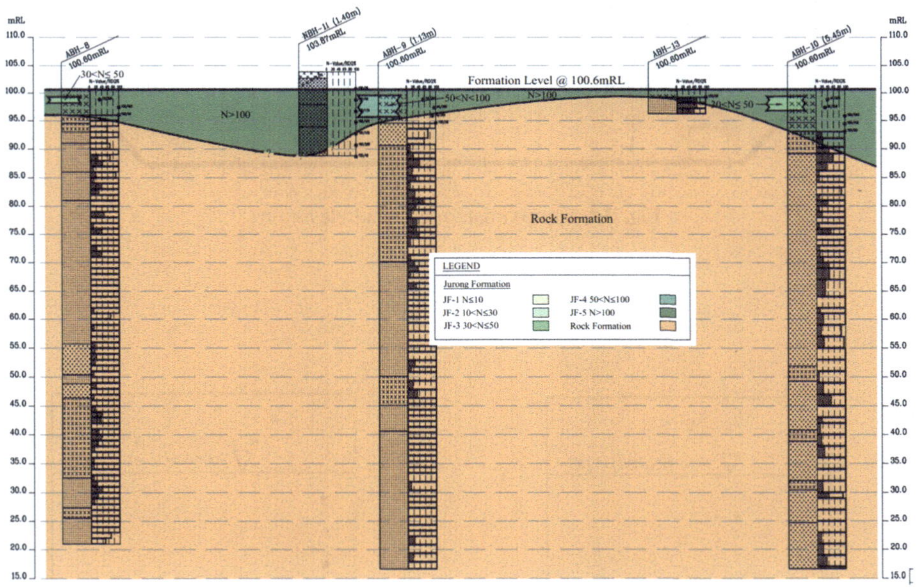

Fig. 2. Typical soil profile across the Tower One's footprint

The topography of the project site is relatively flat with an average ground elevation of +104.0 mRL while the proposed raft soffit was at +100.6 mRL. Based on the soil profile, it can be seen that the raft foundation was directly supported by hard residual soil, predominantly with high standard penetration testing blow counts, SPT-N > 100, with thickness ranging from 1.4 m to 7.6 m, overlying the sedimentary bedrocks. An approximate rock head elevation with the Tower One's footprint is illustrated in Fig. 3.

Figure 4a shows the SPT-N plot with elevation for residual soils with N $\leq$ 100. It can be seen that at relatively shallow depth, localised thin layer of residual soils with SPT-N of about 31 to 56 (35 in average) was encountered in boreholes ABH-8, ABH-9 and ABH-10. At greater depth, the residual soils had a SPT-N close to 100 with an

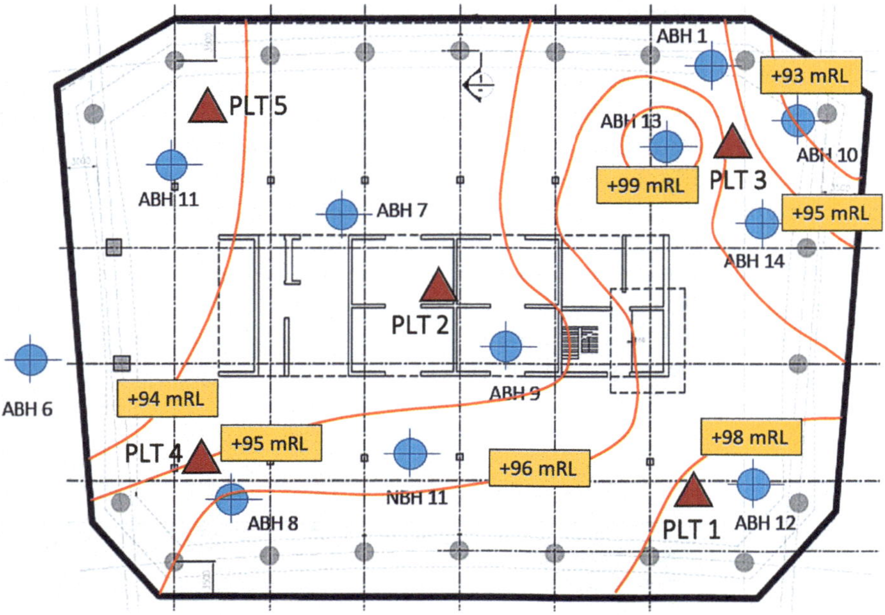

Fig. 3. Approximate rock head elevation

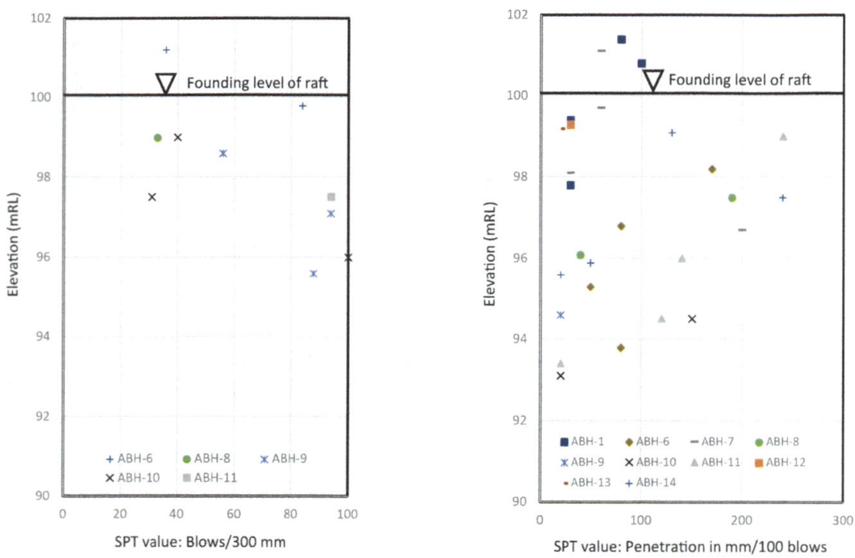

(a) SPT-N plot with elevation for N ≤100

(b) SPT penetration depth for N > 100

Fig. 4. Profile of standard penetration testing with elevation average value of 90. For residual soil with SPT-N > 100, the SPT values in penetration depth over the 100 blows was plotted with elevation (Fig. 4b), in order to reduce the scattered data. Similar plot has been adopted by Leung et al. (1988). The penetration depth was ranging from 20 mm to 240 mm, with an average value of about 100 mm. When extrapolated, this penetration depth corresponds to a SPT-N value of 300.

The basic soil properties of the residual soil is given in Table 1. It can be seen that the water content was ranging from 5.7% to 13.1% and the unit weight of the residual soil is as high as 23.0 kN/m^3 with a void ratio of 0.22. The residual soils are generally classified as hard silt. Although the site investigation revealed that thin pocket lenses of very stiff residual soils were present below Tower One, however, it was judged that the wide, thick and relatively rigid raft system was able to bridge over the slightly "less competent" (30 < SPT-N < 100) hard lenses. Based on our analysis, the hard residual soils have an allowable bearing capacity of more than 1000 kPa, which exceeds the maximum design bearing pressure of 900 kPa. Thus, it is recognised that settlement was a critical factor under the serviceability load of the raft foundation.

Table 1. Basic soil properties of residual soils with SPT-N > 100

Properties	Values
Water content, w (%)	5.7–13.1
Specific gravity	2.68–2.80
Unit weight, γ (kN/m^3)	19.5–23.0
Void ratio	0.22–0.56
Atterberg Limits:	
– Liquid Limit, LL (%)	36–45
– Plastic Limit, PL (%)	18–22
– Plasticity index, PI (%)	18–23
Grain size:	
– Sand	23–25
– Silt	54–60
– Clay	15–23

4 Evaluation of Deformation Parameter

4.1 Plate Load Test

A total of five plate load tests have been carried out near the founding level of raft foundation. The test has been loaded up to 2.5 times of the design allowable bearing capacity. Three loading cycles have been included in this test. Two typical pressure-settlement curves are presented in Figs. 5a and b, representing the indicative

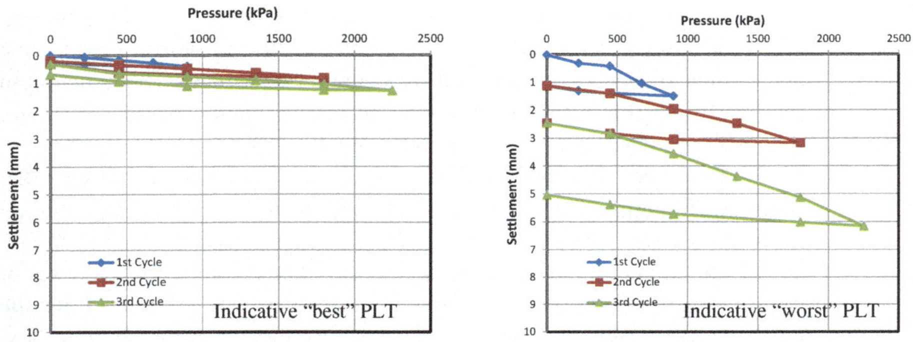

(a) Pressure-settlement curve for PLT 2 (b) Pressure-settlement curve for PLT 5

Fig. 5. Typical pressure-settlement curve using plate loading test

"best" and "worst" ground conditions, respectively. The elastic modulus has been obtained using the solution reported by Poulos and Davis (1974). The equation is given as follows:

$$E = \frac{\pi q r}{2\delta}\left(1 - v^2\right) \tag{1}$$

where q is the applied pressure, r is the radius of the plate, δ is the settlement and υ is the Poisson's ratio of the soil and taken as 0.33. Using the working stress of 900 kPa, the elastic modulus derived from plate load test for the 1st, 2nd and 3rd cycle were ranging from 74 MPa to 482 MPa (189 MPa in average), 161 MPa to 665 MPa (302 MPa in average) and 125 MPa to 1930 MPa (630 MPa in average), respectively.

4.2 Pressuremeter Test

A total of four pressuremeter tests have been carried out on residual soil with SPT-N > 100. An OYO monocell pressuremeter has been used in this site investigation campaign. A test depth ranging from 2.5 m to 7 m has been investigated in boreholes ABH-8, ABH-10 and ABH-11. Figure 6 shows the pressuremeter modulus of the hard residual soil. It can be seen that the pressuremeter modulus was ranging from 40 MPa to 230 MPa (135 MPa in average) and 160 MPa to 700 MPa (435 MPa in average) for 1st cycle and 2nd cycle, respectively. It should be noted that in some tests, the failure pressure cannot be determined, as it was well beyond the pressure capacity measurement (i.e. 5 MPa) of pressuremeter.

4.3 Oedometer Test

A total of four consolidation tests have been carried out using oedometer. The specimens were subject to two loading cycles. A typical oedometer loading curve is shown in Fig. 7. The oedometer modulus, E_{oed}, or known as constrained modulus, has also been determined at the effective vertical stress of interest. Based on theory of elasticity,

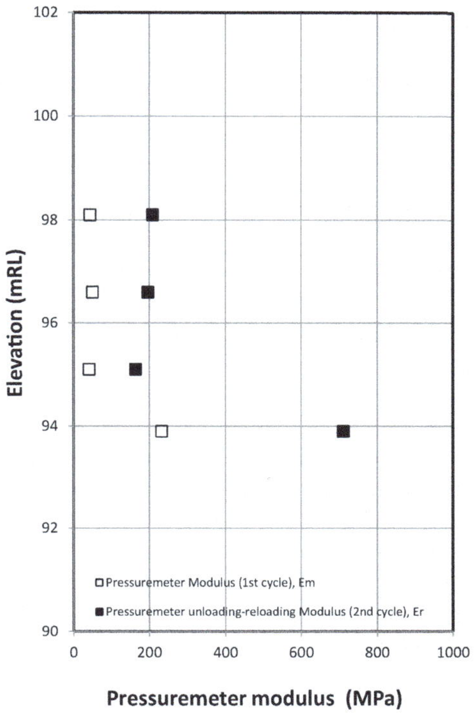

Fig. 6. Pressuremeter modulus with elevation

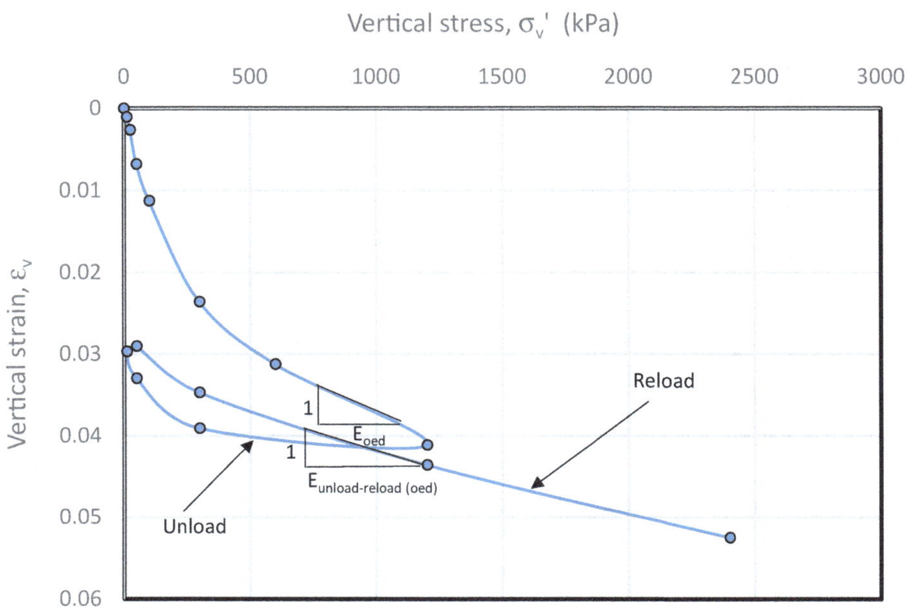

Fig. 7. Typical oedometer loading curves with modulus determination

the relationship between elastic modulus and oedometer modulus can be written as follows:

$$E = \frac{E_{oed}(1 + \nu)(1 - 2\nu)}{(1 - \nu)} \quad (2)$$

where υ is the Poisson's ratio. Under working load condition, the operational υ is in the range from 0.1 to 0.2 (Tatsuoka et al. 1994; Lo Presti et al. 1995). Using a Poisson ratio of 0.15, the elastic modulus of the residual soil was ranging from 27 MPa to 86 MPa while the unload-reload elastic modulus was ranging from 52 MPa to 181 MPa under an average effective vertical stress, σ_v' of 900 kPa.

4.4 Triaxial Test

Three single-stage consolidated drained (CD) triaxial tests have been carried out on the hard residual soil. The soil specimens were loaded at a slow speed with unload-reload cycle up to an axial strain of about 20%. Figure 8 shows the typical triaxial test with unload-reload cycle with elastic modulus determination. The secant elastic modulus derived from triaxial test was ranging from 21 MPa to 99 MPa while the unload-reload modulus was ranging from 151 MPa to 354 MPa, under effective confining pressures ranging from 300 kPa to 1200 kPa.

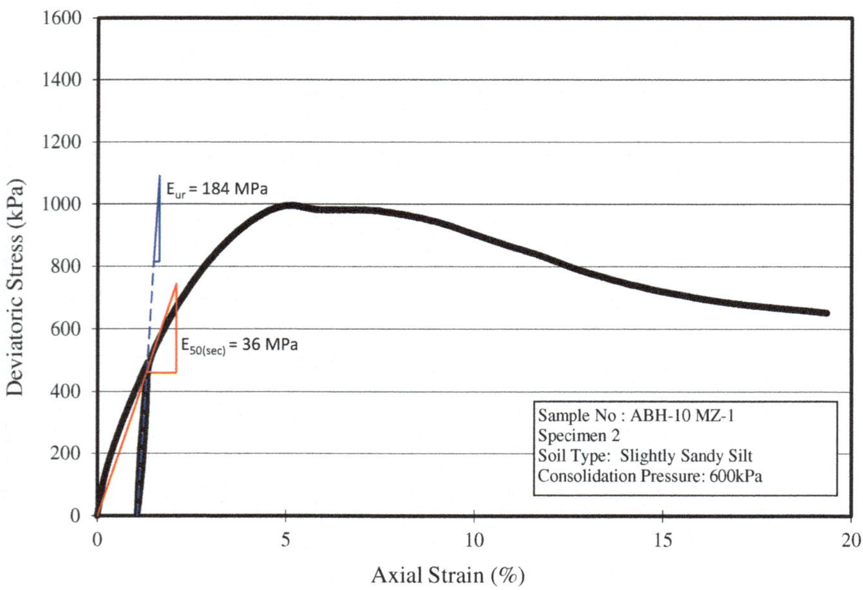

Fig. 8. Typical triaxial test with modulus determination

5 Monitoring and Observation Results

A total of eight building settlement markers have been strategically placed to monitor the settlement behaviour of the raft foundation. The performance of the raft foundation was closely monitored during both the construction and post-construction stages. The settlement readings for all the eight settlement markers are presented in Fig. 9. It can be seen that the settlement increased as the construction progressed, at a rate of approximately 1 mm for every 2 storeys. A final settlement ranging from 14 mm to 18 mm has been recorded with the maximum value being measured at the centre of the raft. Upon completion of construction, most of the settlement readings appear to have stabilised. The instrumentation monitoring was terminated six months post construction and the final settlement reading of each of the markers is presented in Fig. 10.

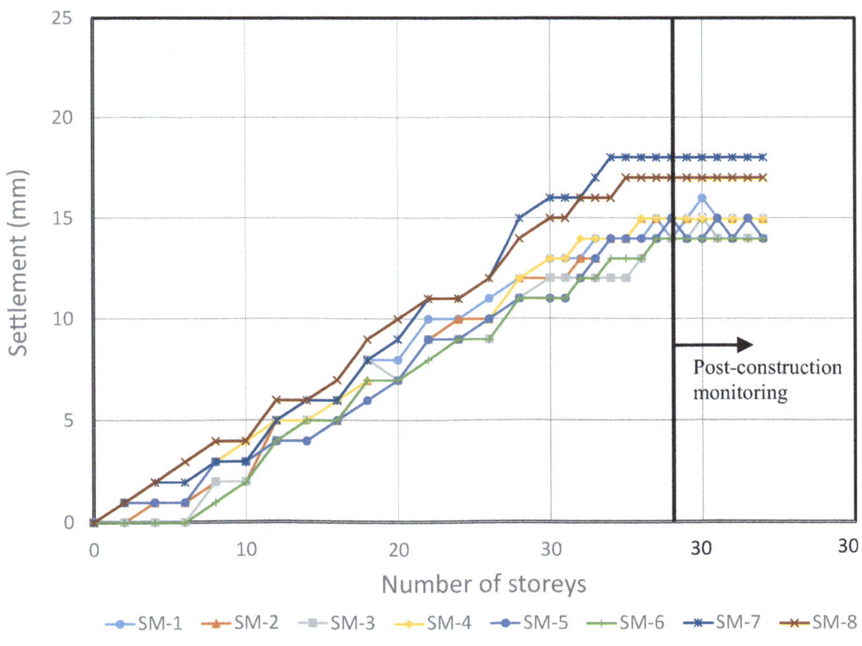

Fig. 9. Settlement readings for Tower One

5.1 Finite Element Model

Back analysis has been carried out using finite element method program, known as Plaxis 3D. Mohr-Coulomb model has been chosen in the analysis due to its simplicity and widely acceptance by the local authority. The raft has been modelled using plate element. The column and corewall loads have been modelled using point load and uniformly distributed load, respectively. A total load of 1970 MN was imposed on the raft with approximately 32% was supported by corewall and the remaining by columns.

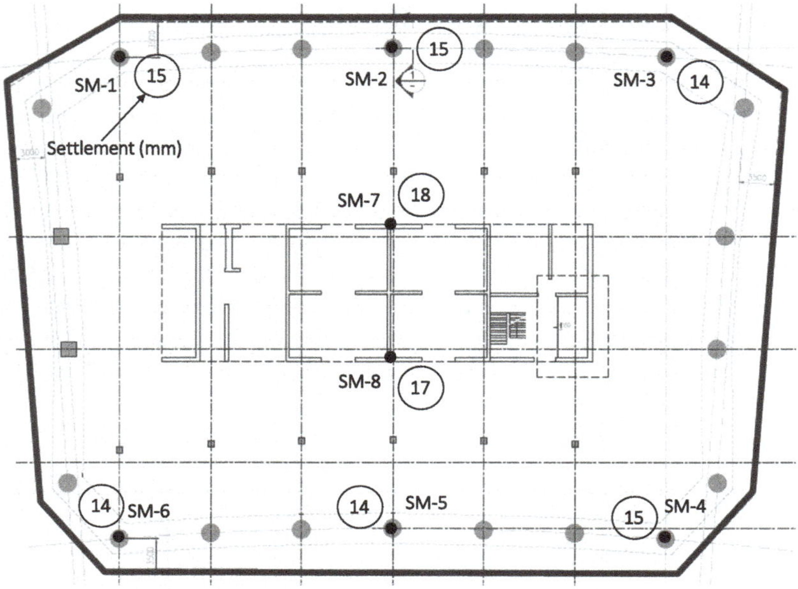

Fig. 10. Final settlement readings across the Tower One

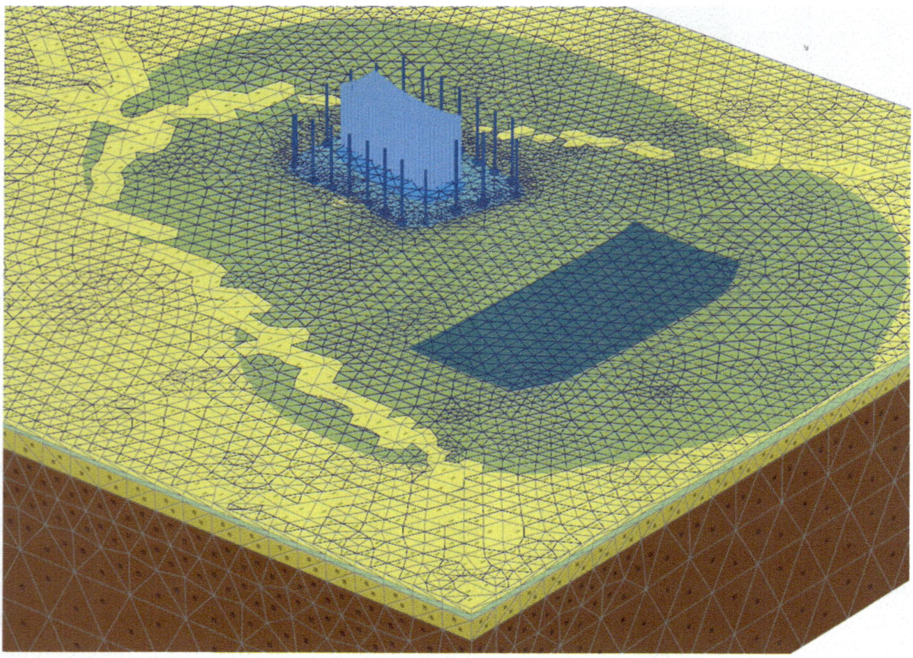

Fig. 11. Finite element model of raft foundation

Based on our past experience on the adjacent building founded on similar rocks, an elastic modulus of 2 GPa was adopted in this study. A typical finite element model of the raft foundation is illustrated in Fig. 11.

Based on the finite element method, the back analysed elastic modulus for the competent residual soil (SPT-N > 100) is on the order of 500 MPa. This value is about the same magnitude reported elsewhere in Singapore where the raft foundations were founded on bouldery clays (Adachi 1987; Ooi 1994; Wong et al. 1996). The resulting settlement contour based on the finite element analysis is presented in Fig. 12. As expected, the maximum settlement can be observed at the corewall centre with a slightly smaller settlement at the perimeter of the raft.

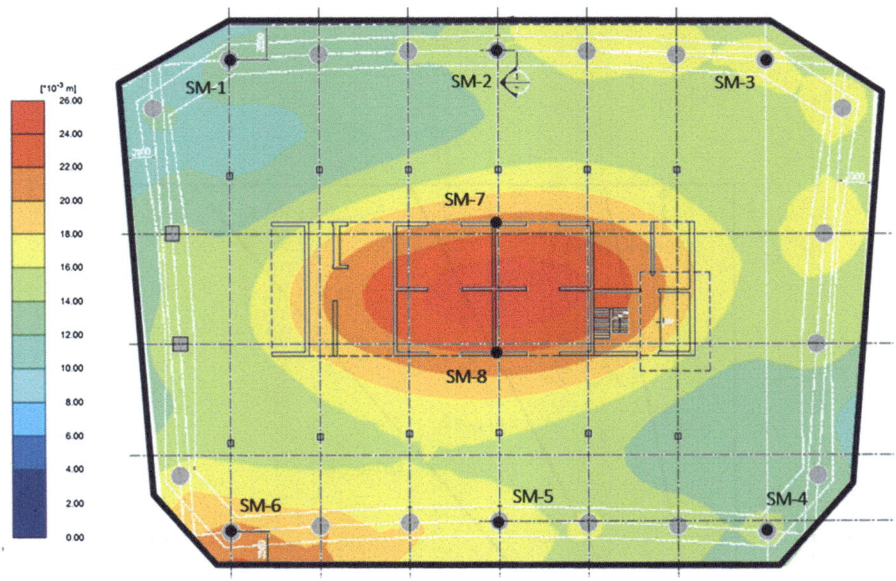

Fig. 12. Settlement contour of raft foundation based on finite element analysis

6 Discussion

Table 2 shows the summary of elastic modulus derived from both the field and laboratory tests. A comparison has been made amongst the field, laboratory and back-analysed elastic modulus. It was found that 2nd loading cycle for plate loading test and pressuremeter test give the more reasonable estimate when compared to the back-analysed value. The 3rd loading cycle of plate load test slightly over estimated the back-analysed elastic modulus. Based on the field test results, it can be postulated that a lower elastic modulus measured in the first loading cycle was likely because of the soil disturbance induced due to drilling of borehole for pressuremeter test and excavation for plate load test.

Table 2. Summary of elastic modulus derived from both the field and laboratory tests

Tests		Elastic modulus (MPa)
Plate load test	1st loading cycle	74–482 (189)
	2nd loading cycle	161–665 (302)
	3rd loading cycle	125–1930 (630)
Pressuremeter test	1st loading cycle	40–230 (92)
	2nd loading cycle	160–700 (320)
Oedometer test	1st loading cycle	26–81 (61)
	2nd loading cycle	49–171 (108)
Triaxial test*	1st loading cycle	21–99 (52)
	2nd loading cycle	151–354 (230)

Notes: (1) Values in parenthesis indicate average values.
(2) *Stress ranging from 300 kPa to 1200 kPa.

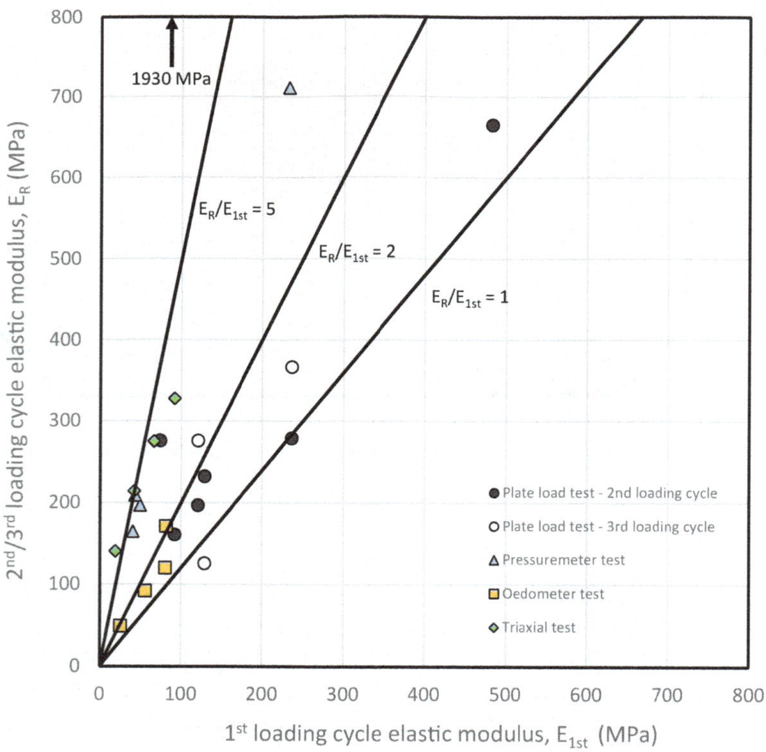

Fig. 13. Relationship between 1st loading and reloading cycles elastic modulus

For laboratory test, the unload-reload elastic modulus derived from triaxial and oedometer tests was much lower than the back-analysed value. Using the laboratory value, the estimated settlement would have been much higher and not realistic. Based on this observation, field tests generally provide a more realistic modulus deformation as compared to laboratory test. This is largely due to the sampling disturbance (i.e., difficult sampling for hard soil and stress relief) for laboratory tests.

From Fig. 13, it can be observed that the average field and laboratory elastic modulus determined from $2^{nd}/3^{rd}$ to 1st loading cycle has a ratio ranging from 1 to 5. This is generally consistent with the values reported in the literature (Adachi 1987; Wong et al. 1996). In the absence of unloading-reloading test cycle, it is suggested that the average elastic modulus determined from single loading cycle can be multiplied by a factor of 2 for practical purpose.

In Singapore, it is widely accepted that the elastic modulus can be correlated to SPT-N with a generalised equation as follows:

$$\frac{E}{p_{atm}} = k \times N \tag{3}$$

where k is the empirical constant and p_{atm} = atmospheric pressure (≈100 kPa).

An attempt has been made to establish a correlation for the elastic modulus with SPT-N under raft foundation loading. Based on the limited data, it appears to suggest

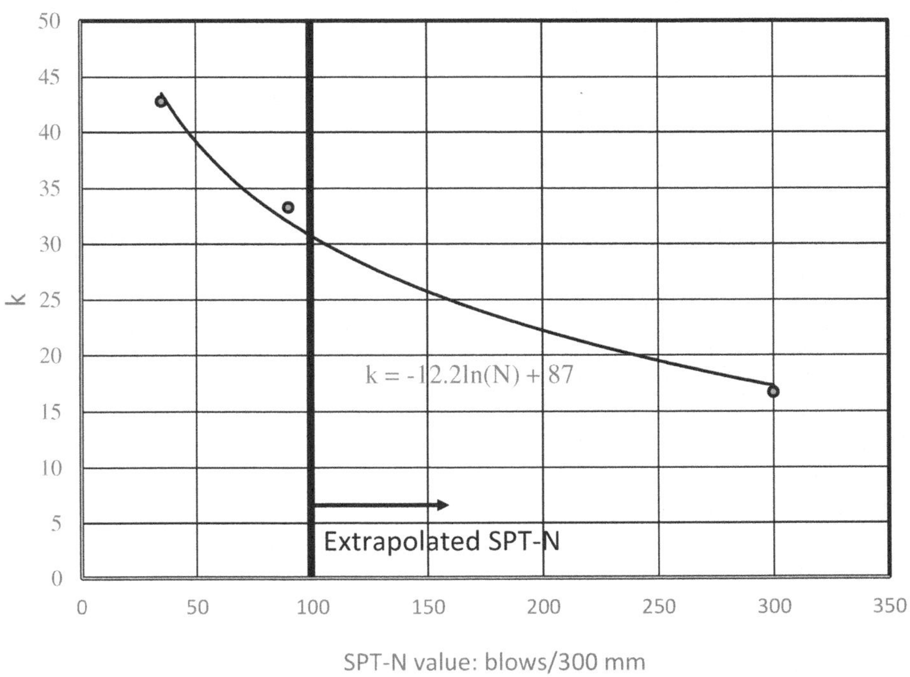

Fig. 14. Variation of k with SPT-N

that elastic modulus of the residual soil is not linearly related to SPT-N. A logarithmic equation is proposed in this study and given below.

$$k = -12.2\ln(N) + 87 \tag{4}$$

where the proposed k generally is ranging from 18 to 45 for 35 < N ≤ 300, as depicted in Fig. 14. This finding is generally in good agreement with the values reported by Goh et al. (2012) where k values were ranging from 25 to 96 under the mobilised radial strains of 0.1% to 1% using pressuremeter test.

Combining Eqs. (3) and (4), the elastic modulus can be re-written as follows:

$$\frac{E}{p_{atm}} = -12.2(N)\ln(N) + 87(N) \tag{5}$$

Using Eq. (5), the increasing elastic modulus with SPT-N is illustrated in Fig. 15. It should be noted that the proposed equation has been developed under foundation loading where the mobilised strain is expected to be in the region of 0.05% to 0.1% (Mair 1993). Further study is required to refine this relationship.

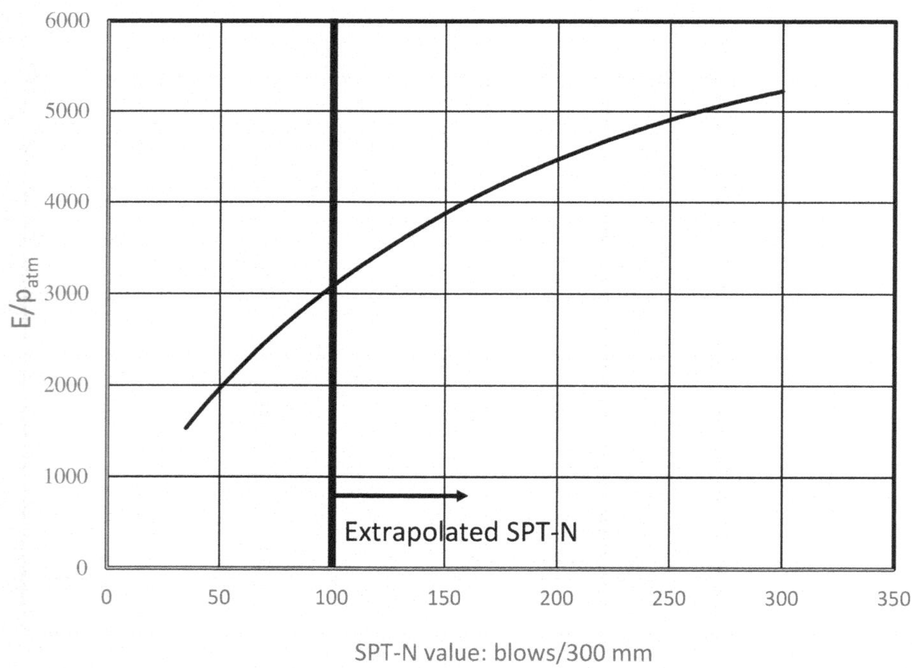

Fig. 15. Proposed relationship between elastic modulus and SPT-N

7 Conclusions

The performance of the 30-storey high-rise building's raft foundation with shallow embedment depth in Jurong Formation residual soils has been evaluated. The recorded maximum settlement of the 30-storey Tower One was on the order of 20 mm. This value was well within the design expectation and found to be very satisfactory. Back analysis has been carried out to determine the elastic modulus of residual soils using finite element method. The back-analysed elastic modulus of the competent residual soils was found to be on the order of 500 MPa. It has also been observed that the second loading cycle of field tests generally give the more reasonable estimate as compared to the laboratory tests. Using limited data, an empirical relationship between elastic modulus and SPT-N has been proposed under small-strain foundation loading. Finally, raft foundation is recommended for future high-rise buildings to be constructed on competent residual soils.

Acknowledgements. The authors would like to thank Mapletree Investments Pte Ltd and Shimizu Corporation for their kind permission to publish this paper and bring this project to a wider audience.

References

Adachi, K.: Settlement of raft foundation of a tall building. In: Proceedings of the 8th Asian Regional Conference on Soil Mechanics and Foundation Engineering, pp. 337–340 (1987)

Goh, K.H., Jeyatharan, K., Wen, D.: Understanding the stiffness of soils in Singapore from pressuremeter testing. Geotech. Eng. J. SEAGS AGSSEA **43**(4), 56–62 (2012)

Leung, C.F., Radhakrishnan, R., Wong, Y.K.: Observations of an instrumented pile-raft foundation in weak rock. Proc. Instn. Civ. Eng. Part 1, 693–711 (1988)

Lo Presti, D.C.F., Pallara, O., Puci, I.: A modified commercial triaxial testing system for small strain measurements. ASTM Geotech.Test. J. **18**(1), 15–31 (1995)

Mair, R.J.: Developments in geotechnical engineering research: applications to tunnels and deep excavations. Unwin Memorial Lecture 1992. Proc. Instn. Civ. Eng. Civ. Eng. **3**(1), 27–41 (1993)

Ooi, I.K.: Raft foundations for high-rise buildings on the bouldery clay in Singapore. Master of Engineering Thesis, Nanyang Technological University, Singapore, 306 pp. (1994)

Poulos, H.G., Davis, E.H.: Elastic Solutions for Soil and Rock Mechanics. John Wiley, New York (1974)

Tatsuoka, F., Teachavorasinskun, S., Dong, J., Kohata, Y., Sato, T.: Importance of measuring local strains in cyclic triaxial tests on granular materials. In: Dynamic Geotechnical Testing II, STP 1213, ASTM, pp. 288–302 (1994)

Wong, I.H., Ooi, I.K., Broms, B.B.: Performance of raft foundations for high-rise buildings on the bouldery clay in Singapore. Can. Geotech. J. **33**(2), 219–236 (1996). doi:10.1139/t96-002

The Application of Large-Scale Caisson in Anchorage Foundation for Suspension Bridge in China

Zheng-zhen Wang[1,2(✉)], Wei-ming Gong[1,2], and Guo-liang Dai[1,2]

[1] Key Laboratory of Concrete and Prestressed Concrete Structure of Ministry of Education, Southeast University, Nanjing 210096, China
765859155@qq.com

[2] Department of Civil Engineering, Southeast University, Nanjing 210096, Jiangsu, China

Abstract. Large-scale caisson has been applied in many long-span bridges in China. Five representative examples of large-scale caisson were described briefly in this study to show application status of large-scale caisson in China. To illuminate the sinking process and development law of the stress and strain of large-scale caisson, the construction monitoring of south anchorage caisson of Maanshan Yangtze River Highway Bridge were introduced and monitored results were analyzed simply. The results show that construction monitoring is a good way to provide important information for designers and constructors and to prevent accidents from occurring. Although there is a gap between monitoring results and expected results, caissons are safe because of the existence of safety factor. In order to overcome drawbacks of traditional large caisson, such as huge volume, high cost, and so on, a rooted-caisson foundation was put forward.

Keywords: Large-scale caisson · Application · Monitored results · Rooted-caisson foundation

1 Introduction

Caisson foundations are tubby structures, which depends on its own weight to overcome side-wall friction to sink to the design elevation and become the foundation of pier or abutment of bridge and other structures through casting base sealing concrete and filling the wellholes with concrete or water. Caisson foundations have so many advantages than other foundations, such as large stiffness, good bearing capacity, well stability, good resistances to earthquake, strong applicability and so on, that they are very popular in many large-scale engineering projects. From London Tower Bridge, the Eiffel Tower, the Brooklyn Bridge to Great belt Bridge, Akashi Keiko Bridge, it is the caisson foundation that supports all the magnificent buildings. Caisson foundation of Great Belt Bridge is shown in Fig. 1.

R. Chen et al. (eds.), *Proceedings of the 2nd International Symposium on Asia Urban GeoEngineering*, Springer Series in Geomechanics and Geoengineering,
https://doi.org/10.1007/978-981-10-6632-0_36

Fig. 1. Caisson foundation of Great Belt Bridge

With the development of Chinese economy, some long bridges have been built and more are under construction or will be constructed. Because of the huge advantages of caisson, the large-scale caisson has been widely used in China, especially in anchorage foundation for suspension bridge. In the study, five practical examples of large-scale caisson are introduced briefly to show application status of large-scale caisson in China.

Although large-scale caissons have been widely applied, it should be noted that many problems, such as cracks in concrete, large displacement and general incline, still arise during the construction and usage stages because of the great volumes, complicated stress conditions and the uncertainty of geotechnical engineering, which cause a considerable worry to designers and constructors. Base on the construction monitoring of south anchorage caisson of Maanshan Yangtze River Highway Bridge, development law of the stress and strain and some related problems during construction are analyzed, which can provide a significant reference and suggestions for the designer and constructors. At last, considering the drawbacks of traditional large caisson, such as huge volume, high cost, and so on, a rooted-caisson foundation was put forward.

2 Application of Large-Scale Caisson in China

As shown in Table 1, there are more than ten large bridges using large-scale caisson as their anchorage foundation from 1997 to now. The size of caisson is becoming bigger and bigger and the depth is increasing too.

Table 1. Some Large-Scale Caisson in Anchorage Foundation for Suspension Bridge in China

Project	Shape	Size (m)	Height (m)	Completion time
North Anchorage Caisson of Jiangyin Yangtze River Highway Bridge	Rectangle	69.0 × 51.0	58.0	1997.05
South Anchorage Caisson of Taizhou Yangtze River Bridge	Rectangle	67.9 × 52.0	41.0	2008.12
North Anchorage Caisson of Taizhou Yangtze River Bridge	Rectangle	67.9 × 52.0	57.0	2009.04
North Anchorage Caisson of Fourth Nanjing Yangtze River Bridge	Rectangle	69.0 × 58.0	52.8	2009.12
North Anchorage Caisson of Maanshan Yangtze River High way Bridge	Rectangle	60.2 × 55.4	41.0	2010.12
South Anchorage Caisson of Maanshan Yangtze River High way Bridge	Rectangle	60.2 × 55.4	48.0	2011.02
North Anchorage Caisson of Yingwuzhou Yangtze River Bridge in Wuhan	Circle	66.0	43.0	2012.05
Caisson of 28# main pier of Shanghai railway Yangtze River Bridge	Rectangle	86.9 × 58.7	105.0	2016.03
North Anchorage Caisson of Wufengshan Yangtze River Bridge	Rectangle	107.0 × 72.1	56.0	Under construction
South Anchorage Caisson of Wenzhou Oujiang Estuary Bridge	Rectangle	70.0 × 63.0	66.0	Upcoming

2.1 North Anchorage Caisson of Jiangyin Yangtze River Highway Bridge

Jiangyin Yangtze River Highway Bridge is a steel suspension bridge with a main span of 1385 m, which is the first bridge whose span exceeds 1 km in China and a large-scale caisson, which was the largest caisson in the world, was used as its north anchorage foundation, shown in Fig. 2. The size of this caisson is 69.2 m × 51.2 m × 58.0 m (length × width × depth).

Fig. 2. North Anchorage Caisson of Jiangyin Yangtze River Highway Bridge

Draining-sinking technology was used when sinking the first 30 m and undraining-sinking was used when sinking the rest part.

2.2 Caissons of Taizhou Yangtze River Bridge

Taizhou Yangtze River Bridge, located in Taizhou, Jiangsu province, China, is the first three-tower two-pan suspension bridge in China and the first bridge in which three large-scale caissons are used. The caisson size of middle tower is 58.0 m × 44.0 m × 76.0 m (Fig. 3), south anchorage is 67.9 m × 52.0 m × 41.0 m and north anchorage is 67.9 m × 52.0 m × 57.0 m.

Fig. 3. Caisson of middle tower of Taizhou Yangtze River Bridge

2.3 North Anchorage Caisson of Yingwuzhou Yangtze River Bridge

The Yingwuzhou Yangtze River Bridge, which is three-tower four-span suspension bridge with the biggest span in the world, is a highway bridge over the Yangtze River in Wuhan, Hubei Province, China. The shape of north anchorage caisson is circle, which is different from other large-scale caissons. There are 16 small hollow well holes and a cross-shaped cross wall in the big circle caisson wall, displayed in Fig. 4. The diameter of the big circle is 66.0 m and the depth is 43.0 m.

Fig. 4. North Anchorage Caisson of Yingwuzhou Yangtze River Bridge

2.4 Caisson of 28# Main Pier of Shanghai Railway Yangtze River Bridge

Shanghai railway Yangtze River Bridge is a rail-road cable stayed bridge with the biggest span in the world. The bridge will use about 480000 tons of steel and 2300000 squares of concrete. Scale and construction difficulties of this bridge create a number of world records, representing the highest level of bridge construction. The size of the caisson of 28# main pier is 86.9 m × 58.7 m × 105.0 m, which is the only one whose depth exceeded 100 m. The photo of this caisson is demonstrated in Fig. 5.

Fig. 5. Caisson of 28# main pier of Shanghai railway Yangtze River Bridge

2.5 North Anchorage Caisson of Wufengshan Yangtze River Bridge

The Wufengshan Yangtze River Bridge, which is a suspension bridge with the biggest span in the world, is the third rail-road Yangtze River Bridge in Jiangsu province, China. The size of the north anchorage caisson is 107.0 m × 72.1 m × 56.0 m. Figure 6 shows a construction stage of this caisson.

Fig. 6. North Anchorage Caisson of Wufengshan Yangtze River Bridge

3 Monitoring and Analysis of South Anchorage Caisson of Maanshan Yangtze River Bridge

3.1 Engineering General Situation

Maanshan Yangtze River Bridge, located in Maanshan, Anhui province, China, is a three-tower double-span suspension bridge. The size of the caisson of the south anchorage is 60.6 m × 55.8 m × 48.0 m, as Fig. 7. 25 small hollow well holes are set in the large-scale caisson. The caisson is divided into 9 parts. The bottom part is 8 m and is steel shell concrete caisson. The top part is 5.5 m, the rest 7 parts are 4.5 m and all the 8 parts are reinforced concrete caissons.

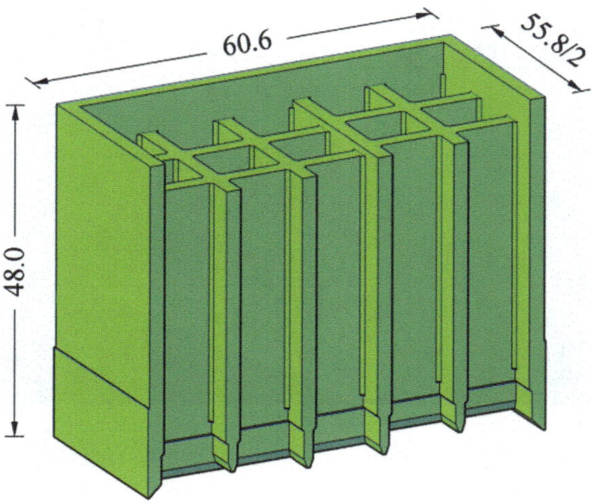

Fig. 7. Semi-structure of the south caisson

3.2 Installment of Monitoring Instruments

3.2.1 Installment of Earth Pressure Gauges for Cross Walls

Eight earth pressure gauges (four for normal cross wall and the others for partition cross wall) were installed at the bottom of the cross walls of the south anchorage caisson to monitor the earth pressure at different position of the cross walls. Positions of eight earth pressure gauges were plotted in Fig. 8.

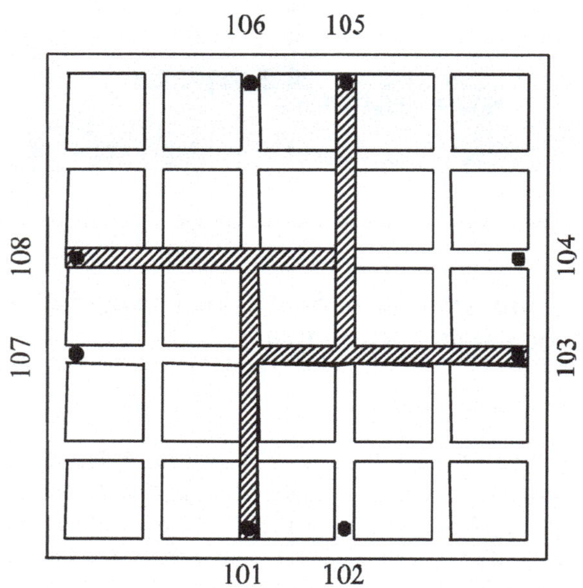

Fig. 8. Layout of earth pressure gauges at the bottom of cross walls

3.2.2 Installment of Steel Plate Strain Gauges for Bottom Part Caisson

There are 48 steel plate strain gauges totally in order to monitor the stress and strain of bottom part caisson, as shown in Fig. 9.

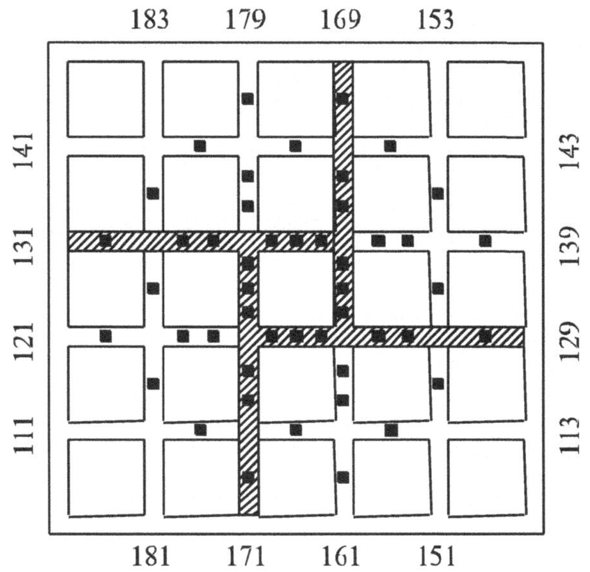

Fig. 9. Layout of steel plate strain gauges

3.2.3 Installment of Concrete Strain Gauges for Bottom Part Caisson

Four concrete strain gauges were installed to monitor the stresses of concrete on the bottom of partition cross walls too. Code of these four concrete strain gauges are 191–194 and positions of them are drawn in Fig. 10.

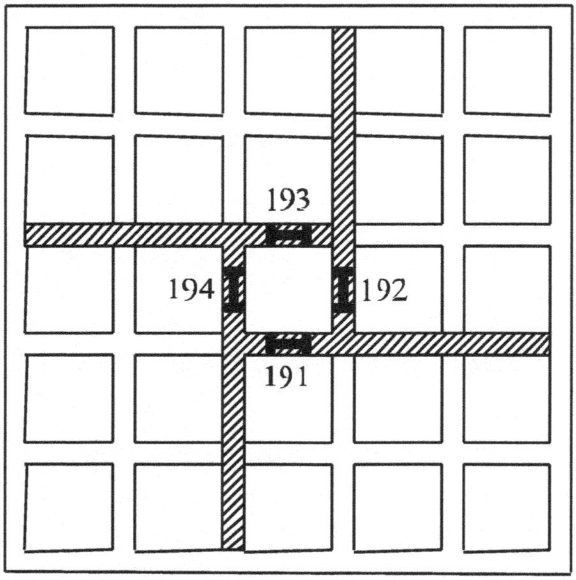

Fig. 10. Layout of concrete strain gauges

3.3 Analysis of Monitoring Results

All the monitoring results were divided into 8 working conditions, just shown in Table 2.

Table 2. Nine Working Conditions of monitoring results

Number	Working conditions
1	Before concrete casting
2	Casting concrete for first part
3	Binding steel bar for second part
4	Casting concrete for second part
5	Binding steel bar for third part
6	Casting concrete for third part
7	Binding steel bar for forth part
8	Casting concrete for forth part

3.3.1 Results of Earth Pressure Gauges for Cross Walls

The results of earth pressure gauges for cross walls are shown in Fig. 11.

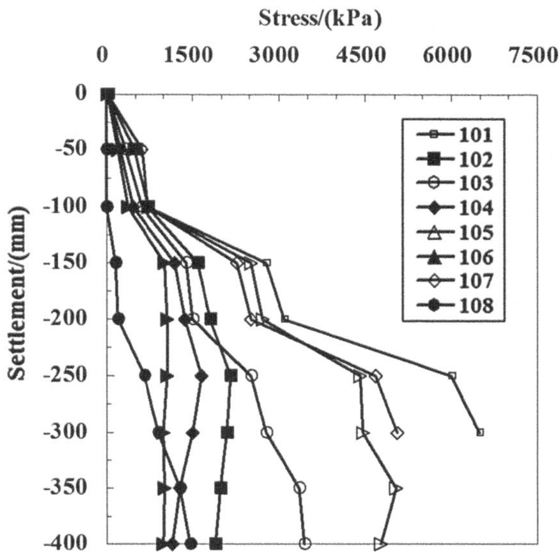

Fig. 11. Curves of reaction forces for cross walls

From Fig. 11, we can know that: (1) During casting concrete for first part, partition cross wall and normal cross wall were under pressure separately, which made the reaction forces very similar because of the similar self-weight. (2) When cast concrete for second part, the first part had suffered stresses as a whole. And because of the bigger stiffness, partition cross wall bore more stresses, so that the difference of stress between partition cross wall and normal cross wall stood out increasingly.

3.3.2 Results of Steel Plate Strain Gauges for Bottom Part Caisson

We chose the results of two typical steel plate strain gauges (135 and 143) to analyse, Change of the stress with working conditions of 135 and 143 were presented in Fig. 12 and two conclusions were reached: (1) The stage of casting concrete for first part determined the level of stresses. (2) Stresses changed little after casting concrete for first part.

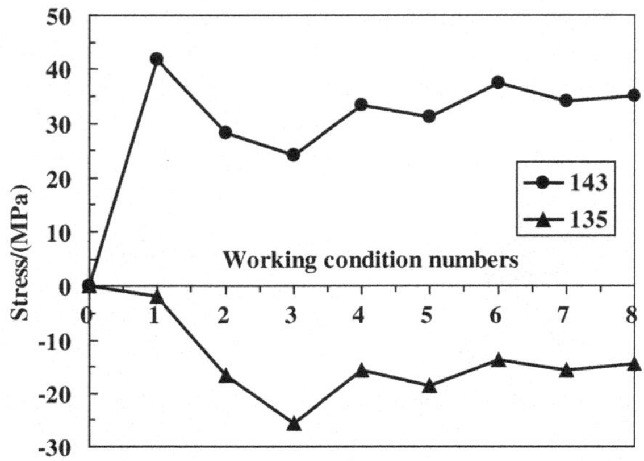

Fig. 12. Curves of stresses of typical steel plate strain gauges

3.3.3 Results of Concrete Strain Gauges for Bottom Part Caisson

The results of four concrete strain gauges, which were compared with the results of steel plate strain gauges at the same position, were shown in Fig. 13.

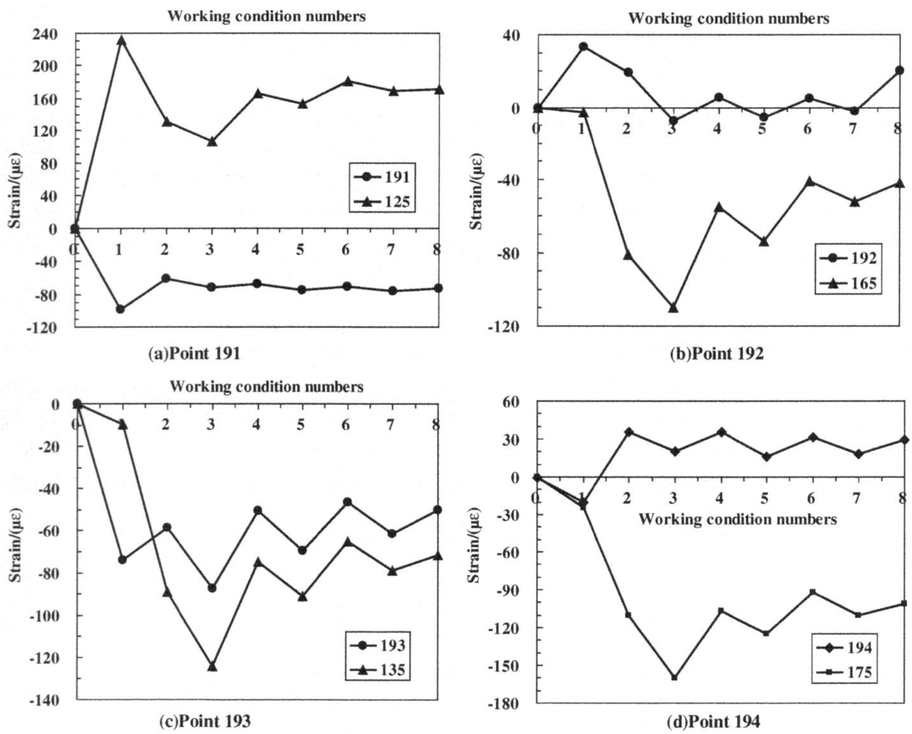

Fig. 13. Curves of stresses of concrete strain gauges and steel plate strain gauges at the same position

As shown in Fig. 13: (1) Numerical values for stresses of concrete strain gauges and steel plate strain gauges at the same position were different. (2) The reason may be that concrete and steel plate are different materials and have different properties.

3.3.4 Results of Sinking Speed

Sinking Process of this large-scale caisson was divided into three times. At the first time, the caisson was sunk to −20 m. At the second time, the caisson was sunk to −31 m and the caisson was sunk to designed elevation at the last time.

We can see from Figs. 14 and 15 that: (1) Sinking speed was slow and uniformed before the depth of 10 m in order to ensure the safety of caisson and the right sinking posture. Another reason to slow down the sinking speed was foundation improvement made the clay had a good bearing capacity. (2) Sinking speed became fast when the caisson entered into sand. (3) Sinking speed of second sinking was uniformed because of the same property of sand layer.

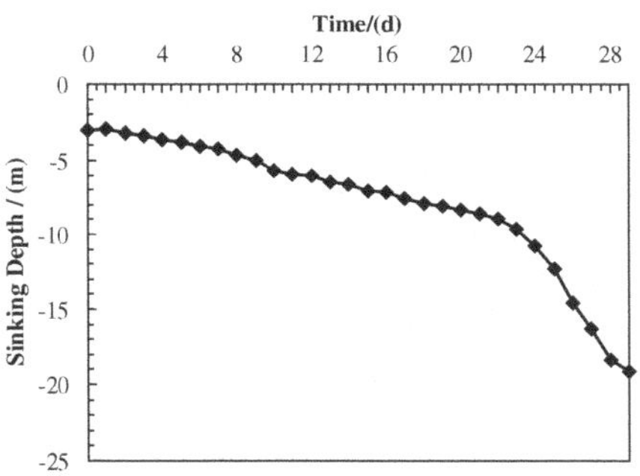

Fig. 14. Curve of first sinking

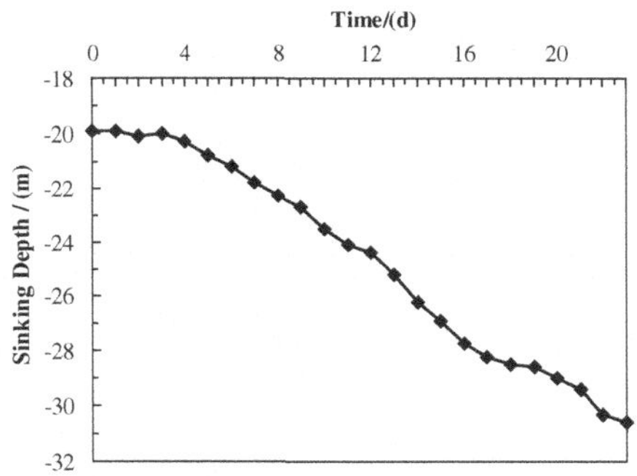

Fig. 15. Curve of second sinking

Because sinking pattern of third sinking was similar with second sinking, the curve of third sinking was not shown in this article.

4 Rooted-Caisson Foundation

Rooted-caisson foundation is a new kind of caisson derived from normal bored pile. Figure 16 exhibited the evolutionary process from normal bored pile to rooted-caisson.

The rooted-caisson consists of some root-bored piles which combine the normal bored pile and roots together.

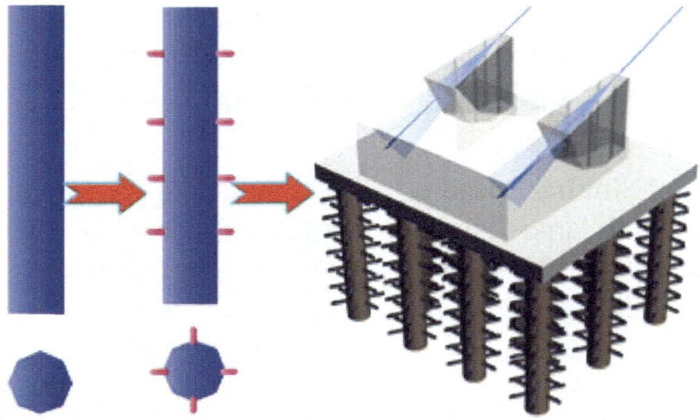

Fig. 16. Evolutionary process of rooted-caisson

Some bearing capacity tests had been conducted on one root-bored pile whose diameter was 6 m and the length was 39.0 m and one normal bored pile whose diameter and length are same with the root-bored pile (Fig. 17).

Fig. 17. Loading box of root-bored pile

We can see from Table 3 that: (1) compared with normal bored pile, the root-bored pile got the larger vertical bearing capacity and lateral bearing capacity. Vertical bearing capacity increased about 75% and the lateral bearing capacity increased about 60%. (2) The roots contributed a lot of power for the bearing capacity of root-bored pile. (3) If same bearing capacity was needed, the diameter and length of root-bored pile can be reduced greatly, thus reduced the cost and construction period. And according to the results of our analysis, the rooted-caisson consisted of some root-bored piles can not only has a better bearing capacity but also achieves good economic benefit.

Table 3. Comparison of bearing behavior of normal bored pile and root-bored pile

Vertical bearing capacity(kN)	Normal	52016
	Rooted	92004
Side friction(kN)	Normal	19046
	Rooted	59074
Lateral bearing capacity (kN)	Normal	10000
	Rooted	16000

5 Conclusions

In this study, five large-scale caissons applied in large bridges are described briefly to show application status of large-scale caisson in China, results of construction monitoring of south anchorage caisson of Maanshan Yangtze River Highway Bridge are analyzed and a new kind of caisson, rooted-caisson, is described simply. In the following, the main conclusions in the study are summarized:

(1) Large-scale caisson, as a kind of safe and reliable foundation, has been widely used in bridges in China and will continue to develop with further studies on it.
(2) Design and construction of large-scale caisson are relatively mature in China, but some problems or accidents still occurred occasionally, so deep studies about caisson are still needed.
(3) Construction monitoring is a good way to provide important information for designers and constructors and to prevent accidents from occurring. Although there is a gap between monitoring results and expected results, caissons are safe because of the existence of safety factor.
(4) Rooted-caisson foundation, which has a good bearing capacity, can provide good economic and social benefits and has good application prospects and popularized value.

References

1. Wand, W.: Caisson sinking technology application in North Anchorage of Taizhou Yangtze River Highway Bridge. Transp. Sci. Technol. **2**, 21–24 (2009). (in Chinese)
2. Li, M., Yang, C.: Design of new type caisson foundation for North Anchorage of Yingwuzhou Changjiang River Bridge in Wuhan. Bridge Constr. **4**(1–4), 24 (2011). (in Chinese)
3. Mu, B., Wang, Y., Zhu, J., et al.: Analysis of large caisson sinking measured resistance. J. Civil Archit. Environ. Eng. **34**(Supplement 2), 107–115 (2012). (in Chinese)
4. Mu, B., Zhu, J., Niu, Y.: Monitoring and analysis of north anchorage caisson of Fourth Nanjing Yangtze River Bridge. Chin. J. Geotech. Eng. **33**(2), 269–274 (2011). (in Chinese)
5. Yu, Y., Qiu, Q., Wang, L.: Scheme of pressure boost and floating aid for steel open caisson of main ship channel bridge of Hutong Changjiang River Bridge. Bridge Constr. **45**(6), 74–78 (2015). (in Chinese)
6. Tong, X., Yin, Y., Xu, M., et al.: Experimental study on vertical bearing capacity of open caisson. Chin. J. Undergr. Space Eng. **27**(11), 897–902, 909 (2009). (in Chinese)

7. Gong, W., Zhu, J., Mu, B., et al.: Dewater sinking of north anchorage caisson of Fourth Nanjing Yangtze River Bridge. Chin. J. Geotech. Eng. **32**(Supplement 2), 537–540 (2010). (in Chinese)
8. Wang, J., Liu, Y., Zhang, Yu.: Model test on sidewall friction of open caisson. Rock Soil Mech. **34**(3), 659–666 (2013). (in Chinese)
9. Zhu, J., Gong, W., Mu, B., et al.: Bottleneck problem study on first stage sinking of super-large caisson. Highway **4**, 13–18 (2011). (in Chinese)

Environmental Geotechnical Engineering

A Modified *p-y* Curve Method for Offshore Pile Foundations Under Cyclic Lateral Loading in Soft Clay

Si-qing Nie, Xin-jun Zou(✉), and Qiong He

College of Civil Engineering, Hunan University, Changsha 410082, Hunan, China
xjzouhd@hnu.edu.cn

Abstract. To discuss the nonlinear pile-soil interaction of offshore pile foundations under cyclic lateral loading in soft clay, the *p-y* curve method proposed by Matlock for the pile shaft under cyclic lateral loads were introduced and modified to reflect the weakening effect of the ultimate lateral bearing capacity of soft clay and the cyclic deformation of offshore pile shaft. In the modified *p-y* curve method, influences by cyclic loads on the lateral bearing behavior of subsoil around the pile shaft were approximately simulated by using a decreased undrained shear strength of the soft clay. And a bigger lateral deflection of the pile shaft was considered to reflect the impact by cyclic loads than the equivalent static load. Based on the formula proposed by Hettler to predict the horizontal displacement of pile shaft after N cycles of loading in sandy soil, a modified formula was established for offshore piles in soft clay to involve the cyclic deformation of pile shaft. The feasibility of modified *p-y* curve method presented in this paper was verified by comparing the more reasonable calculated result with those available achievements from the models of Matlock. Furthermore, a detailed parameter analysis was carried out to find out that, the value of the correlation coefficient ζ used to calculate the Y_{50} in the modified *p-y* curve is 1.5, and the reduction of the ultimate lateral bearing capacity of soft clay increased only 6% when the plasticity index increased from 10 up to 60. Finally, a simplified formula was proposed to calculate the undrained shear strength of soft clay under cyclic loading.

Keywords: Pile foundation · Offshore · Soft clay · Cyclic lateral load · *p-y* curve method

1 Introduction

In recent years, constructions of offshore engineering such as offshore wind turbines, cross-sea bridges, vertical pile-wharf and other maritime structures are being developed rapidly. Because of the complex hydrogeological conditions, the pile foundation was often used to ensure the security and stability of maritime structures. The offshore pile foundations subjected to complex forces, except the axial loads, they often subject to cyclic lateral load resulting from the effects of sea waves and traffic. Such loading induces remarkable deterioration in the interactive performance of the pile-soil system. In addition, the soft clay foundation is widely distributed in Chinese coast. Therefore,

R. Chen et al. (eds.), *Proceedings of the 2nd International Symposium on Asia Urban GeoEngineering*, Springer Series in Geomechanics and Geoengineering,
https://doi.org/10.1007/978-981-10-6632-0_37

it is significant for the development of the marine foundation engineering to study the relationship between the resistance of soft clay and the horizontal displacement of pile foundation under cyclic lateral load in the soft clay foundation.

At present, for the horizontal and nonlinear deformation of pile foundation, the *p-y* curve is adopted by many scholars because of its advantage of reflecting the combination of complexity and non-linearity of pile-soil interaction. Matlock completed a horizontal loading test of steel pipe piles with a diameter of 324 mm and a length of 12.8 m in the underwater saturated soft clay, and presented the *p-y* curves for pile shaft under static loading and cyclic loading in the soft clay, and the *p-y* curves have been used by American Petroleum Institute (API) (Matlock 1970; API RP 2A-WSD 2000). However, the weakening effect of ultimate lateral bearing capacity of soft clay cannot be embodied in the *p-y* curves presented by Matlock for pile shaft under cyclic loading. Jeong et al. (2011) studied the influence on lateral rigidity of offshore piles using proposed *p-y* curves, but the cyclic loading was not included. There were some other research works in this field, Naggar and Bentley (2000) constructed the dynamic *p-y* curves which were a function of the static *p-y* curves and velocity of the soil particles at a given depth and frequency of loading by the damping theory; Rajashree and Sitharam (2001) established a reduction formula of lateral bearing capacity of periodically loaded clay by the finite element analysis method; Cai (2002) presented an empirical formula of *p-y* curves for single pile under reciprocating loading through the concept of reduction factor of reciprocating loading; Dewaikar et al. (2008) employed an iterative scheme in the nonlinear analysis of a pile in soft clay under cyclic lateral loading with secant modulus approach, using the moment area method.

However, the methods of analyzing the effect of cyclic loading on the pile and soil introduced above were mainly obtained by loading tests, and the calculation of these methods which required some complex test parameters were relatively complex and error prone. To ensure the accuracy of the calculation while simplify the calculation process, the influence of cyclic loading on the lateral bearing capacity of soft clay around the offshore pile shaft is externalized as the reduction of undrained shear strength and the influence of cyclic loading on pile displacement is also considered in the modified *p-y* curves, and trying to obtain the modified *p-y* curves, which are simple and more conforms to realistic offshore pile foundation under cyclic lateral load in soft clay.

2 Ultimate Bearing Capacity of Soft Clay Around the Pile Shaft Under Cyclic Loads

According to the *p-y* curves presented by Matlock, the ultimate unit bearing capacity P_u of soft clay increases from $3c_u$ to $9c_u$ as X increases from 0 to X_r (API RP 2A-WSD 2000; Matlock 1970)

$$P_u = \begin{cases} c_u + \gamma' X + \dfrac{c_u}{d} JX & (X \le X_r) \\ 9c_u & (X > X_r) \end{cases} \tag{1}$$

$$X_{\mathrm{r}} = \frac{6d}{\frac{\gamma' d}{c_{\mathrm{u}}} + J} \tag{2}$$

Where P_{u} is the ultimate unit resistance (kPa); c_{u} is the undrained shear strength for undisturbed clay soil samples (kPa); d is the pile diameter (m); γ' is the effective unit weight of soil (kN/m^3); J is the dimensionless empirical constant with values ranging from 0.25 to 0.5 determined by field testing; X is the depth below soil surface (m); X_{r} is the depth below soil surface to the bottom of reduced resistance zone (m).

The shear strength of clay can be calculated by Mohr-coulomb strength theory (Zhao 2014)

$$\tau_{\mathrm{f}} = c + \sigma \tan\varphi \tag{3}$$

Where τ_{f} is the shear strength of clay (kPa); σ is the normal stress acting on the cut plane (kPa); φ is the internal friction angle of clay (°); c is the cohesion of clay (kPa).

Usually, once subjected to load action under operating conditions, the saturated soft clay around the offshore pile foundations would be in an undrained environment, and the internal friction angle $\varphi = 0$. So, the undrained shear strength of the saturated soft clay c_{u} is equal to the cohesion c (Li 2004).

But under a long term cyclic loading, stress concentration phenomenon will be observed in the subsoil surrounding the pile shaft, and some local corners of the irregular particles will be broken by contacting with other particles. Then, the space structure of soil will be changed as soil particles will be rearranged. What is more, the cohesive materials that provide cohesion among soil particles will be disturbed, causing the changes of cohesion. Because of the cyclic loading, the pore pressures will also change and eventually lead to the variation of soil strength (Zheng et al. 2012; Hyde and Ward 1985). Therefore, it is infeasible to replace the undrained shear strength of soft clay under cyclic lateral load with its cohesion.

Zhong et al. (2015) conducted the degradation experiment on pile foundations under cyclic loading, and the experiment results show that the reduction of the lateral bearing capacity of soft clay mainly happened in the first ten cycles of loading, then the attenuation degree reduced dramatically till a relative stable value of the lateral bearing capacity of clay was observed finally after about 100 cycles of loading. In engineering practices, the actual number of loading cycles of wave or car loads subject to pile foundation is far more than 100 times. Therefore, the stable undrained shear strength of saturated soft clay after attenuating is used to calculate the lateral bearing capacity of soft clay around the pile shaft P_{u} in this paper.

Yasuhara et al. (1992) presented the attenuation formula to calculate the undrained shear strength of clay under cyclic loading after the triaxial tests

$$\frac{(c_{\mathrm{u}})_{\mathrm{cy}}}{(c_{\mathrm{u}})_{\mathrm{NC}}} = \left(\frac{1}{1 - \frac{u}{\sigma_c}}\right)^{\frac{A_0}{1 - C_s/C_c} - 1} \tag{4}$$

Where $(c_u)_{cy}$ is the undrained shear strength of clay under cyclic loading (kPa); $(c_u)_{NC}$ is the undrained shear strength of clay (kPa); u is the pore water pressure; σ_c is the initial consolidation stress; Λ_0 is the experiment parameter; C_s, C_c is the coefficient of resilience and compressibility of clay.

According to the definition of quasi-over consolidation (Li 2007; Gao et al. 2004)

$$OCR_{eq} = \frac{\sigma_c}{\sigma_c - u} \tag{5}$$

Substituting Eq. (5) into Eq. (4)

$$(c_u)_{cy} = c_u (OCR_{eq})^{\frac{\Lambda_0}{1 - C_s/C_c} - 1} \tag{6}$$

Where OCR_{eq} is quasi-over consolidation ration determined by testing.

The relationship between Λ_0, C_s, C_c and plasticity index I_p is (Yasuhara 1994)

$$\frac{C_s}{C_c} = 0.185 + 0.002 I_p \tag{7}$$

$$\frac{\Lambda_0}{1 - C_s/C_c} = 0.939 - 0.002 I_p \tag{8}$$

Substituting Eq. (8) into Eq. (6)

$$(c_u)_{cy} = c_u (OCR_{eq})^{-0.061 - 0.002 I_p} \tag{9}$$

Then, the ultimate unit resistance of soft clay under cyclic lateral load $(P_u)_{cy}$ can be expressed as

$$(P_u)_{cy} = \begin{cases} \left(3 + \frac{\gamma'}{(c_u)_{cy}} X + \frac{JX}{d}\right)(c_u)_{cy} & (X \le X_{cy}) \\ 9(c_u)_{cy} & (X > X_{cy}) \end{cases} \tag{10}$$

$$X_{cy} = \frac{6d}{\frac{\gamma' d}{(c_u)_{cy}} + J} \tag{11}$$

Where $(P_u)_{cy}$ is the ultimate unit resistance of soft clay under cyclic lateral load (kPa); $(c_u)_{cy}$ is the undrained shear strength of undisturbed soft clay samples under cyclic loading (kPa); X_{cy} is the turning point depth of ultimate soft clay resistance depth below soil surface to bottom of reduced resistance zone in when the soft clay is forced by cyclic loading (m).

3 Modified p-y Curve Method for Offshore Pile Foundations Under Cyclic Lateral Load in Soft Clay

At present, there are two main methods to predict the horizontal displacement of pile shaft under cyclic lateral load (Zhu et al. 2013): the direct estimation method and *p-y* curve method. In the modified *p-y* curves, the direct estimation method is used to predict the horizontal displacement of offshore pile shaft after *N* cycles, and according to the studies done by Hettler (1981), the Eq. (12) applied to sandy soil showed the relationship between displacement and cycle index

$$Y_N = Y_1\left(1 + C_N \ln N\right) \tag{12}$$

Where N is the number of cycles; Y_N is the horizontal displacement of pile shaft after N cycles; Y_1 is the horizontal displacement of pile shaft after the first cycle; C_N is the weakening coefficient and Hettler suggested that C_N can be 0.17 for sandy soil.

Zhang (2014) did the investigation on lateral bearing capacity of offshore wind turbine pile foundation under cyclic loading, and proposed that the Eq. (12) could be also applied to silty soil. The silty soil in test was from Yellow River Delta, its clay content was about 10%, plasticity index I_p was 9.1 and saturation was 98%. According to the fitting result (Zhang 2014), C_N for embedded pile was ranged from 0.03 to 0.07 and for driving pile was ranged from 0.01 to 0.06, C_N for silty soil was smaller to sandy soil. Generally, the properties of silty soil are between clay and sandy soil. Therefore, the smaller value of the fitting result for silty soil 0.03 is used as the weakening coefficient of soft clay in this paper

$$Y_N = Y_1(1 + 0.03 \ln N) \tag{13}$$

According to the experimental result (Zhang 2014), the lateral deformation on the top of pile foundation in silty soil was significantly increased with the increasing of cycle index in the first 100 cycles, and almost unchanged after 500 cycles. In order to obtain the modified *p-y* curves with the stable displacement of offshore pile shaft under cyclic loading in soft clay, substituting $N = 500$ into Eq. (13)

$$Y_N = Y_1(1 + 0.03 \ln 500) = 1.19 Y_1 \tag{14}$$

$$Y_1 = Y_N / 1.19 \tag{15}$$

According to the *p-y* curves for pile shaft under cyclic loading presented by Matlock When $X \geq X_r$

$$\frac{P}{P_u} = \begin{cases} 0.5\left(\dfrac{Y}{Y_{50}}\right)^{1/3} & (Y < 3Y_{50}) \\ 0.72 & (Y \geq 3Y_{50}) \end{cases} \tag{16}$$

When $X < X_r$

$$\frac{P}{P_u} = \begin{cases} 0.5\left(\frac{Y}{Y_{50}}\right)^{1/3} & (Y < 3Y_{50}) \\ 0.06\frac{X_r - X}{x_r}(15 - \frac{Y}{Y_{50}}) \\ +0.72\frac{X}{X_r} & (3Y_{50} \le Y \le 15Y_{50}) \\ 0.72\frac{X}{X_r} & (Y > 15Y_{50}) \end{cases} \quad (17)$$

$$Y_{50} = 2.5\varepsilon_{50}d \quad (18)$$

Where Y is the actual lateral deflection (m); P is the actual lateral resistance (kPa); Y_{50} is the lateral deflection when actual lateral resistance is the half of ultimate value (m); ε_{50} is the strain which occurs at one-half the maximum stress on laboratory undrained compression tests of undisturbed soil samples, or can be evaluated by standard (JTS167-4-2012 2012).

According to the analysis results mentioned above, based on the *p-y* curves presented by Matlock, considering the weakening effect of the ultimate lateral bearing capacity of soft clay and the cyclic deformation of offshore pile shaft, and then obtaining the modified *p-y* curves for the offshore pile foundation under cyclic lateral load in soft clay, as shown in Fig. 1.

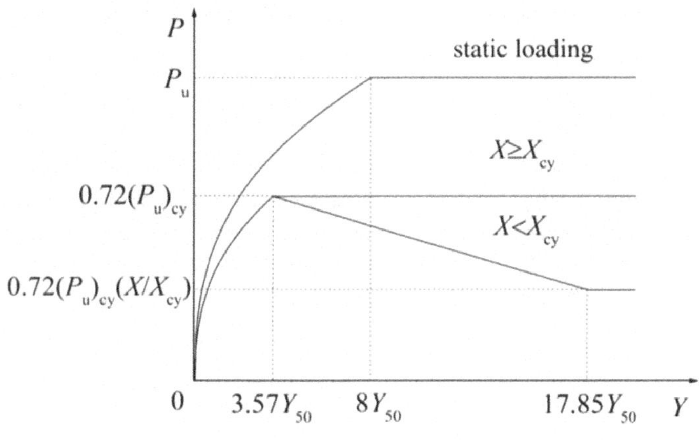

Fig. 1. Modified *p-y* curves for pile foundations under cyclic loading

When $X \geq X_{cy}$

$$\frac{P}{(P_u)_{cy}} = \begin{cases} 0.47\left(\frac{Y_N}{Y_{50}}\right)^{1/3} & (Y_N < 3.57Y_{50}) \\ 0.72 & (Y_N \geq 3.57Y_{50}) \end{cases} \quad (19)$$

When $X < X_{cy}$

$$\frac{P}{(P_u)_{cy}} = \begin{cases} 0.47\left(\frac{Y_N}{Y_{50}}\right)^{1/3} & (Y_N < 3.57Y_{50}) \\ 0.06\frac{X_{cy} - X}{x_{cy}}(15 - \frac{Y_N}{1.19Y_{50}}) & \\ +0.72\frac{X}{X_{cy}} & (3.57Y_{50} \leq Y_N \leq 17.85Y_{50}) \\ 0.72(\frac{X}{X_{cy}}) & (Y_N > 17.85Y_{50}) \end{cases} \tag{20}$$

Where Y_N is the actual stable lateral deflection of offshore pile shaft under cyclic lateral loading (m), can be calculate by Eq. (15); $(P_u)_{cy}$ is the ultimate unit resistance of soft clay under cyclic lateral load, can be calculate by Eq. (10); P is the actual lateral resistance of soft clay (kPa); Y_{50} is the lateral deflection when actual lateral resistance is the half of ultimate value (m), the calculation discussed as below; X is the depth below soil surface (m); X_{cy} is the turning point depth of ultimate soft clay resistance depth below soil surface to bottom of reduced resistance zone in when the soft clay is forced by cyclic loading (m), can be calculate by Eq. (11).

4 Validation of the Modified *p-y* Curve Method for Offshore Pile Foundations Under Cyclic Lateral Load in Soft Clay

Cai organized the data of pile tests in the suburb of Shanghai (Cai 2002), of which the NO. 10D16 test pile with a 0.16 m diameter was subjected to a cyclic lateral load. As for the geological conditions, one meter below the soil surface was yellow grey plastic clay with the cohesion $c = 34$ kPa, effective unit weight $\gamma' = 8.9\text{kN/m}^3$, plasticity index $I_p = 60$, and supposed quasi-over consolidation ration $OCR_{eq} = 2.0$, $J = 0.5$, $\varepsilon_{50} = 0.01$. And the Y_{50} is calculate by Eq. (18). Figure 2 shows the predicted and measured *p-y* curves of the pile for $z = 0.2$ m and $z = 0.6$ m, respectively. The modified *p-y* curves are closer to the general trend of the measured data compared with the *p-y* curves presented by Matlock. The Matlock *p-y* curve for z = 0.2 m has a larger resistance than that by the measured *p-y* curve and the modified *p-y* curves are more reasonable. But the displacement in the peaks of the modified *p-y* curves and Matlock *p-y* curves is larger than the measured data, it shows that the Y_{50} in the modified *p-y* curves and Matlock *p-y* curves is larger than the measured data. The more detailed parameter analysis will be carried out in the below to find out more appropriate *p-y* curves for offshore pile foundations under cyclic lateral loads in soft clay.

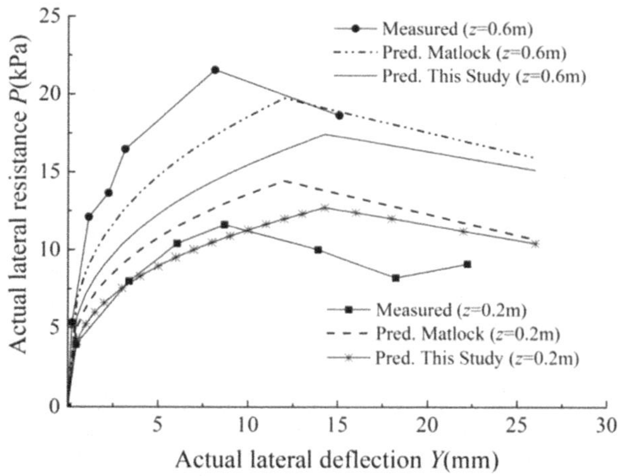

Fig. 2. Comparative analysis of *p-y* curves for pile foundation under cyclic loading

5 Parameter Analysis

The comparison of predicted and measured *p-y* curves for pile foundation under cyclic loading shows that the proposed Y_{50} is larger than the measured data. The ultimate bearing capacity of soft clay will reduce and the lateral deflection of pile shaft will increase because of the cyclic loading, and it is unfaithful to calculate the value of Y_{50} by Eq. (18). So, defining the relation

$$Y_{50} = \zeta \varepsilon_{50} d \tag{21}$$

Where ζ is the coefficient association, it is 2.5 if the pile foundation subject to the static loading; ε_{50} is the strain which occurs at one-half the maximum stress on laboratory undrained compression tests of undisturbed soil samples; d is the pile diameter (m).

Referring to the experimental data (Cai 2002), the pile and the subsoil were respectively described by d = 0.16 m, c = 34 kPa, γ' = 8.9kN/m^3, OCR_{eq} = 2.0, J = 0.5, ε_{50} = 0.01, I_p = 60. Figure 3 (a) and (b) shows the *p-y* curves with different values of ζ for z = 0.2 m and z = 0.6 m, respectively. Comparing the modified *p-y* curve with the measured data, there is an obvious error with prediction if ζ is equal to 2.5, but the modified *p-y* curve is more conforms to the measured data when ζ equal to 1.5.

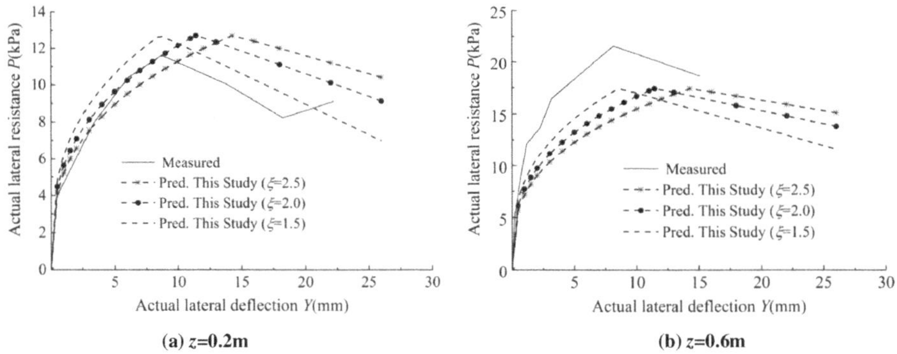

Fig. 3. Modified *p-y* curves with different values of ζ

The cyclic loading will lead to the reduction of the undrained shear strength. And the ultimate bearing capacity of soft clay depends on the undrained shear strength. According to Eq. (9), the reduction of undrained shear strength mainly depends on the plasticity index I_p and quasi-over consolidation ration OCR_{eq}, and it makes sense to analyze the effect of I_p and OCR_{eq} on ultimate bearing capacity of soft clay.

According to the statistics for physical and mechanical parameter of saturated clay distributed in China (Bai et al. 2001), the values of I_p for the saturated clay are generally between 10 and 60, and substituting $I_p = 10$ and $I_p = 60$ in Eq. (8)

$$-0.181 \leq \frac{\Lambda_0}{1 - C_s/C_c} - 1 \leq -0.081 \tag{22}$$

Combining with the engineering practice, do the analysis with $I_p = 10, 30, 60$, and $d = 1.0$ m, $c_{\text{u}} = 20$ kPa, $\gamma' = 8$ kg/m^3, $J = 0.5$, $OCR_{\text{eq}} = 2$. Defining the reduction rate C of ultimate bearing capacity of soft clay by Eq. (23), and Fig. 4 shows the changing law of C with depth and different I_p

$$C = \left(1 - \frac{(P_{\text{u}})_{\text{cy}}}{P_{\text{u}}}\right) \times 100\% \tag{23}$$

Figure 4 shows that C decrease first and then increase and final to be stability with the increasing of depth when I_p is constant, and C increases from about 6% to about 12% as I_p increases from 10 to 60, the plasticity index I_p has little impact on ultimate lateral bearing capacity of soft clay. To make the calculation simple, substituting $I_{\text{p}} = 60$ in Eq. (9), and then obtaining the simplified formula to calculate the undrained shear strength of soft clay under cyclic loading

$$\begin{aligned} \frac{(c_{\text{u}})_{cy}}{c_{\text{u}}} &= \left(OCR_{\text{eq}}\right)^{\frac{\Lambda_0}{1 - C_s/C_c} - 1} \\ &= \left(OCR_{\text{eq}}\right)^{-0.181} \end{aligned} \tag{24}$$

$$\left(c_{\mathrm{u}}\right)_{\mathrm{cy}} = c_{\mathrm{u}}\left(OCR_{\mathrm{eq}}\right)^{-0.181} \tag{25}$$

Andersen et al. (1980) presented that the reduction of undrained shear strength of clay under cyclic loading was less than 25% by the experiment results. According to Eq. (25), the reduction of undrained shear strength of soft clay is less than 25% when OCR_{eq} is less than 5, and that conforms to the experiment results.

Combining with the engineering practice, do the analysis with $OCR_{eq} = 1.5, 2.0, 2.5, 3$, and $d = 1.0$ m, $c_u = 20$ kPa, $\gamma' = 8$ kg/m^3, $J = 0.5$, $I_P = 60$. The changing law of C with depth and different OCR_{eq} is shown in Fig. 5.

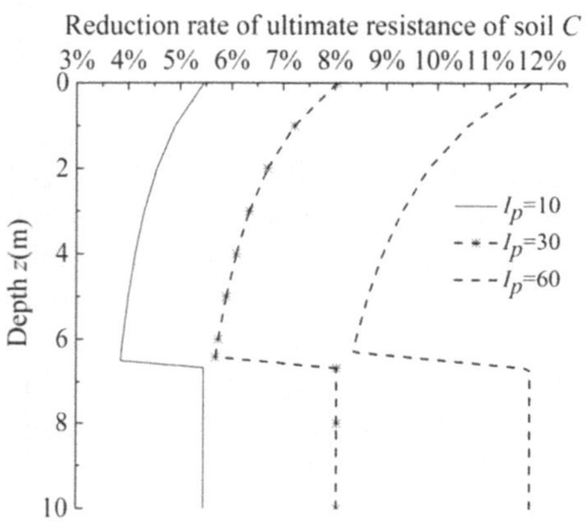

Fig. 4. The effect of plasticity index I_p on the ultimate bearing capacity of soft clay

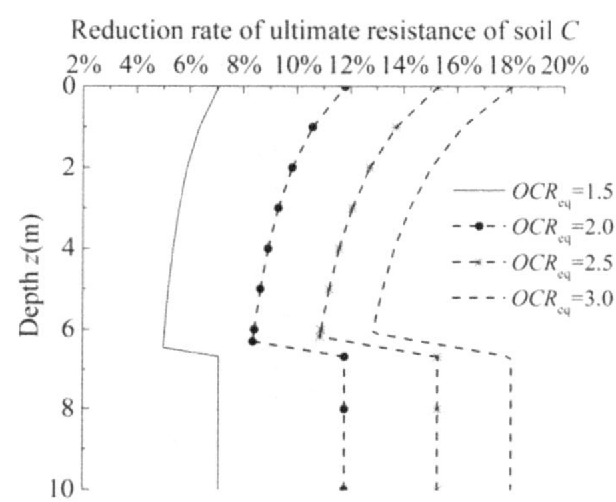

Fig. 5. The effect of quasi-over consolidation OCR_{eq} on the ultimate bearing capacity of soft clay

Figure 5 also shows that, C decreases first and then increases to a final stable value with the increasing of depth when OCR_{eq} keeps constant, and C increases from about 7% to about 18% as OCR_{eq} increases from about 1.5 to about 3.0. This kind of changing rules show that the growth rates of C is larger than the one of OCR_{eq}. As a result, the reduction of the ultimate resistance for soft clay is small when OCR_{eq} is less than 1.5.

6 Conclusions

To find out a more appropriate *p-y* curve method for offshore pile foundations under cyclic lateral loads in soft clay, the *p-y* curve method proposed by Matlock was introduced and modified to reflect the weakening effect of the ultimate lateral bearing capacity of soft clay and the cyclic deformation of offshore pile shaft. The following conclusions are drawn from the analysis by using the modified *p-y* curve method:

(1) The weakening effect of the ultimate lateral bearing capacity of soft clay around the pile shaft was found to depend on the weakening of the undrained shear strength. And based on the analysis by Yasuhara, a simplified formula was established for

saturated clay to calculate the undrained shear strength under cyclic loading, which was then used in the modified *p-y* curve.

(2) The lateral deflection of pile shaft in the modified *p-y* curve was corrected by the theory proposed by Hettler. Compared with those available achievements by using the *p-y* curve method of Matlock, more reasonable results were obtained by the *p-y* curve modified in this paper.

(3) The result by the parameter analysis of the modified *p-y* curve shows that, an obvious error was observed between the predicted *p-y* curve of the pile shaft and the measured one when the valued of coefficient ζ was equal to 2.5. However, the calculated *p-y* curve agreed well with the measured data for $\zeta = 1.5$. Furthermore, the reduction of the ultimate lateral bearing capacity of soft clay increased only 6% when the plasticity index increased from 10 up to 60.

Acknowledgements. The research presented here was undertaken with support from the National Natural Science Foundation of China (Grant No. 51378197 & No. 51578231).

References

Matlock, H.: Correlations for design of laterally loaded piles in soft clay. In: Offshore Technology in Civil Engineering. ASCE, pp. 577–594 (1970). doi:10.4043/1204-MS

American Petroleum Institute: Recommended practice for planning, designing and constructing fixed offshore platformsRP2A-WSD (2000)

Jeong, S., Kim, Y., Kim, J.: Influence on lateral rigidity of offshore piles using proposed *p–y*, curves. Ocean Eng. **38**(2–3), 397–408 (2011). doi:10.1016/j.oceaneng.2010.11.007

Naggar, M.H.E., Bentley, K.J.: Dynamic analysis for laterally loaded piles and dynamic *p-y* curves. Can. Geotech. J. **37**(6), 1166–1183 (2000). doi:10.1139/cgj-37-6-1166

Cai, L.: Study of *p-y* curves of single pile under reciprocating lateral loading. MS Thesis, Hohai University, Nanjing, China (2002). (in Chinese)

Rajashree, S.S., Sitharam, T.G.: Nonlinear finite-element modeling of batter piles under lateral load. J. Geotech. Geoenviron. Eng. **127**(7), 604–612 (2001). doi:10.1061/(ASCE)1090-0241(2001)127:7(604)

Dewaikar, D.M., Padmavathi, S.V., Salimath, S.R.: Ultimate lateral load of a pile in soft clay under cyclic loading. In: The 12th International Conference of International Association for Computer Methods and Advances in Geomechanics, pp. 3498–3507 (2008)

Zhao, M.H.: Soil Mechanics and Foundation Engineering, pp. 99–100. Wuhan University of Technology Publishing House, Wuhan (2014). (in Chinese)

Li, G.X.: Advanced Soil Mechanics, pp. 116–117. Tsinghua University Publishing House, Beijing (2004). (in Chinese)

Zheng, G., Huo, H.F., Lei, H.Y.: Undrained strength characteristics of saturated undisturbed and remolded silty clay after cyclic loading. Chin. J. Geotech. Eng. **34**(3), 400–408 (2012). (in Chinese)

Hyde, A.F.L., Ward, S.J.: The effect of cyclic loading on the undrained shear strength of a silty clay. Mar. Georesour. Geotech. **6**(3), 299–314 (1985). doi:10.1080/10641198609388192

Zhong, C., Mao, D.F., Ni, M.C., Huang, J., Yan, C., Li, X.H., Zhang, T.: Research on pile-soil interaction considering degradation of pile foundation. China Offshore Oil Gas **27**(6), 98–110 (2015). doi:10.11935/j.issn.1673-1506.2015.06.017. (in Chinese)

Yasuhara, K., Hirao, K., Hyde, A.F.L.: Effects of cyclic loading on undrained strength and compressibility of clay. Soils Found. **32**(1), 100–116 (1992). doi:10.3208/sandf1972.32.100

Li, H.: The model of undrained shear strength of saturated clay after cyclic loading and its application. MS Thesis, Tianjin University, Tianjin, China (2007). doi:10.7666/d.y1356444. (in Chinese)

Gao, G.Y., Gu, Z.H., Yang, H.M.: A calculation method for undrained strength of saturated clay under cyclic loads. Rock Soil Mech. **25**(2), 379–382 (2004). doi:10.16285/j.rsm. 2004.s2.079. (in Chinese)

Yasuhara, K.: Post cyclic undrained strength for cohesive soils. J. Geotech. Eng. **120**(11), 1961–1979 (1994). doi:10.1061/(ASCE)0733-9410(1994)120:11(1961)

Zhu, B., Xiong, G., Liu, J.C., Sun, Y.X., Chen, R.P.: Centrifuge modelling of a large-diameter single pile under lateral loads in sand. Chin. J. Geotech. Eng. **35**(10), 1807–1815 (2013). (in Chinese)

Hettler, A.: Verschiebung Starrer und Elastischer Gründungskörper in Sand be Monotoner und Zyklischer Belastung. Veröffentlichungen des Institutes für Bodenmechanik und Felsmechanik der Universität Fridericiana in Karlsruhe, Deutsch-land (1981)

Zhang, D.D.: Investigation on lateral bearing capacity of offshore wind turbine pile foundation under cyclic loading. MS Thesis, Ocean University of China, Shandong, China (2014). (in Chinese)

JTS167-4-2012: Code for Pile Foundation of Harbor Engineering (2012). (in Chinese)

Bai, B., Zhou, J., Zhang, G.: Effect of plasticity index of saturated soft clay on its deformation parameters. ShuiliXuebao **11**, 51–55 (2001). (in Chinese)

Andersen, K.H., Jacobus, H.P., Stephen, F.B., Wim, F.R.: Cyclic and static laboratory tests on Drammen clay. J. Geotech. Eng. Div. **106**(5), 499–529 (1980). doi:10.13243/j.cnki.slxb. 2001.11.009

Digital Survey for Retaining Walls - Using Commercial Cameras as Instrument

P.C. Chi[1(✉)], C.Y. Hsiao[1], C.W. Shen[1], and C.T. Wu[2]

[1] Disaster Prevention Technology Research Center, Sinotech Engineering Consultants, Inc., Taipei City, Taiwan
pcchi@sinotech.org.tw

[2] Department of Geomatics, Chien Hsin University of Science and Technology, Taoyuan City, Taiwan

Abstract. Regular inspection is essential for retaining walls to avoid deterioration or to maintain its function. An innovative technique, named as digital survey, is developed to help engineers obtaining quantitative data of deformation by approachable instrument. Commercial camera is used as the instrument and a masonry wall commonly seen on the urban slopeland is chosen as the specimen. Unconventional Photogrammetry is applied in the development to perform 3D reconstruction from 2D images. The measuring results of digital survey and of the high-precision measuring instruments were compared to verify the feasibility of this developing measurement via digital imaging technique. Furthermore, the accuracy achievement and measuring guidance are suggested in this article.

Keywords: Digital survey · 3D reconstruction · Object sample distance

1 Introduction

Taiwan's slopeland consists about 73% of the island area, and urban areas such as Taipei City, New Taipei City, and other metropolitan areas face the problems, such as dense population, rapid development. Insufficient land space leads highly slopeland development a common sight. Large number of retaining walls was built as the result. Furthermore, erosion, creep or sliding often happened under natural forces. It also needs to be stabilized and reinforced via engineering methods. Engineering methods for slope stabilization can be simply classified as retaining structures and slope face protection. Retaining structures includes ground anchors, retaining piles, and retaining walls to resist overburden pressure, whereas slope face protection use vegetation, surface drainage, shotcrete or stone-pitched surface to prevent surface erosion or shallow slide. Regardless of the type of work, the function will gradually age and deteriorate as time progresses. Hence, regular maintenance is necessary to avoid deterioration or to upkeep their functions. The maintenance inspections can reduce the probability of instability of retaining walls and slope face protection.

Engineer inspection is an effective and simple way in the process of maintenance inspection. The purposes of engineer inspection are to identify all visible changes and signs of distress of the retaining walls, to judge whether these might be significant, and to

R. Chen et al. (eds.), *Proceedings of the 2nd International Symposium on Asia Urban GeoEngineering*, Springer Series in Geomechanics and Geoengineering,
https://doi.org/10.1007/978-981-10-6632-0_38

re-assess the consequence-to-life category of the slope or retaining wall [1]. If necessary, recommend emergency measures (e.g. repair works or detailed investigations). Since 2010, in Taipei City, more than 34,000 retaining walls' spatial information and status assessments have been are recorded and archived so far [8]. According to the risk-level assessed by professional engineers, retaining walls with high-risk (performing immediate danger and important consequences, such as residential or medical institutions) bring to the attention of the owner or party required to maintain the land any immediate and obvious danger noted and, if necessary, recommend emergency measures (e.g. repair works or detailed investigations). For the walls with medium risk, commissioning professional technicians are arranged to carry out the regular inspection. The above-mentioned assessment or periodic inspections are done via visual inspection. Less quantitative data can be obtained via visual inspection. For instance, tape ruler is useful to measure the width of a single or several cracks, however, difficult to obtain surface deformation of entire wall. Moreover, three cases of retaining wall failures on the slopeland in Taipei are taken as examples. It can be seen from Fig. 1 that the commonalities of the rare cases are: (1) all are mortar-stone walls classified as flexible structures; (2) numerous existing cracks were found via visual inspection before failure; (3) destruction due to the local shear force exceeding that the wall allows and thereby causing excess deformation.

Fig. 1. Failure cases of retaining walls in Taipei City. (A) Case A, in May 2013; (B) Case B, in March 2016; (C) Case C, in June 2016

As the visual inspection method only assesses operation status by wall's superficial signs, it is not only difficult to determine its stability; local deformation variances cannot be compared and recorded. Therefore, in practice, for wall with obvious hazard signs, manual or electronic monitoring instrument is also used to supplement. It can be used on monitored and quantified stratum, retaining facility, and groundwater data to monitor internal and external slope stability. It can also set the management standard at the same time, providing information to management with contingency measures. However, the instrument monitoring method is still limited, including the number of monitoring equipment setup points and representativeness, manual measurement possibly containing too much error, recording instrument's power supply and signal transmission. Thus, the more recent Taiwanese researches have been using digital image measurement in side slope monitoring [2, 3]. The common feature is using increasingly mature imaging equipment and processing technology to develop non-contact measurement technology. On the one hand, this reduces the risk of disturbing measured object, and at the same time, we can acquire the two-dimensional and even three-dimensional deformation data that couldn't be obtained by the previous visual or instrument monitoring methods.

It is worth exploring that the deformation of retaining walls is mostly a small change in terms of magnitude. Whether using non-contact image measurement technology can achieve high accuracy, such as at mm level of observation results needing to be verified. In addition, considering the popularity and convenience of application technology, whether the video capturing equipment can use commercial cameras with general specifications is also very worthy of detailed study. This paper will focus on a masonry retaining wall in a slope area located on the mountainside of Taipei City as the experiment site, using consumer camera for image shooting equipment to apply computer visualized unconventional photogrammetry technology to render 3D Reconstruction by Using 2D Images [4], supplemented by popular high-precision measuring instrument: Total Station. The result of 3D Reconstruction by using 2D Images is compared with that of Total Station to verify the feasibility of applying the measurement methods based on the digital imaging technique to quantify observation of retaining walls. This can help to raise the accuracy and follow-up measuring standards, establish digital surveying mechanism to meet technical trend, and become an excellent tool for retaining walls inspection and record for the new generation.

2 Experiment Site

The first consideration in choosing experiment site is significantly abnormal structure. Retaining walls with obvious signs were given priority, so the final result can verify the feasibility of using digital surveying to record quantified changes. The second is having a large space for the field operation because it will be supplemented with auxiliary ground control. The purpose for the experiment site with open space, such as road or plaza, is to avoid environmental interference. Last consideration is the type for man-made slopes. Most of the man-made slopes in Taipei City are masonry retaining walls (including dry-laid and cement) [1] and cantilever retaining wall (including earth anchor), so the selection of the field is expected to include these two types of retaining walls. Based on the above principles, this study selected the dry masonry retaining wall located at Xingyi Road, Beitou District, Taipei City as the experiment site (121.31° east longitude and 25.8° north latitude).

The retaining wall, according to information provided by Taipei City Government's Geotechnical Engineering Office, file 4063-BT-GMS-0131 (4063 is the number for one to one thousandth scale in Taipei City; BT refers to Beitou District; GMS refers to grouting masonry retaining wall; 0131 is the unique serial number). The retaining wall is L-shaped where one side is built along the road (hereinafter referred to as the parallel road side); the other side is almost perpendicular to the road (hereinafter referred to as the vertical road side). The geometry size of the retaining wall are: total length 42.6 m, maximum slope 2.5 m, side slope 85°, the estimated construction time was 1980–1990. The status was evaluated as medium risk level. The main anomaly signs on the wall are tilting, obvious protruding deformation, and cracking. Figure 2 shows the most obvious cracking at the corner of the wall, with maximum crack width of 78 mm. United Geotechnical Consultants Co., Ltd. (2014) conducted the measurements twice with ground three-dimensional scanner. Based on the difference between the two test results, we proposed that the vertical road side has 3–5 mm outward deformation. The cracking at corner is not sustained expanding.

Fig. 2. The masonry retaining wall as the experiment site in this study.

3 Research Methodology and Result

Under normal circumstances, the level of deformation for retaining wall is mm to cm. In order to observe little deformation, we needed to use high-precision Total Station for observation. At the same time, the observation could be used as feasibility assessment benchmark for using consumer camera in restructuring three-dimensional technology for digital surveying as a reference for follow-up shooting project. The flow chart of this paper as shown in Fig. 3.

3.1 Total Station Theodolite Measurement and Inspection Point

This paper used the TOPCON MS05AXII. Total station electronic theodolite is a monitoring level total station, with angle accuracy of 0.5″. It used prism with accuracy ranging up to 0.8 mm + 1 ppm. Prism-free had accuracy ranging 1 mm + 1 ppm. Because the local observation distance was short, all were within 100 m, so the error was less than 1.1 mm, enough to detect a smaller changes. At the same time, this paper provided a more reliable assessment result. When considering the implementation in practice, it was hard to deploy multiple control points on a wall, so we only selected

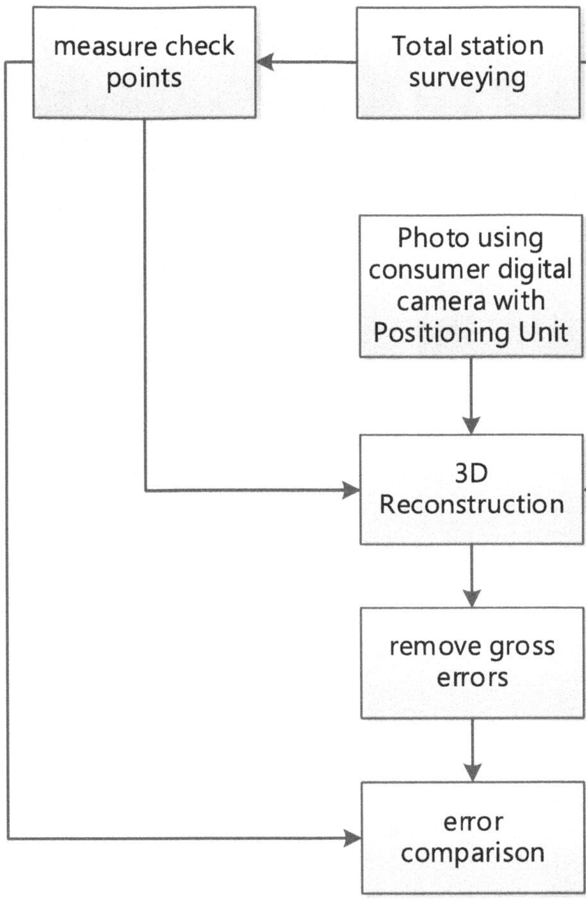

Fig. 3. Flow chart of this study.

control points outside the wall as a reference for reconstructing the three-dimensional image. The checkpoints on the wall were the benchmark for comparing three-dimensional image reconstruction errors.

Therefore, the ground control points were planned to be set up beside the road outside the wall. A total of four control points (numbered L01–L04, as shown in Fig. 4) were arranged on the ground with a white paint markers 10 cm × 10 cm in size and measured the centre steel nail for the three-dimensional coordinates. 5 cm × 5 cm red laser reflection stickers served as 5 check points on the main wall(numbered T01–T05, as shown in Fig. 4) which measure the three-dimensional coordinates of the upper right corner of laser reflection sticker. The results of measurement points and control points of this paper are shown in Table 1.

Fig. 4. Reference point distribution.

Table 1. Electronic total station measuring control points and test results.

Point#	X (m)	Y (m)	Z (m)	Remark
L01	1002.428	1006.917	99.787	Control point
L02	989.969	1006.924	100.805	Control point
L03	971.758	1006.904	102.566	Control point
L04	961.549	1007.537	103.395	Control point
T01	989.647	1009.923	102.752	Control point
T02	989.554	1009.932	103.592	Control point
T03	987.616	1009.931	102.914	Control point
T04	989.953	1013.812	103.697	Control point
T05	991.401	1020.275	104.739	Control point

3.2 Using Commercial Camera with Positioning Function

This paper used Canon EOS 5D Mark II digital camera with 20 mm optical fixed focus lens for effectively 21.1 million pixels to shoot the about 30 meter-long retaining wall. For positioning equipment, we used LEA-6T GPS receiver module and ADIS16405 gyroscope and angle accelerometer. The main mode of shooting considered the resolution of the wall in actual space (object Sample Distance, OSD) and overlap rate. They could be calculated by Eq. 1 (as shown in Fig. 5) and Eq. 2, because of wall-shooting relative distances both less than 3 m. Displacing 70 cm per second to shoot one photograph, the resolution of object was about 1 mm, and the overlap rate was about 80%, with a total of 102 photographs taken.

$$\text{Object Sample Distance} = \frac{\mathit{Object\ Distance} \times \text{Pixel Size}}{\text{Focal Length}} \quad (1)$$

Where max overlap rate (%) = (1 − (Movement rate × continuous shooting's time interval)/(object resolution × shooting image's size)) × 100

$$\text{Max overlap rate } (\%) = \frac{1 - (\mathit{Move\ speed} \times \mathit{Time\ interval})}{\text{Object Sample Distance} \times \text{Pixel Size}} \times 100\% \quad (2)$$

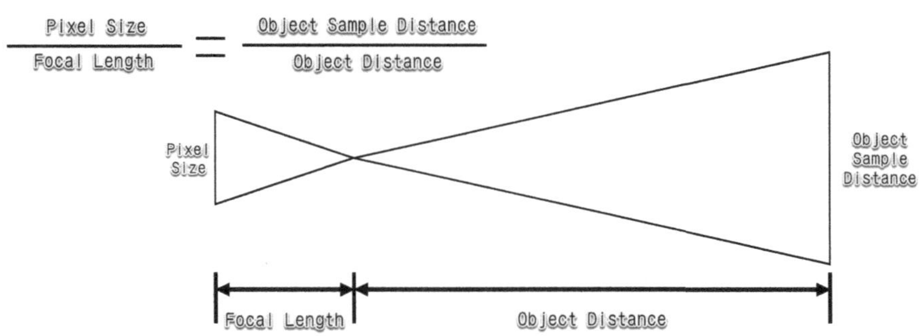

Fig. 5. Calculation of shooting object resolution.

3.3 3D Reconstruction

The three-dimensional reconstruction process is shown in Fig. 6. Consumer camera's photos all contain EXchangeable Image file Format (EXIF), including the shooting focal length, resolution (such as 5616 × 3744 pixels), and camera models. Through the camera model, we could access the original manufacturer's website to check sensor (CMOS) size, and other information., then obtained the coordinates of the main point. This means the interior orientation were known.

At the same time, this paper used SIFT (Scale-invariant feature transform) algorithm [5–7] to conduct image matching. The above algorithm mainly includes "generating eigenvector" and "matching eigenvector" [8], obtaining scale and feature points that are unchanging when rotating as well as description operator. It then achieved good image matching results through the ground control point and the interior orientation parameters. We could use collinear to obtain the exterior orientation parameters containing the error of interior orientation and the correction of the asymmetric lens distortion, and then obtained the three-dimensional point cloud.

Because the number of matching conjugate points obtained by SIFT to reconstruct 3D point cloud is relatively sparse, we further used three-step PMVS (Patch-based Multi-View Stereo, PMVS, [9]), including Matching, Expansion, and Filtering, using the same colinearity to obtain a high density colorful point cloud with three-dimensional coordinates and color values. Finally, we used the Poisson surface reconstruction provided by MeshLab to establish a three-dimensional model.

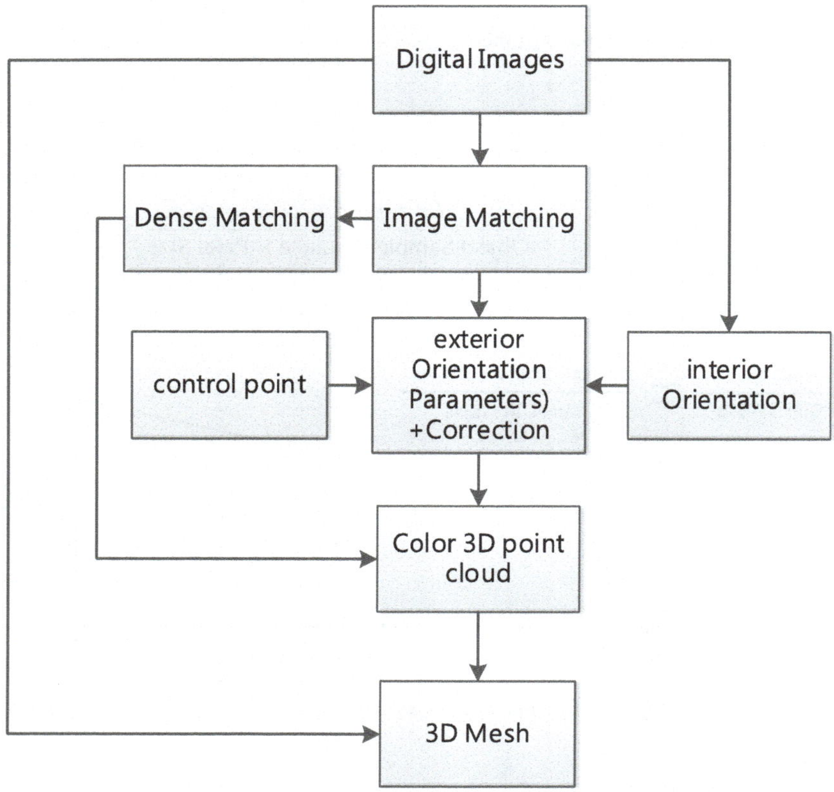

Fig. 6. Three-dimensional reconstruction process.

Fig. 7. Sparse point cloud (a), Encrypted point cloud (b), and Three-dimensional model (c)

The process of sparse point cloud, dense point cloud, and three-dimensional model are shown in Fig. 7.

3.4 Remove Gross Errors

Although obtaining three-dimensional point cloud through collinearity equation was precise, if matching point was wrong, there might be a significant error in the three-dimensional point cloud, and these erroneous three-dimensional point clouds might affect the subsequent future engineers accessing the information for analysis if comparing different periods of data. In this paper, we first established the plane equation of the retaining wall (shown in Eq. 3), and then calculated the distance between each point and the plane. We then found the error of mean squares; if the point has more than three times the error of mean squares (maximum error), annotated with special mark. Because the wall might not be completely flat, it was necessary to be determined by human whether the point was wrong and to be removed.

$$\text{The plane equation } aX + bY + cZ = d \tag{3}$$

Among them: a, b, c, d are coefficients; X, Y, Z are Cartesian coordinates.

3.5 Error Comparison

In general, the distribution of control points should be uniform and cover the entire survey area. In this paper, the control points for monitoring the retaining wall was hard to include the entire area,, so we compared the errors of the control points set on the wall and not on the wall.

1. Control points not on the wall
   The measurement data include the distance between checkpoints (T01–T05) measured by total station and the nearest control point L02, as well as the difference between coordinates of the check points in the three axes and three dimensions (3D) after 3D reconstruction (as shown in Table 2). Moreover, every two points would consist a baseline to compare the baseline lengths between three-dimensional reconstruction and the results of total station baseline (as in Table 3).
2. Control points on the wall
   The three control points T01, T03, and T05 measured by total station were used for three-dimensional reconstruction, checking the difference between the coordinates as T02 and T04 in the three-axes and three-dimensions (3D) (shown in Table 4).

Table 2. Difference comparison for control points not on the wall

Point #	Distance from control point L02 (m)	X Difference (mm)	Y Difference (mm)	Z Difference (mm)	3D Difference (mm)
T01	3.3234	−7.53	−3.32	5.15	9.71
T02	3.4652	−7.43	−4.81	5.97	10.68
T03	4.0851	−3.33	−2.04	6.49	7.57
T04	7.0948	−3.66	−8.84	12.56	15.79
T05	13.5733	2.44	−9.8	34.48	35.93

Table 3. Difference comparison for control points not on the wall.

Baseline combination	Baseline length	3D reconstruction distance			Total station distance			Baseline error		
		X	Y	Z	X	Y	Z	X	Y	Z
T01_T02	0.84518	0.0929	−0.00751	−0.84082	0.093	−0.009	−0.84	−0.0001	0.00149	−0.00082
T01_T03	2.03747	2.0268	−0.00928	−0.16334	2.031	−0.008	−0.162	−0.0042	−0.0013	−0.00134
T01_T04	4.01385	−0.30987	−3.88348	−0.95241	−0.306	−3.889	−0.945	−0.0039	0.00552	−0.00741
T01_T05	10.6859	−1.76397	−10.3455	−2.01633	−1.754	−10.352	−1.987	−0.01	0.00648	−0.02933
T02_T03	2.05318	1.9339	−0.00177	0.67748	1.938	0.001	0.678	−0.0041	−0.0028	−0.00052
T02_T04	3.90187	−0.40277	−3.87597	−0.11159	−0.399	−3.88	−0.105	−0.0038	0.00403	−0.00659
T02_T05	10.569	−1.85687	−10.338	−1.17551	−1.847	−10.343	−1.147	−0.0099	0.00499	−0.02851
T03_T04	4.59748	−2.33667	−3.8742	−0.78907	−2.337	−3.881	−0.783	0.00033	0.0068	−0.00607
T03_T05	11.1649	−3.79077	−10.3362	−1.85299	−3.785	−10.344	−1.825	−0.0058	0.00776	−0.02799
T04_T05	6.70469	−1.4541	−6.46204	−1.06392	−1.448	−6.463	−1.042	−0.0061	0.00096	−0.02192

Note: the baseline error of each axis = each axis' three-dimensional reconstruction distance − each axis' total station distance

Table 4. Result difference between control points on the wall and not on the wall.

Point #	X Difference (mm)	Y Difference (mm)	Z Difference (mm)	3D Difference (mm)
T02	−2.63	0.52	0.85	2.81
T04	−1.30	−2.85	−3.78	4.91

4 Discussion

Regarding the deformation and fracture development in retaining walls, in the past, it was necessary to directly contact the retaining wall using manual recording. The measurement was time-consuming and difficult to record comprehensively, and it was not easy to quantify the spatial relationship of local deformation or fracture development. For new crack or local deformation, direct comparison and accessing past data were impossible. In this paper, we use consumer digital camera that is popular and convenient to shoot the retaining wall. Three-dimensional reconstruction can be created indoor to obtain large number of retaining wall spatial data points and three-dimensional model. This not only saves labour force on-site and time; the data is more complete, better for historical data review, and helping to maintain retaining walls during operation.

In order to verify the feasibility and accessibility of reconstructed three-dimensional model for the retaining wall by consumer digital camera, the error comparison experiment from the above Sects. 3–5 was conducted. The result is discussed below:

1. The control point not on the wall:

(1) Checkpoint error: The error can be seen in Table 2, ground control point range covers only the X axis direction of the inspection point, and only the T03 checkpoint is covered by the two ground control points L02 and L03. The inspection point's Y axis or Z axis direction contains greater error when farther away from control point. It is speculated assumed that control point has a wider distribution in the X axis direction on the experiment site. At the same time,

besides the T03 inspection point, other inspection points also contain greater error along the reconstructed 3D coordinates. The errors in 3D coordinates are in the range of 7.51 mm–35.93 mm, about 8 to 36 times the OSD.

(2) Baseline error: The error is the greatest when baseline length is between 5–7 m, except for baseline associated with the T03 checkpoint. On shorter baselines (less than 4 m) or longer baselines (greater than 10 m), the baseline length errors are small. Baseline length error does not increase as the baseline grows longer (as in Fig. 8). At the same time, the baseline is divided into the directions of three axes X, Y, Z to calculate error. The longer the baseline, the greater the error in each axis. Moreover, X, Z axis errors and the Y axis errors are in the opposite directions (as in Fig. 9). This may be due to inconsistency between the three-axes of three-dimensional reconstruction and total station, so the Y-axis difference is opposite to the X and Z axes, causing the error to be partially offset. Consequently, baseline error does not increase as its length becomes longer.

2. Control points on the wall: From Table 3, when control points were placed on the wall, and they covered all the inspection points, the best control results are obtained, such that errors in T02 and T04's X axis, Y axis, and Z axis directions as well as the 3D reconstruction coordinates were greatly reduced, ranging from 2.81 mm to 4.91 mm, about 3 X–5 X OSD.
3. Comparing the two control point layout methods, placing control points on the wall allows wider distribution in all X, Y, Z axial directions. This not only significantly reduces three-dimensional error and OSD but also the field operation is much simpler, saving hours of work.

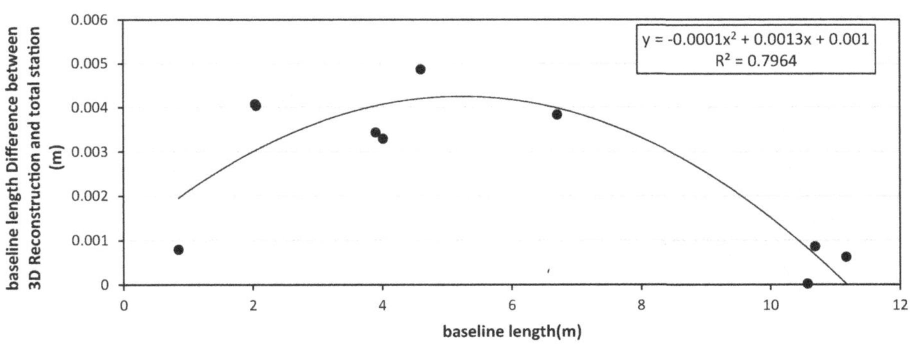

Fig. 8. Baseline length error and correlation.

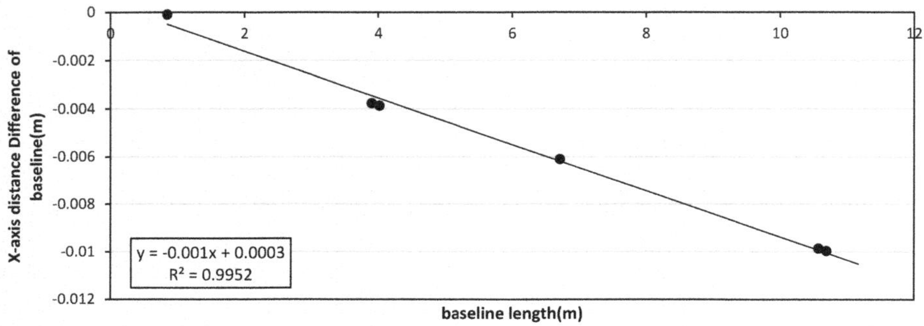

(a)

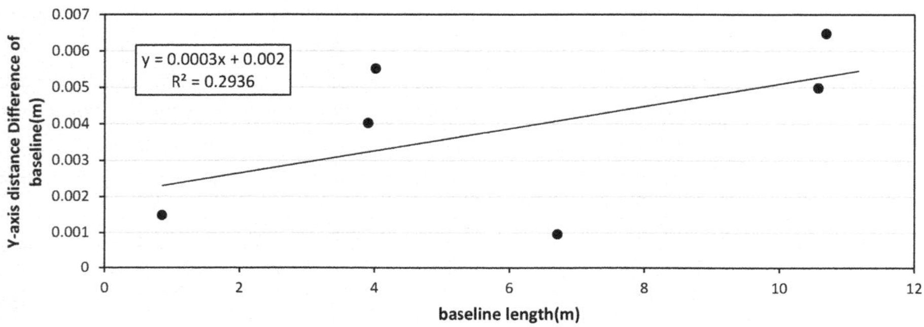

(b)

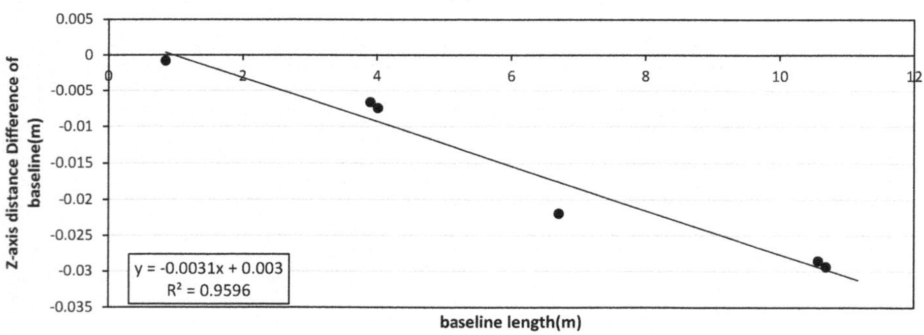

(c)

Fig. 9. Relationship between each axis error and baseline length. (a) X axis; (b) Y axis; (c) Z axis.

5 Conclusions and Recommendations

1. The study proposes using consumer cameras as an instrument for digital survey of retaining walls. It can provide completely 3D record of the retaining wall, whether or not it already has obvious exterior damage, and is conducive to recording the complete quantitative wall data at different times.
2. The results confirmed that using 1 mm OSD digital camera for three-dimensional reconstruction within the shooting distance 13 meters from the control points, if the control points are not on the wall, the farther away from the control point, the greater the error will be. The three-dimensional error is less than 36 times OSD. If the control points are on the wall, the three-dimensional error is reduced and less than 5 times OSD. In this case, when using 1 mm OSD, the X, Y, Z axial direction as well as three-dimensional errors are reduced to within 5 mm. This initially conforms the precision requirement of the three-dimensional point cloud and coordinate reconstruction of the retaining wall.
3. For retaining wall which is not allowed setting up control points, more inspection points may be added to experiment and calculate the rotation relationship from three-axial directions between the 3D reconstruction and total station's measurement, for revising the system error and improve accuracy.
4. For retaining wall surface that allows setting up control points, we may try using photos with higher OSD and accurate instrument to measure control points (such as 0.1 mm). Different periods of information can be compared if this can be used in wall surface monitoring, and establish more suitable shooting plan evaluation method to monitor retaining walls.

Acknowledgement. We would like to extend our gratitude to Geotechnical Engineering Office, Public Works Department, Taipei City Government for their funding and experiment site.

References

1. GEO: Guide to Slope Maintenance, 3rd edn. Geotechnical Engineering Office, Hong Kong (2003)
2. Hu, X.F., Zhao, H.: The Practical Case Selection of Image Processing and Recognition by Using Visual C++/MATLAB. Posts and Telecom Press, Beijing (2004). (in Chinese)
3. Hsiao, C.Y., Hsieh, P.S., Chi, S.Y.: Preliminary study of using unconventional photogrammetry to evaluate the earthwork volume – a case study of the landslide occurred on the Cidu Section of Formosan Freeway. J. Chin. Soil Water Conserv. **42**(2), 120–130 (2011). (in Chinese)
4. Hsiao, C.Y., Hsieh, P.S.: Applied unconventional photogrammetry of computer vision in civil engineering. Sinotech Eng. **131**, 41–49 (2016). (in Chinese)
5. Lowe, D.G.: Object recognition from local scale-invariant features. In: International Conference on Computer Vision, Corfu, Greece, pp. 1150–1157 (1999)
6. Lowe, D.G.: Distinctive image features from scale-invariant key points. Int. J. Comput. Vision **60**, 91–110 (2004)

7. Lowe, D.G.: Automatic panoramic image stitching using invariant features. Int. J. Comput. Vision **74**(1), 59–73 (2006)
8. Sinotech Engineering Consultants, Inc.: Inspection and data maintenance of man-made slopes. Technical report. Geotechnical Engineering Office, Public Works Department, Taipei City Government (2015). (in Chinese)
9. Tung, S.H., Shih, M.H., Sung, W.P., Weng, M.C.: Applying digital image correlation method to detect deformation of abamurus. Disaster Adv. **5**, 496–501 (2012)

Effect of Foreshocks of the 2016 Kumamoto Earthquakes on the Aso-Bridge Slope

C. Hung[1(✉)], B.-C. B. Hsiung[2], K.-H. Yang[3], and L. Ge[3]

[1] Department of Civil Engineering, National Cheng Kung University, Tainan, Taiwan (ROC)
chinghung@mail.ncku.edu.tw

[2] Department of Civil Engineering, National Kaohsiung University of Applied Sciences, Kaohsiung, Taiwan (ROC)
benson@cc.kuas.edu.tw

[3] Department of Civil Engineering, National Taiwan University, Taipei, Taiwan (ROC)
{khyang,louisge}@ntu.edu.tw

Abstract. This paper presents an examination the Aso-bridge slope subjected to the foreshocks of the 2016 Kumamoto earthquakes. While the Aso-bridge slope was marginal stable in the absence of earthquake, it remained stable during the foreshocks that struck on April 14 and 15 2016, with magnitudes of 6.5 and 6.4 M_w, respectively. To validate that the slope did not fail during the foreshocks and the applicability of an enhanced FS method capable of determining the initiation time of co-seismic landslides, relationships between the Aso-bridge slope and the nearest recorded seismic signals, including the main-shock and the foreshocks, were investigated. It was found that the two foreshocks, having PGAs of 43.0 and 129.7 gal, were not strong enough to cause failure of the slope. They were not strong enough mainly because of the smaller magnitude as compared to the main-shock of 7.0 M_w with a PGA of 508.9 gal, leading to a significant difference in the PGAs. The study demonstrates that the enhanced FS method can be one efficient and practical alternative for determining large-scale co-seismic landslides.

Keywords: Co-seismic landslide · Aso-bridge · Earthquake-induced landslide · Initiation time

1 Introduction

Analyzing the mechanism of co-seismic landslide complications are technically challenging undertakings that depend on the geological conditions, seismic loadings, dip angles, prolonged rainfall, seepage, and other factors. In the perspective of geotechnical engineering, studies on earthquake-induced landslides are typically associated with slope stabilities. To that extent, emerging methodologies, considering the Newmark's sliding block concept (Newmark 1965), have been rapidly progressed to tackle landslide problems. Note that understanding the mechanism of failure of seismically induced landslides, such as revealing the initiation time, can be beneficial for hazard mitigation and reconstruction of the progressive behavior of co-seismic landslides; nevertheless, it

R. Chen et al. (eds.), *Proceedings of the 2nd International Symposium on Asia Urban GeoEngineering*, Springer Series in Geomechanics and Geoengineering,
https://doi.org/10.1007/978-981-10-6632-0_39

is nontrivial to both accurately and practically depicts the initiation time of co-seismic landslides.

Despite the fact that numerous studies using various and simpler methods have been practiced for analyzing the co-seismic landslides, large numbers of co-seismic landslides remain unsolved and new ones are inevitable in the future. For the past few decades, when analyzing earthquake-induced landslides, the Newmark's method has been a well-acknowledged approach in evaluating the permanent displacements. Applications and documents based on the Newmark's theory can be found elsewhere (Ling and Cheng 1997; Ling 2001; Jibson 2007; Lin et al. 2017; Hung et al. 2017a). Ling (2001) utilized the sliding block theory to geotechnical design and proposed a procedure that determines the seismic factor of safety against sliding along a joint plane. Lin et al. (2017) indicated that the mechanism of landslides is very complicated and having an appropriate initiation time of a landslide is crucial in understanding the mechanism of an earthquake-induced landslide. These studies address that a defined initiation time of a landslide, with accuracy, is essential for the purpose of understanding the progressive failure behavior of co-seismic landslides. Hung et al. (2017a,b) applied simplified sliding block concepts and numerical codes to evaluate a large-scale landslide in Japan and found that the initiation time of such landslide can be revealed via several indications, including the directional tendency, a factor of safety of the slope, and the Newmark's displacement.

While the studies of the Aso-bridge slope subjected to the foreshocks are not yet documented, and noting that the foreshocks of the 2016 Kumamoto earthquakes reached high magnitudes of 6.5 and 6.4 M_w, this paper studies the effect of these foreshocks on the Aso-bridge slope. An enhanced FS method, considering the Newmark theory and dynamic condition, is utilized to examine the Aso-bridge slope under the foreshocks of the 2016 Kumamoto earthquakes. The main focus of the study is on the relationship between the seismic signals and the occurrence of landslides, and to confirm that the Aso-bridge slope did no fail during the foreshocks. Along with the aforementioned analysis, the permanent displacement of the landslide based on Wilson and Keefer (1983) is also presented.

2 Background

2.1 Aso-Bridge Slope

The Aso-bridge slope was located near to Kumamoto City, Kumamoto Prefecture in the Kyushu Region of Japan. The geometry of the landslide is shown in Fig. 1, and it occupied 700 m in length and 200 m in width.

As suggested by Hung et al. (2017a), the failure mechanism of the Aso-bridge slope was dominated by a rotational slump and further developed into a debris avalanche that covered the foot of hillslope and destroyed the Aso-bridge. The basic geology of the hillslope of the landslide is lava and pyroclastic rocks from eruptions. Lots of voids between particles of the geomaterials suggests that there are prone to adsorb water and be permeated by it, and may be easily weakened subjected to rainfall (Yang et al. 2015). Over characteristics were observed, and they include that the main scarp is flat, the crown exhibited tension cracks, and the failed material was tilting backward of the slope.

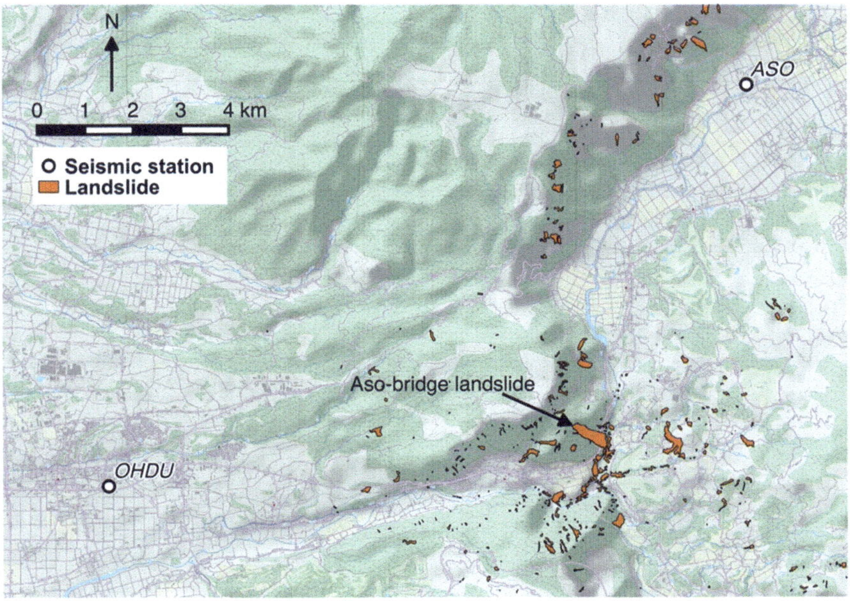

Fig. 1. Aso-bridge landslide (Kumamoto City, Kyushu Region of Japan)

2.2 Strong-Motion Stations

The ground motion records used in this study were obtained from the OHDU strong motion station, with a start level of 6 gal, sampling frequency of 100 Hz, and with a duration of 300 s, and is maintained by the Ministry of Home Affairs, Japan. The station is about 10 km away from the incident slope and is situated at N32.8761, W130.8774, and 124 m in altitude. Figure 2 shows the ground accelerations of the main-shock in the vertical (U-D), north-south (N-S), and east-west (E-W) directions. The recorded maximum accelerations were about 396.89 (N-S), 525.37 (N-S), and 482.33 (E-W) gal. The first acceleration peak occurred at approximately 20.5 s, and between 20.5 and 30 s, most of the accelerations were observed. Figure 3 shows the accelerations in dip-direction.

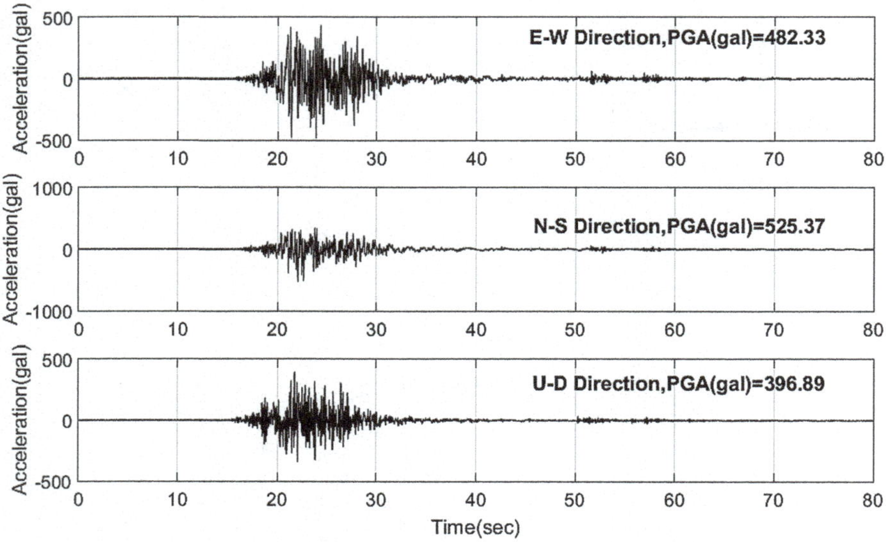

Fig. 2. Seismic signals (OHDU)

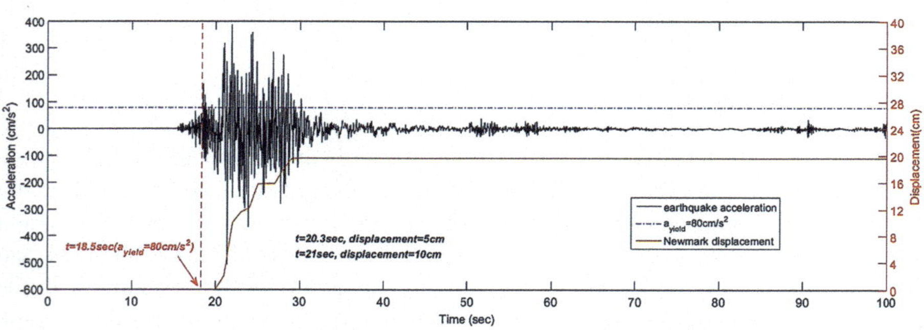

Fig. 3. Newmark's displacement (main-shock)

3 Methodology

Details of the concept of the enhanced FS method can be found in Lin et al. (2017). In brief, the concept related to the sliding block concept proposed by Newmark (1965). The mathematical expression of the Newmark theory is:

$$a_c = (FS - 1)g sin\alpha \tag{1}$$

in the equation, a_c is the critical acceleration, FS, factor of safety of the slope, gravity, and the inclined angle are represented by a_c, FS, g, and α, respectively. According to Wilson and Keefer (1983), it is suggested that the displacement will occur when the seismic acceleration exceeds the yield acceleration. Note that the method is more suitable for a static problem and less applicable considering a dynamic condition.

The detailed formulation and predictive capabilities of the enhanced FS method, which is applicable to a dynamic situation, have been documented in Lin et al. (2017). In brief, the enhanced FS method can be expressed as Eq. (2) as show:

$$FS = \frac{\left[\mu_s\left(gcos\alpha + a_n\right) + \frac{cA}{m}\right] - a_d}{g \sin \alpha} \tag{2}$$

in which, a_n is the direction normal to the slide surface, a_d is the tangential to the slide surface along the dip-direction, g is gravity, α is the slope angle, μ_s is the coefficient of friction of the material, c is the cohesion, A is the area of the interface between the block and sliding surface, and m is the mass of the sliding block. The equation considers simplified sliding and traditional force equilibrium concepts. The dependency of the FS value to the earthquake acceleration a_n and a_d is shown in the equation.

4 Result and Discussion

4.1 The Aso Slope

Figure 3 shows the yield acceleration value obtained using the Newmark's concept, and Fig. 4 demonstrates a yield acceleration of 80 gal subjected to the seismic accelerations. It is shown that the cumulative displacement increases when the acceleration exceeds the yield acceleration of 80 gal. A detailed discussion can be found from Hung et al. (2017a), and the Aso-bridge slope appeared to fail between 20.3–21 s.

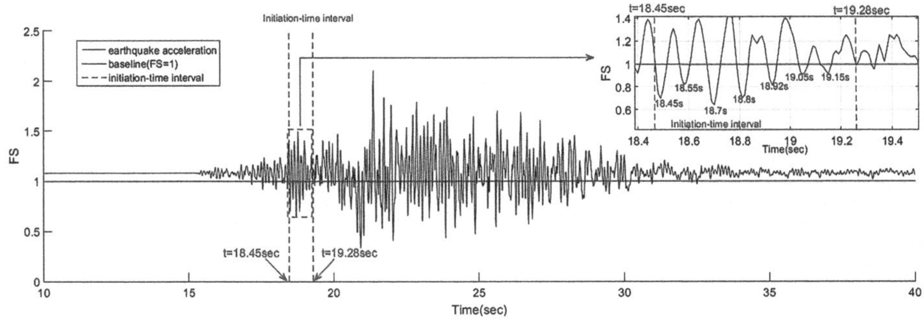

Fig. 4. Enhanced FS values (main-shock)

Results of the two main foreshocks against the slope are shown in Figs. 5 and 6. Based on the results, the two foreshocks, despite their apparent large magnitude of 6.4 and 6.5 M_w, were not strong enough mainly because of the smaller ground accelerations. Note that the accumulated Newmark displacement of the two foreshocks was estimated to be less than 0.01 cm. The results are comparable to the actual situation of the Aso-bridge slope, and the applicability of the enhanced FS method to the Aso-bridge slope under the foreshocks is satisfactory.

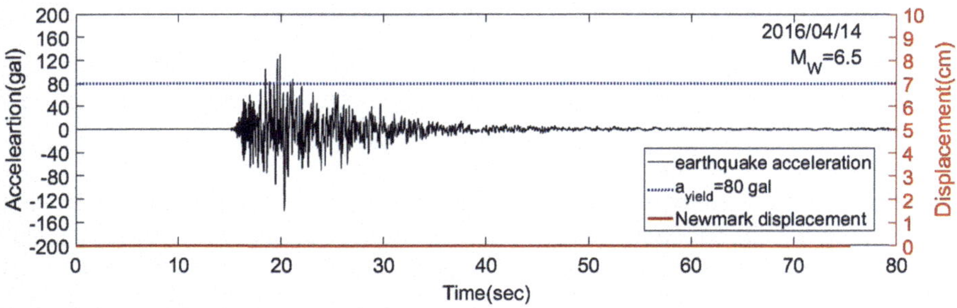

Fig. 5. Enhanced FS values during the foreshock of the 2016 Kumamoto earthquakes: April 14, 2016

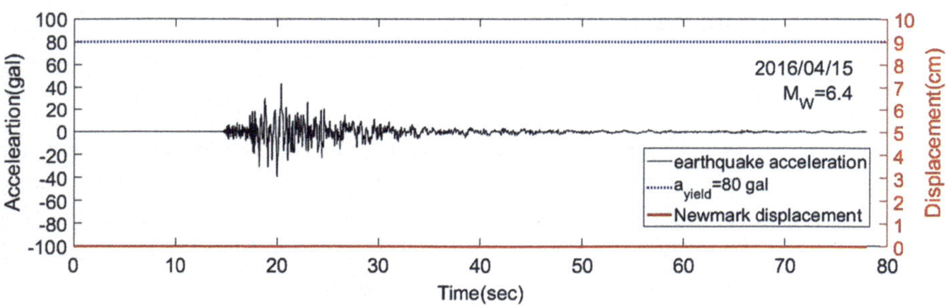

Fig. 6. Enhanced FS values during the foreshock of the 2016 Kumamoto earthquakes: April 15, 2016

5 Conclusion

The relationships between the two foreshocks of the 2016 Kumamoto earthquakes against the Aso-bridge slope were investigated and revealed by both the enhanced FS method and the accumulated Newmark's displacement. The former one showed that the FS values are mostly larger than 1.0 throughout the entire foreshocks, while the latter one resulted in an estimation of merely 0.01 cm. It is evident that the Aso-bridge slope, despite its marginal stable condition in the absence of earthquake, would not fail during the two foreshocks. Based on the satisfactory results as compared to the actual situation of the Aso-bridge slope, the enhanced FS method has demonstrated its potential in engineering application. Although the present study was encouraging, the presented approach should be used to analyze other co-seismic landslides with well-documented monitoring data.

Acknowledgements. This work was supported, in part, by funding from the Ministry of Science and Technology, Taiwan. The authors appreciate the assistance in providing data by the National Research Institute for Earth Science and Disaster Resilience of Japan, the Geospatial Information Authority of Japan, the Geographical Survey Institute of Japan, and the Ministry of Home Affairs, Japan.

References

Newmark, N.M.: Effects of earthquakes on dams and embankments. Geotechnique **15**(2), 139–160 (1965)

Ling, H.I.: Recent applications of sliding block theory to geotechnical design. Soil Dyn. Earthq. Eng. **21**, 189–197 (2001)

Ling, H.I., Cheng, A.H.-D.: Rock sliding induced by seismic force. Int. J. Rock Mech. Min. Sci. **34**(6), 1021–1029 (1997)

Jibson, R.W.: Regression models for estimating coseismic landslide displacement. Eng. Geol. **91**, 209–218 (2007)

Lin, G.W., Hung, C., Syu, H.S.: Evaluation of an enhanced FS method for finding the initiation time of earthquake-induced landslides. Bull. Eng. Geol. Environ. (2017). doi:10.1007/s10064-017-1083-7

Wilson, R.C., Keefer, D.K.: Dynamic analysis of a slope failure from the 6 August 1979 Coyote Lake, California, earthquake. Bull. Seismol. Soc. Am. **73**(3), 863–877 (1983)

Yang, H., Wang, F., Vilimek, V., Araiba, K., Asano, S.: Investigation of rainfall-induced shallow landslides on the northeastern rim of Aso caldera, Japan, in July 2012. Geoenviron. Disasters **2**, 20 (2015)

Hung, C., Lin, G.W., Syu, H.S., Chen, C.W., Yen, H.Y.: Analysis of the Aso-bridge landslide during the 2016 Kumamoto earthquakes, Japan. Bull. Eng. Geol. Environ. (2017a). doi: 10.1007/s10064-017-1103-7

Hung, C., Syu, H.S., Lin, G.W.: Applying simplified sliding block concepts and finite element codes to analyze the Aso-bridge landslide induced by the 2016 Kumamoto earthquakes. In: Geophysical Research Abstracts, vol. 19, EGU2017-5931 (2017b)

Field Measurements and Numerical Analyses of a Pumping-Recharge Test in a Deep Confined Aquifer

Yangqing Zhang, Mingguang Li, and Jianhua Wang(✉)

Shanghai Jiao Tong University, Shanghai, China
wjh417@sjtu.edu.cn

Abstract. Pumping-induced groundwater drawdown and strata deformations can be relieved by artificial recharge. In this study, field measurements and numerical analyses of a single-well pumping with multi-well recharge test are presented to investigate the responses of groundwater and strata to pumping and recharge. A three-dimensional numerical model is developed in a finite-difference software, which considers the fluid-mechanical interaction using the Biot consolidation theory. Both measured data and numerical results indicate that the influence of recharge on the groundwater drawdown caused by pumping correlates with the injection rate. A larger injection rate may result in a greater decrease of drawdown. As artificial recharge can compensate the drawdown by injecting water into the aquifer, the change of effective stress of soils induced by pumping can be reduced. Consequently, pumping-induced strata settlements and ground settlements can be relieved by artificial recharge. Therefore, artificial recharge can be used to reduce the groundwater drawdown and ground settlements during the pumping and protect surrounding facilities from settlement caused by pumping, if an appropriate injection rate and well arrangements are determined.

Keywords: Artificial recharge · Pumping · Deep confined aquifer · Strata deformation · Field test · Numerical analyses

1 Introduction

Deep excavation conducted above confined aquifers calls for dewatering to reduce the artesian water level in order to safeguard the deep excavation (Hsi and Small 1992). Dewatering will increase the effective stress of the soils and lead to soil compression. If the deep excavation is conducted in a congested metropolitan area, the strata deformations and ground settlement caused by dewatering may cause irreparable damage to adjacent facilities (Forth 2004; Zhou et al. 2010, Serrano-Juan et al. 2016). To relieve the dewatering-induced strata deformations and ground settlements in a deep excavation, artificial recharge can be adopted apart from inserting sealing curtains (Wang et al. 2012; Chen et al. 2013; Pujades et al. 2014; Xu et al. 2014; Wu et al. 2015a, b).

The artificial recharge was originally proposed to supply the groundwater and mitigate land subsidence induced by the excessive groundwater extraction (Phien-wej et al. 1998; Bouwer 2002; Missimer et al. 2011; Zhang et al. 2015). In recent years, artificial

R. Chen et al. (eds.), *Proceedings of the 2nd International Symposium on Asia Urban GeoEngineering*, Springer Series in Geomechanics and Geoengineering,
https://doi.org/10.1007/978-981-10-6632-0_40

recharge has been applied to deep excavations for environment protection (Wang et al. 2012; Lu et al. 2015). By carrying out the in-situ pumping and recharge tests in a shallow confined aquifer in Shanghai, and performing the corresponding numerical simulation, Wang et al. (2012) investigated the pumping-recharge effect in shallow confined aquifers and the combined effects of pumping, curtain and recharge. However, the tests only focused on the recharge in shallow confined aquifers as well as the groundwater level control, while no attention was paid to the efficiency on controlling the ground settlement. Lu et al. (2015) investigated the recharge well arrangement and injection rate by field pumping-recharge tests before construction. Most of the existing studies focused on the pumping and recharge in phreatic aquifers and shallow confined aquifers, and few investigated the artificial recharge in deeper confined aquifers during the excavation (Phien-wej et al. 1998; Ni et al. 2011; Wang et al. 2012; Zhou et al. 2014; Wang et al. 2015; Zhang et al. 2017).

In this study, field measurements and numerical analyses of a single-well pumping with multi-well recharge test is presented to investigate the responses of groundwater and strata to pumping and recharge. A three-dimensional numerical model is developed in a finite-difference software (FLAC^{3D}), which considers the fluid-mechanical interaction using the Biot consolidation theory.

2 Background

A pumping-recharge test was conducted for the deep excavation of a metro station in Shanghai. The excavation was adjacent to the Jinfeng Building and the flood control wall of Suzhou River, as shown in Fig. 1 As expected, dewatering will cause adverse impact on the surrounding environment, e.g. land subsidence (Shen and Xu 2011) and settlement of facilities (Shen et al. 2014). To protect the adjacent facilities, artificial

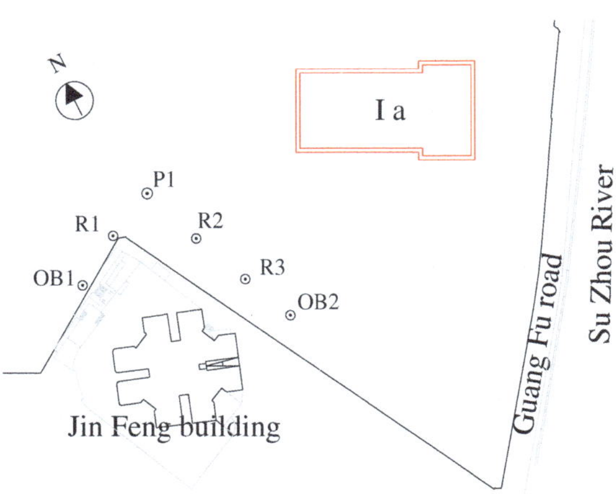

Fig. 1. Plan view of test site

recharge was proposed for the excavation and a field single-well pumping with multi-well recharge test had been carried out for evaluating the recharge efficiency prior to excavation.

According to the geological investigation report, soils at the site are Quaternary deposits and consist of saturated clayey soil and silty soil with the characteristics of stratified distribution. The typical strata profile is shown in Fig. 2. The soft deposits at the site can be defined as a multi-aquifer-aquitard system with a high groundwater head. The groundwater within the depth of 85 m consists of one phreatic aquifer and two confined aquifers (hereinafter labeled as confined aquifer I to Confined aquifer II), and

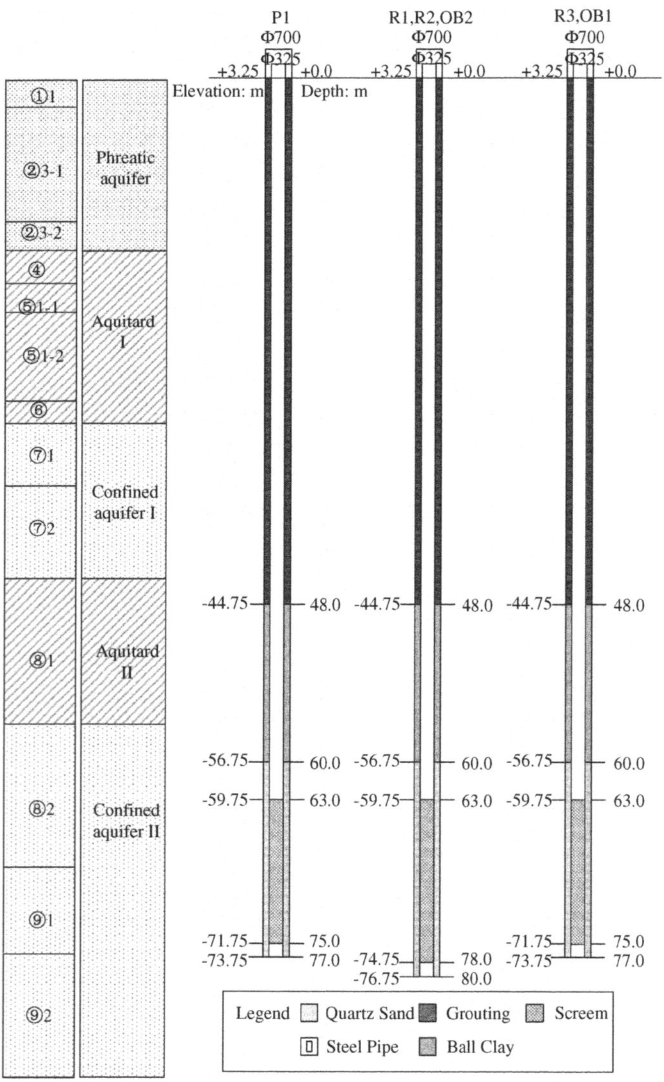

Fig. 2. Typical strata profile and the layout of the wells.

the aquifers are separated by two aquitard (hereinafter labeled as aquitard I to aquitard II). The piezometric head of phreatic water is 0.80 to 1.70 m below the ground surface and changes with precipitation. The confined aquifer I is mainly composed of sandy silt (layer ⑦1) and silt (layer ⑦2) and has a thickness of 10.6~17.2 m. The piezometric head of confined aquifer I is 3 to 11 m below the ground surface and changes with seasons. The confined aquifer II is mainly composed of silty clay with silt (layer ⑧2), silt with silty clay (layer ⑨1) and sand (layer ⑨2). The piezometric head of confined aquifer II is 5.10 to 6.23 m below the ground surface and changes with seasons.

The pumping-recharge test was performed in confined aquifer II. The detailed process of the tests is listed in Table 1. Figure 1 depicts the layout of the wells. One pumping well (labelled P1), three recharge wells (labelled R1 to R3) and two observation wells (labelled OB1 and OB2) were installed in the site. All these wells can be used as observation wells. The structures of these wells are presented in Fig. 2. All screens of these wells were in confined aquifer II. The wells R1, R2 and OB2 had a screen of 15 m in length and a total length of 80 m. The wells P1, R3 and OB1 had a screen of 12 m in length and a total length of 77 m.

Table 1. Process of the pumping-recharge test.

Stage	Pumping well	Recharge well	t_0	t_t (min)	Q_p (m^3/h)	Q_r (m^3/h)
1	P1	–	7:00 6 Oct.	485	21	–
2	P1	R1	15:05 6 Oct.	955	19	13
3	P1	R1, R2	11:00 7 Oct.	360	19	9,7
4	P1	R1, R2, R3	17:00 7 Oct.	480	19	10,9,3

Note: t_0 = starting time; t_t = pumping and recharge time; Q_p = discharge; Q_r = injection.

3 Methodology

A coupled three-dimensional numerical model is developed in a finite-difference software, which considers the fluid-mechanical interaction using the Biot consolidation theory. For the case considered in this study, the pumping-recharge test was conducted for only 38 h, and the deformation of the soils caused by pumping and recharge was within the elastic range. Therefore, the Mohr-Coulomb constitutive model is used to describe the stress and strain behavior of the soils.

3.1 Model Description

Figure 3 presents the meshes of the numerical model and Fig. 4 presents the analysis range and plan view of meshes. The dimensions of the model are set to 1027 m by 1112 m in-plane and 90 m in depth. The total numbers of nodes and elements are 173580 and 165097, respectively. In the vertical direction, the geology of the site includes 13 strata and is represented by 29 layers of elements.

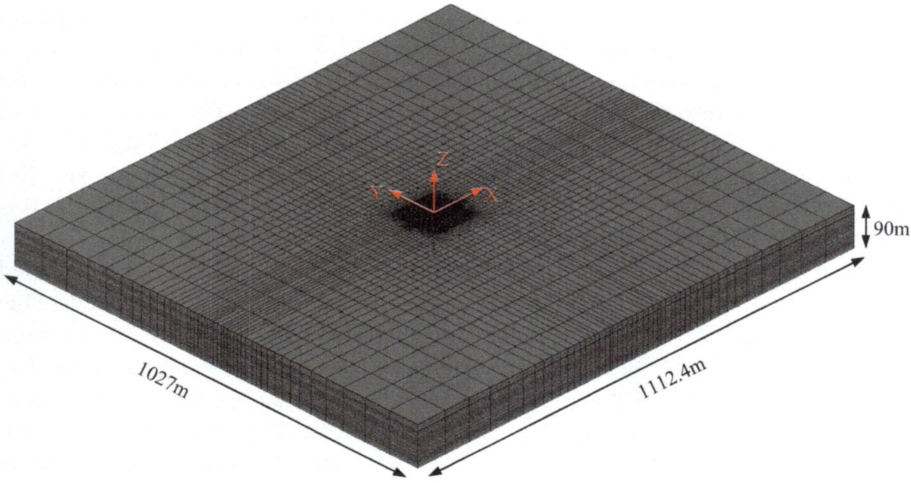

Fig. 3. Meshes of the numerical model.

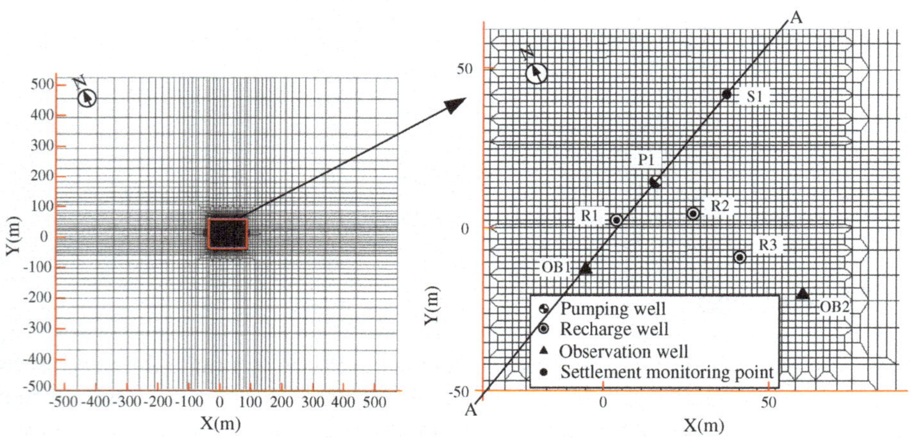

Fig. 4. Plan view of mesh and layout of the wells.

The soil movements at the bottom of the model are restrained in all directions ($u_x = u_y = u_z = 0$). Gridpoints over four vertical sides are only free to move in the vertical direction and restrained from movement in other directions. For convenience of calculation, the initial hydraulic heads of phreatic aquifer and confined aquifer I are set on the ground surface. The initial hydraulic head of confined aquifer II is set as 5.23 m below the ground surface. The hydraulic head at the boundaries of the model remains constant during the simulation. In the numerical analysis, the source and sink term represents the recharge wells and pumping well. A pumping well is simulated by applying a discharge rate of groundwater to a zone where the screen of the pumping well is located. Similarly, a recharge well is simulated by applying an injection rate of groundwater. Figure 4 depicts the layout of the wells.

3.2 Soil Parameters

Table 2 lists the values of the void ratio *e*, elastic modulus *E*, cohesion *c*, friction angle φ and hydraulic conductivity *k* for each layer. The elastic modulus is based on the compression modulus obtained from the oedometer tests. The cohesion and friction angle are determined from the triaxial tests. The vertical hydraulic conductivity and horizontal hydraulic conductivity of soil are determined from laboratory tests.

Table 2. Soil layers and parameters used in analysis

Layer	Depth (m)	γ (kN/m^3)	*e*	*E* (MPa)	*C* (kPa)	φ (°)	k_v (m/day)	k_h (m/day)
① 1 Backfill	2	18.0	0.8	40	10	18	2.68E-02	4.32E-02
②3–1 Sandy silt	11.7	18.5	0.88	62	5	30.5	2.68E-02	4.32E-02
②3–2 Sandy silt	1.8	18.4	0.88	85	5	31	6.48E-02	1.03E-01
④ Silty clay	2.8	16.9	1.36	97	13	11	9.16E-05	1.43E-04
⑤1–1 Clay	3.1	17.9	1.07	107	15	13.5	8.73E-05	1.03E-04
⑤1–2 Silty clay	7.7	18.2	0.97	127	17	18	2.91E-04	7.86E-04
⑥ Silty clay	2.6	19.6	0.69	145	18	17	1.43E-04	3.06E-04
⑦1 Sandy silt	5.6	18.6	0.84	154	4	32	4.20E-01	9.50E-01
⑦2 Silt	8.3	18.9	0.78	178	2	33.5	8.52E-01	3.55E+00
⑧1 Clay	13	18	1.03	227	18	17	3.98E-04	1.71E-03
⑧2 Silty clay with silt	12.4	18.2	0.95	259	22	20.5	1.00E-01	2.00E-01
⑨1 Silt with silty clay	7.6	18.8	0.80	294	9	28.5	7.20E-01	1.02E+00
⑨2 Sand	11.4	19.5	0.65	327	2	34	3.46E+00	5.18E+00

Note: γ = unit weight; *e* = void ratio; *E* = elastic modulus; *c* = cohesion; φ = friction angle; k_v = vertical hydraulic conductivity; k_h = horizontal hydraulic conductivity.

3.3 Simulation Process

The simulation process of the field test includes four stages. In stage 1, only well P1 is activated as a pumping well and works throughout the test. The discharge rate of P1 is 21 m^3/h. In stage 2, the discharge rate of P1 is reduced to 19 m^3/h and remains the same in the next two stages, and well R1 is activated with an injection rate of 13 m^3/h. In stage 3, both wells R1 and R2 are activated, and the total injection rate is 16 m^3/h. In the final stage of the test, wells R1, R2 and R3 are activated, and the total injection rate is 22 m^3/h. The detailed simulation process is listed in Table 1.

4 Results and Analysis

4.1 Groundwater Drawdown

Figure 5 displays the observed and calculated groundwater drawdown in well OB1 during the test. In the first stage, the drawdown increases with time. After the completion of stage 1, both the numerical results and measured data indicate that the drawdown in OB1 decreased with the recharge, and a larger injection rate results in a greater decrease of drawdown in OB1. It can also be seen that numerical results and measured data are in good agreement, which reveals that the numerical analysis is capable to model the groundwater head variation caused by pumping and recharge.

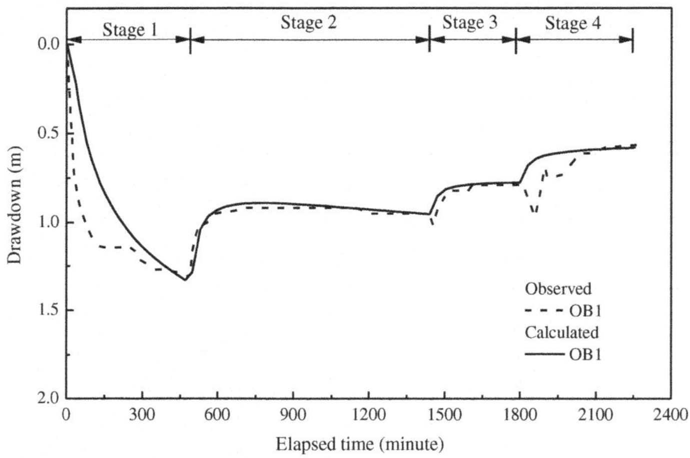

Fig. 5. Observed and calculated groundwater drawdown during the test.

Figure 6 presents the contours of the groundwater drawdown in confined aquifer II during the test. As can be observed in Fig. 6a, the contours of the groundwater drawdown caused by single-well pumping form a circle shape. Figures 6b, c and d present the contours of the groundwater drawdown in the confined aquifer II at the end of stages 2, 3 and 4. It can be seen from these three figures that the drawdown around the recharge wells decreases dramatically as the recharge wells are activated. Note that the well R1 was located far from wells R2 and R3, so the superimposed effect of these recharge is negligible. It can be concluded that artificial recharge can relieve or even eliminate the pumping-induced drawdown and a reasonable well arrangement will lead to a desired recharge effect.

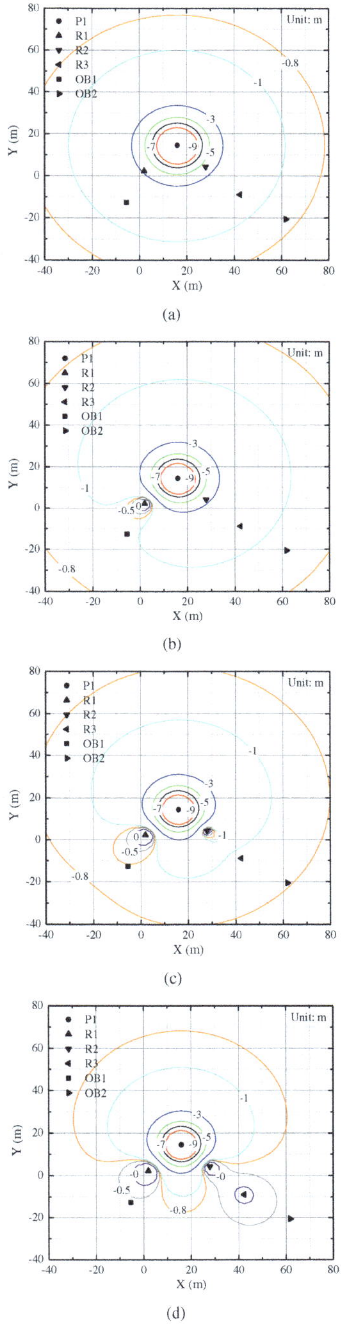

Fig. 6. Contours of drawdown for confined aquifer II during the pumping-recharge test: (a) single-well pumping (stage 1); (b) single-well recharge (stage 2); (c) double-well recharge (stage 3); and (d) triple-well recharge (stage 4).

4.2 Stratum Deformation

Figure 7 illustrates calculated ground settlements at points OB1 and S1 during the test. The layout of OB1 and S1 is shown in Fig. 4. Note that the points OB1 and S1 have a same distance from well P1, but S1 is located far from the recharge wells and OB1 is near recharge wells. In stage 1, only the pumping well is activated. The ground settlements at OB1 and S1 increase with time. After the activation of recharge wells, the tendencies of the ground settlements at these two points differed from each other. The ground surface at S1 settles continuously throughout the test, while the ground settlement at OB1 tends to be stable at the end of stage 2. This discrepancy can be attributed to the different location of these two points. Note that the maximum ground settlement at OB1 is smaller than that at S1.

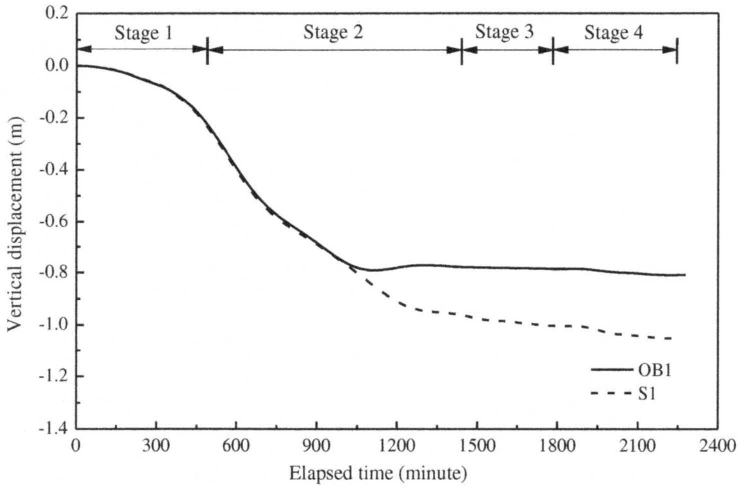

Fig. 7. Calculated ground settlements at points OB1 and S1 during the test.

Figure 8 demonstrates the calculated vertical deformations of strata at points OB1 and S1. As can be seen from Fig. 8, the vertical deformations of strata at OB1 and S1 are the same in stage 1. After the activation of recharge wells in stages 2, 3 and 4, the vertical deformations of strata at OB1 are smaller than that at S1 due to the recharge. It is evident that the recharge can effectively relieve the strata settlement induced by dewatering.

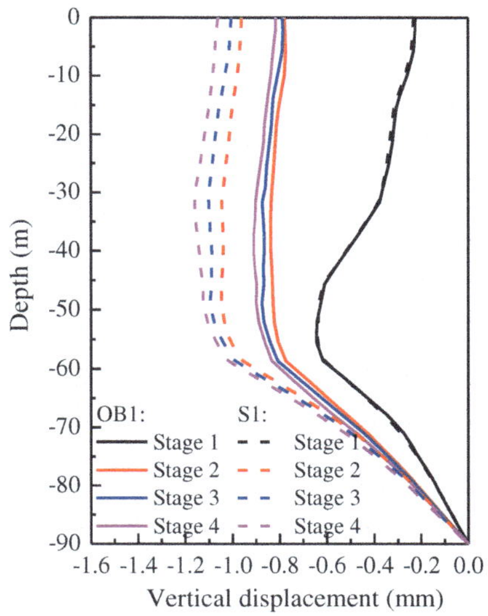

Fig. 8. Vertical deformations of the strata at points OB1 and S1.

5 Conclusions

Through the analyses of measured data and three-dimensional numerical simulation, this study presents field measurements and numerical analyses of a single-well pumping with multi-well recharge test in the confined aquifer II in Shanghai. The responses of groundwater and strata to artificial recharge are investigated. Based on the field measurements and numerical simulation results, the following conclusions can be drawn:

1. Artificial recharge can relieve or even eliminate the pumping-induced drawdown. The influence of recharge on the groundwater drawdown caused by pumping is correlated with the injection rate. A larger injection rate may result in a greater decrease of drawdown. Moreover, a reasonable well arrangement will lead to a desired recharge effect.
2. Artificial recharge can relieve the strata settlements and ground settlements induced by pumping. As artificial recharge can compensate the drawdown by injecting water into the aquifer, the change of effective stress of soils induced by pumping can be reduced. Consequently, pumping-induced strata settlements and ground settlements can be relieved by artificial recharge.
3. Artificial recharge can be used to reduce the groundwater drawdown and ground settlements during the pumping and protect surrounding facilities from settlement caused by pumping, if an appropriate injection rate and well arrangements are determined.

Acknowledgements. The financial support from the National Natural Science Foundation of China (NSFC Grant Nos. 41602283, 41472250, 41330633) is gratefully acknowledged.

References

Bouwer, H.: Artificial recharge of groundwater: hydrogeology and engineering. Hydrogeol. J. **10**(1), 121–142 (2002). doi:10.1007/s10040-001-0182-4

Chen, J.J., Zhang, L., Zhang, J.F., Zhu, Y.F., Wang, J.H.: Field tests, modification, and application of deep soil mixing method in soft clay. J. Geotech. Geoenviron. Eng. **139**(1), 24–34 (2013). doi:10.1061/(ASCE)GT.1943-5606.0000746

Forth, R.A.: Groundwater and geotechnical aspects of deep excavations in Hong Kong. Eng. Geol. **72**(3–4), 253–260 (2004). doi:10.1016/j.enggeo.2003.09.003

His, J.P., Small, J.C.: Ground settlements and drawdown of the water table around an excavation. Can. Geotech. J. **29**(5), 740–756 (1992). doi:10.1139/t92-082

Lu, J.S., Miao, J.F.: Discussion on the integration design of well pumping-recharge of foundation pit in soft soil area. Chin. J. Undergr. Space Eng. **11**(S1), 232–238 (2015). (in Chinese)

Missimer, T.M., Drewes, J.E., Maliva, R.G., Amy, G.: Aquifer recharge and recovery: groundwater recharge systems for treatment, storage, and water reclamation. Ground Water **49**(6), 771 (2011). doi:10.1111/j.1745-6584.2011.00846.x

Ni, J.C., Cheng, W.C., Ge, L.: A case history of field pumping tests in a deep gravel formation in the Taipei Basin, Taiwan. Eng. Geol. **117**, 17–28 (2011). doi:10.1111/j.1745-6584.2011.00846.x

Phien-wej, N., Giao, P.H., Nutalaya, P.: Field experiment of artificial recharge through a well with reference to land subsidence control. Eng. Geol. **50**(1–2), 187–201 (1998). doi:10.1016/S0013-7952(98)00016-7

Pujades, E., Vázquez-Suñé, E., Carrera, J., Vilarrasa, V., De Simone, S., Jurado, A., Ledesma, A., Ramos, G., Lloret, A.: Deep enclosures versus pumping to reduce settlements during shaft excavations. Eng. Geol. **169**, 100–111 (2014). doi:10.1016/j.enggeo.2013.11.017

Serrano-Juan, A., Vázquez-Suñè, E., Monserrat, O., Crosetto, M., Hoffmann, C., Ledesma, A., Criollo, R., Pujades, E., Velasco, V., Garcia-Gil, A., Alcaraz, M.: Gb-sar interferometry displacement measurements during dewatering in construction works. case of la Sagrera railway station in barcelona, spain. Eng. Geol. **205**, 104–115 (2016). doi:10.1016/j.enggeo.2016.02.014

Shen, S.L., Wu, H.N., Cui, Y.J., Yin, Z.Y.: Long-term settlement behavior of the metro tunnel in Shanghai. Tunneling Undergr. Space Technol. **40**(2014), 309–323 (2014). doi:10.1016/j.tust.2013.10.013

Shen, S.L., Xu, Y.S.: Numerical evaluation of land subsidence induced by groundwater pumping in Shanghai. Can. Geotech. J. **48**(9), 1378–1392 (2011). doi:10.1139/T11-049

Wang, J.X., Feng, B., Liu, Y., Wu, L.G., Zhu, Y.F., Zhang, X.S., Tang, Y.Q., Yang, P.: Controlling subsidence caused by de-watering in a deep foundation pit. Bull. Eng. Geol. Env. **71**(3), 545–555 (2012). doi:10.1007/s10064-012-0420-0

Wang, T.T., Zhang, S.S., Huang, T.H.: Determining transmissivity of fracture sets with statistical significance using single-borehole hydraulic tests: Methodology and implementation at Heshe well site in central Taiwan. Eng. Geol. **198**, 1–15 (2015). doi:10.1016/j.enggeo.2015.09.006

Wu, Y.X., Shen, S.L., Xu, Y.S., Yin, Z.Y.: Characteristics of groundwater seepage with cut-off wall in gravel aquifer. I: Field observations. Can. Geotech. J. **52**(10), 1526–1538 (2015a). doi:10.1139/cgj-2014-0285

Wu, Y.X., Shen, S.L., Yin, Z.Y., Xu, Y.S.: Characteristics of groundwater seepage with cut-off wall in gravel aquifer. II: Numerical analysis. Can. Geotech. J. **52**(10), 1539–1549 (2015b). doi:10.1139/cgj-2014-0289

Xu, Y.S., Shen, S.L., Ma, L., Sun, W.J., Yin, Z.Y.: Evaluation of the blocking effect of retaining walls on groundwater seepage in aquifers with different insertion depths. Eng. Geol. **183**, 254–264 (2014). doi:10.1016/j.enggeo.2014.08.023

Zhang, Y.Q., Wang, J.H., Chen, J.J., Li, M.G.: Numerical study on the responses of groundwater and strata to pumping and recharge in a deep confined aquifer. J. Hydrol. **548**, 342–352 (2017). doi:10.1016/j.jhydrol.2017.03.018

Zhang, Y., Wu, J.C., Xue, Y.Q., Wang, Z.C., Yao, Y.G., Yan, X.X., Wang, H.M.: Land subsidence and uplift due to long-term groundwater extraction and artificial recharge in Shanghai, China. Hydrogeol. J. **23**(8), 1851–1866 (2015). doi:10.1007/s10040-015-1302-x

Zhou, N.Q., Vermeer, P.A., Lou, R.X., Tang, Y.Q., Jiang, S.M.: Numerical simulation of deep foundation pit dewatering and optimization of controlling land subsidence. Eng. Geol. **114**(3–4), 251–260 (2010). doi:10.1016/j.enggeo.2010.05.002

Zhou, X.J., Burbey, T.J.: Deformation characteristics of a clayey interbed during fluid injection. Eng. Geol. **183**, 185–192 (2014). doi:10.1016/j.enggeo.2014.10.001

Synthesis and Characterization of Metakaolin Modified Waste Concrete-Based Geopolymer

Wu Chen[1,2(✉)] and Zhiduo Zhu[1,2]

[1] School of Transportation, Southeast University, Nanjing, China
woochan_seu@163.com
[2] Jiangsu Key Laboratory of Urban Underground Engineering and Environmental Safety, Nanjing, China

Abstract. Demolition of construction work generates a significant amount of waste concrete. Most of waste concrete is disposed in landfill, which is costly and occupies vast expanses of land and also causes environmental problems. Currently, the utilization of waste concrete is limited to the utilization of recycled aggregates in non-structure filler and pavement base. This study investigates the feasibility of utilizing waste concrete powder to synthesize geopolymeric binder. The obtained geopolymeric binder can be mixed with recycled aggregates to synthesize new geopolymeric concrete. Due to the high silicon and calcium content for the waste concrete powder, metakaolin is used to optimize the silicon, aluminium and calcium content. The unconfined compressive strengths (UCS) of the geopolymeric binders synthesized at different alkaline activator concentrations and metakaolin contents are evaluated and the optimal values of them are determined. The microstructure and phase compositions of the source materials and the synthesized geopolymeric binders are studied base on SEM/EDS and XRD analyses. The results show that the UCS of geopolymeric binder increases with the addition of metakaolin up to 80% content and then decreases. Therefore, adding proper content of metakaolin to waste concrete powder, the strength of geopolymeric binder can reach the required value.

Keywords: Geopolymer · Metakaolin · Waste concrete · Unconfined compressive strength · Microstructure

1 Introduction

Waste concrete is derived from demolition of the deteriorating construction work. There has been an enormous increase of waste concrete in China during last decades [1]. Disposal of waste concrete has been a great challenge in construction industry [2]. Most of the waste concrete is collected and disposed in landfill [3], which is costly and occupies vast expanses of land and also causes contamination of nearby soil and water [4].

Research on recycling of waste concrete has been conducted by researchers [5, 6]. However, utilization of waste concrete in the current practice is still limited to the use of recycled aggregates in non-structural filler and pavement base [7]. Recycled aggregate contains adhered mortar/paste (from the original concrete) and natural coarse aggregate. The adhered mortar/paste consists of fine aggregate, hydrated and unhydrated cement

R. Chen et al. (eds.), *Proceedings of the 2nd International Symposium on Asia Urban GeoEngineering*, Springer Series in Geomechanics and Geoengineering,
https://doi.org/10.1007/978-981-10-6632-0_41

particles, which causes it has increased water absorption and quantity of dust particles, decreased specific gravity and bulk density, and inferior mechanical properties in comparison to the natural aggregate. These inferior properties hinder the abundant utilization of recycled aggregate in new concrete as it affects the workability, mechanical property and durability of concrete. Some advanced processing methods such as "freeze-thaw", "thermal expansion" and "mechanical grinding" [8–10] have been proposed to produce high quality aggregates by removing the adhered mortar but keeping the integrity of natural coarse aggregate. However, these techniques are high energy consumption with the extra powder still being handled.

Recently, a new inorganic binder, named geopolymer [11], has aroused the interest of researchers from different engineering fields. In addition to comparable performance to Portland cement in many aspects, geopolymer also exhibits additional superiorities such as fast mechanical strength development, excellent adherence to aggregates, and significant reduction of energy consumption and CO_2 emission. Since waste concrete is rich in silica and contains a certain amount of alumina, it is a potential source material to synthesize geopolymers. By utilizing the waste concrete powder fraction as the geopolymer binder and the recycled concrete aggregates as filling material, waste concrete could be completely utilized. The old mortar adhered to the recycled aggregates will incorporate in geopolymeric reaction due to the silicon and aluminium ingredients and thus lead to a strong cementation between the aggregates and the geopolymeric binder. Komnitsas et al. [12] have investigated the effect of synthesis parameters on the quality of waste concrete based geopolymer. However, their study indicated that waste concrete shows limited geopolymerization potential since it reaches a compressive strength of only 13 MPa.

To improve compressive strength of the waste concrete based geopolymer, this study mainly investigates the feasibility of utilizing the waste concrete (WC) powder mixed with metakaolin (MK) to synthesize geopolymeric binder. The silicon (Si), aluminium (Al) and calcium (Ca) contents are optimized by adding MK into the high Si and Ca content WC powder. Sodium hydroxide (NaOH) together with sodium silicate (SS) solution was used as alkaline activator. The unconfined compressive strengths (UCS) of the geopolymeric binders synthesized at different NaOH concentrations and MK contents were measure. To study the microstructure and phases compositions of the synthesized binder, Scanning electron microscopy/energy dispersive spectroscopy (SEM/EDS) imaging and X-ray diffraction (XRD) analyses were also conducted

2 Materials and Methods

2.1 Materials

WC powder, MK, reagent grade 98% NaOH, SS solution, and deionised water were used in this study. The WC powder was derived from the crushed and grinded Portland cement concrete specimens at the pavement and bridge lab of Southeast University. The MK was purchased from China under the brand name of BASF that was produced by thermal activation of high purity kaolin clay within a specific temperature range. The chemical composition of the WC and the MK determined by X-ray fluorescence

(XRF) is presented in Table 1, the WC powder mainly contains SiO_2 and CaO while the MK primarily consists of SiO_2 and Al_2O_3.

Table 1. Chemical components of waste concrete (WC) powder and metakaolin (MK)

Component (%)	WC	MK
SiO_2	38.81	52
CaO	21.91	0.95
Al_2O_3	9.99	40
Fe_2O_3	3.5	2.15
K_2O	1.63	0.4
MgO	1.63	0.8
Na_2O	0.74	0.5

Figure 1 shows the particle size distribution of the WC powder determined by a Hydro 2000Mu Melvin Particle Size Analyzer. The D_{50} of WC is 8.2 μm with about 90% particles finer than 75 μm. The D_{50} of the purchased MK is 4.8 μm with 99.9% particles passing 325 Mesh. The microstructure of raw materials determined by SEM is given in Fig. 2. The WC particles have an irregular shape while the MK particles are platy.

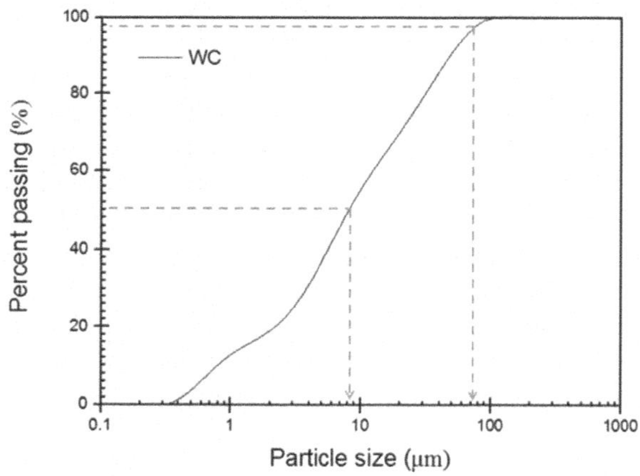

Fig. 1. Particle size distribution of WC powder

The NaOH pellets purchased from ZD chemical company (in China), were dissolved in deionised water to produce NaOH solution. SS solution(Na_2O = 8.52%, SiO_2 = 27.28%, H_2O = 64.2%) was also purchased from ZD chemical company, having a modulus (SiO_2/Na_2O) of 3.1.

Fig. 2. SEM micrographs of metakaolin (MK)-a, b, and waste concrete (WC) powder-c, d

2.2 Sample Preparation and Test Methods

Dry WC powder and dry MK at a given MK content were completely mixed. Then the alkaline activator consists of NaOH and SS solution was mixed with the mixture. The NaOH solution was given enough time to cool down to ambient temperature and then added into the SS solution at a selected SS/NaOH mass ratio. The well stirred activator was mixed with the WC/MK mixture. The resulted mixture was allowed to be stirred completely, leading to dissolution of more Si, Al, and Ca in the liquid phase. Since the addition of MK and higher NaOH concentration resulted in higher viscosity for the paste, various water content were adopted to keep the workability consistent as shown in Table 2. The resulted paste was poured to iron molds measuring 20 × 20 × 20 mm and then shaken for 2 min to release the trapped air bubbles. Next, the mold was capped and cured at 75 °C for 24 h for gel setting and then de-molded. The specimens were placed in plastic bags and cured at ambient temperature for another 6 days' before tested.

Totally, six MK contents (i.e. 0, 20, 40, 60, 80 and 100% by total mass of the dry WC/MK mixture) for the solid powder mixture and three concentrations (i.e. 6, 10, and 14 M) for the NaOH solution with SS at an SS/NaOH mass ratio of 1.5 were used.

Unconfined compression tests were conducted on the cubic sample with a CBR-2 press machine at a constant strain rate of 1.27 mm/min to evaluate the UCS of geopolymers synthesized at different conditions. Three specimens for each condition

Table 2. Combination of variables and conducted tests on WC/MK specimens

Specimen	MK (%)	NaOH conc. (M)	SS/NaOH	Water content (%)	Si/Al	Ca/Si	Na/Al	Na/Si	UCS test	XRD	SEM/EDS
WC	0	NA	NA	NA	3.30	0.60	NA	NA	NA	√	√
MK	100	NA	NA	NA	1.10	0.02	NA	NA	NA	√	√
6-0-1.5[a]	0	6	1.5	24.58	3.83	0.52	0.72	0.19	√		
6-20-1.5	20	6	1.5	29.75	2.60	0.41	0.55	0.21	√		
6-40-1.5	40	6	1.5	34.28	2.04	0.30	0.46	0.22	√		
6-60-1.5	60	6	1.5	37.51	1.71	0.20	0.39	0.23	√		
6-80-1.5	80	6	1.5	40.10	1.49	0.11	0.35	0.23	√		
6-100-1.5	100	6	1.5	47.22	1.36	0.02	0.35	0.26	√		
10-0-1.5	0	10	1.5	26.53	3.86	0.52	0.96	0.25	√		
10-20-1.5	20	10	1.5	33.83	2.65	0.40	0.76	0.29	√		
10-40-1.5	40	10	1.5	37.81	2.06	0.30	0.62	0.30	√		
10-60-1.5	60	10	1.5	40.46	1.72	0.20	0.52	0.30	√		
10-80-1.5	80	10	1.5	45.11	1.51	0.11	0.48	0.32	√		
10-100-1.5	100	10	1.5	50.41	1.37	0.02	0.45	0.33	√		
14-0-1.5	0	14	1.5	27.64	3.87	0.52	1.16	0.30	√	√	√
14-20-1.5	20	14	1.5	35.05	2.66	0.40	0.92	0.35	√		
14-40-1.5	40	14	1.5	41.79	2.10	0.29	0.80	0.38	√		
14-60-1.5	60	14	1.5	45.16	1.75	0.20	0.68	0.39	√		
14-80-1.5	80	14	1.5	48.53	1.53	0.11	0.60	0.39	√	√	√
14-100-1.5	100	14	1.5	54.60	1.39	0.02	0.57	0.41	√	√	√

[a] 6-0-1.5 means specimen at 6 M NaOH concentration, 0% MK and SS/NaOH of 1.5

were measured and the average of the tested values was adopted. To polish the end faces of the samples to make them parallel and flat before tested.

SEM/EDS and XRD were conducted to study the microstructure and phase composition of geopolymers. The SEM imaging/EDS analysis was conducted in a HITACHI SU3500 scanning electron microscope. The XRD analysis was performed with a PANALYTICAL X'Pert PRO MRD/XL diffractometer. WC powder, MK and three geopolymers at different MK contents were investigated. Table 2 presents the combination of variables conducted tests in this study.

3 Results and Discussion

3.1 UCS

Figure 3 shows the effect of MK content and NaOH concentration on UCS of the geopolymeric binder.

3.1.1 Effect of NaOH Concentration

As shown in Fig. 3, increased NaOH concentration leads to higher UCS, which is also reported by other researchers [13, 14]. Two mechanisms have accounted for the effect of NaOH in the geopolymeric reaction: dissolution of Si and Al species; charge-balancing of Al species [15], which can be presented schematically as follows [16]:

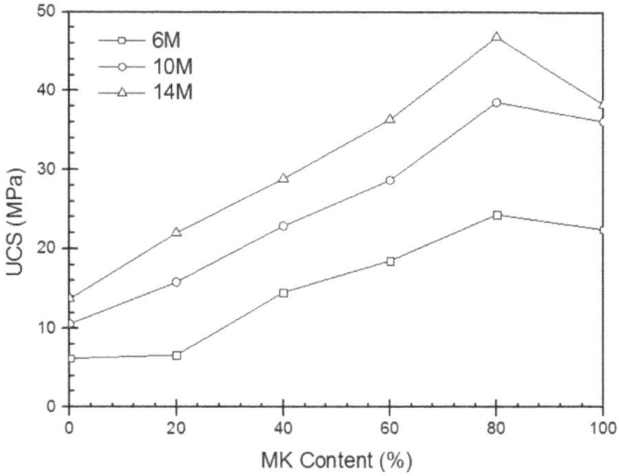

Fig. 3. UCS versus metakaolin (MK) content (percent of total mass of WC and MK) for specimens synthesized at 6, 10, and 14 M NaOH and SS/NaOH of 1.5

$$\begin{aligned} &\mathrm{n}(Si_2O_5, Al_2O_2) + 2\mathrm{nSi}O_2 + 4nH_2O + NaOH \\ &\quad \rightarrow Na^+ + n(OH)_3\mathrm{-Si-O-Al^-(OH)_2-O-Si-(OH)_3} \end{aligned} \tag{1}$$

$$\begin{aligned} &\mathrm{n}(OH)_3\mathrm{-Si-O-Al^-(OH)_2-O-Si-(OH)_3 + NaOH} \\ &\quad \rightarrow \mathrm{Na^+} - \left(\begin{array}{c} \quad | \qquad\quad | \qquad\quad | \\ -\mathrm{Si-O-Al^- -O-Si-O-} \\ \quad | \qquad\quad | \qquad\quad | \end{array} \right) + 4\mathrm{nH_2O} \end{aligned} \tag{2}$$

The increase of UCS with NaOH concentration is primarily due to the fact that at higher NaOH, the more Si and Al dissolve and consequently larger quantities of Si and Al components incorporate in geopolymerization.

It can be seen from the scheme, Na^+ plays the role of an activator on Si and Al species. Therefore, the ratios of Na/Si and Na/Al are important factors affecting the structure of geopolymer. Table 2 shows the Na/Si and Na/Al ratios used in this investigation ranging from 0.35 to 1.16 and 0.19 to 0.41 respectively. According to Table 2 and Fig. 3, despite the two specimens 14-0-1.5 and 14-100-1.5 prepared at highest ratios of Na/Si and Na/Al, they showed lower UCS values, which indicate other factors also play important roles in the geopolymerization such as Ca/Si and Si/Al ratios. However, maintaining the ratios of Ca/Si and Si/Al constant, higher NaOH concentration results in higher UCS because of the large dissolution of Si and Al [17]. Researchers have investigated Na/Al ratios in the range of 0.38 to 2.06 [15, 18–23], a ratio of around 1 is usually recommended [18, 24].

3.1.2 Effect of Initial Si/al

The amount of Si and Al contributes significantly to the synthesis of the geopolymer product as presented in the above scheme. This is probably why researchers have investigated the influence of Si to Al ratios on geopolymerization [22, 24–28]. Duxson et al. [24] indicated that the maximum compressive strength for metakaolin based geopolymer is obtained at a Si/Al ratio of 1.9. Zhang et al. [25] also investigated the effect of Si/Al ratio on the mechanical properties of metakaolin based geopolymer, however they found a Si/Al ratio of 2.75 corresponding to the highest strength, which is close to the value observed by Rowles et al. [22, 26] who synthesized geopolymer from metakaolin activated by sodium silicate solution. Cheng [27] suggested the Si/Al ratio in the range of 1.58–1.73 is optimum for the blended metakaolin and blast furnace slag based geopolymer. Figure 4 presents the Si/Al ratios corresponding to the optimum values of different geopolymeric system reported by the researchers above.

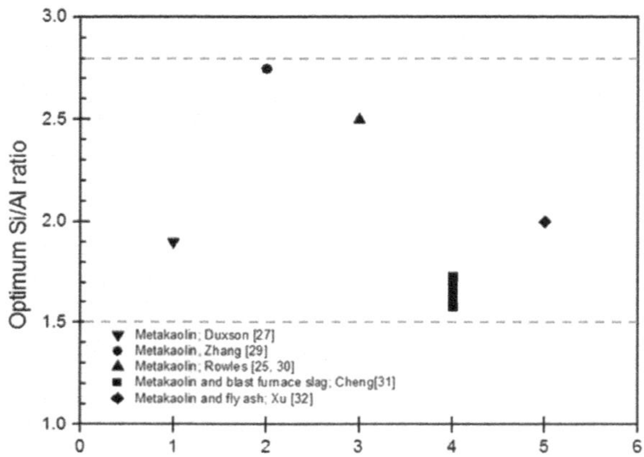

Fig. 4. Optimum Si/Al ratios from different researchers for different source materials

To reach the maximum strength of a geopolymer, generally the Si/Al ratio should be in the range of approximately 1.5 to 2.8 (see Fig. 4). In this study, the Si/Al ratios are in the range of 1.36 to 3.87 (see Table 2) with the addition of MK into the high silica content WC. Figure 5 presents the UCS changes with the initial Si/Al ratio. As the Fig. 5 presents the UCS enhances with the Si/Al ratio up to about 1.53 which is close to the optimum ranges shown in Fig. 4 and then decreases with larger Si/Al.

Besides, MK is regarded as a more reactive source material for geopolymerization because of its high geologic purity and higher reactive phases [24], which may also accounts for the fact that UCS increases with addition of MK. At higher reactivity, the solid phase dissolves more Si and Al species leading to larger amount of Si and Al in reaction phase and then contributing to geopoymerization. Furthermore, Diaz [29] indicated that compared to larger particles, finer ones have larger specific surface area and consequently more reactive due to the fact that the reaction takes place at the interface between particle and liquid. MK particles have a smaller mean size of 4.8 μm and thus larger specific surface area, leading to higher reactivity to alkaline solution.

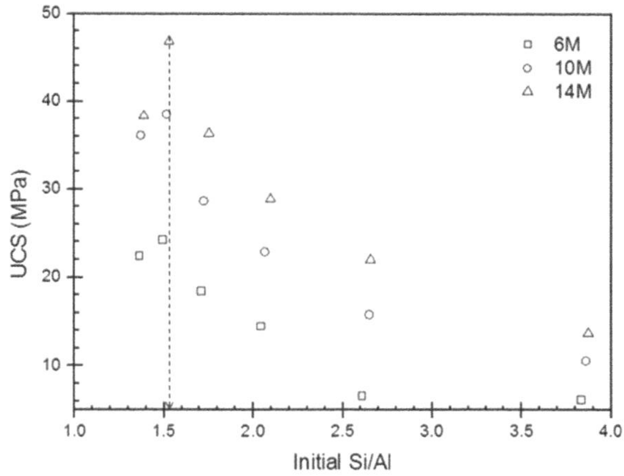

Fig. 5. UCS versus initial Si/Al ratio in the current study

3.1.3 Effect of Calcium

Researchers [30–32] have reported that Ca has an improving effect on the compressive strength of geopolymeric binders. Two mechanisms can be concluded to account for the fact. The first sindicates that Ca^{2+} serves as a charge-balancing component incorporated in the geopolymeric network. The other states that Ca promotes synthesis of calcium silicate hydrate (CSH) gel helping to fill the voids between the reacted phases and unhydrated particles. In the current study the highest UCS was obtained at the MK content of 80%, which is higher than that of the specimen in the absence of the WC. It confirms that Ca originated and diffused from the WC exhibits improving effect on the UCS of the geopolymeric binder.

It should be noted that the positive impact of Ca should be confined to a certain amount of Ca in geopolymerization as researchers reported [30, 33, 34]. In this study, the high content of Ca in WC is decreased by adding the MK into the WC. The UCS increases with the MK content up to content about 80% and then decreases, which can be interpreted as the positive impact of WC is confined to 20% added WC in the WC/MK based geopolymeric system. The adverse impact of WC is primarily attributed to the precipitation of $Ca(OH)_2$. The pH value in the geopolymeric system is approximately 14, which is inclined to cause the precipitation of $Ca(OH)_2$ [35]. $Ca(OH)_2$ is highly unstable, which will further form calcium carbonate as a result of atmospheric carbonation [36]. Furthermore, precipitation of $Ca(OH)_2$ will also have hindered Ca species to be incorporated in the synthesis of CSH, which are the major product in the alkali-activated system of blended MK and $Ca(OH)_2$ [37]. It is therefore expected that a substantial quantity of $Ca(OH)_2$ will be formed in a system with excess WC present, leading to a rapid reduction in UCS, which is in agreement with the observation shown in Fig. 3. Figure 6 is derived from Fig. 3 to present the relationship between Ca/Si ratio and UCS. Corresponding to the highest UCS, the optimum Ca/Si in the WC/MK geopolymeric system is 0.11.

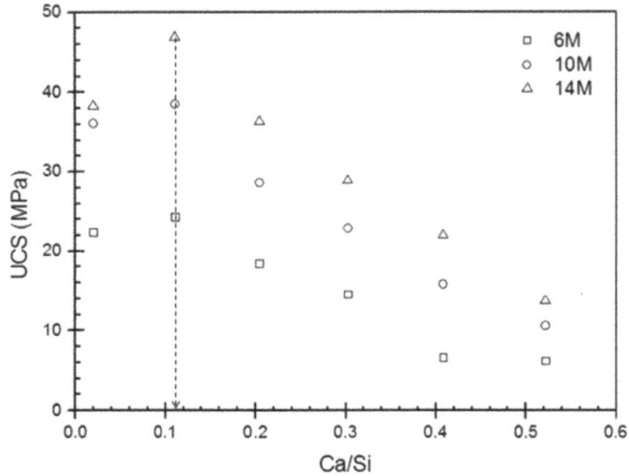

Fig. 6. UCS versus Ca/Si ratio from the current study

3.2 Micro/Nano-scale Properties

3.2.1 XRD Analysis

Figure 7 presents the results of XRD analysis for the MK, WC powders and the three geopolymeric specimens synthesized with 0%, 80%, and 100% MK respectively, prepared at 14 M NaOH and SS/NaOH of 1.5.

The XRD pattern of the MK shows that MK contains semi-crystalline and amorphous phase with a broad hump centered at approximately 28° along with a low-intensity peaks corresponding to muscovite. The WC powder mostly consists of crystalline phases including anorthite [$Ca(Al_2Si_2O_8)$], quartz [SiO2], portlandite [$Ca(OH)_2$], and CSH [$(CaO)_xSiO_2(H_2O)_y$]. Besides, a weak amorphous phase extending from 25° to 40° is observed in the XRD pattern of the WC powder.

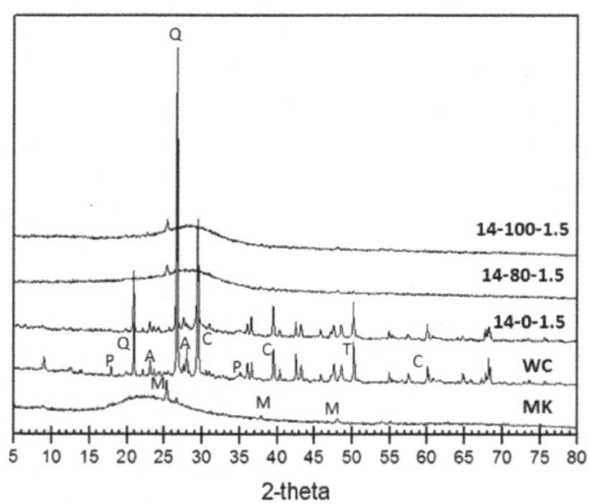

Fig. 7. XRD patterns of MK and WC powders and the geopolymer specimens produced with respectively 0%, 80%, and 100% MK, synthesized at 14 M NaOH and SS/NaOH of 1.5 (M: muscovite, A: anorthite, P: portlandite, Q: quartz, C: CSH, T: tobermorite)

The XRD patterns of the 80% or 100% MK specimens are nearly the same except for some very low-intensity crystalline peaks which could not be definitely identified by JCPDS. The broad hump in both of the two XRD patterns spreads to a larger extent with its center shifting to larger angles, which indicates new amorphous material formed in the specimens after geopolymerization. Such a hump starting at about 21° through to 39° 2-theta is due to the presence of amorphous silica in the binders [35, 38], which is a primary feature of the geopolymeric gel. Besides, it can be found from the two patterns that there are small decreases of the intensity of the crystalline peaks compared to the MK XRD pattern, indicating a portion of the crystalline phase dissolved during the geopolymerization, especially for the MK particles.

No new crystalline phase synthesized as a result of the geopolymerization, which can be interpreted that crystalline CSH is not a main product synthesized as a result of the geopolymerization of the 80% MK specimen under the conditions used in the current study. However, amorphous CSH could be a secondary product together with the geopolymeric gel, as will be presented later.

It should be noted that the crystalline peaks associated with portlandite disappear in the 0% MK specimen, which indicates the corresponding phase in the WC powder dissolves completely after geopolymerization. However, amorphous calcium hydroxide could be part of the product as it will be precipitated when the alkaline solution contains soluble calcium species during the initial hydration step [30]. When comparing the differences between before and after geopolymerization for WC-binder (Fig. 7 specimen 14-0-1.5), the CSH gel only undergoes partial dissolution exhibiting lower reactivity to the alkaline solution than portlandite. Pirssonite is not observed in the WC-binder structure which is the main reason for low compressive strength of the WC-based geopolymer as reported by Komnitsas et al. [14]. It can be explained by considering the materials' chemical composition. The WC Komnitsas et al. used in their study has much higher CaO content (65.42%) compared to the one (21.29%) used in the current study, thus more Ca ions are solubilized in the alkaline solution. It is favored for the formation of pirssonite in that case as a result of atmospheric carbonation.

3.2.2 SEM Imaging/EDS Analysis

Figure 8 shows SEM micrograph of the geopolymer specimens produced with respectively 100% (a), 80% (b), and 0% MK (c), synthesized at 14 M NaOH and SS/NaOH of 1.5.

Figure 8(b) presents a typical SEM micrograph of the specimen containing 80% MK. The phases presented in this micrograph including these appearing in both the 100% (a) and 0% (c) MK specimens. The results of EDS analysis for the specimen are shown in Table 3. According to the Table 3, the major elemental compositions of area A are Si, Al and Na with minor amount of Ca (2.01 wt%) while in area B, Si, Ca and Na are the major and Al is the minor element (2.54 wt%).

Area A is regarded as the geopolymeric gel since it has similar feature of the geopolymeric gel formed in the 100% MK specimen as shown in Fig. 8(a). The minor amount of Ca within the geopolymeric gel is derived from the dissolution of the WC particles. While in the area B, the phase could be some form of CSH. Different types of CSH gel can be obtained from the pozzolanic reaction or geopolymerization depending

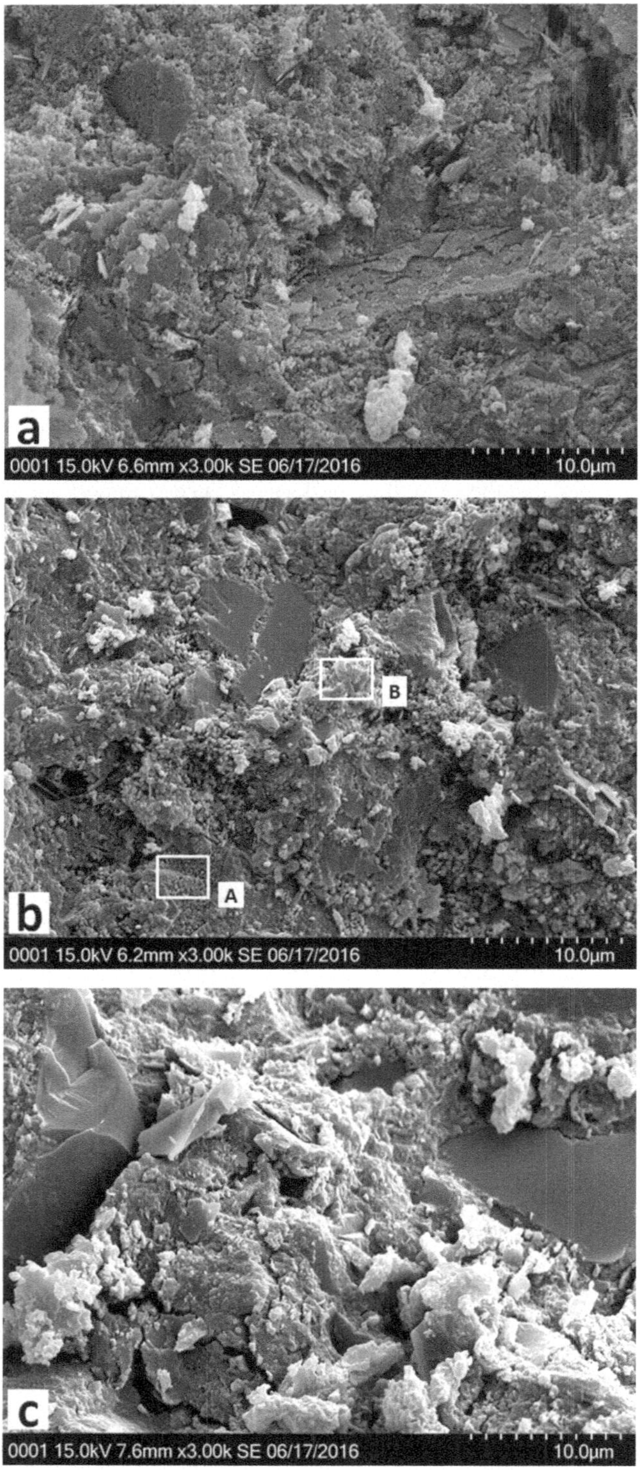

Fig. 8. SEM micrograph of the geopolymer specimens produced with respectively 100% (a), 80% (b), and 0% MK (c), synthesized at 14 M NaOH and SS/NaOH of1.5

Table 3. The elemental compositions of Area A and Area B as shown in Fig. 8 (b)

	Oxygen (O)	Sodium (Na)	Aluminium (Al)	Silicon (Si)	Calcium (Ca)
Area A	59.09	7.30	9.56	22.04	2.01
Area B	61.90	7.51	2.54	14.86	13.20

on the Ca concentration in the reaction phase. The CSH gel derived from the hydration of Portland cement usually has a Ca/Si ratio in the range of 1.5–1.7 [39, 40]. However, the Ca/Si ratio of the CSH gel resulted from the geopolymerization is much lower. Yip et al. [34] found that the CSH formed in a NaOH activated MK in the presence of ground granulated blast furnace slag has a Ca/Si ratio of approximately 1. The Ca/Si ratio in the current geopolymeric system is 0.89, which is close to the previous reported Ca/Si values. Traces of Al incorporated in the CSH phase in this geopolymeric system is probably on account of the substitution with Si atoms [40].

The highest UCS obtained at the MK addition of 80% can be explained by the simultaneous formation of the two phases as shown in Table 3. It is suggested that the coexistence of the two phases result in a binder which is more dense and homogeneous by bridging the voids between the unreacted particles and the formed phases. Consequently, the UCS of the binder will increase.

4 Conclusions

The feasibility of using the WC powder mixed with MK to synthesize geopolymer was studied. To be specific, the effect of NaOH concentrations and MK contents on the UCS was investigated. Microstructure and phase compositions of geopolymeric binder were investigated by SEM/EDS and XRD analyses. According to the study, the following conclusions can be drawn.

1. Increased NaOH concentration helps to dissolve more Si and Al and to form greater amount of geopolymeric gels and thus leads to higher UCS.
2. Addition of MK contributes to improve the UCS of geopolymeric binder and 80% was determined as the optimal MK content at 6, 10 and 14 M NaOH and with SS/NaOH of 1.5. The initial Si/Al ratio is adjusted by adding MK to WC, leading to the optimum ratio (the Si/Al ratio at the highest UCS) of approximately 1.53 for the MK/WC geopolymeric binder.
3. The positive effect of Ca in the geopolymric system should be confined to a certain content of Ca. The improving impact of WC is limited to 20% added WC in the WC/MK based geopolymeric system with the optimal Ca/Si ratio of about 0.11.
4. The SEM/EDS and XRD analyses confirm that the positive effect of Ca on UCS of geopolymeric binders mainly due to the simultaneous formation of geopolymeric gel and CSH gel.

References

1. Chinese Development and Reform Commission: 2014 annual report of Chinese comprehensive resource utilization (2014). http://www.sdpc.gov.cn/gzdt/201410/t20141009_628795.html
2. Formoso, C.T., Soibelman, L., De Cesare, C., Isatto, E.L.: Material waste in building industry: main causes and prevention. J. Constr. Eng. M. **128**(4), 316–325 (2002). doi:10.1061/(ASCE)0733-9364(2002)128:4(316)
3. Kibert, C.J.: Deconstruction as an essential component of sustainable construction. In: Proceedings of the Second Southern African Conference on Sustainable Development in the Built Environment, pp. 1–5 (2000)
4. Kanare, H.K., West, P.B.: Leachability of selected chemical elements from concrete. In: Proceedings of the Symposium on Cement and Concrete in the Global Environment, p. 366 (1993)
5. Xiao, J., Li, J., Zhang, C.: On relationships between the mechanical properties of recycled aggregate concrete: an overview. Smart Mater. Struct. **39**, 655–664 (2006). doi:10.1617/s11527-006-9093-0
6. Poon, C.S., Lam, C.S.: The effect of aggregate-to-cement ratio and types of aggregates on properties of precast concrete blocks. Cem. Concr. Comp. **30**, 283–289 (2008). doi:10.1016/j.cemconcomp.2007.10.005
7. Hack, D.R., Bryan, D.P.: Aggregates. Industrial minerals and rocks. In: Kogel, E.K., Trivedi, N.C., Barker, J.M., Krukowski, S.T. (eds.) Society for Mining. Littleton, Colorado, pp. 1105–1119 (2006)
8. Abbas, A., Fathifazl, G., Isgor, O.B., Razaqpur, A.G., Foo, S.: Proposed method for determining the residual mortar content of recycled concrete aggregates. J. ASTM Inter. **5**(1), 1–12 (2008). doi:10.1520/JAI101087
9. Juan, M.S., Gutierrez, P.A.: Study on the influence of attached mortar content on the properties of recycled concrete aggregate. Constr. Build. Mater. **23**, 872–877 (2009). doi:10.1016/j.conbuildmat.2008.04.012
10. Yanagibashi, K., Inoue, K., Seko, S., Tsuji, D.: A study on cyclic use of aggregate for structural concrete. In: Action for Sustainability-The 2005 World Sustainable Building Conference, pp. 2585–2592 (2005)
11. Davidovits, J.: High-alkali cement for 21st cement concretes. In: Metha, P.K. (eds.) Proceedings of V. Mohan Malhortra Symposium: Concrete Technology, Past, Present and Future, ACI SP, pp. 383–398 (1994)
12. Komnitsas, K., Zaharaki, D., Vlachou, A., Bartzas, G., Galetakis, M.: Effect of synthesis parameters on the quality of construction and demolition wastes (CDW) geopolymers. Adv. Powder Technol. **26**(2), 368–376 (2015). doi:10.1016/j.apt.2014.11.012
13. Hardjito, D., Wallah, S.E., Sumajouw, D.M.J., Rangan, B.V.: On the development of fly ash-based geopolymer concrete. ACI Mater. J. **101**(6), 467–472 (2004)
14. Han, Y., Wang, D., Zhou, W., Lu, H., Wang, L.: Effect of activator and curing mode on fly ash-based geopolymer. J. Wuhan Univ. Technol. Mater Sci. Ed. **24**(5), 711–715 (2009). doi:10.1007/s11595-009-5711-3
15. Stevenson, M., Sagoe-Crentsil, K.: Relationship between composition, structure and strength of inorganic polymer: Part I Metakaolin-derived inorganic polymers. J. Mater. Sci. **40**, 2023–2036 (2005). doi:10.1007/s10853-005-2794-x
16. Xu, H., Van Deventer, J.S.J.: The geopolymerisation of alumino-silicate minerals. Int. J. Mine. Process **59**, 247–266 (2000). doi:10.1016/S0301-7516(99)00074-5

17. Panias, D., Giannopoulou, I.P., Perraki, T.: Effect of synthesis parameters on the mechanical properties of fly ash-based geopolymers. Colliods Surf. A. **301**, 246–254 (2007). doi:10.1016/j.colsurfa.2006.12.064
18. Davidovits, J.: Mineral polymers and methods of making them. US Patent 4349385 (1982)
19. Rattanasak, U., Chindaprasirt, P.: Influence of NaOH solution on the synthesis of fly ash geopolymer. Miner. Eng. **22**, 1072–1078 (2009). doi:10.1016/j.mineng.2009.03.022
20. Silva, P.D., Sagoe-crenstil, K., Sirivivatnanon, V.: Kinetics of geopolymerization: Role of Al_2O_3 and SiO_2. Cem. Concr. Res. **37**, 512–518 (2007). doi:10.1016/j.cemconres.2007.01.003
21. Muniz-Villarreal, M.S., Manzano-Ramirez, A., Sampieri-Bulbarela, S.: The effect of temperature on the geopolymerization process of a metakaolin-based geopolymer. Mater. Lett. **65**(6), 995–998 (2011). doi:10.1016/j.matlet.2010.12.049
22. Rowles, M.R., O'Connor, B.H.: Chemical optimization of the compressive strength of aluminosilicate geopolymers synthesized by sodium silicate activation of metakaolinite. J. Mater. Chem. **13**(5), 1161–1165 (2003). doi:10.1039/B212629J
23. Tippayasam, C., Boonsalee, S., Sajjavanich, S.: Geopolymer development by powders of metakaolin and waste in Thailand. Adv. Sci. Tech. **69**, 63–68 (2010). doi:10.4028/www.scientific.net/AST.69.63
24. Duxson, P., Provis, J.L., Lukey, G.C., Mallicoat, S.W., Kriven, W.M., Van Deventer, J.S.J.: Understanding the relationship between geopolymer composition, microstructure and mechanical properties. Colliods Surf. A. **269**(1–3), 47–58 (2005). doi:10.1016/j.colsurfa.2005.06.060
25. Zhang, Y., Sun, W., Li, Z.: Composition design and microstructural characterization of calcined kaolin-based geopolymer cement. Appl. Clay Sci. **47**, 271–275 (2010). doi:10.1016/j.clay.2009.11.002
26. Rowles, M.R., O'Connor, B.H.: Chemical and structural microanalysis of aluminosilicate geopolymers synthesized by sodium silicate activation of metakaolinite. J. Am. Ceram. Soc. **92**, 2354–2361 (2009). doi:10.1111/j.1551-2916.2009.03191.x
27. Cheng, T.W., Chiu, J.P.: Fire-resistant geopolymer produced by granulated blast furnace slag. Miner. Eng. **16**, 205–210 (2003). doi:10.1016/S0892-6875(03)00008-6
28. Xu, H., Jaarsveld, J.G.S.: The effect of source materials on geopolymerization. Ind. Eng. Chem. Res. **42**, 1689–1706 (2003). doi:10.1021/ie0206958
29. Diaz, E.I., Allouche, E.N., Eklund, S.: Factors affecting the suitability of fly ash as source material for geopolymer. Fuel **89**(5), 992–996 (2010). doi:10.1016/j.fuel.2009.09.012
30. Yip, C.K., Lukey, G.C., Van Deventer, J.S.J.: The coexistence of geopolymeric gel and calcium silicate hydrate at the early stage of alkaline activation. Cem. Concr. Res. **35**, 1688–1697 (2005). doi:10.1016/j.cemconres.2004.10.042
31. Lecomte, I., Henrist, C., Liegeois, M., Rulmont, A., Cloots, R.: Microstructural comparison between geopolymers, alkali-activated slag cement and Portland cement. J. Eur. Ceram. Soc. **26**(16), 3789–3797 (2006). doi:10.1016/j.jeurceramsoc.2005.12.021
32. Fernandez-Jimenez, A., Puertas, F., Sobrados, I., Sanz, J.: Structure of calcium silicate hydrates formed in alkaline-activated slag: influence of the type of alkaline activator. J. Am. Ceram. Soc. **86**(6), 1389–1394 (2003). doi:10.1111/j.1151-2916.2003.tb03481.x
33. Temuujin, J., Van Riessen, A., Williams, R.: Influence of calcium compounds on the mechanical properties of fly ash geopolymer pastes. J. Hazard. Mater. **167**, 82–88 (2009). doi:10.1016/j.jhazmat.2008.12.121
34. Pacheco-Torgal, F., Castro-Gomes, J.P., Jalali, S.: Investigations on mix design of tungsten mine waste geopolymeric binder. Constr. Build. Mater. **22**, 1939–1949 (2008). doi:10.1016/j.conbuildmat.2007.07.015

35. Phair, J.W., Van Deventer, J.S.J., Smith, J.D.: Mechanism of polysialation in the incorporation of zirconia into fly ash-based geopolymers. Ind. Eng. Chem. Res. **39**, 2925–2934 (2000). doi:10.1021/ie990929w
36. Glukhovsky, V.D.: Ancient, modern and future concretes. In: Krivenko, P.V. (ed.) Alkaline Cements and Concrete. VIPOL Stock Company, Kiev (1994)
37. Frias, M., Cabrera, J.: Influence of MK on the reaction kinetics in MK/lime and MK-blended cement systems at 20°. Cem. Concr. Res. **31**, 519–527 (2001). doi:10.1016/S0008-8846(00)00465-8
38. Van Jaarsveld, J.G.S., Van Deventer, J.S.J., Schwartzman, A.: The potential use of geopolymeric materials to immobilize toxic metals: Part II Material and leaching characteristics. Miner. Eng. **12**, 75–91 (1999). doi:10.1016/S0892-6875(97)00046-0
39. Gani, M.S.J.: Cement and Concrete, 1st edn. Chapman and Hall, London (1997)
40. Brough, A.R., Atkinson, A.: Sodium silicate-based, Alkali-activated slag mortars: Part 1. Strength, hydration and microstructure. Cem. Concr. Res. **32**, 865–879 (2002). doi:10.1016/S0008-8846(02)00717-2

Numerical Methods in Geotechnical Engineering

Deformation and Failure Behavior of the Isolated Single Stone Column with and Without Geosynthetic Encasement

Xin Tan(✉) and Minghua Zhao

Hunan University, Changsha, China
xintan@hnu.edu.cn

Abstract. Model tests and corresponding numerical analysis have been conducted on the isolated single stone column to study the failure behavior and bulging deformation behavior of the stone columns with and without geosynthetic encasement. The model test results indicate significant improvement in loading capacity of the stone columns due to geosynthetic encasement. In numerical analysis, the geosynthetic encasement was simply replaced by a cable element inside the stone column, which provides a convenient way to simulate encasement behavior in a 2D condition. The bugling failure behavior involves the interaction between the stone column, the geosynthetic encasement and the typical failure mode for stone columns has been concluded, and the bulging failure mechanism of the isolated single stone column with and without geosynthetic encasement was explained in detail.

Keywords: Stone column · Geosynthetic encasement · Bulging · Numerical simulation · Ground improvement

1 Introduction

Stone column has been used extensively in the ground improvement method for soft soil in various geotechnical engineering, it is preferred because it is economically, easy constructed and environmentally friendly. The stone column can develop the vertical bearing capacity through bulging deformation and thereby inducing the lateral confining stress from the surrounding soils. (Greenwood 1970; Hughes et al. 1975). The deformation and failure behavior of a single stone column is thus controlled by the lateral support provided by the surrounding soil. The application of the stone columns may be limited in extremely soft soils, due to the lack of sufficient lateral confinement for the stone columns. A further development for the stone column application is encasing the individual stone column by geosynthetic encasement, which was firstly proposed by Van Impe in 1985 (Van Impe 1989). The geosynthetic encasement can increase the bearing capacity of stone columns due to the additional confinement from the encasement which prevents the stones from laterally squeezing out of the column into the surrounding soils. The applications of stone columns so were greatly extended by the geosynthetic encasement in extremely soft soil (Murugesan and Rajagopal 2006; Gniel and Bouazza 2009). Model tests have shown that the encasement is needed only where

R. Chen et al. (eds.), *Proceedings of the 2nd International Symposium on Asia Urban GeoEngineering*, Springer Series in Geomechanics and Geoengineering,
https://doi.org/10.1007/978-981-10-6632-0_42

bulging takes place, which were mostly observed to occur at about 2 ~ 3 times the diameter below the top of single stone without encasement (Deb et al. 2011; Castro and Sagaseta 2011; Shivashankar et al. 2010). Lots of analytical solutions to study the bulging failure mechanism were mainly based on the method proposed by Brauns 1978. The encased stone column was also analytically studied by Zhang and Zhao (2015), Castro and Sagaseta (2011), Pulko et al. (2011). Both kinds of stone columns with and without encasement were studied numerically by Elshazly et al. (2008), Chen et al. (2009), the geosynthetic encasement around the stone column was modeled as linear elastic material and discretized as continuum elements around the stone column (Murugesan and Rajagopal 2006; Ghazavi and Afshar 2013). It has been observed that many analytical or numerical studies have been carried out to study the effect of stone column with and without encasement on settlement and bearing capacity, a compilation review of the important findings from physical, numerical, and analytical models of published studies was presented by Najjar (2013) and Badu et al. (2013).

In the present study, model tests and corresponding numerical analysis have been conducted on the isolated single stone column with and without geosynthetic encasement to study the bearing capacity and bulging deformation behavior of the stone columns. In numerical analysis, the geosynthetic encasement was simply replaced by a cable element inside the stone column, which provides a convenient way to simulate encasement behavior in a 2D condition. The bugling failure behavior involves the interaction between the stone column, geosynthetic encasement and surrounding soil.

2 Model Tests

The model tests were conducted in the foundation laboratory of Hunan University. Tests were performed on the single stone column installed in the soil bed which was prepared in a large test tank of size 2200 × 1800 × 1000 mm, four stone columns with 200 mm diameter were installed in the soil bed, and were extended down to the bottom of the tank. In the case of geosynthetic encased stone column, the geosynthetic encasement

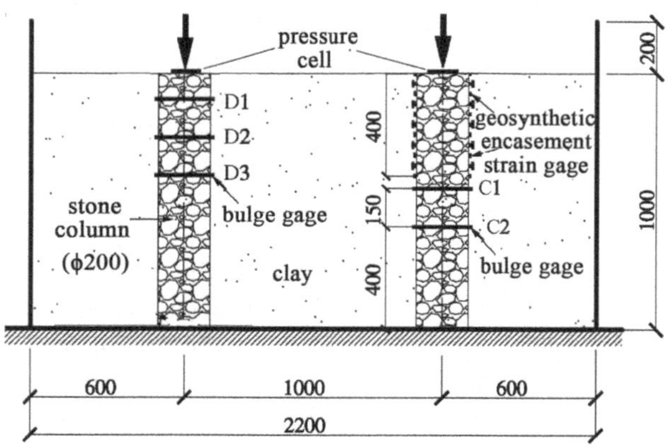

Fig. 1. Load test on single stone column in large test tank

was wrapped on the column at upper 400 mm length range, since most bulging failure happens at the top of column in the field. The bulging deformation of stone column and hoop strain in the geosynthetic encasement at various loading level were measured using bulge gages and strain gages (Fig. 1).

The soft soil used was muck soil, obtained from a flood plain of Xiangjiang River in south China during a low-flow period. The undrained shear strength of undisturbed muck soil was less than 10 kPa, the moisture content was in the range of 30% ~ 60%. The basic parameters of muck soil used in the model tests were summarized in Table 1. Crushed limestone aggregates of sizes mainly between 30 and 50 mm were used to construct stone column. Biaxial polypropylene plastic geogrid was used to form geosynthetic encasement, the physical parameters of geogrid obtained from extension tests were listed in Table 2.

Table 1. Main physical parameters of soil

Liquid limit	Plastic limit	Moisture content	Constrained modulus	Void ratio	Undrained shear strength
%	%	%	MPa		kPa
64.6	35.2	55.6	1	1.5	3.4

Table 2. Physical parameters of geogrid

Grid size	Rib width	Tension at 2% tensile strain	Tension at 5% tensile strain
mm	mm	KN/m	KN/m
40 × 40	5	13.6	20

The results from the model tests indicate significant improvement in loading capacity of the stone columns due to geosynthetic encasement. More details about the deformation behavior of stone columns with and without encasement will be described in next part to explain the failure mode involves the interaction between the stone column, geosynthetic encasement and surrounding soils.

3 Numerical Simulations

The simplified 2D axisymmetric analysis was conducted by FLAC3D code to simulate the model tests numerically. The FLAC3D code has advantages in simulating the nonlinear or large-strain problems, or to situations in which physical instability may occur, it is quite suitable to better investigate the bugling failure mode involves the interaction between the stone column, geosynthetic encasement and surrounding soils (Fig. 2).

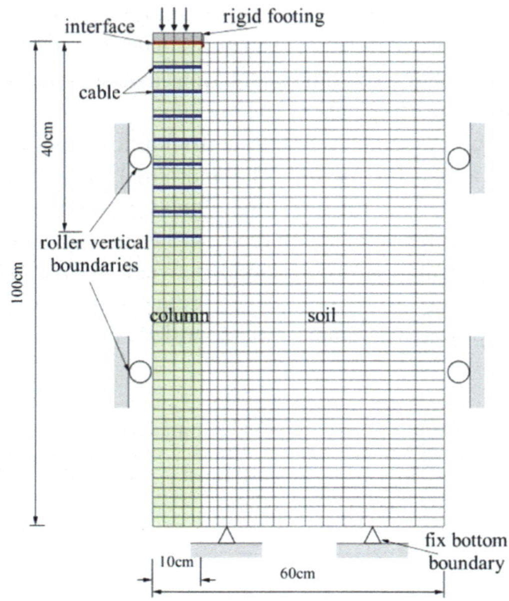

Fig. 2. Numerical model of an isolated single stone column

The numerical grid has a single element thickness (2 cm) along Z-axial direction, all the gridpoints were fixed in Z-axial direction. The unit cell configuration was adopted: a stone column of radius 10 cm and length 100 cm was made at the center of the soil and loaded with a rigid loading plate on its top, the model width was set as 60 cm corresponding to the model test. The top of the model is a free surface. The bottom of the model is fixed in both-direction, roller boundaries are imposed on the sides of the model. The Mohr-Coulomb yield criterion and a non-associated flow rule with a constant dilatancy angle were adopted for both stone column and soil, drained condition and large deformation mode was adopted in the numerical calculations. The material properties used in numerical simulations are given in Table 3.

Table 3. Materials properties in numerical simulation

	E	ν	γ	c	ϕ	ψ	t
	MPa		kN/m^3	kPa	°	°	kPa
Loading plate	2000	0.2	20				
Stone column	20	0.35	20	0	38	5	0
Soil	0.2	0.45	18	5.0	0	0	5.0
	kN	ks	c	ϕ	ψ	t	
	GPa/m	GPa/m	kPa	°	°	kPa	
Interface	100	100	0	5	0	0	

The bulging deformation of stone column causes the geosynthetic encasement to stretch and generate a circumferential tensile stress to provide an additional confining stress on the column. Assuming that the behavior of the geosynthetic encasement is linear elastic, the hoop tensile force T in the geosynthetic encasement, can be transformed into a horizontal stress. Therefore, the confining stress acting on the column therefore consist tow part: the horizontal stress caused by hoop tensile force in the geosynthetic encasement, the lateral confining stress provided by the surrounding soil (Eq. 1). In numerical simulations of encased stone column, the effect of geosynthetic encasement was simply replaced by the axial force N provided by a cable element in the stone column as show in Fig. 3. The cable structural elements provide a shearing resistance along their length. It behaves as an elastic, perfectly plastic material that can yield in tension and compression, but cannot resist a bending moment. A cable may be grouted such that force develops along its length in response to relative motion between the cable and grid, it is suitable for modeling structural-support members in which tensile capacity is important, and for which axially directed frictional interaction with the rock or soil mass occurs.

$$\sigma_{\rm rs} r d\theta + 2T \sin\frac{d\theta}{2} = \sigma_{\rm rc} r d\theta \tag{1}$$

$$\sigma_{\rm rs} r d\theta + \frac{nN}{r d\theta} = \sigma_{\rm rc} r d\theta \tag{2}$$

where T is hoop tensile force in encasement, $\sigma_{\rm rc}$ is confining stress acting on the stone column, $\sigma_{\rm rs}$ is lateral confining stress provided by the surrounding soil, N is axial force in cable element, n is amount of cables per meter.

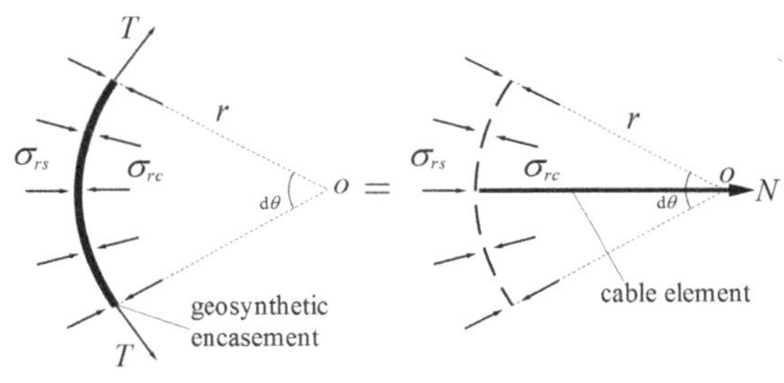

Fig. 3. Lateral stress acting on stone column with geosynthetic encasement

The hoop tensile force T and the axial force N caused by bulging deformation of stone column can be deduced from:

$$T = J_g \varepsilon_\theta \tag{3}$$

$$N = E_a \varepsilon_r a \tag{4}$$

$$\varepsilon_\theta = \varepsilon_r = \frac{dr}{r} \tag{5}$$

Where J_g is tensile modulus of geosynthetic encasement, E_a is elastic modulus of cable material, a is cross section of cable element;

Combing Eqs. (1) – (5), the tensile strength $[N]$ and elastic modulus of cable element E_a can be deduced from the tensile strength and tensile modulus of geosynthetic encasement.

$$[N] = \frac{[T]d\theta}{n} \tag{6}$$

$$E_a = \frac{J_g d\theta}{na} \tag{7}$$

In the model test, the tensile stress in geosynthetic encasement was 20 KN/m at 5% tensile strain, the yield strength of geosynthetic encasement was 40 kN/m, In the numerical model, the section area of cable element was set at 2×10^{-3} m^2, 8 cables were placed in the upper 40 cm length of stone column, the geosynthetic encasement and corresponding cable element parameters were list in Table 4.

Table 4. Cable element parameters used in numerical model

Geosynthetic encasement				Cable element				
Strain	Tension	J_g	$[T]$	$d\theta$	a	n	E_a	$[N]$
%	kN/m	kN/m	kN/m		m^2		MPa	kN
5	20	400		0.2	2×10^{-3}	20	2.0	0.4
12 (yield)	40		40					

The numerical model was firstly brought to an equilibrium stress-state under gravitational loading before the footing on the loading plate. And then a vertical velocity was applied at the loading plate on the top of the stone column. The vertical stress on the top of column was generated by moving the loading plate vertically through the interface between loading plate and top of column. The vertical stress versus axial displacement at the column top was measured during the complete loading process. No interface was used between soil and columns, because stone columns are considered to be tightly interlocked with the surrounding soil. Drained conditions were assumed, no pore pressure was generated during the loading process.

4 Comparison of Model Test and Numerical Results

4.1 Entire Axial Stress-Settlement Curves

The entire load-settlement curves were shown in Fig. 4, compared with the experimental results from model tests, the curves show good agreement with model test. In order to investigate the failure and deformation behavior during the entire loading process, the bulging deformation of stone columns (horizontal displacement) and yielding state of surrounding soil under different settlement were plotted in Fig. 5.

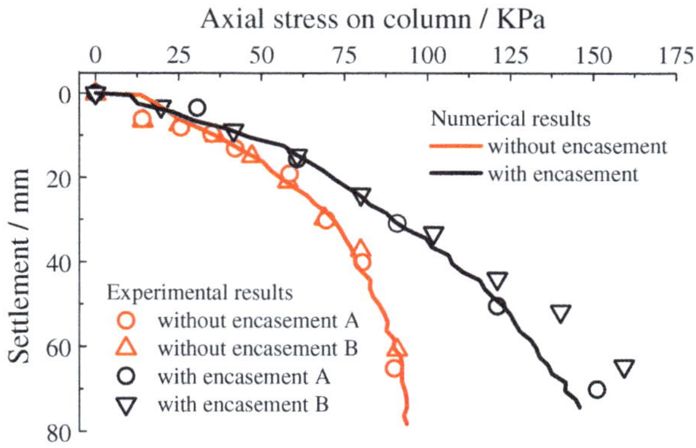

Fig. 4. Entire axial stress – settlement vurves

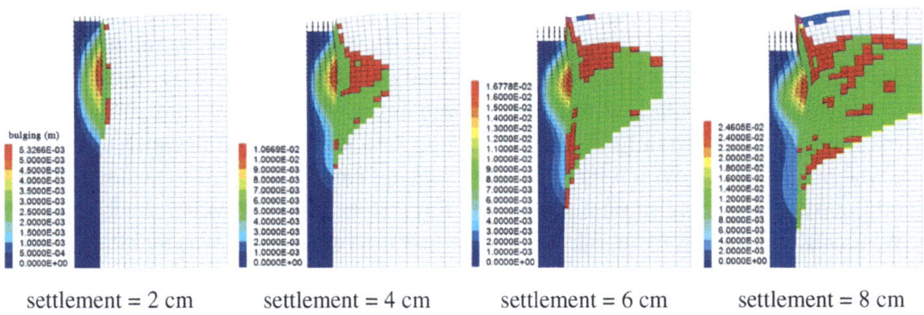

(a) Stone column without geosynthetic encasement

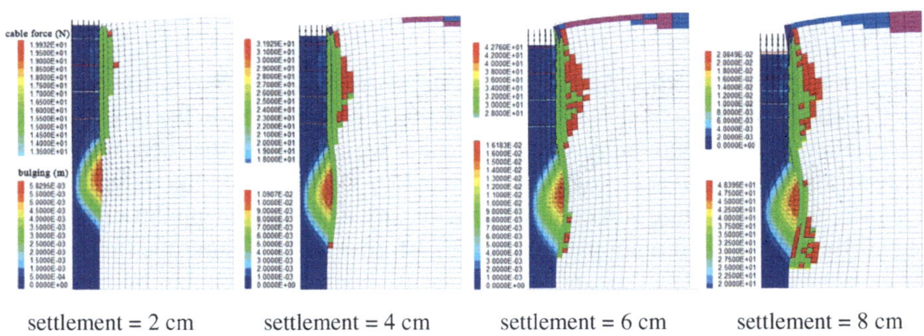

(b) Stone column with geosynthetic encasement

Fig. 5. Deformation of stone columns (horizontal displacement) and yielding state of surrounding soil under different settlement during entire loading process (the scale of grid deformation was exaggerated 2 times)

4.2 Failure Process

When settlement was 2 cm, the axial stress on the stone column was 56 kPa for model without encasement, 72 kPa for model with encasement. The stone columns of both models were in the compression stage, the surrounding soil very close to the upper part of stone column was in shear failure because the shear deformation caused by vertical displacement of stone column. Some part of stone column began to bulge out from the origin profile of column. In the model without encasement, the maximum bulge of stone column was 5.3 mm at 20 cm depth. In the model with encasement, the maximum bulge of stone column was 5.8 mm at 60 cm depth, which located around 20 cm below the encasement. The maximum hoop tensile force (calculated from the axial force of cable element) in encasement was 1.99 kN/m.

When settlement reached 4 cm, the axial stress on the stone column was 79 kPa for model without encasement, 112 kPa for model with encasement. In the model without encasement, the maximum bulge of stone column was 10.6 mm, and the surrounding soil close to bulging zone of stone column began to shear failure because the additional lateral pressure caused by bulging deformation. In the model with encasement, the lateral displacement of upper column was restricted by encasement, however, the maximum bulge of stone column below the encasement increased to 10.9 mm, however, the surrounding soil close to bulging zone didn't go into shear failure because the bulging zone was much deeper, which increased the bearing capacity of soil to sustain a higher lateral pressure. The maximum hoop tensile force in encasement was 3.19 kN/m.

When settlement was 6 cm, the axial stress on the stone column was 89 kPa for model without encasement, 132 kPa for model with encasement. In the model without encasement, the maximum bulge of stone column was 16.7 mm, the shear failure zone of the surrounding soil continuously expanded. In the model with encasement, the maximum bulge of stone column was 16.1 mm. The maximum hoop tensile force in encasement was 4.27 kN/m.

When settlement reached 8 cm, the axial stress on the stone column was 93 kPa for model without encasement, 150 kPa for model with encasement. In the model without encasement, the maximum bulge of stone column was 24.6 mm, the shear failure zone continuously expanded and reached the boundary of unit cell, the general failure happened, the surrounding soil couldn't provide any more lateral support on the stone column, the bugling deformation was going to perform a unstable state, the stone column lost its bearing capacity. In the model with encasement, the maximum bulge of stone column was 20.8 mm, only a small part of surrounding soil close to bulging zone of stone column began to shear failure. The maximum hoop tensile force in encasement was 4.84 kN/m, which was still much smaller than the tensile strength of geosynthetic encasement.

The ultimate vertical bearing capacity of single stone column without encasement was around 90 kPa, whereas for stone column with geosynthetic encasement was more than 150 kPa. The ultimate vertical bearing capacity of the isolated single column could be enhanced more than 60% by installation of geosynthetic encasement with enough tensile strength at the upper length of stone column. The enhancement was also confirmed by lots of laboratory and field tests.

4.3 Bulging Deformation

The bulging deformation of stone column during loading process was measured to study the deformation behavior of the stone column with and without encasement, which is shown in Fig. 6 compared with model test result. In model with encasement, bulging deformation was generally concentrated at a distance of 2 column diameters from the top of the column, the maximum lateral displacement happened at the center of the bulging zone and continually increased with the increasing loading. The maximum bulging deformation in numerical model was closed to the measured value from model test results (Fig. 6a), while it trended to be higher than the model test measured value after the general shear failure happened (settlement = 8 cm in Fig. 6a).

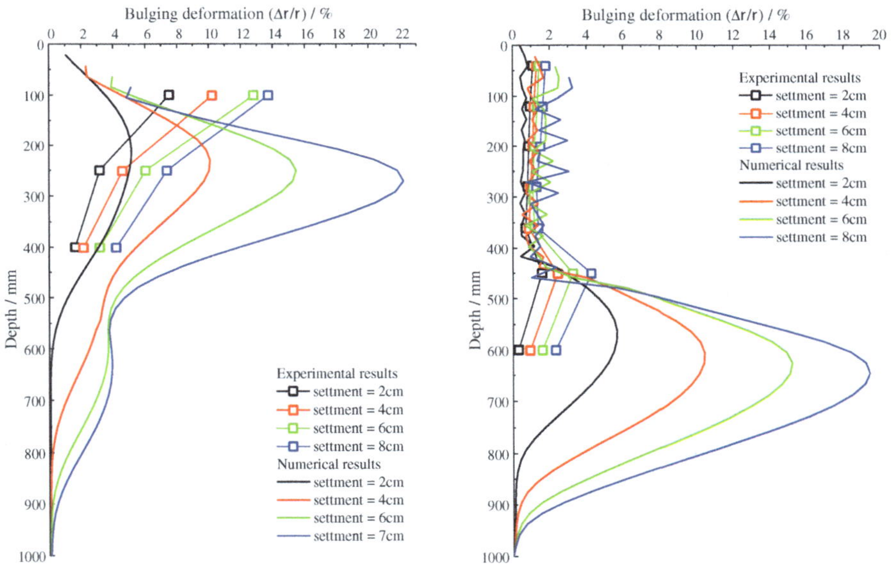

(a) stone column without encasement (b) stone column with encasement

Fig. 6. Bulging deformation along the single column during loading process

In the model with encasement, bulging deformation decreased at the upper region of column and was transmitted to the position under the part wrapped by geosynthetic encasement. In the encased section of stone column, the bulging deformation was reduced significantly and the maximum value was the same as measured in model test ($\Delta r / r < 3\%$), bubt it trended to be much higher than the model test results in the un-encased section of stone column.

The numerical results show very similar bulging trend with model test results, but relatively higher lateral displacement than the model test results when numerical model went into general failure stage or in the un-encased region of model with encasement. This may be ascribed to tow main points: 1. Mohr-Coulomb constitutive model cannot precisely represent the nonlinear behavior of stone column according to various confining pressure, a constant dilatancy angle may overvalue the bulging deformation of stone column especially at deeper depth or after general shear failure happens. 2. The modulus of soil bed was considered as a constant in numerical model, which actually should be increased as increasing in depth.

5 Deformation and Failure Behavior Analysis

The numerical simulations show quit good agreement with model test results and make the failure process to be visualized. According to the failure process revealed from model tests and numerical simulations, the typical failure mode for an isolated single stone column with and without encasement is concluded as follow stages marked in the entire P-S curve (Fig. 7).

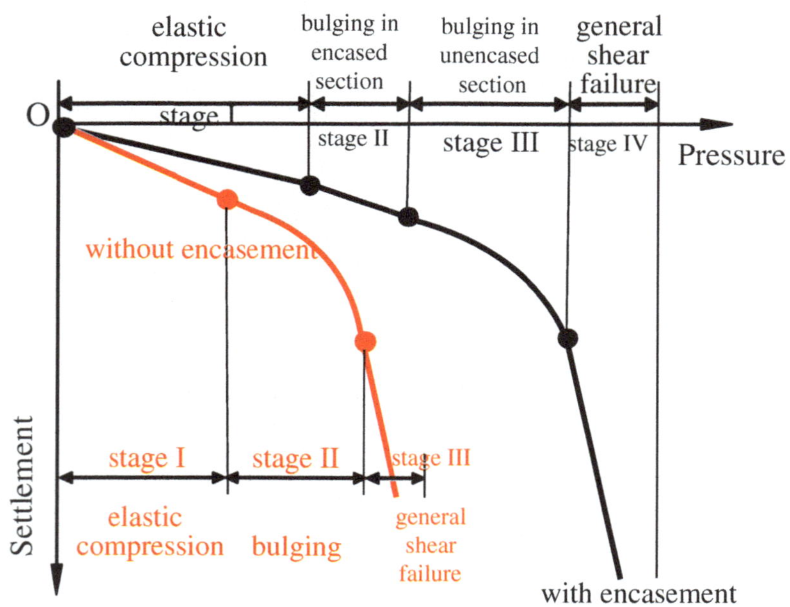

Fig. 7. Failure stages marked in the P-S curve a single stone column

Single stone column without geosynthetic encasement: Stage I, elastic compression stage (Fig. 8a): At the initial loading stage, the stone column is in elastic compression, compression deformation generated along the whole length of stone column. The surrounding soil closed to the stone column deformed following the vertical displacement of stone column, shear failure zone appears when shear stress in surrounding soil exceeded its shear strength around upper column. The P-S curve in stage I is linear. Stage II, bulging deformation stage (Fig. 8b): As the loads increasing, the lateral displacement of stone column is developing rapidly because there is no bond in the aggregates of stone column. The lateral displacement is non-uniform distributed along the depth due to the non-uniform distribution of axial stress of column and lateral earth pressure of soil along the depth. The upper part of stone column began to bulge out from the origin profile of column, and forms a convex shape bulging zone. The surrounding soil closed to bulging zone of stone column begVan to shear failure because the additional lateral pressure caused by bulging deformation of the stone column. The lateral earth pressure caused by horizontal compression of soil acting on stone column provides additional support to prevent the column losing its vertical bearing capacity. The P-S curve in stage II is nonlinear. Stage III, general shear failure stage (Fig. 8c): The shear failure zone in the surrounding soil keeps extending and finally forms a general passive slip surface, and the passive earth pressure zone appears. The bulging deformation cannot be restricted anymore, thus the stone column lost its bearing capacity in this stage.

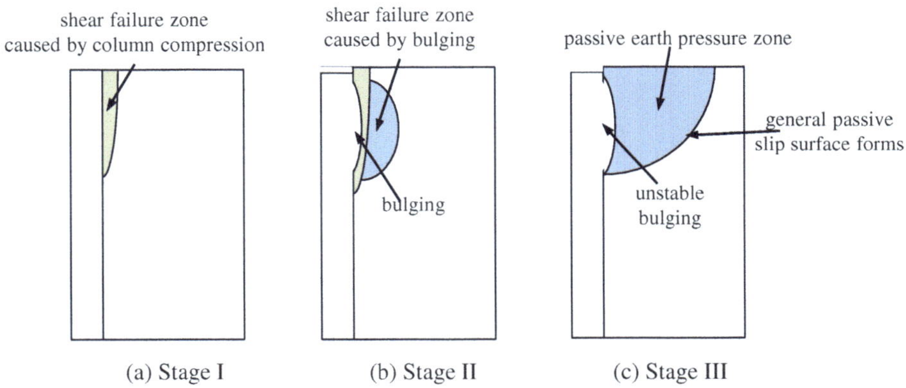

Fig. 8. Failure mode for a single stone column without geosynthetic encasement

For the isolated single stone column with geosynthetic encasement: Stage I, elastic compression stage (Fig. 9a): It is the same stage as the stage I for the isolated single stone column without geosynthetic encasement. The P-S curve in this stage is linear, however, the stiffness of stone column is enhanced due to the restriction provided by encasement, the enhancement of stiffness depends on the tensile modulus of geosynthetic encasement. Stage II, upper bulging deformation stage (Fig. 9b): A certain amount of tensile strain is needed before the enough lateral stress can be generated by geosynthetic encasement. Therefore, the lateral displacement of upper stone column with encasement is still developing in this stage. Then the bugling deformation will be restricted by geosynthetic encasement with enough tensile modulus, only small amount of surrounding soil will go in to shear failure. The P-S curve in stage II is quasi linear, the stone column has higher stiffness than which in stage I. However, if the tensile modulus in geosynthetic encasement is too low to provide enough confining pressure, or the tensile strength is too low and encasement rupture happens, the stone column will then go into general failure stage just as column without encasement. Stage III, deeper bulging deformation stage (Fig. 9c): Suppose that the encasement with enough strength and elastic modulus was used, the upper bugling deformation thus has been restricted by geosynthetic encasement, the vertical stress is transmitted to deeper length of column. The bugling deformation is therefore transmitted to the part of column below the encased column section. Shear failure zone appears and develops closed to the bugling region in the surrounding soil. The P-S curve in stage II is nonlinear.

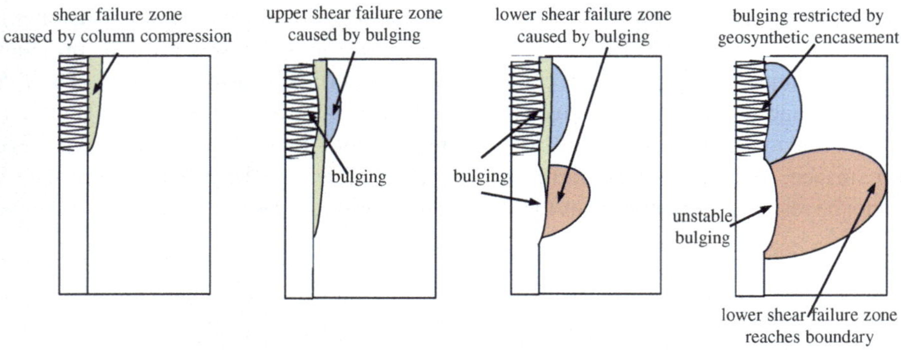

(a) Stage I (b) Stage II (c) Stage III (d) Stage IV

Fig. 9. Failure mode for a single stone column with geosynthetic encasement

Stage IV, general shear failure stage (Fig. 9d): The shear failure zone in the surrounding soil keeps extending and finally reaches the boundary of the unit cell, the bulging deformation cannot be restricted anymore, and the stone column lost its bearing capacity. The general failure happens in a greater depth than the case without encasement, thus the ultimate bearing capacity is much greater than which in the case without encasement.

6 Conclusions

Model tests of the single stone column and corresponding numerical analysis were carried out to study the deformation and failure behavior of the isolated single stone column with and without geosynthetic encasement. Model tests were performed in a large test tank, the results indicate significant improvement in stiffness and loading capacity of the stone columns due to geosynthetic encasement. Numerical simulations were conducted by FLAC3D in the large-strain mode to investigate the bugling failure mechanism involves the interaction between the stone column, geosynthetic encasement and surrounding soils. The geosynthetic encasement was simply replaced by a cable element in the stone column, which provides a convenient way to simulate encasement in a 2D numerical analysis. The numerical simulated failure process of stone column shows good agreement with model tests. According to the failure process revealed from model tests and numerical simulations, the typical failure mode for a single stone column with and without encasement has been concluded, and the bulging failure mechanism of single stone column with and without geosynthetic encasement was explained in detail. The continuum numerical model can be used to predict the failure and deformation behavior of stone column, but a constitutive model which can better represent the non-linear deformation behavior of aggregates depends on the confining pressure should be further studied to simulate the bulging deformation more precisely.

Acknowledgements. This research was supported by the National Natural Science Foundation of China (ID 51508181), and the China Postdoctoral Science Foundation (ID 2015M570679).

References

Babu, M.R.D., Nayak, S., Shivashankar, R.: Acritical review of construction, analysis and behavior of stone columns. Geotech. Geol. Eng. **31**(31), 1–22 (2013)

Brauns, J.: Die anfangstraglast von schottersaulen im bindigen untergrund. Die Bautechnik **55**(8), 263–271 (1978)

Castro, J., Sagaseta, C.: Deformation and consolidation around encased stone columns. Geotext. Geomembr. **29**(3), 268–276 (2011)

Chen, J.F., Han, J., Oztoprak, S., Yang, X.M.: Behavior of single rammed aggregate piers considering installation effects. Comput. Geotech. **36**, 1191–1199 (2009)

Deb, K., Samadhiya, N.K., Namdeo, J.B.: Laboratory model studies on unreinforced and geogrid-reinforced sand bed over stone column-improved soft clay. Geotext. Geomembr. **29**(2), 190–196 (2011)

Elshazly, H.A., Elhafez, D.H., Mossaad, M.E.: Reliability of conventional settlement evaluation for circular foundations on stone columns. Geotech. Geol. Eng. **26**, 323–334 (2008). Engineering, 1–14

Ghazavi, M., Afshar, J.N.: Bearing capacity of geosynthetic encased stone columns. Geotext. Geomembr. **38**(38), 26–36 (2013)

Gniel, J., Bouazza, A.: Improvement of soft soils using geogrid encased stone columns. Geotext. Geomembr. **27**(3), 167–175 (2009)

Greenwood, D.A.: Mechanical improvement of soils below ground surfaces. In: Proceedings of the Ground Engineering Conference, London, pp. 11–22 (1970)

Hughes, J.M.O., Withers, N.J., Greenwood, D.A.: A field trial of the reinforcing effect of a stone column in soil. Geotechnique **25**(1), 31–44 (1975)

Najjar, S.S.: A state of the art review of stone/sand column reinforced clay systems. Geotech. Geol. Eng. J. **31**, 355–386 (2013)

Pulko, B., Majes, B., Logar, J.: Geosynthetic-encased stone columns: Analytical calculation model. Geotext. Geomembr. **29**(1), 29–39 (2011)

Murugesan, S., Rajagopal, K.: Geosynthetic-encased stone columns: numerical evaluation. Geotext. Geomembr. **24**(6), 349–358 (2006)

Shivashankar, R., Babu, M.R.D., Nayak, S., Manjunath, R.: Geotech. Geol. Eng. **28**(5), 695–706 (2010)

Van Impe, W.F.: Soil Improvement Techniques and Their Evolution. A.A. Balkema, Rotterdam (1989)

Zhang, L., Zhao, M.H.: Deformation analysis of geotextile-encased stone columns. Int. J. Geomech. **15**(3), 1–10 (2015)

Numerical Analysis on Strut Responses Due to One-Strut Failure for Braced Excavation in Clays

A.T.C. Goh[3], Zhang Fan[4], Liu Hanlong[1,2], Zhang Wengang[1,2](✉), and Zhou Dong[2]

[1] Key Laboratory of New Technology for Construction of Cities in Mountain Area, Chongqing University, Chongqing, China
cheungwg@126.com
[2] School of Civil Engineering, Chongqing University, Chongqing, China
[3] School of Civil and Environmental Engineering, Nanyang Technological University, Singapore, Singapore
[4] IIBIT-Sydney, Federation University, Ballarat, NSW, Australia

Abstract. In deep braced excavations in clays, struts and walers play an essential role in the whole supporting system. For multi-level strutting systems, accidental strut failure is possible. Once a single strut fails, it is possible for the loads carried from the previously failed strut to be transferred to the adjacent struts and therefore cause one or more struts to fail if these are not of sufficient bearing capacity. Consequently, progressive collapse may occur and cause the whole excavation system to fail. One of the reasons for the Nicoll Highway Collapse in Singapore was attributed to the failure of the struts and walers. Consequently, for the design of braced excavation systems in Singapore, one of the requirements by the building authorities is to perform one-strut failure analyses, in order to ensure that there is no progressive collapse when one strut was damaged due to a construction accident. Therefore, plane strain and three-dimensional finite element analyses of one-strut failure of the braced excavation system were carried out in this study to investigate the effects of one-strut failure on the adjacent struts.

Keywords: Three-dimensional · Braced excavation · One-strut failure · Load transfer percentage · Finite element

1 Introduction

In Singapore, braced retaining wall systems are commonly used to construct cut and cover tunnels/stations for Mass Rapid Transit projects as well as for deep basements for shopping malls. The excavation sides are normally supported by concrete diaphragm walls or secant bored pile walls with two or more levels of struts. The purpose of excavation support system is to provide rigid lateral support for soil surrounding excavation and to limit the surrounding soils movement. The method presented by Peck (1969) for specifying apparent lateral earth pressures forms the basis of design of the excavation support systems and determination of loads on bracing systems.

R. Chen et al. (eds.), *Proceedings of the 2nd International Symposium on Asia Urban GeoEngineering*, Springer Series in Geomechanics and Geoengineering,
https://doi.org/10.1007/978-981-10-6632-0_43

However, the limitation of Peck's method have been highlighted by Liao and Neff (1990), and various enhancements to the method have been proposed by Liao and Neff (1990), Chang and Wong (1997), and Twine and Roscoe (1997). The mechanisms controlling the development of lateral earth pressure around a braced excavation in a deep deposit of soft clay has been presented by Hashash and Whittle (2002).

One concern in the design of these cut and cover excavation projects is the consequence of the failure of one or two struts (due to accidental damage during construction or incidental design/quality problems) in bracing system and whether it would lead to progressive failure and eventual total collapse of bracing systems and the surrounding grounds (Endicott 2013, Saleem 2015). The focus of this paper is on one-strut failure analyses, to investigate how loads from the failed strut are transferred to the adjacent struts and the whole support system.

To date, only limited studies have been carried out to examine this aspect in retaining walls supported by multi-levels of anchors/struts. One widely reported field study was carried out at four sites in Sweden and were described in Stille (1976), and Stille and Broms (1976). Stille (1976) reported that at the four sites, the maximum change of the anchor load expressed as a percentage of the initial load of the failed anchor was 35% for walls with two or more rows of anchors. The field study indicated that the increase in the loads of the adjacent anchors was of the same order of magnitude in the horizontal and vertical directions. As the associated wall movements were small, it was concluded that the transfer of load to the adjacent anchors was not through the arching of the soil behind the wall, but through the wall and wale members. The numerical analysis performed by Goh and Wong (2009) indicated that the failure of one or two struts due to an accident would not result in detrimental failure of the entire excavation system, provided the strut have been adequately against compression failure. Low et al. (2012) presented the design approach and consideration for the temporary removable ground anchor using TR26:2010, based on a Singapore Mass Rapid Transit case. Pong et al. (2012) proposed a simplified procedure to rationally idealise the one-strut failure problem from a 3D analysis to a 2D plain strain analysis.

Unfortunately, in the case of braced strut systems, there is no reported case history detailing the load transfer mechanism due to one-strut failure. Since struts provide passive resistance to wall movement, while anchors rely on stresses in the ground being mobilized to retain the wall, it may not be realistic to compare or equalize the redistribution of forces for anchors and struts. In Singapore, part of the design requirement requires that the braced retaining wall system to be structurally safe, robust and has sufficient redundancy to avoid catastrophic collapse. In the conventional approach for one-strut failure using 2D analysis, the entire level of the failing strut is removed and thus the forces can only be distributed vertically. This generally leads to a more conservative design with heavier strut sections. Thus, 3D analysis of one-strut failure is essential to provide more realistic understanding of the force/stress transfer behaviour of the braced excavation system. This paper describes the both use of 2D and 3D FE analyses to assess the impact of the failure of one strut on the remaining struts.

2 Numerical Schemes

For the numerical simulations carried out in this study, geotechnical software PLAXIS 2D (V9.0) and PLAXIS 3D Foundation were used (Brinkgreve et al. 2011). Figure 1 shows a typical cross-section and plan view for the cases considered. The parameters shown in the figure include: L = excavation length, B = excavation width, D = wall penetration depth, T = clay thickness below the final excavation level (FEL), S_H = horizontal strut spacing, S_V = vertical strut spacing and H_e = depth of final excavation. Vertical retaining walls along the excavation boundary were installed together with a five-level strut and waling system. The compositions of clay layers were varied for the different cases studied.

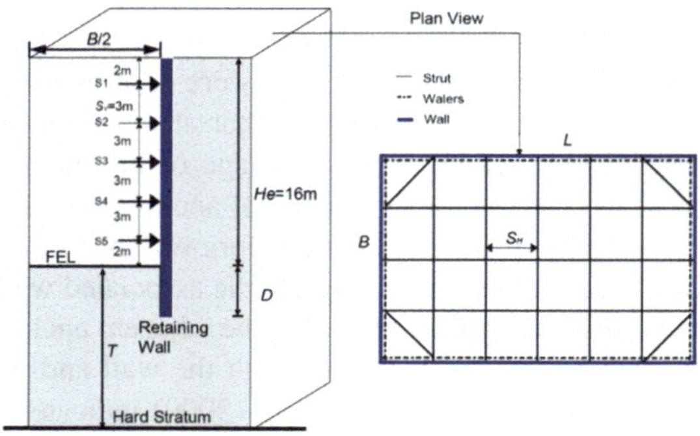

Fig. 1. Cross-section and plan view of the model for braced excavation

For 2D analyses, a half mesh was used due to geometrical symmetry. A very fine mesh size was used for 2D analysis to improve the accuracy of FE calculations. For 3D analyses, a medium mesh size in the horizontal direction and medium coarse mesh size in vertical direction were used to reach a balance between processing time and accuracy. For brevity, only a typical 3D half mesh is shown in Fig. 2, comprising of 15,679 nodes and 4,980 15-node wedge elements.

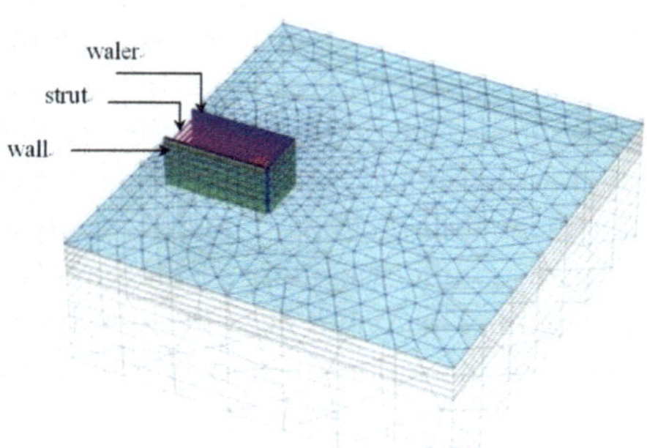

Fig. 2. 3D half mesh of the excavation from PLAXIS 3D foundation

For the 2D simulations, fourth order 15-node triangular elements, which are considered to be very accurate elements, were used to model the soil while the interface elements have 5 integration points. In 3D PLAXIS, the interface elements have 9 point Gauss integration with three translational degrees of freedom for each node. This is described in greater detail in Van Langen (1991). For the 2D analysis, the retaining wall is simulated using 5-node elastic plate elements. The elastic behaviour is defined by the following parameters: *EA*: normal stiffness, *EI*: bending stiffness, v: Poisson's ratio. For the 3D analysis, the wall is simulated using 8-node quadrilateral plate elements with six degrees of freedom per node.

In this study, three different wall stiffness values were considered for each soil type, as listed in Table 1. Based on the approach adopted by Finno et al. (2007), the wall thickness of 0.42 m was set to an arbitrary (constant) value so that the moment of inertia I and area A were kept constant, and only the wall elastic modulus E was varied. A coefficient α was introduced to represent walls with different rigidities (Bryson and Zapata-Medina 2012). The baseline bending stiffness EI for the analysis is 5.04×10^5 kNm^2/m, which refers to a wall of medium stiffness based on the databases of Long (2001) and Moorman (2004). Therefore, $\alpha = 1$ for cases with this wall stiffness, while a smaller $\alpha = 0.1$ represents flexible walls and larger α represents stiff walls. The system stiffness, S is defined as

Table 1. Wall properties for 2D and 3D analyses

Parameters		Wall types				
		Flexible		Medium	Stiff	
		Plane strain (2D) FE parameters				
Wall rigidity α		0.1	0.2	1.0	2.0	10
System stiffness S		32	64	320	320	3200
Wall stiffness EI (kNm^2/m)		5.04×10^4	1.008×10^5	5.04×10^5	1.008×10^6	5.04×10^6
Comp. stiffness EA (kN/m)		3.427×10^6	6.854×10^6	3.427×10^7	6.854×10^7	3.427×10^8
Poisson's ratio v		0	0	0	0	0
		Three-dimensional (3D) FE parameters				
Young's modulus (kPa)	E_1	8.16×10^6	1.632×10^7	8.16×10^7	1.632×10^8	8.16×10^8
	E_2	4.08×10^5	8.16×10^5	4.08×10^6	8.16×10^6	4.08×10^7
	E_3	2.00×10^8	4.00×10^8	2.00×10^9	4.00×10^9	2.00×10^6
Shear modulus (kPa)	G_{12}	4.08×10^5	8.16×10^5	4.08×10^6	8.16×10^6	4.08×10^7
	G_{13}	4.00×10^5	8.00×10^5	4.00×10^6	8.00×10^6	4.00×10^7
	G_{23}	1.333×10^6	2.666×10^7	1.333×10^7	2.666×10^8	1.33×10^8
Poisson's ratio v		0	0	0	0	0

$$S = \frac{EI}{\gamma_w h_{avg}^4} \tag{1}$$

where EI = wall stiffness, γ_w = unit weight of water, and h_{avg} = average vertical strut spacing.

The struts were simulated using node-to-node anchor elements in 2D analysis. For 3D analysis, the struts and walers were modelled as beam elements, which have six degree of freedom per node. This is described in greater details in Brinkgreve et al. (2011). For braced excavations in this paper, the struts were placed horizontally at a spacing of 4 or 5 meters (for different case studies) in two directions to form a frame net. The walings were used to connect the excavation wall and struts. The material properties are tabulated in Table 2.

Table 2. Properties of waling system

Parameter		Struts	Walers
Unit weight γ (kN/m^3)		78.5	78.5
Cross section area A (m^2)		0.007367	0.008682
Young's modulus E (kPa)		2.1×10^8	2.1×10^8
Moment of inertia (m^4)	I_3	5.073×10^{-5}	1.045×10^{-4}
	I_2	5.073×10^{-5}	3.668×10^{-4}
	I_{23}	0	0

The boundary conditions for 2D and 3D cases were: (1) rollers at side boundaries to allow vertical displacements and (2) pinned at the base to restrain any movements. For both 2D and 3D cases, the lateral boundaries in the side directions were set at least 100 m away from the centre of the excavation to eliminate the influence of the boundary restraints on the ground movements. This ensures that the lateral boundaries are beyond the settlement influence zone (which is typically 2 times of the excavation depth) induced by the excavation as proposed by Hsieh and Ou (1998). In this study, the clay thickness below the final excavation level T is assumed as 32 m, which is regarded as fairly large. A typical staged construction simulation is shown in Table 3. The original ground water table was assumed to be 2 m below the ground surface in the retained soil. The water table inside excavation was progressively lowered with soil excavation during each phase.

The properties of three different types of clays which were considered in this parametric study are similar to the properties assumed by Bryson and Zapata-Medina (2012) and are tabulated in Table 4. The soils are assumed to follow the Hardening Soil (HS) model. The three soil types are: soft clay, medium clay and stiff clay. The clays are real soils whose properties have been extensively reported in the literature. The properties of the soft clay with average c_u = 20 kPa are based on the Upper Blodgett soft clay reported by Finno et al. (2002). The medium clay with average c_u = 45 kPa are based on the Taipei silty clay found at the Taipei National Enterprise Centre (TNEC) project (Ou et al. 1998). The Gault clay at Cambridge (Ng 1992; Ng and Yan 1998) with average c_u = 125 kPa was used as the model for the stiff clay.

A series of 2D and 3D analyses covering various cases of wall stiffness α, soil types, excavation geometries and different strut levels and locations using the Hardening Soil (HS) model were conducted. For brevity, only the main findings of the numerical results are presented in the following sections. For all the 2D and 3D cases, the height of wall H_w is fixed at 20 m, so the depth of wall penetration D decreases as H_e increases. Apart from slight differences with regard to the excavation depth H_e, identical construction procedures were applied as described in Table 3.

Table 3. Typical construction sequence for 2D analysis

Phases	Construction details
Phase 1	Install the excavation wall
Phase 2	Reset displacement to zero, excavate to 3 m below ground surface
Phase 3	Install strut system at 2 m below ground surface
Phase 4	Excavate to 6 m below ground surface
Phase 5	Install strut system at 5 m below ground surface
Phase 6	Excavate to 9 m below ground surface
Phase 7	Install strut system at 8 m below ground surface
Phase 8	Excavate to 12 m below ground surface
Phase 9	Install strut system at 11 m below ground surface
Phase 10	Excavate to 15 m below ground surface
Phase 11	Install strut system at 14 m below ground surface
Phase 12	Excavate to 16 m below ground surface
Phase 13	Remove the chosen strut

Table 4. Input HS soil parameters of three clays

Parameter	Unit	A: soft clay (Chicago clay)	B: medium clay (Taipei silty clay)	C: stiff clay (Gault clay)
γ_{unsat}	kN/m^3	18.1	18.1	20
γ_{sat}	kN/m^3	18.1	18.1	20
E_{50}^{ref}	kN/m^2	2350	6550	14847
E_{refoed}	kN/m^2	2350	6550	14847
E_{refur}	kN/m^2	7050	19650	44540
c	kN/m^2	0.05	0.05	0.05
φ	°	24.1	29	33
Ψ	°	0	0	0
ν_{ur}	[−]	0.2	0.2	0.2
p^{ref}	kN/m^2	100	100	100
m	[−]	1.0	1.0	1.0
Knc 0	[−]	0.59	0.55	1.5
R_f	[−]	0.7	0.95	0.96
R_{inter} (interface friction)	[−]	1	1	1

3 Numerical Results and Verification

Before performing the one-strut analysis, the numerical results should be firstly discussed and validated against the previous semi-empirical design chart.

3.1 Typical 2D and 3D Analyses and Results

Figure 3 presents the wall deflection profiles from the 2D analyses at different excavation depths of H_e = 6 m, 9 m, 12 m and 16 m in stiff clays for different wall stiffness before performing the one-strut failure analysis. For brevity, the lateral wall deflections in soft and medium clays are omitted. The profiles reveal a general trend that increasing wall stiffness leads to a smaller maximum wall deflection value. The installation of strut generally results in lateral restraint of the wall movement above the level of the installed strut, especially for the retaining walls of flexible and medium stiffness, so that the wall displacement profiles at different excavation stages almost coincide with each other above the installed struts. This agrees with the previous research by Finno et al. (2007). Generally for the flexible walls, the wall deflection profile has a bulging shape with the maximum wall deflection occurring between the excavation level and the toe of the wall.

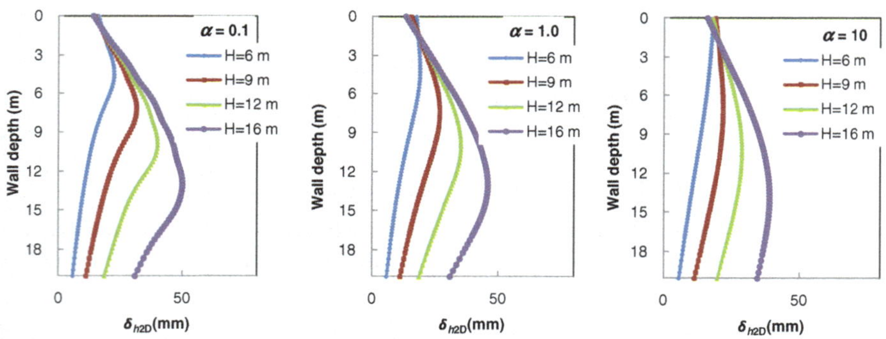

Fig. 3. Wall deflection profiles from 2D analyses for three wall stiffness in stiff clay

For the 3D rectangular braced excavation, the maximum lateral deflection of the wall occurs at the centre line in consideration of the symmetry. Therefore, for 3D analyses, only the profiles of the horizontal wall deflection at the centre line location are presented. For brevity, only wall deflections with L/B = 1.0 are plotted in Fig. 4 for stiff clays. In general, the increase of the wall stiffness leads to the reduction of the horizontal wall deflection. Similar to the trends of the 2D results, the wall deflection patterns from the 3D analyses show that the strut installation generally results in lateral restraint of the wall movement above the level of the installed strut, so that the wall deflection profiles at different excavation stages almost coincide with each other above the installed struts.

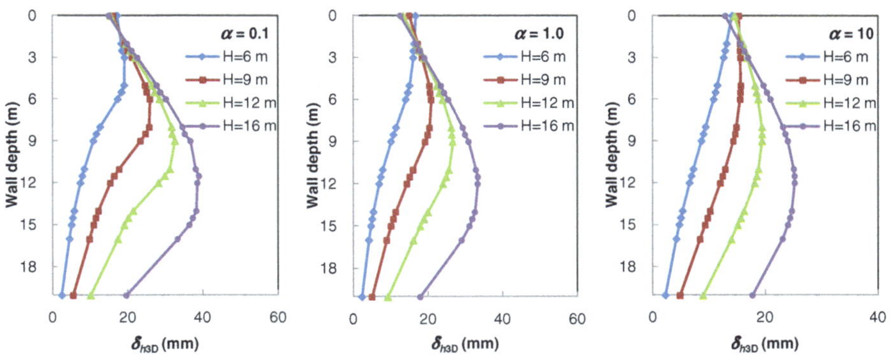

Fig. 4. Wall deflections from 3D analyses at different excavation stages in stiff clay (L/B = 1.0)

3.2 Comparisons with Previous Semi-Empirical Design Chart

In this section, the FE results are compared with two commonly used semi-empirical design charts from Clough and O'Rourke (1990) denoted as the Clough chart. It should be noted that the original Clough chart is based on the basal heave *FS* from Terzaghi's method (Terzaghi 1943) (denoted as FS_{T1} in Eq. 2).

$$\mathrm{FS_{T1}} = \frac{5.7c_u B_1}{\gamma H B_1 + q B_1 - c_u H} \tag{2}$$

In Eq. (2), B_1 = the smaller of 0.7B or T and q = the surcharge load.

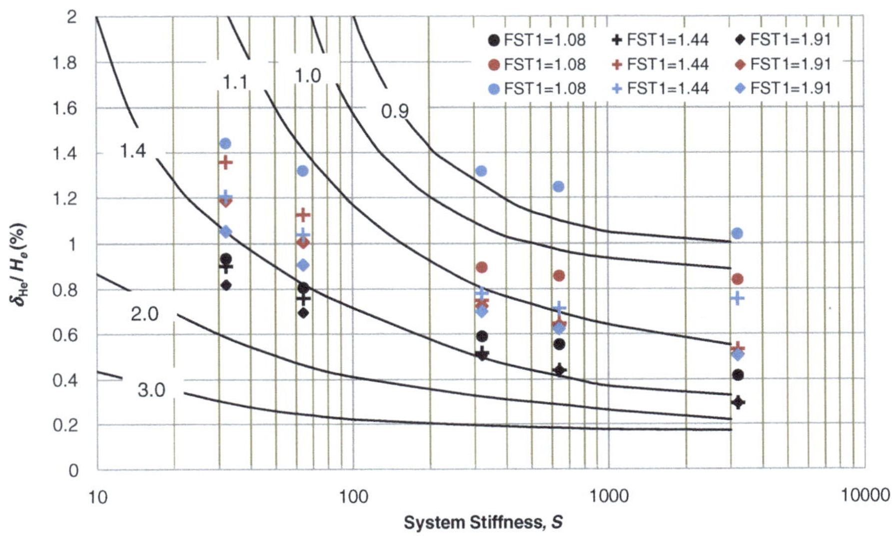

Fig. 5. Comparison of FE results with the Clough chart

In Fig. 5, the data points from this study are plotted on the Clough chart for comparison. The plot shows that generally the numerical results are reasonable. The slight differences between FE results and predictions from the Clough chart may lie in the following facts: (1) The calculated FS_{T1} values did not take into account the wall penetration depth which considerably influences the wall deflection; (2) The Clough chart was compiled based on both the numerical results and field measurements and for the field measurement part, the system stiffness S values were in the lower side, with fewer data sets of S greater than 200 when first compilation in 1990.

4 One-Strut Failure Analysis: Two Hypothetical Cases

In order to investigate the influence of the one-strut failure on the adjacent struts, the strut forces before and after strut failure are examined. By tabulating the load transfer percentage, the influence of the strut failure can be demonstrated. Assume N_{pre} is the load on the strut before strut failure; N_{post} is the load on the strut after strut failure and N_{fail} is the load on the failed strut before failure. Then the load transfer is defined as Eq. (3).

$$\text{Load Transfer } (\%) = \frac{N_{post} - N_{pre}}{N_{fail}} \times 100\% \quad (3)$$

4.1 One-Strut Failure for Soft Clay: H_e = 12 M with 3-Levels of Struts

The load transfer percentages from 2D analyses are tabulated in Table 5, and the 3D results are shown in Table 6. The 2D results indicate that failure of S3 leads to considerable increase in the load for the 2nd level S2 strut. On the other hand, the 3D results indicate that the load of the failed strut is not only transferred vertically upward, but also to the adjacent horizontal and diagonal struts as shown in Fig. 6, where larger arrows denote the larger magnitude of the transferred loads and the red circles denote the struts with more than 10% load increase after the one-strut failure. It should be noted from Table 6 that the load transfer percentage values along the left and right sides of the failed strut (i.e., 9.7 and 9.2 for soft clay, $\alpha = 0.1$, S2 in Table 6, 23.3 and 21.1 for soft clay, $\alpha = 0.1$, S3 in Table 6) are not equal considering that the numerical model is not symmetrical about the failed ($x = 0$) strut. In addition, for the 3D analyses

Table 5. Load transfer percentages from 2D analyses for the one-strut failure of S3 (soft clay)

Struts	Load Transfer %	
	$\alpha = 0.1$	$\alpha = 1.0$
S1	-37.7	-30.4
S2	122.2	120.3
S3	Failed strut	

Table 6. Load transfer percentage from 3D analyses for the one-strut failure of S3 (soft clay, L/B = 2.2)

Struts	Load Transfer %				
	x = -8	x = -4	x = 0	x = 4	x = 8
			soft clay, $\alpha = 0.1$		
S1	-0.1	-2.5	-9.2	-2.5	0.2
S2	-0.5	9.7	**15.4**	9.2	-2.2
S3	2.7	**23.3**	Failed strut	**21.1**	-0.1
			soft clay, $\alpha = 1.0$		
S1	-5.7	-4.4	-3.3	-3.7	-3.0
S2	1.5	**11.9**	**53.6**	**11.2**	-0.7
S3	4.2	**24.4**	Failed strut	**22.5**	3.3

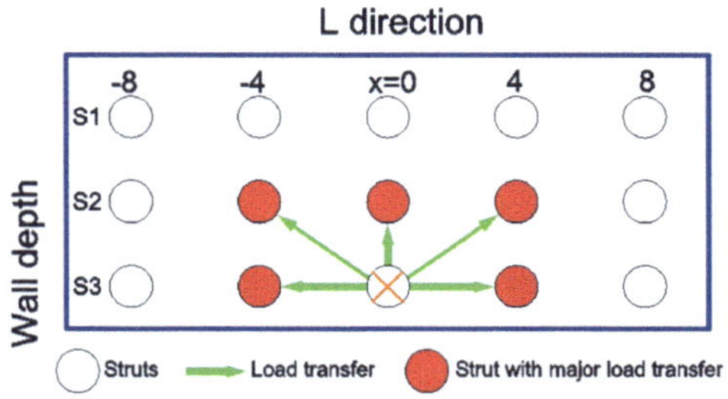

Fig. 6. Influence zone of the one-strut failure of S3 (soft clay)

the magnitude of the load transfer percentages are much smaller, up to approximately 20% for flexible walls and 50% for medium walls compared with the results from the 2D analyses. Therefore, compared to 3D results, the 2D analyses overestimate the possible consequence of the one-strut failure for soft clay, as it ignores the restraining effects of the adjacent horizontal struts.

4.2 One-Strut Failure for Medium and Stiff Clays: H_e = 16 M with 5-Levels of Struts

The load transfer percentages of the one-strut failure from 2D analyses for medium and stiff walls are shown in Table 7. For medium clay, most of the load of the failed strut is transferred to the struts immediately above and below the failed strut, with approximately 30% to 50% of the load transferred to the upper strut and the remainder carried by the lower strut. For stiff clay, the failure of strut S3 leads to the force redistribution for the S2, S4 and S5 struts. The S2 and S4 struts each carry approximately 30% of the load of the failed strut, and S5 carries approximately 10% to 20%. When the wall is also stiff, the top strut is able to carry approximately 10% of the load.

For the 3D analyses, the percentage of the load transfer to adjacent struts of the relevant struts for medium clay with stiff walls are tabulated in Table 8 for *L/B* = 2.2 and Table 9 for *L/B* = 3.4, respectively. The tables show that the struts affected most

Table 7. Load transfer percentage from 2D analyses (medium and stiff clays)

clay type	Struts	Load Transfer (%)		
		$\alpha = 0.1$	$\alpha = 1.0$	$\alpha = 10$
medium clay	S1	-0.9	-0.9	-0.4
	S2	-5.2	**11.2**	**15.5**
	S3	**46.0**	**41.7**	**31.1**
	S4	Strut Failure		
	S5	**53.4**	**47.1**	**46.3**
stiff clay	S1	-0.8	2.4	**15.4**
	S2	**27.6**	**30.2**	**23.1**
	S3	Strut Failure		
	S4	**29.3**	**29.5**	**22.1**
	S5	**10.8**	**11.9**	**14.6**

Table 8. Load transfer percentage from 3D analyses for the one-strut failure of S4 for medium clay (*L/B* = 2.2)

Struts	Load Transfer %				
	$x = -8$	$x = -4$	$x = 0$	$x = 4$	$x = 8$
S1	-1.1	0.2	1.6	0.2	-1.0
S2	-0.3	3.1	7.9	3.2	-0.3
S3	0.5	5.5	**15.3**	5.5	0.5
S4	1.2	7.2	Failed strut	7.2	1.3
S5	2.8	**9.0**	**19.5**	**9.0**	2.7

Table 9. Load transfer percentage from 3D analyses for the one-strut failure of S4 for medium clay (*L/B* = 3.4)

Struts	Load Transfer %				
	$x = -8$	$x = -4$	$x = 0$	$x = 4$	$x = 8$
S1	-0.3	0.8	2.0	0.5	-0.7
S2	-0.1	3.4	8.0	3.2	-0.2
S3	-0.1	5.3	**15.1**	5.4	0.4
S4	0.2	6.5	Failed strut	6.5	0.7
S5	1.4	7.5	**17.9**	7.4	1.5

are located directly above or below the failed strut S4 or diagonally across, as plotted in Fig. 7.

The influence of the one-strut failure from 3D analyses for the stiff clay is shown in Fig. 8. The load transfer percentages for stiff clay is listed in Table 10 for *L/B* = 3.4. The load is mainly transferred to the struts directly above and below the failed strut S3, and the load transfer percentage is approximately 10% to 20%.

The 3D results again highlight that the 2D analyses would result in fairly conservative (i.e., larger) estimates of the loads transferred to the adjacent struts from the failed strut as it ignores the restraining effects of the adjacent horizontal struts.

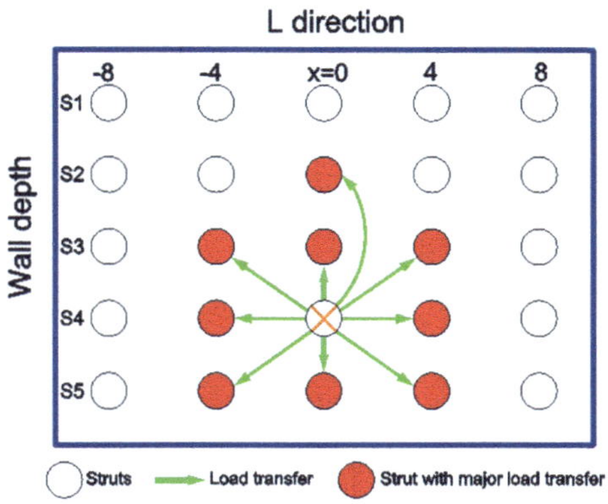

Fig. 7. Influence zone of the failure of S4 (medium clay)

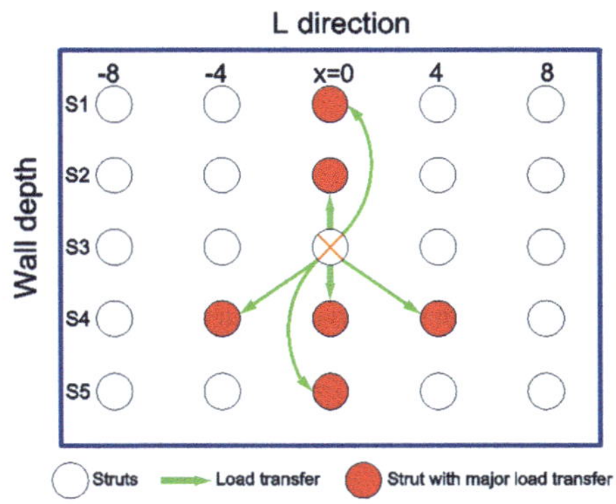

Fig. 8. Influence zone of the failure of S3 (stiff clay)

5 Summary and Conclusions

In this paper, the FE simulation of one-strut failure of multi-level braced excavations was considered. The results were validated against the semi-empirical Clough design chart before performing the one-strut failure analysis. The 3D one-strut failure analysis indicates that the 2D analyses would result in fairly conservative (i.e., larger) estimates of the loads transferred to the adjacent struts from the failed strut as it ignores the restraining effects of the adjacent horizontal struts. Therefore, 2D analysis would result is a more conservative design. This may result in a larger strut size and larger waler size which would increase costs. Actually, the one-strut failure analysis involves an interaction process between neighbouring struts, between struts and wall and is influenced by the strut location, system stiffness, soil types, etc. It should be noted that the load transfer paths and the percentages for the one-strut failure were purely from 3D analyses and the use of them for design of strutting system should be with caution.

Table 10. Load transfer percentage from 3D analyses for the one-strut failure of S3 for stiff clay (L/B = 3.4)

Struts	Load Transfer %				
	x = -8	x = -4	x = 0	x = 4	x = 8
			stiff clay, α = 0.1		
S1	-0.5	-3.7	-5.1	0.0	0.0
S2	0.1	4.8	**12.6**	5.8	1.6
S3	1.4	4.6	Failed strut	5.4	2.6
S4	3.4	6.8	**12.3**	6.6	7.2
S5	4.5	5.0	4.7	4.8	3.8
			stiff clay, α = 1.0		
S1	-0.2	0.6	0.6	0.8	0.1
S2	1.2	4.6	**17.8**	4.9	1.6
S3	1.3	4.1	Failed strut	4.4	1.8
S4	3.2	6.0	**17.3**	6.1	3.0
S5	4.6	5.9	6.6	5.8	4.2
			stiff clay, α = 10		
S1	-0.3	3.0	5.9	3.1	-0.1
S2	0.7	4.4	**12.1**	4.5	0.9
S3	1.6	5.0	Failed strut	5.1	1.7
S4	3.0	4.4	**11.1**	5.9	3.0
S5	4.7	6.6	9.3	6.5	4.4

Acknowledgements. A portion of the work was completed while the corresponding author was on sabbatical leave at Nanyang Technological University, Singapore. The corresponding author is grateful to the support by the National Natural Science Foundation of China (No. 51608071), the Advanced Interdisciplinary Special Cultivation program (No. 106112017CDJQJ208850) and Key Laboratory of Rock Mechanics in Hydraulic Structural Engineering, Ministry of Education (No. RMHSE1601).

References

Brinkgreve, L.B.J., Swolfs, W.M., Engin, E.: Plaxis Manual, PLAXIS bv, Netherlands (2011)

Bryson, L., Zapata-Medina, D.: Method for estimating system stiffness for excavation support walls. J. Geotech. Geoenviron. Eng. **138**, 1104–1115 (2012). doi:10.1061/(ASCE)GT.1943-5606.0000683

Chang, J.D., Wong. K.S.: Apparent pressure diagram for braced excavation in soft clay with diaphragm wall. In: Proceedings of the International Symposium on Geotechnical Aspects of Underground Construction in Soft Ground, London, England, pp. 87–92 (1997)

Clough, G.W., O'Rourke, T.D.: Construction induced movements of in situ walls. In: Design and Performance of Earth Retaining Structures, ASCE Special Conference, Ithaca, New York, pp. 439–470 (1990)

Endicott, J.: Case histories of deep excavation. Examination of where things went wrong: Nicoll Highway Collapse, Singapore. In: International Conference on Case Histories in Geotechnical Engineering, Paper 7 (2013)

Finno, R.J., Blackburn, J.T., Roboski, J.F.: Three-dimensional effects for supported excavations in clay. J. Geotech. Geoenviron. Eng. **133**, 30–36 (2007). doi:10.1061/(ASCE)1090-0241(2007)133:1(30)

Goh, A.T.C., Wong, K.S.: Three-dimensional analysis of strut failure for braced excavation in clay. J. Southeast Asian Geotech. Soc. **40**(2), 137–143 (2009)

Finno, R.J., Bryson, S., Calvello, M.: Performance of a stiff support system in soft clay. J. Geotech. Geoenviron. Eng. **128**, 660–671 (2002). http://dx.doi.org/10.1061/(ASCE)1090-0241(2002)128:8(660

Hashash, Y.M.A., Whittle, A.J.: Mechanisms of load transfer and arching for braced excavations in clay. J. Geotech. Geoenviron. Eng. **128**(3), 187–197 (2002). http://dx.doi.org/10.1061/(ASCE)1090-0241(2002)128:3(187)

Hsieh, P.G., Ou, C.Y.: Shape of ground surface settlement profiles caused by excavation. Can. Geotech. J. **35**, 1004–1017 (1998). doi:10.1139/t98-056

Liao, S.S.C., Neff. T.L.: Estimating lateral earth pressures for design of excavation support. In: Proceedings of the Specialty Conference on Design and Performance of Earth Retaining Structures, ASCE, pp. 489–509 (1990)

Long, M.: Database for retaining wall and ground movements due to deep excavations. J. Geotech. Geoenviron. Eng. **127**, 203–224 (2001). http://dx.doi.org/10.1061/(ASCE)1090-0241(2001)127:3(203)

Low, S.Y.H., Ng, D.C.C., Chin, Y.Y.P., Ting, E.S.K.: A Singapore case history of temporary removable ground anchor design to TR26. The IES J. Part A: Civil Struct. Eng. **5**(3), 181–194 (2010). http://dx.doi.org/10.1080/19373260.2012.696443

Moormann, C.: Analysis of wall and ground movements due to deep excavations in soft soil based on a new worldwide database. Soils Found. **44**, 87–98 (2004). http://dx.doi.org/10.3208/sandf.44.87

Ng, C.W.W.: An evaluation of soil-structure interaction associated with a multi-propped excavation. Ph.D. thesis, University of Bristol, UK (1992)

Ng, C.W.W., Yan, R.W.M.: Stress transfer and deformation mechanisms around a diaphragm wall panel. J. Geotech. Geoenviron. Eng. **124**, 638–648 (1998). http://dx.doi.org/10.1061/(ASCE)1090-0241(1998)124:9(798)

Ou, C.Y., Liao, J.T., Lin, H.D.: Performance of diaphragm wall constructed using top-down method. J. Geotech. Geoenviron. Eng. **124**, 798–808 (1998). http://dx.doi.org/10.1061/(ASCE)1090-0241(1998)124:7(638)

Peck, R.B.: Deep excavation and tunnelling in soft ground. In: Proceedings of the 7th International Conference on Soil Mechanics and Foundation Engineering, Mexico City, Mexico, State-of-the-art Volume, pp. 225–290 (1969)

Pong, K.F., Foo, S.L., Chinnaswamy, C.G., Ng, C.C.D., Chow, W.L.: Design considerations for one-strut failure according to TR26 – a practical approach for practising engineers. The IES Journal Part A: Civil & Structural Engineering **5**(3), 166–180 (2012). http://dx.doi.org/10.1080/19373260.2012.700790

Saleem, M.: Application of numerical simulation for the analysis and interpretation of pile-anchor system failure. Geomechanics and Engineering **9**(6), 689–707 (2015). http://dx.doi.org/10.12989/gae.2015.9.6.689

Stille, H.: Behaviour of anchored sheet pile walls, Ph.D. thesis, Royal Institute of Technology, Stockholm, Sweden (1976)

Stille, H., Broms, B.B.: Load redistribution caused by anchor failures in sheet pile walls. In: Proceedings of the 6th European Conference on Soil Mechanics and Foundation Engineering, Vienna, Austria, vol. 1.2, pp. 197–200 (1976)

Terzaghi, K.: Theoretical Soil Mechanics. Wiley, New York (1943)

TR26: Technical reference for deep excavation. Spring Singapore, Singapore (2010)

Twine, D., Roscoe, H.: Prop loads: Guidance on design. CIRIA Core Programme Funders' Report FR/CP/48, Construction Industry Research and Information Association, London, England (1997)

Van Langen, H.: Numerical analysis of soil-structure interaction. Ph.D. thesis, Delft University of Technology (1991)

Numerical Simulation on the Development of Soil Arching Induced by EPBS Tunnelling

Xingtao Lin[1,2,3], Yuan Liu[1,2,3], Xin Kang[1,2,3], Jun Chen[4], and Renpeng Chen[1,2,3](✉)

[1] MOE Key Laboratory of Building Safety and Energy Efficiency, Hunan University, Changsha 410082, China
chenrp@hnu.edu.cn
[2] National Joint Research Center for Building Safety and Environment, Hunan University, Changsha 410082, China
[3] College of Civil Engineering, Hunan University, Changsha 410082, China
[4] China Construction Fifth Engineering Division Co., Ltd, Changsha 410082, China

Abstract. The development of soil arching around the tunnel is greatly concerned in practice. This paper studies the development of soil arching induced by earth pressure balanced shield (EPBS) tunnelling in dry sand by using finite element method (FEM). The variation of vertical stress in monitoring section during tunnelling is discussed. The arch zone is determined based on the inflection points of the vertical stress σ_{zz}-depth curves, the shear surface and the changes of the porosity ratio in the tunnelling. The area ahead of the tunnel face forms the loosened zone because of the lack of the support pressure. The vertical stress σ_{zz} is significantly reduced in the loosened zone. Due to the influence of the support pressure, the volume loss and the grouting pressure on the soil surrounding tunnel, the arch zone gradually developed upward until segments are installed. The settlement of the arch zone is less than that of the loosened zone.

Keywords: EPBS · Numerical modeling · Soil arching · Stress redistribution

1 Introduction

In the past few decades, EPBS tunnelling technology has been widely used in the construction of the urban subway tunnel. It is important to ensure construction safety during tunnelling. However, there are great challenges due to the complex behaviour of soils. The development of soil arching during tunnelling is a clue to understand the soil deformation and the face stability.

Many researchers have conducted meaningful work in the model tests, numerical simulation and theoretical analysis. In the large-scale model test, Chen et al. (2013) investigated the evolution of soil arching during face failure by monitoring the redistribution of earth pressure in front of the face. Then, a two-stage failure pattern was suggested based on the observation of earth pressure redistribution. Adachi et al. (2003) carried out a series of three-dimensional trapdoor experiments to simulate the tunnel excavations. It is found that the distribution of earth pressure is greatly affected by the

R. Chen et al. (eds.), *Proceedings of the 2nd International Symposium on Asia Urban GeoEngineering*, Springer Series in Geomechanics and Geoengineering,
https://doi.org/10.1007/978-981-10-6632-0_44

process of the excavations. Other than one-gravity model tests, Lee et al. (2006) investigated the tunnel stability and the soil arching effect during tunnelling in soft clayey soil by a series of centrifuge model tests and numerical simulations. In the numerical simulation, Jiang and Yin (2012) analysed the stress redistribution in soil and the earth pressure acting on the tunnel lining due to tunnelling by using discrete element method (DEM). Chen et al. (2011a) investigated the failure mechanism and the limit support pressure of a tunnel face in dry sandy ground by using DEM. Chen et al. (2011b) researched the development of ground arching by stress redistribution during tunnelling using FLAC3D. In the theoretical analysis, Son (2017) deduced a three dimensional formula in which inclined sliding surfaces under general 3D excavation conditions are considered based on the 2D Terzaghi arching equation. However, Most of above researches mainly aimed at the development of soil arching ahead of the tunnel face or the development of soil arching. Few studies consider the development process of soil arching during shield tunnelling.

This paper studies the development of soil arching induced by EPBS tunnelling in dry sand using 3D element method. The change of vertical stress σ_{zz}, the porosity ratio and the settlement during shield tunnelling are investigated. The boundaries of arching zone are determined.

2 Numerical Model

The 3D FEA software, PLAXIS 3D (Brinkgreve et al. 2006), is used in the numerical simulation, as shown in Fig. 1. Considering the symmetry of the research problem, the numerical model employs half of the domain. To eliminate the influence of boundary conditions on the results, lateral boundaries are extended to more than 4*D* (*D* is the diameter of tunnel). Thus, the dimensions of the model are 90 m long, 30 m wide and 40 m high. On vertical boundaries, the horizontal displacements are fixed, and the vertical displacements are allowed. The displacements of the bottom boundary are fixed in three directions. The diameter of the tunnel is 6 m, and the cover depth is 12 m. EPBS

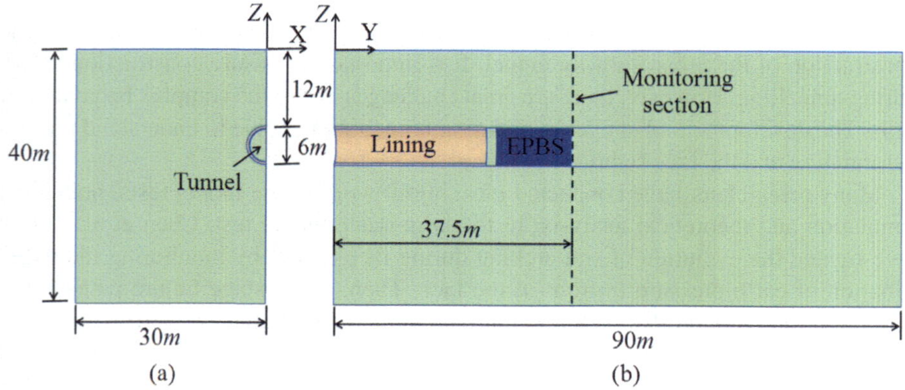

Fig. 1. Model geometry and position of monitoring section: (a) Transversal section (b) Longitudinal section.

is modeled by the shell element. Soil and lining are modeled by solid element. The Hardening Soil model with Small Strain Stiffness (HSS) is used to simulate the behavior of the soil. The tunnel lining and EPBS are modeled as an isotropic linear elastic material. The parameters of the model are shown in Table 1.

Table 1. Material parameters

Parameter type (units)	Constitutive model	*Eref* 50 (MPa)	*Eref* ur (MPa)	*Eref* ur (MPa)	m	e_{init}	c' (kPa)	φ' (°)
sand	HSS	37.8	37.8	113.5	0.5	0.5	2	34
Parameter type (units)	Constitutive model	E (GPa)	ν_{ur}	Ψ (°)	γ (kN/m³)	R_f	$\gamma_{0.7}$	Gref 0 (MPa)
sand	HSS	–	0.2	0	19.5	0.9	1E-4	189.2
tunnel lining	Linear elastic	31	0.1	–	25	–	–	–
EPBS	Linear elastic	230	0	–	120	–	–	–

The monitoring section is located at Y = 37.5 m (see Fig. 1). To eliminate the influence of initial position of EPBS on the monitoring section and to reduce the number of advanced rings, the monitoring section is located on 13.5 m in front of the tunnel face. The excavation of the following 35 rings is simulated ring-by-ring by considering the detailed shield-driving parameters (support pressure, grout pressure, jacking thrust). The cone-shaped EPBS is adopted to simulate the ground loss caused by the shield tunnelling.

3 Results and Discussion

3.1 Soil Arching Zone in Longitudinal Section

The vertical stress σ_{zz} continuously decreases at the monitoring section when the EPBS approaches, as shown in Fig. 2. L stands for the distance between the tunnel face and the monitoring section. $L < 0$ indicates that tunnel face does not pass through monitoring section. $L > 0$ shows that tunnel face has passed through the monitoring section.

When $L < -13.5$ m, the σ_{zz}-depth curves are linear and equivalent to the initial earth stress ($\sigma_{zz} = \gamma H$). So the soil arching is not developed in the monitoring section. When $L = -4.5$ m, the σ_{zz}-depth curves in the location of buried depth of tunnel become nonlinear and are larger than the initial ground stress ($\sigma_{zz} = \gamma H$). The part of monitoring section may be located in the stability zone which supports arch foot. When $L = -3$ m and $L = -1.5$ m, the σ_{zz}-depth curves become obvious nonlinear, indicating that the areas in front of the tunnel face forms soil arching because of the lack of the support pressure, respectively. There are two inflection points with marker ⊕ in the σ_{zz}-depth curves for $L = -1.5$ m, which are represented by points 1 and 2, respectively. The vertical stress is gradually reduced with the increase of depth above the inflection point 1, which indicates that the vertical stress is transferred to surrounding soil through the soil arching. The vertical stress sharply decreases in the range between the inflection points 1 and 2, which shows that the soil in this range is in the loosened state. The vertical stress begins

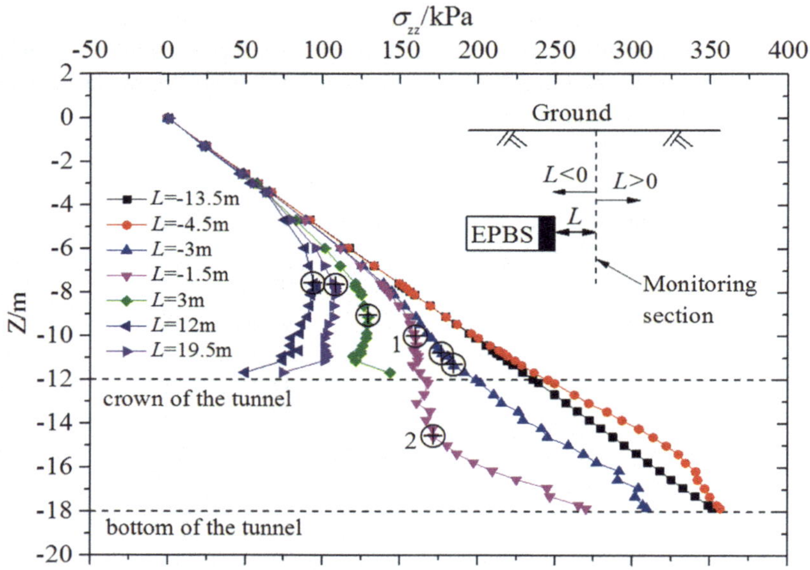

Fig. 2. Variation of σ_{zz}-depth curves in monitoring section during tunnelling.

to increase below the inflection point 2 and the increment is greater than that of the initial vertical stress ($\gamma\Delta H$). It shows that the vertical stress increment includes the self-weight stress and the stress transferred from surrounding soils through the soil arching. The range above the inflection point 1 and the range below the inflection point 2 are defined as soil arching zone. The range between the inflection points 1and 2 is defined as the loosened zone.

The vertical stress σ_{zz} is gradually decreased for the following excavations, $L = 3$ m, $L = 12$ m. At this location, the EPBS is passing through the monitoring section. Due to the increase of the volume loss, these inflection points in the σ_{zz}-depth curves gradually moves up, which implies expansion of the loosened zone.

When L = 19.5 m, The vertical stress σ_{zz} is slightly increased in about 4 m range above the tunnel. Between $L = 12$ m and $L = 19.5$ m, the grouting pressure acts on the surrounding soil and the segments are installed. When $L > 19.5$ m, the σ_{zz}-depth curves only slightly changes, which shows that soil arching is nearly fully developed.

The soil arching zone in longitudinal section is determined based on the vertical stress inflection points in Fig. 2, as shown in Fig. 3. The area above the inflection points represents the soil arching zone. The area below represents the loosened zone. Due to the lack of the support pressure, the areas in front of the tunnel face forms the loosened zone. The same phenomenon has been reported by Kirsch (2010) and Idinger (2011). With the increase of volume loss, the soil arching zone is slowly developing upwards. The grouting pressure is larger than the initial earth stress. The soil arching zone is not further developed when the grout zone passes through the monitoring section. There is no more change in the soil arching zone after the segments are installed.

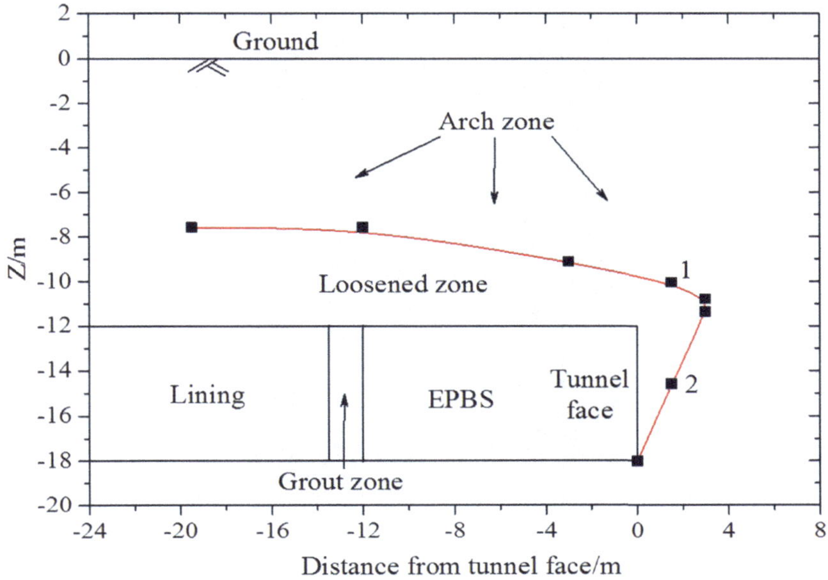

Fig. 3. Approximate soil arching zone distribution around the tunnel.

3.2 Soil Arching Zone in Longitudinal Section

The relative displacement between soil masses produces the shear resistance. The load of the yielded soil is transferred to the adjacent parts by the shear resistance. This load transfer mechanism is called the arching effect. The shear surface occurs where shear strains are large. One side of the shear surface is elastic soil which is defined as stability zone. The other side is the yielded part which is defined as the arch zone. The contours of shear strain in transversal section after tunnel excavation are shown in Fig. 4. The shear strain extends from the shoulder of the tunnel to the ground, which shows that this area is the possible shear surface. The area is defined by the outer boundary of the soil arching (i.e. dash dots in Fig. 5) (Terzaghi 1946). The lower boundary of the soil arching is determined by the inflection points of vertical stress at different positions.

Figure 5 shows the contours of porosity ratio in transversal section after tunnel excavation. The lower and outer boundaries of the soil arching are depicted in the porosity ratio contour in transversal section. According to the outer boundary of the stability zone, the upper boundary of the soil arching is determined. The porosity ratio of the soil arching zone and the stability zone are smaller than that of the other zone, indicating that these zones are compacted. However, the porosity ratio of the loosened zone becomes larger.

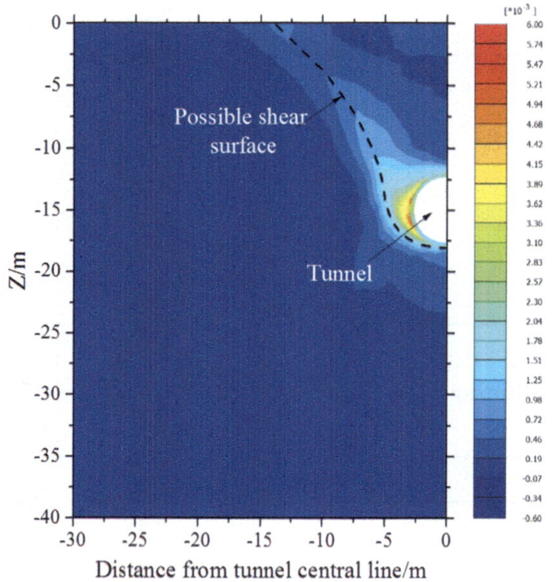

Fig. 4. Shear strain contour of transversal section after tunnel excavation.

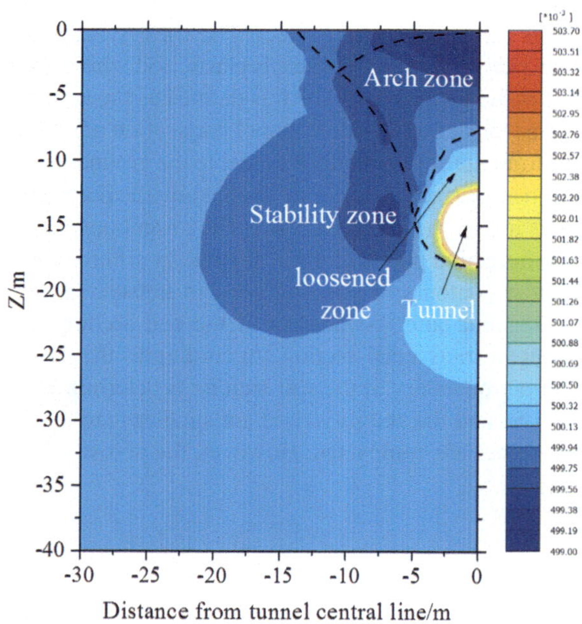

Fig. 5. Porosity ratio contour of transversal section after tunnel excavation.

3.3 Settlement of the Soil Above the Tunnel After Excavation

Figure 6 shows the settlement changes of the four points (i.e., A, B, C and D) in the monitoring section when the tunnel advances. As the depth increases, the settlement decreases before the tunnel face arrives at monitoring section. The support pressure is less than the initial earth pressure. The soil of the area ahead of the tunnel face becomes loosing. Due to the volume loss, the settlement of the soil near the tunnel becomes larger than soil near the ground after the tunnel face passes monitoring section. The settlement of point D is larger than those of other points. The settlement of points A and B is the same. In Fig. 3, point D is located in the loosened zone, and point C is on the edge of the arch zone, and points A and B is located in the arch zone. The settlement of the arch zone is less than that of loosened zone.

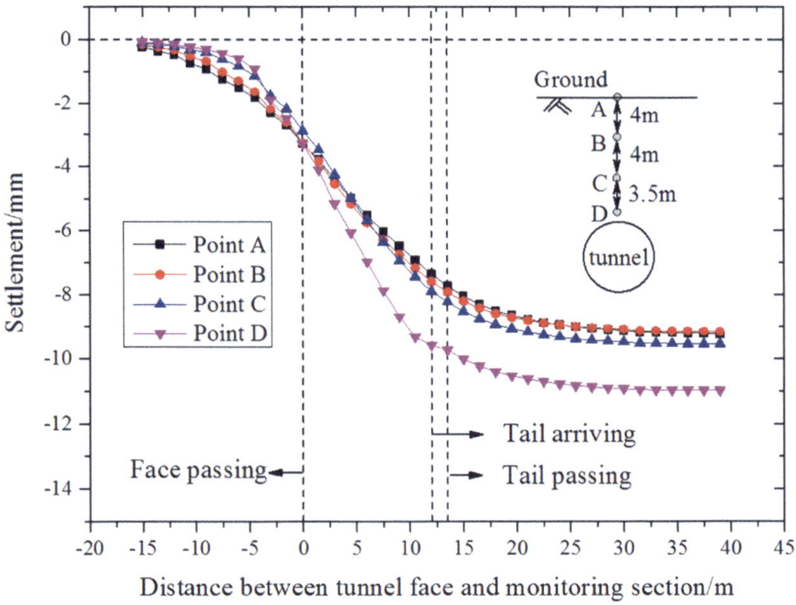

Fig. 6. Variation of vertical displacement of different depths in monitoring section during tunnelling.

4 Conclusion

A 3D FEM model has been developed to analyze soil arching development induced by EPBS tunnelling. The following conclusions can be drawn:

(1) When $L = -4.5$ m, The part of monitoring section is located in the stability zone which supports arch foot. The longitudinal section of arch zone is determined by rising inflection points of the σ_{zz}-depth curves in the tunnelling. Due to the influence of the support pressure, the volume loss and the grouting pressure on the soil

surrounding tunnel, the arch zone gradually develops upward until segments are installed.

(2) The transversal section of arch zone is determined by the flection of the σ_{zz}-depth curves, the possible shear surface and the contours of porosity ratio. The settlement of the arch zone is less than that of loosened zone.

Acknowledgments. The present work is carried out with the support of the National Key Research and Development Program of China (2016YFC0800207), National Natural Science Foundation of China (41472244), the Provincial Key Research and Development Program of Hunan (0105679005), the Industrial Technology and Development Program of Zhongjian Tunnel Construction Co., Ltd. (17430102000417) and the Research Program of Construction Engineering of Shenzhen (20151118003B).

References

Adachi, T., Kimura, M., Kishida, K.: Experimental study on the distribution of earth pressure and surface settlement through three-dimensional trapdoor tests. Tunn. Undergr. Space Technol. Inc. Trenchless Technol. Res. **18**(2), 171–183 (2003)

Brinkgreve, R.B.J., Broere, W., Waterman, D.: Plaxis, Finite Element Code for Soil and Rock Analyses, Users Manual. Plaxis, Rotterdam (2006)

Chen, R.P., Li, J., Kong, L.G., Tang, L.J.: Experimental study on face instability of shield tunnel in sand. Tunn. Undergr. Space Technol. **33**, 12–21 (2013)

Chen, R.P., Tang, L.J., Ling, D.S., Chen, Y.M.: Face stability analysis of shallow shield tunnels in dry sandy ground using the discrete element method. Comput. Geotech. **38**(2), 187–195 (2011a)

Chen, C.N., Huang, W.Y., Tseng, C.T.: Stress redistribution and ground arch development during tunneling. Tunn. Undergr. Space Technol. Inc. Trenchless Technol. Res. **26**(1), 228–235 (2011b)

Idinger, G., Aklik, P., Wu, W., Borja, R.I.: Centrifuge model test on the face stability of shallow tunnel. Acta Geotech. **6**(2), 105–117 (2011)

Jiang, M., Yin, Z.Y.: Analysis of stress redistribution in soil and earth pressure on tunnel lining using the discrete element method. Tunn. Undergr. Space Technol. **32**, 251–259 (2012)

Kirsch, A.: Experimental investigation of the face stability of shallow tunnels in sand. Acta Geotech. **5**(1), 43–62 (2010)

Lee, C.J., Wu, B.R., Chen, H.T., Chiang, K.H.: Tunnel stability and arching effects during tunneling in soft clayey soil. Tunn. Undergr. Space Technol. **21**(2), 119–132 (2006)

Son, M.: Three-dimensional expansion of the terzaghi arching formula considering inclined sliding surfaces and examination of its effects. Int. J. Geomech. **17**(7), 06016043 (2017)

Terzaghi, K.: Rock defects and loads on tunnel supports. In: Proctor, R.V., White, T. (eds.) Rock Tunneling with Steel Supports, pp. 17–99. Commercial Shearing and Stamping Co., Youngstown (1946)

Analyses of Ground Response Induced by Diaphragm Wall Installation

A.-J. Li(✉) and H.-D. Lin

Department of Civil and Construction Engineering,
National Taiwan University of Science and Technology, Taipei, Taiwan
laj871@mail.ntust.edu.tw

Abstract. The purpose of this research is to investigate the ground response due to diaphragm wall construction using three-dimensional numerical modelling. The commercial finite difference method software, FLAC^{3D} is employed in this study. This study investigates the influence of diaphragm wall panel dimensions on soil movement. In addition, a series of diaphragm wall panel constructions are also discussed. The results obtained indicate that the above factors do have influence on ground response in terms of displacements. The discussion in the paper can be utilized as the reference for practical designs.

Keywords: Excavation · Displacement · Three-dimensional

1 Introduction

As the population of global major cities continues to grow, the available space for human inhabitancy is reduced. This is called urbanization. As a consequence, high rise buildings become more popular because they are more efficient at providing the available space. Such buildings generally need larger underground space for car parking or other purposes, and therefore deep excavation, foundations and retaining system are used more frequently. In recent years, the diaphragm wall method of construction has become more popular for excavations in Taipei. The advantages include low vibration and noise, and high stiffness. Moreover, diaphragm walls can provide better control of displacement and water drainage during the excavation, compared to other methods (Ou 2006).

Generally speaking, single diaphragm wall panel construction can probably be divided into three stages, which are (1) Simultaneous trench excavation and bentonite pressure, (2) Place steel cages, (3) Cast the concrete panel. The first stage is often the most critical during the diaphragm wall construction as the unbalanced forces between the trench and soil mass are at their greatest. In fact, the trench installation induced instability and displacement are conventional geotechnical problems, however current evaluations are often based entirely on empiricism.

In engineering practice, often the vertical and horizontal displacements caused by the diaphragm wall construction are not considered. It is because they are generally relatively small, compared with those caused by the main excavation. However, surface settlements induced by diaphragm wall construction have been reported to be as much

R. Chen et al. (eds.), *Proceedings of the 2nd International Symposium on Asia Urban GeoEngineering*, Springer Series in Geomechanics and Geoengineering,
https://doi.org/10.1007/978-981-10-6632-0_45

as 30–50% of final cumulative settlements (Cowland and Thorley 1985; Morton et al. 1980). The measured results from Taipei Rapid Transit system indicated that the surface settlement induced by diaphragm wall installation is significant. Its magnitude also agreed with previous observations. Consequently, the potential to damage adjacent buildings during underground construction could be increased. Based on the discussions above, the importance of understanding ground response caused by diaphragm wall installation is highlighted. In order to understand the ground responses in the Taipei basin, the commercial finite difference method software, FLAC^{3D} (Itasca 2002), is employed in this study.

2 Previous Studies

The construction of a single diaphragm wall panel with the dimensions 1 m × 5 m × 28 m ($W \times L \times D$), was investigated by Dibiagio and Myrvoll (1972) in Oslo. The variables W, L and D are used to define the dimensions of a single trench or panel in this paper, as shown in Fig. 1. Before concreting, the trench stood for 31 days by applying bentonite pressures to the trench faces alone. The maximum surface settlement increased from 4 mm to 8 mm in this period. It was found that the maximum lateral displacement surrounding the trench was recorded as being 7 times the maximum vertical displacement which is significant. Therefore, they concluded that the phenomenon would be due to the suction forces between the excavating machine and soil. The increment of vertical displacements prior to casting concrete in the trench was also observed by Farmer and Attewell (1973). They reported that the ratio of maximum horizontal displacement to maximum vertical displacement is 40%.

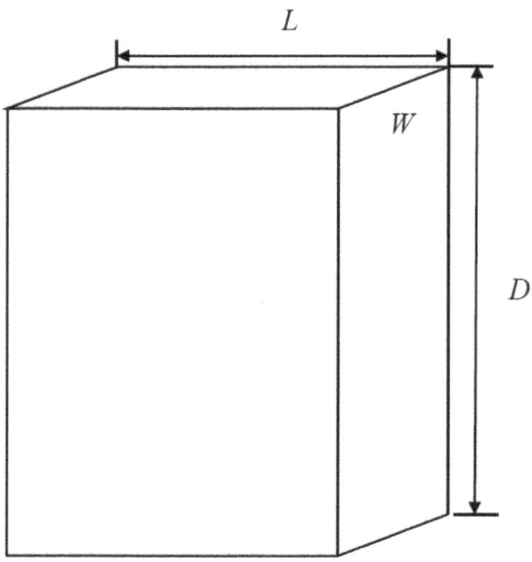

L: The trench length, *W*: The trench width, and *D*: The trench depth

Fig. 1. Dimensions of a single trench (or panel)

A detailed observation and measurement of a diaphragm wall construction was done by Poh and Wong (1998). One of the most important findings in their studies was that the height of bentonite slurry has a significant influence on the soil deformation. It was observed that the maximum vertical and horizontal displacements can increase by up to 50% when the bentonite slurry level is reduced by 1 m. The ground movements induced by the trench installation in Taipei basin have been observed by Yang (2000) and Huang (2003). It was indicated that the diaphragm wall construction will lead to settlements for adjacent buildings. The influences on the spread footing foundations are more significant than those on the mat foundations. As highlighted by Symons and Carder (1993), the construction-induced deformations adjacent to buildings and services could result in serious damages caused by the deformations. It is worth to investigate this issue by simulating the trench installation procedure.

Due to the great progress in computer power for the past two decades, three dimensional (3D) analyses are now possible and have become more commonplace (Li et al. 2014). Based on the 3D numerical modelling, it was found by Ng and Yan (1998, 1999) that the maximum surface settlement occurred when the trench is excavated and only bentonite pressures applied on its surfaces. It demonstrated that the soil behaviour is critical at this stage. In addition, at a distance of one trench depth (1*D*) away, no significant vertical displacement is computed. Gourvenec and Powrie (1999) indicated that the maximum soil lateral displacement will occur on the top of the trench (i.e. ground surface). This horizontal movement decreases with increasing depth. Currently, the study of Arai et al. (2008) is the only study that has investigated the whole process of installing a circular diaphragm wall, including 24 panels. It showed that the soil movements and lateral stresses are not axis-symmetric which changes with the panel construction sequence.

3 Methodology

As mentioned previously, the commercial software, FLAC^{3D}, is employed to simulate the diaphragm wall construction. The Mohr-Coulomb failure criterion with elastic-perfectly plastic model was adopted. The configuration used in the numerical analysis is shown in Fig. 2 where the panel dimensions are 1 m × 4 m × 40 m ($W \times L \times D$). The dimensions will be used for all analyses, unless stated otherwise. Input parameters are shown in Table 1. Groundwater table is at 2.5 m below to the ground surface. In this study, attention will be paid to soil movement due to the change in W and D. In addition, the discussion in this paper focuses on the vertical surface movement. The mesh boundary has been investigated to make sure its effect on the computed settlement is extremely small (less than 1 mm). The same simulation procedures as Ng and Yan (1999) are used which can be divided into three main stages, as following.

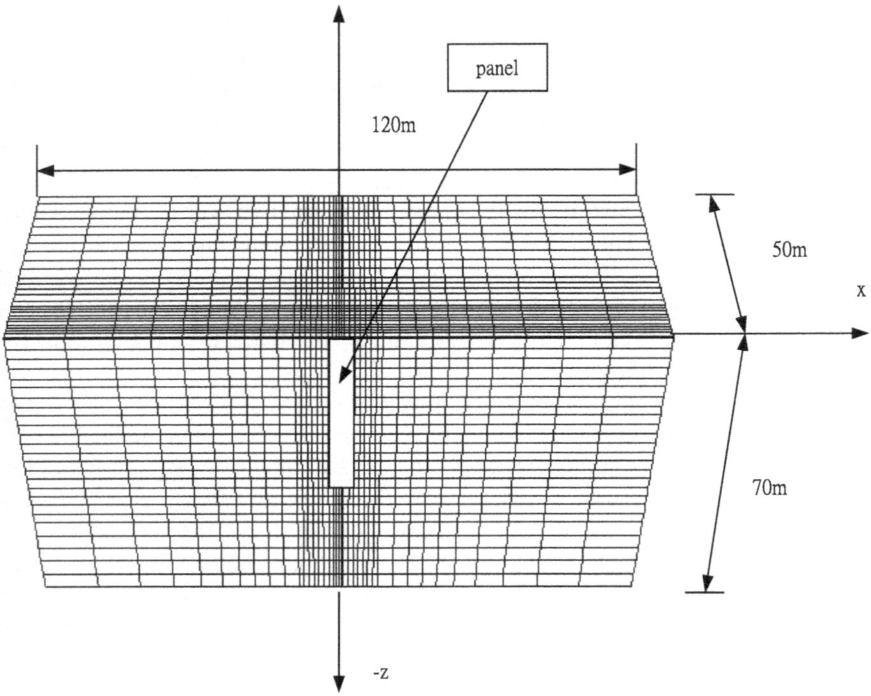

Fig. 2. The mesh configuration

Table 1. Soil input parameters

Layer	Depth(m)	OCR	K_0	s_u(kPa)	E/s_u	ν	ϕ	$\gamma_s(kN/m^3)$
1	0–2.5	4	1	23.8	500	0.49	0	19
2	2.5–5.0	4	1	58.75				
3	5.0–10.0	1.8	0.67	48.8				
4	10.0–15.0	1.34	0.58	57.5				
5	15.0–68.0	1	0.5	$0.33\,\sigma'$				

1. Excavate the trench and apply hydrostatic bentonite pressure on the trench faces simultaneously.
2. Cast the concrete panel by increasing the lateral pressure inside the panel from the bottom using a theoretical bilinear wet concrete pressure envelope proposed by Lings et al. (1994).
3. Cure the concrete panel by replacing the trench with elastic concrete elements and removing the applied bilinear concrete pressures on the trench faces simultaneously.

4 Results and Discussions

Figure 3 presents the obtained results for different diaphragm wall panel widths. The range of panel width is determined based on the reported dimensions in the investigations of Dibiagio and Myrvoll (1972), Farmer and Attewell (1973), Powrie and Kantartzi (1996), Poh and Wong (1998), Yang (2000), Huang (2003), and Arai et al. (2008). It can be observed that the surface movement increases with increasing panel width, as expected. The maximum settlement (S_{max}) appear just next to the excavated trench. The ratio of S_{max}/W ranges from 1.7 to 2.6. Figure 4 shows the vertical displacements parallel to the trench orientation. S_{max} also appears at the center of panel on y-axis and the surface settlement reduce with increasing distance away from the panel. In addition, it can be observed in Fig. 4 that the trench installation induced settlement become less significant when the distance is greater than $1L$.

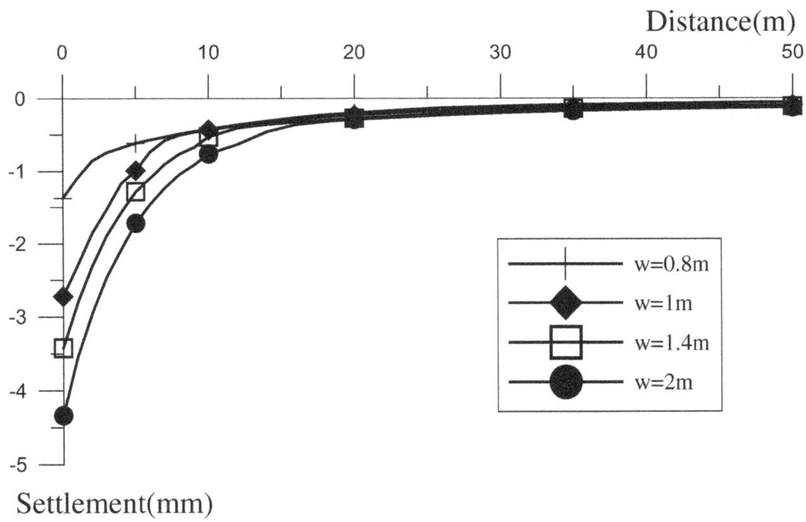

Fig. 3. Surface settlement at the center of panel on y-axis for various panel widths

Figure 5 are the surface settlements for various diaphragm wall panel depths. The settlement trend is similar. S_{max} increases from 0.85 mm to 2.9 mm when D changes from 30 m to 45 m. The increment is 2 times which is significant. Therefore D should be designed carefully due to its influence on soil movement. Figure 6 displays the vertical displacements parallel to the trench orientation for panel depths. Same trend as above was found that the trench installation induced settlement become less significant when the distance is greater than $1L$.

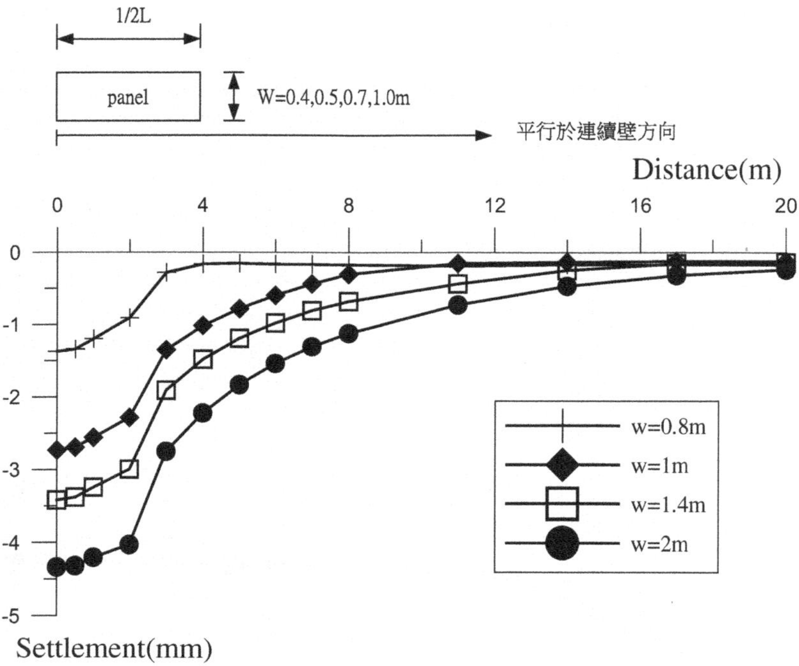

Fig. 4. Surface settlement along diaphragm wall orientation (x-axis) for various panel widths

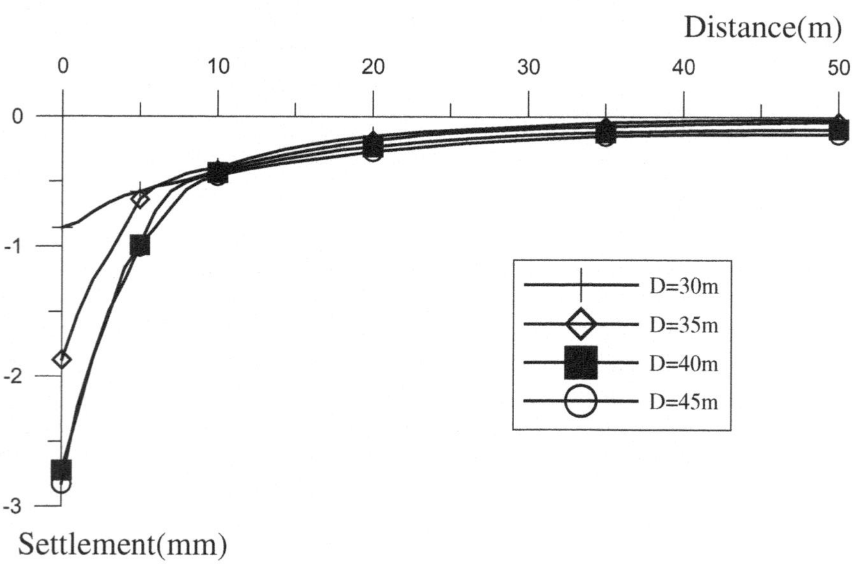

Fig. 5. Surface settlement at the center of panel on y-axis for various panel depths

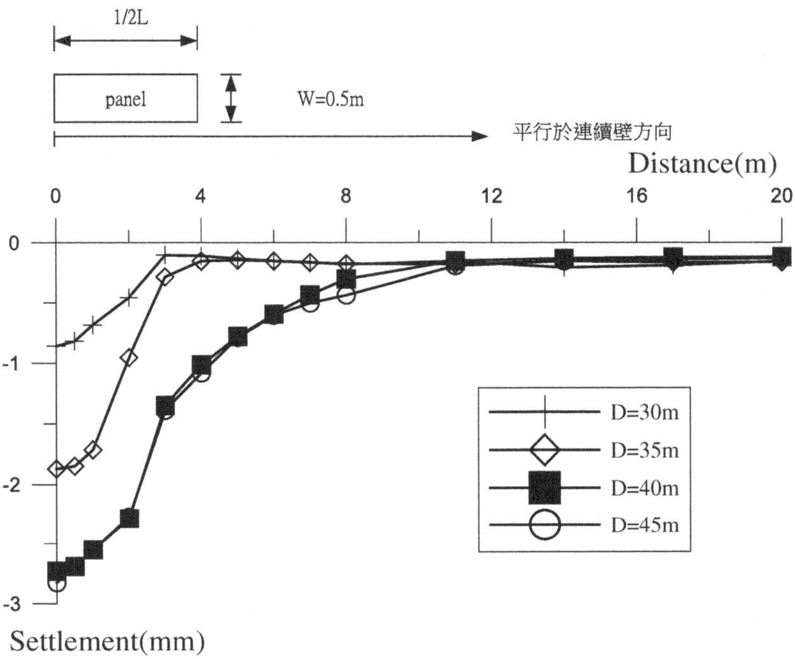

Fig. 6. Surface settlement along diaphragm wall orientation (x-axis) for various panel depths

According to field measurement done by Yang (2000), surface settlement is influenced by multiple 3–4 adjacent diaphragm wall panel construction. In addition, Huang (2003) indicated the surface settlements are mainly induced by right and left hand side 3–4 diaphragm wall panel installations. Therefore, it is worth to analyze multiple diaphragm wall panel construction.

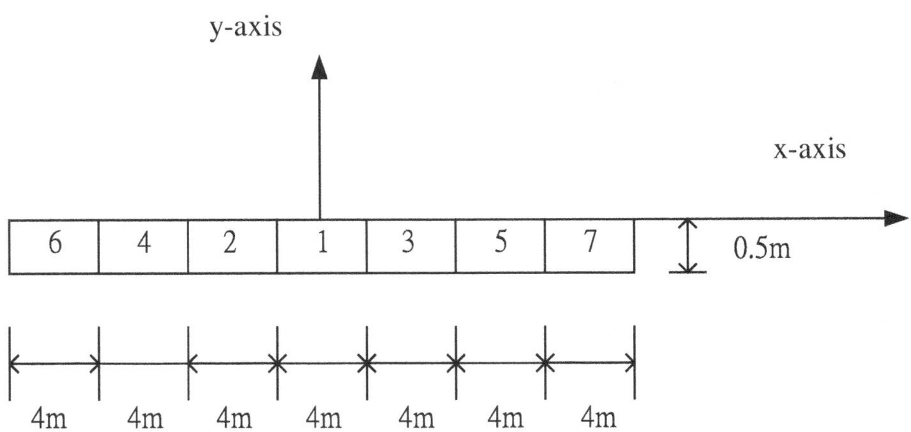

Fig. 7. Surface settlement along diaphragm wall orientation (x-axis) for various panel depths

Seven diaphragm wall construction are simulated. Each panel has dimension of 4 m × 1 m × 40 m. Their installation order is shown in Fig. 7. Figure 8 shows the vertical displacements obtained at the center of Panel 1 along y-axis. It can be seen that the settlement increases gradually with increasing numbers of panel construction. It was also found that the farer panel installation caused less vertical displacement. The results agreed with the findings of Yang (2000) and Huang (2003). However, Fig. 8 also shows the potential that the farer diaphragm wall panel construction could still lead to surface settlement.

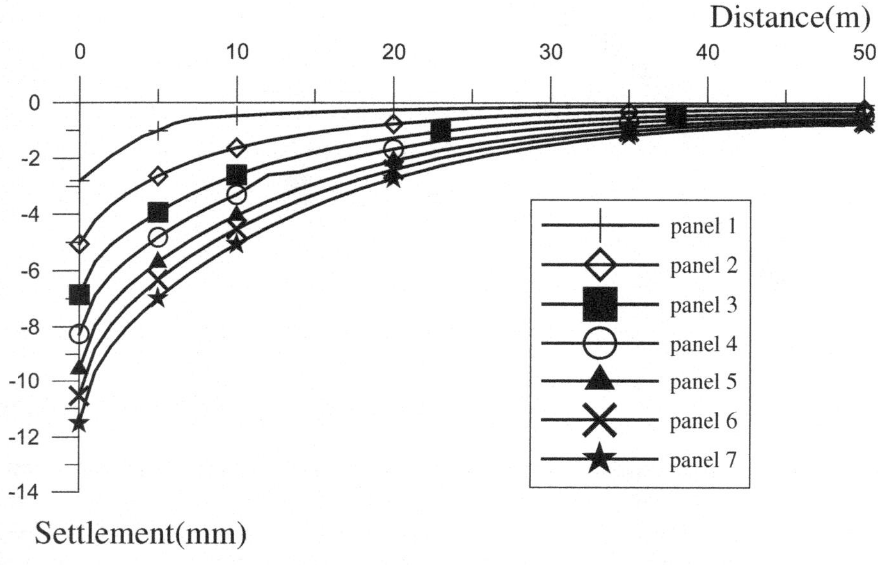

Fig. 8. Surface settlement at the center of panel on y-axis for various panel depths

5 Conclusion

This study investigates three dimensional diaphragm wall panel construction using Taipei basin soil profile based on an assumed case study. The diaphragm installation induced settlement increases with increasing wall dimensions. From the numerical analysis results, surrounding panel installation will definitely cause surface displacement. However, the influence distance is not clear. More detailed investigations are needed in the future.

References

Arai, Y., Kusakabe, O., Murata, O., Konishi, S.: A numerical study on ground displacement and stress during and after the installation of deep circular diaphragm walls and soil excavation. Comput. Geotech. **35**(5), 791–807 (2008)

Cowland, J.W., Thorley, C.B.B.: Ground and building settlement associated with adjacent slurry trench excavation. In: Ground Movements and Structure, pp.723–738. Pentech Press, London (1985)

Dibiagio, E., Myrvoll, F.: Full scale field tests of slurry trench excavation in soft clay. In: Proceedings of the 5th European Conference Soil Mechanics and Foundation Engineering, Madrid, Spain, pp. 461–471 (1972)

Farmer, I.W., Attewell, P.B.: Ground movements caused by a bentonite-supported excavation in London clay. Geotechnique **23**(4), 576–581 (1973)

Gourvenec, S.M., Powrie, W.: Three-dimensional finite-element analysis of diaphragm wall installation. Geotechnique **49**(6), 801–823 (1999)

Huang, T.-Y.: A study of ground deformation and adjacent building's safety due to construction of diaphragm wall. Department of Construction Engineering, National Taiwan University of Science and Technology. Master, Taipei (2003). (In Chinese)

Itasca Consulting Group, Inc.: Fast Lagrangian Analysis of Continua in 3 Dimensions, User manual, Version 2.10. Minneapolis, Minnesota, USA (2002)

Li, A.J., Merifield, R.S., Lin, H.D., Lyamin, A.V.: Trench stability under bentonite pressure in purely cohesive clay. Int. J. Geomech. **14**(1), 151–157 (2014)

Lings, M.L., Ng, C.W.W., Nash, D.F.T.: The lateral pressure of wet concrete in diaphragm wall panels cast under bentonite. In: Proceedings of the Institution of Civil Engineers - Geotechnical Engineering, vol. 107, pp. 63–172 (1994)

Morton, K., Carter, R.W., Linney, L.: Observed settlements of building adjacent to stations constructed for the modified initial system of the mass transit railways. In: Proceedings of the 6th Asian Conference S.M.F.E., Taipei, vol. 1, pp.415–429 (1980)

Ng, C.W.W., Yan, R.W.M.: Stress transfer and deformation mechanisms around a diaphragm wall panel. J. Geotech. Geoenvironmental Eng. **124**(7), 638–648 (1998). ASCE

Ng, C.W.W., Yan, R.W.M.: Three-dimensional modelling of a diaphragm wall construction sequence. Geotechnique **49**(6), 825–834 (1999)

Ou, C.-Y.: Deep Excavation: Theory and Pratice. Taylor & Francis, Oxford, UK (2006)

Poh, T.Y., Wong, I.H.: Effects of construction of diaphragm wall panels on adjacent ground: field trial. J. Geotech. Geoenviromental Eng. **124**(8), 745–756 (1998). ASCE

Powrie, W., Kantartzi, C.: Ground response during diaphragm wall installation in clay: centrifuge model tests. Geotechnique **46**(4), 725–739 (1996)

Symons, I.F., Carder, D.R.: Stress changes in stiff clay caused by the installation of embedded retaining walls. In: Retaining Structures. Proceedings of the Conference Organized by the Institution of Civil Engineers, 20–23 July 1992. Robinson College, Cambridge (1993)

Yang, L.-L.: A study of ground movements due to construction of diaphragm wall. Department of Construction Engineering, National Taiwan University of Science and Technology. Master, Taipei (2000). (in Chinese)

Research on Flow Push Pile Effect of Post-liquefied Sand Based on VOF Method

Enquan Zhou(✉), Wen Yan, and Wang Qiong

Faculty of Civil Engineering and Mechanics, Jiangsu University, Zhenjiang, Jiangsu, China
enquan1986@126.com

Abstract. With regarding liquefied sand as power-law non-Newtonian fluid, the method of volume of fluid (VOF) in computational fluid dynamics is adopted to conduct numerical computation about the flow push pile of post-liquefied sand. This paper analyzes the effect of consistency coefficient K and flow index n of power-law fluid on flow push pile characteristics. The results show that, this method can factually reproduce the flow deformation shape of post-liquefied sand; The displacement and stress of pile at different height both show the development law of the first increases rapidly and then gradually decreases. The variation trend of peak displacement and peak stress response of pile are significantly different along the pile. The peak displacement response of the foundation pile shows a gradually increasing distribution from the bottom to the top of the pile. The peak stress response of the foundation pile shows the rule of decreasing firstly, then increasing, and finally decreasing from bottom to top. Flow slide push pile effect of liquefied sand is simultaneously affected by the consistency coefficient and flow index. With the increase of the consistency coefficient or the flow index, the peak displacement or peak stress of the pile also increases.

Keywords: VOF method · Liquefied sand · Non-Newtonian fluid · Flow push pile

1 Introduction

The dynamic response of pile foundation in liquefiable field under earthquakes is one of the focal points in seismic field of geotechnical engineering. The pile foundation is often used in liquefiable ground because of its high bearing capacity and little settlement. But the seismic data in recent years show that more and more piles were severely damaged in the earthquake liquefaction ground, especially pile foundation failure caused by lateral flow after liquefaction, which is often seen. For example, in 1964 Niigata earthquake, soil liquefaction around pile of Showa bridge caused by site liquefaction triggered continuous five span collapse [1]. In the 1976 Tangshan earthquake, liquefaction caused a large number of river embankment slope flow, which causes the inclination of pile foundations of many highways and railway bridges, abutment tilt, lowering of girder and other seismic hazard [2]. Especially, the four great earthquakes in the past 10 years, 2008 Wenchuan earthquake in China [3], 2011 Chile earthquake [4], 2010–2011

R. Chen et al. (eds.), *Proceedings of the 2nd International Symposium on Asia Urban GeoEngineering*, Springer Series in Geomechanics and Geoengineering,
https://doi.org/10.1007/978-981-10-6632-0_46

New Zealand earthquake [5], 2011 East Japan earthquake [6] all caused damage of pile foundation and structures because of serious liquefaction site lateral flow.

The mechanical property of liquefied sand is the primary problem in the study of pile-soil interaction. Towhata et al. [7] regarded the liquefied soil as a fluid for the first time in the liquefied flow analysis of inclined slope. According to the shaking table test results, the lateral deformation curve of the liquefied soil is assumed to be a sine function curve, the closed solution of the liquefaction displacement of the ground surface is obtained by using the principle of minimum potential energy. Then researches on liquefaction flow slide and push pile gradually developed, in which the liquefied soil was regarded as a fluid. Sato et al. [8] obtained the apparent viscosity magnitude of saturated sand after initial liquefaction through the drag-ball test and the drag-tube test respectively. Kawakami et al. [9] obtained that there is a certain correspondence relationship between the apparent viscosity and the shear strain rate of the soil after the initial liquefaction through the drag-ball test and the viscometer respectively. Towhata et al. [10] obtained the apparent viscosity magnitude of saturated sand after initial liquefaction through the drag-tube test. Sawicki et al. [11] obtained the apparent viscosity magnitude of saturated sand after initial liquefaction through post liquefaction static loading test and drop test respectively. Hwang et al. [12] obtained that there is a certain correlation relationship between the apparent viscosity and the shear strain rate of the soil after liquefaction through the falling ball test and the drag-ball test respectively. Liu et al. [13], Huang et al. [14] studied from the post liquefaction static loading test, Chen et al. [15] studied from drag-ball test, getting that there is a certain correlation between apparent viscosity and shear strain rate of saturated soil, showing shear thinning fluid characteristics. Chen et al. [16] obtained that the liquefied soil has the characteristics of shear thinning fluid according to the post liquefaction static loading test, seen in Fig. 1.

In this paper, the post liquefaction sand is considered as a power law non Newtonian fluid. The mechanical characteristics of the pile foundation in the post liquefaction site are studied based on the VOF method and the fluid structure interaction theory.

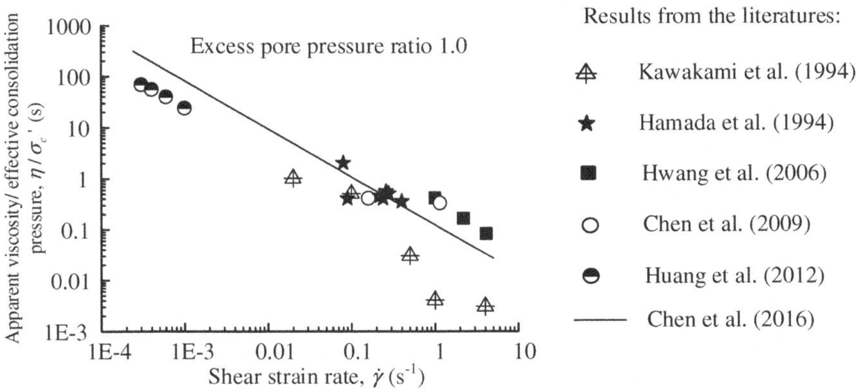

Fig. 1. The relationship between apparent viscosity and shear strain rate of liquefaction sand

2 Fluid Solid Coupling and VOF Method

The fluid solid coupling analysis method is mainly used to study the nonlinear dynamic coupling between solid and fluid. In this paper, the liquefied sand is regarded as fluid, and the pile foundation is regarded as linear elastic solid, so they should respectively meet the fundamental equations of fluid and linear solid.

2.1 Viscous Fluid Equation

In this paper, the liquefied sand is considered as a power-law non Newtonian fluid. The fluid should meet the following basic equations of viscous fluid. The energy conservation equation is not considered in this paper without considering the change of temperature [17]:

① Mass conservation equation

$$\frac{\partial \rho^{(f)}}{\partial t} + \frac{\partial(\rho^{(f)} v_i^{(f)})}{\partial x_i} = 0 \tag{1}$$

In the formula, $\rho^{(f)}$ is the density of the liquefied sand, t is time variable, $v_i^{(f)}$ is the velocity vector of the liquefied sand ($i = 1, 2, 3$). Here, the fluid is incompressible, which means that the density is constant, so the equation can be written:

$$\frac{\partial v_i^{(f)}}{\partial x_i} = 0 \tag{2}$$

② Momentum conservation equation (N-S equation)

The total stress of viscous fluid can be expressed as:

$$\sigma_{ij}^{(f)} = -p\delta_{ij} + \tau_{ij}^{(f)} \tag{3}$$

In the formula, $\sigma_{ij}^{(f)}$ is the total stress tensor of the liquefied sand, p is the static pressure of the liquefied sand, $\tau_{ij}^{(f)}$ is the viscous stress tensor of the liquefied sand, and δ_{ij} is the Kronecker symbol. The momentum conservation equation of liquefied sand can be expressed as:

$$-\frac{1}{\rho^{(f)}}\frac{\partial p}{\partial x_i} + \frac{1}{\rho^{(f)}}\frac{\partial \tau_{ij}^{(f)}}{\partial x_j} + f_i^{(f)} = \frac{\mathrm{d}v_i^{(f)}}{\mathrm{d}t} \tag{4}$$

In the formula, $f_i^{(f)}$ is the volumetric force of mass fluid.

③ Constitutive relation

As power-law non Newtonian fluid, its viscous stress tensor can be expressed as:

$$\tau_{ij}^{(\mathrm{f})} = 2K(S_{ij}^{(\mathrm{f})})^{n} \tag{5}$$

In the formula, K is consistency coefficient, $S_{ij}^{(\mathrm{f})}$ is fluid deformation rate tensor, n is flow index. when $n < 1$ the fluid is shear thinning fluid; while $n > 1$ fluid is shear thickening fluid; especially when $n = 1$ the apparent viscosity is constant, fluid is Newton fluid.

④ Boundary condition [18]

In this paper, a no slip wall boundary is applied. The fluid velocity vector on the wall is equal to zero.

$$v_{i}^{(\mathrm{f})} = 0 \tag{6}$$

While the free boundary of the fluid is a moving boundary, so it is necessary to satisfy the following kinematic and dynamic conditions:

$$(\mathbf{v}_{i}^{(\mathrm{f})} - \dot{\mathbf{d}}) \cdot \mathbf{n} = 0 \tag{7a}$$

$$\boldsymbol{\tau}_{n} = -p_{a}\mathbf{n} + \sigma_{a}\mathbf{H} \tag{7b}$$

Here, $\mathbf{d}$ is the displacement vector of the free interface, $\mathbf{n}$ is the normal vector of the free interface, $\boldsymbol{\tau}_{n}$ is the fluid surface stress, p_{a} is the circumference pressure of the free surface, σ_{a} is the surface tension, $\mathbf{H}$ is the curvature vector of the free interface.

⑤ Fluid finite element equation [19]

The mass and momentum conservation equations of fluid are discretized by finite element method, and the total coefficient matrix is obtained:

$$\mathbf{GV} = \mathbf{0} \tag{8a}$$

$$\mathbf{M}\dot{\mathbf{V}} + \mathbf{CP} + (\boldsymbol{\Gamma} + \mathbf{K}_{\eta})\mathbf{V} = \mathbf{0} \tag{8b}$$

Here, $\mathbf{G}$ is generalized divergence matrix, $\mathbf{M}$ is generalized fluid mass matrix, $\mathbf{C}$ is fluid pressure matrix, $\boldsymbol{\Gamma}$ is a generalized convection matrix, $\mathbf{K}_{\eta}$ is fluid viscosity matrix, $\mathbf{V}$ is total fluid pressure and velocity vector, $\mathbf{V} = \begin{Bmatrix} \mathbf{P} \\ \mathbf{v} \end{Bmatrix}$, here Euler algorithm is used to solve the time term.

The formula (8a) and the formula (8b) are written as a matrix expression:

$$\mathbf{M}^{(f)}\dot{\mathbf{\Phi}}^{(f)} + \mathbf{C}^{(f)}\mathbf{\Phi}^{(f)} = \mathbf{0} \tag{9}$$

In the formula, $\mathbf{M}^{(f)} = \begin{Bmatrix} \mathbf{0} & \mathbf{0} \\ \mathbf{0} & \mathbf{M} \end{Bmatrix}$, $\mathbf{\Phi}^{(f)} = \begin{Bmatrix} \mathbf{P} \\ \mathbf{v} \end{Bmatrix}$, $\mathbf{C}^{(f)} = \begin{Bmatrix} \mathbf{0} & \mathbf{G} \\ \mathbf{C} & \mathbf{\Gamma} + \mathbf{K}_{\eta} \end{Bmatrix}$.

By solving the matrix expression, the convergence speed and pressure value are solved.

2.2 Introduction of VOF Method for Free Interface Solution

The VOF (volume of fluid) method is used to describe the dynamic changes of the free surface in the multiple fluid flow by using the volume ratio function of fluid and mesh in the cell grid. This method is often used to track the flow interface in two kinds of fluid flow. The VOF method is based on the volume ratio function ϕ to construct and capture the free interface, if $\phi = 1$, it means that the cell grid is full of specified fluid; if $\phi = 0$, it means there is no specified fluid in the cell grid; if $0 < \phi < 1$, it means there are fluid interface in the cell grid.

In this paper, the liquefied sand is treated as fluid and the VOF method is used to track the free interface. Therefore, in addition to the basic equation of viscous fluid, the fluid calculation area should also satisfy the following VOF equation [20]:

$$\frac{\partial \phi}{\partial t} + v_x \frac{\partial \phi}{\partial x} + v_y \frac{\partial \phi}{\partial y} = 0 \tag{10}$$

In the formula, ϕ is the fluid volume ratio function, and v_x, v_y are the velocity components of the fluid in the direction of x, y.

Solving the volume fraction function in a specified cell by the differential approximation scheme of donor-acceptor algorithm. As shown in Fig. 2, Hirt and

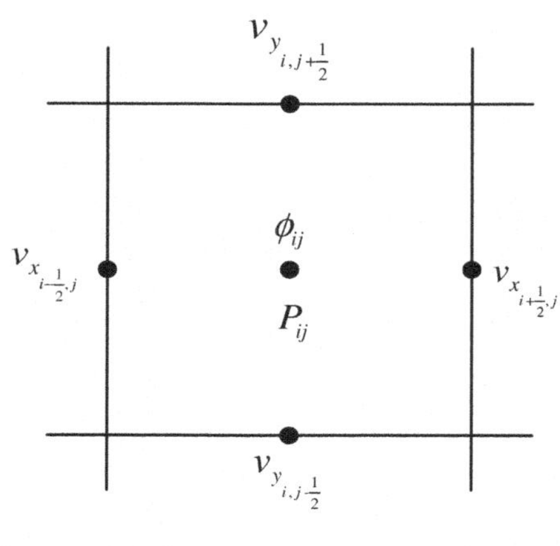

Fig. 2. Staggered lattice of inconsistent variable configurations

Nichols designed the Staggered difference scheme similar to the MAC and PIC method, p_{ij}^n, ϕ_{ij}^n is defined in the center of the grid, choose control body:

$$\Omega = (x_{i-1/2}, x_{i+1/2}) \times (y_{j-1/2}, y_{j+1/2}) \times (t_n, t_{n+1}) \tag{11}$$

The Eq. (7a and 7b) is integrated on the control body:

$$(\phi^{n+1} - \phi^n)\delta x\delta y + \Delta(\phi v_x)\delta t\delta y + \Delta(\phi v_y)\delta t\delta x = 0 \tag{12a}$$

Define:

$$\Delta(\phi v_x)\delta t = \min\{\phi_{AD}|V_x| + \phi_f, \phi_D\delta x_D\} \tag{12b}$$

In the formula, $\phi_f = \max\{(1.0 - \phi_{AD}) \cdot |V_x| - (1.0 - \phi_D) \cdot \delta x_D, 0\}$, $|V_x| = |v_x \cdot \delta t|$. Subscript A , D respectively represents the acceptor and donor unit. The determination of A or D depends on the direction of the free surface in the unit grid and the direction of the velocity on the boundary, and the ϕ_A , ϕ_D represents the fluid volume fraction function of the acceptor and donor unit grid, respectively. Similarly, we can construct $\Delta(\phi v_y)\delta t$.

Based on the above method, we can solve out the volume fraction function ϕ, which can be used to track the free surface of the fluid.

2.3 Solid Basic Equation

In the analysis of liquefied sand flow slide push pile effect, in addition to liquefied sand should meet the basic equations of fluid, as a linear elastic structure, pile foundation should meet the basic equations of linear elastic solid such as [21].

① Geometric equation

$$\varepsilon_{ij}^{(s)} = \frac{1}{2}(u_{i,j}^{(s)} + u_{j,i}^{(s)}) \tag{13a}$$

$$\gamma_{ij}^{(s)} = 2\varepsilon_{ij}^{(s)} \tag{13b}$$

Here, $\varepsilon_{ij}^{(s)}$, $\gamma_{ij}^{(s)}$ are respectively structural strain and engineering shear strain, $u_i^{(s)}$ is structural displacement.

② Stress-strain relationship

$$\sigma_{ij}^{(s)} = \lambda\theta^{(s)}\delta_{ij} + 2\mu\varepsilon_{ij}^{(s)} \tag{14}$$

In the formula, $\sigma_{ij}^{(s)}$ is the structural stress tensor, $\theta^{(s)}$ is the volumetric strain of structure, μ , λ are the lame constants.

③ Momentum conservation equation

$$\sigma_{ji,j}^{(s)} + f_i^{(s)} = \rho^{(s)} \ddot{u}_i^{(s)} \tag{15}$$

In the formula, $\rho^{(s)}$ is the density of the structure, $f_i^{(s)}$ is the volume force.

④ Solid boundary condition

$$\sigma_{ij}^{(s)} \cdot n_i^{(s)} = \bar{T}_i^{(s)} \tag{16}$$

$$u_i^{(s)} = \bar{u}_i^{(s)} \tag{17}$$

Here, $n_i^{(s)}$ is the normal vector of the structure, $\bar{T}_i^{(s)}$ is a known component of surface force on the structure, and $\bar{u}_i^{(s)}$ is a known displacement component of the structure.

In this paper, the bottom of the structure is fixed, so the bottom satisfies the boundary conditions:

$$u_i^{(s)} = 0, \ \ \dot{u}_i^{(s)} = 0 \tag{18}$$

2.4 Introduction of Calculation Model for Fluid Solid Coupling

In the fluid solid coupling problem, the fluid and solid should respectively satisfy the corresponding basic fluid equation and solid equation, and the fluid solid coupling interface should also satisfy the corresponding conditions.

① Kinematic condition:

$$v_i^{(f)} = \dot{u}_i^{(s)} \tag{19}$$

② Dynamic condition:

$$\sigma_{ij}^{(f)} n_i^{(f)} = \sigma_{ij}^{(s)} n_i^{(s)} \tag{20}$$

In the formula, $n_i^{(f)}$ represents the normal vector of the fluid.

For the fluid solid coupling system, the corresponding theoretical model schematic diagram is shown in Fig. 3.

Through discretization to the equation, the unit grid in the fluid model and the structural model can be completely different in the discrete model. As shown in Fig. 4, the node position in the fluid and solid model is different at the coupling interface, and the displacement of node 2 in the fluid model is obtained by the displacement interpolation of the node 1 and note 2 in the structural model, Similarly, the fluid force of

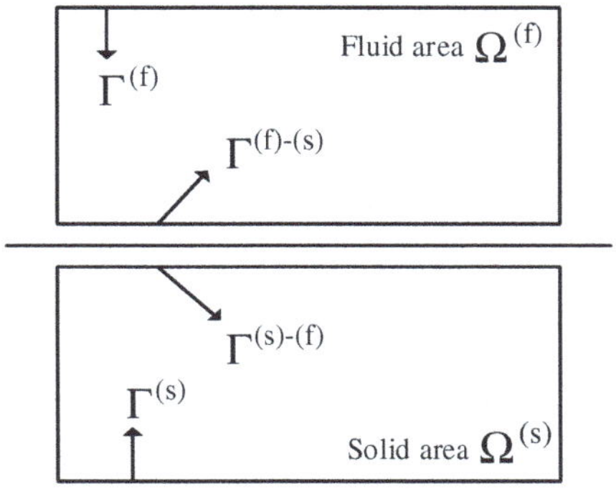

Fig. 3. Schematic diagram of fluid-structure coupling

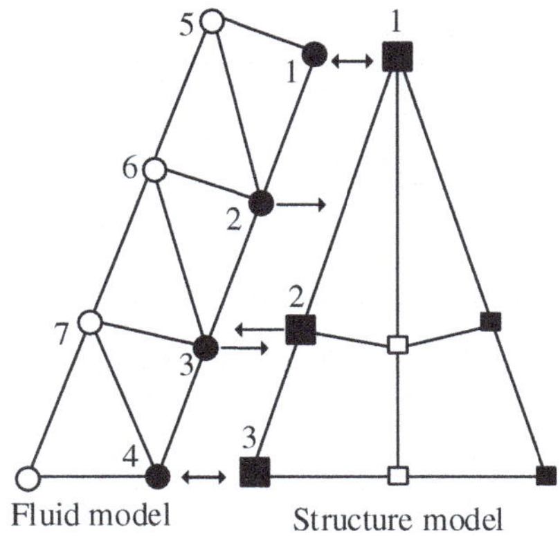

Fig. 4. Fluid-structure interaction interface

node 2 in the structural model is obtained through the fluid stress interpolation of node 2 and node 3 in the fluid model, the stress of node 1 and node 3 in the structural model are respectively equal to the stress of node 1 and node 4 in the fluid model. The forces that fluid acting on node 2 of the structural model include the stress of node 1, node 2 and node 3 in the structural model, the stress of the fluid itself is the sum of the pressure and the shear stress. Therefore, the solution in the node of the fluid model is fully coupled with the solution in the node of the structural model.

3 Computational Model

In this paper, directional coupling method is used to simulate the liquefied sand flow slide push pile, Fig. 5 is the pile group model.

Figure 6 is the Schematic diagram of pile group calculation model. The size of the whole flow field model is 1.2 m × 0.4 m × 0.5 m, and the internal red part is the liquefied sand, the size is 0.4 m × 0.4 m × 0.35 m, and the rest part is air. The flow

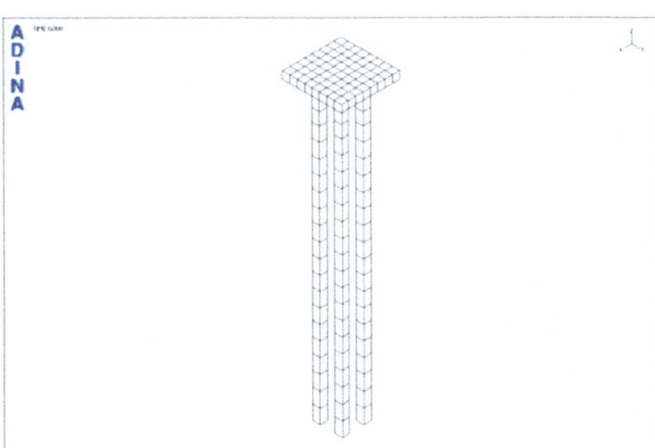

Fig. 5. Finite element model of the pile group

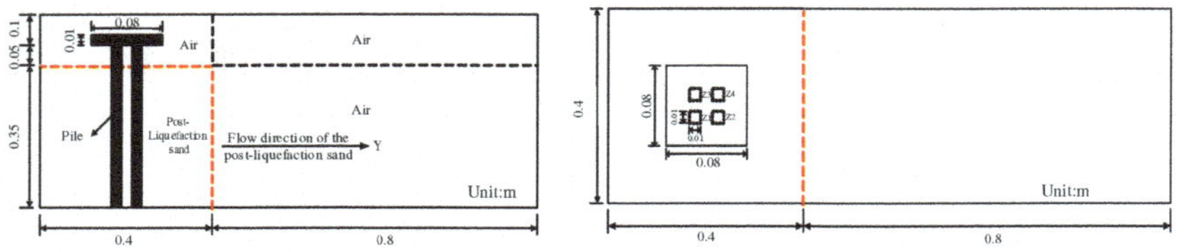

(a)Side view **(b)Top view**

Fig. 6. Schematic diagram of pile group calculation model

field is made up of eight node element, and the density of liquefied sand is $\rho^{(f)} = 1600\ \text{kg/m}^3$, the bottom and side of the flow field are solid walls, and the contact surfaces between the liquefied sand, the air and the pile are fluid solid coupling interfaces, and the fluid parameters are given in Table 1.

Table 1. Fluid model parameters

Number	Consistency coefficient K	Flow index n
1	100	0.2
2		0.4
3		0.8
4	1000	0.2
5		0.4
6		0.8
7	1500	0.2
8		0.4
9		0.8

The pile group are composed of top pile cap and four foundation piles which are fixed at the bottom. The piles are square piles, the length is 0.01 m, the height is 0.4 m, the size of pile cap is 0.08 m × 0.08 m × 0.01 m, the piles top stretch out of the liquefied sand layer. The piles are regarded as linear elastic structure, and the eight node element is used here, the modulus of elasticity is $E = 3.0 \times 10^{10}$ Pa, the density is $\rho^{(s)} = 2500\ \text{kg/m}^3$, the Poisson's ratio is $v = 0.2$. The side surfaces of foundation piles and the surfaces of pile cap are all fluid solid coupling interfaces, except the bottom of the foundation piles and contact surfaces between the piles and the pile cap.

4 Result Analysis

As shown in Fig. 7, the given post-liquefied sand free flow deformation maps at different time show that this method can factually reflect the flow deformation of sand after liquefaction comparing with test in the past research [22].

In the condition of K = 1000, n = 0.8, Fig. 8 shows the dynamic response curves of acceleration and displacement at three height locations (the height distance from pile

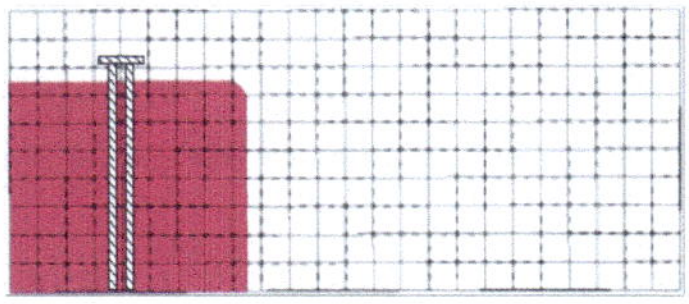

(a) 0s

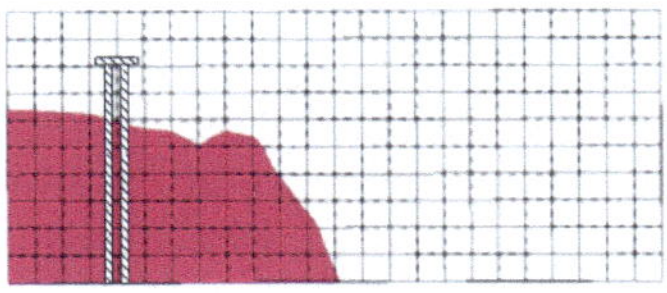

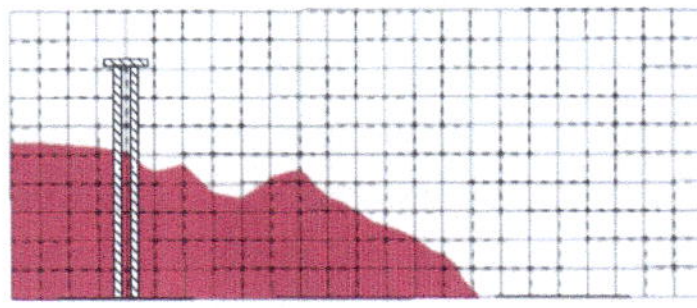

(b) 0.4s **(c) 0.8s**

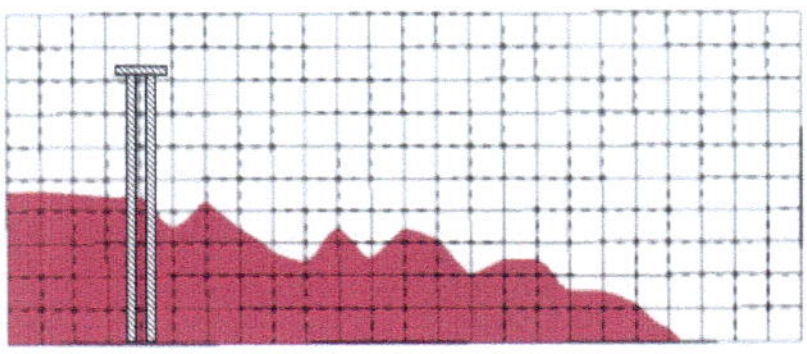

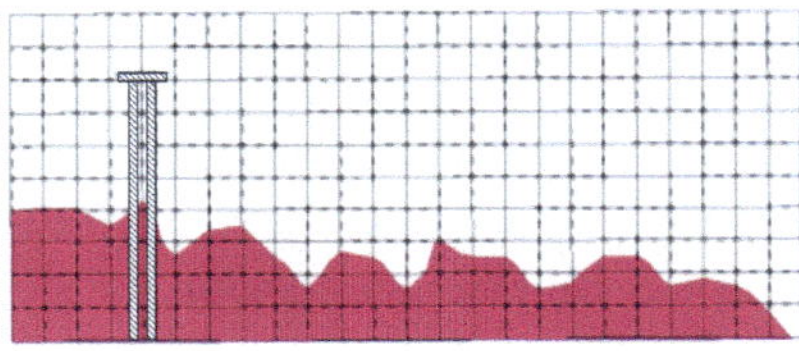

(d) 1.2s **(e) 1.8s**

Fig. 7. Flow deformation of post-liquefied sand under the condition of K = 1000, n = 0.8

bottom are respectively 0.4 m, 0.229 m, 0.057 m) and the pile stress at three height locations (the height distance from pile bottom are respectively 0.4 m, 0.229 m, 0 m) of Z1 pile. The final time in the figures is the ending time of the simulation, it means that fluid has flowed to the right end of the model. It can be seen from Fig. 8(a) that the displacement of the pile increases rapidly to the peak in the early start flow of model, and then with the development of flow, the displacement of the pile decreases gradually. The stress history curve of pile also shows the same law of increasing at first and then decreasing

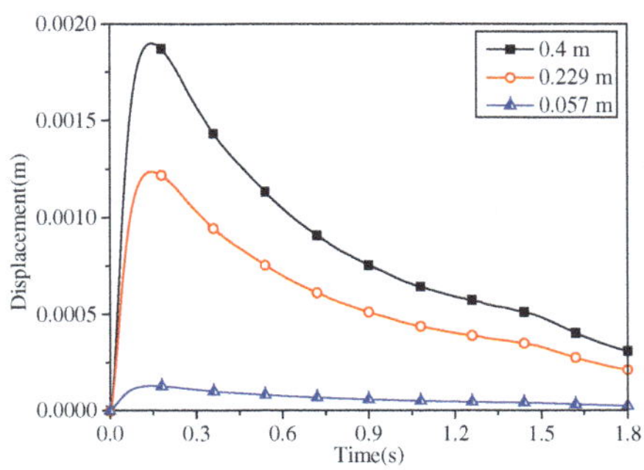

(a)Time history curves of displacement

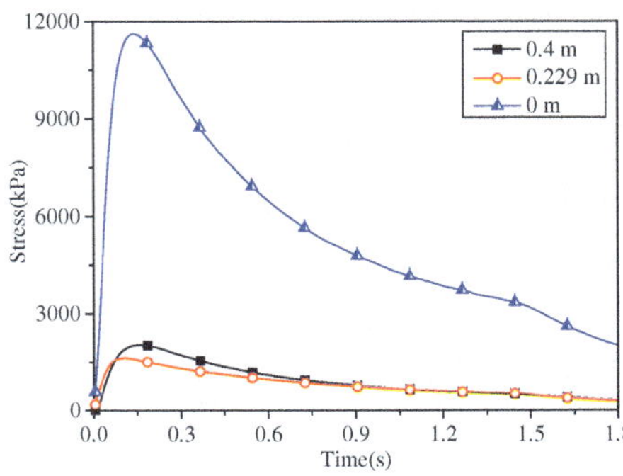

(b)Time history curves of stress

Fig. 8. Time history curves of Z1 pile

In addition, the displacement and stress response of the pile are different obviously at different height of the pile, at the top of the pile, the displacement is the largest, while at the bottom of pile, the stress is the largest. This phenomenon is mainly due to pile were fixed on the bottom.

The mechanical response of the Z1 pile and the Z3 pile, the Z2 pile and the Z4 pile are basically the same because of symmetry of the model. So, Fig. 9 shows the peak displacement and peak stress response of the Z1 pile and the Z2 pile at different positions. The variation trend of peak displacement and peak stress response of pile are significantly different along the pile. The peak displacement response of the foundation pile shows a gradually increasing distribution from the bottom to the top of the pile. At the same height position, the peak displacement of the Z1 pile is larger than that of the Z2 pile. The peak stress response of each pile from bottom to top shows the rule of decreasing first, then increasing, and finally decreasing. The curves show an oblique "2" shape distribution. The difference between Z1 pile and Z2 pile is different at different height position. At the upper part and lower-middle part, the peak stress of the Z1 pile is greater than that of the Z2 pile at the same height position. At the upper-middle part and lower part, the peak stress of the Z1 pile is less than that of the Z2 pile at the same height position.

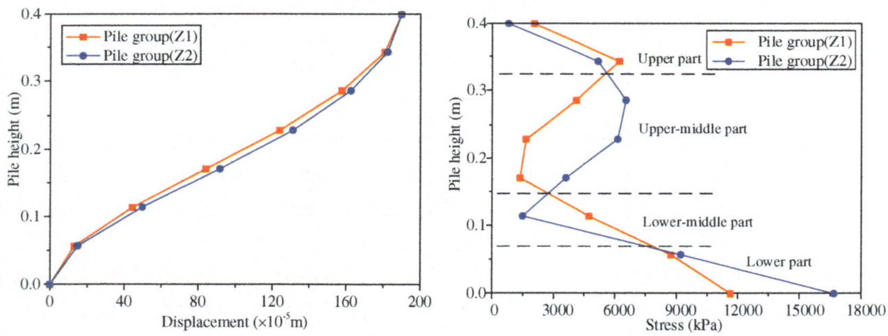

(a)Peak displacement **(b)Peak stress**

Fig. 9. The peak displacement and peak stress response of Z1 piles and Z2 piles at different positions.

4.1 Results Analysis on Effect of Consistency Coefficient K

Figures 10 and 11 respectively shows peak response curves of displacement and stress of pile Z1 in pile group under flow slide push pile effect of liquefied sand with different consistency coefficient K.

The influence of different consistency coefficients on the displacement and stress response of pile is consistent. Under the same flow index n, the displacement and stress peak value of pile both increase with the increase of consistency coefficient. This is because the consistency coefficient K reflects the apparent viscosity characteristics of the fluid, the larger the consistency coefficient K is, the stronger the ability to resist

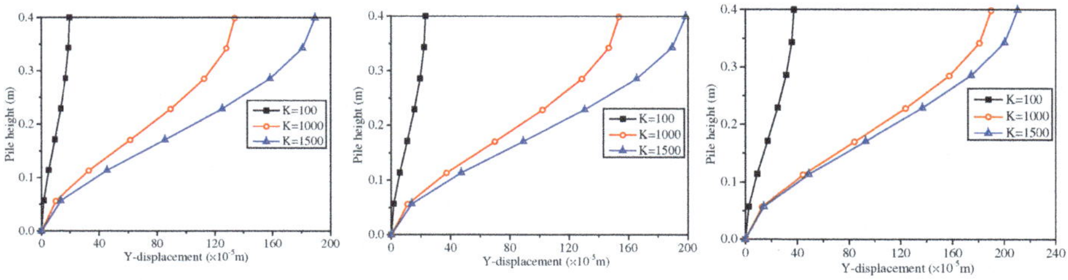

(a)n=0.2 (b)n=0.4 (c)n=0.8

Fig. 10. Peak Y-displacement of foundation pile Z1 at different consistency coefficient

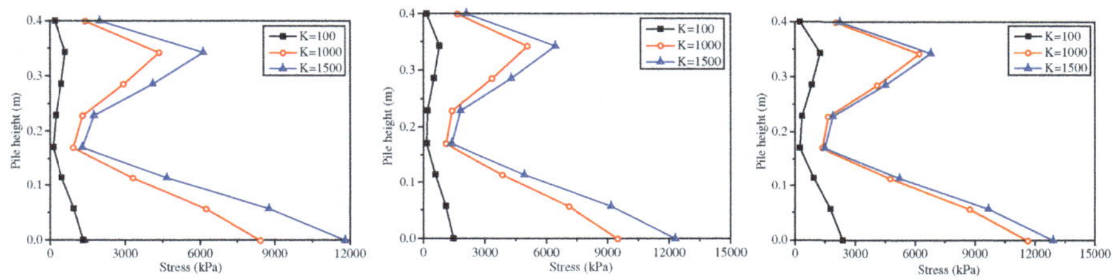

(a)n=0.2 (b)n=0.4 (c)n=0.8

Fig. 11. Peak stress of foundation pile Z1 at different consistency coefficient

deformation of the fluid is, in other words, the larger the consistency coefficient K is, the larger the deformation and stress of the pile induced by the deformation of liquefied sand.

In addition, it can be seen that as the flow index n increases, the gaps between the curves under different consistency coefficients K is becoming smaller and smaller.

4.2 Results Analysis on Effect of Flow Index N

Figures 12 and 13 respectively shows peak response curves of displacement and stress of pile Z1 in pile group under with different flow index under three kinds of consistency coefficient.

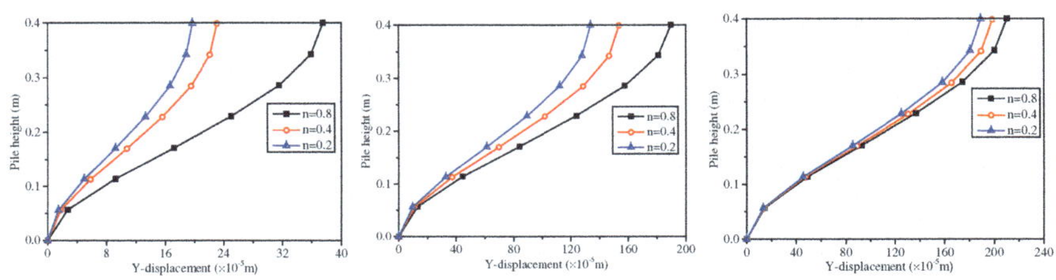

(a)K=100 (b)K=1000 (c)K=1500

Fig. 12. Peak Y-displacement of foundation pile Z1 at different flow index

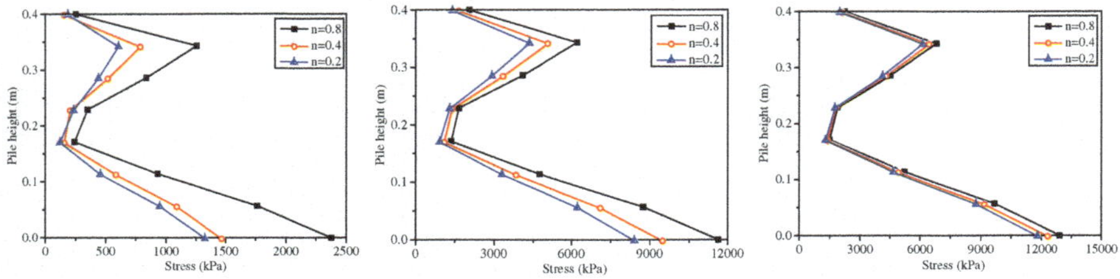

(a)K=100 (b)K=1000 (c)K=1500

Fig. 13. Peak stress of foundation pile Z1 at different flow index

It can be seen from Fig. 12, when the consistency coefficient K is certain, the peak value of the displacement response of the foundation pile Z1 generally increases with the increase of flow index n. From the variation gap between the three figures (a) (b) (c), it can be seen that the gaps between curves under different flow index n decreases gradually with the increase of the consistency coefficient K.

It can be seen from Fig. 13, when the consistency coefficient K is certain, the peak value of the stress response of the foundation pile Z1 generally increases with the increase of flow index n. From the variation gap between the three maps (a) (b) (c), it can be seen that the gaps between curves under different flow index n decreases gradually with the increase of the consistency coefficient K.

5 Conclusions

With regarding liquefied sand as power-law non-Newtonian fluid, the VOF method is adopted to conduct numerical computation about the flow push pile of post-liquefied sand. The following conclusions were drawn.

(1) The method in this paper can factually reproduce the flow deformation shape of post-liquefied sand.

(2) The displacement and stress of pile at different height both show the law of development of the first increases rapidly and then gradually decreases. The variation trend of peak displacement and peak stress response of pile are significantly different along the pile. The peak displacement response of the foundation pile shows a gradually increasing distribution from the bottom to the top of the pile. The peak stress response of the foundation pile from bottom to top shows the rule of decreasing firstly, then increasing, and finally decreasing. The curves show an oblique "2" shape distribution.

(3) Flow slide push pile effect of liquefied sand is simultaneously affected by the consistency coefficient K and flow index n. The influence of the consistency coefficient and the flow index on the displacement and stress of the pile is basically the same. With the increase of the consistency coefficient or the flow index, the displacement or stress of the pile increases. However, the changes of the consistency coefficient and the flow index only affect value of displacement and stress of the piles, but will not affect the deformation mode.

(4) The liquefied sand-piles fluid solid coupling problem is significantly affected by the properties of liquefied sand, so it is worth further study to verify the universality of this method.

Acknowledgments. The authors gratefully acknowledge the financial support for this study from the Project of the National Natural Science Foundation of China (Grant No. 51508236), the Project of the National Natural Science Foundation of Jiangsu Province (Grant No. BK20150519), Project of the Postdoctoral Science Foundation of China (Grant No. 2015M580397).

References

1. Hamada, M.: Large ground deformations and their effects on lifelines: 1964 Niigata earthquake. Technical report No. NCEER-92-0001, Case Studies of Liquefaction and Lifeline Performance during Past Earthquakes, Japanese Case Studies, State University of New York at Buffalo, NY
2. Liu, H.X.: Earthquake Sisaster in Tangshan, vol. 1, pp. 301–436. Seismological Press, Beijing (1985). (in Chinese)
3. Cao, Z.Z.: Study on liquefaction characteristics of Wenchuan earthquake and prediction method of sandy soil liquefaction. Ph.D. thesis, Institute of Engineering Mechanics, China Earthquake Administration, Harbin, China (2010). (in Chinese)
4. Verdugo, R., González, J.: Liquefaction-induced ground damages during the 2010 Chile earthquake. Soil Dyn. Earthq. Eng. **79**, 280–295 (2015)
5. Haskell, J.J.M., Madabhushi, S.P.G., Cubrinovski, M., Winkley, A.: Lateral spreading-induced abutment rotation in the 2011 Christchurch earthquake: observations and analysis. Géotechnique **63**(15), 1310–1327 (2013)
6. Sasaki, Y., Towhata, I., Miyamoto, K., Shirato, M., Narita, A., Sasaki, T., Sako, S.: Reconnaissance report on damage in and around river levees caused by the 2011 off the Pacific coast of Tohoku earthquake. Soils Found. **52**(5), 1016–1032 (2012)
7. Towhata, I., Sasaki, Y., Tokida, K., Matsumoto, H., Tamari, Y., Yamada, K.: Prediction of permanent displacement of liquefied ground by means of minimum energy principle. Soils Found. **3**(32), 97–116 (1992)
8. Sato, H., Hamada, M., Doi, M.: An experimental study of effects of laterally flowing ground on in-ground structures. In: Proceedings from the Fifth U.S.-Japan Workshop on Earthquake Resistant Design of Lifeline Facilities and Countermeasures Against Soil Liquefaction, pp. 405–414 (1994)
9. Kawakami, T., Suemasa, N., Hamada, H., Sato, H., Katada, T.: Experimental study on mechanical properties of liquefied sand. In: Proceedings from the Fifth U.S.-Japan Workshop on Earthquake Resistant Design of Lifeline Facilities and Countermeasures Against Soil Liquefaction, pp. 285–299 (1994)
10. Towhata, I., Vargas-Monge, W., Orense, R.P., Yao, M.: Shaking table tests on subgrade reaction of pipe embedded in sandy liquefied subsoil. Soil Dyn. Earthq. Eng. **18**(5), 347–361 (1999)
11. Sawicki, A., Mierczyński, J.: On the behaviour of liquefied soil. Comput. Geotech. **36**(4), 531–536 (2009)

12. Hwang, J.I., Kim, C.Y., Chung, C.K., Kim, K.K.: Viscous fluid characteristics of liquefied soils and behavior of piles subjected to flow of liquefied soils. Soil Dyn. Earthq. Eng. **26**(2–4), 313–323 (2006)
13. Liu, H.L., Chen, Y.M.: Analysis of flow characteristics of dynamic torsional tests on post liquefied sand. Rock Soil Mech. **30**(6), 1537–1541 (2009). (in Chinese)
14. Huang, Y., Hu, Z., Mao, W.W., Huang, M.S., Li, G.: Triaxial tests on the fluidic behavior of post-liquefaction sand. Environ. Earth Sci. **67**(8), 2325–2330 (2012)
15. Chen, Y.M., Liu, H.L., Shao, G.J., Zhao, N.: Laboratory tests on flow characteristics of liquefied and post-liquefied sand. Chin. J. Geotech. Eng. **31**(9), 1408–1413 (2009). (in Chinese)
16. Chen, G.X., Zhou, E.Q., Wang, Z.H., Wang, B.H., Li, X.J.: Experimental investigation on fluid characteristics of medium dense saturated fine sand in pre- and post-liquefaction. Bull. Earthq. Eng. **14**(8), 2185–2212 (2016)
17. Wang, X.F., Xiong, A.K.: Advanced Fluid Mechanics, pp. 84–139 Huazhong University of Science and Technology Press, Wuhan (2003). (in Chinese)
18. Yue, G.: Advanced Application of Fluid and Fluid Solid Coupling, ADINA, pp. 87–105. China Communications Press, Beijing (2010). (in Chinese)
19. Zhang, Q., Toshiaki, H.: Studies of the strong coupling and weak coupling methods in FSI analysis. Int. J. Numer. Meth. Eng. **60**(12), 2013–2029 (2004)
20. Liu, R.X., Shu, Q.W.: Some New Methods of Computational Fluid Mechanics, pp. 189–209. Science Press, Beijing, China (2003). (in Chinese)
21. Chen, H.F., Sarip, A.F.: Elasticity and Plasticity, pp. 95–167. China Architecture and Building Press, Beijing (2005). (in Chinese)
22. Huang, Y., Mao, W.W., Hu, Z., Li, G.H.: Computational fluid dynamics modeling of post-liquefaction soil flow using the volume of fluid method. Bull. Eng. Geol. Env. **71**(2), 359–366 (2012)

Experimental Investigations on Properties of Soil-Cement Columns: A Case Study

A. Le Van Tuan[1(✉)] and B. Phung Vinh An[2]

[1] College of Civil Engineering, Tongji University, Shanghai, China
levantuan112@gmail.com
[2] Vietnam Academy for Water Resources, Hanoi, Vietnam

Abstract. Ground improvement using soil-cement columns created by the deep mixing method is one of the soil stabilizing techniques that is widely applied in Vietnam recently. However, studies on the soil-cement columns properties conducted in Vietnam geology are limited. At present, the Standard "Vietnam Stabilization of Soft Soil–The Soil-cement Method (TCVN-9403: 2006)" is mainly referred to abroad standards. For this reason, this investigation was carried out to evaluate the properties of soil-cement columns with a case study of Hai Phong site, a city located in the Delta of Northern Vietnam. The tests were conducted with samples created in both laboratory and in-situ using the same type of soil. In laboratory tests, soil was extracted from in-situ at different depth levels, mixed with desired water cement ratios and required cement dosage rates. In in-situ tests, soil-cement columns were created with different cement doses, after desired curing time, the columns were dug to extract samples for tests. The results show that (1) the way to create samples affects the compressive strength of soil-cement mixture; (2) the compressive strength of soil-cement mixture depends mainly on the cement content and the curing time; (3) On the other hand, the compressive strength almost increases within the first month after being created.

Keywords: Viet Nam · Soil-cement column · In-situ · Laboratory
Cement dose · Compressive strength

1 Introduction

The Cement Deep Mixing (CDM) method is one of the wet methods of deep mixing, which is a new geotechnical improvement. Its origins date back to an original patent filling in 1967, but the field application really developed in the 1970s, and has also been frequently applied for on-land works [1] as shown in Fig. 1, a typical system of CMD machine consists of a DM machine and a binder plant. The binder plant consists of a binder silo, water tank, binder-water mixer, agitator tank, pumping unit and control room. The CDM machine consists of a mixing tool and a crawler crane with a leader. The procedure of CDM machine is a "bottom-up" procedure. The grout slurry is injected both during penetration and withdraw, and in general most of the slurry is injected on penetration to ease mixing and penetration [2].

R. Chen et al. (eds.), *Proceedings of the 2nd International Symposium on Asia Urban GeoEngineering,* Springer Series in Geomechanics and Geoengineering,
https://doi.org/10.1007/978-981-10-6632-0_47

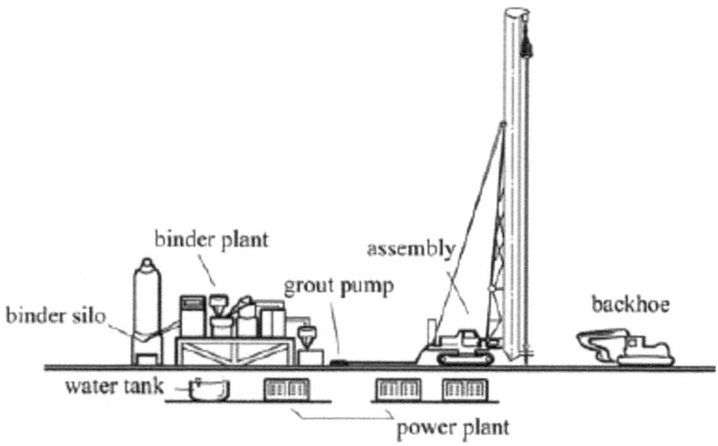

Fig. 1. Typical cement deep mixing machine for on-land work

Viet Nam has two major deltas located in the Northern and the Southern, named Bac Bo and Nam Bo respectively. In these areas, the soft ground layers are quite thick, as the result, the cost for ground improvement occupies a high rate of construction expenses. In recent years, CDM has been introduced in Vietnam, and has become to be a key method to improve the soft ground because of its advantages. Although using CDM for many years, studies on CDM in term of Vietnam conditions have not been paid attention properly. Since CDM firstly introduced in Vietnam in the 1980 by Nguyen Chap [3], some authors such as Ho Chat, Do Minh Toan had their own researches about soil-cement columns properties, but all the tests were carried out in laboratory. Until 2001, a project of soft ground treatment was done in Tra Noc, Can Tho province [4]. In 2002, another project using CDM were executed in Ba Ngoi harbor, Khanh Hoa province. In 2002, some projects such as Vu Ban water factory project (Ha Nam province), Dinh Vu harbor (Hai Phong province) applied CDM to treat the thick soft soil [3].

In the year of 2006, Vietnam Ministry of Construction issued a standard named "Stabilization of soft soil-The soil cement column method" [5]. Since then, soil-cement columns method to treat soft soil ground has applied widely not only in the field of soil improvement but also for anti-seepage purpose in the field of water resources, such as at Cui river (Long An province, 2006), Na Zanh reservoir (Cao Bang province, 2006), Nui Mot basin (Ninh Thuan province, 2009), Bong Canh reservoir (Hoa Binh province, 2010), etc. According to a statistics, more than 100000 m CDM were constructed in Vietnam in the year of 2009, increases three times in comparison with that in 2008 [6].

For the purpose of investigation properties of soil-cement columns using the deep mixing technology method in Viet Nam, this study was conducted with the geology of Hai Phong city that located in the Northern Vietnam.

2 Site Specification

Test site is located in Hai Phong city, about 100 km far from Ha Noi. The ground includes three soft layers under surface: backfill layer (2 ~ 2.4 m thick); silty loam (7.6 ~ 8.0 m thick); and clay layer. The soil-cement columns are mainly used to improve the silty loam layer. The physical properties of soil layers are illustrated in Table 1.

Table 1. Properties of the soil layers

Properties	Symbols	Units	Backfill	Silty loam
Moisture content	W	%	32.33	48.12
Wet density	γ_w	kN/m^3	18.2	18.6
Dry density	γ_d	kN/m^3	12.5	13.1
Void ratio	e_0	–	0.953	1.15
Yield limit	L_L	%	29.59	58.7
Plastic limit	P_L	%	19.5	35.7
Plastic index	I_L	%	10.09	23
Yong's module	E	kPa	1400	11500
Cohesion	C	kN/m^2	7	17
Friction	φ	degree	6.1	10.1
Granular composition:				
Gravel		%	70.14	46
Sand		%	8.27	13
Dust		%	21.59	41

Ground water level under investigation site is 1.5 m lower ground surface. The characteristics of ground water are shown in Table 2.

Table 2. Characteristics of ground water

Properties	Units	Sample M1
pH	–	7
SO_4^{2-}	mg/l	–
CO_2free		0.26
Cl^-	mg/l	976.65
TDS	mg/l	2

3 Cement Properties

Cement used in these tests is PCB 40, chosen according to the Vietnam Standard TCVN 6260–2009 [7]. The properties of PCB 40 are listed in Table 3.

Table 3. Characteristics of cement PCB 40 (Vietnam standard TCVN 6260-2008)

Parameters	Units	Value
Compressive resistance		
3 days	N/mm^2	≥ 18
28 days	N/mm^2	≥ 40
Fineness		
Screening residuary size 0.09 mm	%	≤ 10
Surface area ratio	cm^2/g	≥ 2700
Setting time		
Begin	min	≥ 45
End	hour	≤ 420
Stability volume	mm	≤ 10
Anhydride sulfuric (SO_3) content	%	≤ 3.5

4 Test Equipment

All the tests are axial compression in order to investigate the compressive strength of soil-cement sample extracted from in-situ, as well as mixed in laboratory. The test equipments are shown in Fig. 2.

a) b)

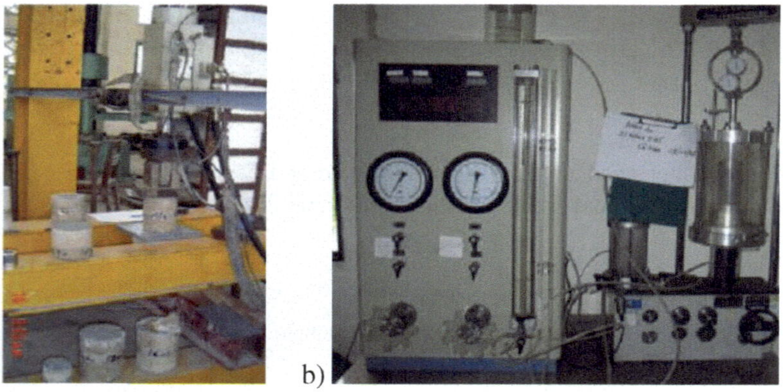

Fig. 2. Testing equipment: (a) axial compression equipment; (b) tri-axial compression equipment

5 Sample Preparation

5.1 Laboratory Test

A laboratory procedure as listed step by step below: soils after extracted from in-situ at different depths were first dried in oven at 105°C for 24 h to ensure having soil with zero initial water content. Then, each soil was sieved using sieve with 2 mm size in order to eliminate any stones and pebbles. Thereafter, the required cement dosage rate of each specimen (100 kg/m^3, 200 kg/m^3, and 300 kg/m^3) was achieved by adding and thoroughly mixing a calculated weight of cement with a specific weight of soil. Finally, according to the desired water cement ratio, a prescribed weight of water was added and mixed for about 3–5 min to make cement–water–soil mixture. The cylindrical cemented soil specimens were prepared by pouring the mixture in 3 layers inside thin-wall UPVC molds. Each layer was compacted by hands using a wooden rod to eliminate air pockets and to unite the layers together. All molds have a constant diameter of 150 mm and a length of 300 mm so as to have samples with a shape factor of 1:2. The molds also have a thin oil layer inside in order to facilitate extracting the samples after hardening. The filled molds were then covered with plastic bags and stored for a specified curing period of 21 days, 28 days, and 56 days in a constant temperature of 25°C. After curing, the specimens were extracted from the molds and left in the air for 1 day before testing. Unconfined compression tests were performed under a testing machine having a maximum load capacity of 250 kN. Each specimen was concentrically loaded until failure. All results were periodically recorded and stored in a computer during the tests by means of a data acquisition system.

5.2 In-Situ Test

With in-situ test, soil-cement columns are constructed in investigated position, with the required cement dosage rate in the range between 100 kg/m^3 and 300 kg/m^3. After a week of curing time in order to soil-cement columns hardening, test samples are extracted in two methods: (1) core drill with drill diameter of 150 mm and the length of samples is controlled more than 300 mm; then package samples with plastic bags; thereafter moving to laboratory. In the laboratory, samples are covered by wet jute sacking, and let in the air until carrying out test. Before a test was carried out, samples were polished at two-side until having samples with a shape factor of 1:2. Experiments are carried out according to Vietnam Standard TCVN 3120:1993 [8]; (2) Digging soil-cement columns, covering by wet jute sacking, then bringing to laboratory. In the laboratory, samples are processed in cylindrical shape, with the height of 140 mm, and diameter of 70 mm, to ensure the shape factor of 1:2. Two-side of samples were also polished before doing tests. The confined compression tests were performed base on Vietnam Standard TCVN 3118: 1993 [9].

6 Analysis the Test Results

6.1 The Effect of Creating Method on Homogeny of Samples

Samples mixed in the laboratory are more homogeneous than the samples extracted from in-situ as shown visually in the Fig. 3. This can be explained by the different creating methods for samples in the laboratory and extracted on-site. In the laboratory, grain bigger than 2 mm was eliminated by using sieve, therefore, samples are more homogeneous.

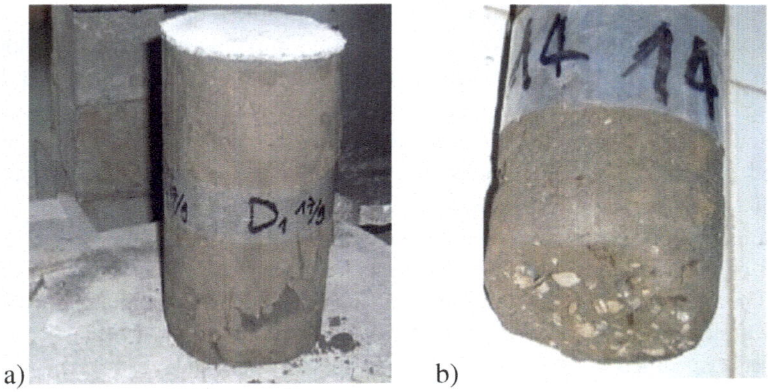

Fig. 3. Viewing of test samples: (a) laboratory test; (b) in-situ test

6.2 The Effect of the Cement Dose on the Compressive Strength

These tests investigate the effect of the cement amount mixed with the soil on the compressive strength of the mixture after 21, 28 and 56 days of curing. There are three different cement dosage rates were investigated in this study. It can be seen from the Fig. 4, the compressive strength of the cemented soil increases with an increase of cement dose. At the maximum cement dose utilized in these tests (300 kg/m^3), the values of compressive strength reach roughly 1800 kN/m^2 after 56 days of curing for both in-situ sample tests and laboratory sample tests.

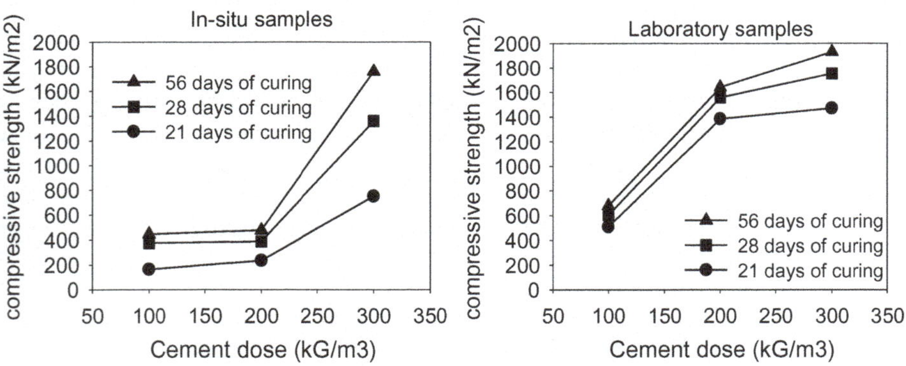

Fig. 4. Variation of compressive strength of the cemented soil with the cement dosage rate

In the tests with in-situ samples, the unconfined compressive strength of samples increases slightly with an increase of cement dose, while to the laboratory samples, the unconfined compressive strength of samples increases significantly with an increase of cement dose. Nevertheless, the compressive strength of in-situ samples increases considerably with an increase of cement dose, and for the laboratory samples, the compressive strength are more stable for the cement dose are higher than 200 kg/m^3.

6.3 The Effect of the Curing Time on the Compressive Strength

The effects of curing time on the compressive strength of soil cement were investigated by testing samples at the age of 21, 28 and 56 days. As shown in Fig. 5, the value of compressive strength increases with an increase of curing time. At 28 days of curing, the compressive strength of soil cement achieved in the range between 140% and 200% in compared with the compressive strength gained after 21 days of curing. Obviously, the value of compressive strength almost increases within the first month, after this period the compressive strength increases slightly and tends to remain constant.

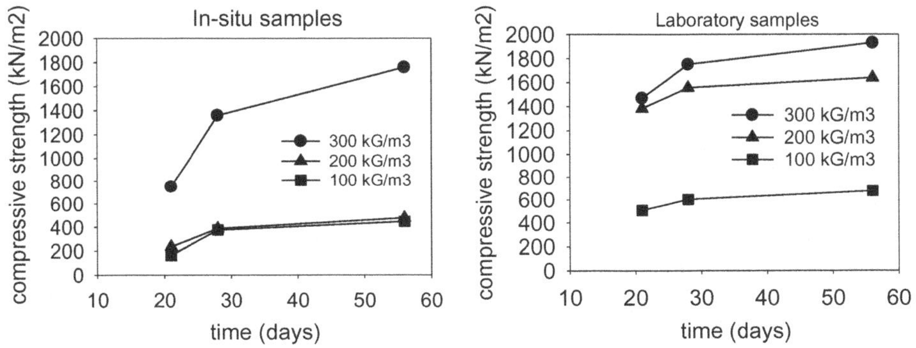

Fig. 5. Effect of the curing time on the compressive strength of soil cement

6.4 The Relationship Between Compressive Strength and Elasticity Modulus of the Soil Cement

There were six in-situ samples extracted from different soil-cement columns for the tests in order to evaluate the relationship between the compressive strength and the elasticity modulus of the soil cement. The tests were carried out after 28 days of curing, cement dose of 300 kg/m^3. The test results were plotted on Fig. 6.

According the test results, the elasticity modulus of soil-cement (in kN/m^2) can be related to the compressive strength (in kN/m^2) by the following equation:

$$E = 275 * q_u^{28} \tag{1}$$

in which q_u^{28} is the compressive strength tested at 28 days of curing. In comparison with Vietnam Standard [5] (that calculated the elasticity modulus by equation $E = (25 \div 50) * q_u^{28}$), the elasticity modulus calculated by the above equation is higher. On the other hand, compared with Coastal Development Institute of Technology [1] calculated by equation $E = (100 \div 300) * q_u$, and U.S Department of Transportation (2000) calculated by equation $E = (100 \div 500) * q_u$ [10], the results calculated by Eq. (1) are the same.

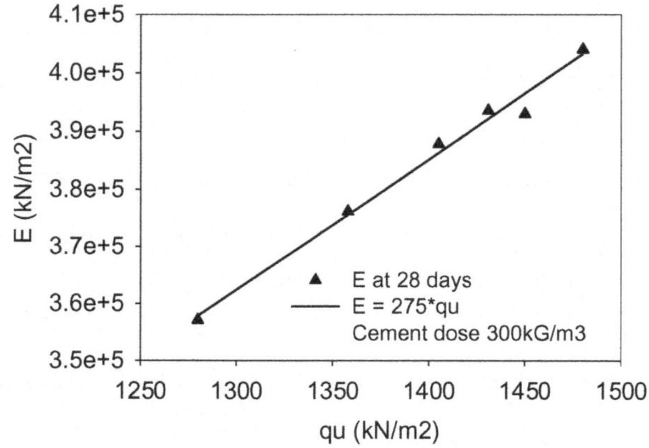

Fig. 6. Variation of elasticity modulus of cemented soil with the compressive strength

7 Conclusion

In this study, in-situ tests and laboratory tests were performed on soils extracted from geology of Hai Phong area - a city located in the Northern Vietnam. The aim of these tests is to investigate the properties of soil-cement column using the mixing technology method. The overall conclusions drawn from this study are followings:

1. There is a clear difference between the homogeny of samples extracted from in-situ columns and mixed in the laboratory.
2. As the cement dosage rate increases, the compressive strength of the mixed soil increases, but there is a different in rate of increase at higher and lower cement dose of 200 kg/m^3. In-situ test samples show that, when cement dosage rate increases, the increase of compressive strength is more significant at a cement dose higher than 200 kg/m^3 rather than at lower ones. However, the samples tests in laboratory show that the compressive strength of soil cement increases considerably with a increase of cement dose from 100 to 200 kg/m^3.
3. The value of compressive strength almost increases within the first month, after this period, the unconfined compressive strength increases slightly and tends to remain constant.
4. The relationship between the elasticity modulus and the compressive strength of the cement soil in Hai Phong area is fitted by regression based on test results of this study can be denoted as $E = 275 * q_u^{28}$.

References

1. Kitazume, M., Terashi, M.: The Deep Mixing Method Principle, Design and Construction, pp. 4–7. Coastal Development Institute of Technology, Balkema, Japan (2002)
2. Kitazume, M., Terashi, M.: The Deep Mixing Method, pp. 201–202. CRC Press, Boca Raton (2013)
3. Dung, N.Q., An, P.V., Huy, N.Q.: The Deep Mixing Technology, pp. 35–37. Agriculture Publishing House, Ha Noi (2005). (in Vietnamese)
4. Dung, N.Q., An, P.V., Huy, N.Q.: Scientific seminar on work construction in the Mekong Delta, pp. 49–55. The Ministry of Transport Hanoi, Viet Nam. (in Vietnamese)
5. TCVN 385:2006. Stabilization of soft soil-The soil Cement column method. (in Vietnamese)
6. An, P.V.: Research on factors affecting the bearing capacity of the soil cement columns by Jet-grouting technology for some soft ground areas in Vietnam. pp. 20–24. Vietnam Academy for Water Resources, Ha Noi, Viet Nam (2012). (in Vietnamese)
7. TCVN 6260:2009. Portland blended cement-Specifications. (in Vietnamese)
8. TCVN 3120:1993. Heavyweight Concrete-Method for determination of direct tensile strength. (in Vietnamese)
9. TCVN 3118:1993. Heavyweight concrete-Method for Determination of compressive strength. (in Vietnamese)
10. U.S Department of Transportation-Federal Highway administration: An Introduction to the Deep Soil Mixing Methods as Used in Geotechnical Applications: FHWA-RD-99-138 (2000)

Design of Full Top-Down Basement Construction with Drain Raft Concept Over Coastal Weathered Sedimentary Formation in Malaysia

Shaw-Shong Liew(✉) and Deng-Ing Ting

G&P Geotechnics Sdn. Bhd., Kuala Lumpur, Malaysia
ssliew@gnpgroup.com.my

Abstract. The paper presents a case study of a full top-down basement construction using Secant Pile Wall (SPW) with top down slabs as both temporary and permanent shoring system in a coastal weathered sedimentary formation in Malaysia with upper highly permeable course grained sandy alluvial deposits within the excavation depth. This full top-down excavation method and wall embedment into the hard weathered sedimentation formation have achieved a well-controlled ground deformation with minimal acceptable ground distresses and excavation seepage quantity as expected. In view of the expected high water table near to the coastal course grained alluvial deposits, design option of using drained raft with passive pumping system to handle the seepage inflow was explored and adopted to avoid the costly tension piles and thick basement slab. The secant pile wall embedded into the weathered sedimentary formation, which is not far from the lowest basement level achieved fairly satisfactory performance in reducing seepage inflow. Recharge wells were provided as mitigation measure for drawdown control during basement excavation construction and operational life time of the basement. During the investigation, there was an expectation of the inclined saline water intrusion interface with depth, which was confirmed from the increasing salinity profile of soil samples with depth. Such saline intrusion, which is not usually a focus of excavation projects near coastal areas, can cause durability and maintenance problems to the planned drained raft system and operational equipment. This paper shares the design performance data of the basement wall behaviours and seepage regime in the unusual sandy alluvial deposits in Malaysia.

Keywords: Full top-down excavation · Drained basement raft
Sandy alluvium · Saline intrusion · Secant pile wall

1 Introduction

The case study in this paper presents the challenges in re-developing an occupied site located in the downtown of capital city of Sabah state in Malaysia. The proposed new development mainly consists of one block of 14-storey office tower with four (4) levels

R. Chen et al. (eds.), *Proceedings of the 2nd International Symposium on Asia Urban GeoEngineering*, Springer Series in Geomechanics and Geoengineering,
https://doi.org/10.1007/978-981-10-6632-0_48

of basements. The maximum depth of excavation below the existing ground level is about 14.1 m to accommodate a 4-storey basement. The main geotechnical concern of this basement excavation construction is that the ground movement associated with the lowering of groundwater during temporary excavation stage and operational stage may cause potential distresses to adjacent existing structures and underground utilities surrounding the development site. Hence, to minimise the ground movement, a full top-down basement construction using Secant Pile Wall with top-down floor slabs as both temporary and permanent shoring system was adopted. On other hand, to resolve the hydrostatic uplift pressure from the high groundwater table to the basement structure during operational life time, design option of using drained raft system with passive pumping system and recharge wells to control the groundwater drawdown was adopted, instead of using costly tension piles and thick basement slab. As the development site is very close to the coast, saline water intrusion profile was also investigated in the durability assessment.

2 General Geology and Subsurface Information

Based on the published geology map, the site is underlain by Crocker formation, which is generally sedimentary rock formed during the Oligocene Epoch consisting of flysch-type sandstone, shale, siltstone with tuff, limestone breccia and agglomerate. Due to intense weathering process of tropical climate and regional tectonic actions, these sedimentary rocks have been subjected to physical and chemical weathering processes in reducing its engineering characteristic, particularly at the upper stratum.

A total of 23 nos. of boreholes were sunken within the 2.2 acres site to investigate and ascertain the subsoil stratigraphy meanwhile four (4) water standpipes were installed at the selected boreholes to monitor the groundwater table regime as shown in Fig. 1. The subsoil stratigraphy for the site generally can be classified into four (4) strata according to the reduce levels: (1) Ground level at RL+2.3m down to RL0.0m: Backfill; (2) RL0.0m to RL–11.0m: Alluvial Deposits; (3) RL–11.0m to RL–19.0m: Weathered Residual Soil; and (4) RL–13.0m to RL–19.0m or deeper: Crocker Formation.

The detailed subsoil stratification interpreted from the borehole logging is illustrated in Fig. 2. Rock head is generally encountered at about RL–19.0m over the entire site, except for the south-east corner area where rock head is somehow shallow at RL–13.0m, which is only slightly deeper than the lowest basement slab level. From standpipe monitoring, the groundwater table is found to be near to ground surface, i.e. at 1 m beneath (RL+1.3m). This is understandable in view of the close proximity (about 600 m) of the site to the coast line where the mean sea level is high and also subject to tidal highs and lows.

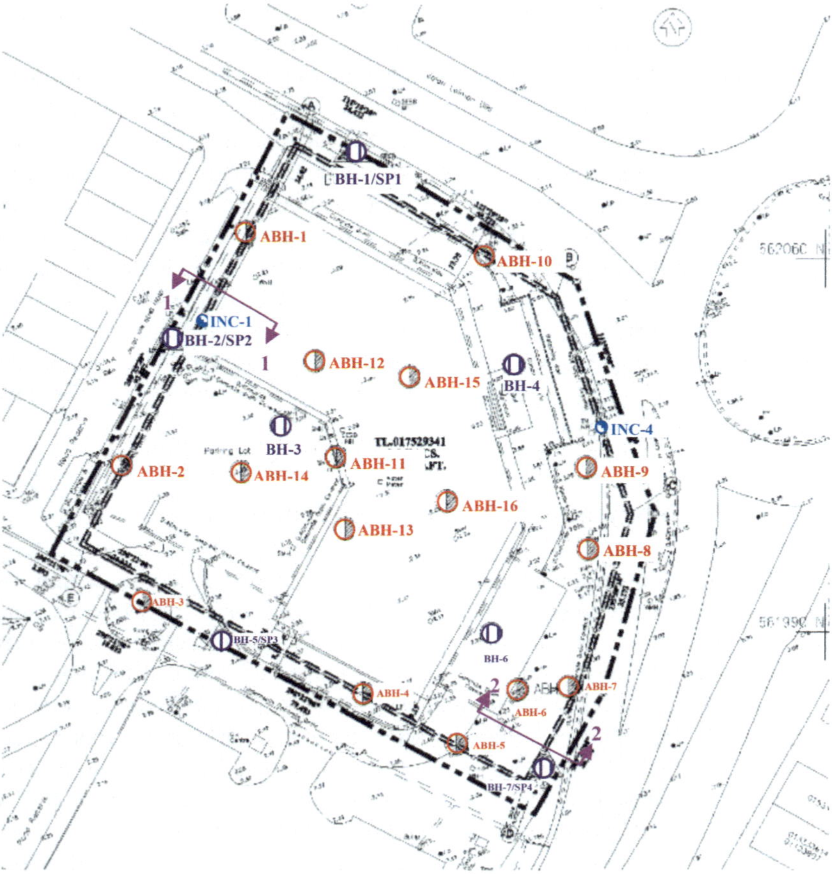

Fig. 1. Site plan with locations of boreholes

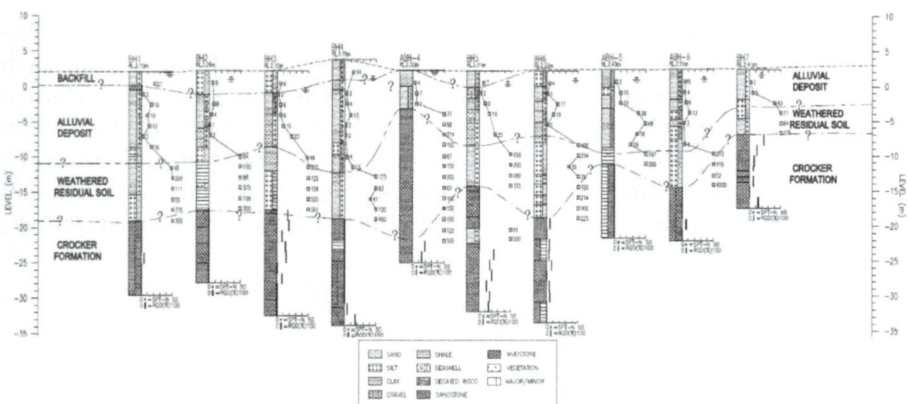

Fig. 2. Subsoil stratification interpreted from borehole logging

3 Full Top-Down Construction Method with Secant Pile Wall

The site has a shape of nearly perfect square with 94 m in length (parallel to north-east direction) by 104 m in width (parallel to south-east direction) and is surrounded by busy road networks with a roundabout at the corner of site, and adjacent buildings (such as hotel, retail shops, government institution, etc.) located immediately across the roads. Without the as-built information of the foundation type and details of those nearby existing structures, the ultimate objective of the basement construction method will have to consider minimizing the potential ground movements in both ground settlement and lateral displacements due to the deep excavation and the associated groundwater drawdown. Hence, full top-down basement construction method using Secant Pile Wall with lateral supports from basement floor slabs was adopted as there was also pressing need to complete the superstructure and handover to the anchored tenant under the agreement. This development is designed with certified LEED Silver Rating Level in the energy efficient and environmental friendly design.

Secant Pile Wall is formed by installing hard cast-in-situ reinforced bored piles (secondary pile with reinforcements) by overcutting the completed soft piles (unreinforced primary piles) along the designated wall alignment. Primary piles without any reinforcing bars are installed first with secondary piles constructed in between primary piles once the latter attain concrete setting without gaining its full strength. The secondary piles are reinforced with steel cage designed to take the design ultimate bending moments and shear forces resulted from soil-structure interaction analysis of staged excavations and constructions. Typical detail of the Secant Pile Wall configuration is shown in Fig. 3.

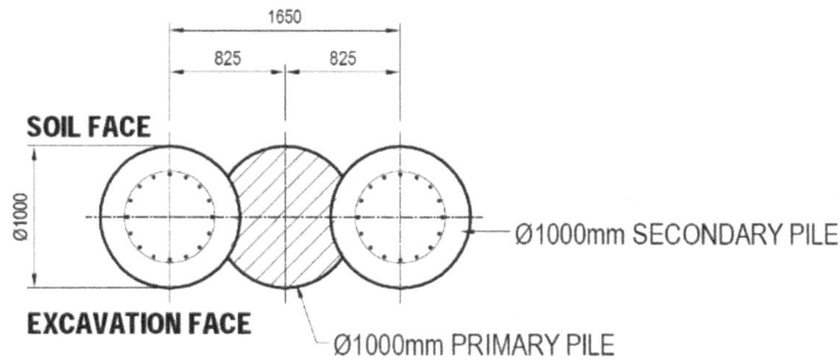

Fig. 3. Typical detail of 1000 mm diameter secant pile wall (overcut 175 mm)

The Secant Pile Wall was generally embedded into the weathered sedimentary rocks with rock socket length ranging from 0.5 m to 1.0 m depending on the rock head level. In some cases, which the rock head level is very deep, the maximum length of primary piles and secondary pile were 19.6 m and 24.0 m respectively. The toe of SPW are down to 10 m below the soffit the lowest basement slab unless the bedrock level is encountered at shallow depth. Two sections (Fig. 1) showing the SPW with toe levels, floor slab levels and excavation stages are shown in Figs. 4 and 5.

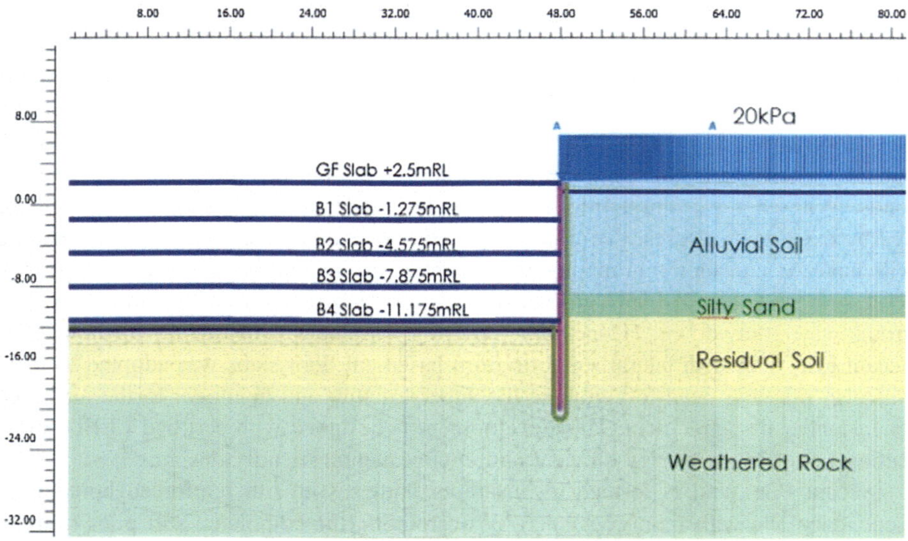

Fig. 4. Section 1-1 showing geotechnical model, the SPW with toe level and floor slab levels.

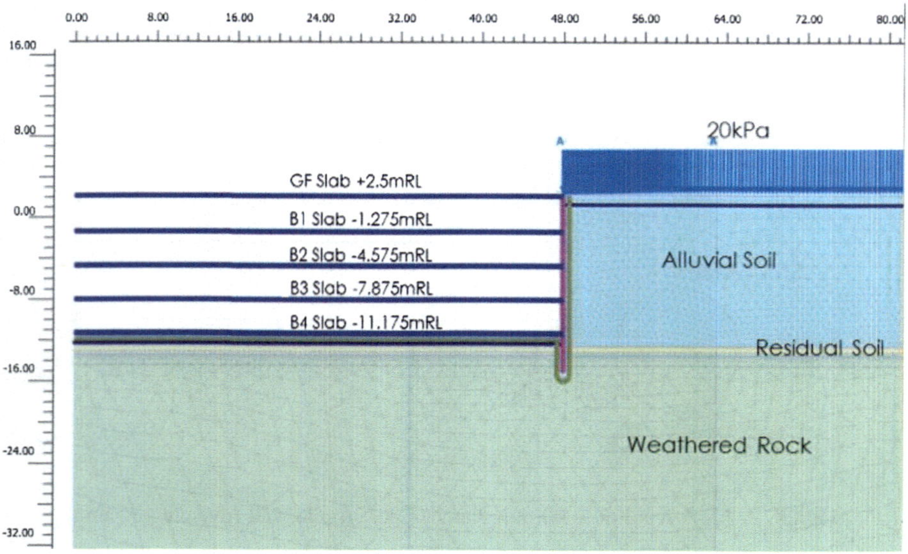

Fig. 5. Section 2-2 showing geotechnical model, the SPW with toe level and floor slab levels

Full top-down construction is a method for deep excavation and basement construction in which the permanent basement floors are constructed simultaneously as the excavation progresses in stages to facilitate the basement floor slab construction. At the same time, the superstructure construction with plunge-in column supports can also commence at the ground floor up more or less simultaneously with the basement excavation in order to expedite the overall construction programme, particularly with short construction duration contractually. For this project site, the overall basement excavation and construction sequence is outlined in Table 1 below.

Table 1. General sequence of full top-down excavation and construction works

Stage	Top-down excavation and construction activities
1	Install Secant Pile Wall (SPW) for basement walls until the pre-determined termination length and foundation piles with steel plunge-in columns embedded to support subsequent basement floor slabs
2	Excavate down 1 m below 'Basement 1' slab top level until RL–2.275m
3	Construct permanent 'Basement 1' floor slab supported by plunge-in columns as shoring support to SPW for further excavation
4	Excavate down 2 m below 'Basement 2' slab top level until RL–6.575m and construct strengthen Ground Floor slab as surface working platform and, also shoring support to SPW
5	Construct permanent 'Basement 2' floor slab supported by plunge-in columns as shoring support to SPW for further excavation
6	Excavate down 2 m below permanent 'Basement 3' slab top level until RL–9.875m
7	Construct 'Basement 3' floor slab supported by plunge-in columns as shoring support to SPW for the final excavation
8	Excavate down to Final Excavation Level (FEL) at RL–12.65m
9	Construct pilecaps and subsoil drainage connecting with collection wells as permanent hydrostatic uplift relief measure
10	Construct pilecaps and permanent 'Basement 4' floor slab as shoring support to SPW
11	Build up all plunge-in-columns with full reinforced concrete section size and reinforcements

The top-down construction begins with basement wall installation and also load-bearing foundation carrying the building structure. In other words, the foundation piles and basement columns are constructed before any basement excavation takes place. When the basement excavation works commence, the construction of the superstructure is also started, in which the self-weight of the superstructure is supported by the plunge-in columns embedded into the foundation piles as described below. These basement columns comprise of plunged-in columns using steel stanchions embedded into the underlying foundation bored piles constructed up to the cut-off level with reasonable concrete overcasting for accommodating the upwards displacing contaminated tremie concrete. First excavation was carried out and the top basement slab was casted with proper connection to the steel plunge-in columns. After the concrete floor has gained sufficient strength for formwork removal, soil under the top basement floor was further excavated to the next designated excavated level. And the similar procedures were repeated for the next basement floor until the Final Excavation Level (FEL). The ground floor slab is usually strengthened structurally as working platform with design loading provision catered for construction machineries and temporary storage of materials. Openings were allowed at the locations of the lift shaft or access ramp areas within the ground floor slab and every basement slabs (except lowest basement) to facilitate transporting construction materials in and also removal of spoil materials. Compact earthwork machineries suitable for very tight working space were used. Figures 6 and 7 show the top-down basement construction activities for this

Fig. 6. Excavation works underneath the completed basement floor slab using top-down method

Fig. 7. Long-arm excavator was removing spoil material through opening within the ground floor slab

Table 2. Advantages & disadvantages of top-down basement construction

Advantages	Disadvantages
a. construction works at every floor are fairly independent with own working platform	a. requires proper ventilation and lighting
b. relatively dry working environment, which is preferred for earthworks if the runoff at ground surface is well managed	b. prompt to damages due to hitting or impact of machineries to the completed structures
c. good isolation of construction noise to ground surface	c. tight working space for installation and removal of formworks for floor slab casting
d. good ground movement control due to stiffer floor strutting before excavation	d. requires post grouting operation to the gaps between the slab or beam soffit and the upwards cast concrete columns encapsulating the plunge-in columns

case study. The advantages and disadvantages of top-down basement construction can be summarised in Table 2.

4 Drained Raft Conceptual Method

Hydrostatic uplift problem arises after complete sealing of the lowest basement slab and the build-up of the recovered hydrostatic groundwater head exceeds the structural capacity of the slab or the dead weight of the structure, the basement floor slab will crack or, more catastrophically, surfer upward punching shear failure at supports or flexural failure at the mid spans. These limit state damage can be resolved structurally by tension anchorage piles tying down the basement slab or/and providing significantly thick slab as counterweight. Another solution is to reduce the hydrostatic uplift action using drained raft, which is primarily a drainage solution with introducing the underlying free draining filtered granular layer beneath the basement slab and half-perforated subsoil collection pipes connecting to several sump pits. The permanent underlying subsoil drainage can only passively reduce building-up of hydrostatic uplift pressure from the recovery of groundwater regime back to the steady state groundwater regime after completion of lowest basement slab. Electrical submersible dewatering pumps with automatic on/off controls by designated triggered water levels. The typical detail of drained raft is shown in Fig. 8. As the basement slab has its own self-weight and structural strengths of the floor slabs with support reactions from the internal columns and the basement wall, the basement floor slab can still withstand fair amount of hydrostatic uplift pressure. In this case, the basement slab can handle about 10 kPa uplift pressure in serviceability limit state. For a more prudent design consideration, the floor slab design has allowed for crack induced design to release the excess uplift pressure in a control manner if the unexpected uplift pressure builds up due to failure of the subsoil drains.

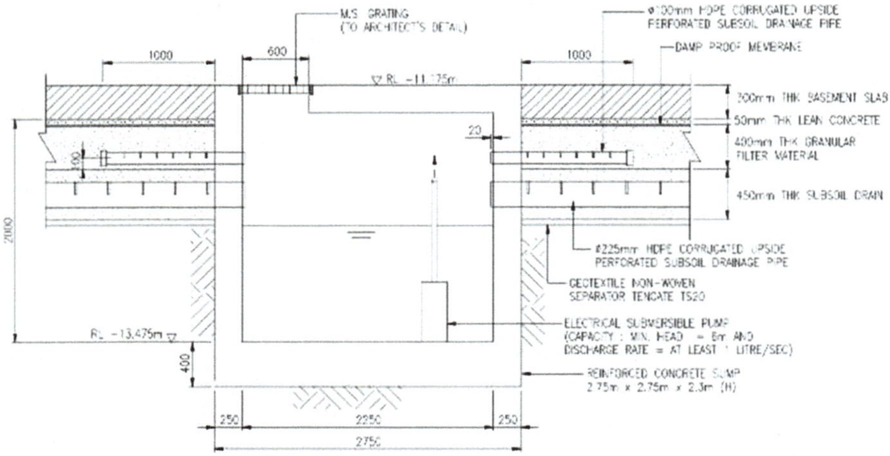

Fig. 8. Typical details of drained raft

The mechanism of the drained raft to reduce uplift pressure is by intercepting any seepage flow reaching the free draining filtered granular drainage layer with phreatic surface at atmospheric pressure, diverting the collected seepage water in subsoil pipes

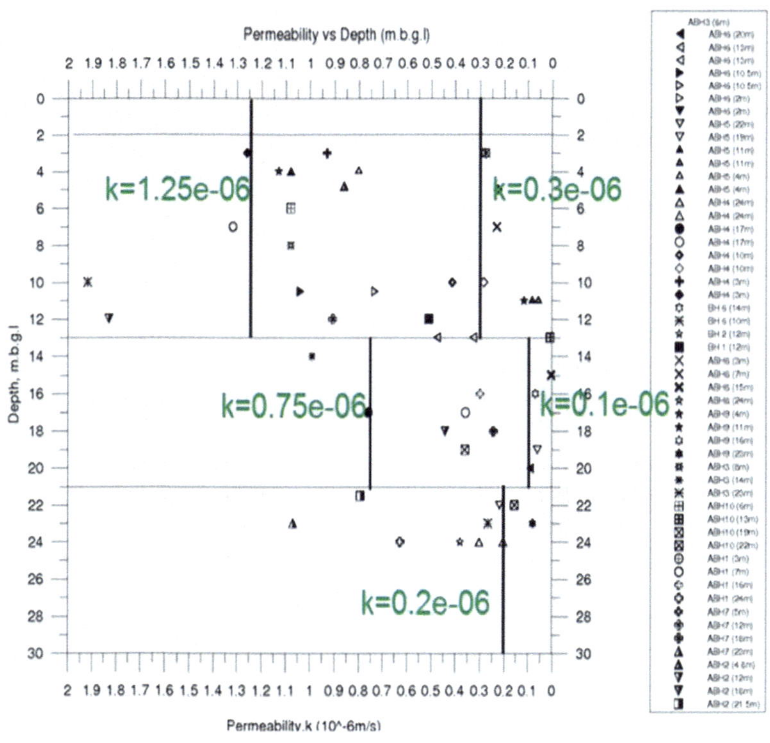

Fig. 9. Soil permeability profile

connecting to circular concrete ring deep sump pits, and lastly discharging the collected seepage water away from the basement to ground surface drainage using high capacity electrical pumps when the full storage threshold is triggered. The design requirements of the filtered granular layer and perforated subsoil pipes have to fulfil two (2) main functionality, namely the high permeable conveyance drainage capacity and the filter retention to prevent internal erosion of the foundation soils. Meanwhile, the volumetric capacity of the sump pits and electrical pumps shall be reasonably configured to more than the rate and anticipated total volume of water ingress into the basement in the seepage analysis and calculation. Given the upper layer of highly permeable course grained sandy alluvial deposits within the excavation depth as shown in Fig. 9, seepage analyses considering upper bound and lower bound soil permeability were cautiously performed and site evaluation of the measured seepage ingress rate and volume was also implemented during the excavation stages (refer to the result and discussion in Sect. 5.2). The estimated seepage rate with the two bound limits ranges from 40 m^3/day to 135 m^3/day.

5 Results of Instrumentation Data During Construction Stage

For any basement excavation project, it is both important and useful to implement instrumentation scheme to monitor the actual magnitude of lateral soil movement, wall deflection profile, settlement of the retained ground, seepage rate with fluctuated groundwater table regime at the retained side due to tidal fluctuation and the changing hydraulic boundary at the excavation side at various excavation stages, etc. The actual performance results with measured magnitudes have been timely reviewed and have the ground models recalibrated at all time for subsequent prediction at corresponding construction stage of analysis. The following two (2) sections will discuss the instrumentation data of wall lateral movement and seepage ingress respectively.

5.1 Actual Wall Lateral Movement Against Prediction

There were total of five (5) inclinometers installed inside the selected secondary piles of the Secant Pile Wall at the critical sections. These inclinometers in the wall were regularly monitored before and after each stage of excavation and by weekly basis during the basement excavation and shoring floor slab construction. Meanwhile, three (3) sets of threshold levels (i.e. Green, Amber and Red Zones) were established from the analytical prediction during the design stage for monitoring and determining the implementation of emergency response where necessary. Appropriate actions will be taken according to the construction specifications if the instrumentation results exceed the designated threshold levels during the review. Preparation of implementing emergency action and caution of following up the developing trend will be exercised when the results fall within Amber zone and shall implement the planned action immediately when entering into Red zone. Figure 10 shows two (2) sets of inclinometer readings (location of INC-1 and INC-4 as shown in Fig. 1) recorded throughout the construction period.

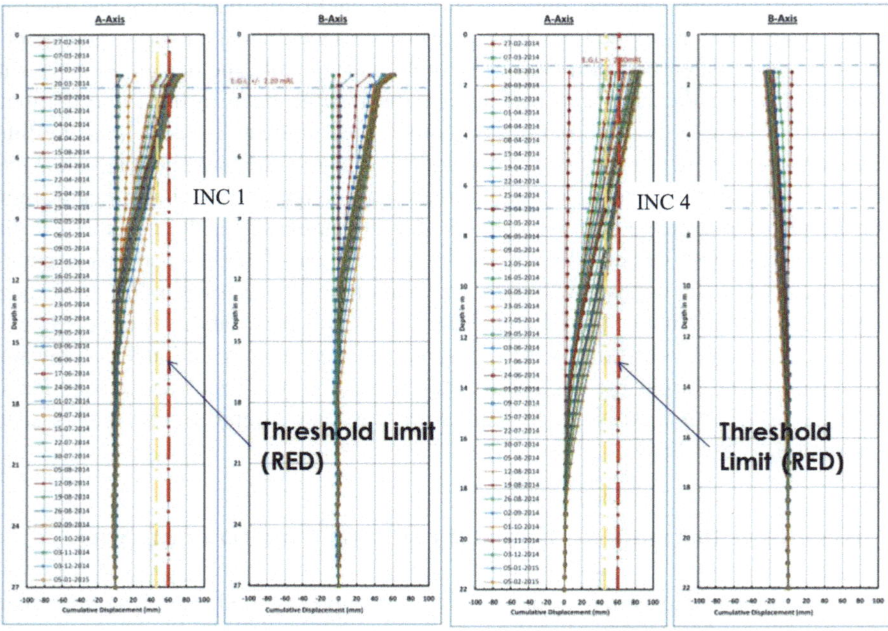

Fig. 10. Inclinometer monitoring results (INC 1 & INC 4)

The threshold level of Red zone is equal to the maximum wall lateral displacement predicted from the soil-structure interaction analysis for the entire stage of excavation and construction works. Meanwhile, the threshold level of Amber zone is taken as about 85% of the expected maximum lateral deflection. It shall be noted that the specified threshold limit for the inclinometer reading is an absolute magnitude, not the deflection profile/trend. The measured wall deflection profile will theoretically match with cautious estimate whilst the exact magnitude of the lateral movement is harder to predict accurately. As such, it is more critical to monitor the measured value against the analytical prediction and observe the developing incremental trends.

As observed from the measured inclinometer readings in the A-axis (towards the excavation face), the wall deflection occurred in a profile where lateral displacement at the top wall level is the largest and gradually reduces with depths until the apparent fixity at the firm ground. Hence, it is surmised to deduce that the most critical wall deflection had occurred during the first stage of excavation down to a meter below Basement 1 slab in which the SBP wall was in cantilever condition with ground floor slab yet to be constructed. Thereafter, all subsequent excavation works had caused the retaining wall to deflect further inwards. The measured magnitudes of wall deflection from the above two inclinometer readings have exceeded the predicted value. When the threshold level was breached, this had prompted the site team to conduct daily inspection on the surrounding retained ground to examine the signs of distress and damages. As shown in Fig. 11, minor ground settlement at the active zone behind the wall was found on the road pavement and rectification works using cement grout to seal

Fig. 11. Ground deformation and cracks on the road pavement

the cracks was carried out immediately. This ground depression in the active zone will be obvious when the wall deflects in the cantilever condition. Inclinometer INC-4 shows larger wall deflection primarily due to low stiffness floor slab support with the existence of opening voids for construction access as shown in Fig. 7.

5.2 Actual Seepage Ingress into Basement Excavation Against Prediction

There were four (4) temporary groundwater extraction wells constructed with installing 500 mm diameter and 6 mm wall thickness mild steel slotted pipe casing until about 18 m depth below ground surface. The main purpose of these temporary extraction wells was to ensure a relatively dry excavation base for the working platform. The water seeping into these open wells were extracted using submersible pumps via temporary hose piping system with flow meter to measure the off-site discharge into the nearby existing storm water drainage. The flow meter was read for recording the daily volume of water pumped out from each well. Figure 12 shows the flow meter readings for three (3) wells whilst the fourth well was not successfully monitored due to technical fault in the measurement.

It was observed that the actual volume of water pumped out for each well ranged from 0 m^3/day (dry) up to 60 m^3/day. Flowmeter No. 2 which was installed in the well located at the north-east side of the site recorded higher inflow of groundwater. In geographical view, the coast line is located parallel with the western boundary of the site meanwhile further east side of the site lays the inland. Generally, this augurs well with the prediction. However, it is important to realise a fact that the SPW has embedded into the hard and weathered Crocker formation and some even into the rock mass of Crocker formation. It was believed that the SPW embedment has somehow effectively cutoff the groundwater seepage regime in the alluvial layer. Figure 9 also evidences that there is a reduction of permeability with depth. If the seepage is effectively cutoff, the groundwater regime shall maintain pretty much in the initial state

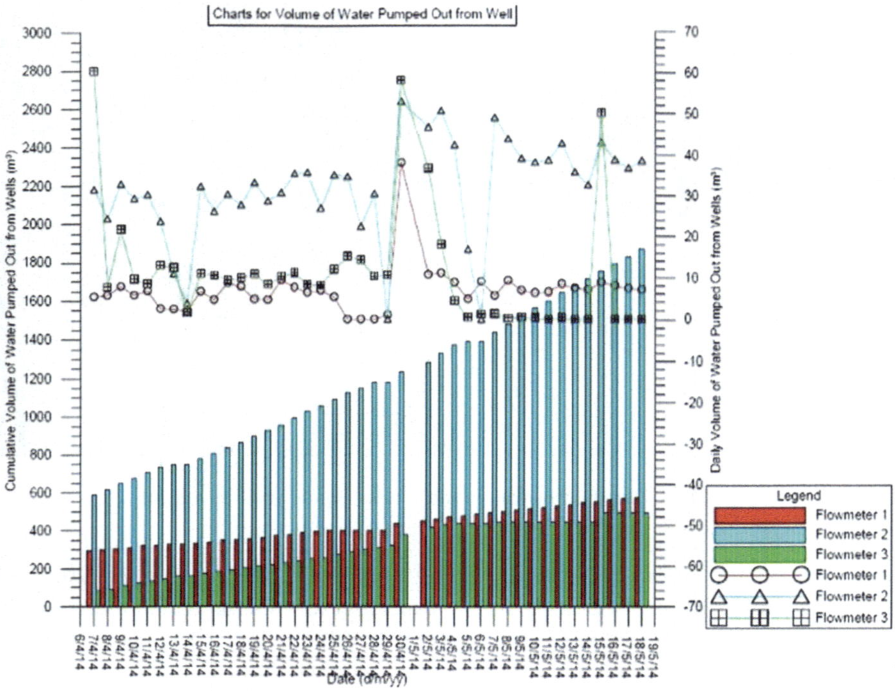

Fig. 12. Daily and cumulative volume of water pumped out from each well

of ambient hydrostatic condition with phreatic line at approximately RL+1.3m. The groundwater condition beneath the excavation will be also of hydrostatic profile, but at lower piezometric head with atmospheric pressure at about RL–13.3m. Thus, there will be hydraulic difference of about 14.6 m along the seepage path of 17 m to 29 m along the SPW.

6 Saline Water Intrusion

Based on the site hydrogeological and geomorphological conditions, it is logical to expect the fresh groundwater flow regime from inland hilly catchment areas towards the coast where terrain elevation and groundwater level is at its lowest. This naturally steady flow movement of fresh water towards sea prevents denser salt water from entering fresh water coastal aquifers. Saltwater has a higher content of dissolved mineral salts, thus resulting its higher density than fresh water. The higher hydrostatic pressure of the heavier salt water, dispersion and diffusion effects form a wedge-shaped intrusion interface into coastal aquifers underlain the fresh water flow body. The salt water and fresh water meet in a transition zone where mixing occurs through dispersion and diffusion.

The distance between the project site and the coastal shoreline is very near, i.e. about 600 m only. Therefore, there would be saline water intrusion occurred within the subsurface region. Based on Ghyben-Herzberg relation by Verrjuit (1968), the saline water intrusion interface can be approximated as shown in Fig. 13. Sampling and laboratory testing of groundwater collected from the site were also conducted to confirm the hypothesis of this salt-water interface. In this case study, the groundwater samples were collected from the recovery of groundwater at the base of casing during performing rising head permeability tests after two times pumping-out of borehole water. However, it is suggested that more appropriate sampling method is to extract the pore water from undisturbed soil samples in order to better preserve the chemical content.

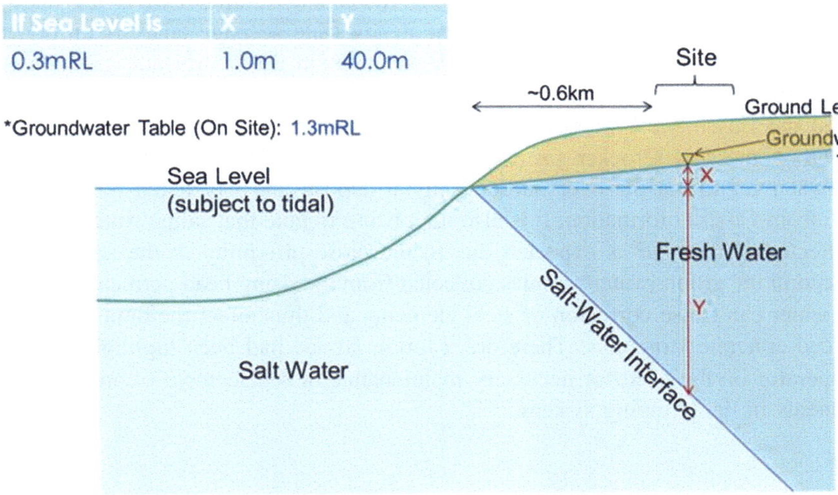

Fig. 13. Approximation of salt-water intrusion interface

As presented in Table 3, it is evident that there is increasing salinity profile of groundwater samples with depth. With the 14 m difference in the piezometric head between the ambient groundwater regime outside the basement and the passive drained raft design inside the basement with wall embedment into formation with higher salinity, the collected water seepage at the drained raft can be of some degree of salinity. Therefore, corrosion of the pumping system and any metallic components along the pipe network can be of shorter life span due to some corrosion. The potential solution is to use less metallic components to lengthen the entire pressure relief system.

Table 3. Laboratory test results of groundwater for salt

Parameters	ABH-5 (6 mbgl)	ABH-3 (8 mbgl)	ABH-10 (10 mbgl)
Total dissolved solid	2,380 mg/L	7,290 mg/L	10,600 mg/L
Chloride (Cl)	1,420 mg/L	4,340 mg/L	5,900 mg/L
Magnesium (Mg)	103 mg/L	226 mg/L	282 mg/L
Sodium (Na)	706 mg/L	1,940 mg/L	2,160 mg/L

7 Conclusion

This paper presents the case study of full top-down basement construction work using Secant Pile Wall (SPW) with permanent top-down slabs as shoring system in coastal weathered sedimentary Crocker formation in the capital city of Sabah, Kota Kinabalu. The actual wall deflection performance as measured by inclinometers in the SPW was slightly more than the prediction established from soil-structure analysis during the design stage. Besides that, drained raft system was successfully constructed as attractive alternative of the costly tension pile option or/and the thick basement slab, to cater for the high hydrostatic uplift pressure. From the measurement of the water quantity pumped out from the collection wells, the actual volume of water ingress was somehow less than the prediction based on steady state seepage analysis during the design stage. The effective cut-off by SPW to maintain the hydrostatic groundwater profile in the highly permeable sandy alluvial layer was observed. The seepage collected at the Basement 4 subsoil drains is likely another groundwater regime in the jointed rock mass of Crocker formation, which can be differentiated from perched groundwater regime in the over burden alluvial deposit and weathered residual soils derived from Crocker formation. It is also interesting to note that saline water intrusion was detected at this site as expected due to the close proximity to the sea shore as evidenced in the groundwater samples collected from the rising head permeability tests. Saline water can cause corrosion of steel elements and thus affect the durability of all reinforced concrete structures. Therefore, caution advice had been highlighted to the asset operator on the need for necessary maintenance or replacement of any corroded components in the pumping system.

Reference

Verrjuit, A.: A note on the Ghyben-Herzberg formula. Bull. Int. Assoc. Sci. Hydrol. **13**(4), 43–46 (1968). Technological University, Delft, Netherlands. Accessed 21 Mar 2009

Revisit of the Surface Surcharge Effect on the Structural Responses of Shield Tunnel in Spatially Varied Soils

Jinzhang Zhang(✉), Dongming Zhang, and Hongwei Huang

Department of Geotechnical Engineering, Tongji University, Shanghai, China
{zhangjz, 09zhang, huanghw}@tongji.edu.cn

Abstract. Surface surcharge is regarded to be one of the main factors that will cause ground settlement and large convergence of tunnel. Although the impact of surcharge on tunnel deformation has been studied in recent years, few studies have been dedicated to investigate the effect of spatial variability in soil strength on the ground surface settlement and convergence of tunnel. Hence, this study focused on the effect of soil spatial variability on the responses of the ground and tunnel linings by using the random finite difference analysis. Random fields are generated and mapped into finite difference analysis to reveal the impact of the variation of soil elastic modulus. The influence of the load thickness of soil and distance from load position to the centerline of the tunnel are discussed. The effect of coefficient of variation (COV) of soil's elastic modulus on surface settlement and tunnel convergence is also investigated. The results indicate that ignoring the spatial variability of soil strength will lead to a lower value of the surface settlement and tunnel convergence compared to the results from deterministic analysis. The maximum tunnel convergence value is 30% (15%) larger than deterministic analysis when the COV of soil's elastic modulus is 0.35. The COV of vertical (horizontal) convergence increases with the COV of elastic modulus. The larger the variability of soil leads to a larger change in the tunnel through soil, and resulting in larger convergence changes. Therefore, it is necessary to consider the variability of soil parameters on analyzing the effect of surface surcharge on the settlement and tunnel convergence.

Keywords: Surface surcharge · Surface settlement · Tunnel convergence
Random finite difference method

1 Introduction

At present, urban rail transit plays an important role in public transport system. Shield tunnel is the main structure of urban rail transit. Based on the site investigation and analysis, it is found that the surface surcharge above the tunnel is one of the significant factors that cause the large deformation of the shield tunnel (Wang and Zhang 2013). Shield-driven tunnel construction typically involves excavation in subsurface regions with spatially variable soil. Under such uncertain soil conditions, it is difficult to obtain accurate soil information in advance, for example, the elastic modulus, E_s, and Poisson's ratio, v_s, i.e., the most significant properties for the deformation of shield tunnel.

R. Chen et al. (eds.), *Proceedings of the 2nd International Symposium on Asia Urban GeoEngineering*, Springer Series in Geomechanics and Geoengineering,
https://doi.org/10.1007/978-981-10-6632-0_49

Without accurate soil parameters' information, conventional analysis thinks these parameters are same in every layer. But in fact, the soil is anisotropic. Thus, it seems that employing random field analysis approach that are able to incorporate soil uncertainties in analyzing the surface settlement and convergence of tunnel under surface surcharge in spatially variable soil is a more sensible analysis.

In the past, a considerable amount of effort has been devoted to analyzing the influence of surface surcharge on shield tunnel. For instance, WANG analyzed the evolution of transverse deformation of the tunnels under surcharge (Wang and Zhang 2013); ZHANG analyzed the influence of the compressibility of soil layers across by tunnel and soil layers overlaying on tunnel structure deformation and surrounding soil pressure under surface surcharge (Zhang et al. 2016). DAI got the effect of load position on tunnel that the impact is weakened when the deviation distance increased to a certain value (in this case 20 m) (Dai et al. 2006). LI did a study on random finite element method for spudcan foundations in spatially variable soils (Li et al. 2016). However, there has been relatively little effort spent in analyzing the spatial variability of soil on the ground settlement and structural behaviors. Therefore, the analysis of the effect of spatial variability of soil properties on tunnel convergence under surface surcharge are quite necessary.

Therefore, the present study will focus on the surface settlement and convergence of shield tunnel under surface surcharge in spatially variable soil. In this study, only the elastic modulus E_s is considered to be a spatially random property, as v_s is believed to have a smaller relative spatial variability (Fenton and Griffiths 2005). The random finite difference analysis is used in the present study. Random fields of soil elastic modulus are generated and mapped into finite difference analysis to reveal the effect of COV of elastic modulus on the surface settlement and tunnel convergence in random fields.

2 Deterministic Finite Difference Analysis

2.1 Finite Difference Modelling

In this paper, a shield tunnel with its outside diameter D = 6.2 m, internal diameter 5.5 m, thickness 0.35 m and depth H = 17 m is considered. The shield tunnel exemplifies the common used shield tunnel in Shanghai metro, and is modelled as elastic homogeneous ring to allow structural deformations. In the deterministic analysis, we set the soil parameters to be consistent. The soil of elastic modulus is 25 MPa, Poisson's ratio 0.31, cohesion 27 kPa and density 1800 kg/m3. Figure 1 shows finite difference model and contour of z-displacement under overloading (deformed factor: 50), as well as the maximum convergence in vertical and horizontal direction. The maximum vertical (horizontal) convergence is denoted by $\Delta dv(\Delta Dh)$.

This section adopts numerical analysis to study the deformation rules and deflections of shield tunnels underground heaped load of different positions and magnitudes. Finite difference analyses are performed using the FLAC3D 5.0 software. The excavation of the soil is simulated by using the convergence-confinement method to account for soil deformation or ground loss caused by time and space effect (Mroueh

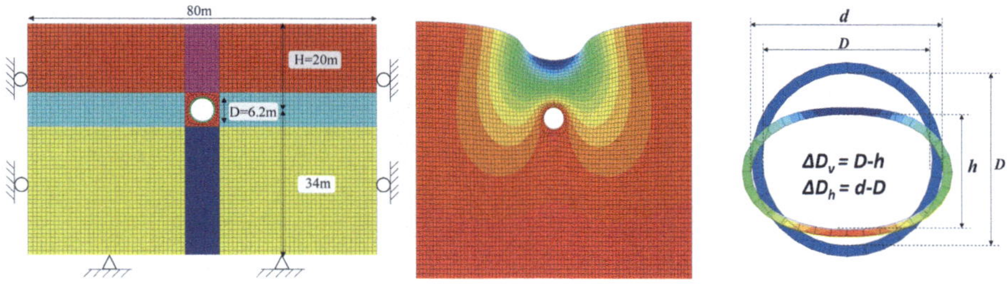

Fig. 1. Finite difference model and the convergence in vertical and horizontal direction of shield tunnel

and Shahrour 2008). The mechanical parameters of the tunnel are shown in Table 1. In the process of simulated the surface surcharge, the loading width is 12 m.

Table 1. Geometric parameters and mechanical parameters of finite element model tunnel

Outside diameter/m	Internal diameter/m	Thickness/m	Elastic modulus/GPa	Poisson's ratio	Density/(kg/m$^{3)}$
6.2	5.5	0.35	34.5	0.2	2500

2.2 Effect of Load Thickness on Surface Settlement and Tunnel Convergence

In this study, we assume that the surface overload is caused by the soil heaped above the tunnel, the surface heap is loaded on a heap of 0.5 m (18 kN/m3) and converted into a uniform load (9 kPa) in the model. In order to get the effect of load thickness on the surface settlement and tunnel convergence, 15 cases with different load thickness are taken into account, as shown in Table 2.

Table 2. Case design on influence of load thickness on the surface settlement and tunnel convergence

Case	Case 1	Case 2	Case 3	Case 4	Case 5	Case 6	Case 7
Load thickness/m	0	0.5	1	1.5	2	2.5	3
Case	Case 8	Case 9	Case 10	Case 11	Case 12	Case 13	Case 14
Load thickness/m	4	5	6	7	8	9	10

Figure 2(a) shows the variation in the tunnel convergence with the load thickness of soil. In the stage of load thickness is 0–1 m, the incremental value of tunnel convergence increases rapidly with the increase of the load thickness of the soil; in the stage of 1–3 m, the incremental value is basically stable; in the stage of 3–4 m, the incremental value increases rapidly with the increase of the load thickness; and in the

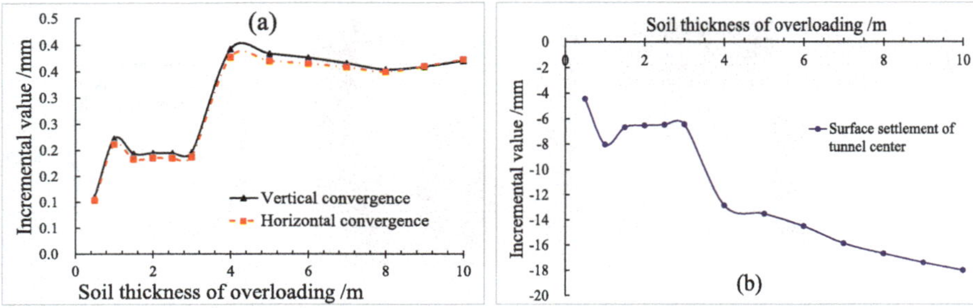

Fig. 2. The effect of load thickness of soil: (a) tunnel convergence; (b) surface settlement of tunnel center

stage of 4–10 m, the incremental value is basically stable. Figure 2(b) shows the variation in the surface settlement of tunnel center with the load thickness. The discipline of the effect is similar to the Fig. 2(a). The load thickness of 3 m the important point that we need pay special attention to.

As shown in Fig. 3, it shows the effect of load thickness of soil on the surface settlement. The incremental value of the surface settlement is relatively small before the thickness is less than 3 m. The incremental value significantly increases when the load thickness of soil lager than 3 m.

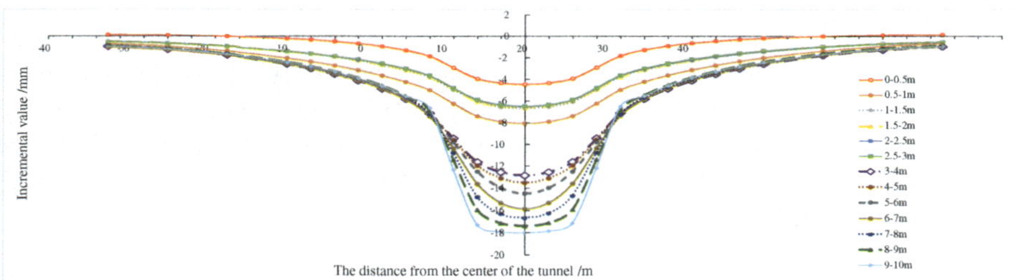

Fig. 3. The effect of load thickness of soil on the surface settlement

To summarize, the 3 m (54 kPa) is a important value that we need pay more attention. In practical engineering, we need to avoid the overloading on the tunnel as far as possible. If we have to be loaded in some emergency special circumstances, the load thickness should not exceed 3 m. Otherwise, the large deformation of tunnel and settlement of surface will occur.

2.3 Effect of the Distance from Load Position to the Centerline of the Tunnel

The location of the overloading is also an important factor. The surface surcharge is not only at the center of the tunnel, so 11 cases with different distance are considered, as shown in Table 3.

Table 3. Case design on influence of deviation distance on the surface settlement and tunnel convergence

Case	1	2	3	4	5	6	7	8	9	10	11
Distance/D	0	0.5	1	1.5	2	2.5	3	3.5	4	4.5	5

Figure 4a (b) shows the variation in ΔDv (ΔDh) with the distance from load position to the centerline of the tunnel. It has a noticeable trend that the larger distance, the smaller tunnel convergence. As can be seen from the graph, there has been a rapid decrease when the distance from 0.5D to 3.5D. A gradually stable period occurs while the distance from load position to the centerline of the tunnel lager than 3.5D.

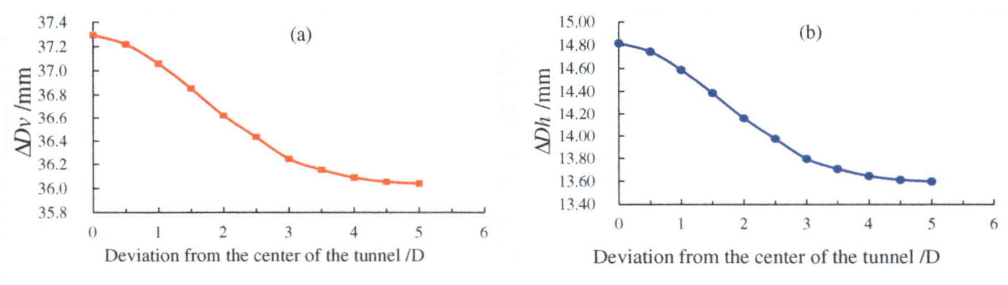

Fig. 4. The effect of distance from load position to the centerline of the tunnel on: (a) ΔDv; (b) ΔD_h

Figure 5 shows the variation in surface settlement with the distance from load position to the centerline of the tunnel. It can be concluded from the figure that there has been a great decline in surface settlement with the distance from 0.5D to 3.5D. The settlement remains a slight change when the distance lager than 3.5D. The incremental value curve has something in common with the settlement curve. The incremental value is obviously small and stable when the distance from load position to the centerline of the tunnel lager than 3.5D (in this case). Which is similar to the conclusion of the literature (Yu et al. 2006; Wang et al. 2008), also conforms to the basic principle of

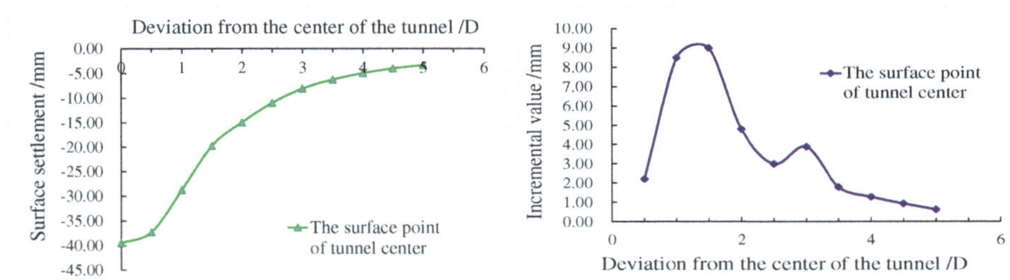

Fig. 5. The effect of the distance from load position to the centerline of the tunnel on the surface settlement

the literature. The farther away the loading position is from the tunnel, the less influence it will have on the tunnel.

In summary, the position of surface surcharge is also a significant factor that we need to consider (Wu 2012). When the distance from load position to the centerline of the tunnel becomes larger, the additional stress of soil around the tunnel becomes smaller, so the effect of tunnel convergence becomes slighter. The impact is rapidly weakening when it becomes from 0.5D to 3.5D. The impact is small while it larger than one particular value. In this case, it is 3.5D. Therefore, we should make the load position as far away from the tunnel as possible to minimize the impact on the tunnel in engineering.

3 Random Finite Difference Analysis (RFDA)

3.1 Random Field Model of Soil's Elastic Modulus

In this section, the model of tunnel also uses the former deterministic model, the load thickness is 3 m (54 kPa). Random fields of elastic modulus are generated using the Karhunen-Loeve expansion. The elastic modulus field is assumed to fit into an exponential covariance function (Huang et al. 2013). The mean value is set as 25 MPa which is consistent with the deterministic analysis, and the scales of COV is shown in Table 4. The COV of Case1 is 0, it was used to compare with other Cases. Other parameters of Case2 to Case7 are exactly same to the deterministic analysis apart from the COV, so that we can get the impact of COV to the tunnel convergence and settlement under the surface loading. For each case, 300 MCS runs are performed for the soil random field and the subsequent finite difference analyses.

Table 4. Scales of COV in anisotropic random fields

Case	Case 1	Case 2	Case 3	Case 4	Case 5	Case 6	Case 7
COV	0	0.1	0.15	0.2	0.25	0.3	0.35

3.2 Spatial Patterns for Max Settlement and Tunnels Convergence

We pay more attention to the worst case rather than the average in practical engineering. It is bound to have different value of E_S in each region because of the COV of E_s, which will result in different settlement and tunnel convergence. Figure 6 shows the worst spatial pattern which have the maximum value of the surface settlement and tunnel convergence. In Fig. 6, the red regions indicate strong soil and blue regions indicate weak soil. From the Fig. 6(a), we can see that the surface settlement is maximum when the surface center soil is very weak. The most interesting thing is that the maximum arch settlement, $\varDelta D_v$ and $\varDelta D_h$ occurs in the same simulation. According to the Fig. 6b, c and d, the soil of tunnel around is so weak that the convergence is very huge. It suggests that the soil of tunnel around play a significant role in the tunnel deformation. The maximum surface (*arch*) settlement is 47.19 *(30.94)* mm, larger than deterministic analysis 21.19 *(17.73)%*. The maximum $\varDelta D_v$ *($\varDelta D_h$)* is 42.77*(19.35)* mm,

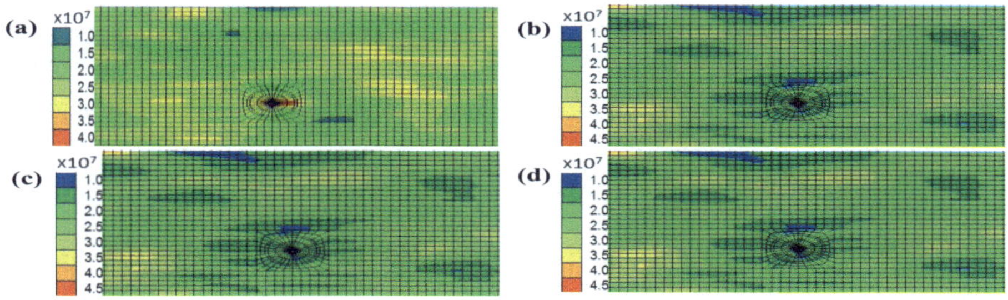

Fig. 6. Realizations with the maximum settlement and convergence for Case7 (COV = 0.35): (a) Maximum surface settlement; (b) Maximum arch settlement; (c) Maximum ΔD_v; (d) Maximum ΔD_h.

larger than deterministic analysis 14.68 *(30.59)%*. It will be disastrous if we design the tunnel using the conventional approach. From another aspect also shows that it is necessary to consider the stratigraphic uncertainty.

3.3 Effect of Elastic Modulus' Variation Coefficient on Surface Settlement

As is demonstrated in Fig. 7(a), we can see clearly that the mean of surface center settlement and arch settlement minor increase with the increase of E_s's COV, but the mean value is basically consistent with the deterministic analysis. However, the maximum settlement value has a considerable increase. From the Fig. 7(b) we can see that the surface center (arch) settlement can lager than deterministic analysis' result. The maximum value is 21.19 (17.73)% larger than deterministic analysis when the COV of E_s reach 0.35. Meanwhile, the percentage of maximum value has a stable increase with the increase of COV which suggests that the bigger E_s's COV, the larger maximum value.

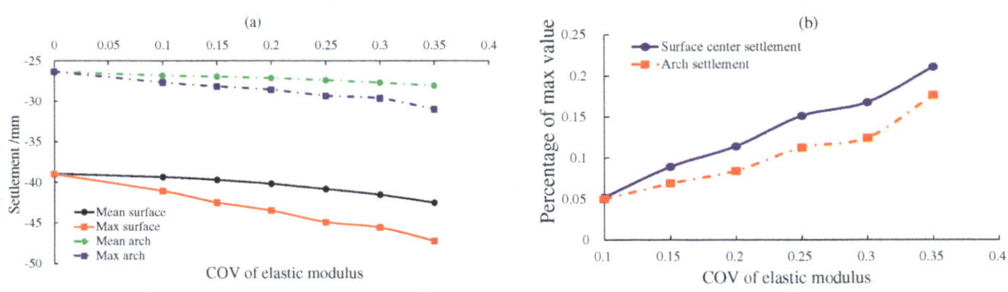

Fig. 7. The impact of COV of E_s on the surface center and arch settlement

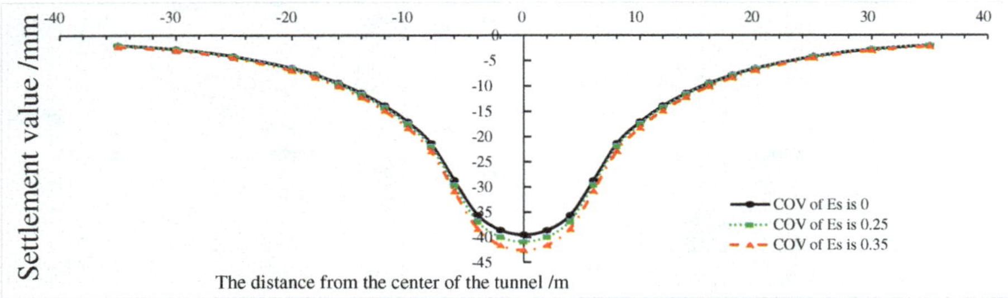

Fig. 8. The effect of COV of E_s on the surface settlement

Figure 8 shows that the change of surface settlement trough with the COV of E_s. In the process of COV from 0 to 0.25 to 0.35, the surface settlement value gradually become large. In a word, the larger the COV value, the greater the spatial variability of the soil. We should pay more attention if the tunnel passes through the soil with large variability of space, otherwise it will have a significant deformation.

3.4 Effect of Elastic Modulus' Variation Coefficient on Tunnel Convergence

Figure 9 shows the variation in the mean and maximum value of ΔD_v and ΔD_h with COV of E_s. The mean value of ΔD_v and ΔD_h is hardly influenced by the COV and keep in line with the deterministic analysis' result. It is clear from the Fig. 9(a) that the maximum value of ΔD_v sharply went up to 42.77 mm while the COV of E_s is 0.35. The value is 14.68% larger than the result of deterministic analysis (37.29 mm). There is not a great deal of difference between ΔD_h and ΔD_v. The maximum value of ΔDh sharply went up to 19.35 mm while the COV of E_s is 0.35. The value is 30.59% greater than the result of deterministic analysis (14.82 mm). The possibility of crossing the weak soil is much greater with the increase of E_s's COV. This is the reason that the maximum value of ΔD_v and ΔD_h is much larger.

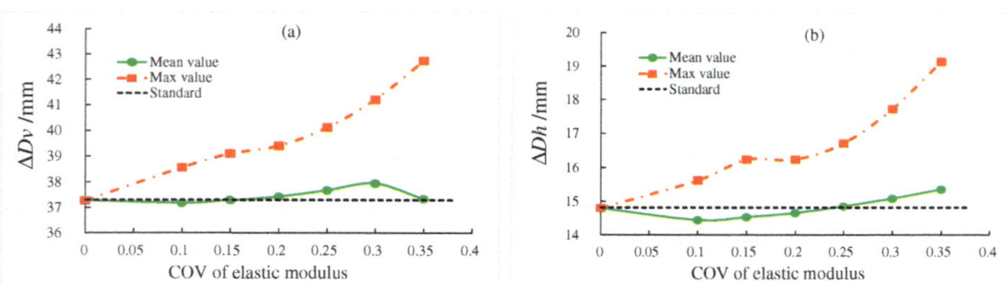

Fig. 9. Mean value of ΔD_v and ΔD_h with COV of elastic modulus

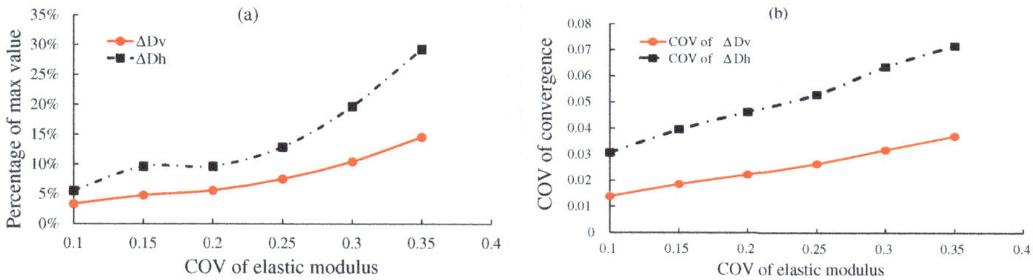

Fig. 10. Variation of convergence statistics with COV of elastic modulus: (a) Percentage of maximum value; (b) COV of ΔDv and ΔDh

As you can see from the Fig. 10(a), this is a cure graph which describes the trend of the percentage of maximum value of ΔD_v and ΔD_h with COV of elastic modulus. The percentage of maximum value of ΔD_v and ΔD_h was on a steady rise along with the increase of COV of E_s. The percentage of maximum value even reach about 30% (15%) of ΔD_h and ΔD_v. It is very dangerous for the stability of the tunnel. As is indicated in the Fig. 10(b), the higher COV of E_s, the higher COV of ΔD_h and ΔD_v. It suggests that the greater the variability of the soil parameters, the greater the variability of the tunnel convergence. It is easy to understand that the greater the variability of soil, the greater the change in the tunnel through soil, resulting in larger convergence changes. Therefore, a larger factor of safety should be adopted while the tunnel passes through the soil with large spatially variability. At the same time, tunnel reinforcement, grouting and other measures should be used to ensure the stability of the tunnel.

4 Conclusions

This study employs a random finite difference analysis for investigating the variation of settlement and tunnel convergence embedded in spatially varied soils. Random fields are generated and mapped into finite difference analysis to reveal the impact of the variation coefficient of elastic modulus. The effects of the load thickness, the load position and the COV of E_s were investigated. Below some findings from the research in this paper is summarized:

(1) The 3 m (54 kPa) is a important value that we need pay more attention to the load thickness of soil. The incremental value of settlement and tunnel convergence significantly increases when the load thickness lager than 3 m (54 kPa).

(2) The impact on surface settlement and tunnel convergence is rapidly weakening when the distance from load position to the centerline of the tunnel becomes from 0.5D to 3.5D. The impact is small while it larger than one particular value. In this case, it is 3.5D. Therefore, we should make the load position as far away from the tunnel as possible to minimize the impact on the tunnel in engineering.

(3) Ignoring the spatial variability of soil strength leads to a lower value of the surface settlement and tunnel convergence. The percentage of maximum value even reach

about 30% (15%) of ΔD_h and ΔD_v. It is very dangerous for the stability of the tunnel.

(4) The COV of ΔD_h and ΔD_v increases with the COV of E_s. It is suggested that the variability of the tunnel convergence increases with the variability of the soil parameters. That means the larger the variability of soil leads to a larger change in the tunnel through soil, and resulting in larger convergence changes.

References

Wang, R.-L., Zhang, D.-M.: Evolution of transverse deformation and assessment index for operation shield tunnel under surface surcharge in soft clay. Chin. J. Geotech. Eng. **35**(6), 1011–1092 (2013)

Zhang, M., Zhou, S., et al.: Analysis of influence of surface surcharge on subway shield tunnel under. Rock Soil Meh. **37**(8), 2271–2278 (2016)

Fenton, G.A., Griffiths, D.V.: Three-dimensional probabilistic foundation settlement. Geotech. Geoenviron. Eng. **131**(2), 232–239 (2005)

Huang, J.S., Lyamin, A.V., Griffiths, D.V.: Quantitative risk assessment of landslide by limit analysis and random fields. Comput. Geotech. **53**, 60–67 (2013)

Wu, Q.: Research on Influence of Shield Tunnel Structure on Ground Heaped Load. Shanghai Jiao Tong University, Shanghai (2012)

Dai, H.-W., Chen, R.-P., Chen, Y.-M.: Study on effect of construction loads on longitudinal deformation of adjacent metro tunnels. Chin. J. Geotech. Eng. **28**(3), 312–316 (2006)

Li, J.H., Zhou, Y., Zhang, L.L., Tian, Y., et al.: Random finite element method for spudcan foundations in spatially variable soils. Eng. Geol. **205**, 146–155 (2016)

Mroueh, H., Shahrour, I.: A simplified 3D model for tunnel construction using tunnel boring machines. Tunn. Undergr. Space Technol. **23**, 38–45 (2008)

Yu, Z., Huang, H., Lou, Y., et al.: Analysis of the shield tunnel constructed in soft soil under locally uniformly distributed load. Spec. Struct. **23**(4), 75–77 (2006)

Wang, T., Li, H., Xu, R.: Analysis of deformation of shield tunnel when vertical loading and unloading. Mod. Transp. Technol. **5**(3), 29–31 (2008)

Chen, X.: Soil Mechanics and Geotechnical Engineering. Qinghua University Press, Beijing (2003)

Influence of Jack Thrust on H&V Shield Behavior

Thi Ngoc Huynh[1,2], Salisa Chaiyaput[3], and Mitsukata Sugimoto[4(✉)]

[1] Graduate School of Nagaoka University of Technology, Nagaoka 940-2188, Japan
[2] Faculty of Civil Engineering, University of Technology, Vietnam National University Hochiminh City, Ho Chi Minh 740128, Vietnam
[3] Faculty of Engineering, King Mongkut's Institute of Technology Ladkrabang, Bangkok 10520, Thailand
[4] Department of Civil and Environmental Engineering, Nagaoka University of Technology, Nagaoka 940-2188, Japan
sugimo@vos.nagaokaut.ac.jp

Abstract. Shield tunnelling technology has been widely used in soft ground conditions to minimize the ground movement and the damage to neighboring structures around a tunnel. Due to limited underground space in urban area and for saving construction cost, multi circular face shield (MF shield) had been innovated to construct a twin tunnel at once. Furthermore, according to the more severe restriction of underground space use, horizontal and vertical variation shield method (H&V shield) was innovated, so that the cross section of a MF shield tunnel is changed from horizontal multi circular shape to vertical one, or vice versa. The H&V shield is manufactured by connecting two articulated shields at their rear bodies and is steered by jack, articulation mechanism and copy cutter, which can be operated individually at each body. The simulation algorithm of H&V shield behavior was developed, based on the kinematic shield model. This paper examines the influence of the jack thrust at the left and right bodies on H&V shield behavior, using the newly developed H&V shield simulation model. As a result, it was found that jack thrust at each body has an effect on shield behavior in term of the machine advance direction during construction process.

Keywords: Shield tunnelling · Simulation · Kinematic model
H&V shield · Jack force

1 Introduction

Due to limited underground space in urban area and for saving construction cost, multi-circular face shield (MF shield) (STA 2011a) had been innovated to construct a twin tunnel at once. Furthermore, according to the more severe restriction of underground space use, horizontal and vertical variation shield method (H&V shield) (Sonoda et al. 1992; STA 2011b) was innovated, so that the cross section of an MF shield tunnel is changed from horizontal multi-circular shape to vertical one or vice versa. The H&V shield is manufactured by connecting two articulated shields (Maidl et al. 2012) at their

R. Chen et al. (eds.), *Proceedings of the 2nd International Symposium on Asia Urban GeoEngineering*, Springer Series in Geomechanics and Geoengineering,
https://doi.org/10.1007/978-981-10-6632-0_50

rear bodies and is steered by jack force, articulation mechanism and copy cutter, which can be operated individually at each body. These steering options can generate rotating force around the shield axis, which can realize the construction of a spiral tunnel. Then the H&V shield enables reducing construction time and saving construction cost. As an example, the H&V shield is shown in Fig. 1. Moreover, the H&V shield behavior was simulated using the kinematic shield model for H&V shield, which has been developed from the one for the single circular shield (Sugimoto and Sramon 2002; Sramon et al. 2002; Sugimoto et al. 2007).

Fig. 1. H&V shield (STA 2011b)

In the previous study, the sensitivity analysis were carried out in term of effective rate of copy cutter length (CCL) in case of H&V shield based on this developed model to examine the influence of a tunnel boring machine operation on its behavior (Chaiyaput et al. 2015). Then, this paper examines the influence of the jack thrust at the left and right bodies on H&V shield behavior, using the newly developed H&V shield simulation model.

2 Methodology

2.1 Kinematic Shield Model

The kinematic shield model for single circular shield had been developed, taking into account shield tunnel engineering practices, i.e. the excavated area, the tail clearance, the rotation direction of the cutter disc, sliding of the shield, ground loosening at the shield crown, and the dynamic equilibrium condition (Sugimoto and Sramon 2002).

The model is composed of five forces: force due to the self-weight of the machine f_1, force on the shield tail f_2, force due to jack thrust f_3, force on the cutter disc f_4, and force on the shield periphery f_5, as shown in Fig. 2. The force due to the self-weight of the machine f_1 and the force on the shield periphery f_5 act on the both sections of two bodies. The force at the shield tail f_2 acts only on the rear section, whereas the force at

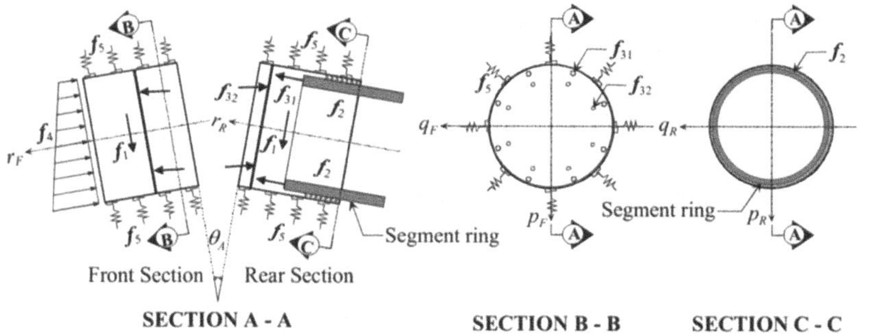

f_1 : self-weight of machine
f_{32} : force due to the articulate jack thrust
θ_A : articulated angle
f_2 : force on the shield tail
f_4 : force acting at the face
f_{31} : force due to the shield jack thrust
f_5 : force acting on the shield periphery

Fig. 2. Model of load acting on both front body and rear body

face f_4 is loaded on the cutter disc of the front section. The force on jack thrust f_3 is composed of the forces due to shield jack and articulated jack. The force on the shield periphery f_5 is due to the ground reaction force acting on the shield skin plate and the dynamic friction force on the shield skin plate.

The shield behavior during excavation can be obtained by solving the following equilibrium conditions of forces and moments acting on the shield.

$$\begin{bmatrix} \sum_{i=1}^{5} \left(F_{LFi}^{M} + F_{LRi}^{M} + F_{RLi}^{M} + F_{RRi}^{M}\right) \\ \sum_{i=1}^{5} \left(M_{LFi}^{M} + M_{LRi}^{M} + M_{RLi}^{M} + M_{RRi}^{M}\right) \end{bmatrix} = 0 \tag{1}$$

where $\boldsymbol{F}$ and $\boldsymbol{M}$ are the force and moment vectors respectively, the first subscripts L and R denote the left and right bodies of H&V shield respectively, the second subscripts F and R denote the front and rear sections of each body respectively, and the superscript M indicates the machine coordinate system. It is noted that the force due to articulated jack f_{32} is not necessary to calculate, since the summation of the force due to f_{32} is always zero.

2.2 Ground Reaction

The interaction force between the shield and the ground can be obtained by considering the coefficient of earth pressure K, which is a function of the distance between the original excavated surface and the shield skin plate U_{n} shown in Fig. 3, where the subscripts h and v are horizontal and vertical directions respectively, the subscripts *min*, *o* and *max* define minimum, initial and maximum respectively, and σ_{vo} is the overburden pressure. In Fig. 3, K at $U_{\mathrm{n}} = 0$ means the coefficient of earth pressure at rest and the gradient of K at $U_{\mathrm{n}} = 0$ represents the coefficient of subgrade reaction k. Furthermore, in the case of soft soil, the ground displacement to shield skin plate can be

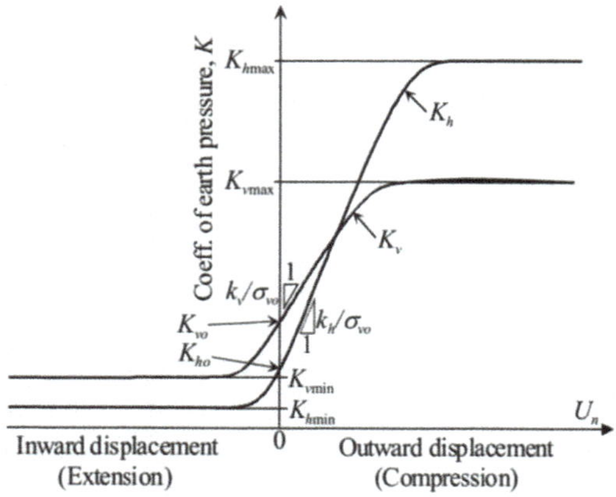

Fig. 3. Ground reaction curve

represented by U_n. On the other hand, for stiff soil, the self-stabilization of the ground can be expected, which causes the appearance of the gap between the shield skin plate and the excavated surface. This consideration is similar to the touching problem in finite element analysis.

2.3 Analysis Condition

The H&V shield machine is divided into left and right bodies. The alignment in the left body is straight on the horizontal and vertical planes, while the right body presents a rotation around the left body. The dimensions of H&V shield in this study are illustrated in Fig. 4. The operations and the ground conditions were assumed as Table 1.

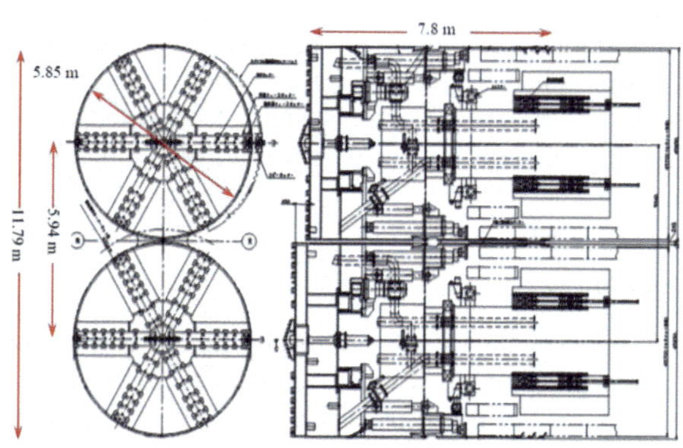

Fig. 4. The dimensions of H&V shield

Table 1. Shield operation and ground condition for simulation

Component	Value
Shield	
Diameter of machine	5.85 m (left and right)
Length of machine	8.25 m (left) and
	7.80 m (right)
Shield operation	
Excavation velocity	2 cm/min
Articulated angle	0° (left) and
	4° (right) during spiral
Ground	
Tunnel center	GL-28.9 m
Ground water level	GL-3.9 m
SPT-N value	30
Earth pressure	Effective earth pressure method

2.4 Sensitivity Analysis

The copy cutter length 200 mm and the articulated angle are applied for the right body to increase the excavation effect and to control the alignment during excavation by ground reaction force (Chaiyaput et al. 2015). The analysis cases were set based on the share of the total jack thrust at each body as shown in Table 2, where F_{3rL} and F_{3rR} are the share of the jack thrust of the left body and right one, respectively. Here the total amount of the jack thrust are same for all of the cases.

Table 2. The share of the jack thrust of each body

Case	1	2	3	4
F_{3rL}	0.8	0.6	0.4	0.2
F_{3rR}	0.2	0.4	0.6	0.8

3 Results and Discussions

In Case 3 and Case 4, the shield moved back during the analysis and finally the analysis was terminated. This is considered as follows:

1. The overcutting at the left body is only 10 mm since the tunnel alignment is straight line, while at the right body, the overcutting is done by copy cutter and the articulation is also applied. Then the friction resistance between the ground and the skin plate at the left body is larger than that at the right body.
2. The jack thrust at the right body is larger than that at the left body. Then since the large horizontal moment to left turn is generated, the H&V shield rotates to leftward.

3. Due to the small overcutting at the left body, the frictional resistance around the skin plate at the left body increases as the shield rotates.
4. Finally the total jack thrust is short against the frictional resistance around the skin plate and the face pressure, then the shield moved back.

Therefore, the results in Case 1 and Case 2 were shown in next section.

3.1 H&V Shield Behaviour

The simulation results are shown in Fig. 5, where "Distance" is the length of the trace of the shield, ϕ_y is yawing angle (+: right turn), ϕ_p is pitching angle (+: downward rotation), ϕ_r is rolling angle (+: clockwise rotation viewed from tail), and v_s is the shield velocity during excavation. The planned and calculated values are plotted for comparison.

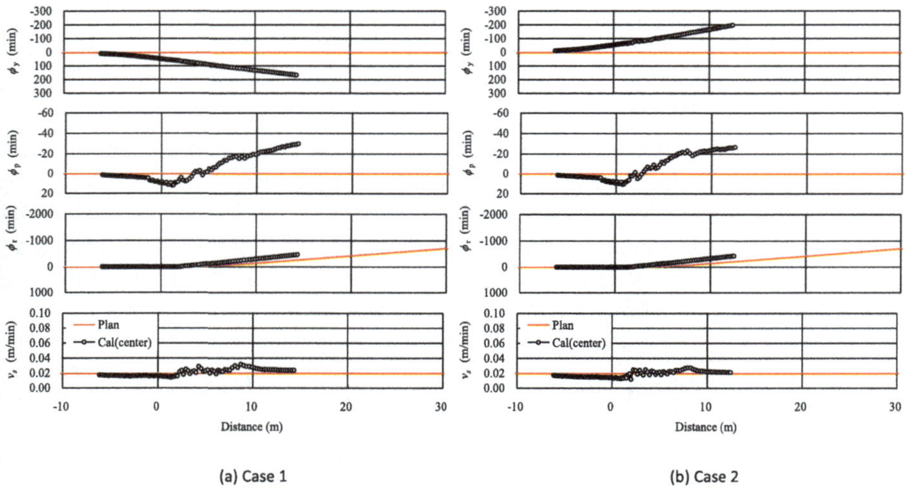

Fig. 5. Results of shield behavior

Comparing both Case 1 and Case 2 in Fig. 5, the followings were found:

1. The total advanced distance in Case 1 is a little bit longer than that in Case 2.
2. The yawing angle ϕ_y in Case 1 shows that the shield takes a right turn, and that in Case 2 shows that the shield takes a left turn, respectively. Furthermore, the rotation angle of the shield at last in Case 1 is a little bit smaller than that in Case 2.
3. The tendency of the pitching angle ϕ_p and the rolling angle ϕ_r is almost same at the both cases. The ϕ_p is close to zero before the shield enters to the spiral part, but increases gradually at the spiral part. The ϕ_r follows the planned ϕ_r.
4. The v_s in Case 1 is a little bit faster than that in Case 2.

These can be explained as follows: in Case 1, the absolute value of the ϕ_y at last is a little bit smaller than that in Case 2; therefore, the frictional resistance around the skin plate is smaller, especially at the left body; the velocity of the shield is larger; and the advanced distance of the shield is longer, compared with those in Case 2.

3.2 Normal Ground Displacement

The distance between the initial excavated surface and the shield skin plate in normal direction against the shield skin plate U_n is shown in Fig. 6, where the left side and the right side of the figure are the left body and right body, respectively, and the shield skin plate is unfolded as a flat plate, that is, the vertical axis shows the length of the shield, and the horizontal axis represents the circumference of the shield. From Fig. 6 the followings were found:

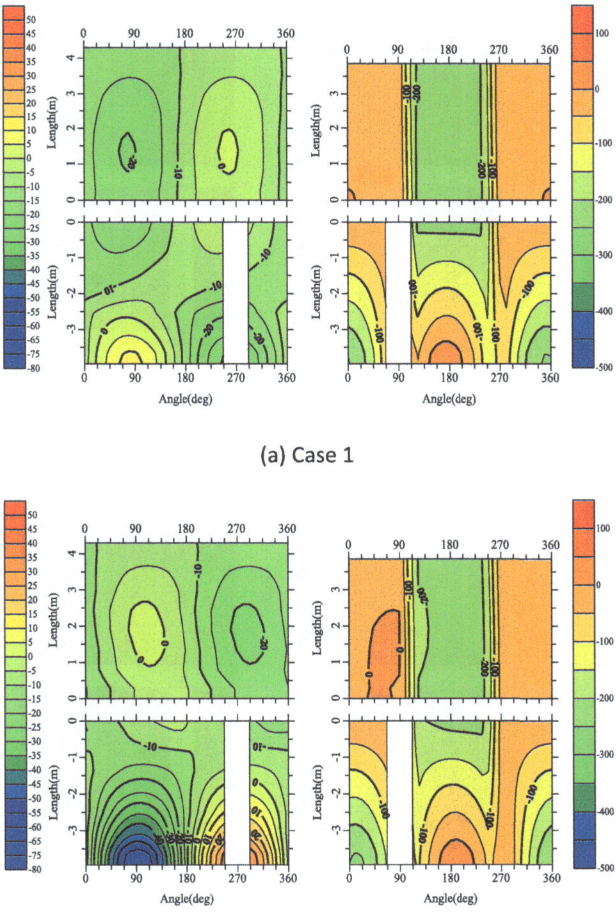

(a) Case 1

(b) Case 2

Fig. 6. Distance from the initial excavated surface to the skin plate (mm)

1. In Case 1, at the left body, the passive state of the ground is developed drastically at the left spring line in the rear part (90°), and the active state occurred at the opposite site.
2. In Case 1, at the right body, the passive state of the ground appears below the spring line of the front body and around the crown at the shield tail of the rear body.
3. In Case 2, the magnitude of the normal distance U_n is a little bit larger than that in Case 1.
4. In case of the left body, the tendency of the distribution of the active state and the passive state in Case 2 is reverse against that in Case 1.
5. The magnitude of the U_n in the left body is much smaller than that of the right body.

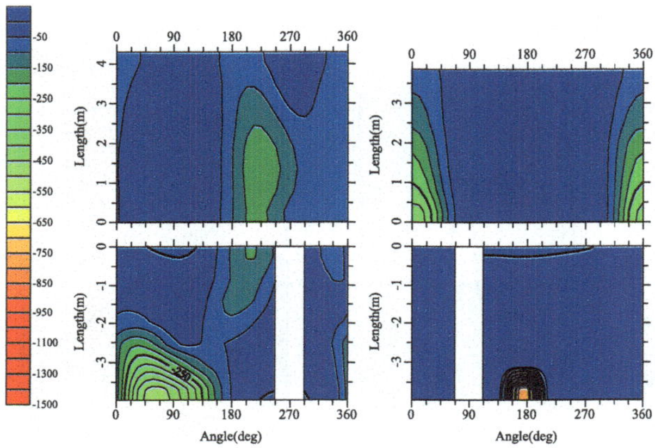

(a) Case 1

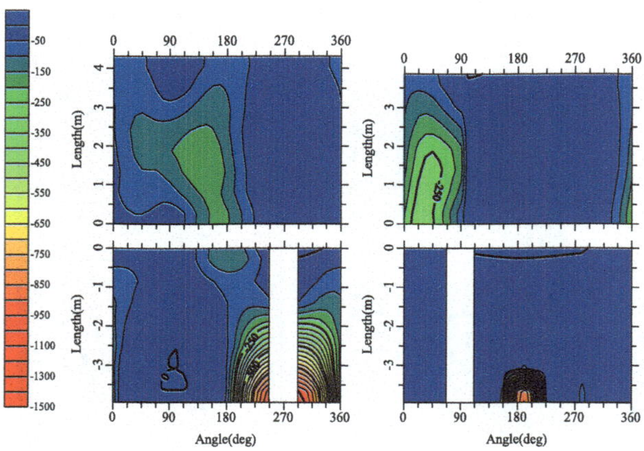

(b) Case 2

Fig. 7. Normal effective earth pressure on the shield periphery (kN/m^2)

There can be consider as follows:

1. The shield rotates to rightward and leftward in Case1 and Case 2, respectively.
2. The overcutting by copy cutter and articulation angle are applied to the right body, while those are not applied to the left body.
3. The overall distribution of the U_n is reasonable from the viewpoint of the geometric condition of the interaction between the shield skin plate and the excavated surface of the ground.

3.3 Normal Effective Earth Pressure

The effective normal earth pressure acting on the shield periphery σ'_n is plotted in Fig. 7. σ'_n is determined by applying U_n to the ground reaction curve in Fig. 3. From Fig. 7, the followings were found:

1. The normal effective earth pressure σ'_n is close to zero at the main part.
2. In the case of the left body, the distribution of the σ'_n in Case 1 is reverse against that in Case 2.
3. In the case of the right body, the distribution of the σ'_n show the same tendency for the both Case 1 and Case 2.
4. There are the reflection of the distribution of the U_n and the stiffness of the ground.

4 Conclusions

This paper describes the results of the simulations of the H&V shield behavior, taking the share of the jack thrust at each body as the analysis parameter. As a result, the following conclusions can be made:

1. In case of H&V shield, the share of jack thrust on each body gives much influence to the shield behavior, especially in yawing angle ϕ_y and velocity v_s, even the total amount of the jack thrust is constant.
2. As the jack thrust at the right body increases, the shield takes a left turn, the frictional resistance around the left body increases, then the shield velocity v_s decreases and finally the shield moved back in the analysis.
3. With the changes of the share of jack thrust on each body, the distribution of the U_n and σ'_n are reasonable from the view point of the geometric condition between the ground and the skin plate. Therefore, it is considered that the proposed model could simulate the H&V shield behavior reasonably.

References

Chaiyaput, S., Huynh, T.N., Sugimoto, M.: Influence of copy cutter length on H&V shield behavior. Jpn. Geotech. Soc. Spec. Publ. **2**(43), 1528–1531 (2015). https://doi.org/10.3208/jgssp.SEA-14

Maidl, B., Herrenknecht, M., Maidl, U., Wehrmeyer, G.: Mechanized shield tunnelling, 2nd edn. Ernst & Sohn, Berlin (2012)

Shield Tunnelling Association in Japan: Technical note on multi-circular face shield method, 6th edn. STA, Japan (2011a). (in Japanese)

Shield Tunnelling Association in Japan: Technical note on Horizontal & vertical variation shield method, 6th edn. STA, Japan (2011b). (in Japanese)

Sonoda, T., Hagiwara, H., Osaki, H., Noguchi, T., Nakamura, M.: Construction of underground space by a new shield tunnelling method: spiral tunnelling and ramification of multi-circular face shield. Tunn. Undergr. Space Technol. **7**(4), 355–361 (1992). https://doi.org/10.1016/0886-7798(92)90065-P

Sramoon, A., Sugimoto, M., Kayukawa, K.: Theoretical model of shield behavior during excavation: II. Application. J. Geotech. Geoenviron. Eng. ASCE **128**(2), 156–165 (2002). https://doi.org/10.1061/(ASCE)1090-0241(2002)128:2(156)

Sugimoto, M., Sramoon, A.: Theoretical model of shield behavior during excavation: I. Theor. J. Geotech. Geoenviron. Eng. ASCE **128**(2), 138–155 (2002). https://doi.org/10.1061/(ASCE)1090-0241(2002)128:2(138)

Sugimoto, M., Sramoon, A., Konishi, S., Sato, Y.: Simulation of shield tunneling behavior along a curved alignment in a multilayered ground. J. Geotech. Geoenviron. Eng. ASCE **133**(6), 684–694 (2007). https://doi.org/10.1061/(ASCE)1090-0241(2007)133:6(684)

A Fuzzy Neural Network System for Architectural Foundation Selection

Bin-Hui Ma[1,2(✉)], Guo Zhong[1], Qiu-Nan Chen[1,2], Cheng-Bin He[1], Yong Lei[1,2], and Bing-Chu Chen[1]

[1] School of Civil Engineering, Hunan University of Science and Technology, Xiangtan, China
bhmahd@126.com

[2] Hunan Provincial Key Laboratory of Geotechnical Engineering for Stability Control and Health Monitoring, Hunan University of Science and Technology, Xiangtan, China

Abstract. Foundation engineering is a comprehensive geotechnical engineering problem, which is costly and time-consuming. And the construction quality is difficult to observe and determine, so is the strengthening and repairing work later. Therefore, people attach greater importance to the selection before building the foundation. This paper, based on the statistical analysis of the forms, characteristics and applicable scopes of the building foundation, establishes the hierarchical factor system and introduces the theory of fuzzy system in order to improve the traditional AHP method and to propose a more comprehensive fuzzy evaluation system combining more accurate FAHP method and interval theory together. Additionally, the basic principles and algorithm of artificial neural network are analyzed and expounded, then combine neural network and fuzzy system to build a fuzzy neural network system for architectural foundation selection. Furthermore, simulated testing and performance analysis for the two models are conducted so as to offer some guidance for the project selection.

Keywords: Architectural foundation selection · Fuzzy evaluation Interval theory · Neural network

1 Introduction

Foundation engineering is the basis of architectural structure, passing the load of the upper structure on to the foundation. According to statistics, the price for building the foundation accounts to 20% to 30% of the total cost, and the construction time up to 30% of the total period. Meanwhile, the foundation bears the major load and has greater underground depth, thus requiring a demanding bearing capacity, compression resistance, and stability. And its construction is more subject to the restriction of the surrounding environment and geological conditions as well. The design of foundation will exert a significant impact on the cost, schedule, quality and safety of the entire engineering. Hence, a reasonable and optimum selection of the foundation, on the premise of meeting the normal function and safety conditions, at the same time to be

R. Chen et al. (eds.), *Proceedings of the 2nd International Symposium on Asia Urban GeoEngineering*, Springer Series in Geomechanics and Geoengineering,
https://doi.org/10.1007/978-981-10-6632-0_51

well-designed, energy-saving and cost-saving, has become an urgent problem in engineering and scientific research.

Three approaches can be applied in selecting the type of foundation and its optimal scheme, and they are qualitative analysis, quantitative analysis, and the combination of qualitative and quantitative analysis. Qualitative analysis mainly relies on the practical experience, subjective judgment and analysis capabilities of the predicting personnel to deduce the nature and trends of certain object. The common methods include expert evaluation and the Delphi method, etc. While quantitative analysis, based on statistical data to build the mathematical model and calculate all indexes and numerical value of the target through certain mathematical model. AHP, and multi-objective decision model are the common methods. The major method for the combination of quantitative and qualitative analysis is expert system.

Foreign schemes and procedures of foundation selection, mainly including expert system GSS, AHP, and logical reasoning mechanism, have been gradually commercialization. For example, Ulshafer developed pile selection module based on expert system FLOPS, using a large number of database and adopting fuzzy comprehensive evaluation to select foundation [1]. Professor Emiroglu in Firat University of Turkey, designed software DTSA ES for dam foundation selection in silt areas [2].

While domestical foundation selection is also developing step by step with a focus on the optimization of pile foundation. For example, He from Tongji University carried out research and discussions on the foundation in Deep-water for FAHP [3]. While Wang from Chang'an University applied general AHP and neural network to run a comprehensive analysis for the selection of building [4]. In addition, Zhang Jinhui of Shanghai Jiaotong University, and Luo Wei of Zhejiang University used FAHP to select an optimum model for the pile foundation.

The selection methods at home and abroad still take the traditional AHP as the theoretical basis and then make summarization and analogy of the collected information. However, due to the different natures of decision-making and forecasting, and the changeable influence factors for their relevance.

In reality, it is not feasible to adopt all the indexes and evaluation factors at present. In view of the above problems, this paper will improve the traditional AHP method and establish a fuzzy optimization model combining qualitative and quantitative indexes for data processing and program decision-making, and the fuzzy neural network is introduced in order to get better option for decision-making, and provide some recommendations to engineering and related scientific research [5].

2 Analysis of Building Foundation

2.1 Characteristic Analysis of Foundation Selection

The scheme for foundation selection should contain both specific index like cost and construction period and those vague ones like construction experience and influence on the surroundings. Combination of the two indexes not only needs a large number of exact data but also rich experience of experts.

2.2 Hierarchical Structure

Therefore, this paper will conduct a comprehensive analysis of such factors as features of upper structure, hydrogeological conditions, construction feasibility and environmental impact so as to make the optimal selection through all-round contrast. First of all, make the optimal foundation as the overall objective. Then choose the construction factors, geological conditions, construction feasibility, economical efficiency, safety concern and environmental adaptability according to the characteristics of this project and take them as the criteria. Finally, further subdivide them into 27 criteria [4].

3 Fuzzy Optimal Decision Model

The building foundation has a complex condition, close relation with geographical conditions, and a great amount of information and empirical data, thus it cannot adapt to the current decision-making by relying merely on the engineering experience and limited judgment. Besides, multi-attribute decision making theory becomes increasingly of ambiguity and uncertainty in information. Therefore, this paper will adopt the fuzzy optimal selection model combining both qualitative and quantitative indexes based on the improved traditional AHP, to process the data and decide the optimizing scheme.

3.1 Hierarchical Structure and Initial Evaluation of Weights

Saaty, an American operational research expert, proposed AHP decision approach in 1970's to make decisions on the complicated problem [6]. It is flexible, systematic and easily operational and concludes solutions through building judgment matrix layer-by-layer (Table 1).

Table 1. Indexes of foundation decision

Foundation selection		
Construction factors	Geological conditions	Construction feasibility
Function Grade Height Stories Structure type Load distribution uniformity Settlement sensibility	Soil layer distribution Soil liquefaction bearing stratum uniformity Substratum condition Underground water condition Regional foundation	Construction facility Transportation condition Construction difficulty Construction experience
Economical efficiency	Safety concern	Environmental adaptability
Cost Period	Integral settlement Integral collapse Differential settlement Anti-seismic property	Surroundings Available transformation Noise and pollution

This paper will combine the hierarchical factor system 1-6-27 listed in the Table 2 and AHP to analyze the foundation selection.

Table 2. Nine-demarcation method

Scale	Description
1	Two elements, equally important
3	Two elements, one slightly important
5	Two elements, one obviously important
7	Two elements, one overwhelmingly important
9	Two elements, one extremely important
2, 4 6, 8	Compromised scale of two adjacent scales
Reciprocal	Plan to plan is a_{ij}, or $1/a_{ij}$ otherwise

Assume a random $n \times n$ matrix $a_{ij} > 0, a_{ij} = 1/a_{ij}$ and call it reciprocal matrix. AHP judgment matrix A is formed to compare in pairs, where a_{ij} stands for the importance of plan i to plan j.

$$\mathrm{A} = (a_{ij})_{n \times n} = \begin{pmatrix} a_{11} & \cdots & a_{1n} \\ \vdots & \ddots & \vdots \\ a_{n1} & \cdots & a_{nn} \end{pmatrix} \tag{1}$$

Psychological research and practice show that the variation of the differential extremity in psychology compared at the same time is 7 ± 2, thus nine-demarcation method is adopted for measurement [7].

Assume A is positive reciprocal matrix, if for any random element, there is $a_{ik} \cdot a_{kj} = a_{ij}$, then we call A is consistent matrix.

For each paired comparison matrix, the original method is applied to calculate its maximum root $\lambda_{\max}$ and feature vector $W = [w_1, w_2, \cdots w_n]^T$ and have them checked, thus the consistent index CI along with its random average RI and the ratio CR are introduced.

$$CI = \frac{\lambda_{\max} - n}{n - 1} \tag{2}$$

$$RI = \frac{CI_1 + CI_2 + \cdots + CI_n}{1000} \tag{3}$$

Satisfactory consistency will be achieved when it meets the follows ing formula, namely, the allocation of the weight coefficient is appropriate.

$$CR = \frac{CI}{RI} < 0.10 \tag{4}$$

Final step is to rank order of the layers and check the total consistency.

$$b_i = \sum_{j=1}^{m} b_{ij}a_j,\ i = 1, \cdots, n \tag{5}$$

$$CR = \sum_{j=1}^{m} CI(j)a_j / \sum_{j=1}^{m} RI(j)a_j \tag{6}$$

3.2 Applying Fuzzy AHP to Finalize the Weights

Nine-demarcation method is commonly used in the traditional AHP for decision-making. However, it is of low accuracy and doesn't work with fuzzy theory. In 1965, Professor Zadeh, a famous American expert on cybernetics, published a thesis called Fuzzy Sets in the international magazine Information and control, and initiated the concept of fuzzy set [8, 9]. And he introduced fuzzy math and carried it out to build the Fuzzy AHP to overcome above shortcomings of traditional AHP. In comparison, Fuzzy AHP is more convenient to determine matrix, and easy to amend the consistency, but more troublesome for evaluation. Therefore, based on the traditional AHP, this paper adopts the Scale Transformation method to establish a fuzzy judgment matrix.

Assume there is a fuzzy matrix R, if $r_{ij} + r_{ji} = 1$, then we call matrix R fuzzy complementary matrix. The Scale Transformation method proposed by Professor Xu will be used here [10].

Assume $A = \left(a_{ij}\right)_{n \times n}$ is the judgment matrix by the traditional nine-demarcation method, then through the following conversion formula:

$$r_{ij} = \frac{a_{ij}}{a_{ij} + 1} \tag{7}$$

The complementary judgment matrix $A' = \left(a'_{ij}\right)$ could be obtained. It means to compare in pairs through the preferential relation matrix R in the Fuzzy AHP.

$$\mathrm{R} = (r_{ij})_{n \times n} = \begin{pmatrix} r_{11} & \cdots & r_{1n} \\ \vdots & \ddots & \vdots \\ r_{n1} & \cdots & r_{nn} \end{pmatrix} \tag{8}$$

The relation between the importance of the two factors and the weight between them can be figured out by the following formula (Table 3):

Table 3. Corresponding indexes

Indexes	Grade 1	Grade 2	Grade 3	Grade 4	Grade 5
Intervals	80–100	60–80	40–60	20–40	0–20

$$r_{ij} = 0.5 + \left(w_i - w_j\right)\beta,\ 0 < \beta \leq 0.5 \tag{9}$$

Assume R is a fuzzy matrix, if for any element, there is $r_{ij} = r_{ik} - r_{jk} + 0.5$, then matrix R is called fuzzy consistent matrix. In this paper, the additive consistency is applied to transform consistency and to strengthen the adjustment of preferential relation matrix into fuzzy consistent matrix.

$$r_i = \sum_{k=1}^{n} r_{ik},\ i = 1, 2, \cdots, n \tag{10}$$

$$f_{ij} = \frac{r_i - r_j}{a} + 0.5,\ a = 2(n-1) \tag{11}$$

Meanwhile, the following two standards for conformance check are presented as follows [11]:

$$M = \max\left\{\left|r_{ij} - f_{ij}\right|\right\} \tag{12}$$

$$e = \frac{1}{n}\sqrt{\sum_{i=1}^{n}\sum_{j=1}^{n}\left(r_{ij} - f_{ij}\right)^2} \tag{13}$$

When $M < 0.2$ and $e < 0.1$ the fuzzy complementary matrix is more reasonable; while when $M > 0.2$ and $e > 0.1$, the consistency of fuzzy complementary matrix is poor, and needs re-evaluating.

The priority of fuzzy consistent matrix can be determined by the following formula [12]:

$$w_i = \frac{1}{n} - \frac{1}{2a} + \frac{1}{na}\sum_{j=1}^{n} r_{ij},\ (i = 1, 2, \cdots, n), a \geq \frac{n-1}{2} \tag{14}$$

Finally, the sequence of matrix R $W = (w_1, w_2, \ldots, w_n)^T$ is calculated to deduce the weights of each factor and convert them into the overall weight in light of the overall objective.

3.3 Evaluation from Interval Theory

Due to the complexity and systematicness of various evaluations, it's rather difficult to achieve accurate measurement and estimation by point value. Russian scholars H. Grell, K. Maruhn, and W. Rinow were among the first to use interval computation to solve the values of traditional expressions and propose the algorithm of interval numbers. Later, Japanese scholar Teruo Sunaga systematically elaborated and extended the algorithm.

This paper introduces the interval theory for property evaluation, thus changing the limitations by the previous point value, increasing the accuracy of decision-making, making the model more in line with actual engineering and enhancing its efficiency.

Assume R is the real number field, then the definition of the interval number is as follows: $a = [a^-, a^+] = \{x|a^- \leq x \leq a^+, a^- a^+ \in R\}$. The midpoint of the interval number can be calculated as in $a = (a^- + a^+)/2$, width as in $l = a^- + a^+$ and algorithms as in $\lambda A = [\lambda a^-, \lambda a^+]$.

The establishment of a evaluation grade domain, and the introduction of interval fuzzy numbers enable each grade to have a corresponding pair of interval fuzzy numbers. The corresponding range of qualitative indexes determined by the five-level partitioning method is offered below:

Three features are showed in the process of judging: one is that the higher the index is, the more optimizing it is; another is the lower the index, the more optimizing, and yet another is that moderation is optimal. However, there are usually two types of index, benefit type and cost-effectiveness type. After marking the qualitative indexes of all factors of the projects and entering the quantitative indexes, the fuzzy relation matrix R will be obtained.

Then the matrix needs to be standardized to obtain standard relational matrix R′. The standard dimensionless method proposed by scholars Xu and Da is adopted and it's also the most widely used method at present [13].

Benefit type:

$$\begin{cases} r_{ij}^- = a_{ij}^- / \sqrt{\sum_{i=1}^{n} (a_{ij}^+)^2} \\ r_{ij}^+ = a_{ij}^+ / \sqrt{\sum_{i=1}^{n} (a_{ij}^-)^2} \end{cases} \tag{15}$$

Cost-effectiveness type:

$$\begin{cases} r_{ij}^- = (1/a_{ij}^+) / \sqrt{\sum_{i=1}^{n} (1/a_{ij}^-)^2} \\ r_{ij}^+ = (1/a_{ij}^-) / \sqrt{\sum_{i=1}^{n} (1/a_{ij}^+)^2} \end{cases} \tag{16}$$

3.4 Comprehensive Priority for Decision

Weighted arithmetic average (WAA) operator is adopted in this paper to address the weighted index of the standard relational matrix R′.

Assume $\tilde{r}_{ij}$ is the index value of a set of interval numbers, then its weighted attribute value can be calculated as follows:

$$\tilde{z}_i = \sum_{i=1}^{m} w_i \tilde{r}_{ij} \tag{17}$$

Finally, the possibility degree matrix put forward by Japanese scholars Nalahara et al. is applied for a comprehensive priority of the interval numbers. Possibility degree formula for comparing the size of the interval number is presented below [14]:

$$P(A > B) = \min\left\{\max\left(\frac{a^+ - b^-}{l(a) + l(b)}, 0\right), 1\right\} \tag{18}$$

While $A = [a^-, a^+], B = [b^-, b^+], l(a) = a^+ - a^-, l(b) = b^+ - b^-$.

When ranking multiple interval numbers, it requires comparing in pairs, and establishing the fuzzy complementary matrix.

$$P = \begin{bmatrix} 0.5 & p_{12} & \cdots & p_{1n} \\ p_{21} & 0.5 & \cdots & p_{2n} \\ \cdots & \cdots & \cdots & \cdots \\ p_{n1} & p_{n2} & \cdots & 0.5 \end{bmatrix} \tag{19}$$

Where p_{ij} represents the possibility degree of $a_i > a_j$. The above formula can be converted to fuzzy consistent matrix for comprehensive priority so that an ultimate decision matrix $[w_1, w_2, w_3, \ldots, w_n]$ can be reached, and then decisions can be made combining with practical cases.

4 Fuzzy BP Neural Network

4.1 BP Neural Network

Artificial Neural Network (ANN) is the simplified model of human central nervous system. BP neural network usually refers to the multilayer feed forward neural network based on the error backpropagation algorithm, namely, BP algorithm. Invented by a research group led by D.E. Rumeiliart and J.L. McCelland, it has currently become the widely used learning algorithm in neural network. Statistics show that about 90% applications in neural network are based on this algorithm [15–17].

BP neural network consists of an input layer, one or more hidden layer and a output layer. Every layer contains numerous nerve cells, with upper and lower ones connecting each other according to their weight. But nerve cells in the same layer have no connection. The structure of a typical three-layer BP neural network is shown in Fig. 1.

BP neural network can well reflect and evaluate the complex non-linear relationship between factors and foundation types for such characteristics as fast training speed, simple structure and strong mapping capacity of non-linear relationship. Furthermore, with its strong self-learning and association ability, it is of high accuracy and requires little manual intervention and expertise.

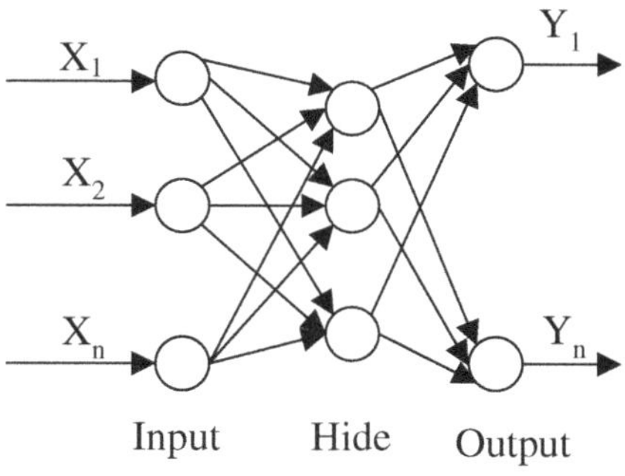

Fig. 1. Neural network structure

Compared with neural network, fuzzy system is superior in its easy understanding of reasoning process, well-applied expertise and lower sample requirements. However, it's inferior in its much manual intervention, slow reasoning and lower accuracy. If the two are combined together, they can be complementary.

4.2 Building Fuzzy BP Neural Network

Fuzzy mathematics applies mathematical methods to research and processes fuzzy phenomena. Based on fuzzy mathematics, this paper sets up the fuzzy BP neural network, a new way to process uncertain and inaccurate problems and a powerful tool to describe brain's processing of fuzzy information.

After initializing the network, the following calculations are needed [18].

Output calculation of hidden layer:

$$H_j = f\left(\sum\nolimits_{i=1}^{n} w_{ij}x_i - a_i\right) \quad j = 1, 2, \cdots, l \tag{20}$$

Output calculation of output layer

$$O_k = \sum\nolimits_{i=1}^{l} H_j w_{jk} - b_k \quad k = 1, 2, \cdots, m \tag{21}$$

Error calculation

$$e_k = Y_k - O_k \quad k = 1, 2, \cdots, m \tag{22}$$

Weights renewal

$$\begin{aligned} w_{ij} &= w_{ij} + \eta H_j(1 - H_j)x(i)\sum\nolimits_{k=1}^{m} w_{jk}e_k \quad i = 1, 2, \cdots, m \\ w_{jk} &= w_{jk} + \eta H_j e_k \quad j = 1, 2, \cdots, l \end{aligned} \tag{23}$$

Threshold renewal

$$\begin{aligned} a_j &= a_j + \eta H_j(1 - H_j)\sum\nolimits_{k=1}^{m} w_{jk}e_k \quad j = 1, 2, \cdots, l \\ b_k &= b_k + e_k \;\; k = 1, 2, \cdots, m \end{aligned} \tag{24}$$

Theoretically, BP learning algorithm is capable of continuous mapping of approaching arbitrary non-linear relationship, but some weaknesses do exist like rather slow rate of convergence, poor numerical stability and difficult adjustments in such parameters as learning rate, momentum coefficient and initial weight value. Whereupon, this paper adopts LM algorithm, a learning algorithm of non-linear neural network, to effectively overcome above weaknesses of BP algorithm.

Network training algorithm is Levenberg-Marquardt, and its formula is:

$$x_{k+1} = x_k - \left[J^T J + \mu I\right]^{-1} J^T e \tag{25}$$

The number of nodes in the hidden layer of BP neural network has a great influence on the accuracy for prediction. With few nodes, the network can't work well, and more training is needed and the training precision is affected. While With too many, the network overfits easily. Therefore, the best number of nodes in the hidden layer can be calculated through the formula below:

$$\begin{cases} l < n - 1 \\ l < \sqrt{(m + n)} + a \\ l = \log_2 n \end{cases} \tag{26}$$

And in this paper, the hidden layer is mainly adopted, which has strong ability in generalization and is of higher precision for prediction.

5 Project Case

5.1 Overview of the Project

The project is planned to build at the junction of the central avenue and the crossroad in Yanghuyuan wetland park of Changsha, about 100 m south to Qi River and 800 m west to Xiang River. The original landscape of this area belongs to the second terrace of Xiang River, with great fluctuation in terrain. The standard height of drilling hole is 29.39–35.26 m, with the largest fall of about 6 m. This route is located in the lowland of Changsha, the secondary syncline of Yong'an multiple syncline, with its axis stretching 40 to 45 degree northeast up to 60 km. Under the constant arch faulting, a flat-folded red basin made up of geodepression deposits came into being. And the stratum is flat with rocks orientation about 300–330°∠10–20.

5.2 Fuzzy Optimum Selection

According to the hierarchical factor system, the weight of each layer has been determined respectively and assessed in accordance with the nine scales, the result is as follows:

A: 0.1687, 0.1720, 0.1467, 0.2020, 0.2020, 0.1087
B1: 0.0941, 0.1615, 0.1060, 0.1100, 0.1100, 0.1094, 0.1534, 0.1556
B2: 0.2173, 0.1280, 0.1867, 0.1867, 0.1533, 0.1280
B3: 0.1898, 0.2269, 0.3730, 0.2103
B4: 0.6667, 0.3333
B5: 0.2556, 0.2556, 0.3426, 0.1463
B6: 0.4375, 0.3333, 0.2292

Choose P1 strip foundation, P2 Pile-supported stand, P3 pile-raft foundation, P4 Flat raft foundation, P5 Beam plate raft foundation, P6 box foundation in turn and evaluate them comprehensively, the fuzzy score of the scheme is:

$$P1(0.3354, 0.4637) \quad P2(0.3658, 0.5020) \quad P3(0.3514, 0.5186)$$
$$P4(0.3306, 0.4467) \quad P5(0.3265, 0.4415) \quad P6(0.3152, 0.4585)$$

Standardize the matrix to obtain the possibility degree matrix, and then rank them to obtain the comprehensive evaluation value.

$$0.1635, \quad 0.1890, \quad 0.1874$$
$$0.1547, \quad 0.1509, \quad 0.1545$$

5.3 Fuzzy BP Neural Network

The neural network used in this case has double hidden layers, with 27 elements in and 1 out. The sample data is taken from the foundation selection report of Yanghu area.

The invoking format for loaded data is

load date input output

The invoking format of fan-in network for program building is

net=newff(input,output,[S1,S2,...SN]

{'tansig','logsig'},'trainlm')

The invoking format for network training program is

[net,tr]=train (net,input,output)

When the network training program maintains stable, the network will be simulated, and the invoking format for checking the predicting ability is

result=sim(net,test)

Related parameters of training are

net.trainParam.epochs=100

net.trainParam.lr=0.1

net.trainParam.goal=0.0004

The nodes in the hidden layer using the formula are 27-15-1-1. And after running the simulation test when the training stabilizes, the comprehensive evaluation value will be attained.

$$\begin{matrix} 0.1653, & 0.1907, & 0.1863 \\ 0.1567, & 0.1567, & 0.1568 \end{matrix}$$

e = –0.0006, –0.0018, –0.0006, 0.0024, 0.0024, perf = ms(e) = 3.1472e–06.

5.4 Comprehensive Analysis

After comprehensive comparison, both fuzzy optimizing model and neural network model take pile-supported stand foundation as the premier scheme, and piled raft foundation as the alternative. And this result matches with the actual engineering, as showed in Table 4.

Table 4. Result comparison

Grade	Fuzzy optimizing scheme	Neural network scheme	Actual scheme
Grade 1	Pile-supported stand	Pile-supported stand	Pile-supported stand
Grade 2	pile-raft foundation	pile-raft foundation	
Grade 3	Flat raft foundation	Raft Combination	

The comprehensive evaluation value, which conforms to check consistency, is obtained combining the weights and assessment of fuzzy optimizing model, thus the value is reasonable and scientific. Meanwhile, the fuzzy neural network system adopts interdisciplinary neural algorithm with high efficiency and expansibility. Domestic expert system and evaluation criteria in foundation selection are not fully established. The two models offered in this paper can provide a practical testing criterion for engineers.

6 Conclusion

The theory of fuzzy system is developed in order to improve the traditional AHP method and to propose a more comprehensive fuzzy evaluation system combining more accurate FAHP method and interval theory together based on the statistical analysis of the forms, characteristics and applicable scopes of the building foundation in this paper. The basic principles and algorithm of artificial neural network are analyzed and expounded, then a fuzzy neural network system for architectural foundation selection is developed. It could draw the main conclusion by comparison and analysis on simulated testing and performance analysis for the two models:

(1) A fuzzy neural network system for architectural foundation selection is developed to improve the traditional AHP method and propose a more comprehensive fuzzy evaluation system combining more accurate FAHP method and interval theory together.
(2) The case study shows that the fuzzy neural network system adopts interdisciplinary neural algorithm with high efficiency and expansibility. Domestic expert system and evaluation criteria in foundation selection are not fully established. The two models offered in this paper can provide a practical testing criterion for engineers.

Acknowledgements. The research is supported by the National Natural Science Foundation of China (Grant No. 51308208), Natural Science Foundation of Hunan (Grant No. 2015JJ3069) and Youth Talent Plan Foundation of Hunan.

References

1. Ulshafer, M.L.: Development of a fuzzy logic expert system for pile selection. Air Force Institute of Technology, Wright-Patterson AFB, OH (1989)
2. Emiroglu, M.E., Tuna, A., Arslan, A.: Development of an expert system for selection of dam type on alluvium foundations. Eng. Comput. **18**(1), 24–37 (2002)
3. He, W., Zhang, Z.: Fuzzy the type selecting model for bridge foundation in deep-water. In: China Civil Engineering Society 12th Annual Cum Tunnel and Underground Engineering Branch of the Fourteenth Annual Meeting Proceedings (2006). (in Chinese)
4. Wang, J.: Selection based on foundations of the construction AHP and neural network. Chang'An University (2004). (in Chinese)
5. Dong-Chao, J.I., Song, B.F., Tian-Xiang, Y.U.: FAHP and its application in the selection of design scheme. Syst. Eng. Electron. **28**(11), 1552–1692 (2006)
6. Saaty, T.L.: The analytic hierarchy process. In: Proceedings of the Second International Seminar on Operational Research in the Basque Provinces, vol. 4, no. 29, pp. 189–234 (1980)
7. Shi, S., Cao, J., Feng, L., et al.: Construction of a technique plan repository and evaluation system based on AHP group decision-making for emergency treatment and disposal in chemical pollution accidents. J. Hazard. Mater. **276c**(9), 200–206 (2014)
8. Zadeh, L.A.: Fuzzy sets. Inf. Control **8**(65), 338–353 (1965)

9. Qin, Z.H., Qin, P.: Evaluation coupling model for high slope stability based on fuzzy analytical hierarchy process-set pair analysis method. Yantu Gongcheng Xuebao/Chin. J. Geotech. Eng. (2010)
10. Xu, Z.-S.: Study on the relation between two classes of scales in AHP. Syst. Eng. Theor. Pract. **19**(7), 97–101 (1999)
11. Lin, J.-C., Xu, Z.-S.: A new scale in fuzzy AHP. Oper. Res. Manag. Sci. **2**, 37–40 (1998)
12. Lu, Y.-J.: Weight calculation method of fuzzy analytical hierarchy process. Fuzzy Syst. Math. **16**(2), 79–85 (2002). https://doi.org/10.3969/j.issn.1001-7402.02.014
13. Xu, Z.-S., Da, Q.-L.: New method for interval multi attribute decision-making. J. Southeast Univ. (Nat. Sci. Edition) **33**(4), 498–501 (2003). https://doi.org/10.3321/j.issn:1001-0505.04.029
14. Nakahara, Y., Sasaki, M., Gen, M.: On the linear programming with interval coefficients. Int. J. Comput. Eng. **23**, 301–304 (1992)
15. Li, W.-G., Yu, Q., Luo, R.-C.: Application of fuzzy analytic hierarchy process and neural network in power transformer risk assessment. J. Cent. South Univ. **19**(4), 982–987 (2012)
16. Zian, T., Yang, X., Sun, Z., et al.: Analysis of command automatic network security simulation based on FAHP and ANN. Fire Control Command Control **38**(12), 108–111 (2013)
17. Wen, J., Zhao, J.L., Luo, S.W., et al.: The improvements of BP neural network learning algorithm. In: Proceedings of the 5th International Conference on Signal Processing (WCCC-ICSP 2000). IEEE, pp. 1647–1649 (2000)
18. Shi, F., Wang, H., Hu, F., Yu, L., et al.: MATLAB Intelligent Algorithm 30 Case Studies. Beihang University Press, Beijing (2011)

Spatial Variation of Soil Suction and Radiant Energy Around a Tree Vicinity in an Urban Green Infrastructure

S. Y. He[1], W. H. Zhou[1(✉)], and A. Garg[2]

[1] Department of Civil and Environmental Engineering, Faculty of Science and Technology, University of Macau, Macau, China
hannahzhou@umac.mo

[2] Department of Civil and Environmental Engineering, Shantou University, Shantou 515063, Guangdong, China

Abstract. Many studies had been conducted to analyse soil moisture/suction distribution in a vegetated soil in context of slopes (geotechnical engineering) or agricultural field (bio-technology). However, such studies are rare in context of urban landscape. Soil suction distribution is essential as it can influence hydraulic conductivity of soil and any subsequent water balance of green infrastructure in an urban landscape. Urban landscape involves complex interaction of vegetation species (tree) with other species, which can influence radiant energy distribution (require for photosynthesis) and hence, soil moisture/suction distribution in green infrastructure. The objective of this study is to investigate soil suction distribution in a green infrastructure in an urban landscape. Mix species cover (presence of grass cover in vicinity of tree) is selected as a green infrastructure site for monitoring. One month field monitoring programme has been designed to monitor the soil (soil suction, moisture), vegetation parameters (tree height, grass leaf area) and environmental parameters (radiant energy distribution; rainfall, humidity and temperature) under drying-wetting cycles. Soil parameters at different distances from tree stem was monitored to understand influence of it. It was found from the field test that suction magnitude is highest near the tree stem and also at shallower depths. Further, the rate of change of suction is also higher near the tree stem and at shallower depth. For drying, this may be because of presence of large number of roots and higher radiant energy interception, which causes higher transpiration induced suction. For wetting, this may be due to higher preferential flow caused by large number of roots near the tree at shallower depths.

Keywords: Soil suction · Plant roots · Meteorological parameters Drying-wetting cycle

1 Introduction

Numerous studies have been conducted to demonstrate hydrological effects induced by vegetation i.e., soil moisture and soil suction from perspective of slope stability (Zhan et al. 2014; Rahardjo et al. 2014; Garg et al. 2015a, b, c; Ng et al. 2016; Gadi et al.

R. Chen et al. (eds.), *Proceedings of the 2nd International Symposium on Asia Urban GeoEngineering,* Springer Series in Geomechanics and Geoengineering,
https://doi.org/10.1007/978-981-10-6632-0_52

2017; Hazra et al. 2017) as well as from agricultural crop yield perspective (Wang et al. 2007; Bhatt et al. 2016) Understanding of soil-plant-atmospheric interaction is important while analyzing such hydrological responses. Recent studies (Garg et al. 2015a, b; Gadi et al. 2016; Gadi et al. 2017) have demonstrated the effects of canopy properties (i.e., leaf area index), root properties (i.e. root area index) and also mix tree-grass interactions in soil-plant-atmospheric continuum. These interactions have found to have significant impact on hydrological responses (such as infiltration; Gadi et al. 2017) from vegetation across seasons.

Understanding of water flow in soil-plant-atmospheric continuum in context of urban green infrastructures (including gardens and lawns) are limited. Water flow mechanism in such urban infrastructure may become complex due to presence of other elements in urban landscape such as buildings, lakes and mixed vegetated species. Presence of these elements may cause shading effects (Gadi et al. 2017) by intercepting incoming radiant energy required for photosynthesis and transpiration of vegetation, which in turn can influence root water uptake. Study by Garg et al. (2015a) and Ng et al. (2016) found that suctions in vegetated soil could vary significantly depending on the % interception of radiant energy by tree canopy. Hence, radiant energy interception may play an important role while analyzing hydrological responses from vegetation in an urban landscape. Understanding of water flow in such urban green infrastructure is necessary for design or analyzing of vegetation in urban landscape. Some studies (Hemmati et al. 2012; Li and Guo 2016) have reported differential settlement caused by water uptake or preferential flow under rain due to presence of vegetation roots nearby around buildings in urban landscape. However, most of them were reported for single tree. Such interactions need to be further investigated in context of soil-plant-atmospheric continuum to better design the infrastructure. For example, understanding of influence zone of suction (spatial variations) induced around tree in an urban landscape could be useful for engineers and architects to improve their planning for infrastructure. In addition, meteorological parameters were found to have significant impact on the activity of vegetation (Iwasaki 2006), such as photosynthesis, respiration and transpiration. These meteorological parameters may vary locally depending on the presence of multiple type of elements in urban landscape (buildings, canopies of another species etc.). Therefore, there is a need to conduct a comprehensive study for investigation of hydrological response from presence mixed vegetation (tree and grass cover) in an urban landscape.

The main objective of this study is to investigate spatial variations of hydrological response from a mixed vegetated site in an urban landscape. In this study, a site containing Elaeocarpus apiculatus tree on a grass cover is selected for investigation. Field monitoring programme for at least one month is designed to investigate spatial variations in soil suction/moisture Atmospheric parameters, such as relative humidity, rainfall amount and air temperature were continuously monitored and interrupted with respect to observed soil suctions/moisture. Further, a comparison was made with other studies reported in literature for a relatively remote landscape (slopes, agricultural fields).

2 Materials and Methods

2.1 Site Description

Site for field monitoring was selected in the new campus of University of Macau (UM), Macau, China. This new campus of UM is located at Hengqin island, Zhuhai, China. The reason for selecting site is the fact that the landscape in University of Macau campus as per campus development plan corresponds to that in urban cities.

Figure 1 shows the location of field monitoring site at university of Macau, Macau. The field monitoring is located just near the artificial river in an urban landscape at University of Macau, Campus. As per the report from the weather bureau (SMG, 2012), the average temperature at Macau in a whole year (2016) was 22.3°. The highest temperature was 36.1°, and the lowest temperature was 5.0°. The average relative humidity was relatively high which is about 80%, and total rainfall in a whole year was 2058.1 mm. The ground water table (GWT) at this site is located at depth of 1.5 m.

Fig. 1. Location of the field test site in University of Macau campus, Macau

2.2 Field Monitoring Plan and Instrumentation

Field monitoring in a selected site (Fig. 2) in an urban landscape of University of Macau was conducted for about one and a half month (i.e., 15th Nov. 2016 to 31st Dec. 2016). This adopted time period was sufficient to achieve a wide range of magnitude of soil suction measurement (i.e., from 0 to 566.5 kPa from Fig. 7) around tree vicinity. Further, it also provide sufficient variation in measurement points around tree vicinity for quantifying spatial distribution of suction (or influence zone).

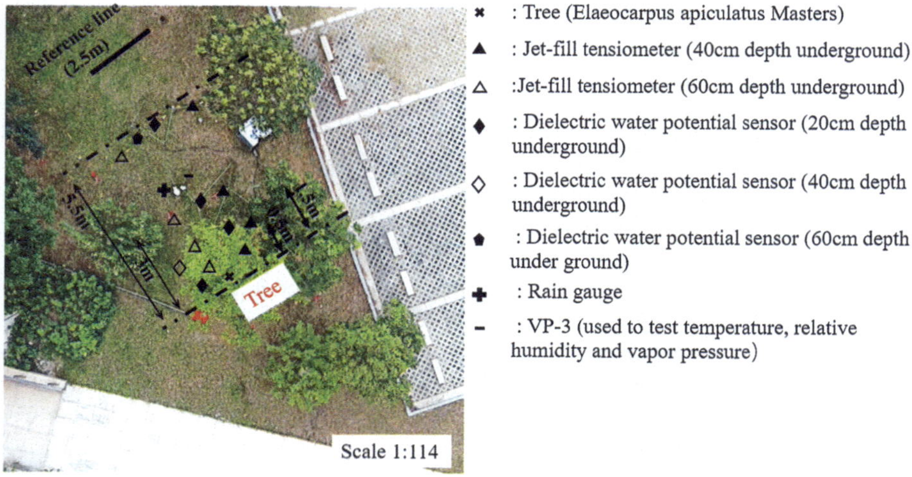

Fig. 2. Top view of selected testing site showing overview of instrumentation for measuring suction and soil moisture around tree

In order to measure suction of wide range i.e., 0 kPa to more than 100 kPa, two different instrumentation were utilized. Jet fill tensiometers (Soil moisture equipment corp., 2010) were adopted for measuring suction below 90 kPa, whereas for suctions above 90 kPa, dielectric water potential sensors (MPS sensors; Decagon services Ltd 2014) were utilized. Figure 2 shows the bird view of selected site (using Drone DJI Phantom) along with the instrumentation for measuring soil suction, soil moisture and atmospheric parameters. In order to further understand the instrumentation, cross sectional view of instrumentation is presented (Fig. 3). It can be seen from Fig. 3 that sensors for measuring suction was installed at three depths (i.e., 0.2 m, 0.4 m and 0.6 m). At shallowest depth (i.e., 0.2 m), dielectric water potential sensors were

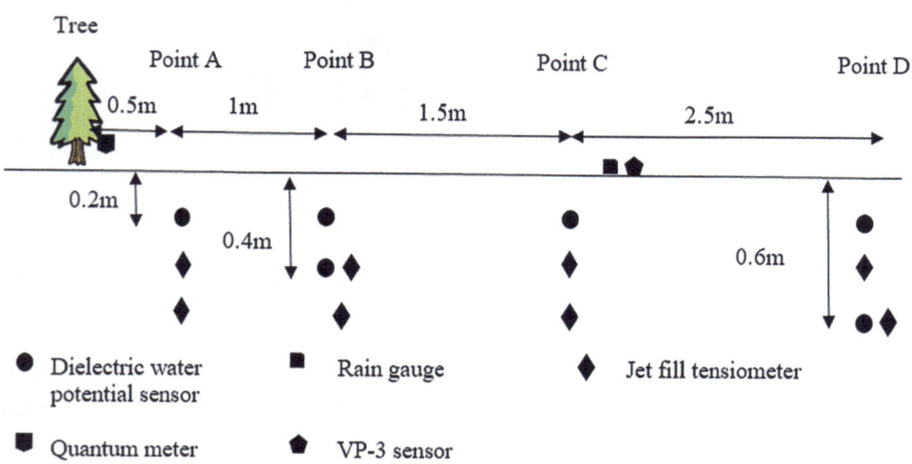

Fig. 3. Layout of instrumentation in field test site

installed as expected soil suction are large due to rapid loss of water near surface by evapotranspiration from grass roots. Whereas, at other deeper depths (i.e., 0.4 m and 0.6 m), where expected suction are lower (< 90 kPa), Jet Fill Tensiometers were installed. In order to quantify influence zone of suction in lateral direction, similar set of suction probes along depth were installed at four different distances from stem (i.e., point A (0.5 m from the tree stem), point B (1.5 m), point C (3 m) and point D (5.5 m)). These points were selected to ensure the coverage includes both within and outside the canopy for quantifying intercepted radiant energy effects.

For interpretation of suction distribution under varying natural conditions, a micro climate monitoring station (Decagon service Ltd 2014) was installed. It can measure rainfall intensity (Rain gauge), air humidity (VP-3 sensor), temperature (VP-3 sensor) and radiant energy (quantum meter) respectively. It should be noted that for measuring spatial distribution of radiant energy (including intercepted energy), measurements using quantum meter were obtained at several points within and outside canopy of tree.

3 Results and Discussion

3.1 Meteorological Parameters at Urban Landscape

Figures 4 (a), (b) and (c) shows the variation of meteorological parameters such as precipitation, air temperature and relative humidity, respectively during the monitoring period. It can be seen from the figure that the total amount of rainfall during the test period was 67.2 mm, which is smaller than the record (84.8 mm) as published on Macao Meteorological and Geophysical Bureau (SMG).

This difference could be due to difference in positioning of weather stations from Macau observatory as compared to our study (within urban landscape). Further, when compared with the precipitation with rainfall records of the last 10 years, the observed value in this period is higher than the average (55.46 mm). This shows that the rainfall this year is higher than usual. As showed in the figure, the rainfall pattern during consisted of mainly multiple rainfall events of light rain and only one major event corresponding to heavy rain and moderate rain.

Figures 4 (b) and (c) shows plot of variation of relative humidity and air temperature during the monitoring period. It can be observed that the air temperature fluctuates during day and night periods. The air temperature drops from around 28°C in beginning of November to around 13°C at end of monitoring period. On the other hand, relative humidity (RH) is close to 90% most of the time during monitoring period except for month of December, where it fluctuates to lower than 60%. Lower value of RH during month of December corresponds to lower precipitation during that period.

3.2 Radiant Energy Characteristics at Urban Landscape

For understanding the effect of canopy of tree in urban landscape, measurements on radiant energy (at 10:00 and 16:00) at four different points (refer to four points in Fig. 3) were plotted with time. Figure 5 shows the variation in received radiant energies on the soil surface at four different points around tree vicinity (at 10:00 am)

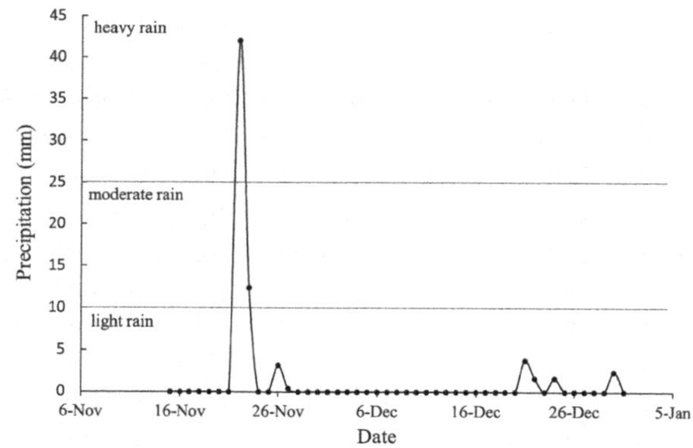

(a)

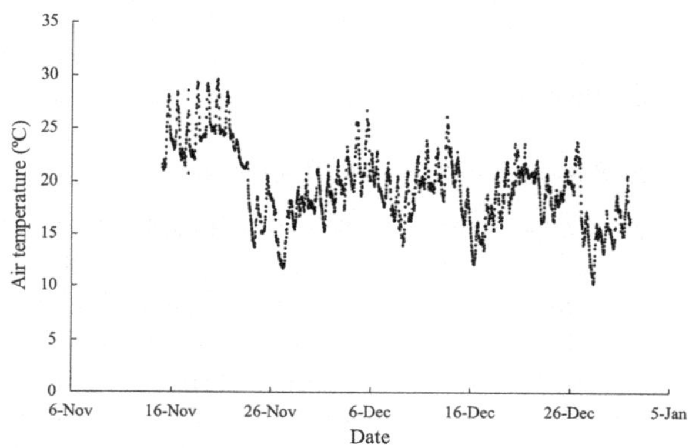

(b)

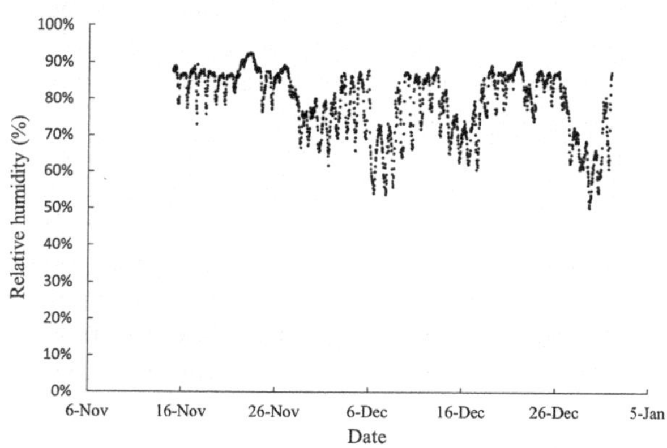

(c)

Fig. 4. Variation of meteorological parameters (a) Rainfall (b) air temperature and (c) relative humidity

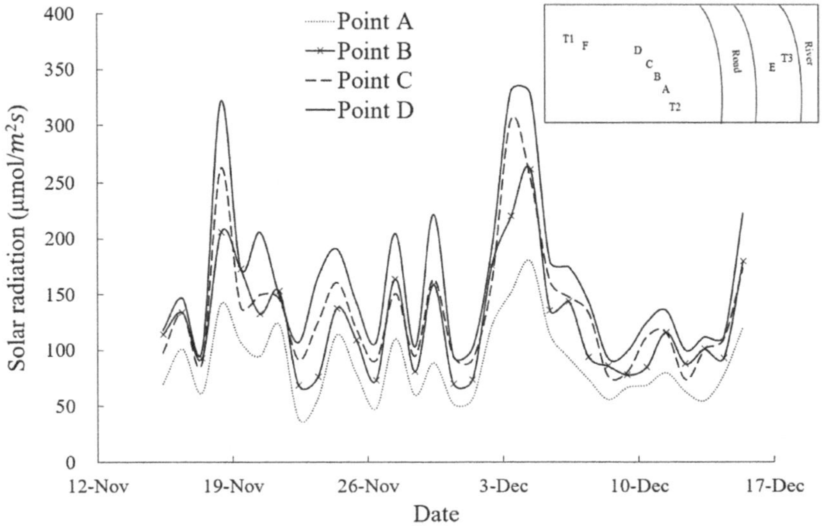

Fig. 5. Variation of solar radiation during monitoring period at different points during day at 10:00 am

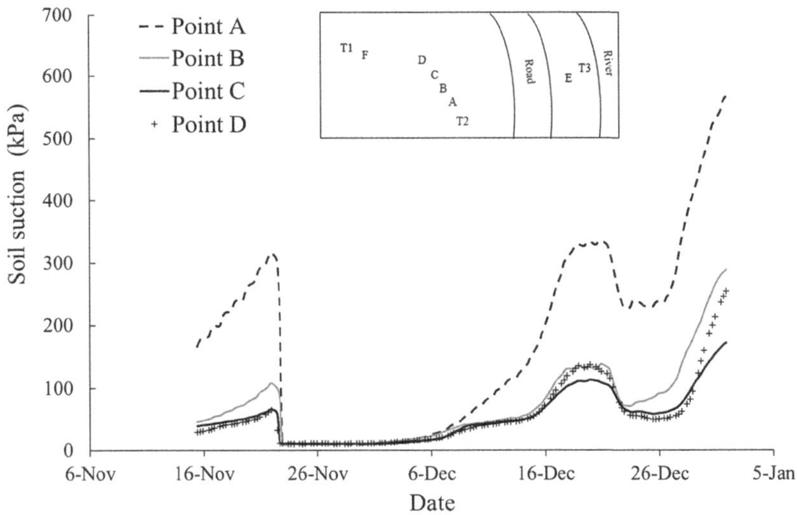

Fig. 6. Comparison of soil suction in different points at 20 cm depth

Generally, there is an increase in measured radiant energy from point A to point D. This is likely because of interception effects of tree canopy that is more dominant for points A and B as compared to C and D.

3.3 Spatial Distribution of Suction Around Tree in an Urban Landscape

3.3.1 Variation in Suction at Different Distances from Tree Stem

Figure 6 shows the variation of soil suction (at shallowest depth, 20 cm) at different distances from tree stem with time. As observed from the figure, suction magnitude for point (A) closest to the tree stem is highest (maximum value of around 566.5 kPa), whereas it is lowest for point (C) that are farther from tree stem. The magnitude of suction (around 566.5 kPa) at 20 cm depth near to tree stem is however consistent with that observed for *Schefflera heptaphylla* grown on recompacted slope in field study under similar climate variations in Hong Kong (Garg et al. 2015a b, c). This trend is opposite to the observed radiant energy distribution (refer to Fig. 5) at different

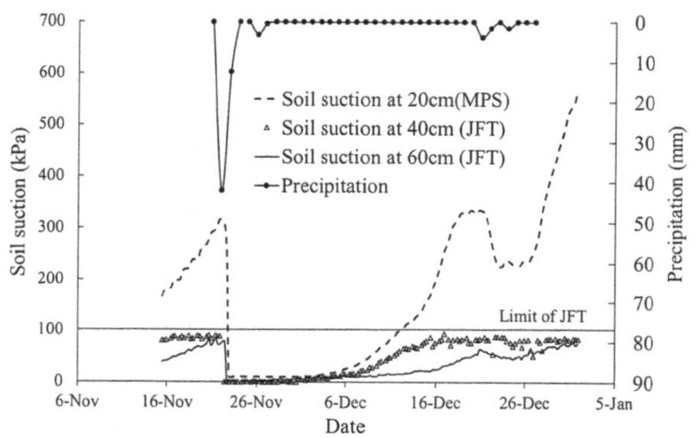

(a)

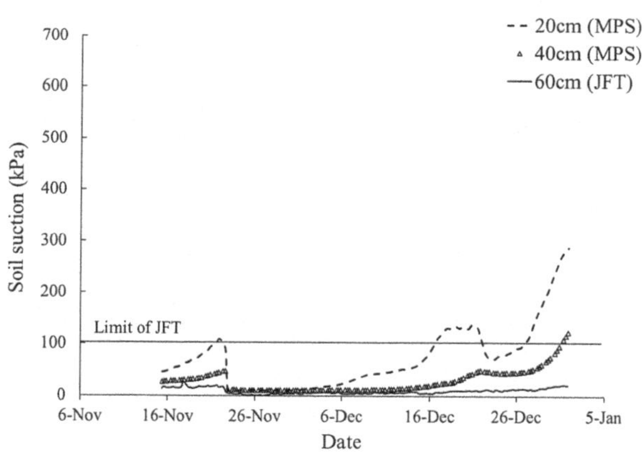

(b)

Fig. 7. Comparison of vertical profile of soil suction at different distances from tree stem i.e., (a) point A, 0.5 m from tree; (b) point B, 1.5 from tree; (c) point C, 3 m from tree; (d) point D, 5.5 m from tree

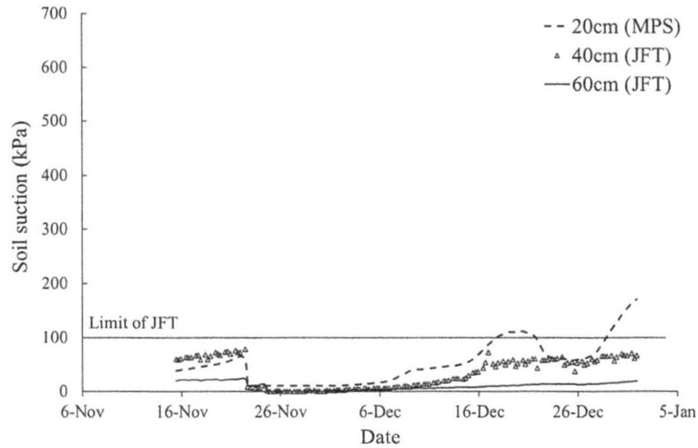

(c)

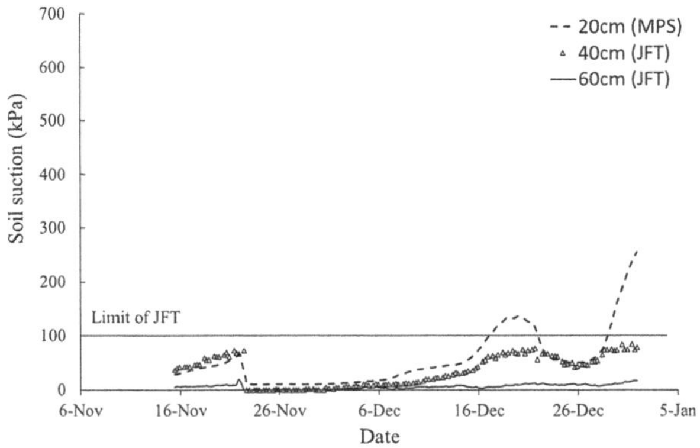

(d)

Fig. 7. (*continued*)

distances from tree stem. Using radiant energy, evapotranspiration rate and hence induced suction must have been higher at distances away from tree stem. This contrasting trend is likely because of dominance of tree root water uptake over evaporation rate. As expected, the dominance is more significant near tree stem as compared to farther locations. Moreover, equation proposed by Penman-Monteith is valid for uniform vegetation cover and ignore any interaction effects from another vegetation species (For example; tree in this study). Though, the trend (for suction magnitude for points from A to C) has been also observed in others study, however, the comparative is not true since their study was for single tree and any interaction affects due to presence of another species was not investigated.

However, the trend do not seem to follow similar for points C and D. The rate of increase in suction under drying is found to be higher at point D than point C on two occasions after wetting. This is very likely because of dominance of evaporation effect from soil surface due to increased radiant energy over root water uptake from tree roots.

3.3.2 Variation in Vertical Suction Profiles with Distances from Tree Stem

In order to fully understand spatial distribution of suction in the vicinity of tree, suction profile at different depths (i.e., 20 cm, 40 am and 60 cm) were plotted for all four points away from tree stem (Figs. 7 (a)–(d)). It can be observed that the magnitude as well as change (under drying or wetting) is found to be much lower than that at 20 cm depth. This difference of suction across depth is much larger for point (A) closest to the stem than away from it. This significant reduction of suction beyond 20 cm depth seems to suggest that the depth and hence, the influence zone of tree root is very shallow. This might because trees were transplanted at the given site of University of Macau campus two years ago before the start of monitoring.

The vertical profile of suction is very contrasting to observed deeper influence zone of suction by tree root water uptake as found in previous studies (Hemmati et al. 2012). This is likely because in an urban landscape, trees are often grown by transplantation and hence the root depth is expected to be shorter during initial few years. On the other hand in their studies, trees were grown in forests whose soil and environmental conditions are more favorable for growth.

4 Conclusions

In this study, spatial distribution of soil suction is investigated in a tree vicinity in an urban landscape. This distribution has been interpreted with measured environmental parameters including radiant energy distribution and compared qualitatively with those observed studied on slope in literature. It can be concluded that presence of tree roots tends to control soil suction near to tree stem and at shallower depths during both drying and wetting period. This is in contrast with those observed for single tree in literature and for natural ground (such as forests), where influence zone of suction of tree was found to be much deeper. The results tempt to conclude that preferential flow due to roots is rapid changes in suction (even despite high initial suction) near to tree stem during wetting. Generally, suction magnitude decreases with increase in distance from tree stem, however, a beyond a certain point it is found to increase slightly. This indicates that the dominance of tree roots seems to vanish at about 1.5 m away from its stem. Radiant energy distribution appears to be the main reason for increase in magnitude of suction beyond such distances.

Acknowledgements. The authors gratefully acknowledge financial support from the University of Macau Research Fund (MYRG2015-00112-FST) and the Macau Science and Technology Development Fund (FDCT) (Code: 125/2014/A3).

References

Bhatt, R., Arora, S., Chew, C.C.: Improving irrigation water productivity using tensiometers. J. Soil Water Conserv. **15**(2), 120–124 (2016)

Operator's Manual for MPS-2 & MPS-6 Dielectric Water Potential Sensors: Decagon Devices. Commercial Publications, Decagon Devices, Inc., Pullman (2014). www.decagon.com

Gadi, V.K., Bordoloi, S., Garg, A., Kobayashi, Y., Sahoo, L.: Improving and correcting unsaturated soil hydraulic properties with plant parameters for agriculture and bioengineered slopes. Rhizosphere **1**, 58–78 (2016)

Gadi, V.K., Tang, Y.R., Das, A., Monga, C., Garg, A., Berretta, C., Sahoo, L.: Spatial and temporal variation of hydraulic conductivity and vegetation growth in green infrastructures using infiltrometer and visual technique. CATENA **155**, 20–29 (2017)

Garg, A., Leung, A.K., Ng, C.W.W.: Comparisons of soil suction induced by evapotranspiration and transpiration of S. heptaphylla. Can. Geotech. J. **52**(12), 2149–2155 (2015a)

Garg, A., Coo, J.L., Ng, C.W.W.: Field study on influence of root characteristics on soil suction distribution in slopes vegetated with cynodon dactylon and schefflera heptaphylla. Earth Surf. Process. Landf. **40**(12), 1631–1643 (2015b)

Garg, A., Leung, A.K., Ng, C.W.W.: Transpiration reduction and root distribution functions for a non-crop species schefflera heptaphylla. CATENA **135**, 78–82 (2015c)

Hazra, B., Gadi, V., Garg, A., Ng, C.W.W., Das, G.K.: Probabilistic analysis of suction in homogeneously vegetated soils. CATENA **149**, 394–401 (2017)

Hemmati, S., Gatmiri, B., Cui, Y.J., Vincent, M.: Thermo-hydro-mechanical modelling of soil settlements induced by soil-vegetation-atmosphere interactions. Eng. Geol. **139**, 1–16 (2012)

Iwasaki, H.: Impact of interannual variability of meteorological parameters on vegetation activity over Mongolia. J. Meteorol. Soc. Jpn. **84**, 745–762 (2006)

Li, J., Guo, L.: Field investigation and numerical analysis of residential building damaged by expansive soil movement caused by tree root drying. J. Perform. Constr. Facil. **31**, 1–10 (2016). D4016003

Ng, C.W., Garg, A., Leung, A.K., Hau, B.C.H.: Relationships between leaf and root area indices and soil suction induced during drying-wetting cycles. Ecol. Eng. **91**, 113–118 (2016)

Rahardjo, H., Satyanaga, A., Leong, E.C., Santoso, V.A., Ng, Y.S.: Performance of an instrumented slope covered with shrubs and deep-rooted grass. Soils Found. **54**(3), 417–425 (2014)

S.M.G.: Annual Meteorological Observation Reports. Macau Meteorologcial and Geophysical Bureau, Macau, China (2012)

Wang, F.X., Kang, Y., Liu, S.P., Hou, X.Y.: Effects of soil matric potential on potato growth under drip irrigation in the North China plain. Agric. Water Manage. **88**(1), 34–42 (2007)

Zhan, T.L.T., Chen, R., Ng, C.W.W.: Wetting-induced softening behavior of an unsaturated expansive clay. Landslides **11**(6), 1051–1061 (2014)

Numerical Investigation of the Efficiency of Buttress Walls in Reducing the Diaphragm Wall Deflection by Deep Excavation

Pio-Go Hsieh[1(✉)] and Shao-Chi Chien[2]

[1] Department of Assets and Property Management, Hwa Hsia University of Technology, New Taipei City 23568, Taiwan
spg@cc.hwh.edu.tw

[2] Department of Information Management, Hsuan Chuang University, Hsinchu 300, Hsinchu, Taiwan
jaky5817@hcu.edu.tw

Abstract. This study performed a series of three-dimensional finite element analyses of diaphragm wall deflections for deep excavations with buttress walls and studied the efficiency on the wall deflection of several parameters, including the buttress wall length, the buttress wall interval, the buttress wall thickness and the buttress wall embedment. Results show that if the buttress wall length was less than 2.0 m, the buttress wall was unable to restrain the wall deflection. Once the buttress wall length was larger than 2.0 m, the efficiency of reducing wall deflection will almost linearly increase with the buttress wall length. The efficiency provided by buttress wall thickness was not significant. By increasing the buttress wall interval, the wall deflections along the diaphragm wall tended to move uniformly. In the same amount of the total length of the buttress walls, the efficiency of increasing the buttress length is better than it of the decreasing the buttress wall interval. The buttress depth only need 2/3 times the final excavation depth, it can yield a diaphragm wall deflection close to that provided by the deeper buttress wall depth in a more economical manner.

Keywords: Deep excavation · Buttress walls · Wall deflection
Three-dimensional finite element analysis · Parametric study

1 Introduction

Deep excavation in soft clay commonly causes large lateral wall deflections and ground settlements. Because the density of buildings in urban areas is relatively high, excessive ground settlement frequently damages adjacent properties. To avoid the damage of adjacent buildings during excavation, it is necessary to adopt effective measures to limit the wall deflection and ground settlement to a tolerable amount. Strengthening of the bracing system and/or installing the cross wall or increasing the passive resistance are commonly adopted auxiliary measures in engineering practice.

Recently, buttress walls have been used as alternative auxiliary measures for protection of adjacent buildings during excavation in Taiwan. Basically, buttress walls are concrete walls constructed from one side of the diaphragm wall to a certain length prior

R. Chen et al. (eds.), *Proceedings of the 2nd International Symposium on Asia Urban GeoEngineering*, Springer Series in Geomechanics and Geoengineering,
https://doi.org/10.1007/978-981-10-6632-0_53

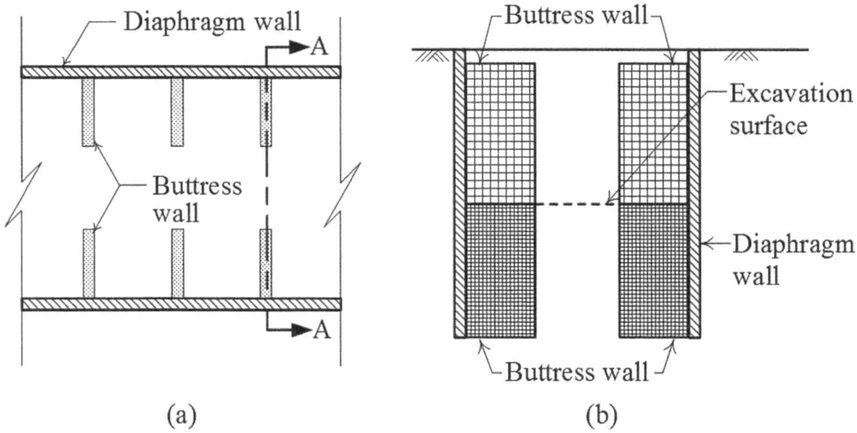

Fig. 1. Schematic diagram of buttress wall (a) plan (b) A-A section.

to excavation, as shown in Fig. 1. They are usually cast with the diaphragm wall at the same time.

In engineering practice, in order to facilitate the construction, the buttress wall in excavations is mainly designed to be demolished along with excavation of soil. Under such conditions, the effect of reduction in the wall deflection mainly come from the frictional resistance between the side surfaces of buttress wall and adjacent soil rather than from the combined bending stiffness contributing from the diaphragm wall and buttress walls (Hsieh et al. 2016).

Although buttress walls have been applied in many excavations in Taiwan. Only a few studies on buttress walls have been found in the literature and indicated that the buttress walls have a certain effect in reducing the lateral wall deflection (Lin and Woo 2007; Hwang et al. 2007; Ou et al. 2008; Chen et al. 2011; Hsieh et al. 2015; Hsieh et al. 2016; Lim et al. 2016). However, there is a lack of specific research results on the effect of the dimensions of buttress walls, so the design of the buttress walls remains stalled at a semi-empirical stage.

In view of this, this study performed a series of parametric studies using a three-dimensional finite element analysis to examine the buttress wall effect factors that reduce the wall deflection, including the buttress wall length, the buttress wall interval, the buttress wall thickness and the buttress wall embedment. It is expected that with a better understanding of the behaviour of buttress walls in reducing the wall deflection, buttress walls in excavations can be designed in a more compact manner and that the protection of adjacent buildings can be ensured by designers.

2 Three-Dimensional Finite Element Analysis

Three-dimensional finite element analyses were conducted using a commercial software package, namely PLAXIS 3D (2013).

As shown in Fig. 2, a basic hypothetical excavation scenario was constructed for the parametric study. This case site is characterised by a rectangular shape with an excavation width (B) of 80 m and a rather large to nearly infinite excavation length (L). The final excavation depth (H_e) is 30 m, and this depth is completed in 10 stages. After Stage 1 to Stage 9 of excavation, steel struts are set up 1 m above the excavation surfaces with a spacing of 4 m, and finally the Stage 10 excavation is carried out. Of which, after Stage 1 to Stage 3 excavations, 2-H400 × 400 × 13 × 21 steel struts are set up; after Stage 4 to Stage 9 excavations, 3-H400 × 400 × 13 × 21 steel struts are set up. Figure 3 presents the temporary struts, excavation levels and excavation sequence of the basic hypothetical excavation. For the diaphragm wall, the embedded diaphragm wall (H_p) is equal to H_e, the thickness (t_{dw}) is 1.3 m, the adopted compressive strength of concrete (f_c') is 27.5 MPa. For the buttress walls, the length (L_{bw}) and the interval (S_{bw}) are 20 m, respectively, the thickness (t_{bw}) is 1.0 m, the embedment (H_{be}), which refers to a measurement taken below the excavation surface to the toe of buttress wall, is 30 m and the adopted f_c' is identical to those of diaphragm wall.

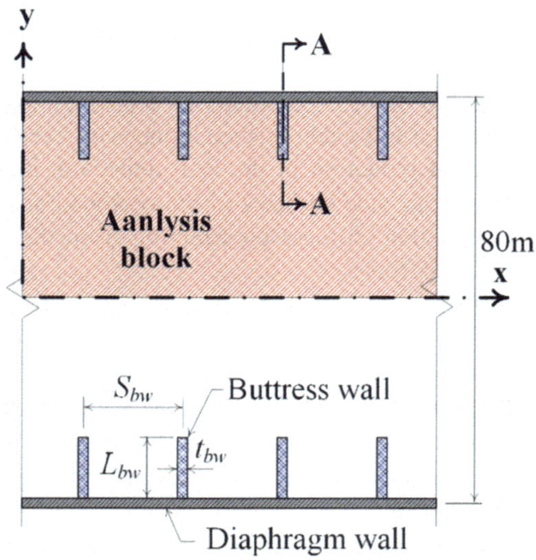

Fig. 2. Plan of the hypothetical excavation.

The subsoil is assumed to be a thick clay layer. The parameters are chosen reference a typical silty clay in Taipei. The groundwater level is assumed to be at GL –3 m, rendering a hydrostatic pressure distribution. Figure 3 illustrates the soil conditions together with their physical properties and model parameters in which the Hardening Soil (HS) model (Schanz et al. 1999) is adopted to simulate the soil behavior under the undrained condition. Of the HS parameters; the secant stiffness (E_{50}^{ref}) corresponding to the reference stress, p^{ref}, the tangent referential stiffness for primary oedometer loading

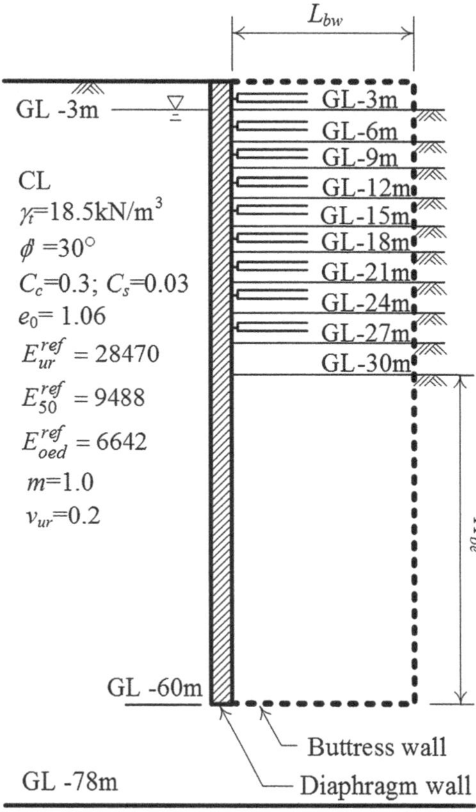

Fig. 3. Schematic diagram for the construction sequence and subsurface soil conditions of the hypothetical excavations.

(E_{oed}^{ref}), the unloading/reloading referential stiffness (E_{ur}^{ref}), and the power for stress-level dependency of stiffness (m); are evaluated according to Lim et al. (2010) and Calvello and Finno (2004). The coefficient of the at-rest earth pressure (K_0) is estimated based on Ladd's equation (Ladd et al. 1977).

The structural members, such as diaphragm walls, buttress walls and temporary steel struts are assumed to behave as the linear-elastic material. Plate elements are used for the diaphragm walls and buttress walls for which the Young's modulus of concrete (E_c) is estimated according to the suggestion by ACI Committee 318 (1995). The temporary steel struts are simulated by node-to-node anchor. The Young's modulus of steel (E_{st}) is 2.0×10^5 MPa.

Table 1 summaries the parameters of the structure used in the analysis, in which the E_c and E_{st} are reduced by 20% and 50%, respectively, considering construction effects (Ou 2006). In analysis, the interface property (R_{int}) between the diaphragm wall and the adjacent soil is set as rigid (R_{int} = 1.0).

Table 1. Structural information and parameters in the analysis

Structure	t(m)	γ (kN/m^3)	E(MPa)	v^*	A(m^2)
Diaphragm	1.5	23.54	19718	0.15	–
Buttress wall	1.0	23.54	19718	0.15	–
Strut of the 1st–3rd layer	–	–	100000	–	0.044
Strut of the 4th–9th layer	–	–	100000	–	0.066

*: Poisson's ratio.

For simplification, only a half of the excavation geometry, marked with hatching and labelled "Analysis block" in Fig. 5, is adopted in the analyses due to symmetry.

When planning the meshes, the meshes in the depth direction (z-direction) are continued down to a depth of 1.3 times the depth of the diaphragm wall, and in the width direction (y-direction) are extended up to four times the excavation depth behind the wall. Since the excavation length is infinite, the meshes in the length direction (x-direction) are extended up to 80 m (four times the buttress wall interval). Figure 4 shows the three-dimensional finite element mesh used in the analyses.

For the parametric studies, the values of L_{bw}, t_{bw}, S_{bw} and H_{be} are varied as shown in Table 2. Of these, $L_{bw} = 0$ implies that the excavation is not installed with buttress walls. When conducting the parametric studies, the variation values for changes in parameters are as listed in Table 2; other parameters maintain the basic values.

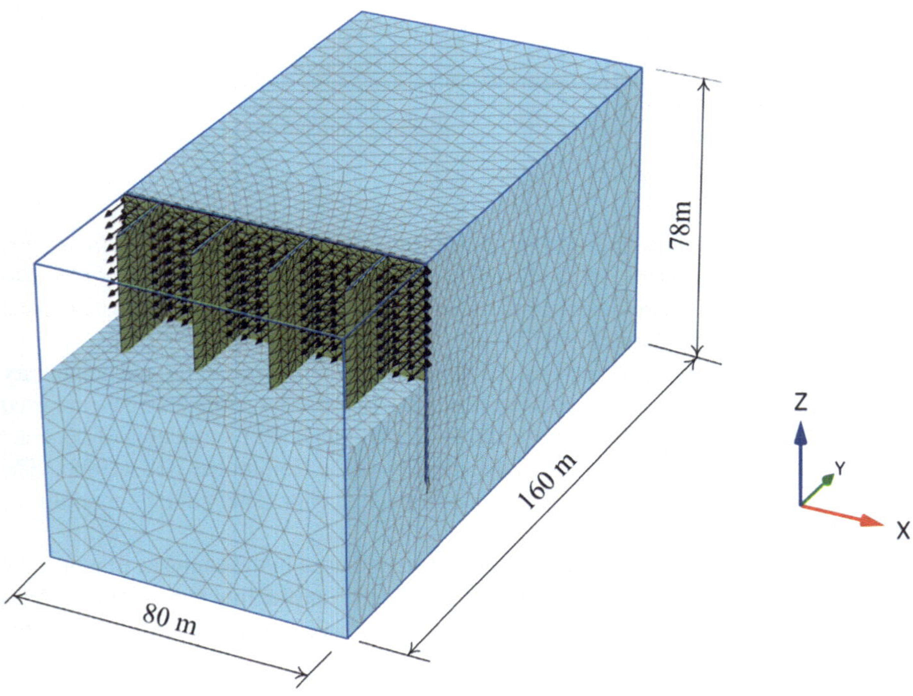

Fig. 4. Three-dimensional finite element mesh used in the analyses.

Table 2. Basic values of parameters and variations applied in the parametric studies

Parameters	Basic value	Variation value
L_{bw} (m)	20	0, 1, 2.5, 5, 10, 15, 35
t_{bw} (m)	1.0	0.5, 1.5
S_{bw} (m)	20	5, 10, 40
H_{be} (m)	30	5, 10, 20, 25, 35

When exploring the restraining effect of buttress walls on the lateral wall deflection, the diaphragm wall's section where the buttress wall is positioned is taken as the object of comparison, as shown in the A-A section in Fig. 2. This is because the lateral deflection of the diaphragm wall for this section is not under the influence of diaphragm wall flexibility in the x direction, and it is easier to clarify the effectiveness of buttress walls.

The performance of buttress walls is assessed quantitatively using the deflection curve of diaphragm wall. Therefore, the deflection area ratio (*DAR*) are proposed and defined as:

$$DAR = \frac{(A_0 - A_i)}{A_0} \times 100\% \tag{8}$$

where:

A_0 = the plane area of lateral wall deflection without buttress walls
A_i = the plane area of lateral wall deflection with buttress walls

Figure 5 illustrates those parameters used for calculating the *DAR*.

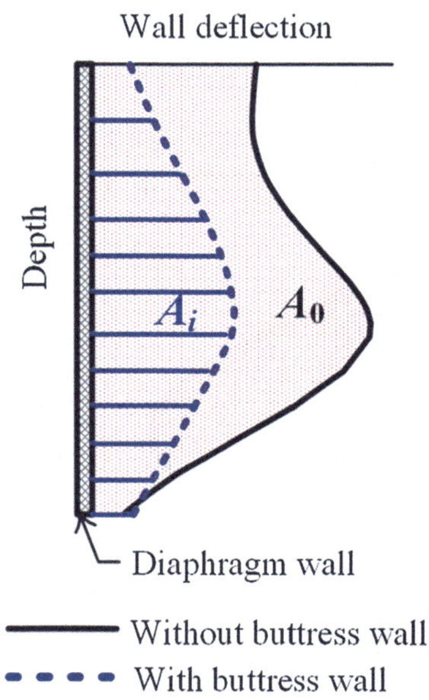

Fig. 5. Illustration of parameters used for calculating DAR values.

3 Results of Parametric Study

3.1 Effect of the Buttress Wall Length

Figure 6 shows the wall deflections under different lengths of buttress wall for the buttress walls being demolished along with the excavation; the result shows that the wall deflection decreases progressively with increasing L_{bw}. When L_{bw} is smaller, the effect of the buttress wall is less significant, but with the increase of L_{bw}, the lateral wall deflection becomes smaller.

Figure 7 shows the variation in the *DAR* with respect to L_{bw}, indicating that L_{bw} is less than 2.0 m, the buttress wall is unable to restrain the wall deflection, while L_{bw} is larger than 2.0 m, the *DAR* increases with increasing L_{bw}, presenting approximately linear relationships.

As described by Hsieh et al. (2016), while the buttress walls are demolished along with excavation of soil, the main mechanism of buttress walls in reducing the wall deflection is due to the frictional resistance between buttress walls and the adjacent soil. The combined bending stiffness of the diaphragm wall from the diaphragm wall and the buttress wall plays insignificant role in the reduction of the wall deflection.

Figure 8 shows the distributions of shear stress and relative shear stress ratio (τ_{rel}) on the surface of the buttress wall for different lengths of buttress wall, where τ_{rel} is defined as the ratio of shear stress to the shear strength of a soil for a given effective normal stress. As shown in Fig. 8(a), the induced shear stress occurred at the diaphragm wall with the smallest value and increased with the increasing distance from the diaphragm wall. The τ_{rel} was also smallest near the diaphragm wall and generally increased with the increasing distance from the diaphragm wall, up to a value of 1.0, as shown in Fig. 8(b). This phenomenon can be explained by the fact that the buttress wall being pushed by the diaphragm wall resembles the movement of a rigid body

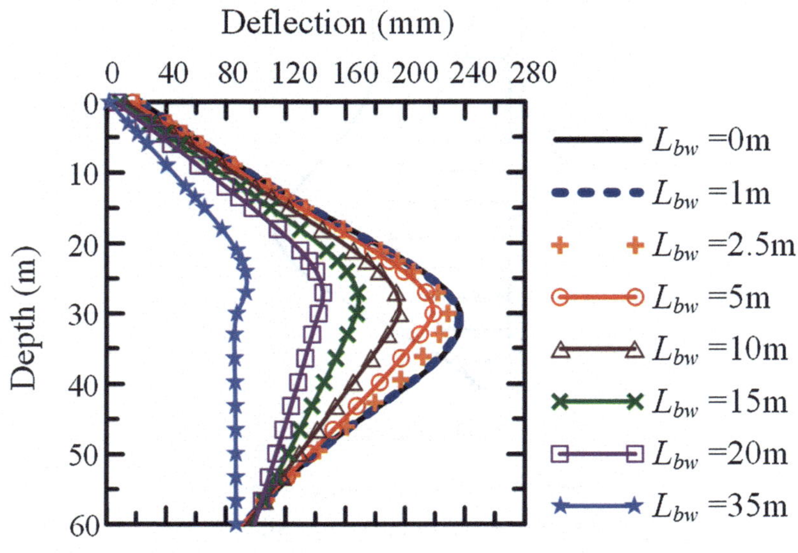

Fig. 6. The wall deflections under different lengths of buttress wall.

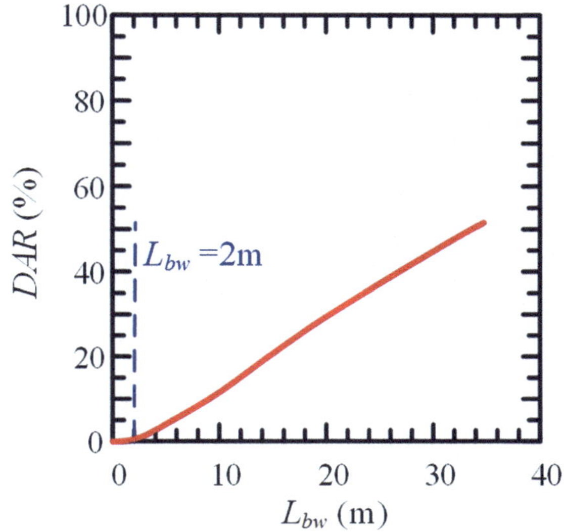

Fig. 7. Relationship between the deflection area ratio and the buttress wall length.

because the buttress wall has a very high axial stiffness, The soil in front of the diaphragm wall are almost pushed together with the buttress wall by the diaphragm wall. Therefore, the relative displacement between the buttress wall and the soil is very small. The driving shear stress and relative shear stress ratio are very small near the diaphragm wall. However, at a more distant position from the diaphragm wall, the displacement of the soils gradually becomes smaller and the relative displacement in between buttress walls is larger. So the part situated at a greater distance to the diaphragm wall has larger driving shear stress and relative shear stress ratio.

Figure 8 also exhibits a relative small driving stress and relative shear stress ratio occurring within the first 2 m from the diaphragm wall, no matter how long the buttress was. Therefore, the L_{bw} that is too short perhaps cannot provide adequate reduction of lateral wall deflection, especially during deeper excavation. Thus, there is a need for longer L_{bw}, so that more frictional resistance can be effectively provided to reduce the lateral wall deflection.

3.2 Effect of the Buttress Wall Thickness

Figure 9 shows the wall deflections obtained in the final excavation stage with different buttress wall thickness at L_{bw} of 5 m, 20 m, and 35 m, respectively. As indicated, as the buttress walls are demolished along with excavations, there is almost no difference in the wall deflections with three types of t_{bw}. Therefore, when the buttress walls are demolished along with the excavations, the efficiency of wall thickness is not significant for restraining the wall deflections. This is because the reason for increasing t_{bw} is mainly to increase the bending stiffness of diaphragm wall, but as described by Hsieh et al. (2016), when the buttress walls are demolished along with excavations, the enhanced bending stiffness provided by the buttress walls to the diaphragm wall cannot

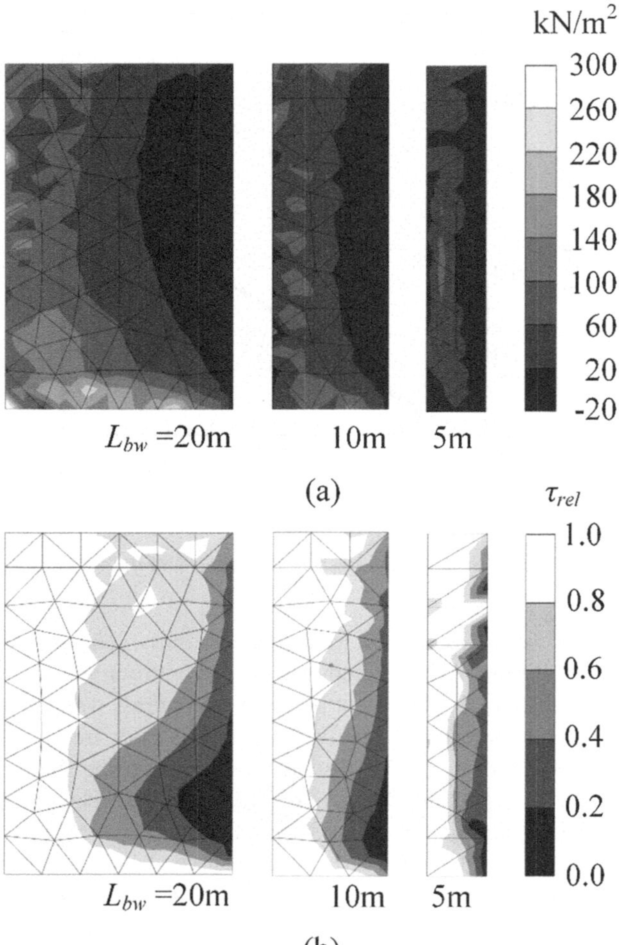

Fig. 8. The distribution of shear stress and relative shear stress ratio under different lengths of buttress wall (a) shear stress (b) relative shear stress ratio.

effectively reduce the lateral deflection of diaphragm wall. Therefore, for the general excavation project, where the buttress walls are demolished along with excavation, the thickness of the buttress wall needs only to resist collapse due to extrusion of the diaphragm wall, in order to ensure that just the action of surface frictions is unleashed.

3.3 Effect of the Buttress Wall Interval

All the previous analyses were conducted under a basic interval between buttress walls of 20 m, and the diaphragm wall section where the buttress walls are positioned were taken as comparison objects, as shown in section A-A in Fig. 2. However, for the wall deflection after installing the buttress walls, on the plan view, at the installation positions of buttress walls, because of the constraints of the buttress walls, the wall

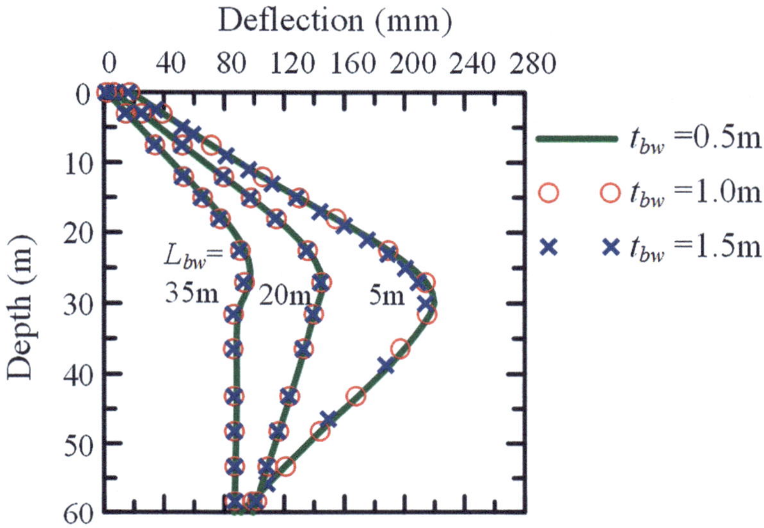

Fig. 9. The wall deflections under different thicknesses of buttress wall.

deflection could be different from the deflection for the middle section between the two buttress walls. Besides, minimizing the interval between buttress walls should be able to reduce the overall wall deflection. Therefore, deciding an ideal interval between buttress walls to achieve the aims of controlling the lateral wall deflection, yet avoiding waste and increasing obstructions to project construction due to an overly dense configuration is an important challenge in the design of buttress walls.

In order to explore the effect of interval between the buttress walls, this study conducted analysis of different S_{bw}. The variation of the S_{bw} parameter is shown in

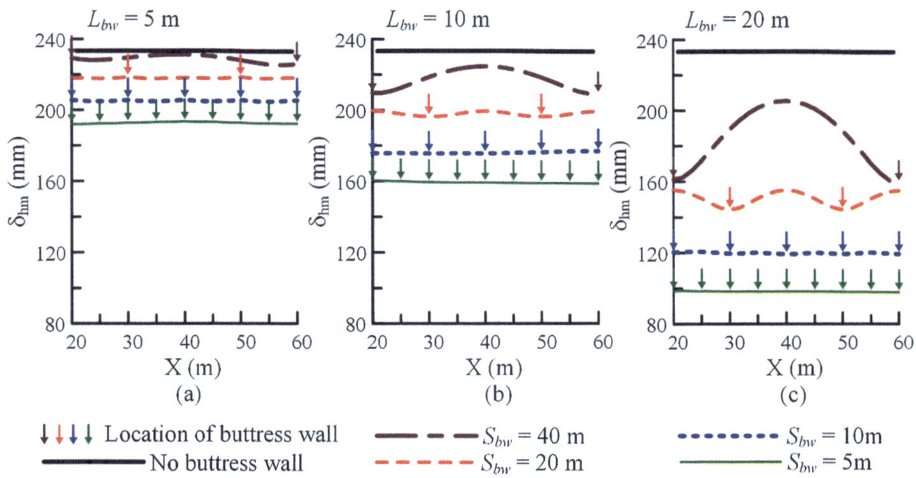

Fig. 10. The lateral wall deflections in plan at the depth of maximum wall deflection for different lengths of buttress wall (a) L_{bw} = 5 m (b) L_{bw} = 10 m (c) L_{bw} = 20 m.

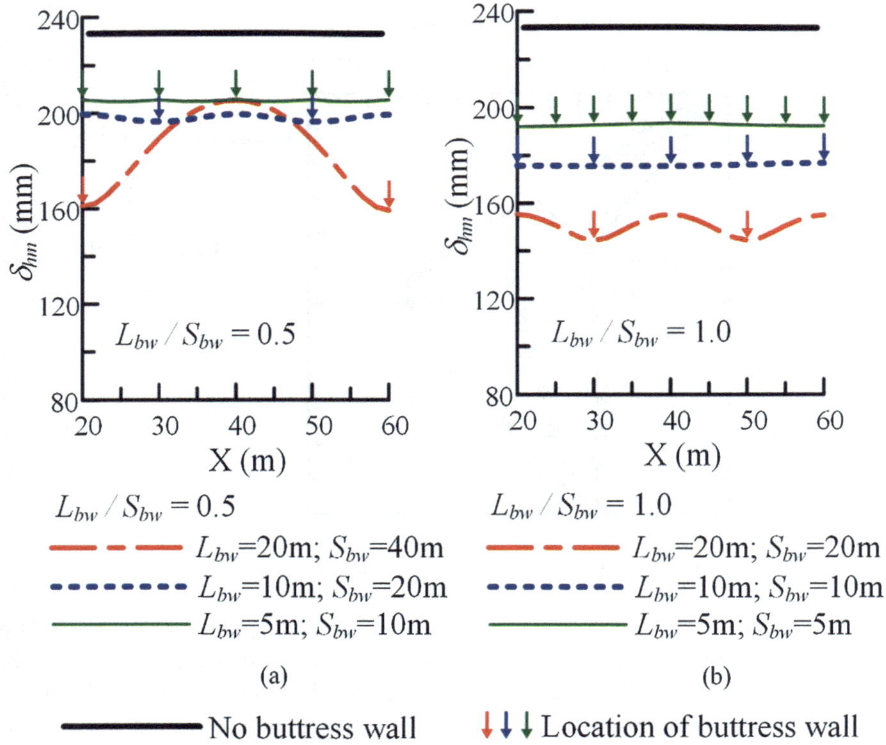

Fig. 11. The lateral wall deflections in plan at the depth of maximum wall deflection for the equivalent total length of buttress wall: (a) L_{bw}/S_{bw} = 0.5 (b) L_{bw}/S_{bw} = 1.0.

Table 2. Figure 10 shows the plan profile of maximum lateral wall deflections (δ_{hm}) obtained by analyzing at H_e = 30 m within a range of X coordinates from 20 m to 60 m (Refer to Fig. 2). As indicated, when the length or the interval of buttress walls are relatively small, the deformations of the diaphragm wall in plan tend to be straight lines; increasing the number of buttress walls can reduce the (δ_{hm}) of the original installation positions of buttress walls, but the reduction amount has a trend of deceleration with the increasing number of buttress walls.

Figure 11 shows the comparisons of lateral wall deflection obtained from analysis of configuring an equivalent total length of buttress wall, but in collocation of different length and interval of buttress walls, including the two matching types of L_{bw}/S_{bw} = 0.5 and 1.0. As indicated, the effect of increasing the length of buttress walls is better than the effect of minimizing the interval between the buttress walls (increasing the number of buttress walls).

3.4 Effect of the Buttress Wall Embedment

When the depth of the hard soil layer is deeper and the buttress walls and diaphragm wall have not penetrated the hard soil layer, whether the buttress wall need to be

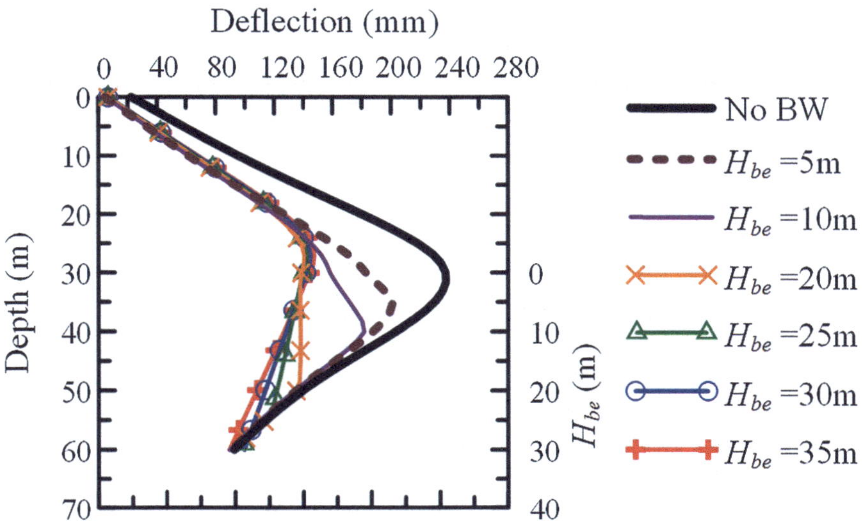

Fig. 12. The wall deflections under different embedments of buttress wall.

installed to the bottom end of the diaphragm wall will influence the decision on the buttress wall embedment (H_{be}).

Figure 12 shows the wall deflections under different of buttress wall embedment. As indicated, increasing H_{be} can mainly minimize the lateral wall deflections in the vicinity and below the bottom of the excavation surface. Figure 13 shows the trend of

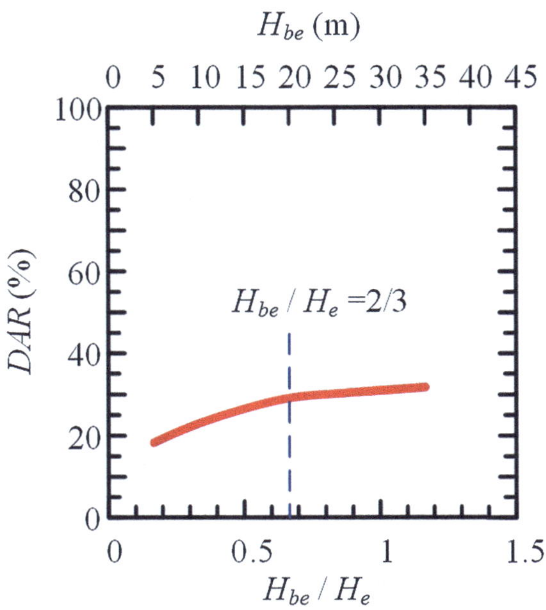

Fig. 13. Relationship between the deflection area ratio and the buttress wall embedment.

increasing *DAR* when H_{be} is increased. Although *DAR* increases with increasing H_{be}, when the H_{be}/H_e exceeds 2/3, the increasing trend gradually slows down. Therefore, during the design of buttress walls, H_{be} do not seem to be as deep as the diaphragm wall and it is adequate to consider adopting $2H_e/3$.

4 Conclusion

Based on the studies in this paper, the following conclusions can be drawn:

1. When the buttress walls are demolished along with excavations, buttress wall with shorter length, for example, 2.0 m, will move together with the surrounding soils, and the driving shear stress is smaller, thus, the effect of buttress walls is less significant. By increasing the length of buttress walls, the driving shear stress is larger at the front end portion of the buttress wall further from the diaphragm wall, and the amount of reduction in wall deflection increased with the increasing length of the buttress wall.
2. The effect of buttress wall thickness on the efficiency of restraining wall deflection is not significant. Therefore, the thickness of the buttress walls only needs to resist collapse due to extrusion of the diaphragm wall.
3. When the buttress wall length is fixed, minimizing the buttress wall interval can reduce the lateral wall deflection, but the reduction amount has a trend of slowing down. Under the condition of equivalent total length of buttress walls, the effect of increasing the length of the buttress walls is greater than the effect of minimizing the buttress wall interval.
4. Increasing the buttress wall embedment can mainly minimize the wall deflection in the vicinity and below the bottom of the excavation surface. During the design of a buttress wall, it is adequate to consider adopting the ratio of buttress wall embedment to excavation depth as 2/3.

Acknowledgements. The authors acknowledge the support provided by the Ministry of Science and Technology in Taiwan via Grant No. MOST 105-2221-E-146-002.

References

ACI Committee 318: Building code requirements for structural concrete (ACI 318-95) and Commentary (ACI 318R-95), American Concrete Institute (ACI), Farmington Hills, Mich (1995)

Calvello, M., Finno, R.: Selecting parameters to optimize in model calibration by inverse analysis. Comput. Geotech. **31**(5), 410–424 (2004). https://doi.org/10.1016/j.compgeo.2004.03.004

Chen, S.L., Ho, C.T., Li, C.D., Gui, M.W.: Efficiency of buttress walls in deep excavations. J. GeoEng. **6**(3), 145–156 (2011)

Hsieh, P.G., Ou, C.Y., Hsieh, W.H.: Efficiency of buttress walls in reducing the deflection of the diaphragm wall in excavation. Acta Geotech. **11**(5), 1087–1102 (2016). https://doi.org/10.1007/s11440-015-0416-6

Hsieh, P.G., Ou, C.Y., Lin, Y.K., Lu, F.C.: Lesson learned in design of an excavation with the installation of buttress walls. J. GeoEng. **10**(2), 63–73 (2015). https://doi.org/10.6310/jog.2015.10(2).4

Hwang, R.N., Moh, Z.C., Wang, C.H.: Performance of wall systems during excavation for Core Pacific City. J. GeoEng. **2**(2), 53–60 (2007)

Ladd, C.C., Foott, R., Ishihara, K., Schlosser, F., Poulos, H.G.: Stress-deformation and strength characteristics. In: Proceedings of the 9th International Conference on Soil Mechanics and Foundation Engineering, vol. 2, pp. 421–494 (1977)

Lim, A., Hsieh, P.G., Ou, C.Y.: Evaluation of buttress wall shapes to limit movements induced by deep excavation. Comput. Geotech. **78**, 155–170 (2016). https://doi.org/10.1016/j.compgeo.2016.05.012

Lim, A., Ou, C.Y., Hsieh, P.G.: Evaluation of clay constitutive models for analysis of deep excavation under undrained conditions. J. GeoEng. **5**(1), 9–20 (2010)

Lin, D.G., Woo, S.M.: Three dimensional analyses of deep excavation in Taipei 101 construction project. J. GeoEng. **2**(1), 29–41 (2007)

Ou, C.Y.: Deep Excavation: Theory and Practice. Taylor & Francis (2006)

Ou, C.Y., Teng, F.C., Seed, R.B., Wang, I.W.: Use of buttress walls to reduce excavation-induced movements. Proc. Inst. Civil Eng. Geotech. Eng. **161**(GE4), 209–222 (2008). https://doi.org/10.1680/geng.2008.161.4.209

PLAXIS 3D: Computer software. PLAXIS, Delft, Netherlands (2013)

Schanz, T., Vermeer, P.A., Bonnier, P.G.: The hardening soil model: formulation and verification. In: Beyond 2000 in Computational Geotechnics, Rotterdam, pp. 281–290 (1999)

Erratum to: Proceedings of the 2nd International Symposium on Asia Urban GeoEngineering

Renpeng Chen, Gang Zheng, and Changyu Ou

Erratum to:
R. Chen et al. (eds.), *Proceedings of the 2nd International Symposium on Asia Urban GeoEngineering*, Springer Series in Geomechanics and Geoengineering, https://doi.org/10.1007/978-981-10-6632-0

The original version of the book was inadvertently published without the following corrections:

New chapters 47–53 have to be inserted.

Misspelt university name "Tianjing University" has to be corrected as "Tianjin University" in Frontmatter.

The erratum book has been updated with the changes.

The updated online version of this book can be found at https://doi.org/10.1007/978-981-10-6632-0

R. Chen et al. (eds.), *Proceedings of the 2nd International Symposium on Asia Urban GeoEngineering*, Springer Series in Geomechanics and Geoengineering,
https://doi.org/10.1007/978-981-10-6632-0_54

Author Index

R. Chen et al. (eds.), *Proceedings of the 2nd International Symposium on Asia Urban GeoEngineering*, Springer Series in Geomechanics and Geoengineering,
https://doi.org/10.1007/978-981-10-6632-0

Keyword Index

R. Chen et al. (eds.), *Proceedings of the 2nd International Symposium on Asia Urban GeoEngineering*, Springer Series in Geomechanics and Geoengineering,
https://doi.org/10.1007/978-981-10-6632-0

The manufacturer's authorised representative in the EU is Springer Nature Customer Service Centre GmbH, Europaplatz 3, 69115 Heidelberg, Germany. If you have any concerns regarding our products, please contact ProductSafety@springernature.com

Printed and bound by CPI Group (UK) Ltd, Croydon, CR0 4YY

25/09/2026

02222853-0018